P9-DTU-005

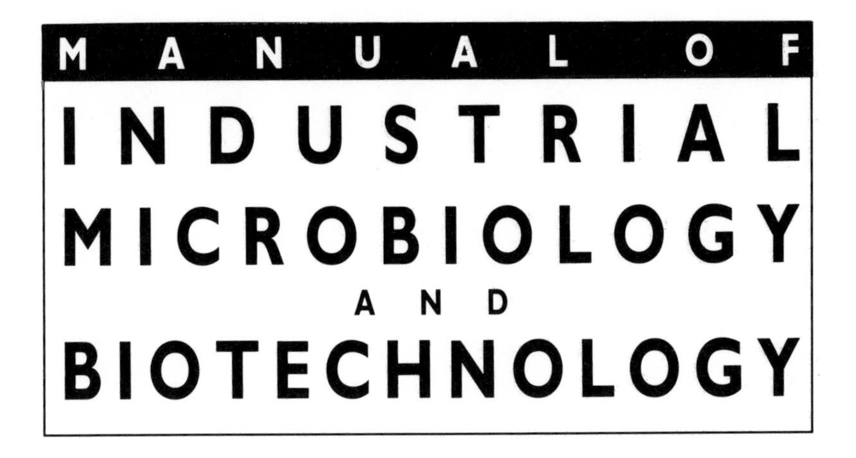
MANUAL OF
INDUSTRIAL
MICROBIOLOGY
AND
BIOTECHNOLOGY

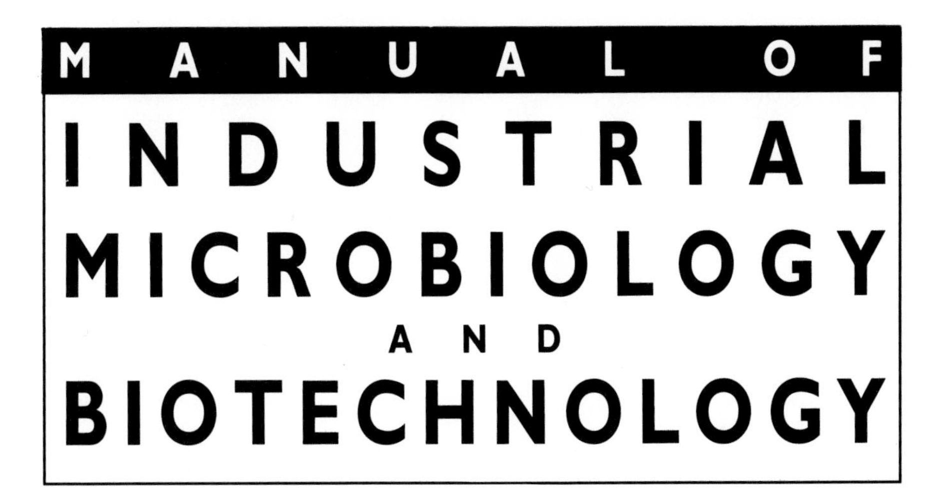

Editors:
Arnold L. Demain and Nadine A. Solomon

Department of Nutrition and Food Science,
Massachusetts Institute of Technology, Cambridge, Massachusetts

American Society for Microbiology
Washington, D.C. 1986

1913 I Street, N.W.
Washington, DC 20006

Library of Congress Cataloging-in-Publication Data

Main entry under title:

Manual of industrial microbiology and biotechnology.

Includes indexes.
1. Industrial microbiology—Handbooks, manuals, etc.
2. Biotechnology—Handbooks, manuals, etc. I.
Demain, A. L. (Arnold L.) II. Solomon, N. A. (Nadine
A.)
QR53.M33 1986 660'.6 85-26678

ISBN 0-914826-72-7
ISBN 0-914826-73-5 (soft)

Printed in the United States of America

Contents

Section VI. SPECIAL TOPICS

Contributors

Bernard J. Abbott
Lilly Research Laboratories, Eli Lilly and Co., Indianapolis, Indiana 46285
Fredric G. Bader
Industrial Division, Bristol-Myers Co., Syracuse, New York 13221-4755
Richard H. Baltz
Lilly Research Laboratories, Eli Lilly and Co., Indianapolis, Indiana 46285
Angela Belt
Cetus Corp., Emeryville, California 94608
F. R. Bernath
Department of Chemical and Biochemical Engineering, Rutgers-The State University of New Jersey, Piscataway, New Jersey 08854
George B. Boder
Lilly Research Laboratories, Eli Lilly and Co., Indianapolis, Indiana 46285
Christopher T. Calam
Liverpool Polytechnic, Liverpool L3 3AF, England
Gary J. Calton
Purification Engineering, Inc., Columbia, Maryland 21046
L. T. Chang
Fermentation Research and Development, Industrial Division, Bristol-Myers Co., Syracuse, New York 13221
Ichiro Chibata
Research and Development Headquarters, Tanabe Seiyaku Co., Ltd., Yodogawa-ku, Osaka, Japan
Bruce W. Churchill
Fermentation Research and Development, The Upjohn Co., Kalamazoo, Michigan 49001
Carrington S. Cobbs
Purification Engineering, Inc., Columbia, Maryland 21046
Arnold L. Demain
Department of Applied Biological Sciences, Massachusetts Institute of Technology, Cambridge, Massachusetts 02139
Steven W. Drew
Merck & Co., Rahway, New Jersey 07065
R. P. Elander
Fermentation Research and Development, Industrial Division, Bristol-Myers Co., Syracuse, New York 13221
Richard Fink
Biohazard Assessment Office, Massachusetts Institute of Technology, Cambridge, Massachusetts 02139
M. E. Fonda
Cetus Corp., Emeryville, California 94608
Oluf L. Gamborg
TCCP, Department of Botany, Colorado State University, Ft. Collins, Colorado 80523
C. T. Goodhue
Research Laboratories, Eastman Kodak, Kodak Park, Rochester, New York 14650
R. Greasham
Merck & Co., Inc., Rahway, New Jersey 07065
Michael Greenstein
Medical Research Division, American Cyanamid Co., Lederle Laboratories, Pearl River, New York 10965
Bruce K. Hamilton
Research Division, W. R. Grace & Co., Columbia, Maryland 21044
John P. Hamman
Purification Engineering, Inc., Columbia, Maryland 21046
David A. Hopwood
John Innes Institute, Norwich NR4 7UH, England
Robert N. Hull
Lilly Research Laboratories, Eli Lilly and Co., Indianapolis, Indiana 46285
G. R. Hunt
Merck Sharp & Dohme Research Laboratories, Rahway, New Jersey 07065
J. C. Hunter-Cevera
Cetus Corp., Emeryville, California 94608
E. Inamine
Merck & Co., Inc., Rahway, New Jersey 07065

Thomas D. Ingolia
Lilly Research Laboratories, Eli Lilly and Co., Indianapolis, Indiana 46285
David M. Isaacson
Research Laboratories, Ortho Pharmaceutical Corp., Raritan, New Jersey 08869
James P. Kalk
Genex Corp., Gaithersburg, Maryland 20877
Joel Kirschbaum
Squibb Institute for Medical Research, New Brunswick, New Jersey 08903
Asger F. Langlykke
Genex Corp., Gaithersburg, Maryland 20877
Allen I. Laskin
New Jersey Center for Advanced Biotechnology and Medicine, University of Medicine and Dentistry of New Jersey-Rutgers Medical School, Piscataway, New Jersey 08854
Daniel F. Liberman
Biohazard Assessment Office, Massachusetts Institute of Technology, Cambridge, Massachusetts 02139
David H. Lively
Biochemical Development Division, Eli Lilly and Co., Indianapolis, Indiana 46285
Lars G. Ljungdahl
Department of Biochemistry, University of Georgia, Athens, Georgia 30602
William M. Maiese
Medical Research Division, American Cyanamid Co., Lederle Laboratories, Pearl River, New York 10965
Patti Matsushima
Lilly Research Laboratories, Eli Lilly and Co., Indianapolis, Indiana 46285
Thomas L. Miller
Fermentation Products Production, The Upjohn Co., Kalamazoo, Michigan 49001
J. Patrick Montgomery
ChemGen Inc., Navarre, Minnesota 55392
Richard E. Mudgett
Department of Food Science & Nutrition, University of Massachusetts, Amherst, Massachusetts 01003
G. P. Peruzzotti
Research Laboratories, Eastman Kodak, Kodak Park, Rochester, New York 14650
Stephen W. Queener
Biochemical Development Division, Eli Lilly and Co., Indianapolis, Indiana 46285
J. P. Rosazza
Department of Pharmacology, University of Iowa, Iowa City, Iowa 52242
Roman Saliwanchik
8753 Merrimac, Richland, Michigan 49083
Tadashi Sato
Research Laboratory of Applied Biochemistry, Tanabe Seiyaku Co., Ltd., Yodogawa-ku, Osaka, Japan
Frederick Schaefer
Toxic Hazards Office, Vanderbilt University, Nashville, Tennessee 37232
Jeffrey J. Schruben
Ajinomoto USA, Inc., Raleigh, North Carolina 27610
Oldrich K. Sebek
Department of Biology and Biomedical Sciences, Western Michigan University, Kalamazoo, Michigan 49008
Nadine A. Solomon
Department of Applied Biological Sciences, Massachusetts Institute of Technology, Cambridge, Massachusetts 02139
Kevin Stafford
Genentech, Inc., South Francisco, California 94060
R. W. Stieber
Merck Chemical Manufacturing Division, Elkton, Virginia 22827
Tetsuya Tosa
Research Laboratory of Applied Biochemistry, Tanabe Seiyaku Co., Ltd., Yodogawa-ku, Osaka, Japan
Antonio Trilli
Microbiology Department, Kings College London, Kensington Campus, London W8 7AH, England
K. Venkatasubramanian
Department of Chemical and Biochemical Engineering, Rutgers-The State University of New Jersey, Piscataway, New Jersey 08854

Henry Y. Wang
Department of Chemical Engineering, The University of Michigan, Ann Arbor, Michigan 48109
Richard J. White
Bristol-Myers Co., Syracuse, New York 13221
Juergen Wiegel
Department of Microbiology and Center for Biological Resource Recovery, University of Georgia, Athens, Georgia 30602
John S. Wood
Lilly Research Laboratories, Eli Lilly and Co., Indianapolis, Indiana 46285

Preface

Our objective in putting together this *Manual of Industrial Microbiology and Biotechnology* was to bring together in one place the biological and engineering methodology required to develop a successful industrial process from the isolation of the culture to the isolation of the product. The structure of the manual indeed resembles the development of a commercial biological process and product.

The first section focuses on the culture: its isolation and preservation, inoculum development, nutritional improvement of its performance, and screening new isolates for new products. Section II emphasizes the process, e.g., shake-flask fermentations, solid-state fermentations, anaerobic processes, continuous fermentations, and substrates for large-scale fermentations. In Section III, emphasis is placed on improving the culture by mutation, screening and selection, protoplast fusion, and cloning. Genetic mapping and its importance are also discussed. The main biological foci in Section III are streptomycetes and yeast. The fourth section features some areas of recent biotechnological interest, i.e., cell and enzyme immobilization and plant and animal cell culture. In Section V, biochemical engineering is highlighted. The very important aspects of scale-up, instrumentation, computerization, large-scale sterilization, and cost estimation are comprehensively covered, as is the construction of a general-purpose pilot plant. In Section VI, special topics such as product recovery, assays, biosafety, and legal protection round out the overall treatment of industrial microbiology and biotechnology.

We are fortunate to have been able to enlist a broad range of experts for this effort, including microbial ecologists, physiologists, geneticists, biochemists, molecular biologists, and biochemical engineers, plus a biohazard expert and a patent lawyer. Some of our authors have had years of experience in the fermentation plant and pilot plant, and most have carried out distinguished research and development in the laboratory. They hail from the chemical and pharmaceutical industries as well as from biotechnology companies and academia. Our section editors also come from these industries, with long experience in practical matters and well-known contributions to the microbiological literature. They were given a list of topics and the freedom to choose the contributors. Many of the contributors had a difficult time with the assignment of writing a "how-to" chapter, as opposed to a review. With sufficient harassment by us and the section editors, however, most came through with useful chapters to guide the users of this manual through successful development of organisms and processes. However, only time will tell whether we have been successful.

To make the manual a more valuable resource, some of our authors have provided lists of commercial sources of supplies, equipment, and expertise. Of course, our intention and that of the American Society for Microbiology (ASM) is not to endorse or exclude individual companies, but only to provide the reader with some examples of commercial outlets. We apologize in advance to those companies that we have not mentioned.

And now, to the audience. We feel that this manual should be useful to advanced undergraduate, graduate, and postgraduate students, to technicians in both academia and industry, and to scientists and engineers in industrial companies, both large and small. All of you will find omissions, but we request that you refer to the ASM *Manual of Methods for General Bacteriology* before you discard our manual in frustration. You will see that certain omitted topics are covered very well in the earlier manual, e.g., genetics of unicellular bacteria.

At this point, after a long effort, we are pleased to thank Helen Whiteley, Walter Peter, Jim Tiedje, and David Schlessinger for their development of the concept of this manual, their insistence that we take on the task, and their advice and encouragement. We are pleased to work with Ellie Tupper and Susan Birch from the ASM Publications Department, who showed us how an efficient operation should be run. Finally, to our authors and section editors, thank you all.

ARNOLD L. DEMAIN
NADINE A. SOLOMON

I. The Culture

Introduction

STEVEN W. DREW

The following chapters deal with the most essential element of any fermentation process, the culture. They describe how to isolate cultures from nature and evaluate their potential, how to provide high-quality inocula, how to examine the effects of various nutrients on product formation, and how to store cultures for long periods.

Chapter 1, "Isolation of Cultures" by Hunter-Cevera et al., describes selective collection and isolation procedures and media for actinomycetes, algae, bacteria, and fungi. After isolation, the culture is tested for its ability to produce new products or activities. This process is addressed in the chapter "Screening for New Products from Microorganisms" by White et al.

Once a culture which makes the desired product has been discovered, reliable inoculum procedures and optimum fermentation media must be devised. Hunt and Stieber, in their chapter "Inoculum Development," describe methods for effective and reproducible inoculum production. In the next chapter, "Nutritional Improvement of Processes," Greasham and Inamine describe statistical approaches to nutrient evaluation as well as biochemical methods.

Preserving microorganisms of industrial importance is not a trivial matter. High levels of product formation, which are almost never advantageous to the culture itself, must be retained. The final chapter in this section, "Long-Term Preservation of Industrially Important Microorganisms" by Chang and Elander, reviews several classes of preservation procedures with detailed protocols for each.

Chapter 1

Isolation of Cultures

J. C. HUNTER-CEVERA, M. E. FONDA, AND ANGELA BELT

Isolating microorganisms from nature is the microbiologist's first step in screening for natural products such as secondary metabolites and enzymes. Unfortunately for industry, no single isolation method will reveal the total number and variety of microorganisms present in a sample (32, 64). It is possible to isolate many different microorganisms by employing enrichment techniques (1, 7, 37, 78) or even single-cell isolation by use of capillary methods (53). However, for industrial screening, such enrichment techniques usually require an inordinate amount of time, labor, and money, since only a few species of a particular genus arise from any one sample. Also, the industrial screen's assay procedures may require modification to "suit" the growth and metabolism of very different genera. Thus, a more scientific and less ad hoc approach for isolation is required (82).

One successful approach for the discovery of new antibiotics (49, 81) and enzymes (J. Geigert, J. C. Hunter, and S. L. Neidleman, Abstr. Int. Conf. Enzyme Eng. VII, White Haven, Pa., 1983, p. 27) involves (i) considering the desired product characteristics and process development, and (ii) using ecological approaches for isolation and screening.

Consideration of the desired product characteristics, process development, or both may aid in answering the first major question: What do I isolate and from where? For example, if an enzyme were needed that functioned optimally at a moderately high pH and salinity, yet at variable temperatures, the odds would certainly favor its discovery among microorganisms found in a desert such as Death Valley, rather than those from a New Jersey hardwood forest. Choice of the microorganisms, in turn, may be influenced by existing capacities for production, biomass and product yield, recovery costs, stability of the microorganism in large-scale fermentors, and ease of genetic manipulation.

Using ecological approaches to isolation can provide a screen with both a large number and a wide variety of microorganisms to examine for the product of interest. Even though microorganisms are highly adapted, specific microbial types are associated with different niches or samples within a variety of ecosystems (J. C. Hunter, Ph.D. thesis, Rutgers, The State University, New Brunswick, N.J., 1978). If one systematically samples different niches within a specific ecosystem, a greater diversity of microbial types can be isolated. Media selection, diluents, incubation conditions, and sample handling dictate the numbers and types of microorganisms isolated from plants, soils, and waters (34). Therefore, what is isolated from the defined ecosystem or habitat is a reflection of the isolation procedures and conditions set by the microbiologist, as well as the fluctuating conditions found in nature. This chapter illustrates different ways of treating and isolating single samples from specific ecosystems to isolate actinomycetes, algae, bacteria, and fungi. The following isolation procedures essentially "milk" the sample for both autochthonous and zymogenous microorganisms that are more representative of the biota associated with that particular sample. In the long run, such "milking" saves money and time and results in the isolation of a large, representative, and diverse microbial population.

Some general rules for applying ecological approaches to isolation are listed below and generally can be applied in the isolation of any one particular group of microorganisms (34).

1. List the groups of microorganisms that are to be isolated.
2. Describe the ecosystem or habitat from which the samples are to be collected.
3. Group samples into types, e.g., plants and plant parts, soils (types and horizons), rocks, waters, insects, etc.
4. List the environmental parameters to be considered and measured, such as pH, salinity, E_h, and temperature.
5. List the available natural substrates in the ecosystem, e.g., chitin in forest soils.
6. Design isolation techniques around data obtained from steps 1 through 5, i.e., diluents, substrates, natural extracts, and incubation conditions.
7. Evaluate "ecological isolation methods" using standard methods as controls.
8. Modify known procedures as required by the ecological parameters of the material to be examined.
9. Employ specific enrichment procedures for microbial groups that may be of screening interest.

1.1 ISOLATION OF ACTINOMYCETES

1.1.1 Sampling and Collection Methods

To isolate a representative actinomycete population from a particular ecosystem, especially those species occurring in unique microenvironmental niches, considerable attention should be paid to sampling. Samples should be collected as aseptically as possible with the aid of sterile spatulas, soil profile samplers, forceps, scalpels, gloves, Nasco Whirlpak bags, and plastic bottles. Samples should be representative of a site, e.g., a particular soil type and its horizons, leaf litter and detritus, rhizosphere plane and zone, marine sand, sediment and muds, plant surfaces and parts, or water column. Samples should be fully labeled with description and date. Seasonal and temporal aspects of collecting should be considered, as a true autochthonous population may occur transiently; e.g., actinomycete numbers decrease after a heavy rainfall. Once the samples are brought into the laboratory, they should be either examined immediately or stored overnight at 4°C in an area separate from actual plating, screening, and culture collection facilities to minimize chances of mite infestations.

1.1.2 Ecological Parameters and Media

Ecological parameters to consider include high or low temperature and pH, ionic strength, E_h potentials, and even substrate concentration. Most of these parameters are measured in the field or in the laboratory (7, 54). The actinomycete isolation media mentioned below are then modified to suit the ecosystem or habitat being examined. For example, seawater is added to isolation agars at different concentrations to match the salt gradient in an estuary (33); pH is lowered to suit the different soil horizons present in a forest (28); temperature of incubation is lowered for psychrophiles (4 to 15°C) and raised for thermophiles (55 to 70°C). A small amount (1 to 5 ml) of "natural extract" (plant, rock, compost) can be added as a growth factor for the initial isolation. Usually, a greater variety of actinomycete types are isolated when isolation methods are thus adjusted than if no modifications are made.

Actinomycetes are bacteria which are most efficient at utilizing substrates available at very low concentrations as well as complex substrates such as chitin. Therefore, most actinomycetes are isolated in lean or complex agars rather than a rich growth medium. Except for thermophilic forms, they are slow colonizers and usually appear on isolation plates within 4 to 20 days. Isolation media have been developed which favor the growth of actinomycetes over other microbial groups. Media such as arginine-glycerol salts (23), AV (44), Benedict (6), collodial chitin (31, 42), GAC (45), M_3 (62), starch-casein (38), and water agar (23) are well suited for the isolation of most actinomycetes. The antifungal agents nystatin (Squibb) and cycloheximide (Calbiochem) are usually incorporated into actinomycete isolation agars to retard fungal development (Table 1; section 1.5.1). In addition, to increase specific genus numbers, other antibiotics are added (18). The isolation agar plates are air dried for 3 days to minimize bacterial development. To the best

TABLE 1. Methods of pretreating soils to enhance isolation of some specific genera

Method of pretreatment	Recommended isolation agars	Antibiotics incorporated (μg/ml)	Days, temp of incubation	Genera	References
1. Air-dry soil, grind, heat for 1 h at 100 to 120°C. Plate out dilutions of distilled water-soil suspensions.	AV, GAC, MGA-SE	Cycloheximide (50), nystatin (50, polymyxin (5), penicillin (1–8)	20-40 days, 30, 32, 40°C	*Actinomadura, Microbispora, Microtetraspora, Streptosporangium*	44–48
2. Mix soil with distilled water, heat for 1 h at 50 to 55°C.	CYC	Cycloheximide (50), novobiocin (25)	1 day, 50–55°C	*Thermoactinomyces*	19
3. Air-dry soil, grind, heat for 2 h at 60 to 65°C. Stamp.	Arginine glycerol salts, starch-casein-nitrate, thin Pablum agar	Cycloheximide (75), nystatin (75)	10–14 days, 26–28°C	*Micromonospora, Nocardia*	Hunter, thesis
4. Mix powdered chitin with soil (1:1). Incubate for 2 to 3 weeks at 26°C. Grind/stamp.	Arginine glycerol salts, starch-casein-nitrate, thin Pablum agar	Cycloheximide (75), nystatin (75)	10–14 days, 28°C	*Micromonospora*	Hunter, thesis
5. Mix $CaCO_3$ with soil (1:1). Incubate for 10 days in an inverted petri plate with saturated disk of filter paper. Dilute in water and plate.	Arginine glycerol salts, water agar	None	10 days, 28°C	Overall numbers of actinomycetes increase	23, 75
6. Mix dead yeast cells with soil (1:3). Adjust pH to 5.0. Incubate for 10 days at 28°C in high humidity. Dilute and plate.	Czapek plus yeast extract	Cycloheximide (75), nystatin (50)	10 days, 26–28°C	*Oerskovia*	Lechevalier (personal communication)
7. Store soil at 4°C; dilute in 0.25× Ringer's solution plus 0.01% (wt/vol) gelatin (pH 7.0) and plate.	Diagnostic sensitivity test agar	Chlortetracycline (50), cycloheximide (45), mycostatin (59), methacycline (10)	7, 14, 21 days, 25°C	*Nocardia*	50, 51

of our knowledge and experience, no single medium will allow the recovery of all the actinomycetes present in a sample. See actinomycete isolation media below (section 1.5.1).

1.1.3 Nonselective Isolation of Actinomycetes from Soil

All muds and soils are air dried in sterile glass petri dishes at room temperature for 3 to 10 days (depending upon moisture content) to aid in reducing the bacterial population. The dried muds and soils are then gently powdered with a sterile pestle and "stamped" (74) onto actinomycete isolation agars. A small circular sponge (Dispo culture plug, 16 mm; Scientific Products) is directly pressed into the dried soil powder and then removed, and excess soil is shaken off. A stack (9 to 12 plates) of alternating different isolation agars are then inoculated by successive "stamping" of the agar surface 13 times (10 stamps on the outside perimeter and 3 in the middle) with the sponge to achieve a dilution effect. Discrete isolated colonies will result after incubation of plates right side up at 4 to 15°C for psychrophiles, 22 to 35°C for mesophiles, and 55°C and higher for thermophiles. Use saturated filter paper disks to prevent drying of agar plates at higher temperatures.

1.1.4 Selective Isolation of Actinomycetes from Soil

There are many ways to pretreat soils to increase overall numbers (count) of actinomycetes isolated or to favor growth of specific genera. Table 1 summarizes some of the more practical pretreatments of soils to aid in isolating actinomycetes for industry. A combination of a pretreatment with a suitable selective

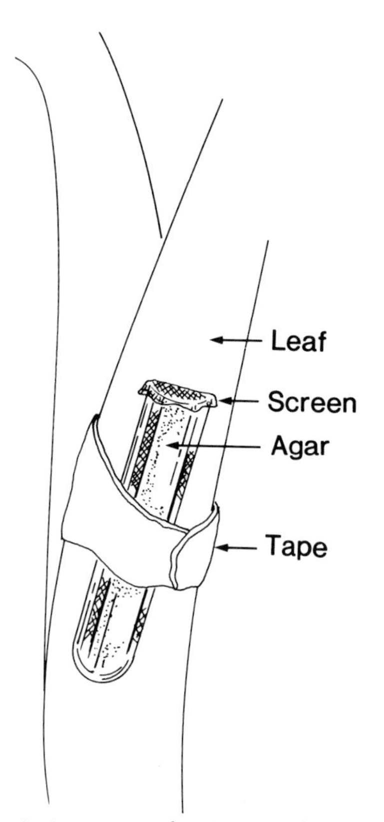

FIG. 1. Actinomycetes baiting trough on a plant.

medium is necessary for the efficient isolation of certain genera (23, 82).

1.1.5 Isolation of Actinomycetes from Plants and Plant Material

In some ecosystems there appears to be a distinct seasonal fluctuation in numbers of actinomycetes isolated from plants, as well as a specific association with plant parts (33; Hunter, thesis). For example, numbers may be low in early spring when plants are young but quite high in late summer when plants begin to senesce and decay. Actinomycetes isolated from the flower usually differ from those isolated from stem or roots.

Plant parts (flowers, leaves, upper stem, middle, base, roots) are chopped aseptically with scissors or scalpel, preferably in a laminar-flow hood, and dried (2 to 7 days) in sterile glass petri dishes. The plant pieces are then either implanted into agar surface or gently rolled over the surface of the agar in a streaking manner. An alternative method is to shake the plant pieces for 20 min (150 rpm) in quarter-strength plant extract or phosphate buffer, serially dilute samples in an appropriate diluent, and spread plate onto isolation agars into which 1 to 5 ml of plant extract has been incorporated. (See section 1.5 for diluents.)

Differential centrifugation works well for separating the bacteroids of *Frankia* species from plant tissue and organelles in homogenates prepared from nodules taken from the host plant (4).

1.1.6 Baiting Technique for Isolation of Actinomycetes Associated with Plants

A simple baiting technique can be used to isolate actinomycetes from standing plants in different ecosystems year round. This method aids in isolating those actinomycetes directly associated with different plant parts and their surfaces.

Glass troughs, made by longitudinally cutting in half a glass tube (12 by 75 mm; Corning), are filled with 1 ml of hot actinomycete agar supplemented with 0.1 ml of plant extract and the antifungal agents cycloheximide and nystatin at levels of 75 μg of each per ml. A Nitex nylon cloth screen (100 μm pore diameter; Tetro, Elmsford, N.Y.) can be glued to the rim of the trough with silicone glue (General Electric) to prevent the entry of insects and large protozoa. The troughs are then attached with time tape (Shamrock), agar side down, to different parts of the plants and trees (Fig. 1).

After a suitable incubation period of 2 to 5 days, the dried agar strips are removed with sterile forceps and are either (i) gently washed with 10 ml of liquid buffer or actinomycete broth, agitated (in a 125-ml flask at 150 rpm) for 15 min, and subsequently serially diluted and plated (0.1 ml) onto appropriate isolation agar or (ii) placed face down (side containing spores and hyphae) onto an actinomycete agar surface and incubated at 28°C overnight. Strips are then removed, and plates are incubated again for 10 to 15 days.

1.1.7 Isolation of Actinomycetes from Waters

Pretreatment of water samples often enhances the number and types of actinomycetes isolated. Centrifugation (6,000 × *g*; 49) or filtration (0.45-μm filter; 2) of water samples, followed by serial dilution and plating, works well for isolation of "aquatic actinomycetes." *Rhodococcus* and *Micromonospora* can be selectively isolated by heating water samples (2.0 ml; previously stored at 4°C) in a glass tube (100 by 12 mm) sealed with silicone rubber bungs for 6 min at 55°C (62). The samples are diluted in quarter-strength Ringer's solution (Oxoid tablets) containing 0.01% (wt/vol) gelatin (pH 7.0) (72) and plated onto M_3 agar (62). Plates are incubated at 30°C. *Rhodococcus* colonies appear within 5 to 7 days, whereas *Micromonospora* will take between 10 and 21 days.

Actinoplanes can be isolated from flowing waters by simply plating water samples on water agar plates made with filtered lake water or chitin agar and incorporating 0.1% (wt/vol) potassium tellurite (83).

1.1.8 Subculturing and Purification

Once colonies develop, they can be tentatively identified by differences in gross morphology such as colony form, aerial spore color, and diffusible and reverse pigments (22). With the aid of a long-working-distance objective (40×), aerial and substrate spore formation can be determined. The success of isolating diverse actinomycete genera, as well as species, depends somewhat upon the sample itself and the agars employed; even more important to success, however, is the ability to recognize the different growth forms initially and tentatively identify them. Therefore, a

TABLE 2. Diluents and general media employed to isolate cyanobacteria and eucaryotic algae from four main sample types[a]

Sample type	Diluent	Media to isolate:	
		Cyanobacteria	Eucaryotic algae
Fresh waters, soils, algae	0.9% saline or 0.5× phosphate buffer	Allen's, BG-11, M2, medium C	CM basal, natural water, proteose, and soil water media
Marine waters, soils, algae	0.25× artificial seawater	ASN 111, marine BG-11, medium A, MN marine	Erdshreiber, M11, natural water, and soil seawater media
Thermophilic waters, soils, algae	Mineral spring water	BG-11, complex hot springs medium, medium D	Allen's medium, complex hot springs medium, and medium D
Other waters, soils, algae	0.5× phosphate buffer	Allen, BG-11, M2, medium C	CM basal, natural water, proteose, and soil water media

[a] See section 1.5 for formulas and instructions for diluents and media.

"trained eye" is of great value in isolating actinomycetes.

Colonies are picked with either a flamed, bent, L-shaped needle or a sterile wood stick and transferred to appropriate maintenance agars or replica plated for screening.

1.2 ISOLATION OF ALGAE

1.2.1 Sampling and Collection Methods

Most methods for the collection of algae have been designed by botanists, whose main interest is the ecology of the algal environment. From their viewpoint, it is important to isolate every representative algal species in a habitat. These studies include numerous samplings of one station, followed by exhaustive (and time-consuming) techniques to isolate all species possible. This approach is inappropriate for industrial screening, where efficiency and simplicity are important. It is better to limit the number of samplings per habitat to only two or three which will yield the greatest diversity. Smith (65) describes the types of algae that may be found in various sections of a body of water and also notes many unusual algal habitats which should not be overlooked.

It is important to record basic information about the samples: date, location, description of sample, any unusual surrounding features, temperature, climate, and sample pH.

Samples can be conveniently divided into four groups: (i) waters; (ii) soils and sludges; (iii) algal blooms, mats, and scrapings; and (iv) miscellaneous, e.g., rocks, twigs, mosses, and other plant material. The samples are sorted within each group according to biophysical characteristics such as pH and salinity. The dividing and sorting of samples, as well as microscopic examination (500× and 1,250× [26]), will determine what techniques and media should be used to isolate algae.

The choice of collection containers depends upon the sample type. Whirlbags (Nasco) may be used for soil, rock, twigs, scrapings, and sometimes more moist samples if the bags are doubled. Generally, sterile polypropylene tubes and bottles are used for collecting water samples, algal blooms, bottom sludge, filamentous mats, and other large samples. It is also advisable to include sterile tubes and bottles which contain a measured amount of diluent (see section 2.5). Diluted samples can be kept for several days at 4°C.

Sterile spatulas and tweezers are used for collecting soils, rocks, twigs, and scrapings of larger objects. A dipcup or "giant pipette" may be used for collecting water samples and sometimes bottom sludge (65).

1.2.2 Ecological Parameters and Media

To effectively isolate large numbers of algae, it is necessary to simulate their environment. Ecological parameters such as temperature, salinity, pH, surface/volume ratios, and nitrogen sources are often varied to isolate a variety of algal types.

In Table 2, both synthetic and natural media are listed according to sample type and purpose. The basic media listed can be varied as needed, as follows.

(i) NaCl or artificial seawater concentration is adjusted based on the salinity content of the sample.

(ii) pH may vary greatly among samples and is adjusted accordingly.

(iii) Incubation temperatures vary according to season sampling. Incubation at 25°C is suitable for most algae, but lower temperatures should be considered (for example, 5°C in the spring). Incubation at 35°C or higher is selective for certain cyanobacteria (34).

(iv) Nitrogen source in the form of $NaNO_3$, KNO_3, etc., is omitted to obtain nitrogen-fixing algae.

(v) Germanium dioxide can be used to inhibit diatoms (29, 70).

1.2.3 Dilution-Based Isolation Methods

Dilution-based methods for isolating algae are well suited for general screening purposes, mainly because they (i) may be applied to all types of samples, (ii) are relatively simple, (iii) are rapid, and (iv) give very good results (34).

A 1.0-ml (water samples) or 1.0-g (soils, sludge, rocks, algae, plants) portion is removed from the original sample and transferred to a tube containing 9 ml of diluent; glass beads (2.0 mm; 77) are added to the tube if the sample is solid. The mixture is vortexed until the sample is well dispersed and then serially diluted until approximately 10^4 cells per ml is achieved.

The diluted samples can now be used in the following ways.

Pour plate method (to isolate filamentous cyanobacteria)

Dispense 20 ml of molten "algae isolation" agar into individual test tubes and place in a water bath at 45°C. Add 1 ml of a dilution sample to each tube of medium. Each tube represents one plate. Use at least the final three dilutions for isolation plates. Work quickly to minimize sample exposure time at 45°C. To ensure equal mixing of sample, always shake or vortex the dilution before dispensing. Roll the melted medium tube gently to mix contents, and pour into a 100-mm plastic petri dish. Allow agar to solidify before moving plates. Incubate in the presence of cyclic light (12 h) at appropriate temperature until growth occurs.

Spread plate method (to isolate filamentous and unicellular cyanobacteria and members of the chlorophyta)

Remove 0.1 ml of dilution to an isolation agar plate and spread evenly with a sterile, bent glass rod. Incubate as described above.

Isolation streak method (to isolate unicellular cyanobacteria and members of the chlorophyta)

Remove a loopful of the dilution and streak onto isolation agar plate. Incubate as described above.

Broth method (to isolate species inhibited on solid substrates)

Add 1 ml of the dilution to 9 ml of broth medium in glass tubes with screw caps. Leave screw caps slightly loosened to allow for gas exchange. Incubate as described above. Save dilution tubes and incubate also.

1.2.4 Filtering/Stamping Method for Isolation of Algae

The filtering/stamping isolation method involves filtering samples through filter paper to concentrate microorganisms from a dilute suspension. This method can be used for any water sample but is especially helpful for dilute water samples (such as alpine lakes) that may have some unusual specimens that would otherwise be missed by the dilution tube method. A 10- to 20-ml water sample is filtered through sterile Whatman no. 1 paper, size 9 cm, using a Büchner funnel and slight vacuum. The filter paper is then stamped on agar medium plates, with each successive transfer equivalent to a 10-fold dilution. When the stamping is complete, the filter is placed right side up on the bottom of a petri dish, covered with melted agar (45°C), and incubated in the presence of cyclic light.

Samples such as rocks, twigs, and soils may be directly inoculated on agar medium plates by stamping, implanting, or rolling. These procedures are outlined in section 1.1.5 above.

1.2.5 Other Methods Used to Isolate Algae

Two other methods used by botanists are very selective and time-consuming. The micromanipulator (58, 70) requires patience and a steady hand, but the results are sometimes worth the effort. Unicellular, small multicellular, or filamentous fragments may be picked up individually and transferred to an appropriate medium. A unialgal culture is immediately established by this method. The disadvantages to this procedure are eyestrain and the potential for missing some species.

Another method involves using a photosynthetic microcapillary pipette and is only useful for those algae with a phototactic response, such as *Euglena* (70).

1.2.6 Subculturing and Purification

Plates and tubes should be examined daily under a dissection microscope until growth appears. Broth cultures are often multialgal, and it is necessary to redilute or use a micromanipulator. The earlier a colony is detected and picked, the more likely a unialgal culture will be established. Colonies are picked with a dissecting needle inserted into a bacterial loop holder. The needle should always be flamed between transfers. If there is overcrowding on a plate, restreak algae onto a new plate and incubate.

A purification method for filamentous cyanobacteria relies on their phototactic response. A block of agar carrying filaments is cut out and implanted near the edge of a new plate. The plate is placed so that the light source is opposite the inoculum. Most filamentous blue-green bacteria exhibit positive phototaxis (68).

Patience and steady hands are important when dealing with algae. Unialgal cultures can be rapidly obtained with some practice.

Cultures are maintained on agar slants, in broth, or in soil tubes at room temperature in dim light. The maintenance medium should reflect the alga's original habitat to ensure viability and stability and also to aid in identification (25, 56, 60, 65, 68, 71).

1.3 ISOLATION OF BACTERIA FROM NATURE

1.3.1 Sampling and Collection Methods

Sampling procedures are essentially the same as those described for actinomycetes (section 1.1.1).

1.3.2 Ecological Parameters and Media

Incorporation of natural extracts and environmental biophysical parameters into media, as well as the diluent employed in plating out the sample, can affect the numbers and variety of bacteria isolated in the laboratory (34).

Many different media are used to isolate bacteria from different samples within an ecosystem. Some agars should contain extracts (see section 1.5.3) in the concentration of 10 to 50% of total liquid volume. Infusions and extracts are usually made from materials (soils, muds, leaves, roots, rocks, compost, detritus, barks, etc.) collected within the ecosystem being examined. Other agars contain multiple carbon and nitrogen sources or complex natural carbon-nitrogen sources, such as chitin, cellulose, or pectin. Agars selective for gram-negative organisms, such as nutrient agar plus crystal violet, red-violet bile agar, brilliant green bile agar, or others manufactured by BBL Microbiology Systems, Difco, or Oxoid, usually have 5 to 15% natural extracts added. All isolation agars have the antifungal agents cycloheximide and nystatin incorporated at 50 μg/ml to retard fungal growth. Agar plates are also air dried for 1 to 2 days before use. Biophysical parameters such as medium pH and salinity are also adjusted to match the sample's ecosystem.

TABLE 3. Isolation media for bacteria[a]

Sample type	Suggested media	Suggested diluents
Soil, muds	Mutiple extract agar, nutrient agar plus crystal violet, soil infusion agar, violet-red bile agar plus extracts	Extract, phosphate buffer, Ringer's solution plus gelatin
Detritus, compost, leaf litter	Compost extract agar, Cellulose agar, multiple carbon-nitrogen source agar, plant extract agar	Distilled water plus artificial seawater (pH adjusted), extract, saline (1.9.5)
Fresh and dried plants and roots	Moss extract agar, nutrient agar plus crystal violet, plant extract agar, root extract agar	Extract, saline
Fresh and marine waters	Cellulose agar, colloidal chitin agar, marine agar, violet-red bile agar plus extracts	Artificial seawater, distilled water plus artificial seawater (pH adjusted), extract

[a] See section 1.5 for formulas and instructions for media and diluents.

1.3.3 Isolation of Bacteria from Soil

Direct isolation of soil bacteria

Mix 5 g (wet weight) of soil with 99 ml of sterile quarter-strength soil infusion or extract contained in a 250-ml flask and shake at 150 to 200 rpm for 25 min at 26°C. Serially dilute the soil-liquid suspension in the appropriate diluent (Table 3), and spread-plate 0.1-ml volumes of three appropriate dilutions, based on turbidity, onto three to seven different agars to which natural extracts have been added. Soil composition will affect turbidity observation; e.g., large clay content may tend to make the suspension more turbid, yet microbial numbers may be quite low. Written records of the soil type and dilutions used to plate out the soil sample should be kept. Incubate plates upside down at between 22 and 26°C for 4 to 10 days. Incubation temperature may vary depending upon original soil temperature. Examine plates every 2 days for new colony formation.

Simple soil enrichment

Some of the 247 genera listed in *Bergey's Manual* (11) require specific enrichment or selection techniques for isolation in pure culture (1, 37). Most enrichments are feasible due to the induction of enzymes, i.e., activation of a specific genome. Thus, many precursors to antibiotics, complex substrates, and growth factors may be employed. Enrichments may also promote antagonism. Therefore, with two microbes possessing different metabolic pathways for utilizing the same substrate, one will probably dominate. The object here is to enrich for the bacteria active in the ecosystem and not dormant microorganisms that are capable of growing rapidly under appropriate artificial selection pressure.

Place 1 g (wet weight) of soil in a sterile 50-ml beaker. Mix the soil with 2 to 10 ml of a sterile substrate suspended in an appropriate quarter-strength soil extract, half-strength Ringer's solution, or distilled water. Substrates can either be those present in the ecosystem or those that will favor development of bacteria with specific or desired enzyme systems, such as chitin, cellulose, pectin, amino acids, metals, antibiotic precursors, inhibitors, alcohols, etc. The pH can be adjusted to the original soil pH or changed to act as a biophysical stress or enrichment. Then overlay the soil-substrate mixture with ethylene-oxide-sterilized, chopped, dried grass (regular lawn grass will do, but preferably a grass from the ecosystem being examined). Cover the beaker and contents with a sterile paper cup. Incubate the enrichment beakers at the appropriate temperature for 2 to 12 days in a sealed vacuum jar to which 200 ml of water has been added. Samples should be withdrawn periodically, diluted, and spread-plated (0.1 ml) onto soil extract plates with and without the enrichment substrate.

Isolation of bacteria from rhizosphere

It is generally accepted that the microbial population associated with plant roots differs in numbers and types from that associated with the surrounding soil (10). The rhizosphere population is affected by the plant species, and therefore, in isolating bacteria from the rhizosphere, it is important to consider the plant species involved. Plant, root, and soil extracts (see section 1.5.3) are incorporated into the isolation agars. Many gram-negative rods and pleomorphic forms can be isolated by direct isolation or enrichment with carbohydrates, cellulose, and various nitrogen sources.

1.3.4 Isolation of Bacteria from Plants

Direct isolation

Cut plant parts with sterile scalpels or scissors, preferably in a laminar-flow hood or clean room. Put 1 g of plant material in a 250-ml flask containing 99 ml of appropriate diluent and glass beads (optional). Shake the flask and contents for 15 to 20 min (150 rpm) at room temperature to loosen the attached surface microbiota. Withdraw a 1-ml sample, serially dilute it in the appropriate diluent (see Table 3), and spread-plate 0.1-ml volumes onto plant isolation agars (see Table 2). Make three or four dilutions.

Simple enrichment for plant bacteria

Mix chopped plant parts with 7.5 ml of a plant polysaccharide, sugar, or protein (5%) suspended in quarter-strength plant extract in a 50-ml beaker. Cover the beaker and contents with a sterile paper cup and incubate in the manner described for soil enrichment. Samples should be periodically withdrawn, diluted in quarter-strength plant extract or half-strength phosphate buffer, and spread-plated (0.1 ml) onto the appropriate plant isolation agars. For additional methodology and general discussion of phyllosphere bacteria, see references 8, 21, and 55.

1.3.5 Isolation of Bacteria from Waters

Collection and storage of water samples

Water samples for bacterial examination should be collected in clean, sterile, wide-mouthed 100-ml plastic bottles. The bottles should be cleaned thoroughly before being used (57). Collect samples from still waters several inches deep because the surface water contains dust particles. Grasp the bottle near the base and plunge it, open mouth downward, to a depth of 12 in. (ca. 30 cm), the hand and bottle taking a wide arc as they quickly pass into and out of the water (59). If a current exists, direct the mouth of the bottle against it. Sample bottles should not be filled to the top. All water samples should be examined immediately after collection or stored at 5°C for no longer than 24 h. As soon as a water sample is collected, its condition of equilibrium is upset and a change in the bacterial population begins (57). For further information concerning sampling techniques and problems, see Collins et al. (13).

Direct isolation of bacteria

Filter 50 ml of a water sample through a 0.22-μm disposable filter. Once the sediment has settled on the filter membrane, add 1 ml of sterile diluent (usually half-strength mud or soil extract made from mud or soil surrounding the body of water). Gently scrape the membrane with a flat, wide-mouth 2-ml sterile pipette. Transfer the silt diluent suspension to 9 ml of corresponding diluent and vortex for 5 min. Next, serially dilute the sample and spread-plate in the manner described for soils (section 1.3.3). For further techniques in aquatic microbiology, see reference 61.

Filter membrane imprints

Filter 50 ml of a water sample through a 0.22-μm disposable filter. Gently cut the filter out. Using sterile forceps, successively lay the filter, collected silt face down, on a stack of six agars to achieve a diminution effect. Agars can be alternatively mixed to isolate a wider variety of types; see Table 2 for media.

Enrichment for isolation of aquatic bacteria

The simplest enrichment is to incubate 50 ml of the water sample in a sterile, 500-ml wide-mouth flask, covered with a cotton gauze plug. Sample the water periodically and plate out onto suitable isolation agars that have the sample water (filter sterilized) incorporated at 50 to 100%. One can also bait for aquatic microorganisms by adding certain substrates such as sugars, polysaccharides, and proteins to different volumes of the sample. It is a good idea to duplicate the enrichment and incubate one flask in the dark and one flask in the light. Low concentrations of antifungal agents (15 to 35 μg/ml) may be added to retard fungal development. Another enrichment involves adding "attachment material," such as sterile plant material or ground rock material, to the flasks. Incubation periods and temperatures can be varied to sort out psychrophiles, mesophiles, and thermophiles.

For additional enrichment procedures for the selective isolation of budding and prosthecate bacteria, see reference 30.

1.3.6 Subculturing and Purification

Once colonies develop, the agar plates are sorted according to types. Similar colonies or duplicates isolated from plating the different dilutions are sorted and then picked with sterile wooden sticks. The colonies are transferred either to master plates for screening or to appropriate maintenance agar slants that have a small amount (50 ml/liter) of natural extracts incorporated. Bacteria can be initially grouped according to plate morphological characteristics, such as pigment and colony forms. Pinpoint colonies are then picked, transferred to additional analogous isolation agars, and reincubated.

1.4 ISOLATION OF FUNGI

1.4.1 Sampling and Collection Methods

Fungi are ubiquitous, heterotrophic organisms occupying nearly every conceivable niche. Before beginning to isolate fungi from nature, it is important to define one's goals (product of interest) and to understand the structure of the ecosystem being sampled.

TABLE 4. Isolation media for lower fungi[a]

Sample type	Suggested media	Additions
Soils: sand, loam, humus, rocks, clay	Soil infusion agar, rose bengal medium, hay infusion agar, PCNB agar, tap water agar	As needed: salts, tetracycline, streptomycin, antifungal agents, etc.
Plant material: fresh or decaying plant matter, detritus	Hay infusion agar, leaf litter extract agar, PCNB agar, tap water agar	As needed: salts, chitin, pectin, cellulose, or sterile plant matter (e.g., carrot disks), inhibitors
Waters: fresh, marine, sewage	Rose bengal medium, PCNB agar, tap water agar, water (baited)	As needed: salts, antibiotics (e.g., chlortetracycline for sewage samples) (15), baits (e.g., seeds)

[a] See section 1.5 for formulas and instructions for media.

This will determine the types of environments sampled.

Samples should be collected using sterile gloves, tweezers, scalpels, or scoops, aseptically placed in Nasco Whirlpak bags or bottles (57), and either used immediately or stored overnight at 5°C.

When collecting soil samples, all horizons of a particular soil, leaf litter and detritus, rhizosphere, sands, gravels, and rocks should be included.

Any part of a plant can be useful for isolating fungi (leaf, stem, bark, root, flower). The shape, size, and physiological state (living, dead, healthy, diseased) of the plant parts, surface conditions (glabrous, pubescent, glandular), and the position or exposure of the sample both on the individual plant and its relationship to the ecosystem as a whole (e.g., the fringes of a forest versus the inner forest) will partially determine the fungal species isolated.

Water samples may be collected from estuaries, sewers, lakes, marine environments, streams, etc. At least 50 ml of water should be collected, using sterile bottles.

1.4.2 Ecological Parameters and Media

Fungal forms and nutritional requirements are as diverse as their habitats. Newhouse and Hunter (43) have demonstrated that changing one constituent of a medium can favor the isolation of one fungus over another. Therefore, there is no one "all-purpose" fungal isolation medium.

The physical parameters (e.g., pH, salinity, nutrient sources, temperature) of the ecosystem being sampled, as well as the "type" of fungi (lignicolous, halophilic, etc.) being isolated, should be considered in developing isolation agars. See section 1.5.4 for fungal isolation agars.

Using media with low C/N ratios yields more discrete countable colonies, resulting in more effective isolation and identification (16). Such lean media may be effective due to the decrease in competition and restriction of spreading fungi (5).

Plates are incubated in stacks and should not be inverted. Low temperatures (5 to 15°C, optimum) are necessary for isolating psychrophiles, 20 to 35°C (optimum) is typical for isolating mesophiles, and temperatures above 45 to 50°C select for thermophiles. Light is also an important factor to be considered as it is sometimes needed to induce fructification (80).

Usually a combination of selective enrichment and inhibition media are used simultaneously (Table 4).

Selective enrichment

Selective enrichments are designed to enhance isolation of one or more species over others; such enrichments may function at a species level (e.g., the separation of *Fusarium* species based on nutritional requirements) or on a more general scale (e.g., separating cellulolytic organisms).

Incorporating sterile substrates (soil extracts, vegetable infusions) from the sampling environment into lean medium is a simple but effective way to enrich for fungi. For example, the number and variety of yeasts isolated from grape leaves on a medium with the same sugars found in vine sap (e.g., grape juice/yeast extract and grape juice/liver extract) were greater than those isolated on standard yeast media (5). In addition, osmotically sensitive yeasts were isolated by incorporating sterile plant parts into water agar (5). Likewise, if isolating alkalophilic fungi with a high salt tolerance, it would be advantageous to include various levels of sea salts in the medium and adjust the pH to an alkaline level (34). There are as many ways to devise enrichment media as there are habitats.

Inhibition of bacteria

Selective inhibition media function by excluding to a degree "unwanted" organisms. Incorporating antibiotics such as chloramphenicol, kanamycin, penicillin, streptomycin, and tetracycline in the medium will inhibit bacterial colony development (76).

Other methods of inhibiting bacteria include drying plates for 3 to 4 days before use, lowering the pH of the medium, or, when a low pH is undesirable, adding rose bengal (1:30,000) to the medium (14, 67).

Inhibition of fungi

Various antifungal antibiotics are available (nystatin, pimaricin, cycloheximide) and can be excellent selective agents. Some nonantibiotic inhibitory agents that have been used include high CO_2 levels, calcium proprionate salts, oxgall, and crystal violet (5). Commercial fungicides include Botran (2,6-dichloro-4-nitroaniline), benlate, pentachloronitrobenzene, and captan. Not all fungi are affected by these agents, and sometimes the effects that do occur may be undesirable. Tsao (76) presents an excellent compilation of various selective inhibitors and their effectiveness.

1.4.3 Isolation of Fungi from Soils

Implant method

Using a sterile spatula or pick, place a small particle of soil near the edge of the agar. Gently press the sample slightly into the agar and incubate.

Dilution plate method

Set up a series of dilution tubes containing 9 ml of sterile distilled water or other diluent. Place 1 g of soil in a dilution tube and agitate for 15 min. Perform serial dilutions and inoculate 0.1 ml of the final three dilutions onto agar plates, spreading with a sterile, bent glass rod. (Dilution should result in 40 to 60 colonies per plate.) Incubate. The isolation plates may be used for replica plating by the same techniques described for bacteria, using a damp velvet.

Warcup method

Place a small amount of soil in a sterile petri dish. Disperse with a little sterile distilled water. Pour cooled molten agar into the plate. This method is useful for isolation of slow growers (79).

Stamping method

Autoclave polyurethane foam cylinders (15 by 40 mm). Moisten the end of a cylinder on the agar surface and press into the pulverized, dried soil sample. Stamp the plug 10 times in a circular pattern around the edge of the petri dish and three times in the center. Continue using the same plug on six to eight plates. This will create a diminution effect. Avoid excessive moisture. Increased numbers and fungal types have been isolated by using this stamping technique (74). The addition of NaCl (4 and 10%) to the medium further increases yields.

Miscellaneous methods

Numerous other methods, such as baiting techniques (79), syringe sampling (24), soil sieving (27), adhesion flotation, sucrose centrifugation, and gelatin column centrifugation (66), are described elsewhere. Although very useful for special circumstances, they are too laborious for effective large-scale isolations.

1.4.4 Isolation of Fungi from Plant Material

Implant method

Using a sterile spatula or pick, place a small piece of plant matter near the edge of the agar plate. Gently press the sample slightly into the agar and incubate. Isolates will spread out from the point of inoculation across the plate. Growing fronts may be isolated.

Impression method

Wearing sterile gloves, gently press a leaf surface onto agar. Remove the leaf and press it onto three or four more plates successively. This will produce a dilution effect. Repeat, using the opposite leaf surface. Parberry et al. (52) found increased frequency of nonsporulators with the increased number of imprints. Originally designed for leaf surfaces, this method will work for many different types of plant material.

Washing method

Place whole plant parts in shake flasks containing 0.05% Tween 80 and water. Shake on a rotary shaker for 2 to 5 min. Filter and wash the inoculum at this point, or immediately dilute and use it in standard spread-plate procedures (see section 1.5.1). Mycelial forms tend to remain attached, while spores are removed more easily (20). The washed plant parts may also be used for the impression method above.

Maceration method

In this method, plant material is first macerated and then handled as for the dilution plate method or implant method. Note that some plants may release inhibitory compounds during the maceration process. In addition, maceration may adversely affect the viability of some forms of fungi or, alternatively, give unreliable counts by increasing the numbers of propagules. Beech and Davenport (5), working with grape leaves, found that washing the sample first yielded greater numbers. The effect may have been one of "drawing organisms out" of the leaves.

1.4.5 Isolation of Fungi from Waters

Dilute samples to yield an optimum of fewer than 50 colonies per plate. Cooke (15) suggests as a general guideline the following dilutions: for liquid, 10^{-1}; rich sludge, 10^{-2}; sludge with 4 to 6% dry matter, 10^{-4}; and rich sludge with 30 to 60% dry matter, 10^{-4}. Prepare dilution plates as described in section 1.2.4.

Other methods for sampling waters have been described (35). Baiting entails placing sterilized "bait," e.g., hair or wool for keratinophilic fungi, hemp for phycomycetes, plant parts, seeds, termite wings, paper, or cellophane, into mesh baskets in the body of water and sampling periodically over several weeks. Other techniques include continuous-flow centrifugation and scrapings. These methods are generally small scale and may be impractical for some industrial screening purposes.

1.4.6 Isolation of Basidiomycetes from Sporocarps

Basidiomycetes may be isolated from fruiting or vegetative structures, resting stages, and substrates. Suitable media include potato dextrose agar, yeast malt agar, and basidiomycete selection medium (see section 1.5.4).

When isolating basidiomycetes from sporocarps, select the freshest specimens; old or excessively wet sporocarps tend to harbor more contaminants. Use a sturdy knife to sever the fruiting body from woody substrate or to dig below the surface of the soil so that the entire structure can be collected. Isolations may be made in the field, using tubes of medium, alcohol, tweezers, scalpel, and a good lighter.

Tissue isolation

To inoculate media from the sporocarp tissue (that will yield dikaryotic mycelia), brush debris from the surface of the fungus. The exterior surfaces may be wiped with alcohol or a weak bleach solution, but this is seldom necessary. Aseptically tear open the sporocarp and expose the interior. Remove a small piece of tramal tissue quickly with a sterile scalpel or tweezers. Take care not to touch the exterior surfaces that have been exposed to the environment. The tissue may originate in the pileus or stipe. Place the tissue on solid agar or, alternatively, float the block of tissue on a piece of sterile nylon mesh or filter paper in liquid medium (80).

Spore isolation

Basidiospores will yield monokaryotic mycelia. Remove the pileus from the stipe and carefully attach it to a petri dish lid by means of a small dab of petroleum jelly. Invert the lid so that the gills or spore surface is facing downwards, and place the lid on a crystallizing dish or deep petri dish. After the spores are deposited on the medium surface, replace the lid with a fresh, sterile one. The medium may be solid or liquid. If liquid medium is used, the spore suspension may be suitably diluted and spread onto plates or inoculated into cooled molten agar and poured into plates. This method should yield more discrete units of germinating spores, with fewer instances of mycelial fusion (Belt, unpublished data).

The same basic inoculating technique can be used with slant tubes of media. Attach the spore surface to the tube side opposite the agar surface with petroleum jelly or a small piece of water agar. Lay the tube horizontally. After the spores are deposited, remove the plug and spore surface.

1.4.7 Isolation of Basidiomycetes from Vegetative Structures

Vegetative structures may also be used as sources of inoculum. Collect rhizomorphs, sclerotia, or even mycorrhizal short roots aseptically in small paper bags. Surface sterilize with dilute bleach, hydrogen peroxide, or mercuric chloride (1:1). Thoroughly rinse several times with sterile distilled water. Aseptically chop, slice, or grind the sample and plate onto medium. Mycelia will slowly grow out of the tissue and can be further isolated.

When collecting wood infected with basidiomycetes, choose areas of profuse, active-growing infection. Surface sterilize with mercuric chloride (1:1) and implant onto agar surface (section 1.4.3).

1.4.8 Subculturing and Purification

Subculturing and purification procedures are usually a necessary prelude to screening fungal isolates. Initial observations may be made with a dissecting microscope; frequency will depend on the fungi being isolated. Some basidiomycetes can take 4 to 6 weeks to emerge from wood samples; yeasts may be apparent in 24 h. Using a dissecting needle (an insect pin attached to a long shank works quite well), make single spore isolations or hyphal tip excision. The propagules can then be removed from the isolation plate and inoculated onto media that will promote growth and the formation of fruiting structures for identification purposes.

1.5 ISOLATION MEDIA AND DILUENTS

1.5.1 Actinomycetes

Arginine glycerol salts medium (23)

Arginine monohydrochloride	1.0	g
Glycerol (specific gravity not less than 1.249 at 25°C)	12.50	g
K_2HPO_4	1.0	g
NaCl	1.0	g
$MgSO_4 \cdot 7H_2O$	0.5	g
$Fe_2(SO_4)_3 \cdot 6H_2O$	0.010	g
$CuSO_4 \cdot 5H_2O$	0.001	g
$ZnSO_4 \cdot 7H_2O$	0.001	g
$MnSO_4 \cdot H_2O$	0.001	g
Agar	15.0	g
Distilled water	1,000	ml

Adjust pH to 6.9–7.1. Autoclave for 15 min at 121 °C. Add sterile additions of cycloheximide (Calbiochem) and nystatin (Squibb) so that the final concentrations are between 50 and 75 μg of each per ml.

Note: Cycloheximide will go into solution in warm water (45°C). Nystatin is not totally soluble in water at pH 7.0. Increase pH to 11 with 1 M NaOH, filter sterilize, and drop the pH down to 7.0 immediately with HCl (nystatin is unstable at high pH).

AV agar (44)

L-Arginine	0.3	g
Glucose	1.0	g
Glycerol	1.0	g
K_2HPO_4	0.3	g
$MgSO_4 \cdot 7H_2O$	0.2	g
NaCl	0.3	g
Agar	15.0	g
Distilled water	1	liter

Autoclave for 15 min at 121°C. Add solutions listed below. Quantities given are final concentration per liter. Filter sterilize.

Vitamin solution

Thiamine hydrochloride	0.5	mg
Riboflavin	0.5	mg
Niacin	0.5	mg
Pyridoxine hydrochloride	0.5	mg
Inositol	0.5	mg
Calcium pantothenate	0.5	mg
para-Aminobenzoic acid	0.5	mg
Biotin	0.25	mg

Mineral solution plus antifungal antibiotics

$Fe_2(SO_4)_3$	10.0 mg
$CuSO_4 \cdot 5H_2O$	1.0 mg
$ZnSO_4 \cdot 7H_2O$	1.0 mg
$MnSO_4 \cdot 7H_2O$	1.0 mg
Nystatin	50.0 mg
Cycloheximide	50.0 mg

Benedict agar (6)

Glycerol	20.0 g
L-Arginine	2.5 g
NaCl	1.0 g
$CaCO_3$	0.1 g
$FeSO_4 \cdot 7H_2O$	0.1 g
$MgSO_4 \cdot 7H_2O$	0.1 g
Agar	20.0 g
Distilled water	1 liter

Autoclave for 15 min at 121°C and adjust pH to 7.0.

Colloidal chitin agar (31, 42)

Colloidal chitin	2.0	g (dry weight)
K_2HPO_4	0.7	g
KH_2PO_4	0.3	g
$MgSO_4 \cdot 5H_2O$	0.5	g
$FeSO_4 \cdot 7H_2O$	0.01	g
$ZnSO_4$	0.001	g
$MnCl_2$	0.001	g
Agar	20.0	g
Distilled water	to 1	liter

Adjust to pH 6.8 to 7.0 before autoclaving.

In flask A, mix colloidal chitin with 500 ml of distilled water. In flask B, mix remaining salts and agar. Autoclave separately for 15 min at 121°C. After autoclaving, swirl both flasks, slowly pour contents of flask A into flask B, and gently mix. Antifungal agents may be added once agar has cooled to 45°C.

Preparation of colloidal chitin

1. At 28°C, wash 100 g of unbleached chitin and shake overnight with 1 N HCl and 1 N NaOH alternatively (24 h), four times with each solvent. Use a Büchner funnel with no. 1 or 4 Whatman paper to dry the chitin between washings.
2. Wash the chitin in cold ethanol four times at 4°C. This washing may be carried out overnight.
3. Moisten the brownish granular material thoroughly with cold acetone in a large beaker (e.g., 4,000 ml).
4. Add cold concentrated HCl to the beaker slowly and carefully in the cold (wear a mask) while stirring with a mechanical stirrer. Continue this process overnight. Within 2 h after adding the HCl, a viscous, dark brown solution should result. If not, add more acid.
5. Prepare several 4-liter or larger side-arm flasks with about half the volume of ice-cold water (distilled).
6. Cover large Büchner funnels with three layers of glass wool and carefully filter the chitin through. This process is slow, and the glass wool must be changed often. This is best accomplished by using several Büchner funnels, each with glass wool. The larger the funnel diameter, the better.
7. When the chitin enters the water, it flocculates into a fluffy white mass. Change cold water flasks when they are full with the flocculent chitin.
8. After the filtration, the chitin chunks in the glass wool may be dissolved in more concentrated HCl and refiltered into cold water.
9. Remove the flocculent chitin by centrifuging at 9,000 × *g* or more for 15 min. Carefully decant the bottles to minimize loss of some chitin.
10. Wash chitin twice with water and centrifuge.
11. Place the chitin (still quite acidic) into large dialysis bags and dialyze, preferably with a continuous flow of tap water. Use a stirring bar to aid in solute/solvent exchange.
12. After continuous flow has been stopped for a few hours, the pH of the solvent is checked until it reaches a pH of 7.0. This may take several days.
13. Cut the bags and check the pH. If the pH is above 3.0, the chitin may be easily adjusted to pH 7.0 with a small volume of 1 N NaOH.
14. The dry weight of the chitin may be determined.
15. Sterilize the chitin with an ethylene oxide sterilizer, though autoclaving is adequate.

CYC (19)

*Czapek-Dox liquid medium powder (Oxoid)	33.4 g
Yeast extract	2.0 g
Vitamin-free Casamino Acids	6.0 g
Agar	16.0 g
Distilled water	1 liter

Final pH 7.2.

Autoclave for 20 min at 121°C. The antibiotics novobiocin (Albamycin; Upjohn) and cycloheximide (Calbiochem) are filter sterilized and added after agar cools to 45°C in final concentrations of 25 and 50 μg/ml, respectively.

*Czapek-Dox liquid medium powder ingredients (per liter)

Sodium nitrate	2.0 g
Potassium chloride	0.5 g
Magnesium glycerol phosphate	0.5 g
Ferrous sulfate	0.01 g
Potassium sulfate	0.35 g
Sucrose	30.0 g

Czapek plus yeast extract (M. P. Lechevalier, personal communication)

Yeast extract	4.0 g
Sucrose	15.0 g
$NaNO_3$	2.0 g
$FeSO_4 \cdot 7H_2O$	0.01 g
K_2HPO_4	0.5 g
KCl	0.5 mg
$MgSO_4$	0.5 g
Agar	15.0 g
Distilled water	1,000 ml

Mix ingredients and autoclave at 121°C for 20 min. Add cycloheximide and nystatin (50 μg/ml final concentration each; filter sterilized) once agar has cooled to 45°C.

Diagnostic Sensitivity Test Agar (Oxoid) (50, 51)

Proteose peptone	10.0	g
Veal infusion solids	10.0	g
Dextrose	2.0	g
Sodium chloride	3.0	g
Disodium phosphate	2.0	g
Sodium acetate	1.0	g
Adenine sulfate	0.01	g
Guanine hydrochloride	0.01	g
Uracil	0.01	g
Xanthine	0.01	g
Thiamin	0.00002	g
Agar no. 1	12.0	g
Distilled water	1	liter

Mix and autoclave for 15 min at 121°C.

Add as sterile additions at indicated final concentration per ml:

Cycloheximide	50 μg
Nystatin	50 μg
Chlortetracycline	45 μg
Methacycline	10 μg

GAC agar (45)

Bottom plate

Glucose	1.0	g
L-Asparagine	1.0	g
K_2HPO_4	0.3	g
$MgSO_4 \cdot 7H_2O$	0.3	g
NaCl	0.3	g
*Trace salts solution	1.0	g
†Antibiotic solution	2.0	ml
Agar	20.0	g
Distilled water	1	liter

Mix and autoclave all ingredients except trace salts solution and antibiotic solution, which are filter sterilized and added once agar is cooled to 45°C. Final pH is 7.4.

*Trace salts solution (per ml):

$FeSO_4$	10.0 mg
$MgSO_4 \cdot 7H_2O$	1.0 mg
$CuSO_4 \cdot H_2O$	1.0 mg
$ZnSO_4 \cdot 7H_2O$	1.0 mg

†Antibiotics solution (per ml):

Nystatin	50.0 mg
Cycloheximide	50.0 mg
Polymyxin B	4.0 mg
Penicillin	0.8 mg

Top plate

Casamino Acids	0.5	g
†Antibiotic solution	4.0	ml
Agar	20.0	g
Distilled water	1	liter

Mix and autoclave all ingredients except antibiotic solution. Pour bottom plate (15 ml) and let solidify. Pour 4 ml of top agar.

M_3 agar (62)

Distilled water	1	liter
KH_2PO_4	0.466	g
Na_2HPO_4	0.732	g
KNO_3	0.10	g
NaCl	0.29	g
$MgSO_4 \cdot 7H_2O$	0.10	g
$CaCO_3$	0.02	g
Sodium propionate	0.02	g
$FeSO_4 \cdot 7H_2O$	200	μg
$ZnSO_4 \cdot 7H_2O$	180	μg
$MnSO_4 \cdot 4H_2O$	20	μg
Agar	18.0	g

Autoclave for 15 min at 121°C. Final pH after autoclaving is 7.0. Once agar is cooled to 45°C, filter sterilize 4.0 mg of thiamine and 50.0 mg of cycloheximide and add.

MGA-SE agar (46)

Glucose	2.0	g
L-Asparagine	1.0	g
K_2HPO_4	0.5	g
$MgSO_4 \cdot 7H_2O$	0.5	g
Soil extract*	200	ml
Agar	15.0	g
Distilled water	1,000	ml

pH 8.0 before autoclaving

*Autoclave 1,000 g of soil with 1 liter of tap water for 30 min. Decant and filter.

Mix ingredients and autoclave at 121°C for 20 min. Filter sterilize the following antibiotic solution and add to agar once it has cooled to 45°C.

Antibiotic solution (final concentration per liter):

Penicillin	1 mg
Polymyxin B	5 mg
Cycloheximide	50 mg
Nystatin	50 mg

Starch-casein-nitrate agar (38)

Starch	10.0	g
Casein (Difco; vitamin free)	0.3	g
KNO_3	2.0	g
NaCl	2.0	g
K_2HPO_4	2.0	g
$MgSO_4 \cdot 7H_2O$	0.05	g
$CaCO_3$	0.02	g
$FeSO_4 \cdot 7H_2O$	0.01	g
Agar (Difco)	18.0	g
Distilled water	1	liter

pH is 7.0 to 7.2 before autoclaving at 121°C for 15 min. Filter-sterilized nystatin and cycloheximide are added at 75 μg/ml each once agar cools to 45°C.

Thin Pablum agar (40, 41)

Pablum	7.5	g
Agar	15.0	g
Tap water	1	liter

Boil in cheesecloth (tea bag effect) for 20 to 30 min. If necessary, adjust pH to 6.8–7.0. Autoclave for 20 min at 121°C. Add antifungal antibiotics.

Water agar (23; Lechevalier, personal communication)

Tap water	1	liter
*Crude agar flakes (USP Flake no. 1)	17.5	g

*Meer Corp., 9500 Railroad Ave., N. Bergen, NJ 07047.

Autoclave for 15 min at 121°C. Add antifungal antibiotics.

1.5.2 Algae

Autoclave all algal media for 20 min at 121°C at 15 lb/in^2, unless otherwise indicated. If solid medium is desired, add 10 g of Noble agar per liter.

Allen's medium (69)

For each 1,000 ml of medium required, add the following to 966 ml of glass-distilled water.

		(Stock solution: g/200 ml of water)
$NaNO_3$	1.5 g	
K_2HPO_4	5.0 ml	(1.5)
$MgSO_4 \cdot 7H_2O$	5.0 ml	(1.5)
Na_2CO_3	5.0 ml	(0.8)
$CaCl_2$	10.0 ml	(0.5)
$Na_2SiO_3 \cdot H_2O$	10.0 ml	(1.16)
Citric acid	1.0 ml	(1.2)
*PIV metals	1.0 ml	

Adjust pH to 7.8.

*PIV metal solution: to 1,000 ml of glass-distilled water, add 0.750 g of disodium EDTA. Dissolve and add:

$FeCl_2 \cdot 6H_2O$	0.097 g
$MnCl_2 \cdot 4H_2O$	0.041 g
$ZnCl_2$	0.005 g
$CoCl_2 \cdot 6H_2O$	0.002 g
Na_2MoO_4	0.004 g

ASN III (60)

NaCl	25.0 g
$MgSO_4 \cdot 7H_2O$	3.5 g
$MgCl_2 \cdot 6H_2O$	2.0 g
$CaCl_2 \cdot 2H_2O$	0.5 g
KCl	0.5 g
Citric acid	3.0 g
Ferric ammonium citrate	3.0 g
Magnesium EDTA	0.5 g
$NaNO_3$	0.75 g
$K_2HPO_4 \cdot 3H_2O$	0.75 g
Na_2CO_3	0.02 g
Glass-distilled water	1,000 ml

Sterile additions:

A-5 trace metal solution (see BG-11, below).	1.0 ml
Vitamin B_{12}	10.0 µg/liter

BG-11 (70)

$NaNO_3$	1.5 g
K_2HPO_4	0.04 g
$MgSO_4 \cdot 7H_2O$	0.075 g
$CaCl_2 \cdot 2H_2O$	0.036 g
Citric acid	0.006 g
Ferric ammonium citrate	0.006 g
EDTA (disodium salt)	0.001 g
Na_2CO_3	0.02 g
Distilled water	1,000 ml

Sterile addition:

*A-5 trace metal mix	1.0 ml

*A-5 trace metal mix

H_3BO_3	2.86 g
$MnCl_2 \cdot 4H_2O$	1.81 g
$ZnSO_4$	0.222 g
$NaMoO_4 \cdot 2H_2O$	0.39 g
$CuSO_4 \cdot 5H_2O$	0.079 g
$Co(NO_3)_2 \cdot 6H_2O$	49.4 mg
Distilled water	1,000 ml

CM basal medium (17)

$(NH_4)_2HPO_4$	1.0 g
KH_2PO_4	1.0 g
$MgSO_4$	0.2 g
$Na_3C_6H_5O_7 \cdot 11H_2O$	0.8 g
$CaCl_2$	0.02 g
$Fe_2(SO_4)_3 \cdot 6H_2O$	3.0 mg
$MnCl_2 \cdot 4H_2O$	1.8 mg
$Co(NO_3)_2 \cdot 6H_2O$	1.3 mg
$ZnSO_4 \cdot 7H_2O$	0.4 mg
H_2MoO_4	0.2 mg
$CuSO_4 \cdot 5H_2O$	0.02 mg
Distilled water	1,000 ml

Sterile additions (per liter):

Thiamine hydrochloride	0.01 mg
Vitamin B_{12}	0.0005 mg

Complex hot springs medium (12)

Hot spring or mineral water	1,000 ml
*Soil extract	50 ml
EDTA-Fe (13% Fe)	0.005 g
$MgSO_4 \cdot 7H_2O$	0.100 g
KNO_3	0.260 g
K_2HPO_4	0.100 g

*Soil extract:

Autoclave ca. 400 g of brown loam in 1,000 ml of water for 40 minutes. Filter the mixed slurry through triple-layered Whatman no. 1 paper while warm.

Erdshreiber medium (69)

Filtered seawater	1,000 ml
Soil extract (supernatant from soil water medium, see below)	50 ml
$NaNO_3$	0.2 g
$Na_2HPO_4 \cdot 12H_2O$	0.03 g
Vitamin B_{12}	1.0 ml (15 µg/100 ml)

Day 1: Filter seawater through no. 1 filter paper; heat water to 73°C.

Day 2: Again, heat seawater to 73°C. Autoclave separate salt and vitamin solution made in distilled water so that 1 ml of each solution gives required amount for 1 liter of culture solution.

Day 3: Add 1 ml of each cold salt solution to cold soil extract; then, add soil extract solution to cold seawater; add vitamin solution.

M2 medium (3)

KNO_3	0.02 g
K_2HPO_4	0.002 g
$MgSO_4 \cdot 7H_2O$	0.002 g

Agar	1.0	g
*Soil extract stock solution	10%	by volume
Water	100	ml

*The soil extract stock solution is made by heating a calcareous garden loam with twice its volume of supernatant water for 2 h in a steamer or by autoclaving for 15 min. It is convenient to make up and sterilize a number of small containers of stock solution, each of a size appropriate to making a batch of medium, as repeated autoclaving is deleterious.

M11 medium (3)

Stock solutions (quantities per liter)

Major salts

*Artificial seawater	38.2	g
Tris	0.5	g

*Instant Ocean, Aquarium Systems, Eastlake, Ohio.

Extra salts

$NaNO_3$	30.0	g
Na_2HPO_4	1.2	g
K_2HPO_4	1.0	g

Vitamins

Biotin	0.2	mg
Calcium pantothenate	20.0	mg
Cyanocobalamine	4.0	mg
Folic acid	0.4	mg
Inositol	1,000	mg
Nicotinic acid	20.0	mg
Thiamine	100.0	mg
Thymine	600.0	mg

Preparation. Add 50 ml of soil extract to 950 ml of major salts stock solution and 3.75 ml of extra salts stock solution. Then adjust the pH to 7.6–7.8 using 1 N HCl. Filter, dispense into appropriate containers, and autoclave. Add vitamin stock solution (0.25% by volume) after sterilizing through a 0.22-μm filter.

Marine BG-11

NaCl	10.0	g
$NaNO_3$	1.5	g
K_2HPO_4	0.04	g
$MgSO_4 \cdot 7H_2O$	0.075	g
$CaCl_2 \cdot 2H_2$	0.036	g
Citric acid	0.006	g
Ferric ammonium citrate	0.006	g
EDTA (disodium salt)	0.001	g
Na_2CO_3	0.02	g
A-5 trace metal mix	1.0	ml
Distilled water	1,000	ml

Add A-5 trace metal solution (see BG-11) sterilely. NaCl may be omitted, with artificial seawater substituted in the following proportions: reduce distilled water to 500 ml and replace with 500 ml of artificial seawater (Instant Ocean, Aquarium Systems, Eastlake, Ohio).

Medium A

1.0 M Trizma base, titrated to pH 8.0 with concentrated HCl	8.5	ml
Disodium EDTA	30	mg
$MgSO_4 \cdot 7H_2O$	5.0	g
NaCl	18.0	g
KCl	0.6	g
$CaCl_2 \cdot 2H_2O$	0.37	g
$NaNO_3$	1.0	g
KH_2PO_4	50	mg
Glass-distilled water	1,000	ml

Sterile additions:

*Trace metal mix	1.0	ml
Vitamin B_{12}	5.0	μg/liter

*Trace metal mix (per 100 ml)

$FeCl_3 \cdot 6H_2O$	389	mg
H_3BO_3	3,430	mg
$MnCl_2 \cdot 4H_2O$	430	mg
$ZnSO_4 \cdot 7H_2O$	63	mg
$Na_2MoO_4 \cdot 2H_2O$	390	mg
$CuSO_4 \cdot 5H_2O$	0.3	mg
$CoCl_2 \cdot 6H_2O$	1,220	mg

Medium C (36)

$MgSO_4 \cdot 7H_2O$	0.25	g
K_2HPO_4	1.0	g
$Ca(NO_3)_2 \cdot 4H_2O$	0.025	g
KNO_3	1.0	g
Sodium citrate $\cdot 2H_2O$	0.165	g
$Fe_2(SO_4)_3 \cdot 6H_2O$	0.004	g
Glass-distilled water	1,000	ml

Sterile addition:

A-5 trace metal mix (see BG-11)	1.0	ml

Medium D (12)

Glass-distilled water	1,000	ml
Nitrilotriacetic acid	0.1	g
$CaSO_4 \cdot 2H_2O$	0.06	g
$MgSO_4 \cdot 7H_2O$	0.10	g
NaCl	0.008	g
KNO_3	0.103	g
$NaNO_3$	0.689	g
Na_2HPO_4	0.111	g

Sterile additions:

$FeCl_3$ solution (0.2905 g/liter)	1.0	ml
Micronutrient solution	0.5	ml

Adjust pH to 7.5–7.7 after autoclaving.

Micronutrient solution:

Distilled water	1,000	ml
H_2SO_4 (concentrated)	0.5	ml
$MnSO_4 \cdot H_2O$	2.28	g
$ZnSO_4 \cdot 7H_2O$	0.50	g
H_3BO_3	0.50	g
$CuSO_4 \cdot 5H_2O$	0.025	g
$NaMoO_4 \cdot 2H_2O$	0.025	g
$CoCl_2 \cdot H_2O$	0.045	g

Prepare medium D as a 10-fold-concentrated stock and store at 4°C unautoclaved. After diluting (glass-distilled water), adjust the pH to 8.2 with NaOH. Final pH is 7.5 to 7.6 after autoclaving.

MN marine medium (60)

Distilled water	250	ml
*Artificial seawater	750	ml
$MgSO_4 \cdot 7H_2O$	0.04	g
$CaCl_2 \cdot 2H_2O$	0.02	g

$NaNO_3$ .... 0.75 g
$K_2HPO_4 \cdot 3H_2O$ .... 0.02 g
Citric acid .... 0.003 g
Ferric ammonium citrate .... 0.003 g
EDTA .... 0.0005 g
Na_2CO_3 .... 0.02 g

*Instant Ocean, Aquarium Systems, Eastlake, Ohio.

Sterile addition:
A-5 metal mix (see BG-11) .... 1.0 ml

After autoclaving, adjust pH to 8.5 with sterile KOH.

Natural water medium

Use collected waters full strength or in combinations with synthetic media. Autoclave waters or filter and store in the cold. Filtering is preferred.

Proteose medium (69)

Proteose peptone (Difco) .... 0.1 g
KNO_3 .... 0.02 g
K_2HPO_4 .... 0.002 g
$MgSO_4 \cdot 7H_2O$ .... 0.002 g
Water .... 100 ml

Since this is a richer medium, contaminants will quickly appear and outgrow the algae. However, it is still a useful medium for isolating more fastidious algae.

Soil water medium and soil tubes (69)

Success with soil water medium depends on selection of a suitable soil. A variety of soil water media can be made by using a basic formula with additional materials. The basic soil water medium is made as follows.

Place 0.25 to 0.5 in. of soil in the bottom of a test tube or bottle; cover with glass-distilled water until container is three-quarters full; cover container and steam (not autoclave) for 1 h on at least 2 consecutive days.

A few algae, such as *Spirogyra* sp., grow well in the basic medium. For most presumptively phototrophic algae which thrive in an alkaline medium, a pinch of powdered $CaCO_3$ is placed in the bottom of the container before the soil and water are added.

Soil seawater medium (69)

Make the soil water medium, except replace the glass-distilled water with the following solution:

*Natural or artificial seawater .... 1,000 ml
$MgCl_2$ .... 5.0 g
NaCl .... 23.0 g
Na_2SO_4 .... 4.0 g

*Instant Ocean, Aquarium Systems, Eastlake, Ohio.

1.5.3 Bacteria

All bacterial isolation agars should have nystatin and cycloheximide incorporated at 50 to 75 μg/ml each to retard fungal growth. See section 1.5.1 (actinomycete media) for instructions for preparing these antifungal agents.

Compost extract agar

Yeast extract .... 5.0 g
Glucose .... 8.0 g
K_2HPO .... 40.5 g
KH_2PO_4 .... 0.5 g
*Compost extract .... 400 ml
Distilled water .... 600 ml

Mix and autoclave for 20 min at 121°C. Adjust pH to match detritus or soil sample. Add filter-sterilized antibiotic solutions when agar has cooled to 45°C.

*Compost extract
Tap water .... 3,000 ml
Compost .... 1,000 g

Bring to boil and steep for 30 min. Decant supernatant. Cool. Filter through cheesecloth. Centrifuge at 3,000 × *g* for 10 min.

Cellulose agar

*Swollen cellulose .... 7.0 ml
Yeast extract .... 1.0 g
Agar .... 15.0 g
Water (distilled, sea, lake, river, etc., or combinations) .... 1,000 ml
K_2HPO_4 .... 0.7 g
KH_2PO_4 .... 0.3 g
$MgSO_4 \cdot 7H_2O$ .... 0.5 g
$FeSO_4 \cdot 7H_2O$ .... 0.01 g
$ZnSO_4$ .... 0.001 g

In flask A, mix swollen cellulose and salts with 500 ml of distilled water. In flask B, mix agar, yeast extract, and remaining water. Follow the same instructions for autoclaving and mixing as for colloidal chitin agar (section 1.5.1).

*Method for preparing swollen cellulose:

1. Slurry 30 g of air-dried cellulose powder in a little acetone and slowly add to 800 ml of 85% phosphoric acid while stirring vigorously with a mechanical stirrer for 2 h to prevent lumping.
2. Add 2 liters of distilled ice water with rapid stirring.
3. Collect the suspended material by centrifugation at 4,000 × *g*.
4. Suspend the pellet in 5 liters of distilled ice water; centrifuge again.
5. Bring product into solution with 1 liter of 2% $NaCO_3$.
6. Homogenize for 5 min at full speed in Waring blender and store for 12 h in refrigerator. *Do not freeze.*
7. Wash product in suction filter with 5 liters of distilled ice water; suspend in 15 liters of distilled water.
8. Pellet by centrifugation at 10,000 × *g* for 5 min and homogenize for 5 min. This should yield a suspension with a pH within 0.1 pH unit of distilled water (pH 6.5). Total yield is calculated based on dry weights of three 1-ml samples of the suspension (41).

Marine agar (Difco) 2216

Bacto-Peptone .... 5.0 g
Yeast extract .... 1.0 g

$FeCl_6H5O \cdot 3H_2O$ . . . 0.1 g
NaCl . . . 19.45 g
$MgCl_2 \cdot 6H_2O$ . . . 8.8 g
Na_2SO_4 . . . 3.24 g
$CaCl_2 \cdot 6H_2O$ . . . 1.8 g
KCl . . . 0.55 g
$NaHCO_3$ . . . 0.16 g
KBr . . . 0.08 g
$SrCl_2 \cdot 6H_2O$ . . . 0.034 g
H_3BO . . . 30.022 g
$Na_2SiO_3 \cdot 9H_2O$ . . . 0.004 g
NaF . . . 0.0024 g
NH_4NO_3 . . . 0.0016 g
$NaHPO_4 \cdot 7H_2O$ . . . 0.008 g
Agar . . . 15.0 g
Final pH 7.6.

Autoclave for 15 min at 121°C.

Moss extract agar

Distilled water . . . 700 ml
*Moss extract . . . 300 ml
Peptone . . . 5.0 g
Yeast extract . . . 5.0 g
Glucose . . . 3.5 g
†Trace mineral solution . . . 1.0 ml
Agar . . . 17.5 g

*Moss extract
Distilled water . . . 3,000 ml
Fresh moss . . . 1,000 g

†Trace mineral solution
$MgSO_4 \cdot 7H_2O$ . . . 0.25 g
$FeSO_4$ . . . 0.01 g
$CuSO_4 \cdot 5H_2O$ . . . 0.001 g
$ZnSO_4 \cdot 7H_2O$ . . . 0.001 g
$MnSO_4 \cdot 4H_2O$ . . . 0.001 g
Distilled water . . . 1 liter

Bring water to boil, and add moss. Steep for 20 min, and decant supernatant. Cool and filter through double-layered cheesecloth. Centrifuge at 3,000 × g for 10 min. Decant supernatant.

Multiple carbon-nitrogen source agar

Asparagine . . . 0.1 g
NH_4Cl . . . 0.1 g
Yeast extract . . . 4.0 g
Glucose . . . 5.0 g
Succinate . . . 0.2 g
K_2HPO_4 . . . 0.1 g
$MgSO_4 \cdot 7H_2O$ . . . 0.1 g
Agar . . . 18.0 g
Distilled water . . . 1,000 ml

Mix ingredients and autoclave for 20 min at 121°C. Final pH should be approximately 6.5.

Multiple extract agar

*Soil extract . . . 100 ml
†Plant extract . . . 100 ml
‡Root extract . . . 100 ml
§Artificial sea salts . . . 0.25 g
¶Vitamin solution . . . 1.0 ml
Peptone . . . 4.0 g
Glucose . . . 2.0 g
Agar . . . 15.0 g
Distilled water . . . 700 ml

Autoclave peptone, glucose, agar, and water for 15 min at 121°C. Add extracts. Final pH will depend on the pH of the extracts.

*Soil extract
Collect garden, farm, forest, marsh, swamp, etc., soil or mud, preferably representative of the ecosystem. Spread soil out to dry under fume hood for at least 6 h.

Steeping tap water . . . 960 ml
Dried soil . . . 400 g

Autoclave soil and water for 1 h at 121°C. Decant liquid and centrifuge to obtain a golden-colored supernatant. Use immediately or filter sterilize and store in the cold until needed.

†Plant extract
Collect plants from ecosystem and cut parts up with scissors.

Plant parts . . . 500 g
Distilled water . . . 1,000 ml

Bring water to boil. Simmer gently for 30 min. Decant liquid and filter through double-layered cheesecloth. Filter sterilize and store in the cold until needed.

‡Root extract
Collect roots from plants. Gently shake soil particles off and simmer roots for 15 min in hot water. Filter sterilize and store in the cold until needed.

§Artificial sea salts: Instant Ocean Aquarium Systems, Eastlake, Ohio
¶Vitamin solution: see ingredients listed for AV agar, section 1.5.1

Nutrient agar plus crystal violet*

Soil or other extract . . . 100 ml
Tap water . . . 900 ml
Beef extract . . . 5.0 g
Yeast extract . . . 5.0 g
Agar . . . 15.0 g

Mix ingredients and autoclave for 3 min at 121°C. Add 1.0 ml of filter-sterilized 0.1% solution of crystal violet.
*Can use Difco Nutrient Agar and add soil extract.

Plant extract agar

*Plant extract . . . 400 ml
Distilled water . . . 600 ml
Glucose . . . 8.5 g
Yeast extract . . . 4.5 g
†Crude agar flakes . . . 17.5 g

*See Multiple extract agar, above, for method to prepare plant extract.
†Meer Corp., 9500 Railway Ave., N. Bergen, NJ 07047.

Final pH should be around 6.6 to 6.8, depending upon the pH of the plant extract used. Mix ingredients and autoclave for 15 min at 121°C.

Root extract agar

*Root extract	375 ml
*Plant extract	25.0 ml
Asparagine	0.5 g
Arginine	0.5 g
Tryptophan	0.5 g
Starch	2.0 g
Yeast extract	1.0 g
*Artificial sea salts	0.5 g
Agar	16.0 g
Distilled water	1,000 ml

*See multiple extract agar, above, for methods to prepare root and plant extract and where to purchase artificial sea salts.

Soil infusion agar

Distilled water	600 ml
*Soil infusion	400 ml
Yeast extract	3.0 g
Glucose	7.5 g
K_2HPO_4	0.5 g
KH_2PO_4	0.5 g
$MgSO_4 \cdot 7H_2O$	0.5 g
Agar	16.0 g

Mix ingredients and autoclave for 15 min at 121°C. Final pH will vary depending upon ecosystem being examined.

*Soil infusion

Fresh Super Soil (any good loam potting soil)	500 g
Distilled water	1,000 ml

Mix soil and water. Gently heat without boiling for 15 min. Strain through double-layered cheesecloth and filter sterilize. Store in the cold until needed.

Violet-red bile agar (Difco) plus extracts

Difco violet-red bile agar	28.0 g
Agar	7.0 g
Natural extract	250 ml
Distilled water	750 ml

Final pH will vary depending upon extract and ecosystem being examined. Mix ingredients and autoclave for 15 min at 121°C.

1.5.4 Fungi

Unless otherwise indicated, autoclave all media for 15 min at 121°C.

Basidiomycete selection medium (63)

Malt extract	3.0 g
Peptone	0.5 g
Agar	2.5 g
ortho-Phenylphenol	0.006 g
Distilled water	100 ml

Lower pH to 3.5 after autoclaving.

Hay infusion agar (73)

Agar

*Infusion filtrate	1,000 ml
K_2HPO_4	2.0 g
Agar	15.0 g

Adjust pH to 6.2 and autoclave.

*Infusion filtrate

Distilled water	1,000 ml
Decomposing hay	50.0 g

Autoclave for 30 min at 121°C; filter.

Leaf litter extract agar

Leaf litter extract	400 ml
Tap water	600 ml
Yeast extract	2.5 g
Agar	17.5 g
Benlate (optional)	15 μg/ml

After autoclaving, addition of penicillin (0.5 g/liter) and streptomycin (0.025 g/liter) is optional.

PCNB agar (pentachloronitrobenzene)

Agar	20.0 g
Peptone	5.0 g
$MgSO_4 \cdot 7H_2O$	0.25 g
KH_2PO_4	0.50 g
Distilled water	1,000 ml

Autoclave. Cool. Add filter-sterilized:

Pentachloronitrobenzene	200 μg/ml
Penicillin G	50 μg/ml
Lactic acid	1.3 ml of 85% solution
Sodium deoxycholate	130 μg/ml

Potato carrot agar (80)

Potatoes (washed, peeled, grated)	20.0 g
Carrots (washed, peeled, grated)	20.0 g
Distilled water	1,000 ml
Agar	20.0 g

Bring carrots and potatoes to a boil in water. Simmer for 1 h. Add agar and distilled water to 1,000 ml. Autoclave.

Potato dextrose agar

Variation no. 1 (39)

Dehydrated potatoes (select without preservatives)	22 g
Distilled water	178 ml

Rehydrate potatoes by heating gently. Add:

Glucose	20.0 g
Agar	17.0 g
Distilled water	to 1,000 ml

Autoclave.

Variation no. 2 (9)

Scrubbed potatoes (peeled and diced)	200 g
Distilled water	1,000 ml

Bring to a boil and simmer for 1 h. Strain through a sieve. Add:

Agar	20.0 g
Water	to 1,000 ml
Glucose	15.0 g

Autoclave.

Rose bengal medium (14)

KH_2PO_4	1.0 g
$MgSO_4$	0.5 g
Soytone or Phytone	5.0 g
Glucose	10.0 g
Rose bengal	0.035 g
Agar	20.0 g
Distilled water	1,000 ml

After autoclaving, add 35 μg of filter-sterilized chlortetracycline per ml.

Soil infusion agar (modified)

Mix equal volumes of Super Soil or sandy loam with distilled water. Let heavy grains settle. Decant supernatant through four layers of cheesecloth.

Solution 1

Soil solution	1.25 liters
Glucose	12.5 g

Solution 2

Distilled water	1.25 liters
Agar	37.5 g

Mix solutions 1 and 2. Autoclave.

Tap water agar

Agar	15.0 g
Tap water	1,000 ml

Modifications: After autoclaving, pieces of sterile plant material, salts, etc., can be added.

Yeast malt agar

Yeast extract	3.0 g
Malt extract	3.0 g
Peptone	5.0 g
Glucose	10.0 g
Distilled water	1,000 ml
Agar	20.0 g

Autoclave.

1.5.5 Diluents

All diluents are filter sterilized and dispensed at 9 ml per tube.

Artificial seawater*

*Instant Ocean Aquarium Systems, Eastlake, Ohio. Prepare according to directions.

Distilled water plus artificial seawater

Distilled water	980.0 ml
Artificial seawater	20.0 ml

Adjust pH to suit sample pH.

Extract diluent

Natural extract	10 to 40%
Distilled water	60 to 90%

Can adjust pH to suit sample pH.

Phosphate buffer diluent

K_2HPO_4	5.62 g
KH_2PO_4	2.13 g
Distilled water	to 1,000 ml

Mix and autoclave for 15 min at 121°C. This will yield a 0.05 M solution at pH 7.0. Molarity and concentrations can be varied to obtain different pH values.

Saline

Distilled water	1,000 ml
NaCl	0.9 g

Ringer's solution plus gelatin

NaCl	8.6 g
KCl	0.3 g
CaCl	0.33 g
Gelatin	0.01 % (wt/vol)
Distilled water	1,000 ml

Mix until clear.

Mineral spring water

Filter sterilize and dispense. Many brands are available, and the composition varies in different areas. Make sure that the water has been collected from a natural spring and has not been distilled.

1.6 LITERATURE CITED

1. **Aaronson, S.** 1970. Experimental microbial ecology, Academic Press, Inc., New York.
2. **Al-Diwany, I. J., B. A. Unsworth, and T. Cross.** 1978. A comparison of membrane filters for counting thermoactinomyces endospores in spore suspensions and river water. J. Appl. Bacteriol. **45:**249–258.
3. **Asher, A., and D. F. Spadling.** 1982. List of strains, 1982. Natural Environment Research Council, Institute of Terrestrial Ecology, Culture Centre of Algae and Protozoa, Cambridge, England.
4. **Baker, D., J. G. Torrey, and G. H. Kidd.** 1979. Isolation by sucrose density fractionation and cultivation in vitro of actinomycetes from nitrogen fixing root nodules. Nature (London) **281:**76–78.
5. **Beech, F. W., and R. R. Davenport.** 1971. A survey of methods for the quantitative examination of the yeast flora of apple and grape leaves, p. 139–157. *In* T. F. Preece and C. H. Dickinson (ed.), Ecology of leaf surface microorganisms. Academic Press, Inc., London.
6. **Benedict, R. G., T. G. Pridham, L. A. Lindenfelser, H. H. Hall, and R. W. Jackson.** 1955. Further studies in the evaluation of carbohydrate utilization tests as aids in the differentiation of species of streptomycetes. Appl. Microbiol. **3:**1–6.
7. **Black, C. A. (ed.).** 1965. Methods of soil analysis, part 2. Chemical and microbiological properties. American Society of Agronomists, Inc., Madison, Wis.
8. **Blakeman, J. P.** 1981. Microbial ecology of the phylloplane. Academic Press, Inc., London.
9. **Booth, C.** 1971. Fungal culture media, p. 84. *In* C. Booth

(ed.), Methods in microbiology, vol. 4. Academic Press, Inc., New York.

10. **Brown, M. E.** 1975. Rhizosphere microorganisms, opportunists, bandits or benefactors, p. 21–38. *In* N. Walker (ed.), Soil microbiology. Butterworth, London.
11. **Buchanan, R. E., and G. C. Gibbons (ed.).** 1974. Bergey's manual of determinative bacteriology, 8th ed. Williams and Wilkins Co., Baltimore, Md.
12. **Castenholz, R. W.** 1969. Thermophilic blue-green algae and the thermal environment. Bacteriol. Rev. **33:**476–504.
13. **Collins, V. G., J. G. Jones, M. S. Hendrie, J. M. Shewan, D. D. Wynn-Williams, and M. E. Rhodes.** 1973. Sampling and estimation of bacterial populations in the aquatic environment, p. 77–110. *In* R. G. Board and D. W. Lovelock (ed.), Sampling—microbial monitoring of environments. Society of Applied Bacteriology Technical Series no. 7. Academic Press, Inc., London.
14. **Cooke, W. B.** 1954. Fungi in polluted water and sewage. II. Isolation technique. Sewage Ind. Waste **26:**661–674.
15. **Cooke, W. B.** 1963. A laboratory guide to fungi in polluted waters, sewage, treatment systems. U.S. Department of Health, Education and Welfare, Cincinnati.
16. **Cooke, W. B.** 1968. Carbon nitrogen relationships of fungus culture media. Mycopathol. Mycol. Appl. **34:** 305–316.
17. **Cramer, M., and J. Myers.** 1952. Growth and photosynthetic characteristics of *Euglena gracilis*. Arch. Mikrobiol. **17:**384–402.
18. **Cross, T.** 1982. Actinomycetes: a continuing source of new metabolites. Dev. Ind. Microbiol. **23:**1–18.
19. **Cross, T., and R. W. Atwell.** 1974. Recovery of viable thermoactinomycete endospores from deepmud cores, p. 11–20. *In* A. N. Barker and G. W. Gould (ed.), Spore research. Academic Press, London.
20. **Dickinson, C. H.** 1971. Cultural studies of leaf saprophytes, p. 129–137. *In* T. F. Preece and C. H. Dickinson (ed.), Ecology of leaf surface microorganisms. Academic Press, London.
21. **Dickinson, C. H., and T. F. Preece.** 1976. Microbiology of aerial plant surfaces. Academic Press, Inc., London.
22. **Dietz, A., and D. W. Thayer.** 1980. Actinomycete taxonomy. SIM Special Publication no. 6. Society for Industrial Microbiology, Arlington, Va.
23. **El-Nakeeb, M. A., and H. A. Lechevalier.** 1963. Selective isolation of aerobic actinomycetes. Appl. Microbiol. **11:**75–77.
24. **Favero, M. S., J. J. McDade, J. A. Robertson, R. K. Hoffman, and R. W. Edwards.** 1968. Microbiological sampling of surfaces. J. Appl. Bacteriol. **31:**336–343.
25. **Fogg, G. E., W. D. P. Stewart, P. Fay, and A. E. Walsby.** 1973. The blue-green algae. Academic Press, Inc., London.
26. **Gantt, E.** 1980. Handbook of phycological methods; developmental and cytological methods. Cambridge University Press, Cambridge.
27. **Gerdemann, J. W., and J. M. Trappe.** 1974. The endogonaceae of the Pacific Northwest. Mycol. Mem. **5:**1–76.
28. **Hagedorn, C.** 1976. Influence of soil acidity on *Streptomyces* populations inhabiting forest soils. Appl. Environ. Microbiol. **32:**368–375.
29. **Hellebust, J. A., and J. S. Craigie.** 1978. Handbook of phycological methods; physiological and biochemical methods. Cambridge University Press, Cambridge.
30. **Hirsch, P., M. Muller, and H. Schlesner.** 1977. New aquatic budding and prosthecate bacteria and their taxonomic position, p. 107–133. *In* F. A. Skinner and J. M. Shewan (ed.), Aquatic microbiology. Academic Press, Inc., London.
31. **Hsu, S. C., and J. L. Lockwood.** 1975. Powdered chitin as a selective medium for enumeration of actinomycetes in water and soil. Appl. Microbiol. **29:**422–426.
32. **Hungate, R. E.** 1962. Ecology of bacteria, p. 95–119. *In* I. C. Gunsalus and R. Y. Stanier (ed.), The bacteria, vol. IV. The physiology of growth. Academic Press, Inc., New York.
33. **Hunter, J. C., D. E. Eveleigh, and G. Casella.** 1981. Actinomycetes of a saltmarsh, p. 195–200. *In* K. P. Schaal and G. Pulverer (ed.), Actinomycetes. Zentralbl. Bakteriol. Parasitenkd. Infektionskr. Hyg. Suppl. 11. Gustav Fischer Verlag, Stuttgart.
34. **Hunter, J. C., M. Fonda, L. Sotos, B. Toso, and A. Belt.** 1984. Ecological approaches to isolation. Dev. Ind. Microbiol. **25:**247–266.
35. **Jones, E., and B. Gareth.** 1971. Aquatic fungi, p. 335–363. *In* J. P. Collins (ed.), Methods in microbiology, vol. 4. Academic Press, Inc., New York.
36. **Kratz, W. A., and J. Myers.** 1955. Nutrition and growth of several blue-green algae. Am. J. Bot. **42:**282–287.
37. **Krieg, N. R.** 1981. Enrichment and isolation, p. 112–142. *In* P. Gerhardt (ed.), Manual of methods for general bacteriology. American Society for Microbiology, Washington, D.C.
38. **Kuster, E., and S. T. Williams.** 1964. Selection of media for isolation of streptomycetes. Nature (London) **202:**928–929.
39. **Lacy, M. L., and G. H. Bridgmon.** 1962. Potato dextrose agar prepared from dehydrated mashed potatoes. Phytopathology **53:**173.
40. **Lechevalier, M. P., and H. A. Lechevalier.** 1957. *Waksmania* gen. nov., a new genus of the actinomycetales. J. Gen. Microbiol. **17:**104–111.
41. **Lechevalier, M. P., H. A. Lechevalier, and P. E. Holbert.** 1968. *Sporicthya*, un nouveau genre de streptomyceteae. Ann. Inst. Pasteur (Paris) **114:**227–286.
42. **Lingappa, Y., and J. L. Lockwood.** 1975. Chitin media for selective isolation and culture of actinomycetes. Phytopathology **52:**317–323.
43. **Newhouse, J. R., and B. B. Hunter.** 1983. Selective media for recovery of *Cylindrocladium* and *Fusarium* species from roots and stems of tree seedlings. Mycologia **75:**228–233.
44. **Nonomura, H., and Y. Ohara.** 1969. Distribution of actinomycetes in soil. VI. A culture method effective for both preferential isolation and enumeration of *Microbispora* and *Streptosporangium* strains in soil. Part 1. J. Ferment. Technol. **47:**463–469.
45. **Nonomura, H., and Y. Ohara.** 1971. Distribution of actinomycetes in soil. VIII. Green-spore group of *Microtetraspora*, its preferential isolation and taxonomic characteristics. J. Ferment. Technol. **49:**1–7.
46. **Nonomura, H., and Y. Ohara.** 1971. Distribution of actinomycetes in soil. IX. New species of the genera *Microbispora* and *Microtetraspora* and their isolation method. J. Ferment. Technol. **49:**887–894.
47. **Nonomura, H., and Y. Ohara.** 1971. Distribution of actinomycetes in soil. X. New genus and species of monosporic actinomycetes. J. Ferment. Technol. **49:**895–903.
48. **Nonomura, H., and Y. Ohara.** 1971. Distribution of actinomycetes in soil. XI. Some new species of the genus *Actinomadura*, Lechevalier, et al. J. Ferment Technol. **49:**904–912.
49. **Okami, Y., and T. Okazaki.** 1972. Studies in marine actinomycetes. I. Isolation from the Japan Sea. J. Antibiot. **25:**456–460.
50. **Orchard, V., and M. Goodfellow.** 1974. The selective isolation of *Nocardia* from soil using antibiotics. J. Gen. Microbiol. **85:**160–162.
51. **Orchard, V. A., M. Goodfellow, and S. T. Williams.** 1977. Selective isolation and occurrence of nocardiae in soil. Soil Biol. Biochem. **9:**233–238.
52. **Parberry, I. H., J. F. Brown, and V. J. Bofinger.** 1981. Statistical methods in the analysis of phylloplane populations, p. 47–65. *In* J. P. Blakeman (ed.), Ecology of the

phylloplane. Academic Press, Inc., London.
53. **Perfil'ev, B. V., and D. R. Gabe.** 1969. Capillary methods of investigating microorganisms. University of Toronto Press, Toronto, Canada.
54. **Pramer, D., and E. L. Schmidt.** 1964. Experimental soil microbiology. Burgess Publishing Co., Minneapolis.
55. **Preece, T. F., and C. H. Dickinson (ed.).** 1971. Ecology of leaf surface microorganisms. Academic Press, Inc., London.
56. **Prescott, G. W.** 1970. How to know the freshwater algae. William C. Brown Co., Dubuque, Iowa.
57. **Prescott, S. C.** 1946. Water bacteriology. John Wiley and Sons, Inc., New York.
58. **Pringsheim, E. G.** 1946. The biphasic or soil-water culture method for growing algae and flagellata. J. Ecol. **33:**193–204.
59. **Rand, M. C., A. E. Greenberg, M. J. Taras, and M. A. Franson.** 1976. Standard methods for the examination of water, 14th ed. American Public Health Association, Washington, D.C.
60. **Rippka, R., J. Deruelles, J. B. Waterbury, M. Herdman, and R. Y. Stanier.** 1979. Generic assignments, strain histories and properties of pure cultures of cyanobacteria. J. Gen. Microbiol. **111:**1–61.
61. **Rodina, A. G.** 1972. Methods in aquatic microbiology. Translated, edited, and revised by R. R. Colwell and M. S. Zambruski. University Park Press, Baltimore.
62. **Rowbotham, T. J., and T. Cross.** 1977. Ecology of *Rhodococcus coprophilus* and associated actinomycetes in fresh water and agricultural habitats. J. Gen. Microbiol. **100:**231–240.
63. **Russell, P.** 1956. A selective medium for the isolation of basidiomycetes. Nature (London) **177:**1038–1039.
64. **Slater, J. H., R. Whittenbury, and J. W. T. Wimpenny (ed.).** 1983. Microbes in their natural environment. Cambridge University Press, Cambridge.
65. **Smith, G. M.** 1950. The fresh-water algae of the United States. McGraw-Hill Book Co., New York.
66. **Smith, G. W., and H. D. Skipper.** 1979. Comparison of methods to extract spores of vesicular-abuscular mycorrhizal fungi. Soil Sci. Soc. Am. J. **43:**722–725.
67. **Smith, N. R., and V. T. Dawson.** 1944. The bacteriostatic action of rose bengal in media used for plate count of soil fungi. Soil Sci. **58:**467–471.
68. **Stanier, R. Y., R. Kunisawa, M. Mandel, and G. Cohen-Bazire.** 1971. Purification and properties of unicellular blue-green algae (order Chroococcales). Bacteriol. Rev. **35:**171–205.
69. **Starr, R. C.** 1978. Culture collection of algae at the University of Texas at Austin: list of strains. J. Phycol. **14:**1–100.
70. **Stein, J. R.** 1973. Handbook of phycological methods; cultural and growth measurements. Cambridge University Press, Cambridge.
71. **Stewart, W. D. P.** 1974. Algal physiology and biochemistry. Botanical Monographs, vol. 10. Blackwell Scientific Publications, Boston.
72. **Straka, R. P., and J. L. Stokes.** 1957. Rapid destruction of bacteria in commonly used diluents and its elimination. Appl. Microbiol. **5:**21–25.
73. **Thom, C., and K. B. Raper.** 1945. The manual of the aspergilli, p. 35. The Williams and Wilkins Co., Baltimore.
74. **Tresner, H. D., and J. A. Hayes.** 1970. Improved methodology for isolating soil microorganisms. Appl. Microbiol. **19:**186–187.
75. **Tsao, P., C. Leben, and G. W. Keitt.** 1960. An enrichment method for isolating actinomycetes that produce diffusible antifungal antibiotics. Phytopathology **50:**88–89.
76. **Tsao, P. H.** 1970. Selective media for isolation of pathogenic fungi. Annu. Rev. Phytopathol. **6:**157–185.
77. **Van Baalen, C.** 1962. Studies on marine blue-green algae. Bot. Mar. **4:**129–139.
78. **Veldkamp, H.** 1970. Enrichment cultures of prokaryotic organisms, p. 305–361. *In* J. R. Norris and D. W. Ribbons (ed.), Methods in microbiology, vol. 6. Academic Press, Inc., New York.
79. **Warcup, J. H.** 1950. The soil-plate method for isolation of fungi from soil. Nature (London)**166:**117–118.
80. **Watling, R.** 1982. How to identify mushrooms to genus. V. Cultural and developmental features. Mad River Press, Eureka, Calif.
81. **Wells, J. S., J. C. Hunter, G. L. Astle, J. C. Sherwood, C. M. Ricca, W. H. Trejo, D. P. Donner, and R. B. Sykes.** 1982. Distribution of β-lactam and β-lactone producing bacteria in nature. J. Antibiot. **35:**814–821.
82. **Williams, S. T., and E. M. H. Wellington.** 1982. Principles and problems of selective isolation of microbes, p. 9–26. *In* J. D. Bu Lock, L. J. Nisbet, and D. J. Winstanley (ed.), Bioactive microbial products: search and discovery. Society for General Microbiology. Academic Press, Inc., London.
83. **Willoughby, L. G.** 1971. Observations on some aquatic actinomycetes of streams and rivers. Freshwater Biol. **1:**23–27.

Chapter 2

Screening for New Products from Microorganisms

RICHARD J. WHITE, WILLIAM M. MAIESE, AND MICHAEL GREENSTEIN

2.1 INTRODUCTION

2.1.1 Microbes as a Source of New Products

Microbes have proved an exceptionally rich source of useful products, and there is every indication that they will continue to be so in the future. Such products vary enormously in terms of both structural complexity and biological activity. It is the purpose of this chapter to examine exactly how one establishes a screening program for new products endowed with particular biological properties. The vast majority of biochemical reactions required to synthesize a new microbial cell have been elucidated, and thus our knowledge of this primary metabolism is fairly complete. In addition to these essential reactions directly associated with the balanced growth of a cell, many microbes have the capacity to synthesize a wide variety of nonessential compounds, referred to as secondary metabolites. Although microbes have played an important role as a source of known primary metabolites, such as amino acids and vitamins, the quest for novel compounds has focused on the products of secondary metabolism. Two major factors governing the success of a screen for such compounds are (i) the producer microorganisms evaluated and (ii) the screening test employed. The emphasis of this chapter will be on the screening tests (detection systems) rather than the isolation of the organisms, which is described in an earlier chapter (chapter 1).

For more than 40 years, the primary emphasis of natural products screening has been the search for antibiotics to treat human disease. Although the most dramatic successes have been achieved in the field of antibacterial therapy with antibiotics, such as the penicillins, cephalosporins, and tetracyclines, significant achievements have also been made in the treatment of cancer and fungal infections. More recently, the trend has been to expand the horizons of natural products screening to compounds having pharmacological activities for other therapeutic uses. Examples of such activities are antihypertensive agents, hypocholesteremic agents, and immunomodulators (7).

Regardless of the type of activity sought, two key

qualities to consider in the choice of a screen are selectivity and sensitivity. The contents of a microbial cell and the medium on which it has been grown present an extraordinarily complex mixture of primary and secondary metabolites, occurring at concentrations ranging from nanograms to milligrams per milliliter; only one of these components may be of interest. Selectivity is necessary to allow detection of a single compound in the presence of many others, while the need for sensitivity stems from the fact that the compound of interest may be present at very low concentrations. It can be noted that this situation differs from the screening of synthetic chemicals, where individual compounds are usually tested at a concentration of choice in the absence of contaminating material.

2.1.2 Nature of Samples To Be Screened

The microorganisms to be evaluated for production of novel natural products can be grown and screened on solid or liquid medium. However, large-scale production of secondary metabolites is carried out by submerged cultivation in liquid medium. For this reason, coupled with the fact that the pattern of secondary metabolites produced on solid medium is frequently different, most groups have chosen to screen samples solely from liquid cultures. Thus, it is assumed that the input samples for the screening systems to be described will be of this type. More detailed discussions on the fermentation process are given later in this volume.

2.1.3 Devising Screening Tests

There are three fundamentally different types of screening tests, based on (i) whole animals, (ii) intact cells, and (iii) subcellular preparations. The most direct approach is to screen for activity in vivo using an animal model. However, the use of in vivo screens is severely restricted due to the limited number of samples that can be handled, excessive cost, and the relative insensitivity of the screen (17). For this reason, only the latter two categories will be considered here.

It is neither practical nor possible (for proprietary reasons) to describe all of the different types of screens that have been and are currently used. Instead, four specific screens, illustrative of the different approaches that can be employed in screening for novel products from microorganisms, will be detailed. It is assumed that the reader is familiar with the general principles of practical bacteriology as described in the *Manual of Methods for General Bacteriology* (5).

2.2 ANTIBACTERIAL SCREEN

2.2.1 Background

A simple and direct screening method commonly used for antibacterial activity is the agar diffusion assay. Over the years, antibiotic agar diffusion assays have played major roles in the search for new compounds, the monitoring of isolation and purification procedures, the determination of antimicrobial susceptibility patterns, and the quantitation of drugs. The conventional antibiotic agar diffusion assay is sensitive, reproducible, and simple to operate on a routine basis. Samples are dispensed on an agar plate inoculated with an indicator microorganism, and after incubation a clear circular zone of growth inhibition surrounding the sample is visible against a partially opaque background of growth. The approach employed for most indicator organisms is very similar, although the details of inoculum, medium, petri dish size, sample application, and zone of inhibition measurement can vary widely. In recent years, many different pieces of equipment have become commercially available to mechanize and automate the agar diffusion assay, thereby greatly increasing the number of samples that can be processed.

2.2.2 Assay Organism

Staphylococcus aureus ATCC 6538P (available from the American Type Culture Collection, 1230 Parklawn Dr., Rockville, MD 20852) serves as the screening organism. Maintain a stock culture by weekly transfers to fresh nutrient agar slants. Alternatively, 1.0-ml samples of an overnight broth culture of *S. aureus*, which can serve as a seed inoculum, can be maintained at −80°C. Aseptically transfer *S. aureus* ATCC 6538P from a nutrient agar slant into a flask or test tube containing nutrient broth or tryptic soy broth. Grow overnight at 37°C with aeration. Dilute the overnight culture in nutrient broth to yield an optical density at 660 nm (OD_{660}) of 0.3 using a spectrophotometer, e.g., Bausch & Lomb Spectronic 20. Use 1 ml of this diluted suspension per 100 ml of assay agar.

2.2.3 Agar Plate Assay

Autoclave the nutrient agar for 20 min; sterilization times will vary depending on the volume of medium per vessel. Allow the sterile molten agar to cool (45 to 50°C), inoculate the diluted *S. aureus* cultures to a final concentration of 1%, mix to obtain a homogeneous suspension, and pour into commercially available petri plates or Pyrex baking dishes. Allow the seeded agar plates to solidify; if they are not to be used immediately, place at 4°C. Inoculated *S. aureus* ATCC 6538P agar plates can be stored for at least 2 days at 4°C before use.

Samples can be applied to filter paper disks (0.25 to 0.5 in.; ca. 6.4 to 12.7 mm) which are placed on the seeded agar plate, in wells that have been cut into the agar, or in cylinders placed on the agar surface. Reference antibiotics such as penicillin, streptomycin, and tetracycline are also included on each assay plate. Incubate the plates overnight at 37°C.

2.2.4 Interpreting Results

The plates can be rapidly scanned for diameters of inhibition zones to determine which samples possess activity (Fig. 1). Examine plates after overnight incubation and record the diameters of clear zones surrounding samples and control antibiotics. Other types of zone may be present: hazy zones, double zones with clear centers and hazy edges, "speckled" zones caused by individual resistant colonies growing within clear

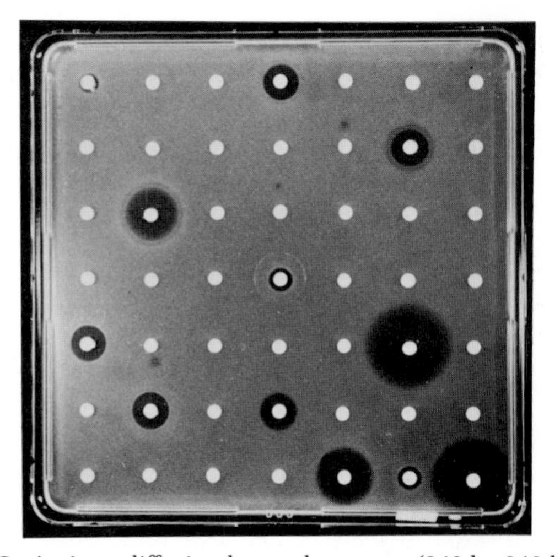

FIG. 1. Agar diffusion large plate assay (243 by 243 by 18 mm; Nunc) showing response of *S. aureus* to fermentation broth samples.

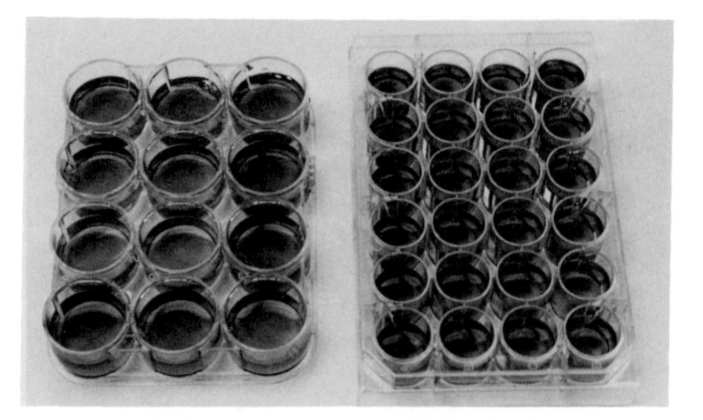

FIG. 2. Multiwell disposable plastic dishes inoculated with chick kidney cells.

areas, and stimulation of growth outside the zone of inhibition. Such qualitative differences may be due to the presence of a particular antibiotic or a mixture of active compounds.

2.3 ANTICOCCIDIAL SCREEN

2.3.1 Background

Coccidiosis is a parasitic disease of economic importance in domestic animals such as poultry, cattle, and sheep. In the United States, coccidiosis in poultry is such a concern that commercially available feeds almost always contain coccidiostats. Among the coccidia, *Eimeria* is probably the most examined and prevalent genus. Many species of *Eimeria* have been identified from domestic poultry in the United States; among the more common are *E. tenella*, *E. necatrix*, *E. acervulina*, *E. brunetti*, *E. maxima*, and *E. mitis*. Screening for anticoccidial compounds in vivo using chickens has been in progress for many years. A major drawback, however, is the requirement for large amounts of relatively pure compound. With the cultivation of *E. tenella* through part of its life cycle in cultured chick kidney tissue cells, an economical, sensitive screening procedure for anticoccidial agents became available (2, 12–14).

2.3.2 Preparation of Primary Chick Kidney Cell Culture

Aseptic techniques and sterile instruments must be used throughout this procedure. Aseptically remove the kidney lobes from 8-day-old chicks that have been sacrificed by cervical dislocation. Place the kidneys in a sterile petri dish containing chilled, sterile Dulbecco phosphate-buffered saline (PBS). Carefully decant PBS and mince kidneys with sharp sterile scissors. Transfer the finely minced pieces with a sterile spatula to a flask containing approximately 25 ml of PBS and a sterile magnetic stir bar. Agitate the minced kidneys slowly on a magnetic stirrer at room temperature for several minutes, and then decant the PBS. Add 20 ml of 0.25% trypsin to the flask and mix slowly for 10 min at room temperature. Allow the cells to settle, and decant the supernatant fluid into a sterile 50-ml centrifuge tube until small clumps of cells begin to appear (keep centrifuge tube on ice). Repeat trypsinization for 5 min. Decant supernatant fluid into another 50-ml centrifuge tube (keep on ice). Repeat the trypsinization procedure one more time for 5 min. Pour this supernatant fluid into a third centrifuge tube. All tubes should be kept in crushed ice until ready to be centrifuged. Pellet the cells by centrifugation at ca. 1,000 × g for 10 min. Decant, and add 20 ml of growth medium 199 to each centrifuge tube. Vortex gently until the cell pellets are resuspended. Place a cell suspension sample in a hemacytometer and determine the cell number. Add a sufficient volume of growth medium 199 to the cell suspension to yield a final concentration of 10^5 cells per ml. Dispense 1 ml into each well of a 24-well cluster plate (Costar 3524), or 2 ml each for a 12-well plate (Costar 3512), using a sterile automatic pipetting syringe, and gently rotate each plate to distribute the cells uniformly (Fig. 2). Incubate the plates in an atmosphere of 5% CO_2–95% air at 41°C. After 48 h of incubation, approximately 40 to 50% confluent cell growth should be observed. The cells are now ready to be inoculated with sporozoites and test samples.

2.3.3 Preparation of Oocysts

E. tenella oocysts can be obtained from Robert Kennett, Agricultural Research Division, American Cyanamid Co., Princeton, N.J. Oocysts are collected from the ceca of infected chickens (1, 14) as follows. Place the contents of cecal pouches that have been removed from sacrificed birds in 500 ml of distilled water, and homogenize for ca. 5 min. Add this homogenate to 2.5 liters of pepsin solution (0.5%, wt/vol, in distilled water) which has been adjusted to pH 2.0, and place in a water bath at 39°C for ca. 4 to 5 h. Remove from the water bath, dilute in a large volume of distilled water, adjust pH to 7.5, and let stand overnight. Decant the supernatant fluid, and wash the oocysts three times by repeated centrifugation in distilled water. Resuspend the oocysts in 1% potassium dichromate solution, and place on a shaker at

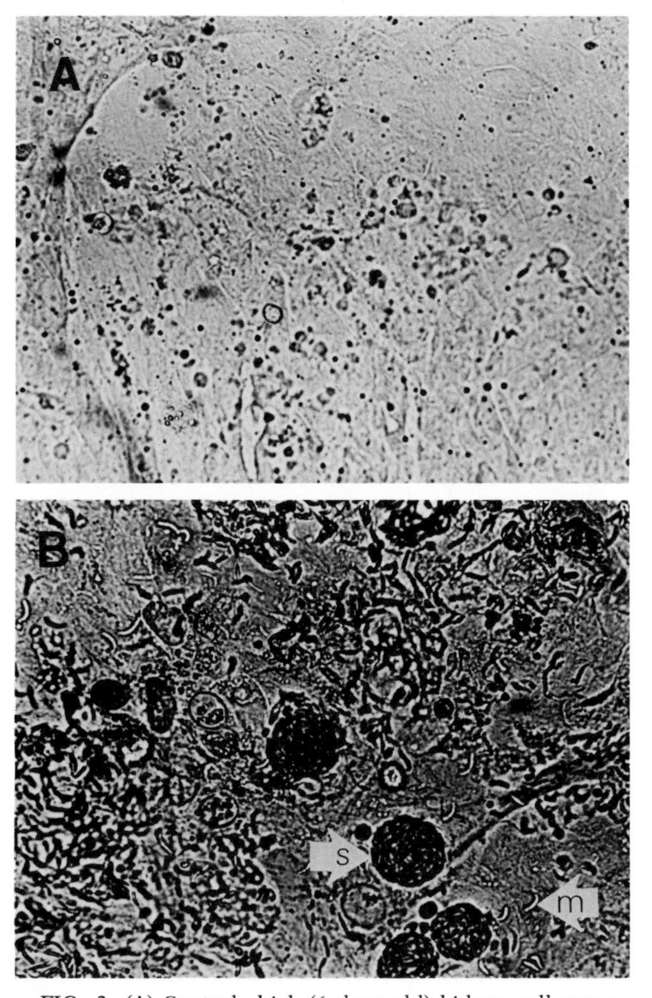

FIG. 3. (A) Control chick (6 days old) kidney cell monolayer. (B) *E. tenella*-infected chick (6 days old) kidney cell monolayer. s, Schizont; m, merozoites.

room temperature for 48 h to sporulate. Centrifuge the sporulated oocysts and wash the oocyst pellet at least four times, each with 500 ml of distilled water. Resuspend the pellet in 20 ml of household-strength Clorox (5%) and hold in an ice bath for 20 min. Add 80 ml of sterile distilled water and wash four times by centrifugation. Resuspend the pellet after the final wash in 100 ml of sterile PBS. These washed oocysts can be held at 4°C for at least 2 months.

2.3.4 Preparation of Sporozoites

Put 5 to 10 ml of sterile *E. tenella* sporulated oocyst stock in a sterile test tube (150 by 25 mm) containing ca. 50 mm of glass beads (2-mm diameter). Add sterile PBS to completely cover the glass beads. Mechanically rupture the oocysts on a rotary agitator (Roto-Torque; Cole-Parmer, Chicago, Ill.) for 20 to 25 min. Transfer the ruptured suspension into a sterile tube, and centrifuge for 10 min at 500 × *g*. Decant, and resuspend the pellet in 10 ml of a 5% solution of sodium deoxycholate and 0.25% trypsin prepared with cold blood Ringer's solution. Vortex to resuspend the pellet, and place in a 41°C water bath for 1.5 h. Centrifuge the freshly excysted sporozoites for 10 min at 500 × *g*, wash once with sterile PBS, and add 15 ml of sterile tissue culture medium 199. Vortex to resuspend the sporozoites. Count the sporozoite suspension with a hemacytometer and dilute with medium 199 to yield 5×10^4 sporozoites per ml.

2.3.5 Infection of Chick Kidney Cell Culture

Aseptically remove growth medium from chick kidney cells which have been incubated at 41°C for 48 h. Using a sterile automatic pipetting syringe, add 2.0 ml of the freshly prepared sporozoite-medium 199 suspension (5×10^4 sporozoites per ml) to each well. Leave a control well without sporozoites. Aseptically add 0.1 ml of fermentation broth sample, diluted at least 1:50 or 1:100 in sterile PBS, to each test well. Leave at least two wells on each plate to serve as positive and negative controls. If toxicity to the cultured cells is noted at 1:50 or 1:100 dilution, prepare additional twofold dilutions in PBS and add 0.1 ml of these. Add 0.1 ml of a monensin solution (Sigma M2152) to the positive control well (final concentration, 0.01 μg/ml) and 0.1 ml of PBS to the negative well. Reincubate the plates at 41°C for 4 days in an atmosphere of 5% CO_2–95% air.

2.3.6 Interpreting Results

After 4 days, examine the plates with an inverted microscope (approximately 200× magnification) for parasite development or drug cytotoxicity. Activity is determined by the presence or absence of second-asexual-generation schizonts and merozoites (Fig. 3) as compared to the infection present in negative control monolayers.

2.4 ANTICANCER SCREEN

2.4.1 Background

Of the various natural products demonstrating clinical utility as cancer chemotherapeutic agents, many function by initiating DNA damage within tumor cells. In recognition of this fact, detection systems have been developed specifically to find compounds with this mechanism of action. The λ prophage induction assay, developed for screening anticancer drugs (8, 10), is one such approach. In this system, DNA damage releases λ prophage from its latent, repressed state in an *Escherichia coli* lysogen. Development of complete virus particles can be detected in an overnight plaque assay. In a more expedient modification of this procedure, called the biochemical induction assay (BIA), a special λ lysogen has been genetically constructed such that the enzyme β-galactosidase, not virus particles, is formed when DNA damage occurs (4).

The BIA, which has been successfully employed in the screening of microbial culture broths for anticancer agents (3, 6, 16), is simple, sensitive, and rapid. The semiquantitative agar plate version of the BIA offers the versatility required for application to all stages of discovery and development of active compounds.

2.4.2 Assay Organism

E. coli BR513 (ATCC 33312) is the λ lysogenic indicator strain employed in the BIA. In this culture, the λ prophage carries the *E. coli lacZ* gene fused to a site adjacent to its leftward promoter under λ repressor control and is deficient in several genes required for phage development. Additional mutations affecting DNA repair functions (*uvrB*) and outer membrane integrity (*envA*) have been included to enhance assay sensitivity to DNA-damaging agents.

The BR513 culture is best stored as a frozen glycerol suspension at −80°C. Stocks of this type can be used for 6 months. Scrape the surface of a frozen suspension of strain BR513 with a sterile bacteriological loop, and inoculate 20 ml of LBE medium (see section 2.7.3) in a 250-ml Erlenmeyer flask. Incubate overnight at 37°C with mild shaking (250 rpm on a gyratory shaker). Aseptically add a sufficient volume of sterile 80% glycerol to the flask to yield a final glycerol concentration of 20%. Mix thoroughly, transfer 0.5 ml of suspension to sterile Nunc cryotubes (distributed by Vanguard International, Inc., Neptune, N.J.), and store at −80°C. After a new group of preservation tubes is prepared, test the response of the culture in the BIA to a set of bleomycin standard solutions (minimum inducing concentration, ≈0.3 to 0.5 μg/ml if a 10-μl sample is applied to the plate), using one of the new tubes as a source of inoculum for the assay.

2.4.3 Assay Procedure

When large numbers of samples are being evaluated, the BIA is most conveniently conducted with Nunc bioassay plates (243 by 243 by 18 mm, distributed by Vanguard International). At least 1 day before the assay, prepare agar base layers by aseptically dispensing 170 ml of LBE-amp agar into each plate on a level surface. After solidification of the agar, these plates may be stored at 4°C and used in the BIA over a period of 5 days.

Grow an overnight culture of strain BR513 as described above. Dilute this culture into fresh LBE broth to an OD_{600} of ca. 0.05. The volume of broth should be calculated assuming 35 ml per assay plate. Incubate this new culture at 37°C with mild shaking. When an OD_{600} of 0.4 has been reached (usually after about 2 to 2.5 h), immediately sediment the cells by centrifugation in the cold (4°C) at 3,000 × *g* for 10 to 15 min. Discard the supernatant, and gently resuspend the pellet (do not vigorously vortex) in 1/25 volume of cold (4°C), sterile distilled water. This cell suspension should be refrigerated immediately at 4°C and can be used over a 4-day period as the BIA inoculum.

For each assay plate, add 1.4 ml of BIA inoculum to 35 ml of melted soft agar that has been cooled to 45 to 47°C. Mix gently, and immediately pour the inoculated soft agar uniformly over a prewarmed (37°C) base layer. Place the plate on a level surface at room temperature, and allow the overlay to solidify for at least 10 min. Apply 20-μl samples of the microbial fermentation broths onto the surface of the agar. Incubate the plates at 37°C for 3 h.

At the end of this incubation period, apply a chromogenic substrate mixture to each plate to detect enzyme (prophage) induction, as follows. For each plate, dissolve 84 mg of fast blue RR and 14 mg of 6-bromo-2-naphthyl-β-D-galactopyranoside (both available from Sigma Chemical Co., St. Louis, Mo.) in 2 ml of dimethyl sulfoxide (gentle trituration with a pipette may be necessary). Add 35 ml of melted soft agar that has been cooled to 45 to 47°C, and pour the solution over the surface of the BIA plate. Let stand for 10 to 15 min at room temperature, and then observe the plate for color development.

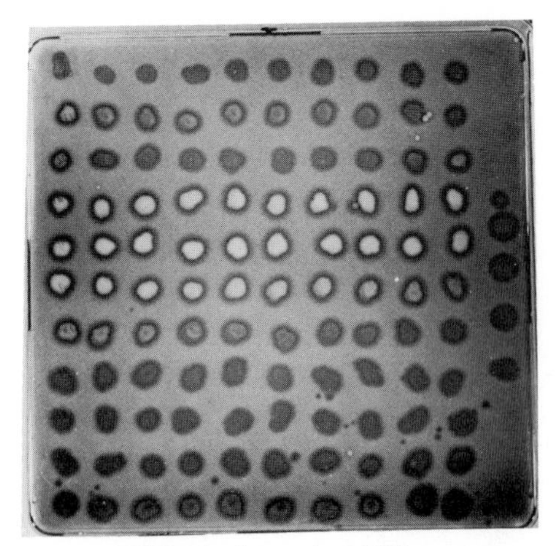

FIG. 4. Agar plate BIA.

2.4.4 Interpreting Results

The appearance of a red-violet zone of enzyme activity surrounding the site of sample application suggests the presence of a compound(s) initiating DNA damage (Fig. 4). An estimate of the relative amount of activity in different samples can be made on a basis of the intensity of color of the zone of induction and, to some extent, the zone size. At very high concentrations of active material, a clear toxic zone surrounded by a colored ring of induction is usually observed.

2.5 β-LACTAMASE INHIBITOR SCREEN

2.5.1 Background

The utility of penicillins and cephalosporins in controlling bacterial infections has been limited by the ability of many pathogens to produce enzymes that destroy their antibacterial activity. Termed β-lactamases, these enzymes are able to hydrolytically cleave the cyclic amide bond of the β-lactam ring, a common structural feature of these antibiotics. Two different approaches have been taken in an attempt to circumvent this limitation: (i) structural modification to make the antibiotic more resistant to enzymatic cleavage and (ii) use of β-lactamase inhibitors in combination with susceptible β-lactam antibiotics. A variety of inhibitors of synthetic, semisynthetic, and natural origin have been reported (18). The test described here allows one to screen microbial fermentation broths for inhibitors of TEM-2 β-lactamase, a clinically important enzyme produced by a variety of gram-negative bacteria and encoded by resistance plasmid RP1 (11). In common with other plasmid-mediated β-lactamases in gram-negative species, it is a cell-bound, constitutive enzyme.

2.5.2 Preparation of TEM-2 β-Lactamase

Grow *E. coli* carrying plasmid RP1 (available from the Culture Curator, Medical Research Division, Lederle Laboratories, Pearl River, NY 10965) to exponential phase with shaking at 37°C in 1% CY medium. Harvest the bacteria by centrifugation (5,000 × *g* for 10 min), wash once by resuspension in 0.1 M NaH_2PO_4–K_2HPO_4 buffer (pH 7.0), and centrifuge again (5,000 × *g* for 10 min). Resuspend the washed cell pellet in 1/20 volume of pH 7.0 buffer at 4°C and disrupt by sonication. Remove cellular debris and surviving intact cells from the sonicated cell suspension by centrifugation at 10,000 × *g* for 30 min at 4°C. This crude cell-free preparation of TEM-2 β-lactamase can be kept for limited periods at 4°C but can be stored for longer periods as a freeze-dried solid at 4°C.

2.5.3 Assay of β-Lactamase Activity

This assay makes use of a commercially available chromogenic cephalosporin, nitrocefin (BBL Microbiology Systems, Cockeysville, Md.), which changes from yellow to red on cleavage of the β-lactam ring (9). This color change can be conveniently monitored spectrophotometrically in the region of 486 nm. The assay described here is a miniaturized and automated version which can be carried out with standard equipment used for enzyme-linked immunoassays. Dispense duplicate samples (up to 10 μl per well) from microbial fermentation broths directly into 96-well, flat-bottomed microtiter plates (e.g., Costar). Add 100 μl of suitably diluted enzyme to one of each pair of samples and 100 μl of 0.05 M phosphate buffer (pH 7.0) to the other. The stock enzyme solution should be diluted such that a 30-min incubation with substrate at 37°C (see below) gives an OD_{490} of 0.5 to 0.7; typically this will require a greater than 10^4-fold dilution in buffer. Preincubate the microtiter plate at 37°C for 15 min before adding substrate. It is known that certain β-lactamase inhibitors, such as clavulanic acid, cause a progressive inhibition of enzyme activity and that incorporation of a preincubation step of enzyme with inhibitor considerably increases the sensitivity to such compounds. After preincubation, add 100 μl of a 100-μg/ml solution of nitrocefin in 0.05 M phosphate buffer (pH 7.0) to each well, using a multichannel pipette. Incubate the plate for a further 30 min at 37°C. Stop the enzyme reaction by adding 50 μl of 0.5 M citrate buffer (pH 3.0), and read the plates in an automated enzyme-linked immunoassay reader, e.g., microELISA Autoreader M580 (Dynatech Laboratories, Inc., Alexandria, Va.). Such instruments are furnished with a series of colored filters permitting measurement of OD at a series of fixed wavelengths. In this case, use a 490-nm filter. Although a manually operated immunoassay reader can be employed to read plates, a major advantage of working with the microtiter system is lost. The automated microELISA apparatus can measure an entire 96-well plate in less than 1 min, analyze the data, and print out the results.

2.5.4 Interpreting Results

The purpose of running a broth sample without enzyme (sample control) is to correct for the presence of interfering pigments or endogenous β-lactamase that may be present in the fermentation. In the absence of such a correction, OD values greater than that of the relevant enzyme control can be obtained. In addition to broth samples run with and without enzyme, an enzyme control (100 μl of buffer plus 100 μl of enzyme) should be routinely included on each plate. After incubation with substrate, this control will provide the 100% activity level.

Results for individual broths can be calculated as follows, using the readings obtained at 490 nm after the 30-min incubation: [(sample + enzyme + substrate) − (sample + buffer + substrate)]/(buffer + enzyme + substrate) × 100. The presence of a β-lactamase inhibitor in a fermentation broth is revealed by a result of <100% of the enzyme control after the sample has been corrected for interfering "pigment" and endogenous enzyme by subtraction of the value obtained for the sample plus buffer control. Bearing in mind the limitations of the assay, only samples giving ≤75% of control values are scored as positive, and these should be rerun at several concentrations to confirm the presence of inhibitory material.

2.6 CONCLUSION

The detection systems discussed in this chapter provide a mere sampling of the types of assay targets and procedures that can be employed in a natural products screening program. The principles presented, however, are generally applicable to the pursuit of a wide variety of biologically active agents. For example, the agar diffusion method, described in assays for antimicrobial and DNA-damaging antitumor compounds, could be employed just as readily for detection of inhibitors of an appropriate target enzyme, such as β-lactamase. Likewise, the BIA or assays for inhibition of microbial growth can be conducted in a liquid medium in microtiter plates in conjunction with automated plate-reading instrumentation. In vitro cell culture in cluster plates can be employed for tumor cells in assays for cytotoxic agents, inducers of cellular differentiation, etc. In fact, the possible screening approaches derivable from the few procedures presented in this chapter should be numerous. The only remaining ingredient is the imagination of the individual investigator.

2.7 MEDIA AND REAGENTS

2.7.1 Antibacterial Screen

Nutrient broth

Nutrient broth (Difco)	8.0	g
Distilled water	1,000	ml

Nutrient agar

Nutrient agar (Difco)	23.0	g
Distilled water	1,000	ml

Tryptic soy broth (soybean-casein digest medium)

Tryptic soy broth (Difco)	30.0	g
Distilled water	1,000	ml

Autoclave at 121°C for 20 min.

2.7.2 Anticoccidial Screen

Most of the media described in this section are available as sterile preparations from several commercial sources. Carry out all operations with sterile material and aseptic techniques.

Medium 199 (with 5% fetal bovine serum)

Medium 199 with Hanks salts (GIBCO no. 330-1181)
Fetal bovine serum (GIBCO no. 230-6140)
Penicillin-streptomycin solution (GIBCO no. 600-5140) Sodium bicarbonate (7.5%) (GIBCO no. 670-5080)

Add 60 ml of medium 199 (10× solution) to 530 ml of sterile, double-distilled water. Mix well. Remove 30 ml and add 30 ml of 5% fetal bovine serum. Mix well. Add 10 ml of 7.5% sodium bicarbonate and 6 ml of penicillin-streptomycin solution. Mix well.

Dulbecco PBS (10×)

Add 100 ml of 10× Dulbecco PBS (GIBCO no. 310-4200) to 900 ml of double-distilled water.

Trypsin

Trypsin (1:300) (GIBCO no. 840-7073)	2.5	g
Double-distilled water	900	ml
Dulbecco PBS (10×)	100	ml

Mix on magnetic stirrer for 1 h. Filter sterilize.

Freeze 20-ml aliquots for trypsinizing cycles.

Cold blood Ringer's solution

NaCl	6.5	g
KCl	0.14	g
CaCl	0.12	g
Distilled water	1,000	ml

2.7.3 Anticancer Screen

LBE medium

Tryptone (Difco)	10.0	g
Yeast extract	5.0	g
NaCl	10.0	g
1 M Tris	5.0	ml
Distilled water	1,000	ml

Autoclave at 121°C for 20 min. After cooling, supplement each 1,000 ml of medium with the following:

50× medium E (below)	4.0	ml
20% glucose (0.22-μm filter sterilized)	10.0	ml

LBE agar

Before autoclaving, supplement LBE medium with 15.0 g of agar (Difco) per 1,000 ml of medium. After autoclaving at 121°C for 20 min, cool medium to 50°C, add 50× medium E and 20% glucose, and then dispense into culture plates.

LBE-amp agar

Prepare as for LBE agar, except also supplement with 1 ml of a 10-mg/ml sodium ampicillin solution (0.22-μm filter sterilized) after autoclaving and cooling to 50°C.

50× medium E (15)

Distilled water	670	ml
$MgSO_4 \cdot 7H_2O$	10.0	g
Citric acid $\cdot H_2O$	100.0	g
K_2HPO_4 (anhydrous)	500.0	g
$Na(NH_4)HPO_4 \cdot 4H_2O$	175.0	g

Add the salts strictly in the order indicated. Each salt must be permitted to dissolve completely before the next one is added. After the last compound has dissolved, adjust the volume to 1,000 ml. Add 2.5 ml of chloroform as a preservative, and store at room temperature in a tightly capped bottle.

Soft agar

Agar (Difco)	10.0	g
Distilled water	1,000	ml

Soft agar preparation does not require asepsis. The agar-water mixture can simply be heated to 100°C to melt the agar and then cooled to 45 to 47°C for use in the assay.

2.7.4 β-Lactamase Inhibitor Screen

1% CY medium

Sodium β-glycerophosphate	0.12	M
$MgSO_4 \cdot 7H_2O$	1.0	mM
Yeast extract	1%	(wt/vol)
Acid-hydrolyzed casein	1%	(wt/vol)
Glucose	0.8%	(wt/vol)
Trace metal solution (below)	0.02	ml/liter

Add glucose and β-glycerophosphate from concentrated solutions after autoclaving at 121°C for 20 min.

Trace metal solution

$CuSO_4 \cdot 5H_2O$	0.5 %
$ZnSO_4 \cdot 7H_2O$	0.5 %
$FeSO_4 \cdot 7H_2O$	0.5 %
$MnCl_2 \cdot 4H_2O$	0.2 %
Concentrated HCl	10 %

Concentrations wt/vol.

2.8 LITERATURE CITED

1. **Davis, L. R.** 1973. Techniques, p. 411–458. *In* D. M. Hammond and P. L. Long (ed.), The coccidia. University Park Press, Baltimore.
2. **Doran, J.** 1970. *Eimeria tenella*: from sporozoite to oocysts in cell culture. Proc. Helminthol. Soc. Wash. **37:**84–92.
3. **Elespuru, R. K., and R. J. White.** 1983. Biochemical prophage induction assay: a rapid test for antitumor agents that interact with DNA. Cancer Res. **43:**2819–2830.
4. **Elespuru, R. K., and M. B. Yarmolinsky.** 1979. A color-

imetric assay of lysogenic induction designed for screening potential carcinogenic and carcinostatic agents. Environ. Mutagen. **1**:65–78.

5. **Gerhardt, P. (ed.).** 1981. Manual of methods for general bacteriology. American Society for Microbiology, Washington, D.C.
6. **Greenstein, M., W. M. Maiese, and R. J. White.** 1984. Prescreens for novel antineoplastic agents. Dev. Ind. Microbiol. **25**:267–275.
7. **Hamill, R. L.** 1982. Screens for pharmacologically active fermentation products, p. 71–105. *In* J. D. Bu'Lock, L. J. Nisbet, and D. J. Winstanely (ed.), Bioactive microbial products: search and discovery. Academic Press, Inc., New York.
   This book in itself is a useful source of screening information.
8. **Heinemann, B., and A. J. Howard.** 1964. Induction of lambda bacteriophage in *Escherichia coli* as a screening test for potential antitumor agents. Appl. Microbiol. **12**:234–239.
9. **O'Callaghan, C. H., A. Morris, S. M. Kirby, and A. H. Shingler.** 1972. Novel method for detection of β-lactamases by using a chromogenic cephalosporin substrate. Antimicrob. Agents Chemother. **1**:283–288.
10. **Price, K. E., R. E. Buck, and J. Lein.** 1964. System for detecting inducers of lysogenic *Escherichia coli* W1709 (λ) and its applicability as a screen for antineoplastic antibiotics. Appl. Microbiol. **12**:428–435.
11. **Richmond, M. H.** 1975. β-Lactamase (*Escherichia coli* $R^{+}{}_{TEM}$). Methods Enzymol. **43**:672–677.
12. **Ryley, J. F.** 1980. Recent developments in coccidian biology: where do we go from here? Parasitology **80**:189–209.
13. **Ryley, J. F., and R. G. Wilson.** 1976. Drug screening in cell culture for the detection of anticoccidial activity. Parasitology **73**:137–148.
14. **Strout, R. G., and C. A. Ouellete.** 1973. *Eimeria tenella*: screening of chemotherapeutic compounds in cell cultures. Exp. Parasitol. **50**:477–485.
15. **Vogel, H. J., and D. M. Bonner.** 1956. Acetylornithinase of *Escherichia coli*: partial purification and some properties. J. Biol. Chem. **218**:97–106.
16. **Wei, T. T., J. A. Chan, P. P. Roller, U. Weiss, R. M. Stroshane, R. J. White, and K. M. Byrne.** 1982. Detection of gilvocarcin antitumor complex by a biochemical induction assay (BIA). J. Antibiot. **35**:529–532.
17. **White, R. J.** 1982. Microbiological models as screening tools for anticancer agents: potentials and limitations. Annu. Rev. Microbiol. **36**:415–433.
18. **Wise, R.** 1982. β-Lactamase inhibitors. J. Antimicrob. Chemother. **9**:32–40.

Chapter 3

Inoculum Development

G. R. HUNT and R. W. STIEBER

The preparation of a population of microorganisms from a dormant stock culture to a state useful for inoculating a final productive stage is called inoculum development. Preparation may range in scale and purpose from a small inoculum for a bioassay to 1,000 liters for production of a vitamin or antibiotic in a 40,000-gal fermentor. The objective of this chapter is to describe methods for inoculum development. These methods aim to (i) minimize the loss of viable microorganisms during recovery from dormancy, (ii) obtain a genotypically identical copy of the population that was stored, (iii) increase biomass, and (iv) develop the culture to a physiological state suitable for performance in the final production stage.

Microbial populations can be obtained from the environment directly or by subculturing (1). To repeat a microbiological task routinely, the population should be developed from a stored stock of the culture. These populations derived from the original stock can have greatly increased capabilities through changes brought about in their physiological makeup.

To obtain a useful inoculum requires the following: (i) a method of transferring the stored culture to the reactivation medium which maximizes viability; (ii) a workplace to do the transfer which minimizes the chance for bulk airflow to carry contaminants into the medium; (iii) a method to quantify the biomass; (iv) determination of a growth curve which indicates a series of physiological states through which the biomass will successively pass; and (v) selection of a growth vessel and environmental condition(s) which allow the expression of these states.

The desired physiological state of inocula varies with each application. For example, actinomycetes and fungi can grow in a pellet form. These pellets begin from a spore center or as agglomerates of mycelia which gradually overlay one another until a clump several hundred micrometers in diameter has been formed (25). A low shear environment, low inoculum levels, or both may enhance the formation of pellets (38). Pellet formation may be greatly magnified as growth of these cultures proceeds to the large biomasses used in industrial processes (24, 40). In this case, therefore, the development of a "useful" inoculum state requires not only recovery within a specified time frame of a biomass that is adequate in terms of quantity, but also an assay to correlate performance in subsequent steps with the pelleting during reactivation. Thus, obtaining information about the factors which cause organisms to take different morphological forms is one area where further research could prove fruitful.

3.1 PRACTICAL IMPLEMENTATION OF INOCULUM DEVELOPMENT

3.1.1 Objectives

A clear understanding of the objectives for inoculum development is not always possible, particularly where the system itself is not well understood. Historically, food processing has been an industry in which inocula are necessary. Yeast cultures for the baking and brewing industries and mold cultures for production of organic acids (17) and foods provided some of the techniques upon which the antibiotic process developments of the 1950s and 1960s depended (3, 35). The application of these early techniques to yeast and penicillin production served as a model for a great deal of subsequent fermentation development. The need for large quantities of pure cultures to fill multi-thousand-liter fermentors imposed new and often difficult limitations on the culture of microorganisms. It was soon apparent that the objectives of inoculum development were rapid growth rates and high biomass concentrations at the outset of the fermentation. These factors were correlated, for example, with the amount of penicillin eventually produced (22).

Another important objective was that of pure culture. This concept applies not only to different species but to variants within a "pure culture." S. J. Pirt, in the introduction to *Principles of Microbe and Cell Cultivation*, noted the historical development of techniques used for culture development. Particularly notable is Pirt's evaluation of Pasteur's work, which was performed without the benefit of pure culture techniques. "Perhaps the reproducibility of Pasteur's results depended on the selectivity of his culture media to ensure sufficient constancy of the microbial population " (33).

3.1.2 Reactivation

Large quantities of microorganisms are routinely grown from very small amounts of stored culture. Depending on the type of storage, as discussed below, the culture may be placed initially on solid or in liquid medium which is selected empirically for its ability to allow the recovery of the majority of the cells in the stored population. On solid media, growth occurs in several weeks for fungal cultures and in 1 to 1.5 days for bacterial cultures (13); in liquid media, fungal cultures may take several days to grow, whereas bacteria can begin to grow overnight.

To start liquid cultures, the stock is added directly to liquid medium in either test tubes or flasks. This procedure effectively enhances the potential for growth from a culture source of low viability. Larger initial volumes of medium can be used if the stock is more viable (20).

3.1.3 Multistaging

A typical multistaging procedure generates a 5% (vol/vol) inoculum for the production stage. When additional stages are needed, flasks are usually used because of their size, convenience, and ease of maintaining sterility. Several stages and flasks of increasing size are sometimes used to maintain a high-percentage inoculum at each successive step (14).

Typically, flasks ranging from 50 ml to 12 liters in size are filled with medium (10 to 20% of the flask volume) and sterilized in an autoclave at 121°C for 15 to 45 min (36). The culture is added to the flask in a controlled environment to prevent contamination. For manufacturing purposes, the scale of these operations demands reproducible methods for inoculation and transfer. Special rooms are designed to provide for these transfers; other rooms are provided with environmental controls for temperature, gas atmosphere, humidity, and light intensity to incubate the cultures.

Oxygen, an important nutrient which readily becomes limiting, is usually supplied to flasks through vigorous shaking on a rotary or reciprocating shaker (2). Constant-temperature rooms operating at different temperatures with shakers operating at a range of speeds are generally used.

Design of facilities for multistage inoculum development must include protection against contamination by extraneous organisms and viruses. While transfer from one flask stage to another can be accomplished without mishap, contamination of only 1 in 100 to 1 in 1,000 would have severe consequences if the growth rates of the organisms were different. To minimize this possibility, inoculation and transfer areas are isolated and provided with filtered air. Each stage, whether flask or fermentor, must be kept free of contaminants. Although small numbers of microorganisms other than the desired one can sometimes be tolerated at the production stage, contamination in early inoculum steps can be overwhelming.

Exclusion measures should be implemented to prevent contaminated materials from entering the inoculum area. All media, flasks, and utensils must enter through a sterilizer. Personnel must put on gowns before entry. Flasks are inoculated in a laminar-flow hood within the area and moved to a connected, environmentally controlled, shaker room for incubation; they are passed out of this latter area only once growth is completed.

Depending on the scale of operation, in addition to the flasks, one or more sparged tanks may be necessary to achieve the required volume of culture. For example, 15,000 liters is required if a 10% inoculum is to be used for a 150,000-liter fermentation. To prepare this volume of inoculum, as many as four stages may be necessary. The initial stages are usually done in flasks. After a desired level of growth has been achieved on the flask level, the contents of the final flask are transferred into the first tank stage. The flask is removed from the shaker and passed through an airlock to the area where it will be used. Alternatively, the flask can be connected to a presterilized vessel in the inoculum area, from which the grown culture can be blown by air or pumped to a larger stirred vessel for growth (4, 32). The following example serves to illustrate important considerations in multistaging.

Frozen vegetative mycelia are rapidly thawed at 37°C. A 1-ml sample is put into a 300-ml Erlenmeyer flask containing 40 ml of a yeast extract medium at pH 6.8 to 7.0. The flask is incubated at 28°C for 24 h on a shaker at 220 rpm with a 2-in. (ca. 5-cm) throw. At the end of 24 h, 10 ml of this cell suspension is transferred to a 2-liter Florence flask containing 400 ml of the same medium. The 2-liter flask is incubated for 24 h at 150 rpm and a 3-in. (ca. 7.5-cm) throw. This grown culture is transferred aseptically into 380 liters of the same medium in a stirred sparged vessel. The airflow is 0.5 vol/vol per min, and the temperature is 27.5°C. After approximately 40 h, this culture is near the end of exponential growth and is transferred into a 10,000-liter vessel containing the same medium at pH 6.7 to 6.9.

Development is monitored by changes in pH, packed cell mass (after centrifugation), and oxygen uptake rate. Transfer to the production vessel is done near the end of exponential growth (Fig. 1).

The sparged tanks are interconnected with lines which can be kept sterile under steam at all times, maintaining temperatures greater than 120°C. When inoculum is to be transferred, the lines are opened into the receiving vessel with steam still coming from a valve at the "sending tank." Immediately before transfer, the steam is shut off and some inoculum is allowed into the line to cool it. The remainder after cooling is rapidly transferred into the "receiving tank."

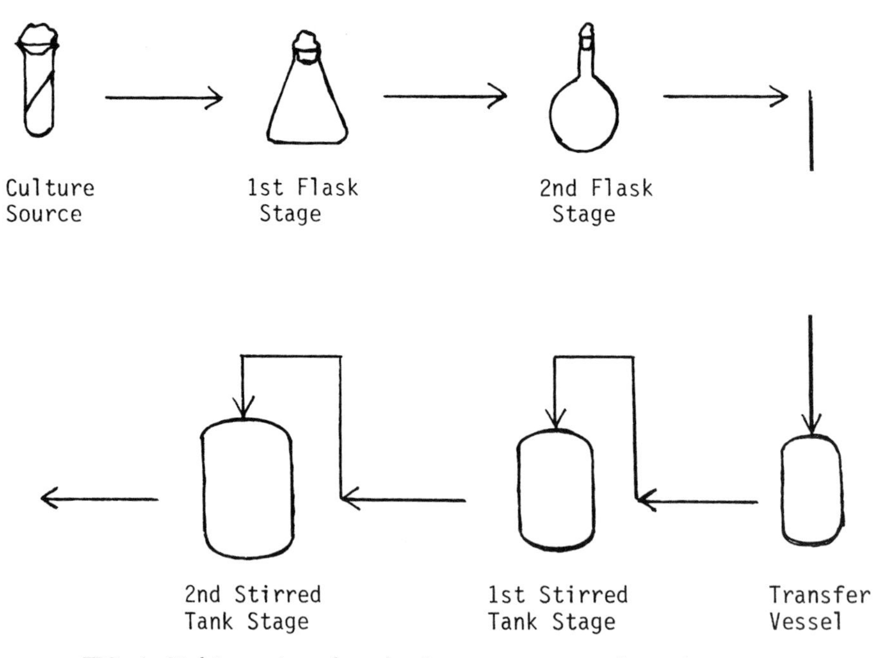

FIG. 1. Multistage inoculum development sequence for cephamycin C.

3.1.4 Media

The development of media for growing inoculum incorporates parameters such as nutrient shiftup, shiftdown, balance, total charge, and type of charge. Inoculum media must also provide for interaction with the production medium. This interaction is crucial in inoculum development. A general rule of thumb is that transfer should be made between identical or similar media (20). This rule establishes growth in the production stage as an extension of growth in the prior seed stage. In batch culture, this may be difficult if high nutrient concentrations are needed for extended biosynthesis of product. When part of the nutrients can be fed during the course of the fermentation, it becomes a simpler task to adjust the production medium composition at inoculation.

Media are broadly divided physically into liquid or solid forms. Solid media result from incorporation of a polysaccharide such as agar into the medium, usually 1.5 to 2% by weight. The medium is heated until dissolved and allowed to cool for solidification. The agar melts and dissolves at boiling temperature and then solidifies near 40°C.

The surface of a solid medium provides a convenient site for temporarily storing aerobic microorganisms as vegetative cells or as spores (after appropriate incubation to generate spores). Culture densities per square millimeter can be as high as 10^8 or 10^9 spores. Although initial culture isolation and culture storage both employ solid media, submerged or liquid culture is generally used for subsequent inoculum growth. The ease of transfer (reducing the potential for contamination), the ability to make simple growth measurements, and the requirement for large volumes of microbial culture to be used as inocula are major reasons to use submerged culture for inoculum development (13).

Media can be divided into chemically defined or complex, depending upon whether or not the exact chemical composition is known (see section 3.5). Defined media form the basis for experimental determination of growth factors. Semidefined, soluble media are routinely used for preparing initial stocks of culture and for reactivating stored culture (spores or cells) (20, 29); these generally contain a protein hydrolysate rather than pure amino acids.

An elemental material balance is used to determine whether sufficient nutrients exist for the amount of biomass desired. For defined media, these calculations are straightforward; quantities of carbon, nitrogen, and phosphorus are directly calculated from the components. For complex media, the calculations are less definitive but nonetheless useful; these quantities must be estimated from the sums of the elements. For example, total carbon is the sum of the carbon from glucose as well as that available from the complex sources, i.e., proteins, lipids, and polymers (such as dextran and starch). An elemental balance on a complex cephamycin inoculum medium is presented in Table 1. Analytical information on complex components can usually be obtained from the manufacturer (43). These analyses primarily are average values rather than assays on a specific lot. However, some manufacturers will supply specific analyses on each lot at additional cost.

Yield coefficients for elements including oxygen are estimated by varying the concentration of the components containing them in batch or continuous culture (2). Although coefficients for elements in complex media are more qualitative than quantitative, they are valuable aids in estimating cell mass. The oxygen yield coefficient (Table 1) can be used to estimate the potential oxygen demand from biomass. Using this estimated oxygen use rate in conjunction with tank oxygen transfer data from historical sources or experimental determination, investigators can prevent pre-

TABLE 1. Cephamycin C fermentation elemental material balance for inoculum medium (43)[a]

Purpose	Raw material	Fraction of element in raw material	Concn of raw material (g/liter)	Amt of element charged (g/liter)	Fraction of element in cell mass[b]	Potential cell mass (g/liter)	Maximum cell mass[c] (g/liter)
Energy source	Glucose	NA	10.0	NA	0.5	20	20
Carbon source	Yeast extract	0.1660	10.0	1.66	0.5	3.3	15.1
	Primary dried yeast	0.1920	10.0	1.92		3.8	
	Glucose	0.4000	10.0	4.00		8.0	
Nitrogen source	Yeast extract	0.0780	10.0	0.78	0.14	5.6	11.1
	Primary dried yeast	0.0770	10.0	0.77		5.5	
Phosphorus source	Yeast extract	0.0959	10.0	0.96	0.03	32.0	39.8
	Primary dried yeast	0.0127	10.0	0.127		4.0	
	KH_2PO_4	0.2276	0.5	0.114		3.8	
Iron source	Yeast extract	0.0003	10.0	0.0028	0.002	1.4	12.1
	Primary dried yeast	0.0001	10.0	0.0012		0.6	
	$FeSO_4 \cdot 7H_2O$	0.2009	0.1	0.0201		10.1	
Potassium source	Yeast extract	0.0004	10.0	0.0041	0.03	0.1	12.1
	Primary dried yeast	0.0224	10.0	0.215		7.2	
	KH_2PO_4	0.2873	0.5	0.144		4.8	
Magnesium source	$MgSO_4 \cdot 7H_2O$	0.0990	0.25	0.025	0.003	8.3	15.0
	Yeast extract	0.0003	10.0	0.003		1.0	
	Primary dried yeast	0.0018	10.0	0.017		5.7	

[a] Oxygen source is calculated as: Air − Y_{O_2} (moles of cell carbon/moles of O_2 absorbed) = 0.85; maximum oxygen uptake (OUR_{max}) = uX/Y_{O_2} for 8 g of cells per liter per liters = 31 mmol/liter h.

[b] Calculated as (moles of ATP per liter) per 10 g of glucose per liter.

[c] Y_{ATP} = 10 g of cells per mol of ATP.

mature cessation of growth due to oxygen limitation. An elemental balance table is a convenient method for estimating any potentially limiting nutrient. Particular uses for these biomass estimates are to ensure reproducibility in biomass from lot to lot of ingredient and to provide a rationale for substituting ingredients or even changing suppliers.

3.1.5 Sterilization

Sterilization techniques have long been recognized to affect the quality of media. The Maillard (browning) reactions have been studied because of their importance in human nutrition as well as in microorganisms. Heat sensitivity of nutrients, vitamin destruction, protein availability, and trace element precipitation are some of the factors which must be considered when preparing media for inoculum.

Heat sterilization can cause interactions between ammonia, amino acids, and glucose, generating factors which can be growth inhibiting or growth stimulating. Medium interactions can reduce the availability of trace metals. Phosphates can complex magnesium and reduce the amount of this critical metal (26).

Because of the low solubility of iron in aqueous media, this element may become limiting if it is complexed during sterilization or pH adjustment. Adequate iron and trace metals must be added to the medium to compensate for those made unavailable by sterilization (34).

Generally, glucose is autoclaved separately from nitrogen sources and vitamins are filter sterilized. Phosphate is often sterilized separately from other inorganic compounds.

On a small scale, the routine autoclaving of flasks and tubes containing liquid media usually yields a homogeneous medium when the constituents are soluble and the autoclave cycle reaches temperature rapidly followed by rapid cooling. However, when solids are present or complex substrates are used, some ingredients must be autoclaved separately. Sterilization of larger volumes of media, especially with solids, presents problems of unequal heat distribution unless sufficient time is allotted for thermal equilibrium at the sterilizing temperature. While some parts of the medium may have sufficient time at this temperature, others may not.

Continuous sterilization with properly sized mixing and charging vessels allows very rigid control of sterilization conditions. If the residence time at the sterilization temperature is appropriate for the bioload and the thermal death kinetics of the organisms in the ingredients, the continuous sterilizer becomes a powerful tool for minimizing the destruction of nutrients and maintaining highly reproducible inoculum growth.

3.1.6 Agitation

Inoculum vessels containing liquid medium are agitated to provide homogeneity. For aerobic fermentations, mixing is especially important because oxygen is a very low-solubility nutrient. The oxygen transfer capability of a flask can limit the amount of biomass that can be grown in the flask. When premature oxygen limitation is imposed on growth, changes in physiological state can occur which result in reduced inoculum effectiveness. This necessity for oxygen transfer capability in inoculum development is sometimes overlooked, especially if only one shaker with a fixed-speed motor of limited horsepower is available for the inoculum vessel.

Agitation, however, also can cause shear. If pellets or shear-sensitive flocs are desired, then alternative methods of obtaining sufficient oxygen must be found to avoid disrupting the pellet or floc. The problem of maintaining these structures increases with scale, and agitation is a major determinant (6).

3.1.7 Temperature

Temperature is a simple but powerful tool in inoculum development. The growth rate of microorganisms is a function of temperature through the reaction sequences which make up the whole of metabolism. Much experimental evidence exists to support the hypothesis that each sequence has its own temperature optimum. The chemical balance for these rates is determined by the medium used for a particular culture. It is important to realize that shifts in temperature can alter the utilization rate of one component as compared to another, thus unbalancing the medium with respect to growth. The early depletion of a critical nutrient can shift the culture from balanced to unbalanced growth and change its performance. Clearly, for reproducible growth rates, temperature must be rigorously controlled.

Temperature can also be used to hold cultures at certain stages of development, thereby increasing their useful life. To maintain fermentor productivity in industrial processes, suitable inocula (at the optimum physiological state) must always be available. Since vegetative inoculum may not be stored long without losing effectiveness, chilled sporulated suspensions are routinely used (20).

For example, a liquid medium which promotes good sporulation is inoculated and sporulation is allowed to occur. A good spore yield is obtained, quantitated, and chilled. Secondary inoculations are made from portions of this chilled spore suspension. The secondary growth cycle is well defined since its inoculum is one of many identical portions (20, 29). Likewise, vegetative cells may be refrigerated for up to 24 h after growth and then used to inoculate secondary sources. Each of these techniques requires thorough testing before being adopted for any experimental or production use. Culture storage is affected by differences in media and how stable the culture is to self-lysis (29).

3.2 TRANSFER CRITERIA

The performance of an inoculum is judged by the job it does in its end use. Reproducible inoculum, whether for experimental purposes or manufacturing, must perform predictably under the conditions selected (10). Transfer criteria are those values of any measured parameter(s) which have been correlated with end use performance. Generally such measurements depend upon a culture's progress along the growth curve with a particular medium and oxygen supply. Measurements such as wet cell volume, pH, dry weight, wet weight, turbidity, respiration, and morphological form have been used with success (9, 19). However, correlations between inoculum parameter and end use performance remain tenuous at best.

There are a number of factors which make correlations difficult. First, in cases where complex substrates are used, the sequential utilization of carbon substrates may change the morphology or enzyme content of a population dramatically (5, 8, 15, 28). Second, the number and cost of the assays necessary to completely analyze a complex system can be overwhelming. Third, if organism physiology or structure changes rapidly, both the cost and the short time available make a detailed analysis difficult. However, as the experience base grows, areas for fruitful effort become defined and study can be focused upon them (37, 38, 42).

3.2.1 Physiological State

Depletion of a major carbon and energy source like glucose brings about changes in parameters (pH, respiration, RQ) which can be measured. These changes are quantitated throughout the standard growth curve (7, 9, 23). Growth or change in the concentration of biomass can be further evaluated by microscopically observing changes in the morphology of the culture as it develops (27). For sporulating cultures, these observations may include changes in refractility of the spores, fragmentation of hyphae, or elongation of hyphae. Physical measurements such as viscosity are linked directly to the presence of mycelia (31).

A further indicator of growth dynamics is macromolecular synthesis. When portions of populations are diluted into new media as in multistaging, these populations respond to the new environment in predictable ways. Studies of RNA, DNA, and protein synthesis indicate that when medium richness changes, regulated responses occur which determine the final growth rate of the culture in the new medium (8, 20).

3.2.2 Measurement

The objective of observation and measurement of population dynamics is to develop a reliable, cheap, and predictive indicator of inoculum performance either in bioassays or in production. Small stirred tanks and shake flasks can be useful for this purpose. When inoculum is needed for bioassays, standard curves can be set up as controls to normalize the efect of a substandard inoculum. In industry, however, the useful state of an inoculum is linked directly to economic performance and must therefore be truly predictable. Therefore, one must identify those characteristics of inoculum which result in (i) rapid initiation of product synthesis and (ii) populations which maintain high rates of synthesis over extended periods.

The performance of inocula is influenced by the chemical environment in which they are growing. The major sources of carbon and energy, nitrogen, and other potentially limiting nutrients can be assessed by chemical assays. Since oxygen can become a critical limiting nutrient, characterization of the oxygen supply environment, whether in tank or flasks, is essential (3). This quantitation is carried out in flasks by observing performance at various medium volumes and shaker speeds; in tanks, dissolved oxygen and oxygen uptake can be followed.

In complex media, substrates can be used sequentially so that the measured consumption of a particular substrate may be associated with satisfactory inoculum performance. Allowing excessive growth in the medium may result in the consumption of a

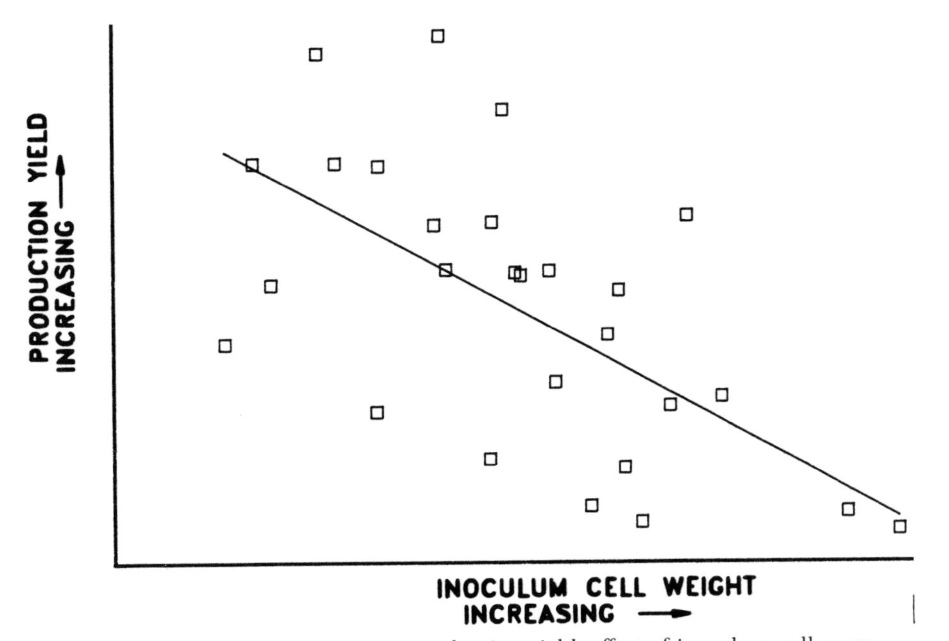

FIG. 2. Vitamin B_{12} fermentation production yield: effect of inoculum cell mass.

secondary substrate, which could cause a significant growth lag when the culture is transferred (5, 33, 37).

Another use for chemical assays is to detect nutrient "carryover." Trace metals and unused carbon and nitrogen sources remaining in the growth medium are diluted by transfer, but they can still interact metabolically through enzyme repression, induction, inhibition, or activation with substrates in the new medium (6, 42).

The time available in which to make decisions about the suitability of a particular population may be limited by the rapidity with which changes occur in batch culture. On-line measurement of respiration data is one way to monitor changes instantaneously (7). Respiratory measurements can reveal not only the rate of the process but also the extent of the reaction, if the recorded rate data can be integrated on line.

3.3 APPLICATIONS TO PARTICULAR FERMENTATIONS

3.3.1 Vitamin B_{12}

For an industrial fermentation process, recommended practice is to routinely examine the effect of measured inoculum parameters on performance of the fermentation. Through continuous examination (11), additional details can be learned of the relation between inoculum parameters and production (12) and small deviations in inoculum growth patterns can be tracked, leading to their control (16). Together these observations may be used for further process improvement.

An investigation made on the vitamin B_{12} fermentation process a few years ago illustrates the methodology and potential benefits to be gained through routine monitoring and analysis. The process involved the culturing of *Pseudomonas denitrificans* in a molasses-based medium (16) through a series of inoculum stages and then into a production stage. The final seed stage consisted of one to three fermentors (depending on their availability) which were used to inoculate one production fermentor. Thus, the percent inoculum varied threefold from 1.3 to 3.9%. Routine measurements of the inoculum included dry cell weight (calculated from optical density) and pH. The final inoculum stages were developed for a specified period. Figure 2 shows the large impact of the average final cell weight of the inoculum on titer in the production fermentor. Although the data are widely scattered, there is 99.9% significance to the correlation. A similar correlation was observed with respect to final inoculum pH, but no effect of percent inoculum was noticed. Apparently, the physiological state of the cells was more important for production performance than were their numbers. As a result of this investigation, the criteria for transfer of the inoculum stages to the production stage were changed from a time basis to a cell weight basis. The variability of the fermentation decreased, and there was substantial overall improvement in titer.

3.3.2 Cephamycin C

Methods for developing inoculum were required for process improvement of the cephamycin C fermentation process in pilot plant stirred tanks. Shake flasks were used to evaluate the inocula at each stage in this multistaging sequence. Parameters of interest were (i) the length of the final inoculum stage, (ii) the time taken for initiation of growth in the production medium, and (iii) the time taken for initiation of cephamycin C synthesis.

The inoculum development protocol was carried out for each stage. In the first stage, growth curves were obtained in flasks used for inoculating the tank. The objective was to achieve growth with the shortest lag in the flask medium without pellet formation. Liquid volume in the flask, sterilization method, and percent inoculum were evaluated.

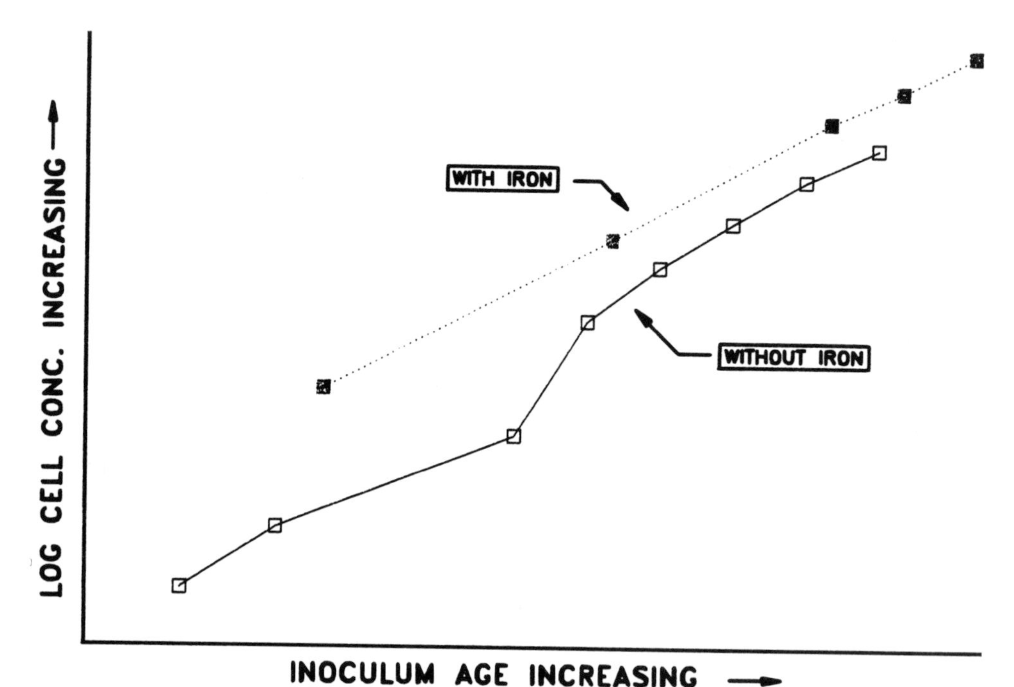

FIG. 3. Cephamycin C fermentation, stirred tank, inoculum growth evaluation.

The culture was observed to grow in pellet form in the production medium. Biosynthesis of product slowed as the formation of biomass in the pellet form increased. Efforts to eliminate the problem by modifying the production stage were unsuccessful. Although the inoculum appeared to grow exponentially, small pellets continued to be formed. When the amount of iron in the seed medium was increased, the pelleting was reduced significantly. When this modified inoculum was used in production, productivity increased. Differences in growth rate were not detected, however (Fig. 3).

These examples point out that inoculum development in industry is often empirical in nature but can reap considerable economic benefits. In many plant operations, the number of fermentations is so high that inoculum data should be assembled on line (e.g., off-gas information) via computer, thereby permitting statistical analysis. Examination may be appropriate weekly or after a change in the fermentation process.

3.4 DIRECTIONS FOR INVESTIGATION

Experimental analysis is needed to enhance the primarily empirical approach used today in developing inocula. Few rigorous studies of a systematic and theoretical nature exist. Developing and refining an inoculum strategy is a continuous process, whether it be for a new fermentation or optimization of a current process (39). Optimization is often necessary because of introduction of improved mutant cultures or use of new ingredients in the inoculum for fermentation media. Improved development approaches based on tested hypotheses would expedite design of a reasonable strategy for optimum product formation and would facilitate maintenance and improvement over time (30).

Calam and Smith studied the relationship of inoculum to final fermentation titer of penicillin and griseofulvin (12). The enzymatic activities of aldolase, glucose 6-phosphate dehydrogenase, and isocitrate dehydrogenase were found to be as important as the morphology of *Penicillium* spp. in determining final titer. Enzymatic methods have practical use because of their ease of implementation and timeliness of results.

The production stage of the xanthan gum fermentation usually requires a 5 to 10% inoculum to ensure active initial growth. However, inoculum volumes of 0.5% or less are sufficient when certain trace elements are at low levels in the production medium, e.g., when iron is below 10 ppm (μg/ml) and zinc and manganese are below 3 ppm (T. R. Jarman, European patent appl. EP.66957A1, November, 1982). This illustrates that a synergistic approach may be best when developing media and conditions for inoculum and production. Also in this example, transfer of inoculum during active growth to the production stage was desirable. The period of inoculum development was minimized by controlling oxygen levels and adding yeast extract to the inoculum medium. Commercially, understanding this behavior may yield economic benefits due to overall reduced processing of inoculum.

Inoculum quality is essential to a reproducible and productive wine fermentation from grape juice (21). The oxidation potential of the inoculum medium, its ergosterol content, and the final yeast concentration are critical and related. A yeast can only produce a given maximum amount of ergosterol, which is later required for yeast growth in the production stage; the amount of ergosterol produced by a yeast is determined by the oxidation potential of the inoculum medium. Thus, to a certain extent, inoculum having a high yeast population and low ergosterol content is as suitable as inoculum of less yeast but more ergosterol.

Knowing such interrelations may allow flexibility in producing inoculum and may even teach one how to upgrade a poor inoculum. Clearly, kinetic analysis is a powerful tool in fermentation development.

Genetic stability. The genetic instability of industrial microorganisms is well known, especially for filamentous organisms (molds and actinomycetes) and recombinant strains. Actinomycetes with or without plasmids can be genetically unstable. The same is true for recombinant strains, where rate of growth, type of replication control, degree of plasmid gene expression, and environmental factors can affect instability. During inoculum development, genetic instability can readily be magnified as the culture proliferates from a stock source to production levels. There are natural selective pressures favoring instability of the microorganism, e.g., a variant may be a faster grower or have an increased affinity for a rate-limiting nutrient.

Procedures should be developed to recognize culture changes due to genetic instability as inoculum growth proceeds. Often plating of samples on agar media can detect changes, indicated, for example, by colony size or pigment formation. Factors affecting the rate of instability should be determined, so that ways may be devised to minimize the problem. Possibly, the frequency of variants from parents, irrespective of time, may be affected by medium ingredients, incubation temperature, or shock of transfer from one stage to the next. Alternatively, selection pressures may be applied to increase stability. For example, a recombinant strain whose plasmid codes for resistance to a metal would be selected for in a medium containing normally toxic concentrations of the metal. Such selectable genetic markers can easily be maintained. It is in dealing with unselected characters that solutions to achieving stability must be devised. Use of strains and plasmids defective in recombination may be a useful approach.

3.5 MEDIA

Examples of complex and defined media for growing inoculum (18, 41) are as follows. Quantities of ingredients are per liter.

Complex medium

Nutrient broth	8 g
Yeast extract	2 g
Glucose	10 g

Adjust pH to 7.0 with NaOH.

Dispense 20 ml each into 250-ml, three-baffled Erlenmeyer flasks. Plug with 26 g of pharmaceutical coil cotton and autoclave for 20 min at 121°C.

Defined medium

Monosodium glutamate	4.25 g
NH_4Cl	1.0 g
$CaCO_3$	0.25 g
K_4HPO_4	2.0 g
Inositol	0.2 g
NaCl	0.5 g
$MgSO_4 \cdot 7H_2O$	0.5 g
$FeSO_4 \cdot 7H_2P$	0.025 g
$MnSO_4 \cdot 7H_2O$	0.005 g
$ZnSO_4 \cdot 7H_2O$	0.01 g
p-Aminobenzoic acid	0.0001 g

Adjust pH to 7.0 with NaOH.

Dispense 40 ml each into 250-ml, unbaffled Erlenmeyer flasks. Plug as before. Autoclave for 20 min at 121°C.

3.6 LITERATURE CITED

1. **Attwell, R. W., and T. Cross.** 1973. Germination of actinomycete spores, p. 197–207. *In* G. Sykes and F. A. Skinner (ed.), Actinomycetales: characteristics and practical importance. Academic Press, Inc., New York.
2. **Bailey, J. E., and D. F. Ollis.** 1977. Biochemical engineering fundamentals. McGraw-Hill Book Co., New York.
3. **Banks, G. T.** 1980. Scale-up of fermentation processes. Top. Enzyme Ferment. Biotechnol. **3:**170–266.
4. **Bartholomew, W. H., and H. B. Reisman.** 1979. Economics of fermentation processes, p. 463–496. *In* H. J. Peppler and D. Perlman (ed.), Microbial technology, vol. 2. Academic Press, Inc., New York.
5. **Borrow, A., E. G. Jefferys, R. H. J. Kessell, E. C. Lloyd, P. B. Lloyd, and I. S. Nixon.** 1961. The metabolism of *Gibberella fujikuroi* in stirred culture. Can. J. Microbiol. **7:**227–276.
6. **Brown, D. E., and M. A. Zainudeen.** 1978. Effect of inoculum size on the aeration pattern of batch cultures of a fungal microorganism. Biotechnol. Bioeng. **20:**1045–1061.
7. **Buckland, B. C.** 1984. The translation of scale in fermentation processes. The impact of computer process control. Biotechnology **1984:**875–883.
8. **Bull, A. T., and A. P. J. Trinci.** 1977. The physiology and metabolic control of fungal growth. Adv. Microb. Physiol. **15:**1–84.
9. **Calam, C. T.** 1969. The culture of microorganisms in liquid medium, p. 567–591. *In* J. R. Norris and D. W. Ribbons (ed.), Methods in microbiology, vol. 1. Academic Press, Inc., New York.
10. **Calam, C. T.** 1976. Starting investigational and production cultures. Process Biochem. **1976:**7–12.
11. **Calam, C. T., and D. W. Russell.** 1973. Microbial aspects of fermentation process development. J. Appl. Chem. Biotechnol. **23:**225–237.
12. **Calam, C. T., and G. M. Smith.** 1981. Regulation of the biochemistry and morphology of *Penicillium chrysogenum* in relation to initial growth. FEMS Microbiol. Lett. **10:**231–234.
13. **Davis, N. D., and W. T. Blevins.** 1979. Methods for laboratory fermentation, p. 303–329. *In* H. J. Peppler and D. Perlman (ed.), Microbial technology, vol. 2. Academic Press, Inc., New York.
14. **Drew, S. W.** 1981. Liquid culture, p. 151–178. *In* P. Gerhardt (ed.), Manual of methods in general bacteriology. American Society for Microbiology, Washington, D.C.
15. **Ensign, J. C.** 1978. Formation, properties, and germination of actinomycete spores. Annu. Rev. Microbiol. **32:**185–219.
16. **Florent, J., and L. Ninet.** 1979. Vitamin B_{12}, p. 497–519. *In* H. J. Peppler and D. Perlman (ed.), Microbial technology, microbial processes. Academic Press, Inc., New York.
17. **Gapes, J. R., V. F. Larsen, and I. S. Maddox.** 1983. A note on procedures for inoculum development for the production of solvents by a strain of *Clostridium butylicum*. J. Appl. Bacteriol. **55:**363–365.
18. **Ginther, C. L.** 1979. Sporulation and the production of serine protease and cephamycin C by *Streptomyces*

lactamdurans. Antimicrob. Agents Chemother. **15:**522–526.

19. **Granade, T. C., M. F. H. Hehmann, and W. M. Artis.** 1985. Monitoring of filamentous fungal growth by in situ microspectrophotometry, fragmented mycelium absorbance density, and ^{14}C incorporation: alternatives to mycelial dry weight. Appl. Environ. Microbiol. **49:**101–108.
20. **Hockenhull, D. J. D.** 1980. Inoculum development with particular reference to *Aspergillus* and *Penicillium*, p. 1–24. *In* J. E. Smith, D. R. Berry, and B. Kristiansen (ed.), Fungal biotechnology, series no. 3. Academic Press, Inc., New York.
21. **Houtman, A. C., J. Marais, and C. S. DuPlessis.** 1980. Factors affecting the reproducibility of fermentation of grape juice and of the aroma composition of wine. I. Grape maturity, sugar, inoculum, concentration, aeration, juice turbidity and ergosterol. Vitis **19:**37–54.
22. **Jarvis, F. G., and M. J. Johnson.** 1947. The role of the constituents of synthetic media for penicillin production. J. Am. Chem. Soc. **69:**3010–3017.
23. **Koch, A. L.** 1981. Growth measurement, p. 179–207. *In* P. Gerhardt (ed.), Manual of methods in general bacteriology. American Society for Microbiology, Washington, D.C.
24. **Konig, B., K. Schugerl, and C. Seewald.** 1982. Strategies for penicillin fermentation in tower-loop reactors. Biotechnol. Bioeng. **24:**259–280.
25. **Metz, B., and N. W. F. Kossen.** 1977. The growth of molds in the form of pellets—a literature review. Biotechnol. Bioeng. **19:**781–799.
26. **Meynell, G. G., and E. Meynell.** 1965. Theory and practice in experimental bacteriology. Cambridge University Press, London.
27. **Meyrath, J.** 1963. Influence of the size of inoculum on various growth phases in *Aspergillus oryzae*. Antonie van Leeuwenhoek J. Microbiol. Serol. **29:**57–78.
28. **Meyrath, J., and A. F. McIntosh.** 1963. Size of inoculum and carbon metabolism in some *Aspergillus* species. J. Gen. Microbiol. **33:**47–56.
29. **Meyrath, J., and J. Suchanek.** 1972. Inoculation techniques—effects due to quality and quantity of inoculum, p. 159–209. *In* Methods in microbiology, vol. 7B. Academic Press, Inc., New York.
30. **Nagamune, T., I. Endo, and I. Inoue.** 1982. Optimal operation for the seed culture process. J. Chem. Eng. Jpn. **15:**481–486.
31. **Nestaas, E., and D. I. C. Wang.** 1981. A new sensor, the "filtration probe," to monitor and control antibiotic fermentations, p. 433–438. *In* M. Moo-Young, C. W. Robinson, and C. Vezina (ed.), Advances in biotechnology, vol. 1. Pergamon Press, New York.
32. **Perlman, D.** 1979. Microbial production of antibiotics, p. 241–280. *In* H. J. Peppler and D. Perlman (ed.), Microbial technology. Academic Press, Inc., New York.
33. **Pirt, S. J.** 1975. Principles of microbe and cell cultivation. John Wiley and Sons, Inc., New York.
34. **Ratledge, C., and P. V. Patel.** 1976. Lipid-soluble iron-binding compounds in *Nocardia* and related organisms, p. 372–385. *In* M. Goodfellow, G. H. Brownell, and J. A. Serrano (ed.), The biology of the nocardiae. Academic Press, Inc., New York.
35. **Rose, A. H. (ed.).** 1979. Economic microbiology, vol. 3 and 4. Academic Press, Inc., New York.
36. **Sehgal, S. N., R. Saucier, and C. Vezina.** 1976. Antimycin A fermentation. II. Fermentation in aerated-agitated fermentators. J. Antibiot. **29:**265–274.
37. **Shida, T., K. Komagata, and K. Mitsugi.** 1975. Reduction of lag time in bacterial growth. 2. Effects of inoculum size and glucose and sodium chloride supplemented in culture media, initial pH of media and cultural temperature. J. Gen. Appl. Microbiol. **21:**293–303.
38. **Smith, G. M., and C. T. Calam.** 1980. Variations in inocula and their influence on the productivity of antibiotic fermentations. Biotechnology Lett. **2:**261–266.
39. **Strehaino, P., M. Mota, and G. Goma.** 1983. Effects of inoculum level on kinetics of alcoholic fermentation. Biotechnology Lett. **5:**135–140.
40. **Vansuijdam, J. C., N. W. F. Kossen, and P. G. Paul.** 1980. An inoculum technique for the production of fungal pellets. Eur. J. Appl. Microbiol. Biotechnol. **10:**211–221.
41. **Wesseling, A. C., and B. D. Lago.** 1981. Strain improvement by genetic recombination of cephamycin producers, *Nocardia lactamdurans* and *Streptomyces griseus*. Dev. Ind. Microbiol. **22:**641–651.
42. **Young, T. B.** 1979. Fermentation scale-up: industrial experience with a total environmental approach, p. 165–180. *In* W. R. Vieth, K. V. Subramanian, and A. Constantinides (ed.), Biochemical engineering, vol. 326. New York Academy of Sciences, New York.
43. **Zabriskie, D. W., W. B. Armiger, D. H. Phillips, and P. A. Albano.** 1980. Traders' guide to fermentation media formulation. Traders' Protein Division, Traders' Oil Mill Co., Fort Worth, Tex.

Chapter 4

Nutritional Improvement of Processes

R. GREASHAM AND E. INAMINE

Major improvements in the productivity of fermentation processes are generally ascribed to the development of superior strains via mutation. However, other parameters such as the nutritional and physical environment to which an organism is exposed are also known to significantly alter product yield. The discussions to follow focus on the nutritional aspects of process improvement.

Media commonly used to screen for industrially important microbial metabolites frequently do not become part of the process definition. Undesirable characteristics of these media include economically unattractive nutrients, support of suboptimal productivity, and support of the synthesis of closely related product components. Therefore, efforts to improve the medium frequently commence early in the development of a successful fermentation process and continue throughout in support of improved culture introduction and fermentation scale-up activities.

Development of an economical production medium requires the selection of a carbon, nitrogen, phosphorus, sulfur, potassium, and trace element source as well as an energy source that will support not only good microbial growth but also maximize product yield, reduce the synthesis of compounds closely related to the product, and enhance product recovery. An equally important consideration is the cost of the candidate nutrients, including not only the purchase price but also transportation, handling, and storage costs, any pretreatment costs (such as enzymatic hydrolysis of starch), and special sterilization costs. Additional information, such as the variability and availability of the nutrient as well as the stability of the purchase price, should also be determined. As medium development proceeds, other economic considerations, such as performance in stirred vessels, product recovery costs, and quality of the product, become important as well.

The formulation of media containing complex nutrients is generally preferred for large-scale fermentations since it leads to the development of cost-effective processes that support maximum product yield. However, since nutrient requirements for growth and product formation vary significantly from culture to culture, empirical procedures should still be employed in the initial screen of complex nutrients. A list of nutrients that have been used successfully in production media is presented in Table 1. Other parameters important in medium optimization include pH, dissolved oxygen level, and cultivation temperature.

4.1 STATISTICAL-MATHEMATICAL METHODS OF OPTIMIZATION

Medium optimization by the classical method involves changing one independent variable (such as nutrient, temperature, pH, etc.) while fixing the others at a certain level. This single-dimensional search is laborious and time-consuming, especially for a large number of variables, and frequently does not guarantee determination of optimal conditions. An alternative method is to institute a full factorial search that would examine every possible combination of independent variables at appropriate levels. This full-dimensional search is useful for a small number of variables (such as three nutrients at two concentrations) but impractical for a large number of variables because of the huge number of experimental runs required to complete the search. For example, the evaluation of six nutrients at three concentrations would require 729 (3^6) experimental trials. Therefore, a more practical method (such as the Plackett-Burman design [27]) is recommended when more than five independent variables are to be investigated.

4.1.1 Plackett-Burman Design

The Plackett-Burman design is a fraction of the two-factorial design and allows the investigation of up to $N - 1$ variables in N experiments. Table 2 shows such a design for conducting 12 experimental trials. Other designs for 8, 16, 20, 24, and up to 100 experiments (at increments of four) may be selected, depending on the number of independent variables of interest (32; L. R. Koupal, Abstr. Annu. Meet. Am. Soc. Microbiol. 1984, O53, p. 197). Each row represents a trial and each column represents an independent or dummy variable. At least three dummy variables

TABLE 1. Nutrients commonly found in fermentation processes[a]

Carbon sources
Dextrose
Glycerol
Molasses (such as beet)
Corn starch
Whey (65% lactose)
Vegetable oil
Alcohols (such as methanol)
n-Paraffins
Cellulose
Nitrogen sources
Cottonseed nutrients
Soybean meal and flour
Fish meal
Corn gluten meal
Dried distillers solubles
Corn steep liquor
Whole yeast
Yeast extract
Malt extract
Casein hydrolysates
Urea
Ammonia gas

[a] Many of the complex carbon and nitrogen sources contain vitamins and trace elements.

should be used to estimate the experimental error. The elements, + and −, represent the two different levels of the independent variable under investigation (such as a nutrient at two concentrations). This design requires that the frequency of each level in a given column be equal and that each row having the same level of one variable must have an equal frequency of each level of the other independent variables. Although the difference between the levels of each variable should be large enough to ensure that the peak area will be included, caution must be exercised when setting the level differential for sensitive variables, since a differential that is too large could mask the results of the other variables.

Once the independent variables and corresponding levels have been selected, the trials are performed in random order to eliminate any bias in experimentation. The response(s) of interest, such as improved productivity, is measured for each trial. The most important variables affecting this response are identified by the following steps.

1. Determine the difference between the average of the + and − responses for each independent and dummy variable. For the independent variables, this value represents the effect that changing each from its + to − levels has on the response. For the dummy variables, this value should be near zero.

2. Estimate the experimental error by averaging the square of the dummy effects. This value is the degree of variability one should expect if any particular combination of variables were run repeatedly at the same experimental condition. The variance of an effect (V_{eff}) is calculated as follows:

$$V_{eff} = \Sigma(E_d)^2/n$$

where E_d is the effect determined for the dummy variable and n is the number of dummy variables and degrees of freedom.

3. Determine the standard error (SE) of an effect by taking the square root of the variance of an effect:

$$SE = \sqrt{V_{eff}}$$

4. Determine the significance level (P value) of each variable effect by using the familiar t test:

$$t_x = E_x/SE \text{ (at } n \text{ degrees of freedom)}$$

where E_x is the effect of variable x.

4.1.2 Response Surface Design

The next stage in medium optimization is to determine the optimum level of each key independent variable as identified by the Plackett-Burman design using response surface optimization techniques (14, 21). A map or a contour plot may be generated by determining the linear, interaction, and quadratic effects of the key variables (i.e., nutrients, medium pH, cultivation temperature, etc.). Although response surface analysis is practical for up to five independent variables, the following quadratic polynomial model represents the relationship fitted for three variables:

TABLE 2. Plackett-Burman design for 12 trials

Trial	Random order[a]	Variables										
		x_1	x_2	x_3	x_4	x_5	x_6	x_7	x_8	x_9	x_{10}	x_{11}
1		+	+	−	+	+	+	−	−	−	+	−
2		−	+	+	−	+	+	+	−	−	−	+
3		+	−	+	+	−	+	+	+	−	−	−
4		−	+	−	+	+	−	+	+	+	−	−
5		−	−	+	−	+	+	−	+	+	+	−
6		−	−	−	+	−	+	+	−	+	+	+
7		+	−	−	−	+	−	+	+	−	+	+
8		+	+	−	−	−	+	−	+	+	−	+
9		+	+	+	−	−	−	+	−	+	+	−
10		−	+	+	+	−	−	−	+	−	+	+
11		+	−	+	+	+	−	−	−	+	−	+
12		−	−	−	−	−	−	−	−	−	−	−

[a] Use random number table.

TABLE 3. Box-Behnken design for three independent variables

Trial	Random order[a]	Variables		
		x_1	x_2	x_3
1		+	+	0
2		+	−	0
3		−	+	0
4		−	−	0
5		+	0	+
6		+	0	−
7		−	0	+
8		−	0	−
9		0	+	+
10		0	+	−
11		0	−	+
12		0	−	−
13		0	0	0

[a] Use random number table.

$$Y = b_0 + b_1x_1 + b_2x_2 + b_3x_3 + b_{12}x_1x_2 + b_{13}x_1x_3 + b_{23}x_2x_3 + b_{11}x_1^2 + b_{22}x_2^2 + b_{33}x_3^2$$

where Y is the dependent variable (such as predicted yield); x_1, x_2, and x_3 are independent variables; b_0 is the regression coefficient at center point; b_1, b_2, and b_3 are linear coefficients; b_{12}, b_{13}, and b_{23} are second-order interaction coefficients; and b_{11}, b_{22}, and b_{33} are quadratic coefficients.

Identification of an optimum requires an estimation of curvature, which in turn requires that each variable be tested at at least three levels. A full three-level factorial design could be used; however, the number of experimental trials would be large (three, four, and five independent variables would require 27, 81, and 243 trials, respectively). Thus, an experimental design that is a fraction of a full factorial is recommended, such as the Box-Behnken design (7). Unlike the Plackett-Burman method, this design evaluates the quadratic effects and two-way interactions among the variables and thus determines the nonlinear nature of the response, if any. An example of this design for three independent variables is shown in Table 3 and for four in Table 4. Each column represents an independent variable, and each row represents an experimental trial. The elements, +, 0, and −, represent the high, middle, and low levels, respectively, of each variable. Generally, values assigned to these levels are equally spaced from the center point (Fig. 1). The trials are performed in a randomized sequence, replicating at least some of the points (triplicates recommended).

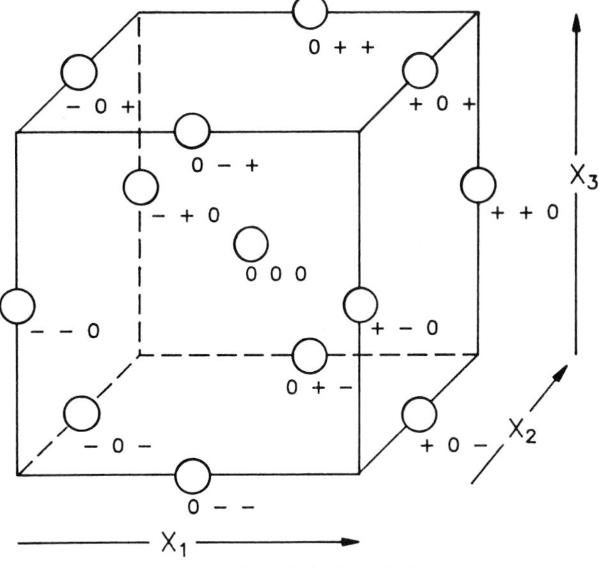

FIG. 1. Box-Behnken design.

Once the trials have been completed, the coefficients of the polynomial model are calculated using standard regression techniques. Several computer programs are available for performing these analyses. (Consult a statistician for a recommendation.) One we have found to be very useful is the PROC RSREG routine in the SAS package (SAS Institute Inc., Cary, N.C.). The calculated equations may be used to generate predicted responses at any combination of factors in or near the region investigated or put into a program which will generate contours of responses versus level of variables (see Fig. 2 for an example). It should be noted that three or more variables may dictate producing several contour plots since any one

TABLE 4. Box-Behnken design for four independent variables

Trial	Random order [a]	Variables x_1	x_2	x_3	x_4
1		+	+	0	0
2		+	−	0	0
3		−	+	0	0
4		−	−	0	0
5		0	0	+	+
6		0	0	+	−
7		0	0	−	+
8		0	0	−	−
9		0	0	0	0
10		+	0	0	+
11		+	0	0	−
12		−	0	0	+
13		−	0	0	−
14		0	+	+	0
15		0	+	−	0
16		0	−	+	0
17		0	−	−	0
18		0	0	0	0
19		+	0	+	0
20		+	0	−	0
21		−	0	+	0
22		−	0	−	0
23		0	+	0	+
24		0	+	0	−
25		0	−	0	+
26		0	−	0	−
27		0	0	0	0

[a] Use random number table.

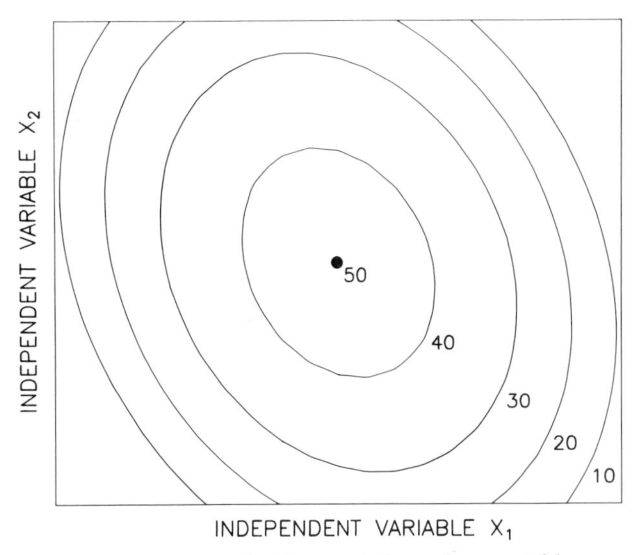

FIG. 2. Contour plot for two independent variables.

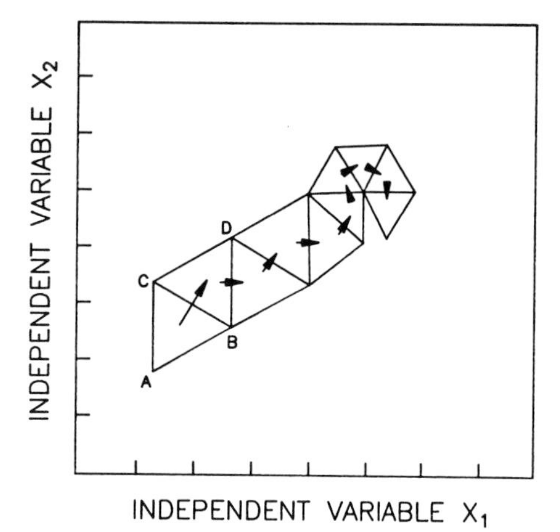

FIG. 3. Simplex optimization for two independent variables.

plot can only show the effect of two variables. These contour plots may indicate that the optimum response has been identified or at least should suggest the region where the optimum is likely to be located. Once the level of each variable for optimal response has been identified, a confirmation shake flask experiment is performed. Positive results indicate that the optimized medium is ready for evaluation in larger vessels.

4.1.3 Simplex Search Technique

Occasionally, conditions optimized in shake flasks will not be transferable to larger-scale vessels. Thus, further optimization must be performed in these larger fermentation vessels (such as fermentors). Because experimental runs at this scale are limited and more expensive than those performed in shake flasks, a simplex optimization procedure is recommended (17, 30, 38). The optimum response is achieved by moving the simplex (a geometric figure defined by experimental points equal to one more than the number of independent variables) in the direction of improved response. Unlike the fractional factorial or Plackett-Burman designs described above, the simplex technique does not require statistical analysis. If an error is made causing the simplex to move in an incorrect direction, subsequent experimental trials should return the simplex toward an optimum response. Although this method is simple, performing the experiments in a sequential manner can be time-consuming.

The key independent variables that have been identified from previous factorial-design experiments are usually selected for this experimental optimization method while the other variables are kept constant. Although several important variables may be employed, only two are used here to describe this method. The first vertex (experimental point) of the simplex usually represents the best current level of key variables. The next two vertices are determined by the level differentials selected for each variable. Initially, the level differentials are usually large so that the optimum response area can be readily approached. Once it appears that this area has been reached, the level differentials are reduced according to a standard procedure (25). Figure 3 is an example of a simplex search for two independent variables (the simplex taking the form of a triangle). The optimization method is started by experimentally determining the response at each vertex of the simplex (ABC). The experimental point with the least desirable response (say vertex A) is discarded, and a new vertex (D) is generated by reflecting point A across the opposite side (BC) of the triangle. The coordinates of the new vertex are easily calculated by using the worksheet shown in Fig. 4 (17). Equivalently, they may also be calculated by using the following formula (23): $R = \bar{P} + (\bar{P} - W)$, where R represents the coordinates of the new vertex, W represents the coordinates of the discarded vertex, and $\bar{P}$ represents the coordinates of the average of all retained vertices.

This new vertex and the two retained vertices define the second simplex. After the response of the new experimental point is determined, it is compared with the previously determined responses of the retained vertices. Again, the vertex with the least desirable response is discarded, and a new one is generated as described above. However, if the new vertex exhibits the lowest response, this simplex would reflect back on the previous one, halting movement toward the optimum. In this situation, the vertex with the second least desirable response is discarded, allowing optimization to continue. Occasionally, the coordinates of a new vertex must be modified to contain the simplex within the constraints (preset high and low levels) of the independent variables. As the sequential simplexes reach the optimal response, they will begin to circle it. As mentioned previously, modifications of this simplex method may be used that include, in addition to reflection, operations of expansion and contraction (12, 29).

The sequence of Plackett-Burman, fractional factorial design, and simplex procedures is attractive

Retained Vertex No.	Coordinates of Independent Variables				
	x_1	x_2	x_3	x_4	x_5
SUM					
$\bar{P}$					
W					
$\bar{P}$ - W					
R					

FIG. 4. Worksheet for calculating the coordinates of the new vertex. SUM, sum of retained coordinates; $\bar{P}$, SUM/n (n = number of retained vertices); W, coordinates of the discarded vertex; $R = \bar{P} + (\bar{P} - W)$, coordinates of the new vertex.

where relatively long fermentation cycles are experienced. However, in instances where results can be achieved more quickly, the simplex procedure should be run before the factorial design to identify the most promising region to run the factorial.

Although the foregoing optimization program can be performed without the assistance of a statistician, the program can be made even more effective by employing such assistance throughout, from the initial design to the final interpretation of the results.

4.2 BIOCHEMICAL METHODS OF OPTIMIZATION

A greater measure of success can be anticipated from studies aimed at the nutritional improvement of a process if certain biochemical data are available. In this respect, information on precursors, biosynthetic pathways, specific requirements of carbon, phosphorus, sulfur, minerals, and trace elements for growth, and product formation is essential. The regulation of the production of many primary and secondary metabolites by carbon, nitrogen, phosphate, and trace elements is well documented (6, 13, 18, 35). Thus, it is also important to determine the extent to which these regulatory mechanisms are operational in a particular organism.

Availability of the kinds of information described is essential for any program aimed at rationally and systematically improving a fermentation process by nutritional means. A chemically defined medium is preferred for studies on the physiology, biochemistry, and regulatory mechanisms of an organism. Such a system allows more precise studies while avoiding the complicated interactions that are inherent in a complex medium fermentation. A useful example that can serve as a guide in the development of a synthetic medium has been described (37).

The cephamycin C fermentation has been selected as a model system for demonstrating how basic studies can contribute significantly to the nutritional improvement of a process. Where possible, explanations of the response by the organism to nutritional factors as they relate to biosynthetic and other metabolic pathways are presented.

4.2.1 Model System: Cephamycin C Fermentation

The cephamycins are β-lactam antibiotics produced by a variety of streptomycetes (24, 31). They differ from cephalosporin C by the presence of a methoxyl group at C-7 in the β-lactam ring. Cefoxitin, the broad-spectrum semisynthetic cephamycin, is derived from cephamycin C.

Medium

The chemically defined and complex media used to investigate cephamycin C biosynthesis by *Nocardia lactamdurans* NRRL 3802 are described below.

The defined basal medium contained (in grams per liter): sodium L-glutamate, 4.25; glucose, 10.0; NH_4Cl, 1.0; inositol, 0.20; K_2HPO_4, 2.0; NaCl, 0.50; $MgSO_4 \cdot 7H_2O$, 0.50; $CaCO_3$, 0.25; $FeSO_4 \cdot 7H_2O$, 0.025; $ZnSO_4 \cdot 7H_2O$, 0.010; $MnS_4 \cdot H_2O$, 0.005; and *p*-aminobenzoic acid, 0.001.

Complex basal media contained the following (in grams per liter). Medium A contained distillers solubles, 30; primary yeast, 10; glycine, 0.5; glycerol, 20; and Mobil Par S defoamer (2.5 ml). Medium B contained corn steep liquor, 23; Cerelose, 40; Proflo, 20; glycerol, 12.5; $MgSO_4 \cdot 7H_2O$, 0.5; and polypropylene glycol P-2000 (2.5 ml). Medium C contained corn steep liquor, 33; meat bone meal, 30; Cerelose, 60; glycine, 1.0; $MgSO_4 \cdot 7H_2O$, 0.5; and polypropylene glycol P-2000 (2.5 ml).

Precursors of cephamycin C

Valine, cysteine, methionine, and α-aminoadipic acid (α-AAA) were established as immediate precursors of cephamycin C by use of radiolabeled substrates (36; E. Inamine and J. Birnbaum, Abstr. Annu. Meet. Am. Soc. Microbiol. 1972, E68, p. 12).

Sulfur precursors

Since sulfur is an integral part of the cephamycin C molecule, initial studies were focused on the sulfur requirements of *N. lactamdurans*. Labeled L-cysteine was found to be rapidly incorporated into the antibiotic, suggesting a direct precursor role. However, culture lysis occurred even at low levels of cysteine. In the search for a substitute sulfur donor, it was found that thiosulfate was not only an excellent source, but it also greatly stimulated antibiotic production (E. Inamine and J. Birnbaum, U.S. patent 3,770,590, November 1973). In contrast, inorganic sulfate was utilized, but no stimulation of antibiotic synthesis was observed. Use of ^{35}S-labeled thiosulfate confirmed that both sulfur atoms were utilized effectively for antibiotic synthesis. Stimulation by thiosulfate indicated that sulfur was rate limiting. Further studies demonstrated that *S*-sulfo-L-cysteine was also stimulatory and that it was an effective competitor of labeled thiosulfate.

The stimulatory nature of thiosulfate was observed in all the complex production media in which it was subsequently examined with cultures of diverse productivity. Stimulation of antibiotic synthesis in both synthetic and complex media ranged from 20 to 40%.

With a wild-type strain, thiosulfate stimulation was best observed when the compound was added after growth phase. Cultures resistant to the somewhat growth-inhibitory nature of thiosulfate were subsequently isolated, which then allowed the addition of the compound at the start of the fermentation.

Although observed only in the wild-type strain, the stimulation of cephamycin C synthesis by thiosulfate was increased further by the concurrent addition of α-AAA, a direct precursor of the antibiotic. The results indicated that with the sulfur requirement satisfied, the supply of α-AAA was now rate limiting. This requirement was overcome with subsequent mutants of improved productivity.

The efficient utilization of thiosulfate for cephamycin C biosynthesis suggested that the sulfate reduction pathway to cysteine was preferred by *N. lactamdurans*. The transsulfuration pathway of cysteine biosynthesis from methionine is the preferred pathway of cephalosporin C biosynthesis by fungi (9). Cystathionine γ-lyase, the pivotal transsulfuration enzyme, was reported to be absent in procaryotes (11). However, the presence and importance of this enzyme for cephamycin C biosynthesis by *N. lactamdurans*, a filamentous bacterium, has been demonstrated conclusively (16). The essential nature of the enzyme was demonstrated by use of propargylglycine, a mechanism-based inhibitor. The compound was a potent inhibitor of both antibiotic production and cystathionine γ-lyase activity but had little or no effect on growth of the organism. The consistent stimulation of product formation by thiosulfate indicates that the organism is also capable of deriving cysteine via the sulfate reduction pathway.

L-α-AAA precursor

Isotope incorporation studies indicated that cephamycin-producing streptomycetes generate α-AAA via the catabolism of L-lysine (36; Inamine and Birnbaum, Abstr. Annu. Meet. Am. Soc. Microbiol. 1972). This contrasts to fungi where α-AAA used for penicillin and cephalosporin C biosynthesis occurs as an intermediate of lysine biosynthesis (26, 34).

D-Lysine is an obligate intermediate of α-AAA synthesis in a strain of *Pseudomonas putida* (10). This pathway was examined in *N. lactamdurans* via isotope competition studies. It was found that unlabeled D-lysine did not dilute the efficient incorporation of L-[^{14}C]lysine into cephamycin C, suggesting that the D-isomer was not an intermediate. However, the titer was increased somewhat in its presence. Subsequent studies showed that at optimum concentration, D-lysine stimulated antibiotic synthesis two- to threefold in the chemically defined medium (E. Inamine and J. Birnbaum, U.S. patent 3,886,044, May 1975). Growth was not affected by the D-amino acid. A regulatory role for D-lysine was suggested by the observation that maximum stimulation was attained only when the compound was added at inoculation. No effect was observed when the addition was made during the period of rapid antibiotic synthesis.

The addition of D-lysine to various complex production media also led to the stimulation of antibiotic production, though not to the extent observed in the chemically defined medium. When cultures of diverse productivity were examined, the stimulation ranged from 20 to 30%. Since the cost of D-lysine precluded its use in large-scale fermentations, the utility of DL-lysine as substitute for the D-isomer was examined. Antibiotic synthesis was stimulated by about 15% with DL-lysine. As with the defined-medium studies, the addition of L-lysine to complex medium failed to bring about a titer increase (Inamine and Birnbaum, U.S. patent 3,886,044, May 1975). These results led to the adoption of DL-lysine along with thiosulfate as additives to complex production media. Subsequent examination of all other D-amino acids showed that D-arginine and D-ornithine were as effective as D-lysine in stimulating antibiotic production (E. Inamine and J. Birnbaum, U.S. patent 3,977, 942, May 1976). The effects of the three D-amino acids were found not to be additive.

In considering possible modes of action of D-lysine, the organism was examined with respect to change in cell wall sensitivity to β-lactam antibiotics. *N. lactamdurans* cells are naturally resistant to the lytic effect of penicillin. However, cultures growing in the presence of D-lysine were found to readily lyse when exposed to penicillin. Sensitivity to cephamycin C was not observed under the same conditions. It was subsequently found that cells grown in the absence of D-lysine contained a penicillinase, while the enzyme could not be detected in D-lysine-grown cells.

The possibility that cadaverine, derived via decarboxylation of D-lysine, was acting as a positive metabolic effector was considered also. Cadaverine was shown to indeed stimulate antibiotic synthesis and to about the same degree as D-lysine. However, the stimulatory effects of the two components were found to be additive, indicating that their modes of action differed. When other naturally occurring polyamines were examined, they were also found to be stimulatory. Spermine, spermidine, agmatine, and 1,3-diaminopropane were all effective in the presence or absence of D- or DL-lysine (Inamine and Birnbaum, U.S. patent 3,977,942, August 1976). The polyamines stimulated production by an average of 50%. Based on utility, cost, and availability, 1,3-diaminopropane was selected for inclusion into the complex production medium along with thiosulfate and DL-lysine. Although the mode of action of 1,3-diaminopropane has not been identified, the compound has been reported to be a potent inhibitor of ornithine decarboxylase in mammals (28).

The first enzyme involved in the conversion of L-lysine to α-AAA has been examined in *N. lactamdurans* (15). The enzyme was classified as L-lysine ε-aminotransferase by identification of the product of the reaction in an in vitro system. Nonsporulating mutants of the culture which do not produce cephamycin C were found to lack this enzyme.

When the enzyme activity was examined in cultures of diverse productivity growing in a synthetic medium, it was found that the productive capacity of the organisms increased coordinately with the enzyme content. The observation indicated that the enzyme plays a pivotal role in antibiotic synthesis. In vitro studies showed that the enzyme is feedback inhibited by α-AAA.

When the intracellular content of α-AAA was examined in these cultures of diverse productivity, a good correlation was found with enzyme content. As the specific activity of the enzyme increased for a given organism, a proportionate increase in the intracellular pool of α-AAA was observed. Interestingly, the contents of other precursors of cephamycin C, lysine, valine, and methionine were high and relatively constant, while cysteine content was uniformly very low for all cultures.

When selected compounds were examined for their ability to inhibit the activity of L-lysine-ε-aminotransferase in vitro, α-amino-ε-hydroxycaproic acid emerged as a potent inhibitor of the reaction. The lysine analog also effectively inhibited antibiotic synthesis in a fermentation without affecting growth.

It has been reported by others (22) that diaminopimelic acid, the precursor of lysine, stimulates cephamycin C production by *Streptomyces clavuligerus* in a synthetic medium fermentation.

Carbamate precursor

Brewer et al. (8) have shown that a specific carbamoyl phosphate transferase enzyme is responsible for the transfer of the carbamoyl group to the C-3′ of cephamycin C in *S. clavuligerus*.

When various compounds containing the carbamate functionality were screened as possible precursor for the carbamoyl moiety, ethylcarbamate was found to significantly stimulate cephamycin C synthesis by *N. lactamdurans* (J. Birnbaum and E. Inamine, U.S. patent 3,769,169, October 1973). However, the *carbonyl*-^{14}C-labeled compound was found not to be incorporated into the antibiotic. Further testing of compounds related structurally to ethylcarbamate led to the finding that dimethylformamide was superior in increasing titer. Feeding studies with *carbonyl*-^{14}C-dimethylformamide, however, indicated that this compound was also not a precursor. The inclusion of dimethylformamide into a variety of complex media resulted in about a 20% stimulation of product formation.

Summary

The complex medium used for early large-scale production (complex medium C, described above) contained thiosulfate, DL-lysine, 1,3-diaminopropane, and dimethylformamide as supplements. The combined addition of these compounds to the basal complex medium improved the titer by over 100%.

4.2.2 Metabolic Deregulation

Carbon catabolite regulation of secondary metabolite biosynthesis is a well-documented occurrence (19). Readily metabolized carbon sources such as glucose decrease the biosynthesis of many metabolites. Production can often be improved in these instances by maintaining a low carbon concentration either by intermittent additions or by slow feeding during the production phase. Alternatively, two carbon sources may be supplied to minimize the problem. A readily utilized carbon source is added to support optimal growth only while a second, less readily metabolized carbon source is added for use by the organism during the production phase.

Inorganic phosphate is also known to adversely affect many fermentations (18). Control of phosphate level is inherently difficult in media containing complex nutrients. Substitution with ingredients low in phosphate content is effective in certain instances, as is the use of MgO or CaO for pH adjustment during media preparation. A recent report suggests that $MgSO_4$ addition may be helpful (39). Soluble phosphate and ammonia concentrations were reduced via the precipitation of magnesium ammonium phosphate. The insoluble salt was found to be slowly utilized by the organism.

Suppression of secondary metabolite production by ammonia is well documented (6). Control of free ammonia level by the addition of trapping agents has been reported (20, 33). Significant titer improvements were obtained in fermentations that were subject to nitrogen regulation. Natural zeolite, tribasic magnesium phosphate, and related compounds were reported to be effective agents. In each instance, the formation of the highly insoluble magnesium ammonium phosphate salt was thought to allow the controlled assimilation of ammonia.

A recent innovation to process improvement is the optimization of product formation by stabilization of a key biosynthetic enzyme. The in vitro stability of gramicidin S synthetase was improved considerably in the presence of a mixture of amino acids that are substrates of the enzyme (5). Subsequent studies showed that the biosynthetic capacity of the cells increased by 50% when the fermentation was supplemented with the same amino acids mixture.

4.3 ACKNOWLEDGMENTS

We thank V. J. Pecore for his advice and critical review of the statistical part of this chapter. Figure 2 was supplied by M. Schnall.

4.4 LITERATURE CITED

4.4.1 General References

1. **Biles, W. E., and J. J. Swain.** 1980. Optimization and industrial experimentation. John Wiley & Sons, Inc., New York.
2. **Box, G. E., W. G. Hunter, and J. S. Hunter.** 1978. Statistics for experimenters. John Wiley & Sons, Inc., New York.
3. **Diamond, W. J.** 1981. Practical experiment designs for engineers and scientists. Lifetime Learning Publications, Belmont, Calif.
4. **Zabriskie, D. W., W. B. Armiger, D. H. Phillips, and P. A. Albano.** 1980. Traders' guide to fermentation media forumlation. Traders Protein, Memphis, Tenn.

4.4.2 Specific References

5. **Agathos, S. N., and A. L. Demain.** 1984. Gramicidin S stabilization *in vivo*. Ann. N.Y. Acad. Sci. **434:**44–47.
6. **Aharonowitz, Y.** 1980. Nitrogen regulation of antibiotic synthesis. Annu. Rev. Microbiol. **34:**209–233.
7. **Box, G. E. P., and D. W. Behnken.** 1960. Some new three level designs for the study of quantitative variables. Technometrics **2:**455–475.
8. **Brewer, S. J., P. M. Taylor, and M. K. Turner.** 1980. An adenosine triphosphate-dependent carbamoylphosphate-3-hydroxymethyl cepham-*O*-carbamoyltransferase from *Streptomyces clavuligerus*. Biochem. J. **185:**555–564.
9. **Caltrider, P. G., and H. F. Niss.** 1966. Role of methionine in cephalosporin synthesis. Appl. Microbiol. **14:**746–753.
10. **Chang, Y.-F., and E. Adams.** 1971. Induction of separate catabolic pathways for L- and D-lysine in *Pseudomonas putida*. Biochem. Biophys. Res. Commun. **45:**570–577.
11. **Delavier-Klutchko, C., and M. Flavin.** 1965. Enzymatic synthesis and cleavage of cystathionine in fungi and bacteria. J. Biol. Chem. **240:**2537–2549.
12. **Deming, S. N., and S. L. Morgan.** 1973. Simplex optimization of variables in analytical chemistry. Anal. Chem. **45:**278–283.
13. **Drew, S. W., and A. L. Demain.** 1977. Effect of primary metabolism on secondary metabolism. Annu. Rev. Microbiol. **31:**343–356.
14. **Hendrix, C.** 1980. Through the response surface with test tube and pipe wrench. Chemtech **10:**488–497.
15. **Kern, B. A., D. Hendlin, and E. Inamine.** 1980. L-Lysine ε-aminotransferase involved in cephamycin C synthesis in *Streptomyces lactamdurans*. Antimicrob. Agents Chemother. **17:**679–685.
16. **Kern, B. A., and E. Inamine.** 1981. Cystathionine γ-lyase in the cephamycin C producer *Streptomyces lactamdurans*. J. Antibiot. **34:**583–589.

17. **Long, D. E.** 1969. Simplex optimization of the response from chemical systems. Anal. Chem. Acta **46:**193–206.
18. **Martin, J. F.** 1977. Control of antibiotic synthesis by phosphate. Adv. Biochem. Eng. **6:**105–127.
19. **Martin, J. F., and A. L. Demain.** 1980. Control of antibiotic synthesis. Microbiol. Rev. **44:**230–251.
20. **Masuma, R., Y. Tanaka, and S. Omura.** 1983. Ammonium ion depressed fermentation of tylosin by the use of a natural zeolite and its significance in the study of biosynthetic regulation of the antibiotic. J. Ferment. Technol. **61:**607–614.
21. **McDaniel, L. E., E. C. Bailey, S. Ethiraj, and H. P. Andrews.** 1976. Application of response surface optimization techniques to polyene macrolide fermentation studies in shake flasks. Dev. Ind. Microbiol. **17:**91–98.
22. **Mendelovitz, S., and Y. Aharonowitz.** 1982. Regulation of cephamycin C synthesis, aspartokinase, dihydrodipicolinic acid synthetase, and homoserine dehydrogenase by aspartic acid family amino acids in *Streptomyces clavuligerus*. Antimicrob. Agents Chemother. **21:**74–84.
23. **Morgan, S. L., and S. N. Deming.** 1974. Simplex optimization of analytical chemical methods. Anal. Chem. **46:**1170–1181.
24. **Nagarajan, R., L. D. Boeck, M. Gorman, A. L. Hamill, C. E. Higgins, M. M. Hoehn, W. M. Stark, and J. G. Whitney.** 1971. β-Lactam antibiotics from *Streptomyces*. J. Am. Chem. Soc. **93:**2308–2310.
25. **Nelder, J. A., and R. Mead.** 1965. A simplex method for function minimization. Comput. J. **7:**308–313.
26. **Neuss, N., C. H. Nash, P. A. Lemke, and J. B. Grutzner.** 1971. Incorporation of carboxyl and methyl carbon-13 labeled acetates into cephalosporin C. J. Am. Chem. Soc. **93:**2337–2339.
27. **Plackett, R. L., and J. P. Burman.** 1946. The design of optimum multifactorial experiments. Biometrika **33:**305–325.
28. **Poso, H., and J. Jane.** 1976. Inhibition of ornithine decarboxylase activity and spermidine accumulation in regenerating rat liver. Biochem. Biophys. Res. Commun. **69:**885–891.
29. **Shavers, C. L., M. L. Parsons, and S. N. Deming.** 1979. Simplex optimization of chemical systems. J. Chem. Educ. **56:**307–309.
30. **Spendley, W., G. R. Hext, and F. R. Himsworth.** 1962. Sequential application of simplex designs in optimization and evolutionary operation. Technometrics **4:** 441–461.
31. **Stapley, E. O., M. Jackson, S. Hernandez, S. B. Zimmerman, S. A. Currie, S. Mochales, J. M. Mata, H. B. Woodruff, and D. Hendlin.** 1972. Cephamycins, a new family of β-lactam antibiotics. I. Production by actinomycetes including *Streptomyces lactamdurans* sp. n. Antimicrob. Agents Chemother. **2:**122–131.
32. **Stowe, R. A., and R. P. Moyer.** 1966. Efficient screening of process variables. Ind. Eng. Chem. **58:**36–40.
33. **Tanaka, Y., R. Masuma, and S. Omura.** 1984. Control of ammonium ion level for efficient nanomycin production. J. Antibiot. **37:**1370–1375.
34. **Trown, P. W., E. P. Abraham, G. G. Newton, C. W. Hale, and G. A. Miller.** 1962. Incorporation of acetate into cephalosporin C. Biochem. J. **84:**157–166.
35. **Weinberg, E. D.** 1970. Biosynthesis of secondary metabolites: roles of trace metals. Adv. Microb. Physiol. **4:**1–44.
36. **Whitney, J. G., D. R. Brannon, J. A. Mabe, and K. J. Wicker.** 1972. Incorporation of labeled precursors into A16886B, a novel β-lactam antibiotic produced by *Streptomyces clavuligerus*. Antimicrob. Agents Chemother. **1:**247–251.
37. **Williams, W. K., and E. Katz.** 1977. Development of a chemically defined medium for the synthesis of actinomycin D by *Streptomyces parvulus*. Antimicrob. Agents Chemother. **11:**281–290.
38. **Yarbro, L. A., and S. N. Deming.** 1974. Selection and preprocessing of factors for simplex optimization. Anal. Chem. Acta **73:**389–391.
39. **Young, M. D., L. L. Kempe, and F. G. Bader.** 1985. Effects of phosphate, glucose and ammonium on cell growth and lincomycin production by *Streptomyces lincolnensis*. Biotechnol. Bioeng. **27:**327–333.

Chapter 5

Long-Term Preservation of Industrially Important Microorganisms

L. T. CHANG AND R. P. ELANDER

The preservation and long-term storage of microorganisms of economic importance, which have the capability to produce high yields of desirable metabolites in large-scale production fermentation, is of prime importance for a successful commercial fermentation process. Ideally, microbial preservation procedures must provide conditions in which highly productive mutant strains are preserved for long periods of time free from phenotypic change, with particular respect to the capability of high production of a primary or secondary metabolic product (5). During recent years, the long-term storage of most microorganisms and higher plant and animal cell lines in the vapor or liquid phase of liquid nitrogen has appeared to be the best available preservation procedure (1–8). Generally, survival levels are significantly higher after storage in liquid nitrogen than after other preservation procedures (18, 23).

This chapter reviews the four major classes of preservation procedures: subculturing or active slant transfer; desiccation in soil or on porcelain beads; freeze-drying or lyophilization; and cryopreservation in ultracold mechanical refrigeration (−20 to −80°C) or in the vapor phase (−156°C) or liquid phase (−196°C) of liquid nitrogen. Detailed protocols for these procedures are outlined. We also include a discussion on the preservation of genetically engineered strains of microorganisms. Finally, we describe some techniques that may be useful in monitoring the viability and eventual productivity of commercially important microorganisms after the preservation process.

5.1 FREEZING

Freezing is the simplest and most commonly used method for preserving microorganisms. For ordinary freezing, no special equipment is required. However, to obtain satisfactory results by freezing, preservatives (cryoprotective agents) must be added to the culture. Also, storage temperatures should be kept lower than −20°C. Care must be taken to ensure that cultures are frozen properly and that storage temperatures are optimal. One disadvantage of freezing is that the culture must be kept frozen or regrown on agar slants during transport.

5.1.1 Ordinary Freezing

Broth cultures or cells harvested from slant culture are dispensed into tubes or vials and stored frozen in the freezing compartment of a refrigerator or in an ordinary freezer with temperatures ranging from −5 to −20°C. The viability of many microorganisms can be maintained for 1 to 2 years by this procedure. However, this method is not suitable for the long-term storage of many microorganisms.

5.1.2 Ultracold-Temperature Freezing (−60 to −80°C)

For long-term preservation, cells should be frozen and stored at temperatures ranging from −50 to −80°C. Such low temperatures can be obtained with mechanical freezers (−80°C) or liquid nitrogen freezers (−156 to −196°C). For storage in mechanical freezers, cells are harvested in the mid- to late logarithmic growth phase by centrifugation and resuspended in a fresh medium containing either 10% (vol/vol) glycerol or 5% (vol/vol) dimethyl sulfoxide (DMSO). Alternatively, a 20% (vol/vol) glycerol solution or 10% (vol/vol) DMSO can be added to an equal volume of sterile broth to achieve a final concentration of 10% (vol/vol) glycerol or 5% (vol/vol) DMSO. Cells grown on agar slants or plates can be harvested by scraping the agar surface with a sterile pipette after flooding with fresh medium containing 10% (vol/vol) glycerol. The cell suspension containing the above cryoprotective agents is then dispensed into cryogenic vials or ampoules and stored in a −70°C freezer. The rate of cooling to −70°C in such a freezer is approximately 1 to 2°C/min. Most bacteria and fungi can be stored at −60°C for up to 5 years without loss of viability (3, 7). To avoid loss of valuable cultures in case of electrical failure, it is advisable to have alarm systems installed on the freezer or to have a backup storage freezer in a different location.

5.1.3 Liquid Nitrogen Freezing and Thawing

Unlike storage at −80°C in a mechanical freezer, liquid nitrogen storage often requires special equipment for controlled-rate freezing before long-term storage in liquid nitrogen. Even though certain organisms can be better preserved by rapid freezing (23), it is generally believed that controlled-rate freezing is superior to rapid freezing for most microorganisms (3). The methods for preparing cells, types of cryoprotective agents, and types of cryogenic containers (vials and ampoules) do not differ from those used in freezing at ultracold temperatures. During the past decade, considerable success has been achieved in preserving fungi and bacteria by storage in liquid nitrogen (−196°C) or the vapor phase of liquid nitrogen (−156°C) (3, 7, 21).

Cryoprotective agents

The most common cryoprotective agents used in liquid nitrogen freezing are glycerol (Difco) and DMSO (Fisher Scientific; 3). These agents are used at final concentrations of 10% (vol/vol) glycerol and 5% (vol/vol) DMSO. Various methods for preparing cells with cryoprotective agents are described above. DMSO is usually sterilized with a Seitz filter and added to the cell suspension to achieve the 5% (vol/vol) final concentration.

Procedure for freezing microorganisms in liquid nitrogen
(Adapted from Shearer [21])

Preparing cultures grown on slants

1. Add 5 ml of nutrient broth (Difco) containing 10% (vol/vol) glycerol to each slant.
2. Scrape surface of the slant with a 1-ml plastic disposable pipette to prepare a dense suspension of spores and vegetative growth. Avoid gouging the agar with the pipette.
3. Using a 2- to 5-ml pipette, dispense 1 ml of cell suspension into a plastic screw-cap ampoule (2-ml capacity; 22). Tighten cap. Alternatively, use glass ampoules (Wheaton Scientific). Seal ampoules with an oxygen/gas burner (Fisher Scientific).
4. Place all ampoules in a refrigerator (5°C) for 30 min to allow equilibration between the cells and suspending medium.

Preparing submerged cultures

1. Add a volume of sterile 20% (vol/vol) glycerol to an equal volume of culture broth to give a final glycerol concentration of 10% (vol/vol).
2. Gently shake flask to mix thoroughly. If the submerged growth is pelleted or in large masses, it may be necessary to break up the masses with the aid of a small, sterile tissue grinder.
3. Using a 2- to 5-ml agar pipette, dispense 1 ml of cell suspension into each 2-ml plastic screw-cap ampoule.
4. Place ampoules in refrigerator (5°C) for 30 min to permit equilibration between the cells and suspending medium.

Controlled-rate freezing. There are many types of commercially available controlled-rate freezers. The general operating procedures for controlled-rate freezers are as follows.

1. Place sealed ampoules in aluminum cans which are contained in larger metal containers. Then place the containers in the freezing chamber of a controlled-rate freezer.
2. Maintain a cooling rate of 1 to 2°C/min until several degrees above the phase change is achieved. The freezing point is usually −30°C.
3. Usually extra liquid nitrogen must be added to the system, either manually or automatically, as the freezing point is approached so that the phase change will occur as rapidly as possible.
4. After the cells are frozen, adjust the cooling rate again to 1°C/min until a temperature of at least −50°C is reached.
5. Quickly transfer ampoules to their final storage space in a liquid nitrogen refrigerator (−156 to −196°C).

Thawing

To thaw the cultures, place the ampoules in a 37 to 40°C water bath. Agitate gently to speed thawing. Swab the exterior of the ampoule with a sterile gauze pad saturated with 70% (vol/vol) ethanol; pay particular attention to the area where the cap and body of the ampoule meet. Do not submerge plastic screw-cap ampoules in alcohol because the ampoules may leak.

If the ampoule is prepared from cultures scraped from the surface of slants, transfer the contents of the ampoule to 2 ml of sterile broth in a test tube, using a 1-ml plastic disposable pipette. Mix by agitating. The use of 0.1 to 0.2 ml of this suspension should normally result in confluent growth on an agar slant.

5.2 FREEZE-DRYING (LYOPHILIZATION)

Freeze-drying involves the removal of water from frozen cell suspensions by sublimation under reduced pressure. It is one of the most effective methods for long-term preservation for many microorganisms even though there are reports that the freeze-drying process may induce mutations in bacteria (9). Cryoprotective agents are required for optimal results when freeze-drying microorganisms for long-term preservation. The most commonly used cryoprotective agents for freeze-drying are skim milk powder (20% [wt/vol] final concentration) or sucrose (12% [wt/vol] final concentration; 3).

Most microbial strains can be preserved in the lyophilized state for periods of up to 10 years (11). Lyophilization is particularly useful when the preserved cultures are constantly in need of transport, since the cultures, once lyophilized, need not be thawed out or regrown on agar slants before being transported. Furthermore, for normal transport there is usually no need to refrigerate lyophilized cultures.

5.2.1 Freeze-Drying Process and Equipment

A typical freeze-drying apparatus consists of the following components: (i) manifold or freeze-drying chamber (shelves), (ii) a condenser, and (iii) a vacuum pump. For manifold drying, bulb or tubular type vials (designed to hold 0.1 to 0.2 ml each) are used. The vials are attached to the manifold while being dried and sealed. Stoppered vials (8.2 ml) are used for batch or shelf drying in commercial freeze-dryers. The American Type Culture Collection uses a "double-vial" method which is a combination of the batch and manifold procedures. The contents of a filled inner vial (0.2 ml per vial) are first lyophilized by the batch method. After lyophilization, the inner vial is placed in an outer vial. After constriction of the outer vial, it is attached to the manifold and sealed while under vacuum (3).

5.2.2 Freeze-Drying Procedures (adapted from Shearer [21])

Equipment and materials (manifold method)

1. Lyophilizer: VirTis 10-145MR-BA Freeze Mobile equipped with a 72-port manifold
2. FTS Systems, Inc., Flexicool with variable temperature control (−40 to 10°C) and a 40-in. (ea. 100-cm) flexible probe
3. Two cooling troughs (VirTis; 41 in. [ca. 102 cm] by 2.25 in. [ca. 5.6 cm] inside diameter), stainless steel with foam plastic insulation
4. Two thermometers (−50 to 50°C temperature range)
5. Ampoules: borosilicate ampoules (9 by 180 mm) prescored 40 mm from bottom, supplied as a special-order item from Bellco Glass, Inc.; the ampoules are loosely plugged with absorbent cotton and sterilized by autoclaving at 121°C for 15 min
6. Skim milk (20% [wt/vol]; Difco), sterilized by autoclaving at 121°C for 15 min
7. Plastic disposable pipettes: 1 ml, graduated in 0.1 ml

Freeze-drying procedures

1. Turn on the refrigeration unit of the freeze-drier. When the condenser temperature reaches −40°C, turn on the vacuum pump. The McLeod gauge should read 100 μm of Hg or less within 20 min. Normal running vacuum is 20 to 30 μm of Hg.
2. Position the cooling troughs beneath the horizontal manifold. With the FTS Systems cooling probe in place, fill the troughs two-thirds full of isopropyl alcohol. Insert thermometers into troughs. Turn on FTS cooling probes and set controls so that the bath reaches −40°C. The bath will require approximately 30 min to reach this temperature.
3. Add 2 ml of 20% (wt/vol) skim milk solution to each slant culture. Scrape the surface of the slant with a 1-ml plastic pipette to effect a homogeneous suspension of spores and vegetative growth to ensure 10^6 CFU/ml. If necessary, combine the suspensions from more than one slant in a screw-cap test tube and mix well. Alternatively, cell suspensions can be prepared by mixing equal volumes of 40% (wt/vol) skim milk and culture broth to effect a final concentration of 20% (wt/vol) skim milk.
4. Remove the cotton plug from each of the ampoules and transfer 0.2 ml of pooled spore or vegetative growth suspension to each with a sterile 1-ml pipette. Reinsert the cotton plugs.
5. Gently agitate ampoules to resuspend cells. Insert ampoule into the Quickseal valve so that the tips containing the cell suspension are suspended in the cooling bath. Allow the cell suspension to fast-freeze in the bath.
6. After 15 min, turn all Quickseal values to the "ON" position, thus placing the ampoules under vacuum.
7. Maintain the temperature of the bath at −40°C for 90 to 120 min. Then reset the FTS cooling probe controls so that the temperature of the bath will rise to −10°C. The bath is maintained at this temperature for 2 h.
8. After 2 h, turn off the FTS cooling probe and allow the temperature of the bath to rise to room temperature (25°C).
9. Throughout the first 4 h, periodically check the vacuum with the Mcleod gauge. The vacuum should reach 100 μm of Hg within 1 h after the ampoules are placed under vacuum and should gradually drop to 20 to 30 μm of Hg as the suspension nears dryness.
10. Allow ampoules to remain on the lyophilizer under vacuum for at least 16 h to effect complete dryness.
11. After drying, seal the ampoules under a vacuum of 20 to 30 μm of Hg, using an oxygen-gas sealing torch.
12. Open the Quickseal valve to break the vacuum after all the ampoules have been sealed.
13. Turn off the vacuum pump.
14. Turn off the condenser refrigeration.

5.2.3 Storage and Recovery

Storage

Freeze-dried cultures are stored at temperatures below 5°C. It is generally believed that lower storage temperatures (−20 to −70°C) are superior for long-term culture stability (13).

Recovery

1. To open the ampoule, score neck with a file. Wipe down ampoule with a sterile gauze pad saturated with 70% (vol/vol) ethanol, paying particular attention to the area around the score.
2. Working under a laminar air-flow hood, wrap the neck of the ampoule in a sterile gauze pad or in a sterile towel. With thumbs placed opposite the score, break open the ampoule.
3. Add 0.5 to 1.0 ml of nutrient broth to the ampoule and rotate to rehydrate the contents. Transfer to a sterile test tube containing suitable recovery medium. Alternatively, reconstituted broth can be transferred directly onto agar slants or petri dishes.
4. The contents of small tubular ampoules are usually in a pelleted form and can be shaken directly from the ampoule into a test tube containing 1 to 2 ml of broth. Allow the pellet to stand for 5 to 10 min in the broth to rehydrate. Gently mix. Use the suspension to inoculate suitable recovery media.

5.3 MISCELLANEOUS PRESERVATION METHODS

Methods other than freezing and freeze-drying have also been used widely for preserving various microorganisms, with varying degrees of success. Some of the more important methods which are still practiced in many microbiology laboratories are described below.

5.3.1 Subculturing

Many laboratories preserve cultures simply by periodically transferring them onto fresh agar media, followed by incubation at a suitable growth temperature. This method is convenient and inexpensive and requires no special equipment. To avoid the selection of culture variants, transfers should be made on a minimal medium (17), and the number of transfers should be kept to a minimum. The slant cultures should be stored in a refrigerator (5°C) in a closed container to avoid desiccation and also to minimize metabolic activity. This procedure is not recommended and, in our opinion, is not suitable for maintaining strains of microorganisms over long periods of time.

5.3.2 Immersion in Mineral Oil

A simple procedure to preserve a microbial strain is to immerse the culture (agar slants or broth culture) in mineral oil and store in an upright position at 25°C (3). The oil can be sterilized in an oven at 170°C for 1 to 2 h. The culture is submerged under oil and can be easily subcultured onto fresh media with an inoculating needle. This procedure is also not recommended for the long-term storage of microorganisms.

5.3.3 Drying

Many spore-forming fungi and streptomycetes can be preserved by drying the spores on the surface of various inert solid substrates, such as soil, silica gel, or glass beads. Soils or silica gels are washed and dispensed into screw-capped tubes. The tubes are sterilized by autoclaving and then dried at 25°C.

Drying on soil

In preparing soil preparations for sporulating fungi, a modification of the technique described by Greene and Fred (14) may be used. The procedure involves pipetting 1 ml of a heavy conidial suspension of a given strain onto 5 g of dry, sterilized soil (a mixture of equal parts of quartz sand and well-screened, pulverized garden loam) in a culture tube stoppered with a gauze-wrapped cotton plug. The inoculated soil tube is then allowed to dry out slowly at 25°C, covered by a porous paper cap. Once prepared and dried, the soil stocks require no further attention. In the University of Wisconsin botanical laboratories, soil tubes of strains of *Penicillium chrysogenum* were not refrigerated and were given no special care except to keep them protected from dust. Such soil stocks not only were very easy to prepare but served to maintain the various strains in a dormant condition over a long period of time (10). To recover the fungus in an active condition, it is necessary only to remove aseptically a small quantity of the soil and dust it over the surface of a fresh agar slant. The soil preservation procedure can be used for preparation of a variety of sporulating bacteria, actinomycetes, and fungi.

Drying on silica gel

A method (20) widely practiced by *Neurospora* researchers is described as follows.

1. Half fill screw-capped tubes (13 by 100 mm) with silica gel (6- to 12-mesh, grade 40, desiccant activated; Davison Chemical Corp., Baltimore, Md.).
2. Dry sterilize the tubes at 180°C for 1.5 h and store in tightly sealed containers.
3. Prepare a dense conidial or vegetative suspension from a fresh culture slant, using 1 to 2 ml of 10% (vol/vol) nonfat skim milk (Carnation).
4. Add the suspension dropwise to each silica gel tube (0.5 ml per tube). To minimize the heat generated by absorption of the introduced culture fluid, submerge the tube in an ice bath.
5. Agitate or vortex each tube to loosen the silica gel granules.
6. After drying the tubes at 25°C, store them in a closed container in the presence of desiccants (Tel-Tale silica gel; Davison Chemical Corp.).

Drying on porcelain beads (15)

Another convenient drying method for preservation involves the use of porcelain beads. The procedure was originally developed by J. Lederberg and is recommended for most species of bacteria.

1. Place 10 to 12 porcelain beads ("Fishspine" beads, no. 2; Taylor, Tunnicliffe and Co., London, WC1, England) in a small (10-ml) screw-capped glass vial.
2. Autoclave the vials containing the beads for 15 min at 121°C.
3. Prepare cell suspensions by flooding a 24- to 48-h slant culture with 1 to 2 ml of a 20% (wt/vol)

sucrose solution. Scrape the culture surface with a sterile 1-ml pipette.
4. Aseptically transfer the beads to a sterile petri dish and inoculate each bead with 1 drop of cell suspension (0.2 to 0.3 ml per bead).
5. Transfer inoculated beads with sterile forceps back into the vial. Each vial usually contains 10 to 12 beads.
6. Place the sterile screw caps loosely on the vials.
7. Place each vial in a vacuum desiccator and dry under vacuum for 72 to 96 h.
8. Remove the beads from the vacuum jar and break them up with a sterile spatula.
9. Store the vials in a closed metal cabinet containing the desiccant Drierite (Fisher Scientific Co.) at 25°C.

5.4 PRESERVATION OF GENETICALLY ENGINEERED MICROORGANISMS

The preservation of genetically engineered strains containing hybrid plasmids annealed with foreign DNA segments presents a new series of problems for culture preservation laboratories. Strains containing foreign segments of DNA in vector plasmids, etc., are usually less stable genetically and are more likely to lose their foreign plasmid replicons (16). Plasmids usually carry nonessential genes, and when cells lose their plasmid(s), they not only maintain their viability but generally exhibit a more rapid growth habit (12, 16). Since plasmid genes are generally not essential for growth, cells containing them carry a biosynthetic burden. Therefore, precautions must be taken during the growth period to ensure that plasmid genes are not lost, since non-plasmid-containing cells usually outgrow their plasmid-containing counterparts.

Plasmid-encoded antibiotic resistance allows for a useful technique to enrich for plasmid-containing cell populations. Propagation in the presence of antibiotics, etc., provides a useful selection pressure in favor of the plasmid-containing cell population (12). Two commonly employed plasmids used as vectors for carrying foreign genes into genetically engineered strains of *Escherichia coli* are pSC101 and pBR322. These particular plasmids convey resistance to the antibiotics tetracycline (pSC101) and tetracycline and ampicillin (pBR322). Therefore, culturing strains containing these plasmids in the presence of these antibiotics provides a selective environment for plasmid-containing cells, since cells which lose their plasmid elements become susceptible to the antibiotics and ultimately are eliminated in the selective antibiotic environment. It becomes essential, therefore, in a microbial fermentation process dependent upon a genetically engineered culture, that the cloned foreign plasmid-encoded genes are kept in situ in the vector plasmid during replication in coordination with the host cells. Growth in the presence of antibiotics will help maintain this coordination of plasmid duplication together with chromosomal replication (12).

We recommend that genetically engineered strains be preserved in a solution containing low concentrations of the selective agent to maintain the selective pressure even in the preserved state. Imanaka and Aiba (16) have published an important review on this subject.

5.5 MONITORING OF VIABILITY AND STABILITY OF MICROORGANISMS

Regardless of the many methods available for preserving microbial cultures, all preserved strains should be checked for viability before and after preservation to determine the degree of cell death occurring during the actual preservation process. In addition, we recommend periodic viability checks during storage to ascertain the shelf life of the preserved cultures. Whenever possible, we recommend that both morphological and, preferably, discrete biochemical characteristics, such as metabolite yields, enzyme levels, and other well-defined genetic or biochemical criteria, be examined after preservation storage. This is especially important with industrial microorganisms (R. P. Elander, Abstr. XII Int. Cong. Microbiol., Munich, 1978, abstr. no. 528.1, p. 39).

There are other methods that can be used to predict stability of the preserved culture during storage. One of these methods, the "accelerated storage test," predicts the stability of a strain of *Lactobacillus acidophilus* at a given temperature by extrapolating from short-term loss of viability at higher temperatures (19). Of course, the applicability of these kinds of monitoring procedures should be examined on a case-by-case basis.

Criteria for the successful long-term preservation of valuable industrial strains include percent recovery (viable units) after 6 months and 1 or more years of storage; uniformity in colony population pattern or some particular biochemical characteristic before and after preservation; and, most important, stability of product titer in laboratory, pilot, and production-scale fermentations (Elander, Abstr. XII Int. Cong. Microbiol., Munich, 1978, abstr. no. 528.1, p. 39).

5.6 SUPPLIERS OF CULTURE PRESERVATION EQUIPMENT

Item	Supplier
Vials and ampoules	
Ampoules and vials for freeze-drying	VWR Scientific, Philadelphia, Pa.
	Bellco Glass, Vineland, N.J.
	Kontes Glass, Vineland, N.J.
Polypropylene vials for freezing	Vangard International, Neptune, N.J.
Glass ampoules for freezing	Wheaton Scientific, Millville, N.J.
Reagents	
Skim milk	Difco Laboratories, Detroit, Mich.
Silica gel	Fisher Scientific, Pittsburgh, Pa.
DMSO	Fisher Scientific, Pittsburgh, Pa.
Glycerol	Difco Laboratories, Detroit, Mich.
Programmable freezers	Cryo-Med, Mt. Clemens, Mich.
Mechanical freezers	Revco, West Columbia, S.C.
	Kelvinator, Manitowoc, Wis.

Liquid nitrogen freezing chamber	Linde, Div. Union Carbide New York, N.Y.
Liquid nitrogen storage boxes	Murray and Heister Beltsville, Md.
Freeze-driers	The VirTis Co., Inc. Philadelphia, Pa. FTS Systems, Inc. Stone Ridge, N.Y. Edwards High Vacuum Grand Island, N.Y. LSL Biolafitte Princeton, N.J.

5.7 CULTURE REPOSITORIES FOR PRESERVED MICROBIAL STRAINS (2)

ATCC [a, b(1,2,3)]
The American Type Culture Collection
12301 Parklawn Dr.
Rockville, Maryland 20852

CBS [a, b(4,5,6)]
Centraalbureau voor Schimmelcultures
Baarn, Netherlands

CMI [a, b(4,5,6)]
Commonwealth Mycological Institute
Kew, Surrey, England

CCM [b(1,5,6)]
Czechoslovak Collection of Microorganisms
University J. E. Purkyne
Brno, Czechoslovakia

DSM [a, b(1,2,3)]
Deutsche Sammlung von Mikroorganismen
Grisebachstrasse 8
D-3400 Göttingen, Federal Republic of Germany

FERM[a] [a]
Fermentation Research Institute
(Biseibutsu Kogyo Gijitsu Kenkushyo)
Higashi 1-chome, Yatabe-cho
Tsukuba-gun, Ibaragi-ken
T 300-22 Tokyo, Japan

IFO [b]
Institute for Fermentation
Osaka, Japan

IMET [a]
Institute for Microbiology and Experimental Therapy
Jena, German Democratic Republic

JFCC [b]
Japanese Federation of Culture Collections of Microorganisms
Institute of Applied Microbiology
University of Tokyo
Tokyo, Japan

NCIB [a, b(7)]
National Collection of Industrial Bacteria
Torry Reserach Station
P.O. Box 31, 35 Abbey Road
Aberdeen, Scotland

NRRL [a]
Northern Regional Research Center
U.S. Department of Agriculture
Science and Education Administration
Peoria, Illinois 61604

RIA [a]
USSR Research Institute for Antibiotics
Moscow, USSR

Notes: a, repository for patent strains; b, culture catalog available. Codes in parentheses indicate information on cultures provided in catalog: 1, growth conditions; 2, reference to patents and publications; 3, utility data; 4, growth conditions available on request; 5, culture source; 6, limited utility data; 7, information available on request.

[a]FERM-P identifies patent strains.

5.8 LITERATURE CITED

5.8.1 General References

1. **Clark, A. W.** 1968. The American Type Culture Collection: experiences in freezing and freeze-drying microorganisms, viruses, and cell lines, p. 309–318. *In* H. Iizuka and T. Hasegawa (ed.), Proceedings of the International Conference on Culture Collections, Tokyo.
2. **Dietz, A.** 1981. Pure culture methods for industrial microorganisms, p. 412–434. *In* H. J. Rehm and G. Reed (ed.), Biotechnology, vol. 1. Verlag Chemie, Weinheim.
3. **Gherna, R. L.** 1981. Preservation, p. 208–217. *In* P. Gerhardt (ed.), Manual of methods for general bacteriology. American Society for Microbiology, Washington, D.C.
4. **Heckly, R. J.** 1978. Preservation of microorganisms. Adv. Appl. Microbiol. **24:**1–53.
5. **Hesseltine, C. W., and W. C. Hayes.** 1973. Sources and management of microorganisms for development of a fermentation industry. Prog. Ind. Microbiol. **12:**3–46.
6. **Lapage, S. P., J. E. Shelton, T. G. Mitchell, and A. R. Mackenzie.** 1970. Culture collections and the preservation of bacteria, p. 135–228. *In* J. R. Norris and D. W. Ribbons (ed.), Methods in microbiology, vol. 3A. Academic Press, London.
7. **Onions, A. H. S.** 1971. Preservation of fungi, p. 113–159. *In* C. Booth (ed.), Methods in microbiology, vol. 4. Academic Press, London.
8. **Perlman, D., and M. Kikuchi.** 1977. Culture maintenance. Annu. Rep. Ferm. Proc. **1:**41–48.

5.8.2 Specific References

9. **Ashwood-Smith, M. J., and E. Grant.** 1976. Mutation induction in bacteria by freeze-drying. Cryobiology **13:**206–213.
10. **Backus, M. P., and J. F. Stauffer.** 1955. The production and selection of a family of strains in *Penicillium chrysogenum.* Mycologia **470:**429–463.
11. **Cabasso, V. J., and R. H. Regamy (ed.).** 1977. International Symposium on Freeze-Drying of Biological Products. S. Karger AG, Basel.
12. **Clewell, D. B.** 1972. Nature of Col E_1 plasmid replication in *Escherichia coli* in the presence of chloramphenicol. J. Bacteriol. **110:**667–676.
13. **Fortney, K. F., and R. W. Thoma.** 1977. Stabilization of culture productivity. Dev. Ind. Microbiol. **18:**319–325.
14. **Greene, H. C., and E. B. Fred.** 1934. Maintenance of vigorous mold stock cultures. Ind. Eng. Chem. **26:** 1297–1299.
15. **Hunt, G. A., A. Gourevitch, and J. Lein.** 1958. Preservation of cultures by drying on porcelain beads. J. Bacteriol. **76:**453–454.
16. **Imanaka, T., and S. Aiba.** 1981. A perspective on the application of genetic engineering: stability of recombinant plasmids. Ann. N.Y. Acad. Sci. **369:**1–14.
17. **Kirsop, B.** 1983. Culture collections—their services to biotechnology. Trends Biotechnol. **1:**4–11.
18. **McDaniel, L. E., and E. G. Bailey.** 1968. Liquid nitrogen preservation of standard inoculum gas-phase storage.

Appl. Microbiol. **16**:912–916.
19. **Mitic, S., and I. Otenhajmer.** 1974. Predicting the stabilities of freeze-dried suspensions of *Lactobacillus acidophilus* by the accelerated storage test. Cryobiology **11**:116–120.
20. **Perkins, D. D.** 1962. Preservation of *Neurospora* stock cultures with anhydrous silica gel. Can. J. Microbiol. **8**:591–594.
21. **Shearer, M. C.** 1979. Actinomycetes: permanent preservation, p. 54–67. *In* Workshop on preservation of microorganisms by freezing and freeze-drying. Society for Industrial Microbiology, Arlington, Va.
22. **Simione, F. P., P. Daggett, M. S. McGrath, and M. T. Alexander.** 1977. The use of plastic ampoules for freeze preservation of microorganisms. Cryobiology **14:** 500–502.
23. **Sokolski, W. T., E. M. Stapert, and E. B. Ferrer.** 1964. Liquid nitrogen freezing in microbiology assay systems. I. Preservation of *Lactobacillus leichmanii* for direct use in the vitamin B_{12} assay. Appl. Microbiol. **12**:327–329.

II. The Process

The Process: Introduction

OLDRICH K. SEBEK

This section describes techniques used to develop processes by which products of interest can be obtained in maximal yields.

The first chapter, by Calam, describes shaken-flask methods, which have been shown to be effective, practical, and useful for carrying out strain selection and examining parameters (such as nutritional requirements, aeration, agitation, or temperature) which affect the growth of the culture and the formation of the product. The application of these techniques is illustrated by examples of product formation, mutant selection, and oxygen transfer.

Next, Mudgett describes solid-state fermentations, which are characterized by microbial growth and product formation occurring on the surfaces of solid substrates. Such methods have been used to make fermented foods in the Orient for centuries; other products include mold-ripened cheeses, starter cultures, and mushroom cultivation. More recently, solid-state fermentations have been used for the production of extracellular enzymes, specialty chemicals, and mycotoxins. Some representative processes are described, and variables and controls are discussed. Such fermentations have been used also for mass production of spores, which in turn have been applied to transform various substrates of interest.

Anaerobic fermentation technology is discussed in the chapter by Ljungdahl and Wiegel. In view of the considerable industrial and ecological importance of anaerobic bacteria (production of alcohols and fatty acids, denitrification in soils, methanogenesis in anaerobic treatment processes), the methods of their cultivation are described in detail.

The chapter by Goodhue et al. deals with the biotransformations of different organic compounds, which can be carried out not only by spores but also by growing, resting, or dried whole cells as well as by cell-free preparations either in batch processes or immobilized systems. Since the reactions are enzymatic in nature, there are considerable advantages to the use of such processes over chemical ones: microbial enzymes react under mild physiological conditions, functionalize nonactivated carbons, display high reaction specificity, resolve racemic mixtures, and introduce centers of chirality into optically inactive molecules.

To produce the desired compounds economically on the industrial scale, high levels of enzymes involved in the production of such compounds are crucial. Thus, the fermentation media selected for the manufacturing process must not only generate good growth and, in specific cases, provide substrates, intermediates, and precursors of the desired product, but they must also support an effective synthesis of the enzymes involved. In the chapter by Miller and Churchill, a variety of relatively inexpensive raw materials used successfully in large-scale fermentations is presented. The methods for developing suitable media, the selection of additional nutrients, their nutritional balance, lists of commercially available raw materials and their suppliers, etc., are also given in this chapter.

In the chapter on continuous fermentation, Stafford describes ways to separate and define parameters (such as the microbial growth rate, nutritional and physical environment, cell density) which are interdependent in batch fermentation. Study of these parameters allows one to examine microbial form, function, and population dynamics, which in turn may lead to improvements in large-scale production. It also allows the selection of different genotypes or mutants with altered enzyme expression from an originally homogeneous population.

Chapter 6

Shake-Flask Fermentations

CHRISTOPHER T. CALAM

Shaken culture represents an important microbiological technique, providing a convenient method of growing microorganisms in submerged culture. The method began to emerge in the 1930s but developed as a major technique with the rise of the antibiotic industry, as a large-scale testing method for laboratory cultures. In many industrial processes, also, shaken cultures are used for the initial stages of inoculum production, and the technique has a wide range of other applications.

Shaken cultures, as part of the general field of submerged culture, have been discussed previously (1, 3), and the present chapter is based on industrial and academic experience.

While submerged shaken cultures may be aimed at cell production, the objective is usually the formation of a product obtained by the growth of mycelial microbes. Production cultures are started from spores or from specially grown inocula. Growth is rapid at first, becoming apparent after 24 h. It then slows down, giving a concentration of 10 to 30 g of cellular mass per liter after 5 to 7 days, when the stationary phase is usually reached. Growth is then usually thick, consisting of large or small pellets or some kind of sludgy growth with a characteristic appearance and odor. Product formation begins 1 to 3 days after inoculation and continues for 3 to 4 days or more. The extent of production depends on the quantity of cells present, their specific activity, and the span of their product-forming capacity.

Since shaken cultures are usually aimed at screening or yield improvement, rather than at theoretical studies, complex media are often used, based on carbohydrates and various proteinaceous materials (corn steep liquor, soybean meal, peanut meal, etc.) and vegetable oil. The components of such media are metabolized at different rates, providing a long period during which conditions are suitable for optimal growth and metabolism. Solid materials, such as chalk ($CaCO_3$), are often added, which can help with pH control and which favor particular types of pellet formation for best results. Optimal results depend a great deal on the use of an optimal inoculum, and in industrial work much attention is given to this. Production in shaken culture can approach that obtained in stirred culture.

Chemically defined media may also be used for shaken cultures, but need careful design to avoid fluctuations in pH or the sudden exhaustion of important components. Such media usually have a basis of sucrose plus ammonium tartrate, phosphate, metal salts, and additional growth factors such as yeast extract (1 g/liter) as needed. All kinds of media are used, and the examples given below illustrate a number of media intended for different purposes.

A wide range of shaken-culture apparatus has been used and found suitable for different types of work. The main types of shaken culture are based on either rotary or reciprocating shaking machines; the former are the most often used. With rotary shakers, in which the flasks move in orbits of usually 50 mm at 200 to 250 rpm (but may vary between 100 and 500 rpm), the culture moves smoothly around the inside of the flasks. In reciprocating shakers the culture is thrown back and forth, which may cause serious splashing. The fermentation vessels are usually Erlenmeyer flasks, but may be flasks of special types (e.g., Sakaguchi flasks [11]) or test tubes.

In general, filamentous microorganisms are grown for the production of secondary metabolic products, such as antibiotics, and require good growth and aeration, whereas unicellular bacteria are grown for enzyme production and similar processes, for tests of culture purity, or for assays where rapid growth is essential. Shaken culture, because of the mixing effect, can often be made to produce rapid and uniform growth which is very suitable for biochemical and genetic work.

The types of flasks used, and the type of shaking, depend on the nature of the fermentation under investigation. Shaken culture is usually applied to aerobic processes, which tend to be of two types: (i) those in which relatively large amounts of oxygen are needed so as to produce large amounts of cells, as is often the case with filamentous organisms, and (ii) those in which aerobic conditions are required but the quantity of oxygen needed is much less, as often happens with bacteria. For the first type of culture, small volumes (25 to 50 ml) of medium are used in large flasks (250- to 500-ml Erlenmeyers, respectively), giving good rates of oxygen transfer. For the second type, larger volumes of medium may be used with a slower shaking rate. Test tubes or other small vessels may be used instead of flasks when large numbers have to be handled. Such small vessels offer both advantages and disadvantages.

Although shaken culture appears to be a very simple technique, it can present a number of difficulties. These are, mainly, that special care is often necessary to obtain the correct type of growth, and that shakers require considerable care and attention if breakdowns are to be avoided. These subjects will be considered later. The methods used in shaken culture may be illustrated by a number of examples.

6.1 CONDUCT OF SHAKEN-FLASK CULTURES

The purpose of shaking in submerged culture is to supply oxygen and nutrients to the growing cells. In shaken culture the medium in the fermentation flasks is inoculated with spores or cells. The strain used as inoculum is held as a master culture, in the freeze-dried state or at −70°C, and then grown on slants by standard techniques. In some cases the master culture is inoculated directly into a rich medium, and after incubation on the shaker, the culture is used to inoculate slants.

Production flasks may be inoculated with spores, but are usually started from specially grown inoculum cultures. The details vary (see section 6.3 for examples and section 6.2 for discussion). In most cases, there is an optimal spore concentration for the inoculum, which has to be determined experimentally. Aseptic conditions should be used for the inoculation of the flasks. If large numbers of flasks are being used, a safe and convenient system must be developed, and a sterile working area may be necessary.

Shake-culture machines involve, essentially, a platform to hold the flasks, which is driven by a motor and transmission system. It is convenient to consider the subject of shaker design and use under several headings: (i) the apparatus used; (ii) incubation conditions and temperature control; (iii) maintenance requirements; (iv) makes of shakers; and (v) shaken cultures in tubes or small fermentors.

6.1.1 Apparatus Used for Shaken Culture

The shaker platform consists of an aluminum plate with spring clips to hold the flasks. The most common types used are 500- and 250-ml Erylenmeyer flasks, but smaller flasks (50 ml) may also be used to allow greater numbers. Large flasks (e.g., 2 liters) are used for the production of inocula or spore suspensions. As a rule, standard flasks are used, but sometimes the flasks are provided with some form of baffling (see section 6.3). The platform may hold 50 to 100 flasks or more. Frequently, with large installations, the shaker will have several platforms arranged in a rack, each holding 50 or more flasks. In many cases the flasks are carried on trays which can be directly attached to the shaker. This is important because if the shaker is stopped for more than about 5 min, the cultures may be seriously damaged and will give poor results. Therefore the rapid loading and unloading of large shakers, by means of trays, should avoid lengthy stoppages.

Flasks are often closed by cotton plugs, though different sorts of reusable plugs may be used (e.g., Steristoppers from Exogen Co.). Lighter closures, e.g., a layer of surgical lint or a thin respirator pad, held by a rubber ring or wire spring, may be used to give improved diffusion of oxygen, as illustrated in an example given below (11).

A support mechanism is provided to hold the platform or the framework supporting several layers of flask trays. Various systems are used. In some cases the platform rests on a central column or shaft which provides the motion, or it may rest on supporting rotary columns. The platform also may be supported on an independent crank mechanism which allows free, but controlled, movement. When the platform is supported by the central drive shaft, the corners are steadied by various reinforcements.

It is important that the support system be firm and stable. The main mechanism usually depends on large ball bearings, but if any kind of oscillation develops, excessive strains can be produced which can cause the failure of the bearings or some other part of the system.

The drive mechanism usually consists of a horizontally mounted electric motor, connected to the vertical shaft by gearing. Power is transferred to the shaker by belt or gearboxes. In some cases a shaft drives three or four gearboxes, the vertical shafts of which support and drive the platforms holding the flasks. The motion of the shaker, which is usually imperfectly balanced, puts a considerable strain on the drive mechanism which, in turn, puts a strain on the motor. Frequently, the motor is small, which makes it likely to burn out. An oversized motor, on the other hand, may damage the drive mechanism.

The drive mechanism may be based on an AC electric motor, giving a constant rotational speed, but more commonly a variable speed is provided by an electronically controlled DC motor. In this case a knob is provided which, when rotated, sets the speed of the machine. The carbon brushes in these DC motors need regular inspection, as they wear out and cause stoppages.

6.1.2 Incubation Conditions: the Need for Cooling

Shakers are usually operated in the range of 25 to 37°C. It is important to realize that the motor and drive give off heat. The shakers are often stated to work at "2°C above ambient temperature." It is easy to have problems with temperature control, since the ambient temperature in many laboratories in temperate climates is in the region of 25 to 28°C. For many actinomycetes and fungi, a temperature of 30°C can seriously disturb metabolism, and such temperatures can easily develop in shaken flasks.

It is therefore nearly always necessary to provide cooling sufficient to hold the temperature in the correct range. It is also necessary to provide a good flow of air around the shaker units, so that all the flasks are at the same temperature. Careful checking is necessary. In incubation rooms, cooling may be provided by small cooling units.

Shakers may also be mounted in an incubator box fitted with a temperature-control unit. Unless cooled, these shakers usually operate above the desired temperature, which may make them unsuitable for the intended task.

6.1.3 Maintenance

Maintenance is always important, and shakers must be checked every 2 to 3 months. In particular, lubrication must be attended to, and carbon brushes must be checked for wear and replaced in good time. The effect of missed maintenance usually is that, after expensive repairs, the shaker is never as reliable as it was before. It is advisable to have a spare motor in stock. The temperature control system should also be checked from time to time, as well as shaker speed and other conditions necessary for good results.

6.1.4 Makes of Shakers and the Choice of Machines

The choice of shakers is not easy. It has to take into account the object of the purchase and the physical conditions where it is to be placed. Shakers of the large, industrial type are usually rugged pieces of machinery, of considerable cost, which are expected to work regularly for many years. Suppliers of small and large shakers which have been suggested to the writer are Adolf Kühner AG, B. Braun Melsungen AG, Infors AG, Marubishi Co., New Brunswick Scientific Co., and Queue Systems (addresses are given in section 6.5). Most of these firms have agencies in other countries.

Large machines must be set up on firm foundations in incubation rooms of ample size, with good temperature control and air distribution and cooling machinery sufficient to allow the desired working temperature to be maintained. Although the above companies refer to temperature control, only heating systems are mentioned as a rule, so the cooling equipment is left to the purchaser to provide.

6.1.5 Shaken Culture in Small-Tube Fermentors

The use of shaken test tubes for microbial culture offers considerable possibilities. Test tubes, e.g., 25 by 150 mm or small vials, may be used with either rotary or reciprocal action. The tubes are usually held upright in racks or in a sloping position, e.g., 15° to the vertical. The advantage of small, parallel-sided fermentation units is that they can be packed together in large numbers in a minimum of space, in contrast to flasks, which have to be fitted into holders. On the whole, however, tube fermentors are not as effective as small flasks (50 ml, wide mouth), which also can be used in large numbers.

Small-tube fermentors may be used with a small amount of medium to provide a well-aerated system for filamentous organisms or with a large volume of medium for rapid bacterial growth such as may be required in purity testing. In the former case, rotary shaking is suitable, at, e.g., 250 rpm with 50-mm orbits. For purity tests and similar work, reciprocating shaking (250 strokes per min, 25-mm stroke) may be better, with the tubes incubated in a water bath to give rapid heating to the required temperature. Tubes may be plugged with cotton, but it is more convenient to use plastic or metal closures.

When growing filamentous organisms in tube cultures, a variety of difficulties may be encountered. The vigorous shaking of a small quantity of medium can throw medium and culture up the walls, with consequent loss and possible infection. It may be possible, by experimentation, to find a system that works well, but it is likely, with different isolates, that there will be failures and irregular behavior. It is understood that the first improved penicillin-producing mutants were obtained in 12.5-by-150-mm test tubes containing 1 to 2 ml of medium. These were incubated in batches of 100 to 150 in baskets attached to a reciprocating shaker and were then tested for penicillin production. Attempts to repeat this procedure were rather unsuccessful due to splashing of the culture up the walls and erratic behavior. Other trials with soil isolates also gave variable results and made it desirable to reconsider the real suitability of tube fermentations as a test system. The screening process requires a number of steps: (i) inoculation of the tubes (it is desirable to use several different media for each culture); (ii) incubation on the shaker; (iii) culture sampling, filtration, and dilution, a step that may prove rather inconvenient if the culture is thick and spongy; (iv) dilution and testing of the samples.

It is evident that growth in the shaken tubes is only a part of the system. If this step is unreliable and must be repeated because of the variable responses of different isolates, a good deal of time and effort can be lost. It is worthwhile to consider carrying out this preliminary evaluation in test tubes in surface culture and verifying the results in small Erlenmeyer flasks (50 ml) on a shaker. Some isolates which work in surface culture may fail in submerged culture, but there is likely to be a net gain in such an operation because of the larger numbers that could be handled.

A valuable new device for surface culture screening is any of the available plastic block systems, which provide about 100 0.5-ml culture cups that can be integrated with a system of automatic dilution, colorimetric assay, and recording (R. D. Nolan, *in* Z. Vaněk and Z. Hošťálek, ed., *Excessive Metabolites: Strategy of Strain Construction and Process*, in press), thus enabling large numbers of cultures to be screened with minimum labor, time, and space. The "Titertek Multiscan" system (Flow Laboratories) provides a range of equipment for this purpose, with the availabilty of computer recording and analysis of data.

The objective of the first screening step is usually to produce a crop of isolates likely to be of interest, representing a small proportion of the main population of isolates. It is better to collect a large number of possible isolates and then to use greater care to test them in small shaken flasks, which will give reliable results. There are situations where screening in shaken test tube cultures is prudent, but in such cases careful standardization of procedures and a degree of replication are essential.

6.2 PRACTICAL PROBLEMS IN SHAKEN-CULTURE FERMENTATIONS

Practical problems arise in two areas, in the microbiological process and in the mechanical and electrical operation of the shakers themselves. Mechanical problems are discussed above. For many workers, shaken culture is an introduction to work in submerged culture, and it will be found that more care is usually needed than when working with surface culture or cultures on agar.

The preparation of media and flasks, with their necessary closures, presents no particular problem. The optimum volume of medium must be used, and this may require practical trials. For filamentous organisms, typical volumes are 50 to 100 ml in 500-ml Erlenmeyer flasks or 25 to 50 ml in 250-ml flasks. As a rule, the smaller volumes give better results. This effect is associated with oxygen transfer, which can be improved by the use of baffled flasks and thin closures instead of cotton plugs. The shaker speed may be increased to 400 rpm, though this can lead to wear, and careful placing of the flasks is needed to improve balance. It is important that the plug or closure remains dry; otherwise it will become impervious to oxygen diffusion and the cultures will soon suffocate.

The flasks are inoculated with spores or pregrown inoculum and incubated on the shaker. It is essential that the temperature remain within the desired range and that shaking be continuous. Stoppages of more than a few minutes often have a very harmful effect on the results, although growth may not appear to be affected. Examples of typical shaken cultures are given below, but it should be remembered that microorganisms behave differently and that development work is usually needed to get the best results.

After incubation for 24 to 48 h, the flasks may be examined to judge the extent and type of growth. With aerobic growth, 1 g of carbohydrate usually gives about one-quarter to one-half its weight of cells. With unicellular bacteria, a dense or fairly dense growth of cells will appear. With filamentous species, in media with, e.g., 5% of carbohydrate, a thick growth of cells should develop, reaching 20 to 30 g of cells per liter after 4 to 5 days. If the culture is producing toxic substances or large quantities of by-products, growth will be correspondingly reduced.

It is important to make some quantitative or semiquantitative estimate of cell growth, either by optical density (4), by centrifuging the culture and noting the percentage volume of cells, or by filtering, drying, and weighing the cells. A crude estimate may be made by placing some of the culture in a small measuring cylinder and allowing it to settle. Other changes may also be noted, such as pH, residual sugar, color, appearance, and odor. If these checks are not made and recorded, experimental variations may pass unnoticed, making the interpretation of results difficult. Microscopic examination is also of assistance; the state of the hyphal tips, branching, pellet formation, and infection all provide useful information as to the state of the culture.

The cell or spore concentration used to seed the cultures is of great importance, not only in securing good growth, but also in establishing the quality of the culture. The factors involved in this are not understood but are of great practical importance (10).

A problem which often arises is poor growth. This effect is usually caused by using too few cells to seed the medium. The spore concentration should be in the range of 5×10^4 to 1×10^6 per ml of medium. From experience, I prefer the inoculum to mature in 2 days, but some workers prefer a longer time to obtain the best results in their processes. The smooth growth of filamentous cultures is enhanced by using rich, complex media and by adding chalk or other particulate matter. In some cases when growth has started off in lumps, inocula have been made by disintegrating the lumps in a blender, but this does not always lead to the best results.

The strain or mutant used has an important effect on the mode of growth, and strain selection is important in obtaining both good growth and maximum formation of the desired product.

If foaming occurs and an antifoam agent is needed, vegetable oils or polypropylene glycol 2000 (Dow Chemicals) (1 g/liter or less) may be used, based on experience. Alternatively, the large chemical companies supply a variety of antifoam agents, which should be evaluated before use.

Infection can give rise to problems, though shaken cultures are not particularly subject to it. It manifests itself as a change in the appearance of the culture, an unusual odor, cloudiness of the filtrate, loss of product, and other effects. Infection can be checked by observing the culture under a microscope (500×). It is better to examine an unstained culture, as motility can be seen more readily. Care is needed, because the media used often contain solid particles or oil, and some substrates, such as corn steep liquor, contain dead bacteria. Infection can also be detected by adding 0.5 ml of culture to about 100 ml of nutrient medium in a bottle and incubating, or by plating on agar. As a rule, the contaminant organisms grow out after 24 to 48 h. It is advisable to carry out the incubation at 25 to 30°C, as some common contaminants do not grow well at 37°C. In some cases, it may be thought that the culture is contaminated, but no foreign organisms can be detected. In this case, a careful study of the techniques used may suggest whether the culture has really been infected or not.

6.3 EXAMPLES

These examples serve to illustrate the use of shaken cultures for different purposes, using typical media and typical conditions.

6.3.1 Citric Acid

Erlenmeyer flasks (500 ml, with 50 ml of medium) were incubated on a rotary shaker (25-mm orbits at 270 rpm [9]). The medium contained, per liter, 140 g of deionized glucose, 2.5 g of ammonium nitrate, 2.5 g of KH_2PO_4, 2.5 g of $MgSO_4 \cdot 7H_2O$, 0.06 mg of Cu, 25 mg of Zn, 1.3 mg of Fe, and 1.0 mg of Mn (as sulfates), at an initial pH of 3.8. The flasks were inoculated with a high concentration of spores of *Aspergillus niger* 72-4, and a 72% conversion of sugar to citric acid was obtained after 9 days. There are many references dealing with the need to use a high concentration of spores to produce an optimal form of pellet growth. For example, Miles Laboratories (British patent 672,128, 1952) used 80,000 spores per ml to produce small pellets, 0.1 to 0.5 mm in diameter, consisting of stubby mycelia with many swellings and with a granulated and vacuolated structure.

6.3.2 Penicillin

Furylmethylpenicillin (M. Cole, G. N. Rolinson, and M. J. Soulal, British patent, 1,118,566, 1968), for conversion to 6-aminopenicillanic acid, was made by growing a mutant of *Penicillium chrysogenum* in a medium containing corn steep liquor to provide an inoculum for the production medium, which consisted of (per liter): 20 g of cottonseed meal (Pharmamedia [Traders Oil Mill Co.]), 20 g of peanut meal, 70 g of lactose, 10 g of $CaCO_3$, 5 g of $CaSO_4$, 4 g of Na_2SO_4, and 2.5 g of ground-nut oil. 2-Furylacetic acid (5% in water, pH 7) was added daily (1 ml). The 500-ml conical flasks, having three single glass spikes as baffles, each contained 100 ml of solution and were closed with gauze pads. After 7 days of growth at 25°C, 5.2 g of sodium 2-furylmethylpenicillin was obtained per liter. For a more recent account of the production of penicillin in shaken flasks, see Queener and Swartz (7).

6.3.3 Effect of Spore Concentration on Penicillin Production

Cultures were grown in 250-ml Erylenmeyer flasks in 50 ml of medium, shaken on a rotary shaker (220 50-mm orbits per min) at 25°C (10). Different concentrations of *P. chrysogenum* JV101 spores were added to a series of seed flasks containing glucose plus corn steep liquor. After 42 h, 5-ml volumes of the inocula were transferred to production flasks which contained 45 ml of medium containing (per liter): 14.7 g of corn steep solids, 38.5 g of lactose, 6.2 g of $CaCO_3$, 0.5 g of KH_2PO_4, 0.25 g of $MgSO_4 \cdot 7H_2O$, 1.0 g of $Na_2S_2O_3 \cdot 10H_2O$, and 2 g of phenoxyacetic acid. To each flask were added 1 ml of peanut oil and 0.5 ml of White oil (Shell, London). The flasks were plugged with cotton wool. Typical cell yields and penicillin production are shown in Table 1. The spore concentration for optimal results was 10^4/ml in the inoculum stage, when the yield of cells was 0.5 g/g of sugar used.

TABLE 1. Penicillin production and growth with different spore concentrations in the inoculum

Spore concn in inoculum (per ml)	Cell concn (g/ml) at time:		Penicillin yield (U/ml) at time:		Growth type
	48 h	120 h	48 h	120 h	
2×10^2	4.7	11	30	750	Smooth, compact pellets
1×10^3	9.0	17	50	900	
5×10^3	11.0	28	140	4,800	Small, feathery pellets, 0.4 mm in diameter
1×10^4	15.5	28	400	5,000	
5×10^4	17.5	28	470	5,000	

6.3.4 Penicillinase

Penicillinase was produced in 5-liter conical flasks containing 1 liter of a tryptone medium, shaken at 100 strokes per min with a 50-mm throw on a reciprocating shaker at 37°C. The flasks were seeded with a suitable strain of *Bacillus subtilis*, and penicillinase production was induced by the addition of penicillin solution (6).

6.3.5 Rifamycin

An evaluation of potential precursors of rifamycin was carried out as follows (5). An inoculum of *Nocardia mediterranei*, using 40 ml of a complex medium, was grown in 200-ml flasks incubated for 4 days at 250 rpm at 28°C (5). The inoculum was added to a complex medium containing, per liter: 70 g of glucose, 20 g of glycerol, 30 g of Protanimal, 10 g of soya meal, 8 g of $CaCO_3$, 3 g of ammonium sulfate, 1 g of KH_2PO_4, 0.003 g of $CuSO_4 \cdot 5H_2O$, 0.05 g of $ZnSO_4 \cdot 7H_2O$, 0.004 g of $MnSO_4 \cdot 4H_2O$, 0.002 g of $CoCl_2 \cdot 6H_2O$, and 0.001 g of $(NH_4)_2Mo_7O_{24} \cdot 4H_2O$, adjusted to pH 7.1. Single-baffled, 200-ml flasks containing 40 ml of medium were used, shaken at 250 rpm, and the fermentation was continued for 5 days.

6.3.6 Selection of Griseofulvin Mutants

Spores of *Penicillium patulum* were mutated with ^{35}S and plated, and colonies were picked off onto malt agar slopes and cultured for testing (R. S. Aytoun and R. W. McWilliam, British patent 788,188, 1957). In the preliminary tests, 7.5 ml of Nitrochalk (a commercial fertilizer; I.C.I. Ltd.) plus brown sugar medium was used because it gave the most consistent results. This medium was distributed in 30-ml glass vials and incubated on a rotary shaker (250 rpm, 50-mm-diameter orbits) at 24°C for 7 days. After assay, the highest yielding mutants were tested in 500-ml Erlenmeyer flasks containing 60 ml of a complex medium comprising, per liter, 28.5 g of corn steep solids, 70 g of lactose, 1 g of KCl, 4 g of KH_2PO_4, and 8 g of ground limestone, on a rotary shaker at 250 rpm. The screening of 750 to 800 isolates gave four markedly improved mutants.

6.3.7 Oxygen Transfer in Shaken Flasks

Yamada et al. (11) tested oxygen transfer rates in Erlenmeyer flasks, with or without a baffle or indentations, and the special Sakaguchi flasks by the sulfite method (2) or by the rate of conversion of sorbitol by *Aerobacter suboxydans*. Erlenmeyer flasks with cotton plugs were used on a rotary shaker (190 rpm, 35-mm orbits). Oxygen transfer rates were 0.92×10^{-5} mol/ml per h; these were increased to 4.37×10^{-5} and 5.29×10^{-5} mol/ml per h in flasks having one or three indentations. With three indentations, the oxygen transfer rate was still further increased to 8.9×10^{-5} mol/ml per h when a polyfluorocarbon fiber closure was used. With Sakaguchi flasks (500-ml capacity, cylindrical, with a long neck) on a reciprocating shaker with 140 strokes (85 mm) per min and with 50 ml of medium, the oxygen transfer rate was 4.33×10^{-5} mol/ml per h with a cotton plug and 5.85×10^{-5} mol/ml per h with a polyfluorocarbon closure. Rates of oxygen transfer with the bacterial culture were equal to those obtained by the sulfite method (2).

6.4 DISCUSSION: THE SHAKEN FLASK AS A BIOREACTOR

The shaken-culture technique is one form of the general technique of submerged culture, which also includes stirred fermentor culture. Both of these much-used methods involve aeration and agitation and are extensively used with filamentous microorganisms, but the two methods tend to give different metabolic patterns. The growth of the cells and the production of metabolites involve a number of cellular processes, such as the dissimilation of substrate carbohydrates, the formation of intermediates, and the generation of energy. The results are affected by the pattern of aeration and agitation, the characteristics of the cells themselves, their efficiency and productivity, and their response to the conditions in the fermentor. In stirred fermentor cultures, the agitation system stirs and mixes the medium and enables the uptake of oxygen, mainly by diffusion from the medium. In shaken culture, on the other hand, the cells seem to obtain oxygen largely by direct absorption from the air, as the culture swirls around the walls of the flask in a thin layer, enabling good growth to take place. Altogether, shaken flasks give much better growth and production than would be expected from the rate of oxygen uptake indicated by the sulfite oxidation method (2). This is illustrated by the cited example dealing with the effect of spore concentration in the inoculum medium, where excellent levels of cell production were achieved, and by the high penicillin titers described by Queener and Swartz (7). Attempts are made to increase the degree of mixing in shaken flasks, as illustrated by the rifamycin example above in which a wedge-shaped indentation was made in the flask, or in the penicillin production example, where spikes were used. These procedures, however, can lead to foaming.

The differences in agitation and aeration tend to create different metabolic patterns in shaken and stirred cultures, which in turn can cause the two systems to yield different amounts of the desired product. Mutants can be selected which respond well in both systems.

The above example of penicillin production, in which different spore concentrations were used to prepare the inoculum, emphasizes the importance of the regulatory pattern in the cells as they begin to grow in the production stage. This effect is often of practical significance, and the optimal way of preparing starter cultures is of considerable importance. Spore concentration is also important in citric acid production; the literature gives many references to the desirability of particular forms of pellets, to suit particular mutants (Miles Laboratories Inc., British patent 672,128, 1952).

Another factor brought out by the examples above is the frequent use of complex organic media, often with added $CaCO_3$ and vegetable oil. This is in contrast to the theoretical objective, which favors chemically defined media. As mentioned above, the metabolic patterns of growth and production are complicated, and most microorganisms grow better on complex media. This is especially so in shaken cultures, where it is difficult to apply the degree of control of pH or sugar feed which is possible in stirred cultures. Chalk ($CaCO_3$) is added to control pH, but it also has an effect on the growth pattern. This effect may extend to the particular type of $CaCO_3$ used, as it does in the case of oxytetracycline production (8).

In spite of its complexity, shaken culture is essentially a very practical and useful fermentation system. It is usually fairly easy to adjust medium and inoculum conditions to obtain good results. At the same time, it is advisable to realize that the system is relatively crude. If closely controlled fermentations are required, with automatic pH control, specialized sugar feed rates, and other factors, stirred fermentor culture is better in all respects. Although the problem of mechanical failure has been stressed, it can be overcome by careful attention to maintenance and other requirements, and it should not be regarded as too serious an obstacle in the way of successfully conducting shake-flask fermentations.

This discussion has related mainly to the use of filamentous microorganisms, but the shaken-culture method is often applied to other types of fermentations. It has the advantage of producing rapidly growing, efficient cells at high concentrations and is useful for many unicellular bacterial and yeast culture processes. Shaken culture can also be applied to test tubes or other small fermentation vessels, although this is not always the best approach to mass screening in the search for new microorganisms.

6.5 COMMERCIAL SOURCES

B. Braun Melsungen AG, P.O.B. 346, D-3508 Melsungen, Federal Republic of Germany
Dow Chemical Co., Midland, Michigan 48640
Exogen Co., Glasgow, Scotland
Flow Laboratories, Irvine, Ayrshire, Scotland
I.C.I. Ltd., Agriculture Division, Billingham, Teesside, England
Infors AG, Rittergasse 27, CH-4103, Bottmingen, Switzerland
Adolf Kühner AG, Argensteinstrasse 46, CH-4053 Basel, Switzerland
Marubishi Trading Co. Ltd., Koko Building, 8-7, 2-chome, Kaji-cho, Tokyo, Japan
New Brunswick Scientific Co. Inc., P.O. Box 986, Edison, New Jersey 08817
Queue Systems, Box 1901, Parkersburg, West Virginia 26102
Traders Protein, Buckeye Cellulose Corp., Memphis, Tennessee 38108

6.6 LITERATURE CITED

1. **Calam, C. T.** 1969. The culture of microorganisms in liquid medium, p. 225–326. *In* J. R. Norris and D. W. Ribbons (ed.), Methods in Microbiology, vol. 1. Academic Press, London.
2. **Cooper, C. M., G. A. Fernstrom, and S. A. Miller.** 1944. Performance of agitated gas-liquid contactors. Ind. Eng. Chem. **36:**504–509.
3. **Drew, S. W.** 1981. Liquid culture, p. 151–179. *In* P. Gerhardt (ed.), Manual of methods for general bacteriology. American Society for Microbiology, Washington, D.C.
4. **Flowers, T. H., and S. T. Williams.** 1977. Measurement of growth rates of streptomycetes: comparison of turbidimetric and gravimetric techniques. J. Gen. Microbiol. **98:**285–289.
5. **Ghisalba, O., P. Traxler, and J. Nüesch.** 1978. Early intermediates in the biosynthesis of ansamycins. I. Isolation and identification of protorifamycin. J. Antibiot. **31:**1124–1131.
6. **Pollock, M. R.** 1950. Penicillinase adaptation in *B. cereus* adaptive enzyme formation in absence of free substrate. Br. J. Exp. Pathol. **31:**739–753.
7. **Queener, S., and R. Swartz.** 1979. Penicillins: biosynthetic and semi-synthetic, p. 35–122. *In* A. H. Rose (ed.), Secondary products of metabolism. Academic Press, London.
8. **Riviere, J.** 1977. Industrial applications of microbiology, p. 206. (Translated and edited by M. O. Moss and J. E. Smith.) Surrey University Press, London.
9. **Shu, P., and M. J. Johnson.** 1948. Citric acid. Production by submerged fermentation with *Aspergillus niger*. Ind. Eng. Chem. **40:**1202–1205.
10. **Smith, G. M., and C. T. Calam.** 1980. Variations in inocula and their influence on the productivity of antibiotic fermentations. Biotechnol. Lett. **2:**261–266.
11. **Yamada, S., M. Wada, and I. Chibata.** 1978. Oxygen transfer in shake flask cultures and the conversion of sorbose by *Acetobacter suboxydans*. J. Ferment. Technol. **56:**20–28.

Chapter 7

Solid-State Fermentations

RICHARD E. MUDGETT

Solid state fermentations may be briefly defined as those in which microbial growth and product formation occur on the surfaces of solid substrates. Although such fermentations include a number of well-known microbial processes, such as soil growth, surface culture, composting, wood rotting, mushroom cultivation, and the production of familiar Western foods, e.g., bread, mold-ripened cheese, and sausage, this chapter will focus on solid-state fermentation methods traditionally used to manufacture oriental foods. Commonly known in the East as koji and in the West as solid-substrate fermentations, such processes are well established in the Orient and may have significant advantages over submerged-culture methods for the manufacture of nontraditional products of interest to the food, pharmaceutical, and chemical industries.

7.1 GENERAL CONSIDERATIONS

7.1.1 Solid-State Characteristics

Solid-substrate fermentations are distinguished from submerged cultures by the fact that microbial growth and product formation occur at or near the surfaces of solid materials with low moisture contents. Substrates traditionally fermented in the solid state include a variety of agricultural products, such as rice, wheat, millet, barley, corn, and soybeans. However, nontraditional substrates which may also be of interest in industrial process development include an abundant supply of agricultural, forest, and food-processing wastes. Solid substrates may be viewed as gas-liquid-solid mixtures in which an aqueous phase is intimately associated with solid surfaces in various states of sorption and is in contact with a gas phase continuous with the external gas environment. Depending on the moisture content of the solid, some of the water is tightly bound to solid surfaces, some is less tightly bound, and some may exist in a free state in capillary regions of the solid. The gas-liquid interface provides a boundary for oxygen-carbon dioxide exchange and for heat transfer. The solids phase provides a rich and complex source of nutrients which may be complete or incomplete with respect to the nutritional requirements of an organism to be cultured. The chemical composition of agricultural solids is generally classified by proximate analysis in terms of protein, carbohydrate, lipid, and ash contents and by elemental analysis. Although the solids generally contain some small carbon compounds, the bulk of dry weight is in complex polymeric forms requiring enzymatic hydrolysis for use as carbon energy sources in microbial metabolism. In comparison with submerged cultures, which generally employ less complex carbon energy sources, solid substrates provide mixed substrates of high-molecular-weight carbon compounds which may involve induction, inhibition, or represssion mechanisms in microbial metabolism. The unique characteristic of solid-substrate fermentations is their ability to provide a selective environment at low moistures for mycelial organisms that produce a variety of extracellular enzymes, surface bound or free, and can grow at high nutrient concentrations near solid surfaces. These organisms include a large number of filamentous fungi and a few bacteria, such as the actinomycetes, in addition to at least one strain of *Bacillus*.

Solid-state methods are not as well characterized on a fundamental scientific or engineering basis as are the submerged-culture methods used almost exclusively in the West for the industrial production of microbial metabolites. They are, however, widely used in the Orient, and traditional methods used in food processing for more than 2,000 years have been modernized and extended to nontraditional products. Solid-state fermentations have been classified by Hesseltine (27) according to the physical state of the substrate as follows. (i) Low-moisture solids are fermented without agitation, e.g., tempeh and natto; with occasional stirring, e.g., miso and soy sauce; or with continuous agitation, e.g., aflatoxin. (ii) Suspended solids are fermented in packed columns through which liquid is circulated, e.g., rice wine, or in stationary or agitated liquid media, e.g., Kaffir beer.

Solid-state processes have also been classified by Ralph (50) according to the nutritional state of the substrate, i.e., (i) the solid is the major nutrient source, or (ii) the solid is nutritionally inert. Processes in the latter category are uncommon, but include the production of citric acid on inert supports impregnated with nutrients (1, 36). These processes would, however, permit the use of defined or complex media in natural physical environments. Most solid-state processes to date are in the former category and differ from submerged cultures in that major nutrients are in complex polymeric mixtures in low-moisture environments which favor extracellular enzyme synthesis in growth-associated metabolism. Although physiological effects of the aqueous environment on biomass and product formation by the filamentous fungi have been studied intensively in submerged cultures (7, 11), little information is available on effects of the microbial environment in the solid state. Presumably, such effects are similar to those in submerged culture, as modified by chemical composition and physical structure of the substrate and by local variations in temperature, pH, nutrients, and dissolved-gas concentrations. Additional areas of physiological interest for mycelial organisms in the solid state include phenomena of solute transport, translocation (31), and differentiation (70). Such effects are not easily investigated in natural solid-state environments because the liquid phase in low-moisture solids is generally inaccessible to in vivo measurements for major parameters of interest. Tsuchiya (70) has made the interesting point that fungal growth and differentiation with time occur in various morphological stages which may favor the production of specific secondary metabolites. For example, *Aspergillus awamori* cultured on semisolid medium containing 1.0% glucose, 0.5% yeast extract, and 0.5% peptone exists in four morphological states and produces α-amylase, glucamylase, protease, and alkaline phosphatase in an overlapping time sequence. Tsuchiya's point is that residence time in plug-flow continuous fermentors may be controlled to favor the production of specific metabolites associated with the morphological state (age) of the culture. Such effects may be evaluated in the solid state by screening batch cultures for specific extracellular metabolites extracted from the solids as a function of time.

The kinetics of fungal growth in submerged cultures with a single carbon energy source may be linear or exponential (14, 56), depending on the physiological state of the organism. However, useful kinetic models for fungal growth in the solid state are not easy to obtain because of the complex nutritional nature of the substrate and the difficulty in estimating rates of biomass formation in the presence of the substrate. Similar difficulties are seen in modeling heat and mass transfer effects in microenvironments influenced by the following general characteristics.

- Traditional solid-substrate fermentations may involve mixed cultures of indigenous microbial flora, seed inocula, or both.
- Solid substrates provide selective environments for a large number of filamentous fungi and a few bacteria, such as actinomycetes, which grow in mycelial form.
- Natural solid substrates provide mixed carbon energy sources and a diverse and complex source of nutrients which may or may not be complete with respect to nutritional requirements of the organism(s) to be cultured.
- Traditional substrates contain some small carbon compounds. However, the bulk of the dry weight of these substrates is in high-molecular-weight biopolymers, e.g., starch, cellulose, hemicellulose, pectin, protein, and lipid, which require enzymatic hydrolysis for assimilation in growth-associated primary metabolism.
- Hydrolytic enzymes for the assimilation of high-molecular-weight compounds by mycelial organisms generally employed in the solid state are extracellular and may be surface bound or free.
- Mixtures of high- and low-molecular-weight carbon compounds lead to complex patterns of induction, repression, and inhibition in the regulation of microbial metabolism.
- Apical growth of mycelia on solid surfaces may permit primary and secondary metabolism to occur simultaneously in different parts of the mycelia.
- Microbial growth and product formation occur at or near solid surfaces in a liquid phase which interfaces with both the solid phase and the external gas environment.
- Some substrate moisture is tightly bound, some is less tightly bound, and some exists in a free state on internal and external solid surfaces in proportions determined by moisture isotherm characteristics.
- The gas-liquid interface provides a boundary for oxygen-carbon dioxide exchange and heat transfer at high liquid surface-to-volume ratios. It also provides a large surface area for the exchange of moisture between the liquid and gas phases, which may influence biological activity with respect to sorption-desorption effects on substrate moisture content.
- Fungi commonly employed in solid-substrate fermentations are obligate aerobes and need to obtain oxygen from the gas phase under relatively stagnant conditions for gas transfer.

- Biomass densities in the liquid phase may be very high, leading to extreme oxygen demand and carbon dioxide evolution rates in late fermentation stages. Since chemoheterotrophic processes are exothermic, rates of heat generation in microbial biomass formation may also be stringent under relatively stagnant conditions for heat transfer.
- Growth-associated enzyme synthesis or product formation in some solid-state fermentations is extremely shear sensitive; solids may not be agitated greatly, or at all, in such cases.

7.1.2 Advantages and Disadvantages

Choice of the solid-state method for a new fermentation process must be based on consideration of the advantages and disadvantages of the method as compared with those of submerged culture. Advantages of the solid state method cited by Hesseltine (23) include the following.

- Solid substrates may require only the addition of water; other nutrients may also be added.
- Fermentation vessels may be small relative to product yield, since little water is used and the substrate is concentrated.
- Seed tanks are unnecessary, and spore inocula may be used.
- Low moisture reduces the problem of contamination.
- Conditions for fungal growth are similar to those in natural habitats.
- Culture agitation inhibits mold sporulation, reducing the risk of laboratory contamination.
- Aeration is facilitated by spaces between substrate particles and particle mixing.
- Product yields may be much higher than those in liquid media and are reproducible.
- Fermented solids may be extracted immediately by the direct addition of solvents or maintained in frozen storage before extraction.
- Products may be incorporated directly into animal feeds.

In addition, solid-state methods can be used to provide low-shear environments for shear-sensitive mycelial organisms.

Disadvantages of the method cited by Hesseltine include the following.

- Fermentations with continuous agitation or rotation of fermenting solids may involve high power requirements.
- The addition of water in early fermentation stages may increase the risk of bacterial contamination.
- Spore inocula may be quite large and may need to be produced and harvested aseptically.
- Agricultural substrates may require some kind of pretreatment such as cracking or surface abrasion.
- Considerable developmental work may be required to make a process feasible on a large scale.

7.1.3 Substrate Selection

The selection of a solid substrate for a fermentation process may involve screening a large number of agricultural materials for microbial growth and product formation. Substrates suggested by Hesseltine (23) include whole rice kernels, polished rice, pearled barley, pearled wheat, cracked corn, and dehulled, cracked soybeans. Wheat and barley were pearled by lightly abrading the seed coat; corn and soybeans were cracked into five or six pieces. Hesseltine noted that particles or kernels of grain must (i) be within a limited size range to prevent particle agglomeration, (ii) be maintained at relatively low moistures to prevent contamination, and (iii) occupy a relatively small total volume compared with the size of the fermentation vessel.

In several fermentations for mycotoxin production, glutinous rice was used as a static (stationary) control for comparison of yields on other grains under static or agitated culture conditions. For example, the production of cytochalasin E and the tremorgenic toxins tryptoquivaline and tryptoquivalone by *Aspergillus clavatus* was tested (16) on grains suggested by Hesseltine. Crude toxin production on agitated pearled barley was three times greater than on static glutinous rice and 50% greater than on static pearled barley. Procedures employed for mycotoxin production were as follows (16).

Spores from a soil culture maintained at 4°C were inoculated onto Czapek solution agar, enriched with 0.5% yeast extract and 0.5% Casamino Acids (Difco Laboratories), in a 120-ml dilution bottle. The incubation lasted 4 to 6 days at 30°C. Spores were suspended in 100 ml of 0.01% (wt/vol) sodium lauryl sulfate, and 10 ml of the spore suspension was added to each 2.8-liter, wide-mouthed Fernbach flask. The Fernbach flasks had received 300 g of pearled barley and 150 ml of warm water; they were then autoclaved, cooled, and incubated on a rotary shaker at 200 rpm (2-in. [ca. 5.08-cm]-diameter orbit) for 12 days at 30°C. The barley was extracted twice, each time with 500 ml of methylene chloride per flask. After the first addition of methylene chloride, the suspension was allowed to soak overnight. The flask contents were then filtered through glass wool, and the grains were blended in a Waring blender for 60 s with a second 500-ml portion of methylene chloride. The homogenate was filtered through cheesecloth, and the solids were discarded. After the two extracts were combined, anhydrous magnesium sulfate was added to remove traces of water; the suspension was then filtered on a Büchner funnel with no. 595 filter paper (Schleicher & Schuell) or Hyflo Super Cel. The filtrate was evaporated to dryness in a flask evaporator, and the oily residue was then precipitated by pouring it slowly into ice-cooled petroleum ether and placing it overnight in the ice bath. The crude toxin fraction was recovered by filtration under vacuum on a medium-pore, sintered-glass filter and allowed to dry in the hood before being weighed. To remove traces of petroleum ether before the weight of the fraction was determined, the vial was stored overnight under negative pressure.

Similar procedures have been employed in the production of cyclopiazonic acid by *Aspergillus flavus* (39), the production of malformin C, a cyclic pentapeptide, by *Aspergillus niger* (34), and the production of cyclochlorotine, a hepatotoxin, by *Penicillium islandicum* (19). In a preliminary study of spore production by *Penicillium roqueforti* (40), wheat bran, hulled wheat, whole wheat, buckwheat, oats, sorghum, barley, rice, and potato solids were screened as potential substrates at moistures up to 33% by weight

TABLE 1. Typical characteristics of some traditional food fermentations (22, 76)

Product	Primary genus	Common substrate	Thermal processing		Initial moisture (%)	Incubation		Further processing
			Temp (°C)	Time (min)		Time (h)	Temp (°C)	
Soy sauce	*Aspergillus*	Soybean, wheat	110	30	45	72	30	Yes
Miso	*Aspergillus*	Rice, soybean	100	40	35	44	30	Yes
Tempeh	*Rhizopus*	Soybean	100	30	40	22	32	No
Hamanatto	*Aspergillus*	Soybean, wheat				36		Yes
Sufu	*Actinomucor*	Tofu	100	10	74		15	Yes

in static petri dishes incubated for 10 days at 25°C. Spore counts on these substrates ranged from 0.8×10^9 to 1.8×10^9 spores per g of dry solids. Buckwheat was selected as the substrate of choice because of its excellent mechanical properties (structure retention and lack of particle agglomeration) and high spore yields. Spore production on buckwheat seeds with a specific surface area of 750 cm^2/cm^3 (a particle size roughly equivalent to a sphere 80 μm in diameter), soaked in water for 24 h to a final moisture content of 0.48 g/g of dry solids, was then investigated in a series of factorial experiments to determine optimal cultivation conditions of temperature, moisture content, and aeration. These optimal conditions were found to be, respectively, 23.5°C, 0.48 g/g of dry solids, and 4.42 vvh (liters of gas per liter of liquid per hour).

7.2 SOLID-STATE PROCESSES

7.2.1 Traditional Foods

Much of the detailed information on traditional oriental food fermentations is written in Japanese or is proprietary in nature. There are, however, a few comprehensive reviews of such processes in the Western literature (22, 76, 77, 80). These are essentially descriptive and provide little information on microbial-substrate interactions, metabolic regulation, growth kinetics, or heat and mass transfer in the solid state. They do provide considerable insight as to the practical nature of such processes and parameters of interest in fermentation process development. For example, Table 1 illustrates some characteristics of the traditional koji process for oriental food products of general interest, several of which are becoming more popular in the West (22, 76). While the values indicated are generally typical of processing conditions for each product, some variations in these conditions are observed in the literature from one process description to another for the same product. Inoculum densities of 2.5×10^7 and 10^6 spores per 100 g of wet solids are reported for shoyu (soy sauce) and tempeh (fermented soybeans), respectively. Wang and Hesseltine (76) observed that although a content of 60% for cooked soybeans used in soy sauce manufacture is ideal for bacterial growth, combining this mixture with roasted wheat reduces the total moisture content to 45%, which is adequate for fungal but not bacterial growth. Soaking the soybeans used in tempeh fermentation in 0.85% lactic or acetic acid, dehulling the beans, and boiling them in the acid solution has been recommended to provide a selective environment for fungal growth (62). Similar procedures have been employed in other fermentations. For example, the use of 0.1 to 0.3 N hydrochloric acid at pH 3.5 to 4.5 is reported to inhibit the growth of undesirable organisms in solid-state fermentations for amylase (73) and soy sauce (81) production. A recent patent (81) is also based on the addition of 0.2 to 0.8% by weight of acetate to inhibit the growth of bacterial contaminants in koji processing. Lactic acid (0.5% by volume) has also been used in rice wine production (48). As suggested in Table 1, many traditional food fermentations involve two major processing stages. The primary stage, an aerobic solid-state fermentation by one or more fungal species, is known as the "koji." The secondary stage, an anaerobic submerged-culture fermentation with mixed bacterial populations, is designated the "moromi." Conditions of aeration and agitation for the primary, koji phase of these processes are not reported in the literature. Since most fungi are obligate aerobes, these conditions are important considerations in the design and operation of the koji-making apparatus to be discussed below. The products shown in Table 1 are in the solid state during the koji process. Four of these products are subject to further processing. For example, the koji in soy sauce manufacture is transferred to a stirred concrete or wooden tank and mixed in an approximately 1:1 ratio with 17 to 19% (wt/vol) aqueous sodium chloride to make a mash (moromi) which is agitated and aerated frequently to prevent the growth of undesirable anaerobic organisms, maintain uniform temperature, and facilitate carbon dioxide removal. Fermentation in this second stage may be accomplished with indigenous microbial flora, since the koji is not prepared aseptically, or by adding pure cultures of yeasts and bacteria. Strains of *Saccharomyces*, *Torulopsis*, and *Pediococcus* are reported to be important flavor producers in soy sauce manufacture. The initial pH of the mash (6.5 to 7.0) gradually decreases as the lactic acid fermentation by *Pediococcus* sp. proceeds, and yeast fermentation begins at about pH 5.5. The moromi is generally fermented for 8 to 12 months, although Wang et al. (78) indicate that quality soy sauce can be produced in 6 months by temperature profiling, i.e., 15°C for 1 month followed by 28°C for 4 months and finishing at 15°C for 1 month. The final product is pressed to recover the liquid soy sauce, soy oil is removed from the liquid, and the product is pasteurized, filtered, and bottled. Miso koji is subjected to a similar process, except that the final product is a moist (48%) or dehydrated solid.

Hamanatto and sufu are brined in 18 and 12% (wt/vol) sodium chloride, respectively, and aged for 6 to 12 months (hamanatto) or 1 to 3 months (sufu). Hamanatto is produced in koji by *Aspergillus* spp. on a mixture of soybeans and wheat flour and has been classified by Hesseltine (26) as one of two kinds of natto; the other kind is called itohikinatto and involves solid-substrate fermentation of soybeans by the bacterium *Bacillus subtilis* (also called *Bacillus natto*). Hesseltine noted (22) that the order *Mucorales* is well adapted to food manufacture for the following reasons: (i) many genera such as *Rhizopus*, *Mucor*, and *Actinomucor* grow rapidly; (ii) growth from a single spore soon invades large masses of substrate; and (iii) members of these genera produce an abundance of (mycelial-bound or free extracellular) saccharolytic, proteolytic, or lipolytic enzymes. Hesseltine also stated that only one or two mycotoxins have been reported for any member of this order, but that when tempeh, usually made from soybeans by a *Rhizopus* species, is made from copra (coconut meal), food poisoning has occasionally resulted due to *Pseudomonas* contamination. Wang and Hesseltine (76) concluded that the preparation of foods by traditional fermentation methods has a number of advantages. Such fermentation (i) produces desirable enzymes, (ii) destroys or masks undesirable flavors and odors, (iii) adds (desirable) flavors and odors, (iv) preserves, (v) synthesizes desirable constituents such as vitamins and antibiotics, (vi) increases digestibility, (vii) changes physical state (e.g., softens), (viii) produces color, and (ix) reduces cooking time. They observed that the microorganisms used in these fermentations are not toxin producers and that, properly fermented, such foods are not hazardous to health.

7.2.2 Rice Wine

The rice koji process provides an interesting and useful model for aerobic solid-substrate fermentations. While rice koji is used as a starter culture in a variety of traditional food fermentations, it is also used as a starter culture in rice wine (sake) production. The process is described in great detail by Nunokawa (48), with extensive references to the Japanese literature. In the rice koji process, the primary substrate, brown rice, is milled and polished, washed, and steeped in water to solubilize some of the rice nutrients and swell the grains. Steeping time is critical, since the absoprtion of too much water causes stickiness and particle agglomeration. Polished rice absorbs sufficient water in 3 to 4 h. The steeped rice is then steamed for 30 to 60 min, during which time the rice is pasteurized, its starch is converted to alpha form, and its proteins are denatured, making these constituents more susceptible to enzymatic hydrolysis. Normally, the final moisture content is about 30 to 35% of the original rice weight. The steamed rice is then inoculated with spores of *Aspergillus oryzae* and incubated on trays in environments maintained at 25 to 30°C and 80 to 90% relative humidity. After 20 to 24 h, the rice is transferred to smaller trays for improved temperature control and aeration. After a total fermentation time of 44 to 48 h, during which the temperature of the moist solids may rise to 38 to 42°C, the koji is harvested for use in sake or food production.

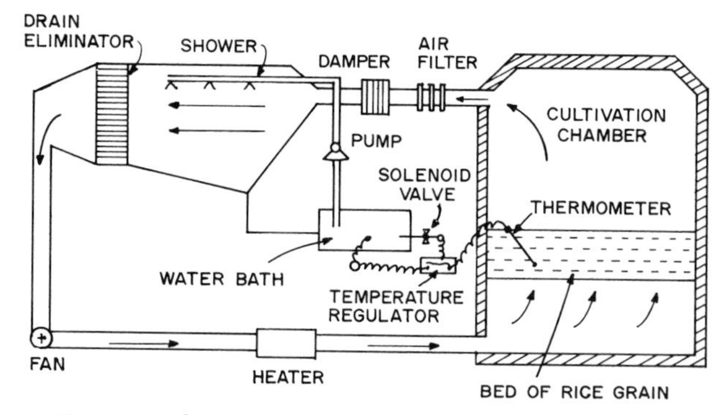

FIG. 1. Schematic representation of a koji-making apparatus. From Nunokawa (48).

A schematic diagram of a koji-making apparatus described by Nunokawa (48) is shown in Fig. 1. Washed, filtered air is circulated through the grain bed, humidified by a spray shower, and heated to control the temperature and humidity of the fermenting solids. The temperature of the solids is monitored in a feedback control system which regulates the temperature of a water bath that feeds the shower. When the temperature of the solids is too low, a heater is switched on to warm the water. The airflow rate is regulated by a damper, apparently manual for the unit shown. Nunokawa commented that many kinds of koji-making apparatus have been devised based on similar principles. He classified these in two major groups, rotary drum systems and airflow systems with thick, medium, or thin beds in single or multiple layers. In drum systems, mixing and temperature control of the solids are accomplished by rotating the drum with circulation of conditioned air. In airflow systems, Nunokawa noted that mixing should be done by hand and is sometimes omitted, with temperature and moisture controlled by surface or penetration forced-convection systems on a continuous or intermittent basis. In both systems, the air is conditioned by spraying with temperature-controlled water. Hesseltine (personal communication) has commented that koji is now made in large rotary tray fermentors.

For alcohol production in the second stage, some of the koji is mixed with unfermented steamed rice and unsterilized water and inoculated with a pure culture of wine yeast to produce yeast starter culture (moto), which is cultivated anaerobically. Yeast propagation begins after nitrate-nitrite reduction and lactic acid fermentation by indigenous bacteria, which prevent contamination by bacterial pathogens and wild yeasts. Alternatively, commercial lactic acid may be added to about 0.5% by volume. The moto is then mixed with the remaining koji and additional water to obtain a main mash (moromi), which is fermented for about 20 days at 25°C. The moromi is dense and mushy and provides a solid matrix which retains the bulk of the yeast population in suspension throughout the fermentation. The moromi is then filtered, and the filtrate is settled, pasteurized, and stored before blending and bottling. Nunokawa reported the process to yield about 20 gal (ca. 76 liters) of sake, at 16% alcohol content, per 150 kg of brown rice.

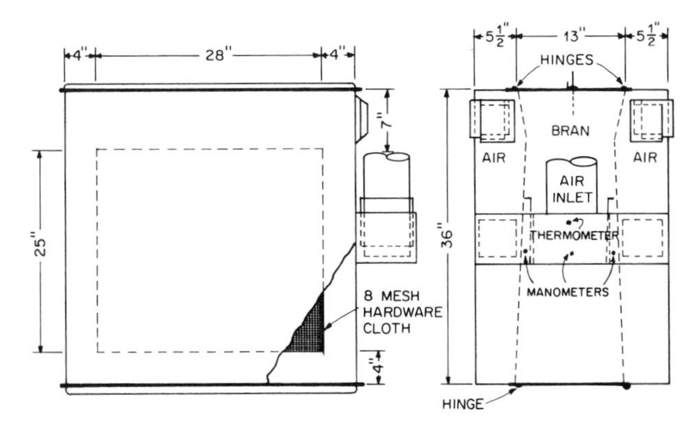

FIG. 2. Vertical incubation cell. From Underkofler et al. (73).

7.2.3 Enzymes

The major industrial use of the koji process, apart from food and beverage manufacture, is for the production of fungal enzymes. Koji is probably the largest enzyme product on a worldwide basis (5). Such processes are widely used in Japan (67) for enzyme manufacture, but are not well described in the literature in terms of current research or levels of manufacturing productivity. In 1896, Takamine marketed the enzyme Takadiastase, a fungal amylase preparation, and subsequently introduced the process to the West to replace barley malt for the production of alcohol by fermentation of grains and potatoes (63). The process involved wheat bran, pretreated with formaldehyde to reduce contamination, inoculated with *A. oryzae* spores, and incubated in a slowly rotating drum through which conditioned air was passed. This led to investigations by Underkofler and co-workers (71–73), which showed that acidifying the bran to a pH of 3.5 to 4.5 with dilute hydrochloric acid in place of water inhibited the growth of undesirable organisms. These studies involved laboratory, pilot, and plant-scale fermentations using perforated pans, trays, rotating drums, and cells (packed beds) with forced ventilation. Semicommercial production of mold bran in two units showed that an incubator with hinged trays (Fig. 2), in a ventilated room, was easier to operate and gave a more uniform product than one designed in accordance with a two-stage inclined packed-bed system (Fig. 3), developed in pilot plant studies. The tray system was limited to bed depths of 2 in. (ca. 5 cm). The packed-bed system (inclined to reduce compacting of the solids) involved primary and secondary fermentors of similar design operated at bed depths of up to 14 in. (ca. 35.5 cm). Mold bran was removed from the primary incubator at the maximum point of heat generation in 15 h, broken into pieces about 1 in. (ca. 2.5 cm) in diameter, loosely packed in the secondary incubator, and fermented for an additional 18 to 20 h. A semicommercial tray system of 1-ton/day capacity (Fig. 4) was operated successfully for 9 months, after which a commercial plant of 10-ton/day capacity was constructed.

Solid-state methods have been used for the commercial production of cellulase, amylase, and pectinase (66), protease (33), and lipase (82). An increase in cellulase production from 400 to 3,000 filter paper units per mg of solids was obtained during the period from 1962 to 1974 in extracts of *Trichoderma viride* (now *reesei*) cultured on wheat bran (67). Activities for the commercial cellulases Onozuka (Kinki Yakult) and Meicelase (Meiji Seika) on wheat bran in 1976 were ca. 3,750 filter paper units per mg. The former enzyme is produced in trays, and the latter is produced in an automatic rotary tray process (68) on wheat bran, the usual substrate for enzyme production in Japan. Typical koji-making apparatus used in such fermentations (67) is shown in Fig. 5 through 7. Toyama reported (67) that *Trichoderma* cultured on sieved (6-mesh) wheat bran produced twice as much filter paper activity as the culture grown on unsieved wheat bran; culture conditions were not indicated. It is not clear whether this result was due to a difference in chemical composition or particle size of the residue, although the latter is suspected. In the same study, the organism was cultured on natural and delignified rice straw, newspaper, and cardboard. Similar activities were obtained for the natural materials and were not significantly lower than on wheat bran. Toyama concluded that saccharification of cellulosic wastes with enzymes from cultures grown on wheat bran is impractical, due to the cost of wheat bran, and that cellulase-producing organisms should be cultured on natural cellulosic wastes. He also noted that cellulase production by the koji method is incompatible with waste saccharification, since although

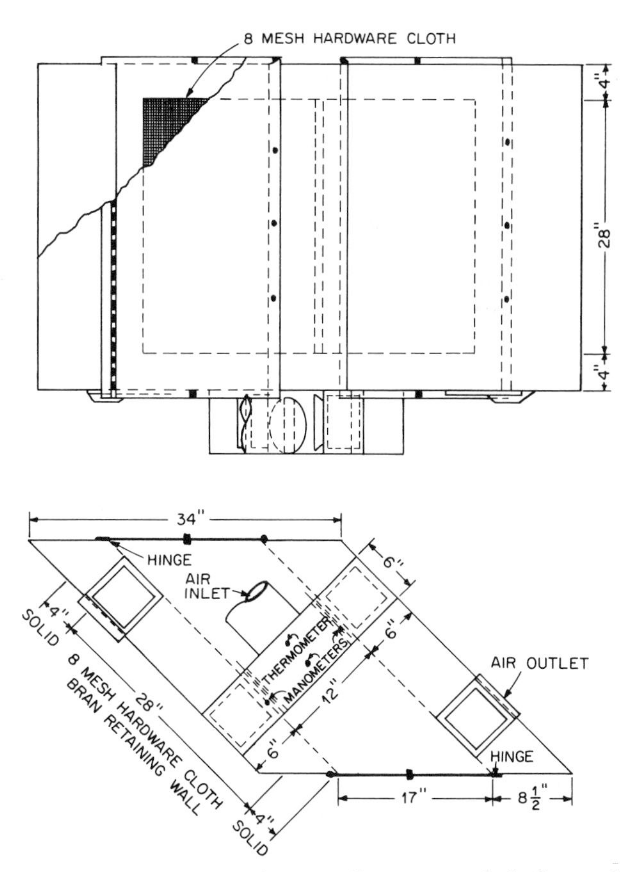

FIG. 3. Inclined incubation cells. From Underkofler et al. (73).

delignified wastes are required for saccharification, they support production of only weak filter paper activity during cultivation. It is difficult to compare enzyme activities reported for the koji process with those obtained by submerged culture in the United States because of differences in assay methods employed. For example, *T. reesei* cellulase activities in submerged cultures were characterized at the U.S. Army Natick Laboratories by filter paper activities based on reducing sugar release in International Units (IU) of 1 mol/min (41). The cellulase assay reported by Toyama is based on the time to disintegrate 2 cm of filter paper into fine particles in 5 ml of a 1% (wt/vol) enzyme solution and is expressed as units per milligram of solids. Preliminary studies in our laboratory showed that stationary cultures of *T. reesei* QM9414 on a 1:1 mixture of wheat bran and soybeans at 50% moisture for 5 days produced twice as much activity as cultures on wheat bran alone (46). In addition, agitated cultures produced significantly lower activities than stationary cultures. The fungus *Pestalotiopsis versicolor* was cultured in the solid state and in submerged cultures on various lignocellulosic substrates (52). Cellulase activities were two to three times higher in the solid state, with maximum filter paper activity on bagasse at about 90% moisture. As in previous studies by Toyama (67), delignified substrates gave lower activities than natural substrates. Finally, α-galactosidase and invertase have been produced by *A. awamori* on wheat bran at 50% moisture in aerated roller bottles (61). Maximum activities were 20 and 80 IU/kg of dry solids, respectively, for 150- and 300-kg batches of wet bran. Activities at aeration rates of 0.2 liter of gas per kg per min were considerably higher than those at 0.05 liter/kg per min. α-Galactosidase production was relatively insensitive to rotation rates of 0.15, 1.5, and 15 rpm, whereas invertase formation was highest at 15 rpm. There was significantly less growth at higher agitation rates for both aeration rates. In recent studies in our laboratory, 16,000 U of *p*-nitrophenyl galactoside activity per kg of dry soybeans was produced by *A. oryzae* in agitated solid cultures (unpublished data).

7.2.4 Primary Metabolites

While solid-state methods have been used for gluconic and citric acid production (1), current industrial production of organic acids is largely by submerged-culture methods. Solid-state methods have been applied in citric acid production by *A. niger* on inert solids (8), based on earlier work in which sugar cane or beet pulp solids were impregnated with sucrose or molasses. In this work, unsterilized cane mixtures were inoculated with *A. niger* and fermented at 20 to 35°C for about 4 days. Citric acid yields were 55% on sucrose and 45% on molasses sugars. The method also gave 0.45 kg of citrate from a mixture of 2.56 kg of molasses and 1.24 kg of beet pulp. The cane pulp was reusable as an inert support. A thick-bed process (64, 65), with water-cooling pipes imbedded on the solids, used diluted molasses (heated at 90 to 95°C for 30 min) mixed with sterilized sawdust. The mixture was inoculated with an *A. niger* spore preparation (1% wt/vol) and cultured at 30°C for 4 to 5 days. Moisture-saturated air at 30°C was used to aerate the

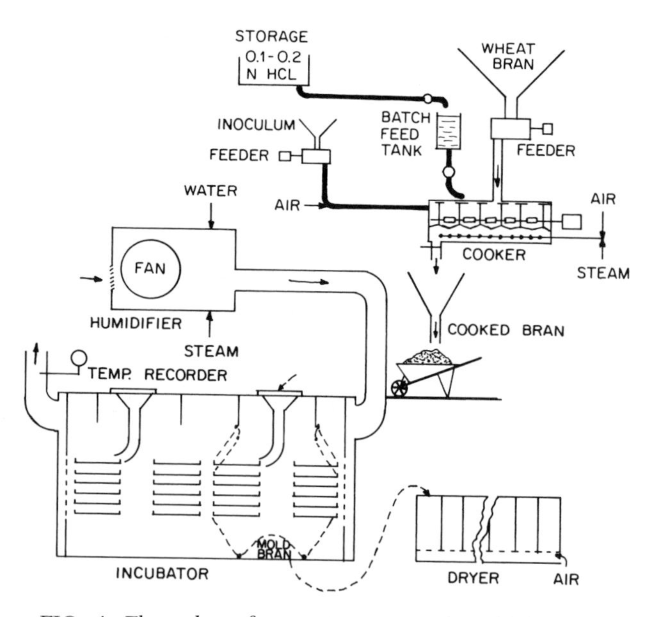

FIG. 4. Flow sheet for semicommercial mold bran productions. From Underkofler et al. (73).

bed, which was cooled by water at 33°C (64, 65). Although yields for the process were not reported, submerged-culture yields were more affected by raw material impurities, and rates of power consumption were higher than for surface-culture or solid-state methods. Laboratory cultures in flasks and trays used sucrose or molasses mixed with bagasse, inoculated with *A. niger* spores, and incubated at 30°C for 6 days without agitation (36). Citrate was recovered by extraction with water. Methanol (3%, vol/vol) stimulated productivity considerably, with maximum yields of about 60% (flasks) and 70% (trays) of original sugar concentrations. Yields were higher in the solid state than in submerged cultures for comparable cultivation conditions. Citrate has also been produced in Japan by cultivating *A. niger* on cooked vegetable residues in tray fermentations (12), but is more generally produced in submerged culture on beet molasses (83).

7.2.5 Secondary Metabolites

Antibacterial agents have been produced by strains of *Rhizopus* and *Actinomucor* used in traditional food fermentation, although their significance in such products is uncertain (75). A compound produced by *Rhizopus oligosporus*, used in tempeh production, is active against some gram-positive bacteria, including microaerophilic and anaerobic strains of *Streptococcus*, *Bacillus*, *Staphylococcus*, and *Clostridium* (77). There are few reports on the production of antibiotics in the solid state; Hesseltine has reported (24) a preparation from the USSR in which *Streptomyces aureofaciens* was grown on millet seeds to be used in swine feed mixtures. Tetracyclines have also been reported in Nubian bones (c. 350 A.D.) discovered by archaeologists in the Sudan (6). Bassett et al. (6) postulated that food grains of the period, such as barley, wheat, and millet, had been stored in warm, dry mud bins with environments similar to those of

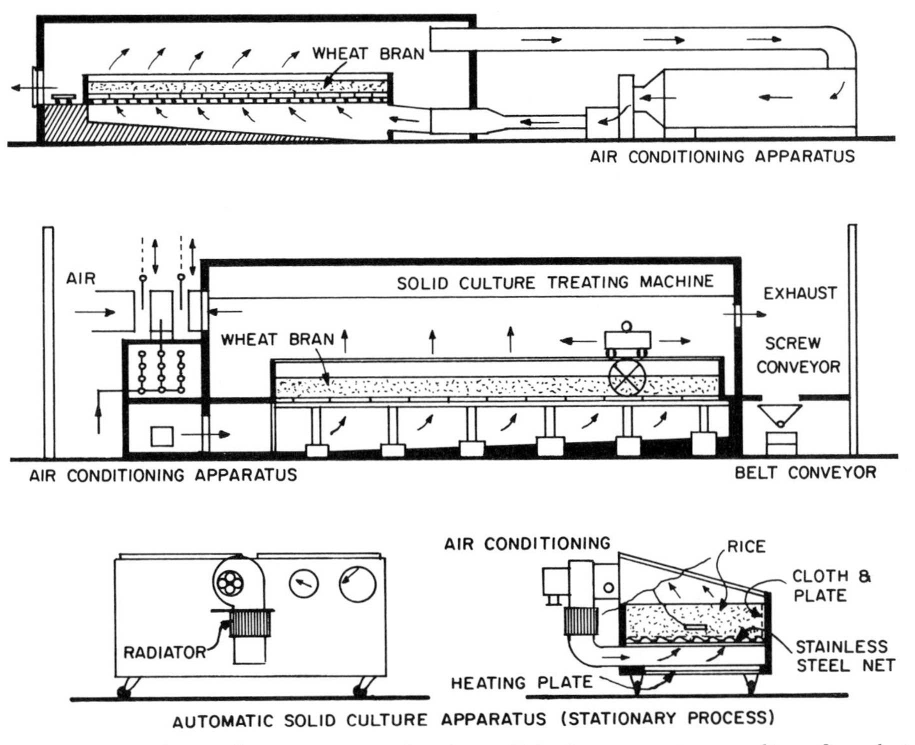

FIG. 5. Stationary automatic koji-making apparatus. An air-conditioning apparatus supplies a forced-air current through the wheat bran layer. A movable machine supplies wheat bran moistened with water and loosens the solid culture solidified with mycelium. The finished solid culture is exhausted by the machine after 3 to 5 days of incubation. From Toyama (67).

solid-state fermentations and that antibiotics were produced by strains of *Streptomyces* contaminating the grains. *Streptomyces badius* and *Streptomyces flavovirens* have been cultured in our laboratory on various solid substrates. Although antibiotic activity was not detected in preliminary screening (46), more recent experiments have discovered antimicrobial activity in extracts of both species cultivated on oats, millet, barley, or soy (unpublished data).

Hesseltine and co-workers at the U.S. Department of Agriculture Northern Regional Research Laboratories have produced mycotoxins in the solid state at concentrations substantially higher than in submerged cultures (24). Aflatoxin titers of up to 1.5 g/kg of solids were obtained from rice cultured at 40% moisture (40 g of moisture per 100 g of wet solids) by *Aspergillus parasiticus* in agitated Fernbach flasks. Yields were much lower in stationary cultures (60). The method has been adopted for producing mycotoxins from various genera and species with extremely good results. Similar levels were obtained for ochratoxin production by *Aspergillus* cultured on wheat (23). Ochratoxin has also been regularly produced at titers of up to 2.4 g/kg of solids on wheat by

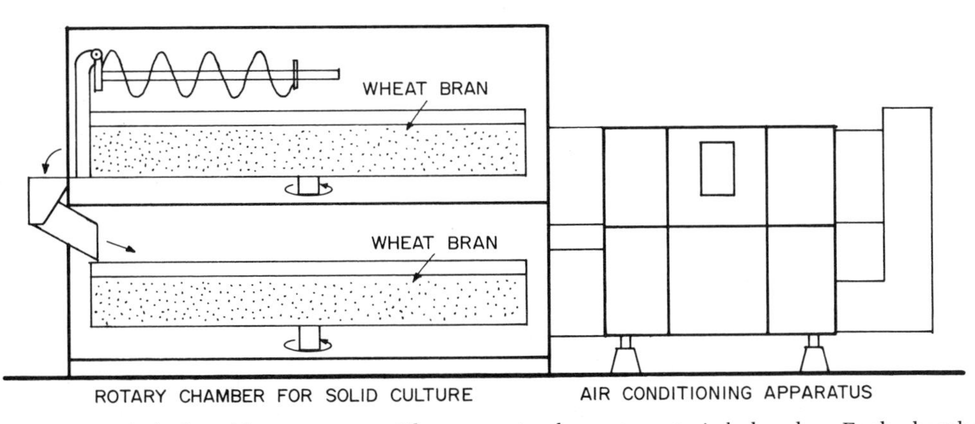

FIG. 6. Rotary automatic koji-making apparatus. The apparatus has a two-storied chamber. Each chamber has a large rotary tray on which wheat bran is heaped evenly. After inoculated fungus has grown sufficiently, solid culture is transferred by a screw conveyor to the lower rotary tray. Solid culture is loosened by a fixed stirrer in the rotary tray and crushed in the hopper. From Toyama (67).

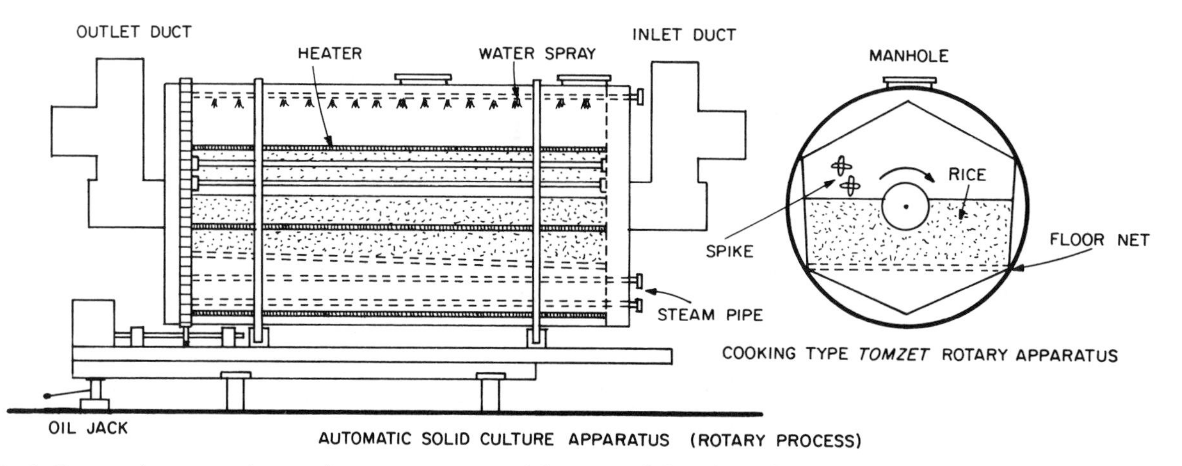

FIG. 7. Rotary drum-type koji-making apparatus used for rice solids culture by *A. oryzae*. All operations, i.e., washing, cooking, inoculation, loosening of solids, water spraying, cooling, air circulation, filling, and exhausting, can be done in this apparatus. From Toyama (67).

using a four-chambered rotating fermentor with baffles to elevate and drop the grain during each revolution (38). The fermentor was designed to pass sterile air through each chamber. Occasional titers of 3 to 4 g/kg of solids were obtained. The highest agitation rate (16 rpm) gave the highest production but required 12 to 19 days to reach peak levels. At 1 rpm, normal titers of 2.3 to 2.5 g/kg of solids were obtained in 8 to 9 days; titers in stationary cultures were only 0.1 to 0.2 g/kg of solids. Hesseltine (24) summarized the advantages of agitation in mycotoxin production as (i) effective distribution of spore inoculum, (ii) maintenance of homogeneity and prevention of mycelial pellet formation, (iii) improved aeration by exposure of individual particles to the gas environment, and (iv) facilitation of heat transfer to prevent localized overheating of the substrate. As previously described, a number of other mycotoxins have been produced on natural grains by strains of *Aspergillus* and *Penicillium* (16, 19, 34, 39, 40).

7.2.6 Lignocellulosic Wastes

The microbial conversion of an abundant supply of agricultural and forest wastes has become a subject of considerable interest as a renewable source of raw materials in chemical, food, and pharmaceutical manufacture (10, 17, 69). The current use of solid-state processes for such conversions is well documented in a review by Aidoo et al. (1) and will not be described in detail here, since the methods used in such processes are similar to those discussed above. The state of the art in this area is indicated by references on upgrading of agricultural (20, 21, 35, 43, 49, 59) and animal (25, 30, 54, 55) wastes in protein content and digestibility for use as animal feeds, enhanced plant protein recovery for human foods (3), and detoxification of cassava (9). Lignin, which acts as a barrier to microbial utilization of cellulose and hemicellulose in such materials, may be removed by chemical methods such as treatment with sodium hydroxide and peracetic acid or by selective solvents which spare the hemicellulose (2). It may also be degraded to phenolic residues by microbial conversion in the solid state. This has been demonstrated in our laboratory in solid-state fermentations of natural birch wood lignin by *Phanerochaete chrysosporium* (47).

7.2.7 Transformations by Spores

Solid-substrate methods have been used for mass production of spores which can be used to transform organic compounds such as steroids, antibiotics, fatty acids, and carbohydrates (1). Fungal spores have also been shown to produce food flavors (37) and insecticides (50). The advantages of solid-state methods include simplicity, yield, and homogeneity of spore preparations (74). On a dry-weight basis, spores are said to have from 3 to 10 times the transforming activity of fungal mycelia and to offer some interesting possibilities as catalysts for the production of novel biochemical compounds (50).

7.3 PROCESS VARIABLES

It is important to recognize that each microbe-substrate system is unique and must be considered in terms of the chemical composition and physical properties of the substrate, the growth characteristics and physiology of the organism to be cultured, and the nature of the product; i.e., such processes may involve the production of primary or secondary metabolites based on the synthesis of extracellular enzymes in growth-associated metabolism. The preceding sections suggest a number of key variables which should be considered in process development of solid-substrate fermentations. These are discussed in the following sections.

7.3.1 Pretreatment

Natural solid substrates generally require some kind of pretreatment to make their chemical constituents more accessible and their physical structure more susceptible to mycelial penetration. As seen in the rice koji process (48), the rice is dehulled and milled to remove the bran and some of the starchy endosperm, soaked in water, and then steamed. This softens the rice, solubilizes nutrients for initial growth, swells the grain structure, and pasteurizes the

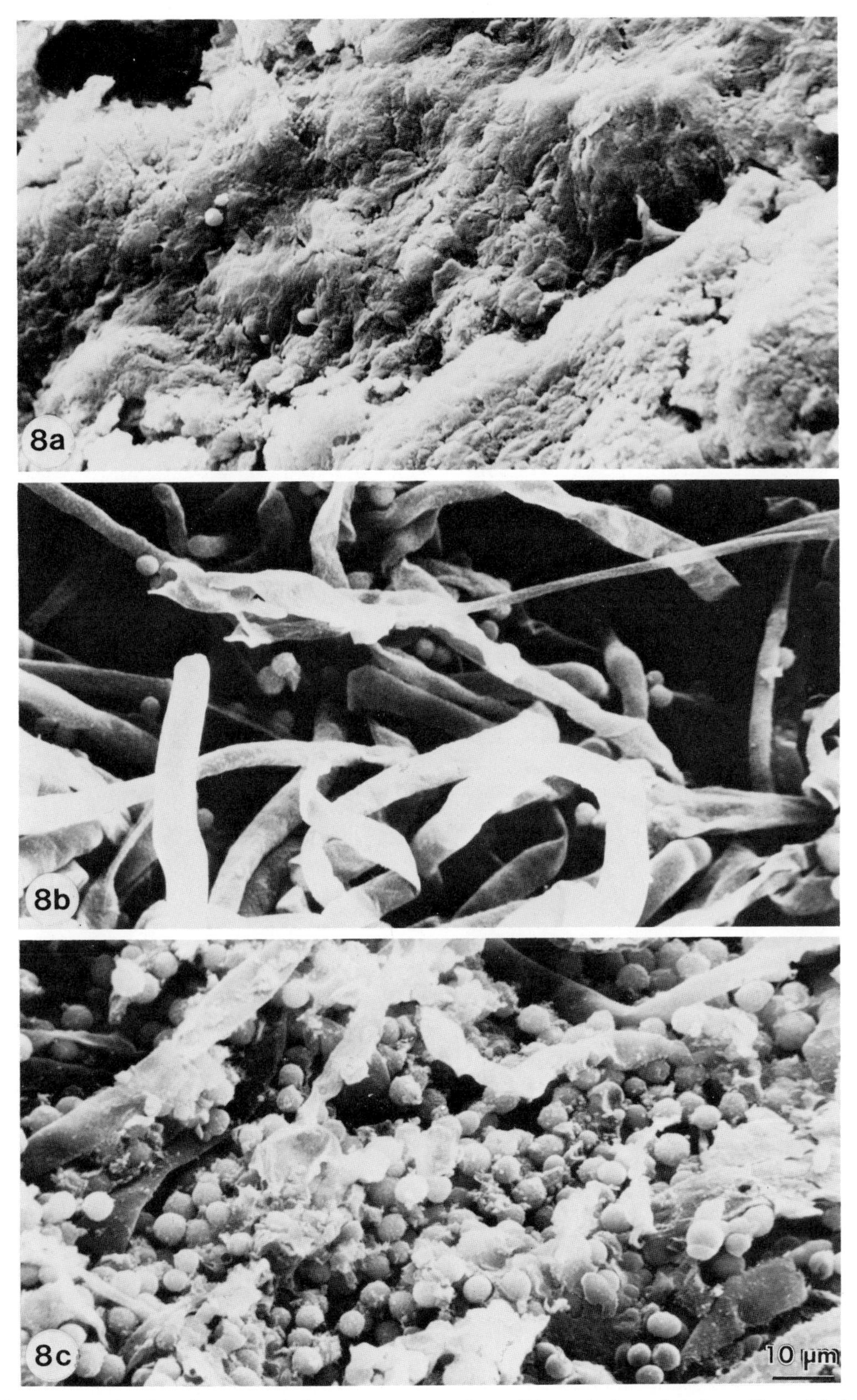

FIG. 8. Characteristics of surface growth by *A. oryzae* on rice solids at 0, 2, and 4 days. From Mudgett et al. (46). Magnification, ×500.

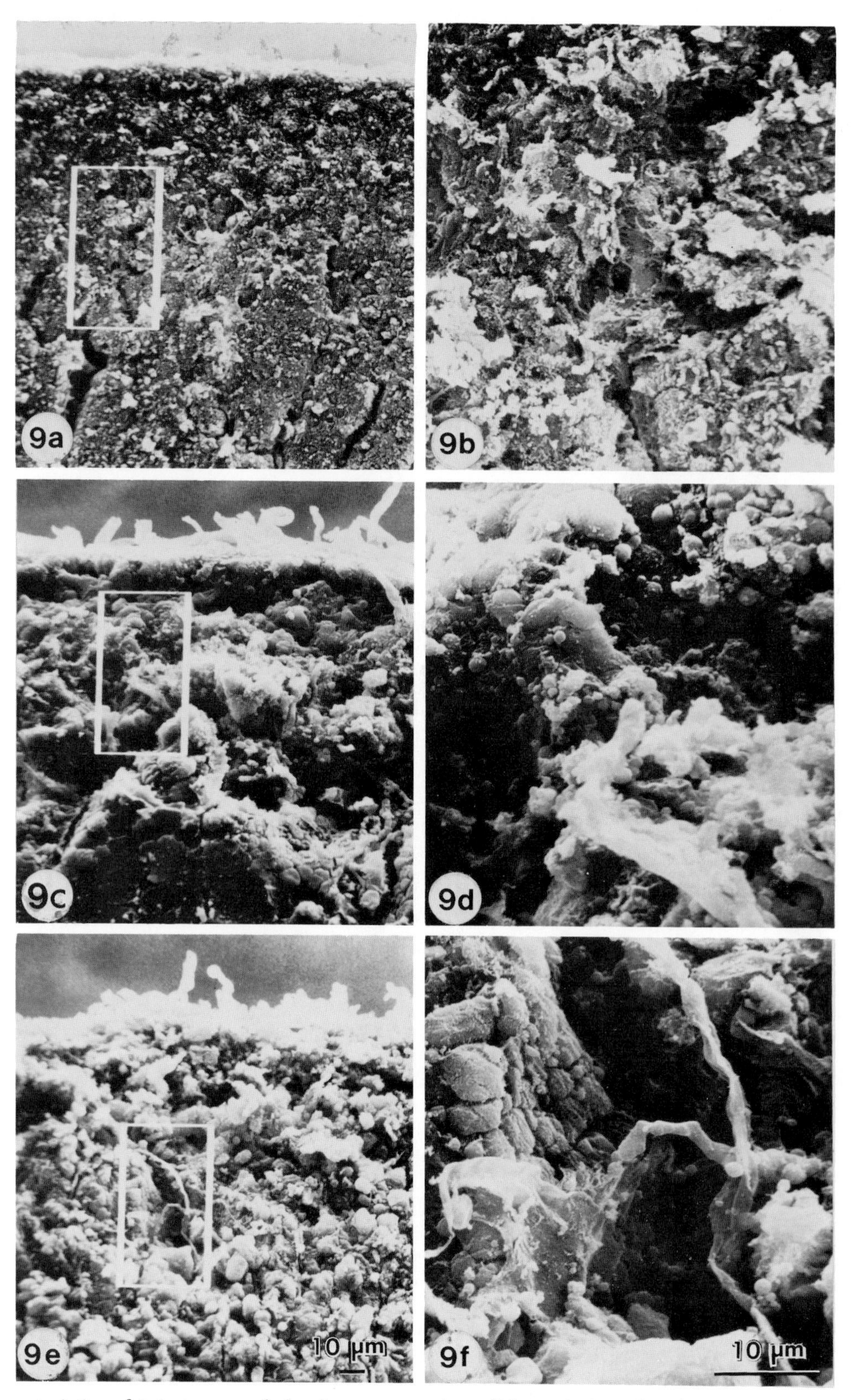

FIG. 9. Characteristics of interior growth by *A. oryzae* on rice solids at 0, 2, and 4 days. From Mudgett et al. (46). Magnification: (a, c, e) ×200; (b, d, f) ×1,000.

moist solids. In amylase production on rice solids processed in this manner, extensive surface growth and microbial penetration of interior solids have been observed (Fig. 8 and 9) (46).

Surface and interior structures of the rice solids were compared at days 0, 2, and 4. Surface characteristics of the fermented rice solids are shown in Fig. 8. The surface structure of the autoclaved solids before inoculation (day 0; Fig. 8a) was rough in texture, with irregular cracks and fissures. Growth at 2 days (Fig. 8b) was largely mycelial, with little spore formation. In contrast, growth at 4 days (Fig. 8c) was characterized by extensive spore formation, with few mycelia.

Interior characteristics of the fermented rice solids are shown in Fig. 9. The structure of the autoclaved solids before inoculation (day 0; Fig. 9a) was relatively smooth, with extensive internal cracks and fissures. Growth at 2 days (Fig. 9c and d) consisted of surface mycelia with penetration of the rice solids to a depth of nearly 0.15 mm. The minor axis of the rice solids used was approximately 1 mm. Growth at 4 days (Fig. 9e and f) also involved surface mycelia, but with greater penetration and degradation of the rice solids. At the higher magnification, mycelial growth at 2 days and spore formation at 4 days appeared to be more characteristic, as seen in Fig. 8b and c (dense mycelial structures are also visible in Fig. 9e, suggesting that surface mycelia at 4 days are covered by spores). The interior of the rice solids appears to be highly accessible to the gas phase during the course of fermentation, thus providing favorable conditions for gas transfer and control. No mycelial pellet formation was observed in the specimens examined.

Similar procedures are employed in traditional and nontraditional processes. For example, in soy sauce manufacture, soybeans are dehulled, washed, soaked from 12 to 24 h with several water changes to prevent bacterial contamination, and steamed for from 1 to several hours at temperatures from 100 to 120°C (64, 65). The wheat used in the process is first roasted and then coarsely crushed. Other pretreatments include pearling (abrasion of the surface to remove some, but not all, of the seed coat) and cracking of grains (25).

7.3.2 Supplementation

Although most traditional food fermentations do not require nutritional supplementation, it may be beneficial in nontraditional fermentations to provide supplemental media to initiate biomass production, induce enzyme synthesis, provide balanced growth conditions, or prolong secondary metabolite production. For some brown rot fungi, lignin degradation does not provide enough energy for enzyme induction and utilization of cellulosic constituents and may require a supplemental carbon source for growth and further lignin degradation, i.e., cometabolism (29). For example, *Poria monticola*, *Lentinus lapidus*, and *Lenzites trabea* are unable to grow on natural woods such as aspen, pine, and spruce, which contain significant amounts of lignin and cellulose (28). However, these organisms grow well in cellulosic media supplemented with 0.5% (wt/vol) glucose or cellobiose and, to a considerably lesser extent, in media supplemented with 0.5% peptone, asparagine, or yeast extract. Negligible growth was seen in unsupplemented media. Similarly, the white rot fungi *Phanerochaete chrysosporium* and *Coriolus versicolor* are unable to degrade spruce lignin to a significant extent without a cometabolite such as cellulose or glucose for growth (32). For these organisms, mannose, xylose, cellobiose, glycerol, and succinate were much less effective as cosubstrates.

Supplementary nutritional requirements for a developmental microbe-substrate system of interest may be determined by preliminary experiments in either submerged or solid-state cultures at the laboratory scale. Culture of the organism and the solid substrate in separate compartments divided by a membrane with a molecular-weight cutoff permitting the passage of enzymes and small molecules and restricting microbial and substrate solids would provide a means for evaluating the effects of nutritional supplements on microbial growth and product formation (3). One of the major problems in the development of solid-state fermentations has been the difficulty in separating microbial biomass from residual substrate solids, since the mycelia are firmly attached to the substrate. Methods generally used to estimate biomass yields and growth rates have included measurements of glucosamine, DNA, RNA, protein, oxygen consumption, and carbon dioxide or heat evolution. Such methods are of limited value in stoichiometric analysis of solid-state fermentations because the organism and substrate are intimately associated. In contrast, biochemical analyses of solids and fluids recovered from dialysis cultures would permit estimation of stoichiometric coefficients such as substrate, oxygen, carbon dioxide, and thermal yields, in addition to levels of product formation, as affected by nutritional supplements. Dialysis culture, however, is of little value in obtaining kinetic data, because transport of nutrients and products from one compartment to another is limited by low molecular diffusivities in liquid media. Such studies also require that the enzymes involved in growth-associated metabolism be completely extracellular, i.e., diffusible.

With respect to secondary metabolites, solid substrates may present a different problem in that the nutrient whose deficiency triggers the pathway leading to a secondary metabolite of interest (42) may be available in excess when growth becomes limited by some other nutrient. This suggests that the choice of a solid substrate and its supplementation may be somewhat more critical for a process to be designed for antibiotic production than one for enzyme or organic acid production.

7.3.3 Particle Size

Generally, smaller particle sizes provide a larger surface area for heat transfer and gas exchange. Given the same void volume fraction (porosity) and pore size distribution, smaller particle sizes also distribute equivalent moisture concentrations in thinner films on external surfaces exposed to the gas environment, although internal pores maintain the same surface-to-volume ratios with respect to solid surfaces, based on geometric considerations of spherical particles. This results in higher surface nutrient concentrations and shorter pathways for diffusion of nutrients, both at the surface and in the pores of substrates having the same tortuosity. On the other hand, smaller particle sizes result in closer packing densities of the substrate and reduction of the void space between particles, which tends to reduce the area for heat transfer and gas exchange with the surrounding atmosphere, unless the particles can be sufficiently agitated to provide a high degree of particle separation. There may therefore be a lower limit in particle size at which heat transfer or gas exchange becomes rate limiting and an upper limit at which nutrient transfer becomes limiting. Whatever the basis, particle size has been suggested by Hesseltine (23) as one of the major variables in process development. Particle size reduction methods include milling, grinding, and chopping. For lignocellulosic materials, ball-milling is the method of choice (33).

7.3.4 Moisture Content

Water in biological materials generally exists in three states, as determined by moisture isotherm measurements in which the solids sorb or desorb water vapor in equilibrium with relative humidities (water activities) in a gas phase which can be maintained by saturated salt solutions at a constant temperature (57). In a monolayer region at the surface, water is tightly bound to solid surfaces. Monolayer binding is generally 5 to 10 g/100 g of dry solids for agricultural materials. In a multilayer region beyond the surface monolayer, water is less tightly bound in additional layers at progressively decreasing energy levels. Beyond the multilayer region, free water exists in a region of capillary condensation. The distinction between the multilayer and capillary regions, in terms of the relationship between water activity and moisture content, is ambiguous. The dividing line between the two regions has been estimated from electrical measurements of an agricultural substrate of high starch content. The line was defined by a moisture content of about 25 to 30% by weight at a water activity of 80 to 85%, the lower limit for microbial growth except for some halophilic or osmophilic microbes. Although sorption isotherms vary from one product to another and hysteresis is seen in sorption and desorption isotherms, the point is that water may exist in the free state at moisture levels of interest in solid-state fermentation, in contrast with a general perception in the literature that such processes do not involve free water. Hesseltine noted that moisture is a critical factor in aflatoxin production on rice (24) and growth of lactic acid bacteria on feedlot waste liquids mixed with cracked corn. In the former case, aflatoxin yields decreased rapidly at moistures above 40%. In the latter, moistures less than 35% limited growth and acid production, while moistures above 42% caused the mixtures to become gummy and aggregate. Nunokawa (48) noted the stickiness of rice particles at moistures above 30 to 35%. Hesseltine (22) also commented that one of the secrets to successful solid-state fermentations is having the substrate moist enough for mold growth, but not so wet that bacterial growth is promoted. In view of such factors, the optimum moisture content for each microbe-substrate system should be determined based on the desired product and the conditions for cultivation.

7.3.5 Sterilization

Since most bacteria require high moisture levels, generally above 40%, solid-state fermentations may exclude or greatly reduce the problem of bacterial contamination; many processes require no sterilization whatsoever (13). In fermentations at higher moisture levels, mild thermal processing conditions may be employed to pasteurize the solids. This is done in traditional fermentations (Table 1) by preparative steaming of the solids at 100°C for relatively short times; this also opens the structure for mycelial penetration. Selective conditions for fungal growth may also be obtained by moistening the solids with buffered or unbuffered medium at low pH (62, 72) and by using high densities of spore inoculum.

7.3.6 Inoculum Density

It is generally necessary to optimize inoculum density in solid-state fermentations. Too low a density may give insufficient biomass and permit the growth of undesirable organisms; too high densities may produce too much biomass and deplete the substrate of nutrients necessary for product formation. At the laboratory scale, spores are generally prepared by inoculating agar slants of suitable media from a broth culture of the organism, incubating the slants, and harvesting spores by adding 3 to 5 ml of water or sodium lauryl sulfate (0.01%) and aseptically scraping or shaking the slants to free the spores from the agar surface. Spore densities may be determined by colorimeter measurements at 540 nm, calibrated against hemacytometer counts; the spores are then diluted aseptically to obtain fixed densities (4). Similar procedures may be employed for agar plates. Disadvantages of the method are that it is unsuitable at higher scales and that spore viability is not generally determined.

A method developed for mass production of spores at the U.S. Department of Agriculture Northern Regional Research Laboratories (78) involves solid-state fermentation of grains for spore production and freeze-drying of the fermented grains for use as inocula. Since inocula from previous fermentations under normal processing conditions led to bacterial contamination, the process was developed aseptically. Viable spore counts before and after freeze-drying are estimated by plate count. A 1-g sample of the fermented solids is aseptically weighed and then transferred to a sterilized blender containing 99 ml of sterile water. After blending for 2 min at high speed, serial dilutions are made, and 1 ml of each dilution is mixed with plate count agar (0.5% peptone, 0.25% yeast extract, 0.1% dextrose, and 1.5% agar) at 45°C in petri plates. The plates are incubated at 32°C for 20 to 24 h, and the colonies are counted. Aerobic bacterial counts are made in dishes of plate count agar containing 100 μg of cycloheximide per ml, previously sterilized using a 0.45-μm Millipore filter and added to the agar at 45°C. The plates are incubated at 32°C for 3 days, and colonies are counted daily. High-viability spore counts, from about 10^8 to 10^9 per g of dry solids, have been obtained from *R. oligosporus* on several grains with typical loss of viability in freeze-drying of about 1 log unit. Preparations are generally stable for 6 months. Inocula from such preparations may then be used in subsequent fermentations at calibrated spore densities on a weight basis, i.e., grams of inoculum per 100 g of wet solids. It is noted that moisture levels and adequate aeration levels are critical for spore production by this method. Bacterial contamination is not usually a problem in spore preparation or storage.

7.3.7 Temperature and pH

Temperature and pH of the fermenting solids are significant variables in solid-state processes and are generally specific to the organism and the product to be cultured, as in submerged fermentations. Both, however, are more difficult to control in the solid state, since direct measurements in the liquid phase are impractical because of their association with the solids at low moistures. Optimal temperatures for growth may not be the same as for product formation, suggesting a possible need for temperature shifts (profiling) in later stages of fermentation. The rate of heat generation in the bed at high levels of biological activity may also induce thermal gradients within the bed which lead to heat transfer limitations and suboptimal conditions for biomass or product formation.

The bulk pH of the liquid phase may also be considerably different from local pH levels at solid surfaces near which major biological activity occurs, due to surface charge effects and ionic equilibria modified by solute transport effects. While there is no widely accepted procedure for measuring the pH of agricultural products, a general procedure for forest products involves mixing 1 part (by weight) of dried solids in 3 parts of freshly boiled and cooled water and measuring the pH with a glass electrode after 5 min. Minimum sample size is about 1 g of solids. Such procedures may be used to monitor pH changes during fermentation on a sampling basis.

Temperatures may be measured at various depths below the bed surface by thermistor or thermocouple probes. Total heat generation of up to 600 kcal/kg of fermenting solids has been reported in several koji processes for biomass, enzyme, or organic acid production (1). Such heat loads may lead to rapid temperature rise of the fermenting solids in fermentations limited by heat transfer, as indicated by a study on composting of agricultural and animal wastes (18). The study also showed that biological activity near the surface of the compost pile was significantly higher than in the interior at oxygen pressures somewhat above atmospheric. This was attributed to a decrease in interior oxygen concentrations in the bed under previously rate-limiting conditions for oxygen transfer. Thus, the overall rate of heat generation in solid-state fermentations may be coupled to conditions for both heat and mass transfer.

7.3.8 Agitation and Aeration

Agitation and aeration conditions are critical in providing an adequate supply of oxygen to maintain aerobic conditions and in ventilating excess carbon dioxide evolved during fermentation (44). However, many solid-state fermentations are shear sensitive and cannot be agitated vigorously or, in some cases, at all. This is especially important in early fermentation stages (73). Shear sensitivity is attributed to disruption of mycelial-substrate contact, particularly for organisms with mycelial-bound enzymes for the hydrolysis of solid polymers. Hesseltine (24) commented that in most koji processes for enzyme manufacture in Japan, the fermenting solids are not greatly agitated. Instead, the substrate is turned periodically to bring the bottom of the koji to the top. He also noted that such processes have been developed in highly controlled environments, with automatic inoculation, turning of the molding substrate, and harvesting of the finished product. Similarly, most traditional food fermentations in Japan appear to involve koji preparation by the rotary tray method, with circulation of humidified air under conditions suitable for heat transfer and gas exchange (48, 76). However, agitation on shakers or in rotating vessels with circulating conditioned air to provide mixing and particle separation is used in some enzyme processes (67) and has been particularly useful in secondary metabolite production, e.g., aflatoxin and ochratoxin (24, 38, 61). Maximum rotation rates are generally low and decrease with the size of the vessel (73). Solid-state fermentations appear well suited to production of microbial products by mycelial organisms with extreme sensitivity to shear rates at the impeller speeds required for stringent oxygen demand rates in submerged culture. A potential advantage is also seen in the reduction of mycelial pellet formation in the solid state by microbe-substrate attachment.

Aeration is also of major importance in solid-state processes and is implicated not only in gas transfer, as in submerged cultures, but also in heat and moisture transfer between the fermenting solids and the gas environment. The temperature of the gas phase can be used to supply or remove heat and to maintain the relative humidity in equilibrium with the liquid phase. With respect to moisture transfer, moisture loss or gain during fermentation may be extremely sensitive to the water activity of the gas phase, unlike submerged cultures, in which the substrate is dissolved at low substrate concentrations in large fluid volumes. That is, large changes of moisture content in the solid state may result from small changes in the relative humidity of the gas phase in equilibrium with the solids, depending on the sorption-desorption characteristics of the substrate. Although the primary function of the gas phase in the solid state is to supply oxygen and remove carbon dioxide, its secondary functions in heat and moisture transfer may be critical when oxygen demand and carbon dioxide evolution rates are not limiting. The gas environment is also involved in the control of solid-state fermentations, since low-moisture solids are not accessible to direct measurements of dissolved oxygen or carbon dioxide concentrations during fermentation on either a continuous or sampling basis (44). In addition, conditions for gas transfer may be relatively stagnant, depending on the method of aeration, and may lead to oxygen limitation at small penetration depths or to inhibitory carbon dioxide concentrations in normal atmospheric environments. Oxygen and carbon dioxide pressures in the gas phase are generally measured by analyzers based on thermal conductivity, paramagnetism, or infrared absorption. They may also be monitored by gas chromatography (51).

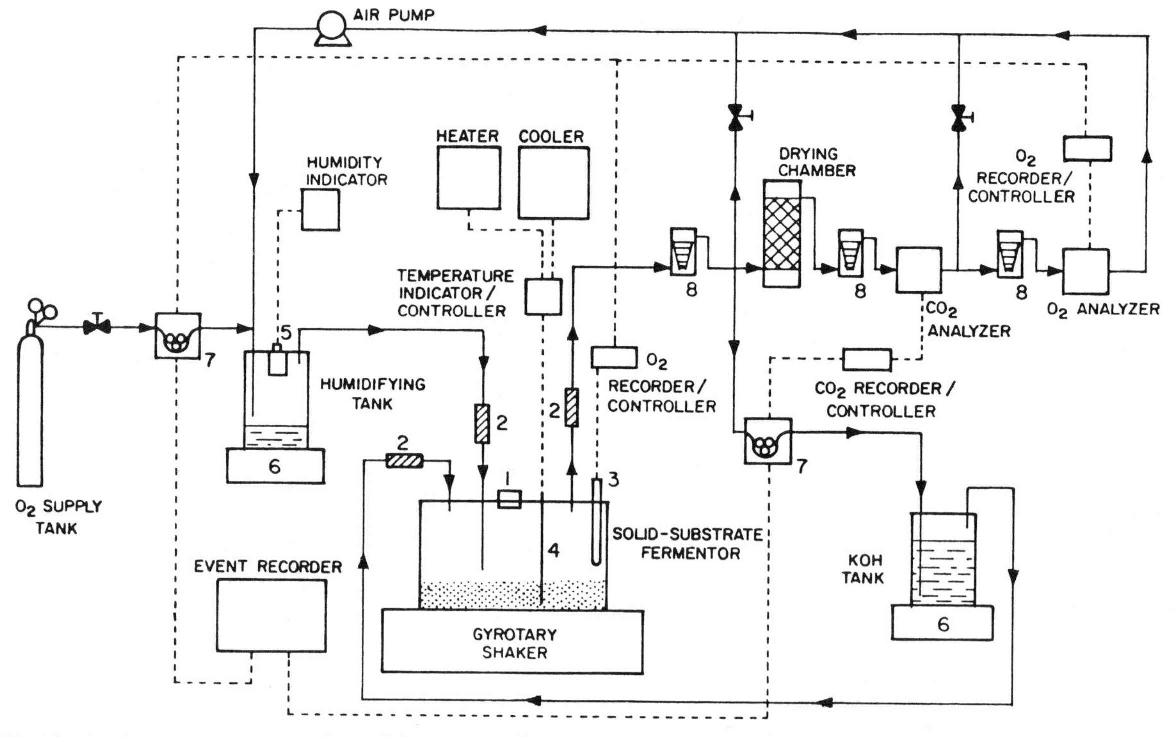

FIG. 10. Static aeration system for solid-substrate fermentations. (1) Inoculum probe; (2) filter; (3) oxygen probe; (4) thermistor probe; (5) humidity sensor; (6) magnetic stirrer; (7) tube pump; (8) flow meter. From Bajracharya and Mudgett (4).

7.4 PROCESS CONTROL

The disadvantages of large-scale solid-state fermentations center on problems of heat buildup, process control, and scale-up (53). Despite these disadvantages, the Japanese have been successful in developing large-scale processes for the manufacture of traditional food products, enzymes, and organic acids. As suggested, these processes are primarily based on stationary or rotary tray methods in which temperature- and humidity-controlled air is circulated through stacked beds of fermenting solids. Rotary drum-type fermentors are employed less frequently, presumably due to shear sensitivity in some of these processes and because tray methods have been used for centuries in traditional food manufacturing processes. Temperature and humidity control within practical limits appear to be based on water temperatures used to humidify the circulating air stream at flow rates consistent with heat and mass transfer requirements, although little information is available in the Western literature on the details of modern control systems in large-scale solid-state processes.

In recent laboratory studies, the gas environment has been found to significantly affect the rate and extent of biomass and product formation in the solid state. For example, in amylase production on rice by *A. oryzae*, oxygen pressures above atmospheric significantly stimulated amylase productivity, suggesting oxygen limitation at normal atmospheric pressures, but had little effect on biomass formation as estimated by DNA measurements (4). Carbon dioxide pressures above 0.01 atm (10^{-3} Pa) severely inhibited amylase productivity but stimulated biomass formation. By contrast, in plant protein recovery from pressed alfalfa residues with a different *Aspergillus* species, oxygen or carbon dioxide pressures above atmospheric levels stimulated both cellulase and pectinase activities (45) without significantly affecting biomass formation. These studies were conducted in controlled gas environments at constant partial pressures maintained by admitting pure oxygen on demand at pressures below a set point and purging carbon dioxide in 30% KOH at pressures above a set point in a closed aeration system (Fig. 10). In a more recent study, high oxygen pressures were found to stimulate, and high carbon dioxide pressures were found to inhibit, the degradation of natural birch lignin by *Phanerochaete chrysosporium* (47). The stimulatory effect of oxygen on breakdown of labeled synthetic lignins and natural wood lignins had previously been reported by a number of other researchers.

The point of the present discussion is to emphasize the potential of the gas environment for control of gas transfer and regulation of microbial metabolism in the solid state. Although the liquid and solid phases in the solid state are not accessible to direct measurement of substrate consumption, growth, or product formation, kinetic information for modelling such processes can be obtained from chemical analysis of fermented solids and extracted fluids sampled during fermentation. Analysis of solids and fluids from dialysis cultures of solid substrates can also be used to obtain stoichiometric data for modeling biological processes in the liquid phase. The gas environment is of particular interest in solid-state fermentations, since it is able to provide (i) a means of monitoring biological activity in the liquid phase by real-time measurement of parameters such as oxygen, carbon dioxide, temperature, and humidity, and (ii) a potential basis for development of improved process control systems to obtain more stable microbial environments in solid-state fermentations, e.g., model reference adaptive control systems.

Given the present state of the art, the most promising approach in solid-state fermentation process development appears to be measurement and control of temperature, humidity, oxygen, and carbon dioxide levels in the gas phase, coupled with biochemical analysis of fermented and unfermented solids and their extracts. Demain (15) has observed that the manufacturing productivities of certain industrial submerged-culture fermentations have increased logarithmically over a period of several decades, e.g., penicillin and streptomycin. Such increases have been due partly to research in microbial biochemistry, physiology, and genetics and partly to engineering research based on concepts of stoichiometry, kinetics, thermodynamics, and heat and mass transfer in control of the microbial environment.

Potential economic advantages of solid-state methods for suitable microbe-substrate systems may include reduced thermal processing requirements (many processes are not aseptic), reduced energy requirements for agitation (surface-to-volume ratios for gas transfer are high and many processes do not require agitation or may be shear sensitive), and high extracellular product concentrations, which may be recovered by efficient leaching (58) or supercritical extraction (79) methods. While direct economic comparisons of solid-state and submerged culture fermentations are not generally available, it is apparent that the Japanese have developed large-scale solid-state processes on an economic basis (76, 82, 83).

7.5 LITERATURE CITED

1. **Aidoo, K. E., R. Hendry, and B. J. B. Wood.** 1982. Solid substrate fermentations. Adv. Appl. Microbiol. **28:** 201–237.
2. **Avgerinos, G. C., and D. I. C. Wang.** 1983. Selective solvent delignification for fermentation enhancement. Biotechnol. Bioeng. **22:**2219–2235.
3. **Bajracharya, R., and R. E. Mudgett.** 1979. Solid substrate fermentation of alfalfa for enhanced protein recovery. Biotechnol. Bioeng. **21:**551–560.
4. **Bajracharya, R., and R. E. Mudgett.** 1980. Effects of controlled gas environments in solid-state fermentations of rice. Biotechnol. Bioeng. **22:**2219–2235.
5. **Barbesgaard, P.** 1977. Industrial enzymes produced by members of the genus *Aspergillus*, p. 391–404. *In* J. E. Smith and J. A. Pateman (ed.), Genetics and physiology of *Aspergillus*. Academic Press, London.
6. **Bassett, E. J., M. S. Keith, G. J. Armelagos, D. L. Martin, and A. R. Villaneuva.** 1980. Tetracycline-labelled human bone from ancient Sudanese Nubia (A.D. 350). Science **209:**1532–1534.
7. **Berry, D. R.** 1975. Environmental control of the physiology of filamentous fungi, p. 16–32. *In* J. E. Smith and D. R. Berry (ed.), The filamentous fungi, vol. 1. John Wiley & Sons, Inc., New York.
8. **Berry, D. R., A. Chmiel, and Z. Al Obaidi.** 1977. Citric acid production by *Aspergillus niger*, p. 405–426. *In* J. E. Smith and J. A. Pateman (ed.), Genetics and physiology of *Aspergillus*. Academic Press, London.
9. **Brook, E. J., W. R. Stanton, and A. Wallbridge.** 1969. Fermentation methods for protein enrichment of cassava. Biotechnol. Bioeng. **11:**1271–1284.
10. **Buchholz, K., J. Puls, B. Godelmann, and H. H. Dietrichs.** 1980. Hydrolysis of cellulosic wastes. Process Biochem. **16:**37–43.
11. **Bull, A. T., and M. E. Bushnell.** 1976. Environmental control of fungal growth, p. 1–31. *In* J. E. Smith and D. R. Berry (ed.), The filamentous fungi, vol. 2. John Wiley & Sons, Inc., New York.
12. **Cannel, E., and M. Moo-Young.** 1980. Solid state fermentation systems. Process Biochem. **15:**24–28.
13. **Cannel, E., and M. Moo-Young.** 1980. Solid state fermentation systems. Process Biochem. **15:**2–7.
14. **Cocker, E., and R. N. Greenshields.** 1977. Fermentor cultivation of *Aspergillus*, p. 361–390. *In* J. E. Smith and J. A. Pateman (ed.), Genetics and physiology of *Aspergillus*. Academic Press, London.
15. **Demain, A. L.** 1977. The health of the fermentation industry: a prescription for the future. Dev. Ind. Microbiol. **18:**72–77.
16. **Demain, A. L., N. A. Hunt, V. Malik, B. Kobbe, H. Hawkins, K. Matsuo, and G. N. Wogan.** 1977. Improved procedure for production of cytochalasin E and tremorgenic toxins by *Aspergillus clavatus*. Appl. Environ. Microbiol. **31:**138–140.
17. **Detroy, R. W., and C. W. Hesseltine.** 1978. Availability and utilisation of agricultural and agro-industrial wastes. Process Biochem. **13:**2–8.
18. **Finger, S. M., R. T. Hatch, and T. M. Regan.** 1976. Aerobic microbial growth in semi-solid matrices. Biotechnol. Bioeng. **18:**1193–1218.
19. **Ghosh, A. C., A. Manmade, J. M. Townsend, A. Bousquet, J. F. Howes, and A. L. Demain.** 1978. Production of cyclochlorotine and a new metabolite, simatoxin, by *Penicillium islandicum* Sopp. Appl. Environ. Microbiol. **35:**1074–1078.
20. **Han, Y. W., and A. W. Anderson.** 1975. Semisolid fermentation of ryegrass straw. Appl. Microbiol. **30:**930–934.
21. **Han, Y. W., G. A. Grant, A. W. Anderson, and P. L. Yu.** 1976. Fermented straw for animal feed. Feedstuffs **48:**17–20.
22. **Hesseltine, C. W.** 1965. A millenium of fungi: food and fermentation. Mycologia **57:**149–197.
23. **Hesseltine, C. W.** 1972. Solid-state fermentation. Biotechnol. Bioeng. **14:**517–532.
24. **Hesseltine, C. W.** 1977. Solid-state fermentation. Process Biochem. **12:**24–27.
25. **Hesseltine, C. W.** 1977. Solid-state fermentation. Process Biochem. **12:**29, 30, 32.
26. **Hesseltine, C. W.** 1981. Future of fermented foods. Process Biochem. **16:**2–6, 13.
27. **Hesseltine, C. W.** 1983. Microbiology of oriental foods. Annu. Rev. Microbiol. **37:**575–601.
28. **Highley, T. L.** 1973. Influence of carbon source on cellulase activity of white-rot and brown-rot fungi. Wood Fiber **5:**50–58.
29. **Horvath, R. S.** 1972. Microbial co-metabolism and the degradation of organic compounds in nature. Bacteriol. Rev. **36:**146–155.
30. **Hrubant, G. R.** 1975. Changes in microbial population during fermentation of feedlot waste with corn. Appl. Microbiol. **30:**113–119.
31. **Jennings, D. H.** 1976. Transport and translocation in filamentous fungi, p. 32–64. *In* J. E. Smith and D. R. Berry (ed.), The filamentous fungi, vol. 2. John Wiley & Sons, Inc. New York.
32. **Kirk, T. K.** 1980. Studies on the physiology of lignin metabolism by white-rot fungi, p. 51–64. *In* T. K. Kirk, T. Higuchi, and H. M. Chang (ed.), Lignin biodegradation: microbiology, chemistry and potential applications, vol. 2. CRC Press, Boca Raton, Fla.
33. **Knapp, J. S., and J. D. Howell.** 1980. Solid substrate fermentation. Enzyme Ferment. Biotechnol. **4:**85–143.
34. **Kobbe, B., M. Cushman, G. N. Wogan, and A. L. Demain.** 1977. Production and antibacterial activity of malformin

C, a toxic metabolite of *Aspergillus niger*. Appl. Environ. Microbiol. **33**:996–997.
35. **Kokke, R.** 1977. Improvement of carob pods as feed by solid substrate fermentation. J. Appl. Bacteriol. **43**: 303–307.
36. **Lakshminayara, K., K. Chaudhary, S. Ethiraj, and P. Tauro.** 1975. A solid-state fermentation method for citric acid production using sugar cane bagasse. Biotechnol. Bioeng. **178**:291–293.
37. **Lawrence, R. C.** 1966. The oxidation of fatty acid by spores of *Penicillium roqueforti*. J. Gen. Microbiol. **44**:393–405.
38. **Lindenfelser, L. A., and A. Ciegler.** 1975. Solid substrate fermentation for ochratoxin A production. Appl. Microbiol. **29**:323–327.
39. **Luk, K. C., B. Kobbe, and J. M. Townsend.** 1977. Production of cyclopiazonic acid by *Aspergillus flavus* Link. Appl. Environ. Microbiol. **33**:211–212.
40. **Maheva, E., G. Djelveh, C. Larroche, and J. B. Gros.** 1984. Sporulation of *Penicillium roqueforti* in solid-state fermentation. Biotechnol. Lett. **6**:97–102.
41. **Mandels, M., R. Andreotti, and C. Roche.** 1976. Measurement of saccharifying cellulase. Biotechnol. Bioeng. Symp. **6**:21–34.
42. **Martin, J. F., and A. L. Demain.** 1980. Control of antibiotic synthesis. Microbiol. Rev. **44**:230–251.
43. **Moo-Young, M., A. J. Daigulis, D. S. Chahal, and P. G. MacDonald.** 1979. The Waterloo process for SCP production from waste biomass. Process Biochem. **14**:38–40.
44. **Mudgett, R. E.** 1980. Controlled gas environments in industrial fermentations. Enzyme Microb. Technol. **3**:273–280.
45. **Mudgett, R. E., and R. Bajracharya.** 1979. Effects of controlled gas environments in microbial enhancement of plant protein recovery. J. Food Biochem. **3**:135–149.
46. **Mudgett, R. E., J. Nash, and R. Rufner.** 1982. Controlled gas environments in solid substrate fermentations. Dev. Ind. Microbiol. **23**:397–405.
47. **Mudgett, R. E., and A. J. Paradis.** 1985. Solid-state fermentation of natural birch lignin by *Phanerochaete chrysosporium*. Enzyme Microb. Technol. **7**:150–154.
48. **Nunokawa, Y.** 1972. Sake, p. 449–487. *In* D. F. Huston (ed.), Rice: chemistry and technology. American Society of Cereal Chemists, St. Paul, Minn.
49. **Pamment, N., C. Robinson, J. Hilton, and M. Moo-Young.** 1978. Solid-state cultivation of *Chaetomium cellulolyticum* on alkali pretreated sawdust. Biotechnol. Bioeng. **20**:1735–1744.
50. **Ralph, B. J.** 1976. Solid substrate fermentations. Food Technol. Aust. **28**:247–251.
51. **Ramstack, J. M., E. R. B. Lancaster, and R. J. Bothast.** 1979. Gas chromatographic headspace analysis of solid substrate fermentations. Process Biochem. **14**:2–4.
52. **Rao, M. N. A., B. M. Mithal, R. N. Thakkur, and K. S. M. Sastry.** 1983. Solid-state fermentation for cellulase production by *Pestalotiopsis versicolor*. Biotechnol. Bioeng. **25**:869–872.
53. **Rathbun, B. L., and M. L. Shuler.** 1983. Heat and mass transfer effects in static solid substrate fermentations: design of fermentation chambers. Biotechnol. Bioeng. **25**:929–938.
54. **Rhodes, R. A., and G. R. Hrubant.** 1972. Microbial population of feedlot waste and associated sites. Appl. Microbiol. **24**:369–377.
55. **Rhodes, R. A., and W. L. Orton.** 1975. Solid substrate fermentation of feedlot waste combined with feed grain. Trans. ASAE (Am. Soc. Agric. Eng.) **18**:728–733.
56. **Righelato, R. C.** 1975. Growth kinetics of mycelial fungi, p. 79–103. *In* J. E. Smith and D. R. Berry (ed.), The filamentous fungi, vol. 1. John Wiley & Sons, Inc., New York.
57. **Rockland, L. B.** 1960. Saturated salt solutions for static control of relative humidity between 5 and 40 C. Anal. Chem. **32**:1375–1376.
58. **Schwartzberg, H. G.** 1980. Continuous counter-current extraction in the food industry. Chem. Eng. Progr. **76**:67–85.
59. **Senez, J. C.** 1979. Solid fermentation of starchy wastes. Food Nutr. Bull. **1**:18–20.
60. **Shotwell, O. L., C. W. Hesseltine, R. D. Stubblefield, and W. G. Sorenson.** 1966. Productioin of aflatoxin on rice. Appl. Microbiol. **14**:425–428.
61. **Silman, R. W.** 1980. Enzyme formation during solid substrate fermentation in rotating vessels. Biotechnol. Bioeng. **22**:411–420.
62. **Steinkraus, K. H., Y. B. Hwa, J. P. van Buren, M. I. Provvidenti, and D. B. Hand.** 1960. Studies on tempeh—an Indonesian fermented soybean food. Food Res. **25**:777–778.
63. **Takamine, J.** 1914. Enzymes of *Aspergillus oryzae* and the application of its amyloclastic enzyme to the fermentation industry. Ind. Eng. Chem. **6**:824–828.
64. **Terui, G., I. Shibasaki, and T. Mochizuki.** 1959. I. Studies on the high heap process as applied to some industrial fermentations. J. Ferment. Technol. **37**:479–483.
65. **Terui, G., I. Shibasaki, and T. Mochizuki.** 1960. II. Studies on the high heap process as applied to some industrial fermentations. J. Ferment. Technol. **38**:29–39.
66. **Toyama, N.** 1963. Degradation of foodstuffs by cellulase and related enzymes, p. 235–253. *In* E. T. Reese (ed.), Advances in enzymatic hydrolysis of cellulose and related materials. Macmillan Publishing Co., Inc., New York.
67. **Toyama, N.** 1976. Feasibility of sugar production from agricultural and urban cellulosic wastes with *Trichoderma viride* cellulase. Biotechnol. Bioeng. Symp. **6**:207–219.
68. **Toyama, N., and K. Ogawa.** 1975. Sugar production from agricultural woody wastes by saccharification with *Trichoderma viride* cellulase. Biotechnol. Bioeng. Symp. **5**:225–244.
69. **Tsao, G.** 1978. Cellulosic material as a renewable resource. Process Biochem. **13**:12–14.
70. **Tsuchiya, H. M.** 1983. The holding time in pure and mixed culture fermentations. Ann. N.Y. Acad. Sci. **413**:184–192.
71. **Underkofler, L. A.** 1954. Fungal amylolytic enzymes, p. 97–121. *In* L. A. Underkofler and R. Hickey (ed.), Industrial fermentation. Chemical Publications, New York.
72. **Underkofler, L. A., E. I. Fulmer, and L. Schoene.** 1939. Saccharification of starchy grain mashes for the alcoholic fermentation industry. Ind. Eng. Chem. **31**:734–738.
73. **Underkofler, L. A., A. M. Severson, K. J. Goering, and L. M. Christensen.** 1947. Commercial production and use of mold bran. Cereal Chem. **24**:1–22.
74. **Vezina, C., and K. Singh.** 1975. Transformation of organic compounds by fungal spores, p. 158–192. *In* J. E. Smith and D. R. Berry (ed.), The filamentous fungi, vol. 1. John Wiley & Sons, Inc., New York.
75. **Wang, H. L., J. J. Ellis, and C. W. Hesseltine.** 1972. Antibacterial activity produced by molds commonly used in oriental food fermentations. Mycologia **64**: 218–221.
76. **Wang, H. L., and C. W. Hesseltine.** 1979. Mold-modified foods, p. 96–131. *In* H. J. Peppler and D. Perlmann (ed.), Microbial technology, 2nd ed., vol. 2. Academic Press, Inc., New York.
77. **Wang, H. L., and C. W. Hesseltine.** 1982. Oriental foods, p. 492–538. *In* G. Reed (ed.), Prescott and Dunn's industrial microbiology, 4th ed. Avi, Westport, Conn.
78. **Wang, H. L., E. W. Swain, and C. W. Hesseltine.** 1975. Mass production of *Rhizopus oligosporus* spores and their application in tempeh fermentation. J. Food Sci. **40**:168–170.
79. **Williams, D. F.** 1981. Extraction with supercritical gases. Chem. Eng. Sci. **36**:1769–1788.
80. **Wood, B. J. B.** 1977. Oriental food uses of *Aspergillus*, p.

481–498. *In* J. E. Smith and J. A. Pateman (ed.), Genetics and physiology of *Aspergillus*. Academic Press, London.

81. **Wood, B. J. B.** 1982. Soy sauce and miso, p. 39–87. *In* A. H. Rose (ed.), Fermented foods. Academic Press, London.
82. **Yamada, K.** 1977. Recent advances in industrial fermentation. Biotechnol. Bioeng. **19:**1563–1621.
83. **Yamada, K.** 1977. Statistics of fermentation industry in Japan, p. 3.1–3.14. *In* Japan's most advanced industrial fermentation technology and industry. The International Technical Information Institute, Tokyo.

Chapter 8

Working with Anaerobic Bacteria

LARS G. LJUNGDAHL AND JUERGEN WIEGEL

Studies of anaerobic bacteria and fungi have intensified tremendously over the past 10 to 15 years. Such microorganisms, in addition to being of clinical interest, have considerable ecological importance as well as potential for use in industrial processes. Consequently, methods for studying anaerobic microorganisms and other microorganisms under anaerobic conditions have been improved. Components and enzymes of anaerobic microorganisms in many cases necessitate work under strictly anaerobic conditions. Methods for such studies are now also available. In this chapter we describe the more commonly used methods for studies of anaerobic bacteria and refer to more specific applications of such methods. We do not include methods which are used in clinical laboratories and involve pathogenic microorganisms; such methods are described in references 20 and 52.

The successful handling and cultivation of an anaerobic microorganism requires techniques to effectively remove oxygen (air) from both the medium and the gas phase in contact with the medium. Some microorganisms also require medium with a low redox potential (E'_0), in some cases as low as -0.33 V. This is accomplished by adding reducing agents to the medium. Current methods for the removal of oxygen and for preparation of reduced media are based on techniques described by Hungate (24, 25), which are now referred to as the "Hungate technique" (15). Modifications of the Hungate technique have been described by several workers. Most of these modifications are minor and often reflect a special application or an individual preference.

8.1 REMOVAL OF O_2 FROM GASES

Gases used in anaerobic work generally are CO_2, H_2, N_2, or mixtures of these. Normally, they are supplied in cylinders. Premixed gases are available but expensive, and it is more economical in the long run to use a gas-mixing system such as that supplied by Matheson Gas Products. (For names and addresses of suppliers mentioned in this chapter, see section 8.9.)

Cylinder gases contain small amounts of O_2, which must be removed. This can be done by passing the gas or the gas mixture through a column containing copper wire electrically heated to about 350°C. Directions for constructing the copper column are given by Hungate (25) and Bryant (11). A heating oven with a glass insert containing copper oxide wire is sold by Sargent-Welch Scientific Co. (catalog no. S-36517 and S-36518). The copper column is prepared for use by first passing CO_2 or N_2 through it to remove oxygen, which otherwise may cause an explosion if it is mixed with H_2. The copper oxide is then reduced to metallic copper by passing H_2 through the heated column. Water formed in this process may condense in the tubing connected after the copper column, and it should be removed before gases from the column are let into media or sterile equipment. The copper column is now ready to use. Small amounts of oxygen present in the cylinder gases react with the copper to yield copper oxide, turning the copper black as it is oxidized. The copper is then regenerated by passing H_2 through the column. For some applications we use gas mixtures containing from 3 to 5% H_2. The copper column is then automatically regenerated and always ready to use. Of course, this practice cannot be used when H_2 interferes with the microbiological work that is to be performed.

Microbial work almost always involves processing a large number of bacterial cultures; thus it is convenient to dispense the O_2-free gas through a gassing manifold to several tubes or bottles at the same time. The construction of a gassing manifold has been described by Balch and Wolfe and their associates (5, 6). From the gassing manifold the gas is led through thick-walled plastic or rubber tubing which ends with a 3- to 4-in. (ca. 75- to 100-mm), 18-gauge, stainless-

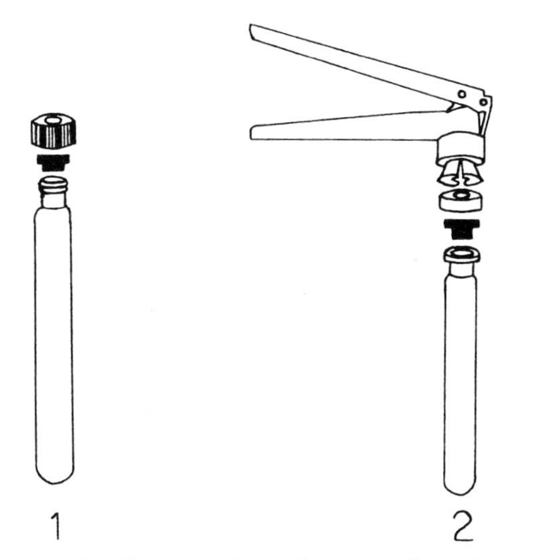

FIG. 1. Tubes for anaerobic cultures (Bellco Glass, Inc.). (1) No. 2047 (16 by 125 mm) Hungate tube (39) with butyl rubber stopper and screw cap. (2) No. 2048 (18 by 150 mm) Balch tube (56) with butyl rubber stopper and aluminum seal to be crimped.

steel syringe needle bent into a hook so that it can be hung conveniently over the rim of a culture tube. For flexible tubing we prefer butyl rubber. Very little gas dissolves in this material, and gases do not penetrate through the walls of tubing made of butyl rubber, in contrast to tubing made of most plastics or natural rubber. Butyl rubber tubing is sold as Viton tubing (many suppliers) or Iso-Versinic tubing (M.G.C. Verneret). The Iso-Versinic tubing is available with narrow bore suitable for connections to carry eluants in chromatographic separations (37).

8.2 TUBES AND BOTTLES FOR ANAEROBIC CULTURES

Tubes are now especially manufactured for use with the Hungate technique (Fig. 1) and are available from Bellco Glass, Inc. (catalog no. 2047 and 2048). This company supplies other equipment needed for anaerobic work, such as a spinner (no. 7790) to prepare roll tubes, as well as the Anaerobic Culture System described in the *Anaerobe Laboratory Manual* of Virginia Polytechnic Institute (23). The Bellco tube no. 2047 (also called a Hungate tube) (16 by 125 mm) was described by Macy et al. (39). It is sealed with a butyl rubber septum (no. 2047-11600), which is held by a screw cap (no. 2047-16000) that is open in the center to allow inoculation with syringes through the rubber septum. The Bellco tube no. 2048 (18 by 150 mm) was used by Balch et al. (5, 6). This tube is constructed with a flange and is sealed with a special black rubber stopper (Bellco no. 2048-11800). An aluminum seal (Bellco no. 2040-11020; Pierce Chemical Co., no. 18182) is crimped over the stopper and the flange of the tube. This type of tube is particularly useful in studies of bacteria such as methanogens (5) or acetogens (36), which use gaseous substrates. The tubes can be pressurized to 2 to 3 atm (200 to 300 kPa) and repressurized as the gases are consumed by using the gassing manifold described above (5, 6).

In addition to the Hungate tube, serum bottles are excellent for use with the Hungate technique (40). We use Hypo-Vials, obtainable from Pierce Chemical Co., in sizes from 1 ml (no. 12901) to 125 ml (no. 12995). These bottles are sealed with the same rubber stoppers (Bellco no. 2048-11800) and aluminum seals (Bellco no. 2040-11020 or Pierce no. 12182) as are used with the Bellco no. 2048 tubes described above. To deal with the aluminum seals, special hand crimpers are available from Pierce (no. 13212) and Bellco (no. 2048-10020), as well as a DeKapitator to remove the seals (Pierce no. 13210).

Larger cultivation vessels of up to 6,000-ml capacity can be constructed using Pyrex glass bottles (Bellco no. 5636) or Erlenmeyer flasks (Bellco no. 2514) which are closed with 38-mm screw caps. Modification of the screw caps of the bottles and the flasks by inserting cut-off Bellco no. 2048 tubes through the screw caps allows the use of rubber stoppers and aluminum seals (5). A further modification of this bottle system is to fuse a second Bellco tube at a right angle to the first. The second tube can be inserted into a Bausch & Lomb or similar spectrometer to determine the optical density of the culture (7).

For larger cultures, 10 to 20 liters, carboys of Pyrex glass can be used. We have successfully grown several anaerobic bacteria in 20-liter carboys closed with black butyl rubber stoppers. For the growth of acetogens and other gas-utilizing organisms the stoppers have been equipped with an inlet tube, reaching the bottom of the carboy, and an outlet tube. Sterile oxygen-free CO_2 or other gases are bubbled through the medium throughout the culturing (37). Similarly, by bubbling sterile CO_2 through the medium, acetogens have been cultured in 400-liter New Brunswick fermentors with yields of over 3 kg of cells per run.

8.3 REDOX POTENTIAL OF MEDIA

8.3.1 Reducing Agents

Removal of O_2 lowers the redox potential of the medium. However, as pointed out by Hungate (25), it is impossible simply by removal of O_2 to obtain a redox potential as low as -0.33 V, which is apparently required for methanogens. Furthermore, although O_2 is removed as completely as possible, some will still be present because it adheres to glassware and other materials and is dissolved in rubber and plastic tubing and stoppers. Therefore, it is necessary to add reducing agents to the medium to remove small traces of O_2 and to obtain the low redox potentials needed for growth of many anaerobic bacteria. However, it has been pointed out by several investigators (see Walden and Hentges [53] and references therein) that oxygen or products from its reactions such as H_2O_2 are toxic for anaerobic bacteria (13) and that a high redox potential is not necessarily a growth-limiting factor per se.

Some reducing substances which are added to lower the redox potential in anaerobic media are listed in Table 1. The E'_0 values are standard redox potentials of the 50% reduced substances at pH 7 based on measurements against a standard hydrogen electrode.

TABLE 1. Reducing agents used for lowering the redox potential (E_0') of media

Compound	E_0' (mV)	Concn normally used in media	References
Ascorbate	+80	0.1%	21, 33
Sodium thioglycolate	<−100	0.05%	24, 37, 48
Yeast extract	−50 to −150	0.1–1%	
Cysteine hydrochloride	−210	0.02–0.08%	12, 13, 24, 40, 48
Dithiothreitol	−330	0.05%	14, 53
FeS, amorphous	<−350	0.05 mM	9
H_2 and palladium chloride	−420		4, 25, 42
E. coli membrane fraction	<−300		2, 3
Titanium(III) citrate	−480	0.5–2 mM	27, 44, 60
Sodium sulfide or hydrogen sulfide	−571	0.01–0.025%	25
Sodium sulfide plus cysteine hydrochloride	−571	0.025% each	
Dithionite	<−600	0.001%	8, 25, 27

In redox reactions involving uptake or liberation of protons, the redox potentials increase as the pH is lowered and decrease when it rises. Thus, the redox potential may increase during fermentations in which acids are products. For discussions of the redox potential, its relation to pH, and its importance for microorganisms see Jacob (26) and Rabotnowa (45).

In general, solutions containing reducing substances are prepared and autoclaved separately from the rest of the medium and added to the medium just before inoculation. Since reducing agents react with oxygen to form toxic substances (13, 53), reducing solutions are prepared by first boiling the water to rid it of dissolved O_2. The water is bubbled with an O_2-free gas such as CO_2, N_2, or Ar while it is allowed to cool. The reducing substance is then dissolved in the water, and the solution is sterilized, preferably in a Hypo-Vial serum bottle stoppered with a butyl rubber stopper and an aluminum seal before being autoclaved. It is often convenient to autoclave solutions of reducing substances in batches of the exact amount required for the medium.

Recipes for the preparation of reducing solutions are found in the references listed in Table 1. In addition, some of the more commonly used reducing solutions are prepared as follows.

Cysteine hydrochloride, sodium thioglycolate, and dithiothreitol are normally used at a concentration of 0.05% in the final medium. Stock solutions can be made up by dissolving 2.5 g of the thiols in 100 ml of O_2-free water. Autoclaved solutions are kept under CO_2 in butyl rubber-stoppered vials. A 2-ml portion of the solutions is included in each 100 ml of ready medium. Cysteine hydrochloride is acid, and it can lower the pH of the medium. To neutralize it, a sterile solution of sodium carbonate, also stored under CO_2, is often added to the medium.

Perhaps the most commonly used reducing solution is a mixture of sodium sulfide and cysteine hydrochloride (8, 25, 32). In our laboratory, we prepare a stock solution by heating 200 ml of 0.2 N NaOH to a boil to remove oxygen. The sodium hydroxide solution is then bubbled with N_2. After some cooling, 2.5 g of $Na_2S \cdot 9H_2O$ and 2.5 g of cysteine · HCl · H_2O are added. Convenient volumes (4 ml is used to prepare 100 ml of reduced medium) are then transferred into vials that have been preflushed with N_2. The vials are sealed with butyl rubber stoppers, crimped with aluminum seals, and autoclaved for 15 min at 121°C. It should be noted that reducing solutions containing sodium sulfide should be stored under H_2, N_2, or argon and not under CO_2 (25), because sodium sulfide is alkaline and CO_2 is rapidly absorbed. This will lead to a vacuum which in turn may allow entrance of air.

Titanium(III) citrate was first used as a reducing agent in bacterial cultures by Plotz and Geloso (44). It reacts rapidly with oxygen and has been used successfully in media for several strictly anaerobic bacteria (27, 41, 44, 60). However, it inhibits the growth of some facultative anaerobic bacteria (60). Some strict anaerobes do not grow well in the presence of titanium citrate; in these cases it has been demonstrated that growth is dependent on cysteine (27). Thus, titanium(III) citrate can be used as a reducing agent in studies of growth requirements for cysteine and perhaps other sulfur compounds. The titanium(III) citrate is prepared from $TiCl_3$, available in 15 or 20% solutions. An amount of titanium chloride solution corresponding to 3 g of $TiCl_3$ is added to 200 ml of oxygen-free 0.2 M sodium citrate. The pH is adjusted to 7 by the addition of a saturated O_2-free solution of sodium carbonate. The solution is divided into suitable volumes and autoclaved anaerobically under N_2 or CO_2. One advantage with titanium citrate is that the complex is blue to violet in solution; when oxidized, the solution becomes colorless. Thus, the complex serves as a redox indicator (60).

Amorphous ferrous sulfide appears to be an efficient reducing agent which has been used for clostridia and methanogenic bacteria (9). It is prepared by mixing 0.1 mol of ferrous ammonium sulfate (39.2 g) with 0.1 mol of $Na_2S \cdot 9H_2O$ (24 g) in about 500 ml of oxygen-free boiling water. The precipitate of amorphous ferrous sulfide is allowed to settle, and the clear solution is decanted. It is then washed several times with boiling anaerobic distilled water, and finally it is suspended in O_2-free water and made up to a volume of 1 liter in an anaerobic flask which, after being stoppered and crimped, is autoclaved. Theoretically, the suspension should contain 8.8 g of FeS. However, apparently a loss occurs in the preparation and there is in fact somewhat less (9). The reaction of FeS with oxygen involves several steps which are best summarized in the following overall equation (9):

$$FeS + 2.25\ O_2 + 2.5\ H_2O \rightarrow Fe(OH)_3 + SO_4^{2-} + 2H^+$$

TABLE 2. Stock solution of vitamins according to Balch et al. (5)[a]

Vitamin	Concn (mg/100 ml)
Biotin	4
p-Aminobenzoic acid	10
Folic acid	4
Pantothenic acid, calcium salt	10
Nicotinic acid	10
Vitamin B_{12}	0.2
Thiamine hydrochloride	10
Pyridoxine hydrochloride	20
Thioctic acid	10
Riboflavin	1

[a] Vitamins are dissolved in 100 ml of water; 0.05 ml of this solution is added to 100 ml of medium. The vitamin solution is best stored frozen or in a refrigerator in the absence of light.

Ferric hydroxide, one of the products of the reaction of FeS with O_2, is red and is less soluble than FeS. The reaction is therefore pulled toward the right. The red color of the hydroxide is a good indicator; if a substantial amount of it is formed in the medium, the medium is probably not anaerobic. One disadvantage with FeS as a reducing agent is the formation of H^+. However, considering that only about 0.05 mmol is added per liter of culture medium, this acidification is small.

Anaerobic conditions in a medium can also be achieved by first allowing a facultative anaerobe, e.g., *Escherichia coli*, to grow in the medium, thus consuming the oxygen. The culture, in an anaerobic vial, is then autoclaved and inoculated with the anaerobic bacterium. A more novel technique along this line has been developed by Adler et al. (2, 3), who used a sterile membrane preparation for aerobically grown *E. coli* (1) to remove the oxygen. This preparation, containing a cytochrome system, reduces oxygen with either sodium lactate or sodium succinate as hydrogen donor. The membrane preparation, sterilized by filtration through a 0.22-μm membrane filter, is stable for a month when frozen at −10 or −70°C. From 20 to 100 μg of the membrane fraction is added per ml of medium, which contains 0.25 M sodium lactate or succinate. The medium is rendered oxygen free within 7 or 2 min with 20 or 100 μg of the preparation per ml, respectively.

8.3.2 Determining Redox State of Medium

It is not within the scope of this article to give details of how to measure the redox potentials of media. Such details are described by Jacob (26). For most anaerobic work, it is sufficient to know whether or not oxygen has been removed from or incidentally reintroduced into an anaerobic medium. With some of the reducing agents discussed above, e.g., amorphous ferrous sulfide and titanium citrate, the agents themselves become colored when oxidized, and thus media prepared with these agents have a built-in redox indicator. With thiols and sulfides as reducing agents, it is common to add a suitable redox indicator which should not be toxic for the bacterium to be cultured. The redox indicator of choice seems to be resazurin (6, 12, 15, 25), which has an E'_0 of −51 mV. It goes from blue to pink to completely colorless when the redox potential is lowered to about −110 mV or lower. When reoxidized, resazurin becomes pink and does not turn blue. It is practical to make a stock solution of resazurin by dissolving 100 mg in 100 ml of water and adding 0.1 ml of this solution per 100 ml of medium. Other redox indicators, including phenosafranine (10), which has an E'_0 of −252 mV, have been used. A list of redox indicators is given by Jacob (26). The redox indicator is not required in the medium to obtain bacterial growth. With practice, one should be able to prepare reduced media without the addition of the indicator. Nevertheless, a medium containing resazurin without pink color is a very good assurance of anaerobicity, although it does not indicate the low redox potential of −330 mV needed for growth of some anaerobes.

8.4 MEDIA

8.4.1 Liquid Growth Media

It is impossible to list here the compositions of all media for anaerobic bacteria. Many organisms require quite rich media which in addition to carbon and energy sources also contain yeast extract, tryptone, vitamins, minerals, and reducing agents. However, attempts to develop chemically defined media for several heterotrophic bacteria have been successful (38). In some instances, bacteria formerly regarded as strict heterotrophs have been found to be able to grow autotrophically. A prominent example is the acetogen *Clostridium thermoaceticum* which, when growing heterotrophically on glucose, converts glucose almost quantitatively to acetate. This bacterium has now been shown to grow on CO_2-H_2 or CO as sole carbon and energy source (30). The fixation of CO_2 or CO occurs via a unique new pathway of autotrophic CO_2 fixation, yielding acetyl coenzyme A (36, 47).

Although many types of media can be prepared as one single solution which is autoclaved after being freed of O_2, it is often better to prepare medium in three separate solutions that are autoclaved separately. This is especially so for media containing sugars, which when heated caramelize with amino acids or other ingredients in the medium to form inhibitory substances. Typically (37), the three solutions contain (i) the carbon source (carbohydrates), (ii) growth factors such as yeast extract, tryptone, amino acids, vitamins, and trace elements, and (iii) buffer substances. The solutions, freed of O_2, are autoclaved separately and are combined anaerobically, just before inoculation, with the reducing solution prepared as described earlier. Large carboys with buffer solution, especially if they contain bicarbonate, should be autoclaved with cotton plugs as used with aerobic solutions. After some cooling (to about 90°C) the carboys are removed from the autoclave and immediately bubbled with CO_2. It should be noted that many saccharophilic anaerobic bacteria do not grow well in media containing more than 1% of a carbohydrate.

Vitamins and trace elements which are needed in minute amounts are most conveniently prepared as stock solutions (5, 55, 57, 61). Tables 2 and 3 show examples of vitamin and trace element solutions, respectively. In rich media containing yeast extract, tryptone, or a similar complex ingredient, it is difficultto demonstrate a requirement for vitamins and trace elements. However, with a chemically defined medium such requirements are often detected, although many bacteria can extract their needs for trace elements from stainless steel and even glass (51; K. O. Stetter, personal communication).

A convenient method for preparing an anaerobic solution is described by Bryant (11). The heat-stable ingredients are dissolved in the appropriate volume of water and heated in a round-bottomed flask to a boil. The size of the flask should be about 1.5 times the volume of the solution. As the solution starts to boil, it should be gassed rather briskly (500 to 800 ml/min for a 1-liter flask) through a stainless-steel needle with an O_2-free gas, usually CO_2. The flask, still being gassed, is stoppered with a butyl rubber stopper which is applied rapidly as the gassing needle is withdrawn. The stopper is wired or crimped onto the flask, which is then autoclaved. Before autoclaving, the solutions may also be divided (10-ml portions) and transferred into culture tubes or serum bottles which are pregassed with O_2-free gas (Fig. 2), preferably by using the gassing manifold described earlier. The transfer can be made with a syringe or a serological pipette manipulated with a pipette device (Pelleus ball or similar). The syringe or pipette should also have been pregassed and have an internal gas phase free of O_2. The transfer is made while both the original flask and the tube or serum bottle are gassed. Immediately after the transfer, the tube or serum bottle is closed with a butyl rubber stopper and crimped before being autoclaved. Alternatively, the flask with the anaerobic solution is inserted into a glove box or so-called Freter box (4; Coy Laboratory Products, Inc.), to be discussed below, with an essentially O_2-free gas atmosphere. We use a mixture of 5% H_2 and 95% N_2. The culture tubes or serum bottles already in the glove box and free of O_2 can now be easily prepared with the anaerobic medium solution, stoppered and crimped before being removed from the glove box, and autoclaved.

In preparing a bicarbonate-containing medium for *C. thermoaceticum* (37) for a 400-liter fermentor, we have adopted the following procedure. The buffer ingredients, bicarbonate, yeast extract, vitamins, and trace metals are dissolved in 340 liters of water. The solution is heated to a boil and bubbled rather briskly with CO_2, and the reducing solution is added. Cysteine hydrochloride, neutralized with NaOH, or sodium thioglycolate is used as reducing agent; sodium sulfide is not as effective because when the pH of the medium drops, H_2S is formed, and it is removed with the CO_2 bubbling through the medium. The fermentor vessel is closed, and the solution is autoclaved for 2 h at 121°C in the fermentor. When the temperature of the solution decreases to below 100°C, the solution is again bubbled with CO_2, but less briskly. The temperature is adjusted to 60°C, the temperature at which the fermentation is performed. Glucose (8 kg dissolved in 20 liters of water) is autoclaved separately and added aseptically, as is 40 liters of a 24-h culture of *C. thermoaceticum* in two 20-liter carboys. The fermentation is performed with sterile CO_2 slowly bubbling through the medium and is complete at between 24 and 36 h. A cell yield of between 2.5 and 3.5 kg (wet cells) is obtained. The cells are harvested with a Sharples centrifuge. For further detailed information regarding liquid cultivation of bacteria, the excellent article by Drew (17) should be consulted.

TABLE 3. Stock solution of trace elements according to Balch et al. (5)[a]

Chemical[b]	Concn (mg/liter)
Nitriloacetic acid	1,500
$MgSO_4 \cdot 7H_2O$	3,000
$MnSO_4 \cdot H_2O$	500
NaCl	1,000
$FeSO_4 \cdot 7H_2O$	100
$Co(NO_3)_2 \cdot 6H_2O$	100
$CaCl_2$, anhydrous	100
$ZnSO_4 \cdot 7H_2O$	100
$CaSO_4 \cdot 5H_2O$	10
$AlK_2(SO_4)_3$, anhydrous	10
Boric acid	10
$Na_2MoO_4 \cdot 2H_2O$	10

[a] The nitriloacetic acid is suspended in about 500 ml of water and dissolved by titration with 2 to 3 N KOH until the pH is stabilized at 6.5. The chemicals are then added and dissolved in the order listed. The volume is finally adjusted to 1 liter. For 100 ml of medium, 0.5 ml of the trace element solution is used.

[b] In addition to the chemicals listed it is often necessary to include $NiCl_2 \cdot 6H_2O$ (20 mg), anhydrous Na_2SeO_3 (1 mg), and $Na_2WO_4 \cdot 2H_2O$ (10 mg) in the stock solution. Nickel, selenium, and tungsten have recently been found to be parts of several microbial enzymes including hydrogenase (43), carbon monoxide dehydrogenase (46), formate dehydrogenase (59), and methyl reductase (18).

8.4.2 Solid Media

The standard techniques for the estimation of the number of viable bacteria in an aerobic culture and for purifying aerobes as single-cell colonies require the use of solid medium (31). Similar methods are now available for use with anaerobes.

In general, solid media have the same or similar compositions as liquid media except for the addition of a solidifying agent. The most widely used agents are agar, carrageenan, silica gel, and Gelrite. All of these can be used to solidify media for anaerobic bacteria.

The choice of solidifying agent depends on the growth requirements. Agar is undoubedly the best known and also the most used. It is obtained from certain red marine algae, and it consists of two polysaccharides, agarose and agaropectin, which are not degraded by most bacterial species. It is commercially available in different grades from several suppliers of bacteriological agents and chemicals. To prepare anaerobic agar-solidified medium, agar can be added to the medium at the same time it is made anaerobic, according to the procedure of Bryant (11) described above. The medium is then autoclaved and dispensed into culture tubes, serum bottles, or petri dishes while still molten. Normally the concentration of agar in the medium is between 1.5 and 2%. Agar gels solidify at about 40°C, but after solidification they will remain solid up to 65°C. Bacteriological-grade agar or better does not influence the pH of the medium. However, when media are prepared at pH values below 5 or above 8.5, the agar should be autoclaved

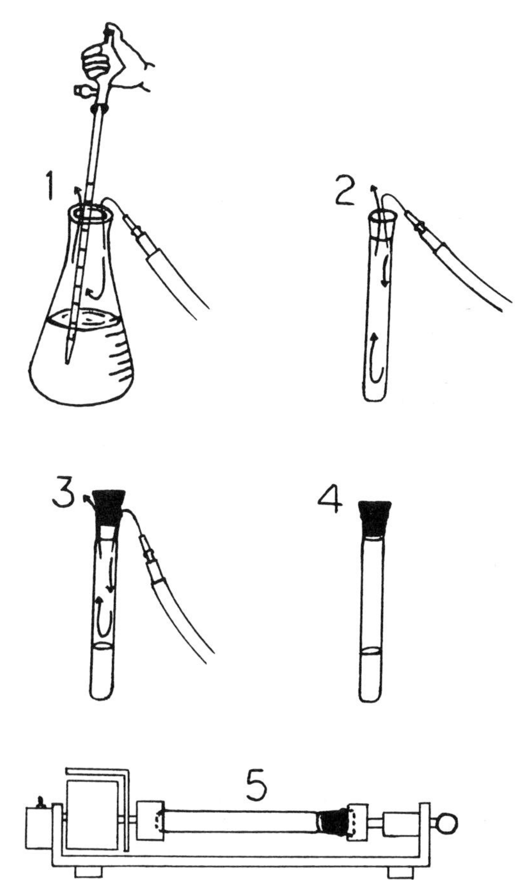

FIG. 2. "Hungate technique" (11, 15, 25, 40) for preparing anaerobic medium and dispensing it into anaerobic tubes.

separately, because at the more extreme pH values heating will cause breakdown of the agar.

Carrageenan, like agar, is obtained from certain red marine algae. Available from Sigma Chemical Co, it is cheaper than agar and is a good substitute (35, 54). It is used in the same concentration as agar, and it is stable (solid) at 60°C. Carrageenan may cause some changes in medium pH and a correction may be necessary.

Silica gel is inorganic and is particularly suitable as a solid medium in studies of autotrophic bacteria in the absence of organic compounds. It is also used for growing anaerobic, extremely thermophilic bacteria such as *Methanothermus fervidus* (51) at 85°C. For the latter purpose, polysilicate plates are prepared in an anaerobic chamber by first mixing 180 ml of a medium suitable for methanogens (5) and containing resazurin with 20 ml of sodium silicate solution (DAB6; 1.37 kg/liter; Merck). Solid sodium dithionite is added until the resazurin becomes colorless. The pH is quickly adjusted to 7 with H_2SO_4, using vigorous stirring, and the solution is poured immediately into glass petri dishes. The plates solidify within 5 min. They are then flooded with the methanogenic medium containing an antibiotic (vancomycin, 20 mg/liter) to avoid contamination problems. The plates are allowed to equilibrate with the medium for 24 h, with the liquid medium changed about five times. The plates are then placed in a pressure cylinder (5) containing 80% H_2 and 20% CO_2 and autoclaved for 2 h at 120°C.

A second method of making silica gel plates is described by Funk and Krulwich (19) and referred to by Krieg and Gerhardt (31). Though the preparation is described for aerobic bacteria, it probably can be used with anaerobes, if the plates are prepared in an anaerobic box and with prereduced medium. However, we have no experience with the method. The preparation is as follows. Double-strength, prereduced medium is prepared and sterilized. A potassium silicate solution is made by dissolving 10 g of powdered silica gel (grade 923, 100 to 200 mesh; Fisher Scientific Co.) or silicic acid (J. T. Baker Chemical Co.) in 100 ml of 7% (wt/vol) KOH. The solution is sterilized by autoclaving. Equal volumes of the double-strength medium and the silicate solution are mixed and neutralized (pH 7) aseptically with 20% *o*-phosphoric acid. Plates must be poured immediately since the medium starts to solidify within 1 min and is firm in 15 min.

Gelrite is an agarlike polysaccharide that is produced by a *Pseudomonas* species (29). An acetylated form is sold by Kelco. With the aid of a cation such as Ca^{2+} or Mg^{2+} it forms clear gels which have good thermal stability and also the ability to withstand low pH. Consequently it has been used for studies of acidophilic and thermophilic bacteria, including *Bacillus acidocaldarius*, *Bacillus stearothermophilus*, *Thermus thermophilus*, and *Thermus aquaticus* (34), as well as bacteria from deep-sea hydrothermal vents (J. W. Deming, personal communication) at 120°C under pressure. A Gelrite medium is prepared by mixing a sterile double-strength medium, containing extra $CaCl_2$ or $MgCl_2$ as gelling aid (from 0.114 to 0.2%), with a sterile solution of 1.1 to 2.2% Gelrite (34). The mixture is immediately poured into petri dishes. Bacteria appear to respond differently to the high calcium and magnesium concentrations, and therefore the best medium composition for each species must be found. Again, there is no detailed literature available yet for the use of Gelrite with strictly anaerobic bacteria. One of us (J.W., unpublished data) has used Gelrite in shake-roll tubes successfully with thermophilic clostridia and other anaerobes (J. Wiegel, *in* T. D. Brock and J. G. Zeikus, ed., *Thermophiles: General, Molecular and Applied Microbiology*, in press). The medium was prepared with a cysteine hydrochloride-sodium sulfide mixture (see above) as reducing agent.

It has been emphasized that solid plates to be used for anaerobic bacteria should be prepared with the exclusion of O_2 and with the addition of a reducing agent to the medium. An alternative would be to use the membrane fraction from *E. coli* (2, 3), which apparently is an efficient method for removal of O_2 from a medium.

Even if solid plates have been prepared under strictly anaerobic conditions, inoculating or streaking the plates may cause exposure to oxygen. With the use of a glove box or anaerobic chamber (4), such exposure can be avoided since the work is performed in an anaerobic gas mixture normally consisting of H_2-N_2 (5:95%) or H_2-CO_2 (5:95%). Several anaerobic chambers are available. However, we have found the flexible Coy chamber (Coy Laboratory Products, Inc.) to be the most practical and least expensive. It comes in several sizes and is equipped with an airlock system which permits transfers of samples and equipment between the outside and the inside of the chamber essentially without the introduction of any O_2 into the chamber. Any O_2 that is introduced or penetrates the flexible vinyl film used in constructing the chamber is rapidly reacted with the H_2 present in the gas atmosphere. This is facilitated by a palladium catalyst placed in a gas-circulating catalyst box. The chamber can be equipped with several accessories, including electrical outlets, gas analyzer, temperature control, microscope sock, temperature-controlled incubator, and most importantly an incandescent flaming device for test tubes and bacteriological loops.

To function well, the Coy chamber should be operated almost daily. Some O_2 penetrates the vinyl film, and normally it reacts with H_2 in the gas atmosphere inside the chamber. However, after the chamber originally is made anaerobic, H_2 is introduced only through the airlock system. If this system is not operated, the H_2 in the chamber will eventually be consumed and O_2 will be present. This will happen if the chamber is not used within 5 to 6 days. After such a period, it is thus advisable to partly evacuate the chamber and introduce fresh H_2-N_2 or H_2-CO_2 gas mixture. Another problem which sometimes arises is a buildup of moisture, as water is formed in the reaction between H_2 and O_2. This can be eliminated by placing a box with Drierite (Thomas Scientific) or a similar drying agent inside the chamber. Drierite is regenerated by heating at 200 to 225°C for 1 to 2 h. The palladium catalyst over a period of time becomes inactive. It too can be regenerated by heating at about 200°C for a couple of hours. Eventually, it is rendered inactive and must be replaced; however, we have found that it lasts about 3 years.

After plates have been inoculated, they can be placed either in a temperature-controlled incubator inside the chamber or in the pressure cylinder described by Balch et al. (5), which is taken out of the glove box and placed in an incubator outside the glove box. As an alternative to the pressure cylinder, an incubator jar for a GasPak system (BBL Microbiology Systems) can be used (49).

The GasPak system is described as being a complete anaerobic system. However, when the system is used without a glove box, plates are exposed to air during preparation and inoculation. Furthermore, about 1 h is required to establish anaerobic conditions inside the GasPak jar and to obtain a low redox potential of the medium after the GasPak system has been activated.

The principle of the GasPak system is the generation of H_2 and CO_2 from sodium borohydride- and CO_2-generating tablets with water inside the cylinder containing the plates. Oxygen is then removed by reacting with H_2, which is catalyzed by a palladium catalyst. Although the GasPak system is widely used, it appears best suited for bacteria that tolerate some exposure to oxygen.

With the Hungate technique (24, 25) and modifications thereof (5, 6, 11, 15, 23, 39, 40), roll tubes with solidified media are used for the isolation of single colonies of anaerobic bacteria without the benefit of an anaerobic glove box or GasPak supply. The anaerobic agar-containing medium is prepared as described and distributed into anaerobic tubes or serum bottles. These are stoppered with butyl rubber stoppers, crimped with aluminum seals, and sterilized by autoclaving. The tubes or bottles are then incubated at 45°C to prevent the agar medium from solidifying. The medium is thereafter allowed to solidify as a film around the walls of the containers by means of gentle rotation while the containers are chilled by flushing with cold tap water or immersion in ice. This is best done with a mechanical spinner (Fig. 2) (Bellco Glass, Inc) which evenly rotates the tubes and bottles as they are chilled.

To inoculate and streak out organisms on the agar film, about 0.5 ml is withdrawn from a liquid culture and injected into an agar tube, using a sterile syringe pregassed with O_2-free gas. The tube is then slowly rotated as the fluid washes down over the agar film. The tube is left upside down to accumulate the liquid at the stopper. The excess liquid can then be withdrawn with a sterile O_2-free gassed syringe. By using a dilution row (1:10 or 1:100 dilutions) and injecting several tubes, good single-cell colonies can be obtained on the agar film.

Streaking can also be done with a regular loop. However, this requires that the stopper be removed from the agar tube and replaced, which involves a risk for contamination and the introduction of oxygen. The procedure involves gently removing the rubber stopper (using aseptic technique, since the rubber stopper is used for restoppering) and at the same time inserting a gassing needle carrying O_2-free gas into the tube. The bacterial culture is then introduced with a platinum loop and is streaked out from the bottom to the top of the tube as the tube is slowly rotated. The tube is then reclosed with the butyl rubber stopper as the gassing needle is removed.

An alternative to the above methods is to use the so-called agar shake method. This technique has been successfully used in our laboratory for the isolation of thermophilic anaerobic bacteria and can be used also for cryophilic, mesophilic, and thermotolerant bacteria. Serum bottles are inoculated through the rubber stoppers, using an O_2-free gassed syringe, before the agar solidifies. After inoculation, the agar is allowed to solidify and the bottles are incubated at an appropriate temperature (up to 65°C). As described above, by using inoculations from a dilution row, well-isolated colonies, scattered throughout the solid agar medium, may be obtained. Colonies can be picked without opening the serum bottle by inserting a 3-in. (ca. 7.5-cm) pregassed O_2-free syringe with an 18-gauge needle through the stopper and manipulating the needle close to a colony. The colony is then gently sucked into the needle together with a small additional plug of agar. The latter protects the colony as the tip of the needle passes through the rubber stopper when it is withdrawn. It also protects the colony from direct exposure to air during the transfer.

An anaerobic technique using solid medium in flat, 100-ml glass bottles has been described by Braun et al. (8) (Fig. 3). It was developed for isolating bacteria which form methane or acetate from a 67% H_2 and 33% CO_2 gas mixture. The technique combines the advantages of the Hungate technique and those of petri dishes. The bottles are gassed with the H_2-CO_2 mixture as 10 ml of liquid, sterile, anaerobically

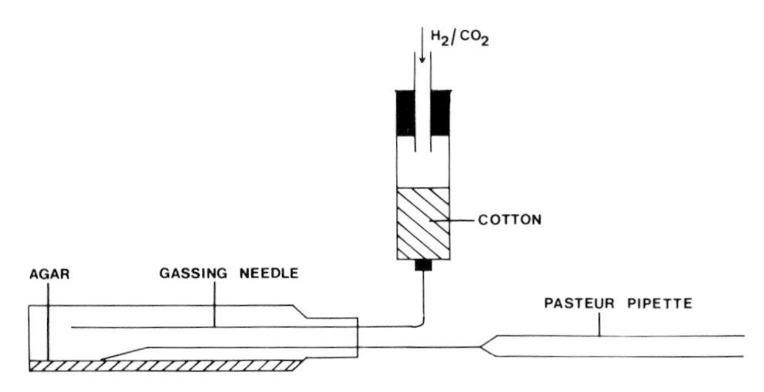

FIG. 3. Anaerobic agar bottle technique; method of Braun et al. (8).

prepared agar medium is introduced per bottle. The bottles are closed with butyl rubber stoppers, and the agar medium is allowed to distribute evenly and to solidify with the bottles lying on their flat sides. The bottles are then opened and again gassed to maintain anaerobicity while about 0.1 ml of culture (again, several bottles may be inoculated from a dilution series) is introduced and streaked out. The streaking can be done using a Pasteur pipette bent into a hook. Colonies can be picked with a Pasteur pipette with a tip drawn out to a capillary and slightly bent. This tool can also be used to pick colonies from the Hungate roll tubes.

With solid medium in closed tubes and bottles, syneresis (the contraction of a gel with the expulsion of water) may cause a problem. To minimize smearing of individual colonies on the surface of the medium, the bottles and tubes should be incubated upside down and in a tilted position. Water accumulating at the rubber stopper can then be removed with a syringe.

8.5 STORAGE OF ANAEROBIC BACTERIA

The viability of anaerobic bacteria in culture media varies between extremes. This is exemplified by the extreme thermophile *Methanothermus fervidus* (51), which dies within a few hours after exhaustion of its energy source, and by *Clostridium thermoautotrophicum* (56), which is a poor sporeformer in liquid medium and whose vegetative cells die within a day after the substrate is depleted. On the other hand, many clostridial species, including *Clostridium thermocellum* and *Clostridium thermohydrosulfuricum*, sporulate easily and survive for months in liquid growth medium when stored at 4°C. Many anaerobes, including those clostridial species that do not readily sporulate, survive quite well in agar medium when stored at 4°C. An example is *C. thermoaceticum*, which we have found viable after more than 10 years of such storage.

A general method for storing anaerobes is to add glycerol to the liquid medium and then to store the culture in a freezer. We have adopted the following procedure. A 6-ml serum bottle sealed with a butyl rubber stopper and crimped, and containing from 1 to 2 ml of prereduced medium, is inoculated. During mid-logarithmic growth, an equal volume of sterile, anaerobically prepared glycerol is injected with a syringe flushed with O_2-free gas. The contents of the bottle are thoroughly mixed, and the bottle is stored at −20°C. This 50% glycerol mixture does not freeze, so aliquots can be withdrawn with a gas-flushed syringe without changing the temperature of the stock culture. To prepare a new culture, 0.1 ml is withdrawn from the stock culture and directly injected into 2 to 6 ml of fresh medium. It is our experience that growth begins without a long lag period. We have stored *Thermoanaerobacter ethanolicus*, *Methanobacterium thermoautotrophicum*, and *Acetobacterium woodii* in this way for over 6 years. These three bacteria do not form spores. Hippe (22) has used the same method for storing methanogens at −70°C for more than 5 years. Occasionally, it is desirable to keep a dense stock culture. This is achieved by centrifuging a 10-ml mid-log-phase culture in a Hungate tube. The supernatant is withdrawn aseptically and anaerobically. The cells are suspended in 1 ml of fresh medium, and 1 ml of glycerol is added. The concentrated stock culture is stored in the freezer.

Bacteria can also be preserved in liquid nitrogen, which is the method of choice for long-term storage. Bacterial cultures growing in the mid-log phase are centrifuged and resuspended in prereduced medium (0.05% cysteine plus 0.05% sodium sulfide). A cryoprotectant is added, and the culture is placed in a glass ampoule which has been pregassed with N_2, CO_2, or argon. The ampoule is closed, frozen in a stream of cold nitrogen, and stored in liquid nitrogen. The following are used as cryoprotectants: 10% glycerol, 5% dimethyl sulfoxide, and 10 to 20% sucrose (16, 28). Cryoprotectants should be sterilized and made anaerobic separately from the medium.

Hippe (22) has described a method for storing bacterial cultures in glass capillaries kept in liquid nitrogen. Several capillaries are prepared at the same time, and a single capillary is used to inoculate fresh medium, leaving the remaining capillaries undisturbed. The only equipment required is a micropipettor to be used with sterilized capillaries (no. 0658-33; Difco Laboratories). The bacterial culture is prepared with reducing agent and cryoprotectant as described above, and as it is flushed with O_2-free gas, about 0.025 ml of culture (one-third of the capillary tube) is drawn into the capillary with the aid of a micropipettor. The capillary is taken out of the bacterial suspension but still held inside the O_2-free tube; the suspension is then drawn further into the capillary, together with some gas at the end of the capillary. This end is now quickly sealed in a small flame, as is the other end at about 1 cm below the attachement in the pipettor. The capillary is washed with 75% ethanol, and the seal is inspected with a stereo microscope before the capillary is frozen in the gas phase above liquid nitrogen, in which it is later stored. To inoculate with the bacterial culture, the capillary is quickly warmed to thaw the contents. The tip of the capillary is then scratched with a diamond pen, dipped in ethanol, and flamed. The tip is broken off, and the bacterial suspension is removed with an O_2-free gassed syringe with a 26-gauge hypodermic needle and directly injected into a Hungate tube or serum vial with prereduced medium.

Anaerobic bacteria can also be stored freeze-dried, at room temperature or at 4°C, with methods normally used for aerobic microorganisms. The only precau-

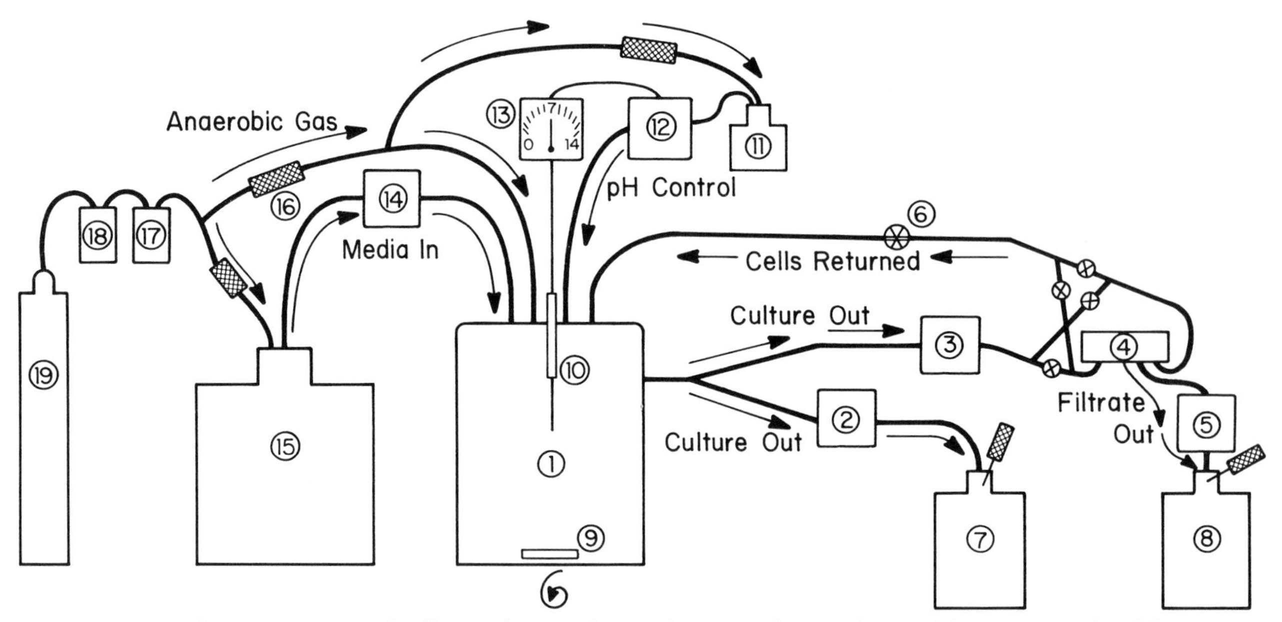

FIG. 4. Setup for continuous and cell recycling studies under anaerobic conditions (1) Fermentor (model C30, New Brunswick Scientific Co.); (2) cassette pump (junior model, catalog no. 72-510-000 [115 V]) with pumping cassette (catalog no. 72-550-000, Manostat); (3) peristaltic pump (model 1203, Harvard Apparatus); (4) hollow-fiber cartridge filter (type H1MP01-43) with hollow-fiber cartridge adapter (model DH2, no. 54077, Amicon Corp., Scientific Systems Div.); (5) same as (2) above; (6) regulating valve (e.g., Nupro part no. B-4JR, Nupro Co.); (7 and 8) outflow container (glass carboy); (9) magnetic stir bar; (10) pH electrode (catalog no. 5573705-DL, Phoenix Electrode Co.); (11) container for NaOH (glass flask); (12 and 13) pH controller (model pH-40, New Brunswick Scientific Co.); (14) peristaltic pump (type 4912-A, LKB Instruments, Inc.); (15) medium container (glass carboy); (16) sterile cotton gas filter; (17) backflow trap (glass flask); (18) gas-purifying furnance (model S-36517) with copper granule-filled gas-purifying tube (model S-36518, Sargent-Welch Scientific Co.); (19) gas cylinder (CO_2); (20) connection tubing (butyl rubber tubing, Viton or Versinic).

tion needed is to keep the cultures in an O_2-free gas phase. For very oxygen-sensitive bacteria, some amorphous ferrous sulfide (9), prepared as described previously, can be added to the culture medium before the freeze-drying. We have not tested this method.

8.6 CONTINUOUS CULTURES WITH AND WITHOUT CELL RECYCLING UNDER ANAEROBIC CONDITIONS

Continuous-culture techniques allow studies of bacterial growth yields, production rates, maintenance energy requirements, survival at extreme conditions, and effects of concentrations of substrates, products, and growth factors under strictly controlled and reproducible conditions. Continuous culture is also a good technique for studying interrelations between bacteria in cocultures, for establishing enrichment cultures, and for selecting mutants. The principles and general procedures of continuous culture have been discussed in several articles, including some in this book and a series of monographs edited by Dean et al. (16). Here we will discuss only some details pertinent to performing continuous cultures under anaerobic conditions.

Figure 4 is a diagram of a setup for continuous anaerobic fermentations with the possibility of cell recycling. It is used in our laboratory for studies of acetate production by fermentation of glucose with themophilic acetogenic bacteria and for ethanol production with *Thermoanaerobacter ethanolicus* (L. G. Ljungdahl, L. H. Carreira, R. J. Garrison, N. E. Rabek, and J. Wiegel, *in* Proceedings of the 7th Symposium on Biotechnology for Fuels and Chemicals, in press; J. Wiegel, C. P. Mothershed, and J. Puls, *in* Proceedings of the 6th International Biodeterioration Symposium, in press).

The fermentation unit (labeled 1) in Fig. 4 is designed for a medium volume of 350 ml. Other sizes of fermentors from 50 to 1,000 ml can easily be manufactured in a glass-blowing shop. We use double-walled containers to be able to control the temperature by circulating water from a constant-temperature bath. To provide for inserts and attachments, the fermentors are equipped with side pieces such as Wheaton necks for screw caps (no. 807808 and 240506, Wheaton Scientific) or screw-thread tubes (no. 0,190,006, Witeg Scientific). The smallest fermentors (50 ml) are very economical since they do not need large volumes of medium. However, for cell recycling a fermentor of 350 ml or larger is preferred because of the dead volume of the recycling unit.

The entire fermentor setup must be kept anaerobic. Gases from gas tanks (no. 19, Fig. 4) are led through a heated oven (no. 18) with copper granules to remove any O_2 and then through a backflow trap (no. 17). The gas line then branches, and the gas is passed through separate sterile cotton filters to a 10- to 20-liter medium reservoir (no. 15), the pH control unit (no. 11 through 13), and the fermentor (no. 10). The prereduced medium in the reservoir is pumped into the fermentor with a peristaltic pump (no. 14; model 1203, Harvard Apparatus; type 4912-A, LKB Instruments; or no. 72-510-000 cassette pump, Manostat) which is able to work with relatively thick-walled black rubber tubing. The peristaltic pumps (no. 2, 3, and 5) are also of the same quality and makes.

For continuous fermentations without cell recycling, the outflow from the fermentor is transferred via pump (no. 2, Fig. 4) into the sterile reservoir (no. 7), which has a gas outlet through a 25-gauge hypodermic needle and a syringe with sterile cotton as filter. We have found it necessary to keep the outflow reservoir sterile since pressure differences may occur, leading to backflow and the possible introduction of contaminating bacteria. If gas pressure is desired during the fermentation, the gas outlet from the reservoir (no. 7) may be equipped with a check valve (no. B-4c PA 2-3, Nupro Co.).

For cell recycling, the culture outflow is led through pump no. 3 (Fig. 4) to an Amicon hollow fiber cartridge (no. 4) (model DH2, no. 54077). This hollow-fiber system does not plug up as easily as some other filter systems, and it allows the flow to be reversed, which we do every 12 h. The cartridge holding the hollow fiber is made of plastic and is permeable to oxygen in the air; therefore, we have enclosed the cartridge in a glass tube. The filtrate is discharged from the hollow-fiber system via a peristaltic pump (no. 5) into a sterile reservoir (no. 8). The degree of recycling is regulated by the peristaltic pump (no. 5) and the Nupro valve (no. 6).

Success or failure with the setup described depends, to a large extent, on the choice of tubing connecting the different parts. Most plastic tubing, natural rubber tubing, and valves are permeable to air and consequently also oxygen. Therefore, we keep all connections short, and where possible we use glass or stainless steel. Flexible connections are made with black butyl rubber tubing with an outer diameter of 12 mm and a 1.5-mm-thick wall. This tubing has very low permeability for gases, and it can be handled by the model 1203 Harvard Apparatus peristaltic pumps mentioned earlier.

8.7 HARVEST OF CELLS UNDER ANAEROBIC CONDITIONS

Most fermentations are probably conducted to convert a substrate to a desirable product or products. The products in general are not sensitive to oxygen and can be recovered without exclusion of oxygen or maintenance of sterility. If, on the other hand, the bacteria are going to be used for physiological studies or as an inoculum, sterile and anaerobic conditions may still be required during harvest or concentration of the cells. Bacterial cells are also often the source of enzymes. Cells grown for this purpose generally do not need to be harvested under sterile conditions and perhaps not even anaerobically. However, it is now becoming quite clear that many enzymes and proteins in anaerobic bacteria are extremely oxygen sensitive. This is especially so for metal-containing enzymes isolated from acetogenic and methanogenic bacteria (18, 46, 59) and for membrane-bound proteins which are constituents of electron transport or energy generation mechanisms in autotrophic anaerobic bacteria (M. Ivey, personal communication). Cells to be used in studies and isolation of such proteins should be kept as anaerobic as possible, although the proteins generally are relatively stable as long as the cells are intact.

Harvesting cells under anaerobic and sterile conditions is easy only when volumes are small. The simplest method is to grow the bacteria in serum bottles or Hungate tubes which can be centrifuged at low speed (1,000 to 2,000 $\times$ g). If the tubes contain pressurized gas, the pressure should be released before centrifugation. For cultures of larger volumes (1 to 3 liters), the best way is to bring the culture inside an anaerobic chamber and to transfer the culture medium into sterilized centrifuge tubes of stainless steel (e.g., no. 303380 for use in a Beckman Preparative Ultracentrifuge [Beckman Instruments, Inc.], or no. 00522 for Sorvall centrifuges [DuPont Co.]) or of heavy glass with tightly fitting caps. After centrifugation the tubes are brought back into the anaerobic chamber before they are opened. If a chamber is not available, the centrifuge tubes can be flushed with O_2-free gas, and the cell culture is then transferred by the regular Hungate technique (11, 25).

Harvest of large volumes (5 to 1,000 liters) under anaerobic and sterile conditions can be done by employing the appropriate size of the "through flow filtration" ceramic tube system (Ceraflo, Norton). The filtration tubes are made of ceramic (sintered aluminum oxide) and are contained in stainless-steel tubes. The system is available in four different sizes from 0.3 to 10 ft^2 (0.028 to 0.93 m^2). When the filtration system is used under anaerobic conditions, the culture fluid of the fermentor should be passed to the system via heavy-wall black butyl rubber, copper, or stainless-steel tubings to exclude air. The Amicon hollow fiber cartridge system, as described above in connection with fermentation with cell recycling, can also be used to harvest cells from large volumes of media. However, with the Amicon system less pressure is used, and thus the bacterial slurry that is obtained is less concentrated than that with the ceramic tube system. With both systems, if a cell pellet is desired instead of a slurry, a final concentration step has to be done using centrifugation.

When the cells are going to be used for enzyme work and sterile conditions during the harvest are not required, the best harvest method uses a Sharples centrifuge (Pennwalt Corp.). The fermentor, still under anaerobic conditions, is connected to the centrifuge by tubing, as described above. A slightly positive pressure with O_2-free gas in the fermentors pushes the medium through the Sharples centrifuge. At the completion of the centrifugation, the cells are removed from the centrifuge rotor as it is flushed with O_2-free gas and stored frozen under anaerobic conditions. Cells of *C. thermoaceticum*, when harvested as described, wrapped tightly with Saran wrap, immediately frozen, and stored at $-70°C$, maintain a high activity of formate dehydrogenase for several months; such cells stored at $-20°C$ lost activity rather rapidly. This formate dehydrogenase has a K_i for O_2 of 7.6 μM and is one of the most O_2-sensitive enzymes that has been isolated (59).

8.8 CARBON RECOVERY AND REDOX BALANCE AS CONTROL OF FERMENTATIONS

In anaerobic fermentations the substrate is converted to organic products which differ depending on the substrate that is utilized, the bacterial species, and the conditions of the fermentation. Therefore, a stoichiometric analysis of the consumption of substrate and formation of products of the fermentation is of extreme value. Such an analysis shows whether the substrate is converted to products in yields as predicted, whether or not there are side products, the utilization of nonsubstrate carbon (e.g., of amino acids in carbohydrate fermentations), and whether or not a culture is contaminated. The analysis involves the quantitative determination of amounts of substrate utilized, the products formed, and the yield of cells (dried). Calculations of carbon recovery and redox balance are then performed. The perfect fermentation would have a 100% carbon recovery and a redox balance or quotient of 1. Methods of calculation of carbon recovery and redox balance have been discussed by Wood (58).

8.9 SUPPLIERS

Amicon Corp.
Scientific System Div.
17 Cherry Hill Dr.
Danvers, MA 01923

J. T. Baker Chemical Co.
222 Red School Lane
Phillipsburg, NJ 08865

BBL Microbiology Systems
Cockeysville, MD 21030

Beckman Instruments, Inc.
2500 Harbor Blvd.
Palo Alto, CA 92634

Bellco Glass Inc.
340 Edrudo Rd., P.O. Box B
Vineland, NJ 08360

Coy Laboratory Products, Inc.
P.O. Box 1108
Ann Arbor, MI 48106

Difco Laboratories
P.O. Box 1058A
Detroit, MI 48232

DuPont Co.
Quillen Bldg.
Concord Plaza
Wilmington, DE 19898

Fisher Scientific Co.
585 Alpha Dr.
Pittsburgh, PA 15238

Harvard Apparatus
22 Pleasant St.
South Natick, MA 01760

Kelco
Div. of Merck & Co., Inc.
8355 Aero Dr.
San Diego, CA 92123

LKB Instruments Inc.
9319 Gaither Rd.
Gaithersburg, MD 20877

Manostat
519 Eighth Ave.
New York, NY 10018

Matheson Gas Products
932 Paterson Plank Rd., P.O. Box 85
East Rutherford, NJ 07073

Merck GmbH
Darmstadt, F.R.G.

M.G.C. Verneret
BP 211
F-94203 Ivvy-Cedex, France

New Brunswick Scientific Co., Inc.
44 Talmadge Rd., P.O. Box 986
Edison, NJ 08818

Norton
P.O. Box 350
Akron, OH 44309

Nupro Co.
4800 East 345th St.
Willoughby, OH 44994

Pennwalt Corp.
99 Business Park Dr.
Armonk, NY 10504

Phoenix Electrode Co.
6103 Glenmont St.
Houston, TX 77081

Pierce Chemical Co.
P.O. Box 117
Rockford, IL 61105

Sargent-Welch Scientific Co.
7300 North Linder Ave.
P.O. Box 1026
Skokie, IL 60077

Sigma Chemical Co.
P.O. Box 14508
St. Louis, MO 63178

Thomas Scientific
Vine St. at Third
P.O. Box 779
Philadelphia, PA 19105

Wheaton Scientific
1000 N. Tenth St.
Millville, NJ 08332

Witeg Scientific
700 HN Valley St.
Anaheim, CA 92801

8.10 ACKNOWLEDGMENTS

Our work with anaerobic bacteria is supported by grant no. DE-FG09-84ER13248 and contract DEA-S09-79ER10449, both from the U.S. Department of Energy, and by Public Health Service grant AM27323 from the National Institute of Arthritis, Diabetes, and Digestive and Kidney Diseases.

8.11 LITERATURE CITED

1. **Adler, H. I., A. Carrasco, W. D. Crow, and J. S. Gill.** 1981. Cytoplasmic membrane fraction that promotes septation in an *Escherichia coli lon* mutant. J. Bacteriol. **147:**326–332.
2. **Adler, H. I., and W. D. Crow.** 1981. A novel approach to the growth of anaerobic microorganisms. Biotechnol. Bioeng. Symp. **11:**533–540.
3. **Adler, H. I., W. D. Crow, C. T. Hadden, S. Hall, and R. Machanoff.** 1984. New techniques for growing anaerobic bacteria: experiments with *Clostridium butyricum* and *Clostridium acetobutylicum*. Biotechnol. Bioeng. Symp. **13:**153–161.
4. **Aranki, A., and R. Freter.** 1972. Use of anaerobic glove boxes for the cultivation of strictly anaerobic bacteria. Am. J. Clin. Nutr. **25:**1329–1334.
5. **Balch, W. E., G. E. Fox, L. J. Magrum, C. R. Woese, and R. S. Wolfe.** 1979. Methanogens: reevaluation of a unique biological group. Microbiol. Rev. **43:**260–296.
6. **Balch, W. E., and R. S. Wolfe.** 1976. New approach to the cultivation of methanogenic bacteria: 2-mercaptoethanesulfonic acid (HS-CoM)-dependent growth of *Methanobacterium ruminantium* in a pressurized atmosphere. Appl. Environ. Microbiol. **32:**781–791.
7. **Bhatnagar, L., M. Henriquet, and R. Longin.** 1983. Development of an improved bottle-system for cultivation of strict anaerobes (methanogens). Biotechnol. Lett. **5:**39–42.
8. **Braun, M., S. Schoberth, and G. Gottschalk.** 1979. Enumeration of bacteria forming acetate from H_2 and CO_2 in anaerobic habitats. Arch. Microbiol. **120:**201–204.
9. **Brock, T. D., and K. O'Dea.** 1977. Amorphous ferrous sulfide as a reducing agent for culture of anaerobes. Appl. Environ. Microbiol. **33:**254–256.
10. **Bryant, M. P.** 1963. Symposium on microbial digestion in ruminants: identification of groups of anaerobic bacteria active in the rumen. J. Anim. Sci. **22:**801–813.
11. **Bryant, M. P.** 1972. Commentary on the Hungate technique for culture of anaerobic bacteria. Am. J. Clin. Nutr. **25:**1324–1328.
12. **Bryant, M. P., and L. A. Burkey.** 1953. Cultural methods and some characteristics of some of the more numerous groups of bacteria in the bovine rumen. J. Dairy Sci. **36:**205–217.
13. **Carlsson, J., G. P. D. Granberg, G. K. Nyberg, and M.-B. K. Edlund.** 1979. Bactericidal effect of cysteine exposed to atmospheric oxygen. Appl. Environ. Microbiol. **37:**383–390.
14. **Cleland, W. W.** 1964. Dithiothreitol, a new protective reagent for SH groups. Biochemistry **3:**480–482.
15. **Costilow, R. N.** 1981. Biophysical factors in growth, p. 66–78. *In* P. Gerhardt (ed.), Manual of methods for general bacteriology. American Society for Microbiology, Washington, D.C.
16. **Dean, A. C. R., D. C. Ellwood, G. G. T. Evans, and J. Melling (ed.).** 1976. The action of antibacterial agents on bacteria grown in continuous culture. Continuous cultures: applications and new fields. Society of Chemical Industry, London. Ellis Horwood Ltd., Chichester, England.
17. **Drew, S. W.** 1981. Liquid culture, p. 151–178. *In* P. Gerhardt (ed.), Manual of methods for general bacteriology. American Society for Microbiology, Washington, D.C.
18. **Ellefson, W. L., W. B. Whitman, and R. S. Wolfe.** 1982. Nickel-containing factor F_{430}: chromophore of the methylreductase of *Methanobacterium*. Proc. Natl. Acad. Sci. USA **79:**3707–3710.
19. **Funk, H. B., and T. A. Krulwich.** 1964. Preparation of clear silica gels that can be streaked. J. Bacteriol. **88:**1200–1201.
20. **Gall, L. S., and P. E. Riely.** 1981. Manual for the determination of the clinical role of anaerobic microbiology. CRC Press, Inc., Boca Raton, Fla.
21. **Gill, J. W., and K. W. King.** 1958. Nutritional characteristics of a *Butyrivibrio*. J. Bacteriol. **75:**666–673.
22. **Hippe, H.** 1984. Maintenance of methanogenic bacteria, p. 69–184. *In* B. E. Kirsop and J. Snell (ed.), Maintenance of microorganisms. Academic Press, London.
23. **Holdeman, L. V., E. P. Cato, and W. E. C. Moore (ed.).** 1978. Anaerobe laboratory manual, 4th ed. Virginia Polytechnic Institute, Blacksburg.
24. **Hungate, R. E.** 1950. The anaerobic mesophilic cellulolytic bacteria. Bacteriol. Rev. **14:**1–49.
25. **Hungate, R. E.** 1969. A roll tube method for cultivation of strict anaerobes, p. 117–132. *In* J. R. Norris and D. W. Ribbons (ed.), Methods in microbiology, vol. 3B. Academic Press, Inc., New York.
26. **Jacob, H.-E.** 1970. Redox potential, p. 91–123. *In* J. R. Norris and D. W. Ribbons (ed.), Methods in microbiology, vol. 2. Academic Press, Inc., New York.
27. **Jones, G. A., and M. D. Pickard.** 1980. Effect of titanium(III) citrate as reducing agent on growth of rumen bacteria. Appl. Environ. Microbiol. **39:**1144–1147.
28. **Jones, J. B., and T. C. Stadtman.** 1977. *Methanococcus vannielii*: culture and effects of selenium and tungsten on growth. J. Bacteriol. **130:**1404–1406.
29. **Kang, K. S., G. T. Veeder, P. J. Mirrasoul, T. Kaneko, and I. W. Cottrell.** 1982. Agar-like polysaccharide produced by a *Pseudomonas* species: production and basic properties. Appl. Environ. Microbiol. **43:**1086–1091.
30. **Kerby, R., and J. G. Zeikus.** 1983. Growth of *Clostridium thermoaceticum* on H_2/CO_2 or CO as energy source. Curr. Microbiol. **8:**27–30.
31. **Krieg, N. R., and P. Gerhardt.** 1981. Solid culture, p. 143–150. *In* P. Gerhardt (ed.), Manual of methods for general bacteriology. American Society for Microbiology, Washington, D.C.
32. **Leedle, J. A. Z., and R. B. Hespell.** 1980. Differential carboyhydrate media and anaerobic replica plating techniques in delineating carbohydrate-utilizing subgroups in rumen bacterial populations. Appl. Environ. Microbiol. **39:**709–719.
33. **Lehmberg, C.** 1956. Untersuchungen über die Wirkung von Ascorbinsaure, Stoffwechselgiften und anderen Faktoren auf den Stoffwechsel von *Clostridium butyricum* Prazm. Arch. Mikrobiol. **24:**323–346.
34. **Lin, C. C., and L. E. Casida, Jr.** 1984. Gelrite as a gelling agent in media for the growth of thermophilic microorganisms. Appl. Environ. Microbiol. **47:**427–429.
35. **Lines, A. D.** 1977. Value of the K^+ salt of carageenan as an agar substitute in routine bacteriological media. Appl. Environ. Microbiol. **34:**637–639.
36. **Ljungdahl, L. G.** 1983. Formation of acetate using homoacetate fermenting anaerobic bacteria, p. 219–248. *In* D. L. Wise (ed.), Organic chemicals from biomass. Benjamin/Cummings Publishing Co., Inc., Menlo Park, Calif.
37. **Ljungdahl, L. G., and J. R. Andreesen.** 1978. Formate dehydrogenase, a selenium-tungsten enzyme from *Clostridium thermoaceticum*. Methods Enzymol. **53:**360–372.
38. **Lundie, L. L., Jr., and H. L. Drake.** 1984. Development of a minimally defined medium for the acetogen *Clostridium thermoaceticum*. J. Bacteriol. **159:**700–703.
39. **Macy, J. M., J. E. Snellen, and R. E. Hungate.** Use of syringe methods for anaerobiosis. J. Clin. Nutr. **25:** 1318–1323.
40. **Miller, T. L., and M. J. Wolin.** 1974. A serum bottle modification of the Hungate technique for cultivating obligate anaerobes. Appl. Microbiol. **27:**985–987.
41. **Moensch, T. T., and J. G. Zeikus.** 1983. An improved preparation method for a titanium (III) media reductant.

J. Microbiol. Methods **1**:193–202.

42. **Mylroie, R. L., and R. E. Hungate.** 1954. Experiments on the methane bacteria in sludge. Can. J. Microbiol. **1**:55–64.
43. **Odom, J. M., and H. D. Peck, Jr.** 1984. Hydrogenase, electron transfer proteins and energy coupling in the sulfate reducing bacteria *Desulfovibrio*. Annu. Rev. Microbiol. **38**:551–592.
44. **Plotz, H., and J. Geloso.** 1930. Relations entre la croissance des microorganismes anaerobies et le potentiel du milieu de culture. Ann. Inst. Pasteur (Paris) **45**:613–640.
45. **Rabotnowa, I. L.** 1963. Die Bedeutung physikalisch-chemischer Faktoren (pH und rH_2) für die Lebenstätigkeit der Mikroorganismen. Gustav Fischer Verlag, Jena, D.D.R.
46. **Ragsdale, S. W., L. G. Ljungdahl, and D. V. DerVartanian.** 1983. Isolation of carbon monoxide dehydrogenase from *Acetobacterium woodii* and comparison of its properties with those of the *Clostridium thermoaceticum* enzyme. J. Bacteriol. **155**:1224–1237.
47. **Ragsdale, S. W., and H. G. Wood.** 1985. Acetate biosynthesis by acetogenic bacteria. Evidence that carbon monoxide dehydrogenase is the condensing enzyme that catalyzes the final steps of the synthesis. J. Biol. Chem. **260**:3970–3977.
48. **Reed, G. B., and J. H. Orr.** 1943. Cultivation of anaerobes and oxidation-reduction potentials. J. Bacteriol. **45**:309–320.
49. **Seip, W. F., and G. L. Evans.** 1980. Atmospheric analysis and redox potentials of culture media in the GasPak system. J. Clin. Microbiol. **11**:226–233.
50. **Smith, P. H., and R. E. Hungate.** 1958. Isolation and characterization of *Methanobacterium ruminantium* n. sp. J. Bacteriol. **75**:713–718.
51. **Stetter, K. O., M. Thomm, J. Winter, G. Wildgruber, H. Huber, W. Zillig, D. Jane-Covic, H. Konig, P. Palm, and S. Wunderl.** 1981. *Methanothermus fervidus* sp. nov., a novel, extremely thermophilic methanogen isolated from an Icelandic hot spring. Zentralbl. Bakteriol. Parasitenkd. Infektionskr. Hyg. Abt. 1 Orig. Reihe C **2**:166–178.
52. **Sutter, V. L., D. M. Citron, S. M. Finegold, and K. S. Bricknell.** 1980. Wadsworth anaerobic bacteriology manual, 3rd ed. C. V. Mosby Co., St. Louis.
53. **Walden, W. C., and D. J. Hentges.** 1975. Differential effects of oxygen and oxidation-reduction potential on the multiplication of three species of anaerobic intestinal bacteria. Appl. Microbiol. **30**:781–785.
54. **Watson, N., and D. Apirion.** 1976. Substitute for agar in solid media for common usages in microbiology. Appl. Environ. Microbiol. **31**:509–513.
55. **Widdell, F., and N. Pfennig.** 1981. Studies on dissimilatory sulfate-reducing bacteria that decompose fatty acids. Arch. Microbiol. **129**:395– 400.
56. **Wiegel, J., M. Braun, and G. Gottschalk.** 1981. *Clostridium thermoautotrophicum* species novum, a thermophile producing acetate from molecular hydrogen and carbon dioxide. Curr. Microbiol. **5**:255–260.
57. **Wolin, E. A., M. J. Wolin, and R. S. Wolfe.** 1963. Formation of methane by bacterial extracts. J. Biol. Chem. **238**:2882–2886.
58. **Wood, W. A.** 1961. Fermentation of carbohydrates and related compounds, p. 59–149. *In* I. C. Gunsalus and R. Y. Stanier (ed.), The bacteria, vol. II: Metabolism. Academic Press, Inc., New York.
59. **Yamamoto, I., T. Saiki, S.-M. Liu, and L. G. Ljungdahl.** 1983. Purification and properties of NADP-dependent formate dehydrogenase from *Clostridium thermoaceticum*, a tungsten-selenium-iron protein. J. Biol. Chem. **258**:1826–1832.
60. **Zehnder, A. J. B., and K. Wuhrman.** 1976. Titanium(III) citrate as a nontoxic oxidation-reduction buffering system for the culture of obligate anaerobes. Science **194**:1165–1166.
61. **Zeikus, J. G., P. W. Hegge, and M. A. Anderson.** 1979. *Thermoanaerobium brockii* gen. nov. and sp. nov., a new chemoorganotrophic, caldoactive, anaerobic bacterium. Arch. Microbiol. **122**:41–48.

Chapter 9

Methods for Transformation of Organic Compounds

C. T. GOODHUE, J. P. ROSAZZA, AND G. P. PERUZZOTTI

The use of biocatalysts in reactions of organic compounds has advanced significantly during the past three decades. Factors contributing to this advance include the discovery of many new enzymes from microbial cells and the development of methods for their isolation and stabilization. In addition, various techniques for immobilizing cells and enzymes for use in bioreactors have been discovered.

Traditional fermentation technology has matured so that most bench-top reactions can be scaled up to industrial volumes without great difficulty. This period of advancement in biocatalysts coincided with the development of sensitive analytical methods needed for the convenient measurement of chemical products and intermediates in complex biological media. Of major importance were high-pressure liquid and thin-layer chromatographic techniques (HPLC and TLC), which permit the isolation and structure elucidation of small amounts of reaction products.

Further advances in biocatalysis will develop as a result of the close association of traditional research skills from the diverse areas of organic chemistry, analytical chemistry, biochemistry, microbiology, and engineering.

Biotransformation reactions are catalyzed by enzymes and have several features that make them attractive for organic synthesis. Enzymes are chiral catalysts with high regio- and stereoselectivity. Catalysis occurs under mild reaction conditions requiring no strong acids or bases, heavy metals, or other materials commonly associated with chemical catalysts. Enzyme reactions occur optimally between 20 and 70°C, so only a modest energy input is required. They can be conducted in solvent mixtures of relatively low water content, thus extending the traditional utility of these reactions to the more lipophilic starting materials and making new reactions possible, such as the use of esterases to catalyze esterifications rather than the more usual ester hydrolysis. Because of the high selectivities achievable with enzyme-catalyzed reactions, the use of troublesome protecting groups, which is frequently required in organic synthesis, is not necessary.

9.1 GENERAL FEATURES OF MICROBIAL TRANSFORMATIONS

Microbial transformations (also known as biotransformations and bioconversions) are reactions of organic compounds catalyzed by microorganisms. Of course, the true catalysts of these reactions are the enzymes synthesized by the microbial cells. The anabolic and catabolic reactions necessary to life processes such as bioenergetics, growth, and replication are catalyzed, and indeed controlled, by the enzyme network of the cell.

Anabolic enzymes are usually substrate specific, but catabolic enzymes seem to have evolved broader ranges of specificity. Thus, most organic compounds (excluding unstable and highly reactive compounds) can serve as substrates for microbial transformations. This has been demonstrated convincingly by the natural degradation of organic compounds in the environment and by the many isolations of microorganisms capable of total degradation of organic compounds. Indeed, it is possible to select experimental conditions that favor the production of desired enzymes that can be used to perform single and highly specific reactions. This is done by controlling the growth and transformation environment of the culture and the physical form of the organic substrate and by establishing reproducible experimental protocols.

Often the natural functions of enzymes necessary for catalyzing organic reactions are not known, since most screening is done with growing cells. Of course, enzymes of primary or secondary metabolism may accept unusual substrates that fall within the boundaries of their specificity requirements. Few enzymes are specific enough to catalyze reactions with only their natural substrates. The substrate specificities of some enzymes are remarkably broad. For example, cytochrome P-450 (122) can accept substrates with a broad range of different structures. Furthermore, many organic compounds can induce microbial cells to form enzymes that catalyze specific degradative reactions of the inducer compound and its structural analogs (114).

Biotransformation enzymes occur both intracellularly and extracellularly (123). The location of these enzymes varies among the different types of microorganisms used in biocatalytic experiments. Bacteria contain soluble enzymes within the cytosol or particulate enzymes bound to membrane structures. Yeasts and fungi are more complicated; their enzymes are often associated with the various organelles, including mitochondria, nuclei, vacuoles, membranes, and cell walls.

Frequently, the location of the useful enzymes is not known. Therefore, experimental methods are designed to allow the transport of reactants to catalytic sites, wherever they might be, by favoring the highest possible solubility and dispersion in the reaction medium and by favoring permeability of the cells to the reactants.

The basic tenets of microbial transformations are as follows.

(i) A large number of enzymes in microorganisms can be used to catalyze chemical reactions. These are constitutive enzymes. Culture collections can be screened for probable candidates (more often molds and yeasts than bacteria) which perform desirable transformations based on these enzymes.

(ii) If enzymes are not constitutive, they frequently can be induced in microorganisms (usually bacteria) by organic compounds. Inducible enzymes can perform a variety of reactions on the inducers and their analogs. Frequently, multienzyme pathways are induced. By careful manipulation of environmental conditions, these enzymes can be used singly or in concert to catalyze useful reactions.

The rest of this chapter discusses the methods required to perform experiments in microbial transformation of organic compounds. Complete discussions of the possible types of enzyme-catalyzed reactions have been published (13, 23, 60, 63, 65, 76, 83, 123, 147).

9.2 THE MICROBIAL BIOCATALYST

Large numbers of microbes coexist in almost all natural environments, particularly soils, waters, and sewage. It has been estimated that as many as 10^9 microbial cells occur in a single gram of rich garden soil. The makeup of the flora in these ecosystems is determined by the availability of oxygen and water, the degree of exposure to light (55), the temperature ranges, and the nutrients present. Experience has shown that a wide variety of bacteria, molds, yeasts, and algae can be isolated from nature by using the rich natural ecosystems as a source of inocula for enrichment to produce enzymes as well as useful secondary natural products (112, 118).

The mixed populations of microorganisms found in nature function synergistically in their use of available nutrients, where the waste products of one species may feed another species. Through these complex biological interactions, microorganisms, even those with highly specific metabolic needs, can grow and replicate in the mixed-culture environment. Some cultures are difficult to work with in the laboratory when removed from their natural environments. On the other hand, many others can be isolated in pure cultures, manipulated, and maintained in a viable state. Generally, it is preferable to work with pure cultures since they are easier to maintain and control.

9.2.1 Taxonomy

A detailed exposition of taxonomy is beyond the scope of this chapter. Nevertheless, it is necessary to understand the basic organizational framework of classification of the microorganisms commonly used as biocatalysts. This is naturally more useful to readers whose main scientific experience is in the physical rather than the natural sciences.

Classifications are constructed from individual organisms that occur within populations of species. The species grouping is the fundamental level of organization in taxonomy. Individuals that closely resemble one another in many different characteristics are considered to be of one species. Species that share many common characteristics are placed in a group termed the genus. Families consist of still more inclusive groups, and so on to higher organizational levels. The taxonomic classification for a given organism, *Mycobacterium fortuitum*, for example, is as follows: kingdom *Procaryotae*, division *Bacteria*, order *Actinomycetales*, family *Mycobacteriaceae*, genus *Mycobacterium*, species *fortuitum* (20). Where possible, all microorganisms are placed in a similar taxonomic framework. Detailed descriptions and the taxonomic classifications of microorganisms are available (6, 8, 20, 132, 150).

An understanding of taxonomic organization is useful when one is searching for a better microbial catalyst for specific organic reactions. For example, when a culture such as *Mycobacterium fortuitum* performs an interesting biotransformation but in low yield, it is logical to examine taxonomically related cultures to find other candidates that might provide a higher yield or that might perform related reactions on similar starting materials. To do so, other *Mycobacterium* spp., other genera of the class *Actinomyces*, and members of closely related families such as the *Nocardiaceae* would also be examined. This approach is reasonable, since there are often similarities in enzymatic makeup among members of related genera and families. Similarities should be even more striking among strains of the same species.

9.2.2 Bacteria

Bacteria are single-celled microorganisms that occur as rods, cocci, or spirals. They are the smallest of the cells used for biotransformations, with sizes at the micrometer level. Cocci are often 0.5 to 4 μm in diameter; rods and spiral-shaped organisms have the same diameter but can be up to 20 μm long or longer (5, 57, 118). Aiba et al. (5) list various components of the bacterial cell with descriptions of their chemical composition and function. Belonging to the kingdom *Procaryotae*, bacteria lack a defined nucleus (20). Bacterial cells contain a variety of organelles in the cytosol, such as storage granules, ribosomes, spores, and membraneous structures. Enzymes useful for bioconversions are found in all of these structures.

The membrane surrounding the cell is a semipermeable barrier to nutrients and waste products entering and leaving the cell. This membrane must be penetrated by organic substrates for reactions to occur with intracellular enzymes. If enzymes are located in the periplasmic space between the membrane and the rigid cell wall, the organic substrates need only to penetrate this outer wall. In addition to the rigid cell wall, the shells of bacteria may consist of an extra membrane ("outer membrane" of gram-negative bacteria), slime layers, and capsules adhering to the wall. These can present barriers to the penetration of organic compounds.

Bergey's Manual of Determinative Bacteriology (20) classifies bacteria taxonomically into 5 orders, 42 families, and 225 genera. Another useful reference, Skerman's guide (130), describes bacterial genera and species. Also see *Bergey's Manual of Systematic Bacteriology* (80a). Not all bacteria are listed in these works; many useful bacteria are not included because of incomplete or uncertain taxonomic description. The three major classes of bacteria are the *Eubacteria*, *Myxobacteria*, and *Spirochetes*. Of these, the *Eubacteria* have been used for most biotransformation work (53).

Bacterial classification is based on a combination of characteristics including the following.

(i) Dye reactions. Cell wall reactions with dyes show some bacteria to be gram positive or gram negative.

(ii) Sources of energy. Photosynthetic bacteria derive energy from light and use carbon dioxide and mineral salts for nutrients. Autotrophic bacteria derive energy from the oxidation of inorganic substances such as sulfur, hydrogen, and ammonia; they also assimilate carbon dioxide. Heterotrophic bacteria derive energy from the oxidation of organic compounds.

(iii) Oxygen requirement. Aerobic organisms require oxygen, anaerobic organisms do not require oxygen, and facultatively anaerobic organisms can live under either condition.

At present, biotransformations are carried out using bacteria that are largely aerobic and heterotrophic. However, current research on anaerobes and autotrophs is resulting in more use of other types of bacteria in biocatalysis.

9.2.3 Fungi

Examples of eucaryotic organisms used most frequently in biotransformation work are the yeasts and the molds. Most fungi are aerobic organisms that form long, filamentous, nucleated cells known as hyphae. The cells are large compared to bacteria, being 4 to 20 μm wide and possibly hundreds of micrometers long. Hyphae grow intertwined to form mycelia. Fungal classification is based more on morphological characteristics than on dye staining and biochemical reactions.

Based on the nature of their life cycles, fungi are classified into four categories: (i) *Zygomycetes* (also known as *Phycomycetes*), (ii) *Ascomycetes*, (iii) *Basidiomycetes*, and (iv) Fungi Imperfecti. The life cycles of the fungi have been described (130, 132, 150).

In general, *Zygomycetes* reproduce from small asexual spores borne within sporangia. They lack septal division in their hyphal strands and may form sexual spores in an unprotected fashion. Downy mildews and black molds belong to this class. The *Ascomycetes*, on the other hand, have septate hyphae and form sexual spores within sacs called asci, though they may also reproduce asexually. Spores are dispersed upon rupture of the sacs. This class includes bread molds, citrus molds, and most yeasts. The *Basidiomycetes* are septate and carry spores on club-shaped appendages known as basidia. Macroscopic mushrooms, including the common edible varieties, belong to this class, as do puffballs and wood-rotting fungi. Fungi Imperfecti include the fungi for which a sexual reproductive stage has not been observed. Microorganisms from each of these four classes are commonly described in the literature of microbial transformations.

Yeasts are grouped with the fungi even though they usually have distinctive growth habits. Yeast cells are elliptical, ranging from 8 to 15 μm in length and from 3 to 5 μm in width. Although they form colonies similar to those formed by bacteria on solid media, they are much larger and reproduce generally by budding. They also may form sexual spores. Many yeasts, especially the common *Saccharomyces cerevisiae* and related species, grow anaerobically. Yeasts are commonly used as reductive biocatalysts.

9.2.4 Pathogens

Microorganisms pathogenic to humans are listed in textbooks of microbiology and mycology (7, 20, 132, 138, 146, 150). In general, these organisms are not recommended for routine biotransformation experiments. Few of these pathogens are listed in the biotransformation literature. The organisms identified as plant pathogens are also of concern. Although not harmful to humans directly, they can cause great agricultural damage. To obtain these cultures, specific permission from state agriculture departments must be obtained (118). Upon request, investigators will receive a form from the agriculture department that will ask for the specific ways the organism will be used and the planned manner of disposal of the culture. It is good standard practice to dispose of all pathogens by autoclaving when experiments are complete. It also is advisable to consider all new, unidentified cultures as potentially pathogenic to some life form; thus, they should be handled with care and disposed of by autoclaving until shown to be harmless.

9.2.5 Culture Acquisition

Selection of the microbial biocatalyst is the most critical of all the operations in a biotransformation experiment. Cultures with desirable properties are obtained from other investigators in the biotransformation field, from standard culture collections, or by isolation from natural habitats (53, 106).

Before cultures are obtained from collections or other sources, the literature should be reviewed for leads as to the types of organisms with the desired enzymatic capabilities. It is advisable to concentrate on the more recent literature because the cultures described in these publications are likely to be readily available.

Catalogs of chemical reactions catalyzed by microorganisms have been referred to earlier in this chapter. These are assembled with specific attention to groups such as the alkaloids (63, 65, 83, 123), the steroids (23, 83, 123, 147), and nonsteroidal cyclic compounds including various drugs (13, 76). Such resources are critical to serious research in this field. New chemical reactions continue to appear in the primary literature.

A comprehensive list of most of the significant culture collections around the world is available (91), and smaller lists of culture collections have also been published (23, 76, 147). In many collections strains are maintained in lyophilized form or on agar slants suitable for mailing. Investigators often find it convenient to maintain their own culture collections of 100 to 400 strains. In addition, microbiology and mycology departments on many college campuses maintain culture collections. Investigators are usually willing to share their cultures with researchers in other laboratories. Table 1 gives a partial listing of culture collections.

Enrichment techniques are used to isolate cultures from nature (see chapter 1 in this manual; 1, 12, 143).

9.2.6 Maintenance and Preservation

When new cultures are procured, it is important to establish their purity and to transfer the organism to a fresh culture medium for propagation in the laboratory. Upon acquisition, the following information should be recorded: culture name and number, source, acquisition and lyophilization dates, lyophilization medium, growth medium and temperature, literature source, reactions known to be catalyzed, and unusual properties and comments. This information may be kept conveniently on file cards with appropriate cross-indexing. For large collections computer cataloging is recommended.

Culture purity is determined by plating on appropriate media. For short-term storage and routine culturing, slants in screw-top vials should be used. After significant growth, many strains can be stored in a refrigerator. For longer-term storage, lyophilization or ultralow-temperature freezing is recommended (10, 93). Dietz (39) has extensively surveyed the methods used for storage of industrial microorganisms. The advantages of various methods have been tabulated along with a compilation of methods used for storage of strains producing antibiotics, enzymes, other fermentation products, and steroid bioconversion products (110). The preservation of cultures is discussed in detail in chapter 5 of this manual. Additional information regarding maintenance and preservation can be found in the literature (34, 36, 54, 81, 82, 121).

9.2.7 Growth Fundamentals

Growth is the cumulative process resulting in the orderly increase of all chemical components of the living cell (53, 138). Different groups of organisms behave quite differently in liquid culture. As with enzyme-catalyzed reactions, an approximately linear

TABLE 1. Selected culture collections

Abbreviation	Address
ATCC	American Type Culture Collection, 12301 Parklawn Dr., Rockville, MD 20825
CBS	Centraalbureau voor Schimmelcultures, Baarn, The Netherlands
CMI	Commonwealth Mycological Institute, Kew, Surrey, England
DSM	German Collection of Microorganisms, Grisebachstr. 8, D-34 Goettingen, Federal Republic of Germany
IFO	Institute for Fermentation, 4-54 Juso Nishimomachi, Higashi-Yodogawa-ku, Osaka, Japan
NCIB	National Collection of Industrial Bacteria, Chemical Research Laboratory, Teddington, Middlesex, England
NCTC	National Collection of Type Cultures, Central Public Health Laboratory, London, NW9, England
NRRL	Northern Regional Research Laboratories, Northern Utilization Branch, ARS, USDA, 1815 North University Ave., Peoria, IL 61604
QM	Quartermaster Culture Collection, Quartermaster Research and Engineering Command, U.S. Army, Natick, MA 01760
WC	Waksman Collection, Institute of Microbiology, Rutgers, The State University, New Brunswick, NJ 08901
WISC	School of Pharmacy, University of Wisconsin, Madison, WI 53706
UI	College of Pharmacy, The University of Iowa, Iowa City, IA 52240
UT	College of Pharmacy, The University of Texas at Austin, Austin, TX 78712

relationship exists between the amount of microbial cells present and the rate of reaction in the incubation mixture. Therefore, when biotransformations are conducted with growing cultures, conditions that favor enhanced growth usually result in greater yields of metabolites.

Growth rates of the unicellular bacteria and yeasts are regular and can be determined mathematically in a simple manner (51, 53, 132, 138). Growth of multicellular fungi and filamentous procaryotes (actinomycetes) such as *Streptomyces* spp. is more difficult to define mathematically (132).

When cells are inoculated into a fresh medium they display a lag phase until they are able to grow exponentially at the maximum rate allowable in the new physical and nutritional environment. Rapid growth continues until a limiting condition is reached, such as the depletion of a required nutrient or the accumulation of an inhibitory substance. In the resulting stationary-growth phase, cells can survive in what is essentially a resting state for various times, depending on the types of cells present and their environmental history.

Different organisms have significantly different growth and death rates. Growth should be measured turbidimetrically at about 650 nm. It also is convenient to make the direct correlation between the dry weight of cells and turbidity. Correlation with viable cells is obtained by plating diluted suspensions on appropriate agar media (53).

Molds and actinomycetes have less predictable growth habits than unicellular bacteria and yeasts (132). In liquid culture, cells of these filamentous microorganisms may occur as homogeneous suspensions, but more often they grow in pellets of various sizes and shapes, in large clumps, or in stringy masses. The growth kinetics of these organisms are difficult to generalize. Although they usually display a lag phase in newly inoculated cultures, they may never reach exponential growth rates, and the actual times necessary to achieve maximum growth are considerably longer than for unicellular cultures.

Doubling times for unicellular bacterial cells are typically ~1 h; those for fungi are 2 to 4 h. Macroscopic fungal growth can be measured by obtaining mycelial dry weights of filtered samples. Blending the culture immediately before sampling yields homogeneous samples, but the blending can affect growth adversely. Fungi of the *Penicillium* and *Aspergillus* genera produce mycelial dry weights of about 2 g/100 ml of medium. Similar dry weights are common for yeasts and unicellular bacteria.

The complement of enzymes varies greatly at given times during the life cycle of the cell. The desired enzyme may be present from the start of the growth cycle, or it may not appear until the late exponential phase. Thus, the optimum time for adding organic reactants or for harvesting cells must be established by experimentation. The changes in enzyme activities during growth reflect the changes occurring within the cell and the culture medium as the organism grows and metabolizes nutrients.

9.3 FORMS OF THE BIOCATALYST

The use of live, growing microbial cells as biocatalysts for biotransformations of organic compounds has been extensively documented. Pure cultures are grown to a point where desired enzyme activities are maximal, at which time organic chemical substrates are added to the incubation mixture in which the transformations take place. However, much experimental latitude is now possible in the use and form of the biocatalyst. The possibilities are outlined in Fig. 1, which illustrates the relationships of the forms of the biocatalyst.

9.3.1 Growing Cultures

Both batch and continuous cultures are used in bioconversion experiments. In the batch culture technique, a pure culture is grown in a suitable medium. At an experimentally determined time the substrate is added, and incubation is continued until transformation of the substrate ceases or additional reactions seriously begin to affect yields.

1. Continuous Culture
2. Batch Culture

 a. Clear Supernatants → Extracellular Enzymes → Purified Enzymes

 b. Cells and Solids
 i. Immobilized cells
 ii. Dried cells
 iii. Permeabilized cells
 iv. Resting cells
 v. Cell extracts - intracellular enzymes → purified enzymes

FIG. 1. Relationship of different forms of biocatalysts in growing cultures.

In the batch process, the biocatalyst is used only once and then discarded. This procedure is straightforward and useful for screening procedures. However, it requires the repetitive production of cells for each experiment, and the isolation of reaction products from complex fermentation media can be complicated.

The physiological state of cells in batch culture varies continuously throughout the growth cycle. This is not true of cells in continuous culture, in which cells can be maintained in a steady physiological state for long periods of time by means of continuous addition of fresh nutrient medium and simultaneous withdrawal of equal amounts of spent medium. Details of the continuous culture method are given in *Methods in Microbiology* (141), in the *Manual of Methods for General Bacteriology* (51), and in chapter 11 of this volume. Although it would seem that water-soluble substrates in particular would be amenable to transformation by this procedure, continuous culture has not been used widely in biotransformation work, perhaps because of its inherent limitations in this area.

9.3.2 Resting Cells

When biotransformations are conducted with growing cultures in complex fermentation media, the isolation and purification of products can be difficult due to the presence of organic acids, phenols, amino acids, lipids, and other metabolites produced by the culture. The use of resting cells avoids these problems. Resting cells are nongrowing, live cells that retain most of the enzyme activities of growing cells.

Resting cells are obtained by removing cells from the culture medium at a time in the growth cycle when enzyme activities are highest or at least present at useful levels. Mycelial growth can be removed by filtration, but yeasts and bacteria are best harvested by centrifugation. Concentrated cells are then suspended in appropriate buffers, modified culture media (usually without some required nutrient for growth), distilled water, or even nonaqueous solvent mixtures for use as biocatalysts.

As biotransformation catalysts, resting cells have several advantages over growing cells or isolated enzymes. Whole resting cells can accomplish multistep enzymatic reactions without the need for expensive coenzymes. Furthermore, one must assume that for metabolic pathways, enzymes present in their native state are more efficient than isolated enzymes that are simply recombined in incubation medium. For multistep transformations by enzymes that are part of a metabolic pathway, resting cells are the catalyst of choice. For single-step reactions, the direct use of resting cells minimizes loss of enzyme activity, which is unavoidable during isolation and purification of enzymes. Furthermore, enzymes in intact cells usually are more stable than their isolated counterparts. The incubation media used for resting-cell biocatalysis may contain substances that are inimical to growth but nevertheless useful for inhibiting undesirable side reactions.

Cometabolism and enzyme induction increase the usefulness of resting-cell biocatalysis. For some reactions to occur, it is necessary to use a cosubstrate such as glucose, glycerol, succinic acid, or another oxidizable metabolite along with the organic compound to be transformed. These cosubstrates drive reactions to completion by providing the necessary energy during their utilization. They also provide energy for recycling of coenzymes for the enzymatic reactions.

In some cells, desired enzyme activities are dramatically increased by cultivating the organism in the presence of the organic compound to be transformed (enzyme induction). Conversely, the use of resting cells in which enzyme activities have not been induced can result in poor yields.

Many examples of the use of resting cells in biotransformations have been described (44, 96, 113, 133, 159). Cells of a *Nocardia* sp. catalyze the oxidation of cholesterol in organic solvents to cholestenone. *Septomyxa affinis* ATCC 13425 dehydrogenates the A ring of a variety of steroids. The dehydrogenase activity of the cells is induced by progesterone or 3-oxobisnor-4-cholen-22-al added to the growth medium. *Cunninghamella bainieri* ATCC 9244 transforms an anthelminthic agent into two of its mammalian metabolites (133). *Sepedonium ampullosporum* carries out 16α-hydroxylations of steroids on pilot-plant scale in fermentor tanks (96). Because of the greater cell densities that can be used, bioconversion times are significantly less than with conventional batch cultures. Up to 12 sequential batches of steroid substrate can be converted completely by a single batch of resting cells. The level of oxygen required for these conversions is critical.

Nearly all reductions of ketones by yeasts are done in the resting-cell mode. A recent example is the asymmetric synthesis of L-carnitine (159). Commercial yeast cake (12 g) suspended in 20 ml of tap water was incubated with chloroacetoacetic esters (2.7 mmol) to achieve highly selective reductions of ketone functional groups.

9.3.3 Dried Cells

In some cases the enzyme activities of microbial cells can withstand drying treatment. The resulting powders are convenient biocatalysts. Esterase, amidase, oxidoreductase, and dehydrogenase are some of the enzymes that can survive the cell-drying procedures (44, 113). The two most common methods of drying are lyophilization and acetone dehydration.

Lyophilization is relatively simple, since it does not require aseptic handling. After harvesting by centrifugation or filtration, cells are resuspended in distilled water or very dilute buffer (no more than 0.05 M). The suspension should be as thick as can be readily frozen in a thin shell inside a round-bottom flask or one of the more convenient breakaway flasks. Freezing can be carried out in a dry ice-acetone (or ethanol) bath. The frozen shell should be no more than 5 mm thick; thicker masses tend to thaw during prolonged lyophilization. When the process is done correctly, water sublimes from the frozen cells, which form a fluffy, dry powder. Storage of these powders in a freezer will preserve enzyme activities for many years. A variety of lyophilized yeasts and bacteria can be obtained from biochemical supply houses (e.g., Sigma Chemical Co.). These products should be tested for surviving enzyme activities before use.

The treatment of cell pastes or cakes with cold (−20°C) acetone is a reliable method for preparing dried cells. Water should be removed from the cells by centrifugation or filtration before slurrying in cold acetone. The cells are then collected by suction filtration. The treatment is performed twice more, and then a cold ether wash is used to remove residual acetone, which can be detrimental to enzyme activities. Acetone powders thus produced are most stable when stored frozen (56).

Dried cells offer many of the same advantages of resting cells compared to growing cultures. Both lyophilized and acetone-treated cells are easy to prepare. An adequate supply of powdered cells helps to ensure experimental reproducibility, since many experiments can be performed with one batch and since it is not necessary to establish rigorous fermentation protocols for each experiment. Incubations with dried cells do not require asepsis. Examples of the use of dried cells are given by Perlman (113) and Fischli (44).

9.3.4 Permeabilized Cells

A variety of substances alter the permeability of microbial cells to facilitate the contact of substrates with enzymes and the excretion of products. Permeabilizing agents include surfactants (90) and solvents (52, 99). These agents are applied after growth stops. Alternatively, the addition of inhibitors of cell wall synthesis during growth enhances permeability. Antibiotics (e.g., penicillin) and glycine (a D-alanine analog) are effective in sublethal doses (37). The solvents dimethylformamide (DMF) and dimethyl sulfoxide, commonly used to disperse steroids and other lipophilic compounds, also increase the permeability of the cells. These solvents should be used with care since at higher concentrations they may adversely affect the viability of the cells.

Ghisalba and co-workers provide a typical example of the use of solvents to prepare permeabilized cells (52). The mycelium from 40 ml of a culture of *Nocardia mediterranei* N813 is suspended in 12 ml of 0.066 M phosphate buffer (pH 7.0). After the addition of 12 ml of cold ether, the suspension is mixed by hand agitation for 1 min and then centrifuged at 12,000 × *g* at 0°C. The ether phase is removed, and the permeabilized mycelium is resuspended in the aqueous supernatant. Measurement of the hexokinase-glucose 6-phosphate dehydrogenase reaction provides control values for the extent of permeabilization. This enzyme reaction, dependent on the permeability of glucose, ATP, and NADP, does not occur with untreated, intact cells.

Permeabilized *Cephalosporium acremonium* cells are used to transform the antibiotic rifamycin S. A 10-ml reaction mixture containing 2.5 ml of cells, 0.5 mM rifamycin S, 0.5 mM NADH or NADPH, 1.5 mM MgCl, and 0.05 M phosphate buffer (pH 7.6) is incubated with shaking at 250 rpm and 28°C in an Erlenmeyer flask. The antibiotic is converted to the related rifamycins B and L by the permeabilized cells. The results compare favorably with those obtained with resting cells and cell-free preparations of *N. mediterranei* (52).

9.3.5 Spores

Microbial spores are excellent biocatalysts, although they frequently are difficult to use. Thus spores have not found a wide application for biotransformations (see review by Vézina and Singh [144]). The main problems are the maintenance of both spore integrity and enzyme activities within the spore. Washing the surface growth of fungal cultures with dilute Tween 80 surfactant (about 0.1%) is a convenient way to harvest spores, which can either be used immediately or stored at −20°C.

Generally, spores suspended in buffer media containing glucose as an energy source perform biotransformation reactions without germinating. Other nutrients, particularly those containing organic nitrogen, favor germination and outgrowth of cells. The spore titers used in incubation mixtures range from 10^8 to 10^9/ml.

9.3.6 Cell-Free Preparations

An ultimate goal of some microbial-transformation researchers is to obtain stable, pure enzymes with all of the desirable catalytic properties of whole cells. Enzymes catalyzing specific chemical reactions would be ideally suited for use in the organic laboratory, if the only requisite for use were the addition of an aqueous solvent and the organic substrates to be transformed. Indeed, it is now possible to do this with certain stable enzymes readily available from biochemical supply houses.

Examples of such enzymes are many of the hydrolases (various esterases and proteases) and some oxidoreductases (e.g., horse liver oxidoreductase) that require pyridine nucleotides as cofactors. Many other important enzymes, including the monooxygenases, are not stable enough in purified forms. In such cases, one should explore the possibilities of using crude cell-free preparations instead (56). Such mixtures are obtained by breaking cells under mild conditions so that the contents are released in an active state to the buffer medium.

Several reviews have described methods for the disruption of microbial cells (127, 133). Common techniques are pressure shearing, enzymatic digestion of cell walls, osmotic lysis, autolysis, and freezing-thawing (127). Ultrasonic disintegration and pressure shearing are used most widely in exploratory research. The French press (American Instrument Co.) is a reliable device for pressure shearing on a small scale. It consists of a solid steel cylinder containing a well that holds 5 to 10 ml of a thick cell suspension. The cylinder is fitted with a solid stainless-steel piston that can be rapidly forced into the well under high pressure (4,000 to 20,000 lb/in^2). Cells are broken by the high shear forces as they are squeezed through a small release orifice in the well.

Ultrasonic oscillators also break cells by shearing. Rapidly moving bubbles in the sonic field at the probe tip cause high shear forces capable of breaking the toughest cell walls. Since high-frequency sonic oscillations generate much heat, the operation must be conducted in an ice bath. The tip of the sonic probe should be inserted just below the surface of the cell suspension. Short bursts of 15 to 20 s break cell suspensions of 30 to 40 g (wet weight) per 100 ml. The short-burst method minimizes the heat buildup. The operator of a sonic oscillator should wear protective ear covering.

Crude broken-cell suspensions, even without purification, catalyze useful organic reactions. Appropriate volumes of these suspensions are added to reaction vessels along with organic substrates in buffer. If coenzymes are required, these are added in stoichiometric amounts; otherwise a coenzyme regeneration system may be used. (For more information on coenzyme regeneration, see the study by Jones and Taylor [72].)

It may be advantageous to have most of the enzymes present in the broken-cell mixture in an active state, but sometimes interfering enzymes must be removed. This is done by simple fractionation. The simplest procedure is high-speed centrifugation, during which solid debris including cell walls and unbroken cells is removed. Centrifugation at 5,000 × *g* for 10 min is sufficient to produce a murky supernatant mixture containing most of the soluble enzymes and cofactors of the microbial cell. Portions of this mixture can be used to catalyze biotransformations under proper conditions. The highly sophisticated procedures necessary to obtain pure enzymes from these crude mixtures are described elsewhere (66, 127).

Although it is conceptually appealing to use soluble, purified enzymes in biotransformation experiments, the technique is of limited applicability, especially because the desired enzyme frequently does not have the required stability. Furthermore, proteolytic enzymes released along with desired enzymes cause enzyme degradation. Crude preparations have short lifetimes compared to enzymes in unbroken cells. Cofactors may be depleted or degraded. A requirement for cofactors could become expensive in the absence of a practical regeneration system. The disruption of important linkages between enzyme systems may substantially reduce overall activities. The optimum conditions for transformations should be determined since they could be different from the conditions for growing cultures. One advantage of cell-free preparations is that permeability barriers are absent.

Naturally occurring, exocellular enzymes are usually stable and thus can be useful biocatalysts. For convenience of use and storage, the enzyme may be concentrated by ultrafiltration, precipitation by solvents or salts, or adsorption to a carrier.

An example of adsorption is given for laccase, a copper-containing oxidase formed by a variety of organisms including fungi. *Polyporus anceps* grown in a defined medium secretes laccase at predictable times (115). The time of maximum excretion of the enzyme is established, and at that point cells are filtered from the fermentation broth. Dry DEAE-cellulose (H^+ form, 3 g/liter) is added to the broth with stirring for 30 min at 4°C. By then most enzyme activity is bound tightly to the resin and can be removed by simple filtration. After two washes with distilled water, bound enzyme is stored in small portions in a freezer. Elution of the DEAE-cellulose with 0.2 M phosphate buffer (pH 5.0) gives quantitative recovery of the enzyme. This enzyme catalyzes the oxidation of the alkaloid vindoline in good yields (40).

9.3.7 Immobilized Systems

The forms of biocatalysts discussed above are normally used only once in batch processes and then discarded. The exception is the resting-cell suspension, from which cells can be recovered for reuse. Methods developed mostly in the last 15 years have been devised for immobilization of cells and enzymes that may then be recovered and used again. These immobilized biocatalysts have potential for wide use in chemical analysis and synthetic organic chemistry (2, 26, 38, 47, 107, 108, 125, 126, 142).

Bioreactors containing immobilized materials are valuable for continuous processing. Under ideal conditions, solutions of organic reactants continuously passed over immobilized catalyst emerge as products in the effluent. Such column reactors are more efficient and economical than batch reactors. Biocatalysts often are remarkably stable when immobilized in the polymer matrix (2, 24, 25, 47, 49, 62, 77, 80, 101, 102, 129, 136, 140, 145; F. Leuschner, British patent 953,414, Chem. Abstr. **62:**760c, 1964). Both living and dead microbial cells can be immobilized. Living cells in the polymer matrix sometimes can be rejuvenated after prolonged use by passing fresh nutrients through the reactor (Leuschner, British patent 953,414, 1964). Living cells also tend to multiply to pack voids in the polymer, giving increased catalytic activity.

Methods for immobilizing cells and enzymes are discussed in chapters 18 and 19 of this volume (see also references 46, 48, 50, 74, 75, 86, 92, 99, 156).

9.4 EQUIPMENT

Biotransformation is an interdisciplinary field involving microbiology, biochemistry, and organic chemistry. Therefore, some of the necessary equipment can be found in laboratories devoted to these three disciplines. The following discussion will emphasize microbiological aspects. A general list of equipment is given in Table 2.

9.4.1 Culture Vessels

There are many ways to culture microorganisms. Laboratories tend to develop their own routines that are convenient but not necessarily superior to alternative methods. However, certain items are indispensable, such as culture tubes and flasks of various forms. Culture tubes are simply test tubes with flush rims or screw-cap tops and are especially useful for culturing in liquid or on agar slants. Screw-cap tubes are used for storing agar slants in a refrigerator. They also are used for growing small amounts of cells. Flush-rim tubes are closed with special metal or plastic tops that allow sterile exchange of atmospheric and metabolic gases. Disposable tubes and tops are available from commercial sources. Common sizes of tubes range from 13 by 100 mm to 18 by 150 mm or larger sizes. Tube sizes must be compatible with racks and shakers. No more than a fourth of the total volume of the tube should be used for medium.

Erlenmeyer flasks are ideal for aerobic culturing on shakers. For screening, the smaller sizes (50 to 250 ml) are recommended. Larger flasks (up to 2 liters) are convenient for scaled-up experiments. At this size, 2.8-liter Fernbach flasks are used frequently. Normally only 10 to 20% of the volume of the flask is used for medium, since this allows for maximum agitation and aeration (see below) without splashing or excessive evaporation.

Sterile closure of the flasks is essential. Cotton and plastic foam plugs are widely used, but cotton-gauze filter disks are better; these are held in place by stainless-steel springs or rubber bands. Flasks with flush necks (Delong flasks) are closed with special stainless-steel or plastic caps.

Petri dishes about 100 mm in diameter and sterile, cotton-plugged pipettes are routinely used. From the standpoint of housekeeping in the laboratory, sterile, disposable labware is most convenient.

TABLE 2. Equipment for biotransformation experiments

Category	Equipment
Glass and plastic ware	Pipettes, 1 to 10 ml, disposable plastic, serological type
	Erlenmeyer (or DeLong) flasks, 50 to 1,000 ml
	Fernbach flasks, 2.8 liter
	Culture tubes, flush rims or screw capped, 13 by 100 mm to 18 by 150 mm or larger
	Petri dishes, 100 mm, sterile, plastic, disposable
	Closures for flasks and tubes, plastic, disposable
	(Note: many of these items are available commercially in presterilized packages)
Incubators, sterilizers, shakers	Temperature-controlled room
	Incubator, desktop
	Shakers, rotary or reciprocal
	Fermentors, stirred, bench-top
	Autoclave
	Filter sterilizing apparatus
	Filters
Other	Centrifuge, refrigerated
	Centrifuge, clinical, desk-top
	Balances
	Burner, for flaming
	Inoculating needles and loops (or toothpicks)
	Hood, sterile air, laminar flow
	pH meter
	Chromatographic equipment
	TLC
	GC
	HPLC
	Spectrophotometer, UV-visual
	Pipettes, automatic, in a range of sizes

9.4.2 Incubators and Shakers

Bioconversion experiments require some form of temperature-controlled environment. Incubators that control temperatures from below ambient temperature to 50°C or higher are available in bench-top sizes up to full-room size. Some incubators have integrated shakers.

The value of shakers in fermentation experimentation is well known (45). It is important to have the capability for shaking aerobic cultures, as this is the only economical and practical way to do screening. The two common types of shakers are the rotary (or orbital) shaker and the reciprocal shaker. The shaking speed is continuously adjustable from 0 to about 350 oscillations or rotations per min. Shakers range from desk-top size to ones that can accommodate hundreds of 250-ml flasks. Reciprocal shakers are best for tube cultures; orbital shakers are best for flasks. With either type, different platforms are available for holding the various sizes of vessels.

9.4.3 Sterilizers

Steam sterilizers are essential for microbiological work. Culturing vessels and media must be sterilized. Automatic autoclaves with capacity for several hundred 250-ml flasks are generally used. Commercial sources are listed in Table 3. For small-scale work, a desk-top size or even a pressure cooker can be used.

9.4.4 Fermentors

Fermentors are useful for larger-scale biotransformations. Although the general sophistication of the devices has increased, the basic design of the vessel has not changed for about 30 years.

With a fermentor it is possible to control culture parameters in ways and degrees not possible in flasks

TABLE 3. Commercial sources for equipment

ABEC Inc. (Associated Bioengineers & Consultants), Airport Rd., Commercial Park, Box 2606, Lehigh Valley, PA 18801
American Sterilizer Co., 2424 W. 23 St., Erie, PA 16514
LSL Biolafitte, Inc., Alexander Commerce Park, 715 Alexander Rd., Princeton, NJ 08540
Biotech, Inc., 12221 Parklawn Dr., Rockville, MD 20852
B. Braun Instruments, 805 Grandview Dr., South San Francisco, CA 94080
Castle Co., 1777 E. Henrietta Rd., Rochester, NY 14623
Chemap AG, Alte Landstrasse 415, CH-8708 Mannedorf, Switzerland
Chemapec, Inc., 230 Crossways Park Dr., Woodbury, NY 11797
Marubishi Laboratory Equipment Co., Ltd., Koki Building, 14-24 2-chome, Kadakaji-cho, Chiyoda-ku, Tokyo, Japan
New Brunswick Scientific Co., Box 986, 44 Talmadge Rd., Edison, NJ 08818
Queue Systems, Box 1901, Parkersburg, WV 26102
The Virtis Co., Rt. 208, Gardiner, NY 12525

and tubes. Stirring and air-sparging devices allow the maximum possible aeration. Many parameters (e.g., pH) can be measured and controlled continuously. Therefore, it is useful to have access to one or two bench-top fermentors for experiments that cannot be done conveniently in flasks and for scaling-up processes. Larger scale fermentor studies can be done in cooperation with bioengineers. Fermentor monitoring and computer control of fermentations are discussed in chapter 23 of this manual. Some suppliers of fermentation equipment are listed in Table 3. Several excellent textbooks are available that address the principles of biochemical engineering with specific regard to fermentors (5, 11, 135, 148).

9.5 TECHNIQUES

Success in the application of microbial cultures as catalysts for organic reactions requires a working knowledge of several microbiological laboratory techniques. In particular, one must be able to transfer and cultivate microorganisms aseptically. Also, one must know how to sterilize materials, how to prepare media, and how to use the light microscope. The use of the light microscope is discussed in the *Manual of Methods for General Microbiology* (51). The following discussion describes workable techniques in the other areas mentioned above.

9.5.1 Aseptic Manipulations

Contaminants cannot be tolerated. Aseptic technique is essential to the maintenance of pure cultures and can be simplified to two requirements: (i) the use of sterile equipment, vessels, substrates, media, etc., and (ii) the exclusion of airborne dust when making additions or transfers to medium.

Cultures are transferred to and from solid medium with sterile platinum (or nichrome) loops and needles. Presterilized disposable needles and loops of plastic are available. Sterile toothpicks are useful for picking and streaking colonies on agar plates. Needles are used for stab-inoculating and removing inocula from tubes of agar medium. Loops can be used for transferring small amounts of liquid culture. To transfer larger amounts of liquid, sterile pipettes are used. Of course, the contents of one flask can be simply poured into another flask, if care is taken to exclude contaminants. Loop and needle techniques are discussed thoroughly elsewhere (51, 53).

The streaking of plates to obtain isolated, presumably pure colonies is one of the more important techniques that should be strictly adhered to (see Fig. 2; reference 51). This method is used mostly for aerobic cultures, but it can be used for anaerobes and facultative anaerobes, if incubation is carried out in anaerobic jars. New varieties of these jars are available and minimize the difficulties of culturing anaerobes.

9.5.2 Sterilization

Sterilization is the complete removal or destruction of all living entities from materials by using heat, filtration, radiation, or chemicals. There are many excellent discourses on the practical and theoretical aspects of sterilization (5, 30, 53, 135, 139). Four commonly used procedures are described briefly here.

Steam sterilization

Autoclaving with steam under pressure (15 lb/in^2, 121°C) is the best and most convenient sterilization method for heat-resistant materials likely to be used in bioconversion research. Small amounts of materials are completely sterilized after 15 min under these conditions. Up to 10 flasks, each containing 25 to 100 ml (or a similar quantity of dry material), is considered a small amount. For larger amounts the time of autoclaving must be increased, but care should be taken not to destroy medium components by prolonged autoclaving. If longer exposures are necessary, heat-sensitive components should be sterilized separately for a shorter time by autoclaving or some other method (filtration), and added after cooling. Common problems are the precipitation of organic salts and the destruction of sugars when heated in the presence of nitrogenous organic nutrients.

Dry heat

Ovens heated to 180°C sterilize dry goods such as glassware, pipettes, and other utensils in 1 h. Slow cooling for about 2 h should minimize breakage of glassware.

Filtration

Filtration is a reliable method for obtaining sterile solutions of heat-labile materials or of materials dissolved in nonaqueous solvents. For this purpose, asbestos and silica filters, as well as cellulose ester membrane filters (0.2-μm pore size), are generally used.

Chemical sterilization

Chemical sterilization is usually carried out with β-propiolactone and ethylene oxide but is not used as frequently as the other methods described above. β-Propiolactone (155°C bp) in 0.5 to 0.75% concentrations sterilizes not only liquid culture media (17) but also heat-sensitive equipment made of rubber or plastic material, which it permeates. Sterilization of solids with ethylene oxide is carried out effectively in special vessels or bags and is easy to handle.

9.5.3 Analytical Techniques

A few analytical techniques are well suited for use in biotransformation screening. Foremost is thin-layer chromatography (TLC). Bioassays are also used where applicable, but automated high-pressure liquid chromatography (HPLC) and gas chromatography (GC) systems are faster and so can increase analytical capabilities considerably.

TLC

TLC plates with various absorbents can be prepared directly in the laboratory and are also available commercially. Theory and techniques are discussed in detail in several excellent texts (15, 137).

Silica gel, alumina, kieselguhr, and cellulose are the most commonly used absorbents and have a wide range of properties which can be altered to suit the particular need by pretreatment with acids, bases, buffers, or specific reagents (e.g., $AgNO_3$). The selection of the suitable solvent system depends on the nature of the compounds to be separated (see the references given above for selection of suitable systems). The samples to be examined are dissolved in a moderately volatile solvent and applied approximately 1.5 cm from the base of the plate and 1.0 cm from each other. Disposable calibrated micropipettes may be used for easy application. The chromatograms are then run in the selected system(s) until the solvent front has reached the proximity of the upper end of the plate. For better separation of compounds with similar R_f values, two-dimensional chromatograms are preferred.

After the plates have been removed from the chamber and dried, the position of the compounds may be located by several methods described in the above references. When applicable, UV light may be used since it is noncorrosive and usually will not damage the material. Exposure to iodine vapors, another relatively mild visualization method, is done conveniently in a TLC tank containing iodine crystals. In most cases, however, the plates are sprayed with selected reagents.

GC

GC is a versatile tool for analysis of biotransformation mixtures. Separation is effected by the distribution of components between two phases which may be either gas-liquid or gas-solid. Any compound that is volatile or that can be derivatized to a volatile substance can be analyzed. The retention times of compounds on particular columns at specified temperatures and gas flow are characteristic of the compounds.

The automated capillary instruments are recommended. When these instruments are used in combination with a mass spectrometer, detailed structural analyses of many samples can be obtained.

The advantages of GC are that it permits rapid analysis; microliter quantities of sample are sufficient; there is a high degree of resolution; the method can be quantitative or qualitative; it is highly sensitive (parts per billion); it is simple to operate; and it can be combined conveniently with mass spectrome-

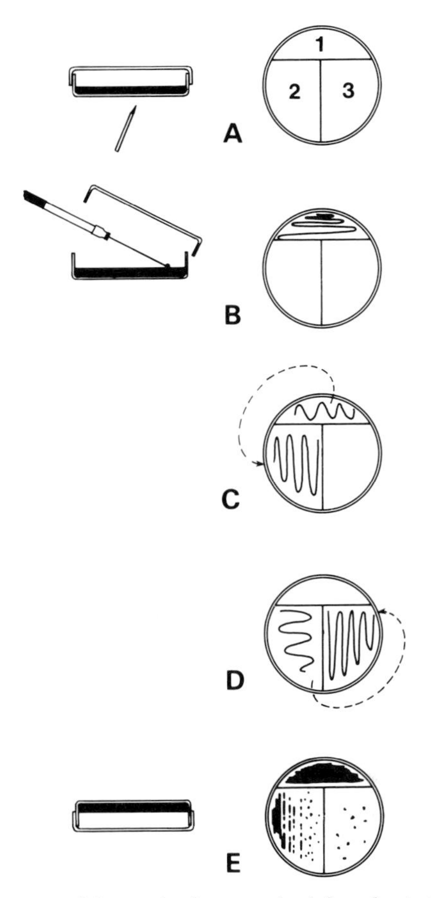

FIG. 2. A useful streak-plate method for obtaining well-isolated colonies. (A) With a glass marker pencil, draw a "T" on the bottom of the petri dish to divide the plate into three sections. (B) Streak a loopful of culture lightly back and forth on the surface of the agar over section 1 as shown. Raise the lid of the dish just enough to allow the streaking to be done, then replace it. Flame-sterilize the loop and allow it to cool (15 s). (C) Draw the loop over section 1 as shown, and immediately streak back and forth over section 2. Flame the needle and allow it to cool. (D) Draw the loop over section 2 as shown, and then streak back and forth over section 3. (E) Incubate the dish in an inverted position as shown to prevent drops of condensed water on lid from falling onto the agar surface. Section 1 will develop the heaviest amount of growth, while section 2 or 3 will usually have well-isolated colonies.

try. (For a comprehensive treatment of the theory and practice of GC, see McNair and Bonelli [97] or Mitruka [98].)

HPLC

HPLC is useful for both analytical and preparative work. In contrast to GC, HPLC analyses are not limited by the volatility and thermal stability of compounds. HPLC is suited for rapid separation of labile natural products and high-molecular-weight compounds and for routine analyses. Automated instruments allow the convenient processing of many samples.

Components in the column effluent are detected with refractive-index monitors. UV-visible and fluorescent monitors, and electrochemical detectors. Postcolumn reactors can be used to make derivatives of separated compounds that are detectable. (For discussion of theory, instrumentation, columns, and solvent selection, see the works by Snyder and Kirkland [134], Brown [19], Parris [111], and Hamilton and Sewell [59].)

9.5.4 Measurement of Cell Mass

For quantitative work it is necessary to determine the cell mass.

Dry-weight estimation

To determine dry weight, homogeneous samples of a suspension of microorganisms are weighed in duplicate or triplicate. Since this simple and direct procedure is rather tedious, an easier way is to calibrate the dry weight of a suspension against its turbidity under standard conditions. Cells are collected by centrifugation or filtration and then resuspended thickly so that 1-ml portions have about 10 to 50 mg (dry weight) of cells. Several portions are removed for drying and weighing. A series of dilutions is then made of the suspension so that six to eight different densities between 0.1 and 1.0 optical density are obtained at 650 or 660 nm. A plot of these densities should yield a straight line falling off somewhat in the more turbid region. A separate calibration curve is constructed for each cultured strain. This procedure works well for nonfilamentous bacteria and yeasts.

The dry weight of filamentous bacteria (actinomycetes) and fungi usually cannot be measured by estimation of turbidity, so it may be necessary to dry and weigh samples. Other methods for measuring growth are discussed by Calam (22).

Viable counts

The plating of serial dilutions is a simple and direct way to obtain viable-cell numbers. A portion (1 ml) of a cell suspension is placed in a flask containing 99 ml of sterile water or 0.9% saline. Saline is preferred because it lessens osmotic shock to the microorganisms. One milliliter of the diluted suspension is further diluted the same way, and the process is repeated three times to obtain dilutions of 10^{-2}, 10^{-4}, 10^{-6}, 10^{-8}, and 10^{-10}. The last three dilutions are plated twice on an appropriate nutrient agar medium in 0.1- and 1.0-ml portions, using a glass "hockey stick" spreader. The resulting six plates should grow colonies in numbers ranging from 30 to 300 per plate.

A handy rule of thumb for estimating cell numbers in suspensions is that a suspension with just barely visible turbidity contains about 10^6 cells per ml. This estimate is fairly accurate for bacteria of average size such as *Escherichia coli*.

9.6 MEDIA

9.6.1 Defined, Semidefined, and Complex Media (16, 82)

Catalytically active microbial cells are obtained by growth in balanced nutrient media, especially those containing inducers of the desired enzyme activities. A comprehensive discussion of microbial nutrition and medium design is given by B. M. Guirard and E. E. Snell in *Manual of Methods for General Bacteriology* (51).

Different microorganisms have special requirements for optimal growth. In addition to environmental factors (temperature and pH), the ratios and amounts of carbon, nitrogen, phosphorus, trace minerals, and special growth factors are important in proper nutrition (119). The elemental composition of a dried cell provides insight into the nutritional requirements of that cell. Water makes up 80 to 90% of the cell weight. For dried *E. coli* cells, the relative concentrations of the various elements are as follows: carbon (50%), nitrogen (15%), phosphorus (3.2%), sulfur (1.1%), sodium (1.3%), potassium (1.5%), magnesium (0.5%), calcium (1%), and iron (0.24%), with trace amounts of manganese and copper. Hydrogen and oxygen account for the balance. Cell composition, however, is dependent on the growth medium used. For instance, the high amount of sodium in this analysis reflects the fact that the *E. coli* was grown in nutrient medium containing sodium chloride, which is customarily added even though there is no requirement for it.

Culture media support growth until a critical nutrient is depleted, and then growth ceases. Cell yields differ widely among various organisms grown on the same carbon and energy sources (119). The important factors that determine cell yield are the oxidation-reduction potential of the carbon and energy sources, the degree of polymerization of these sources, the operative metabolic pathways, the intrinsic growth rate of the cell itself, and temperature, pH, and aeration. Some elements must be provided; for example, elemental nitrogen and sulfur usually are required at 6 to 13% and 0.7 to 1.3% of total medium ingredients, respectively. These values reflect the percentages of these elements found in proteins.

The composition and preparation of selected liquid media are given below for illustration. To make solid media, about 1.5 to 2% agar or other solidifying agents are added (29).

Chemically defined media

Chemically defined media are made by the addition of known ingredients to distilled water. Although more expensive, these media offer the advantages of reproducibility and greater simplicity in the analysis of biotransformation end products.

Typical defined medium (53)

Carbon source	2 g
Ammonium sulfate	1 g
Dipotassium phosphate	1 g
Salt solution A	10 ml
Distilled water	990 ml

Adjust medium to pH 7.0 before autoclaving.

Salts solution A (53)

Magnesium sulfate · $7H_2O$	25.00 g
Ferrous sulfate · $7H_2O$	2.80 g
Manganous sulfate · H_2O	1.70 g
Sodium chloride	0.60 g
Calcium chloride · $2H_2O$	0.10 g
Sodium molybdate · $2H_2O$	0.10 g
Zinc sulfate · $7H_2O$	0.06 g
Hydrochloric acid, 0.1 M	1.00 liter

A variety of carbohydrates, organic acids, alcohols, hydrocarbons, and lipids serve as carbon sources. Nitrogen sources can be salts other than ammonium sulfate, such as sodium or ammonium nitrate. Urea is another good nitrogen source, as are certain amino acids (e.g., asparagine). Vitamins, purines and pyrimidines, and amino acids are added to the defined media to stimulate growth of more fastidious organisms.

Semidefined media

When a small amount of a single vegetable or meat extract or preparation is added to the chemically defined medium, the resulting mixture is termed semidefined. These ingredients are added in amounts varying from 0.05 to 0.5%. Small amounts of these organic nutrients supply growth factors and vitamins that often enhance growth significantly compared to the completely defined medium. Yeast extract, meat peptones, soy peptones, malt extract, casein hydrolysates, and corn steep liquor are some of the most useful additives.

Complex media (53)

Most of the nutrients in complex media are provided by extracts or enzyme digests of plant and animal products. A few examples of media generally suitable for growth and maintenance of a wide variety of microorganisms are given below.

Nutrient broth

Beef extract	3.0 g
Peptone	5.0 g
Distilled water	1.0 liter

For a solid medium, add 15 g of agar. The pH after autoclaving is 6.8.

Glucose-yeast extract broth

Glucose	10.0 g
Yeast extract	10.0 g
Dipotassium phosphate	1.0 g
Salt solution A (see Defined media)	10.0 ml
Distilled water	990.0 ml

Adjust pH to 7.0 before autoclaving.

Yeast extract-tryptone broth

Yeast extract	2.5 g
Tryptone	5.0 g
Glucose	1.0 g
Distilled water	1.0 liter

Sabouraud dextrose broth

Neopeptone	10 g
Glucose	40 g
Distilled water	1 liter

The pH before autoclaving is 5.7.

Sabouraud maltose broth

Peptone	10 g
Maltose	40 g
Distilled water	1 liter

The pH before autoclaving is 5.6.

***Streptomyces* sporulation agar (8)**

Yeast extract	1 g
Beef extract	1 g
Tryptose	2 g
Ferrous sulfate	1 mg
Glucose	10 g
Agar	15 g
Distilled water	1 liter

Adjust pH to 7.2 before autoclaving. For broth, eliminate agar and reduce other nutrients to one-third of the value given.

Mannitol broth

Yeast extract	5 g
Peptone	3 g
Mannitol	25 g
Distilled water	1 liter

9.6.2 Specialized Media for Biotransformation Studies

Conversion of thalicarpine by *Streptomyces* (103)

(i) Seed medium

Glucose	20 g
Soybean meal	5 g
Yeast extract	5 g
Sodium chloride	5 g
Potassium phosphate, dibasic	5 g
Distilled water	1 liter

Adjust to pH 7 with 6 N HCl.

(ii) Transformation medium

Glucose	10.0 g
Corn steep liquor	6.0 g
Potassium phosphate, monobasic	3.0 g
Calcium carbonate	3.5 g
Soybean oil	2.2 ml
Yeast extract	2.5 g
Distilled water	1.0 liter

Hydroxylation of benzoylalkylpiperidines by *Sporotrichum sulfurescens* (68)

Corn steep liquor (60% solids)	20 g
Glucose	10 g
Tap water	1 liter

Adjust to pH 4.9.

Microbial transformation of maridomycin by *Streptomyces* sp. (128)

(i) Seed medium

Glucose	20 g
Soluble starch	30 g
Soybean flour	10 g
Corn steep liquor	10 g
Polypeptone	5.0 g
NaCl	3.0 g
$CaCO_3$	5.0 g
Distilled water	1.0 liter

pH 7.0.

(ii) Transformation medium

Glucose	50 g
Polypeptone	5 g
Corn steep liquor	20 g
NaCl	3 g
$CaCO_3$	5 g
Distilled water	1 liter

pH 7.0.

Transformation of 14β-bromocodeinone (3)

Glucose	1 g
Peptone	2 g
Meat extract	1 g
Yeast extract	1 g
Corn steep liquor	3 g
Distilled water	1 liter

Transformation of mycophenolic acid (69)

Cerelose	30.0 g
Ammonium tartrate	7.5 g
Dipotassium phosphate	2.0 g
Salts solution A (see Defined Media)	10.0 ml
Yeast extract	1.0 g
Distilled water	1.0 liter

Adjust pH to 5.5 for fungi and *Streptomyces* and to 7.0 for unicellular bacteria.

Hydroxylation of estrogenic compounds by *Curvularia lunata* (85)

Sucrose	10.0 g
Tryptone medium (Difco)	10.0 g
$NaNO_3$	2.0 g
K_2HPO_4	1.0 g
$MgSO_4 \cdot 7H_2O$	0.5 g
KCl	0.5 g
$FeSO_4 \cdot 7H_2O$	10.0 mg
Distilled water	1.0 liter

Adjust pH to 7.0 with H_2SO_4 and add 2.5 g of $CaCO_3$ before sterilizing.

Transformation of trichothecenes (28)

(i) Seed medium

Glucose	30.0 g
Lactose	10.0 g
Ammonium sulfate	2.0 g
Potassium phosphate, monobasic	0.5 g
Corn steep liquor	30.0 ml
Deionized water	1.0 liter

(ii) Transformation medium

Glycerol	40.0 g
Cornstarch	20.0 g
Pharmamedia	15.0 g
Washed brewers' yeast	10.0 g
$MgSO_4 \cdot 7H_2O$	0.5 g
$CaCl_2$	0.5 g
$ZnSO_4 \cdot 7H_2O$	10.0 mg
Deionized water	1.0 liter

Bioconversions of tylosin-related antibiotics by *Streptomyces fradiae* (109)

Glucose	10 g
Starch	20 g
Peptone	5 g
Yeast extract	5 g
L-Asparagine	3 g
$CaCO_3$	4 g
Distilled water	1 liter

Adjust pH to 7.4.

Hydroxylation of steroids with *Aspergillus* and *Penicillium* (14)

Malt extract	1.0 g
Beef extract	1.0 g
Yeast extract	1.0 g
Corn steep liquor	1.0 ml
Glucose	5.0 g
Deionized water	1.0 liter

pH adjusted to 5.5 with 2 N HCl.

Biotransformation of colchicine (64)

Glucose	10 g
Corn Steep liquor	10 g
Soy Flour	10 g
Dry malt extract	5 g
$CaCO_3$	1 g
NaCl	5 g
Distilled water	1 liter

(Five other media are described in this reference.)

9.7 ADDITION OF ORGANIC COMPOUNDS TO REACTION MIXTURES

Most organic reactions are run in nonaqueous solvents. Since microbial growth and biological reactions take place in aqueous environments, there is a natural tendency to restrict biotransformation reactions to water-soluble organic substrates. Actually, biotransformations occur with both lipophilic and hydrophilic substrates.

An example of the biotransformation of a lipophilic substance is the hydroxylation of progesterone to 11α-hydroxyprogesterone by *Aspergillus ochraceus* (149). The use of a micronizing technique with wetting agent to suspend the starting material greatly improved the yield of product. Starting with progesterone at 20 g/liter, the yield of hydroxylated product was 90%; at 50 g/liter the yield was 65%.

The key to success with biotransformations of lipophilic compounds is the enhancement of substrate availability to the active site of the enzyme. It is generally assumed that access to the active site of an enzyme is available only to compounds dissolved or dispersed in the reaction medium. Thus, methods have been developed to improve the solubility and dispersion of reactants in water (53, 88, 113, 133). As the accessibility to the cell surface and ultimately to the enzyme increases, the yield of biotransformation products also increases. Once contact with the cell occurs, substrates can penetrate the cell wall and membrane by passive or active transport. Cell surfaces and membranes, as well as enzymes themselves, have hydrophobic domains that facilitate reactions with lipids. In addition, microorganisms produce a variety of endogenous emulsifiers which promote these reactions (157).

9.7.1 Physical Forms of Organic Substrates

Nearly every class of organic compound is amenable to microbial and enzymatic transformation. Water-soluble compounds with ionizable or hydrophilic functional groups, such as carboxylic acids, amines, alkaloids (especially when used as salts), alcohols, sugars, phenols, and polyfunctional substances are easy to handle and can be added directly to incubation mixtures at the appropriate time. Some problems may arise, however, from the toxicity of compounds, ionic strength and concentration, and pH changes.

Lipophilic substrates may also be toxic, but the major problem is their accessibility to the biocatalyst. An examination of growth rates of bacteria cultivated in media containing water-insoluble hydrocarbons illustrates this point. Bacteria grown on naphthalene, phenanthrene, or anthracene have generation times of 1.5, 10.5, and 29 h, respectively (153). These growth rates are directly proportional to the water solubilities of the hydrocarbons but independent of the total amount of solid substrate present (133, 153). The bacteria grow at the interfaces between the solids and the aqueous medium. Thus, it appears that generation times are subject to the rate of substrate dissolution, if the rates of hydrocarbon uptake and metabolism are not the limiting steps.

9.7.2 Dispersion and Solubility

Jones and Baskevitch (70) demonstrated that steroid aggregation results in deviation from pseudo-first-order transformation kinetics. Lee and co-workers examined in detail the aggregation and solubilization phenomena of steroid substrates (84). A mixed culture of *Arthrobacter simplex* and *Curvularia lunata*

TABLE 4. Effect of method of substrate dispersion on bioconversion yields (84)

Vehicle	Yield of steroid (%)
Hot borate-ethanol-water	90
Hot ethanol	65
Hot ethanol-acetone	67
Cold DMF	60
Cold ethanol-acetone suspension	30
Cold ethanol suspension	20
Cold aqueous 0.1% Tween 80 surfactant	10

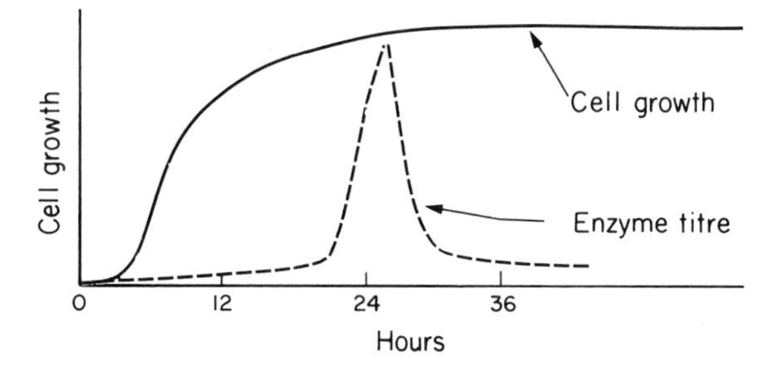

FIG. 3. Hypothetical time course for the production of a biotransformation enzyme.

catalyzed the simultaneous 1-dehydrogenation and 11β-hydroxylation of 16α-hydroxycortexolone-16,17-acetonide. Substrate was prepared (i) as a suspension in cold solvent, (ii) in 0.1% (wt/vol) aqueous Tween 80 surfactant, or (iii) as solutions in hot and cold solvents. The substrates were added to the cultures immediately after preparation and again 25 h later.

The best yields were obtained with hot solvents and cold DMF (Table 4). Yields were related to the particle size of the substrate. Hot solvents gave dispersions with 0.5- to 2-μm particles; those from cold solvents ranged from 10 to 100 μm. Apparently, ultrafine, amorphous particles are more accessible to enzyme active sites than crystalline forms. This may be due to improved compound solubility. Also, the solubilizing vehicle can improve cell permeability.

Chien and Rosazza (27) used solubilization with polyvinylpyrrolidones (PVPs) to enhance the hydroxylation of ellipticine by *Aspergillus alliaceus*. This brilliant yellow alkaloid is barely soluble in water (<5 mg/liter), but solutions of 30% PVP (40,000 average molecular weight) gave the highest initial water solubilities of ellipticine, and the solubility increased proportionately with the concentration of PVP (60 g of PVP per liter solubilized 100 mg of ellipticine per liter). At that concentration, yields of hydroxylated ellipticines were doubled.

Nakamatsu et al. (105) evaluated the influence of surfactants on the oxidation of sterols by *Nocardia corallina*. The bioconversions of substrates dispersed with 1 of 16 surfactants were compared with the performance of substrates sonicated to reduce their particle size. Cationic, nonionic, and anionic surfactants were used at 0.01% concentration. Two cationic detergents (Sanisol C and Amiet 106) and one nonionic detergent (Nonal 106) significantly inhibited cell growth. Most nonionic surfactants did not inhibit growth and provided good emulsification. Emal 10C, Emulbon T-83, Sorbon T-40, and Tween 80 surfactants stimulated the oxidation of soy sterols.

9.7.3 Timing of Additions

When growing cells are used for biotransformations, the time of addition of the organic substrate profoundly influences the yield of product. Toxic substances such as many antibiotics and antitumor compounds often inhibit growth and enzyme production if they are added early in the growth cycle. In other cases, it is advantageous to add at least small amounts of substrate at the beginning of the growth phase to promote enzyme induction. The addition of substrate during the late logarithmic growth phase avoids toxicity effects and also promotes enzyme induction. At this point in their growth cycle, cells are still capable of enzyme synthesis, while the proliferation of the biomass is less inhibitable by toxic substrates. The same reasoning applies to substrates added in toxic solvents.

Toxic substrates or substrates in toxic solvents should be added incrementally. Karow and Petsiavas (73) added progesterone semicontinuously to growing *Aspergillus ochraceus*. This dosing technique minimized toxicity as well as the mechanical loss of starting material through aggregation. Unexpectedly, fewer side reactions occurred. Dosing techniques also sidestep the undesirable phenomenon of substrate inhibition, which almost invariably occurs when large amounts of substrate are added at a single time. The effect is most prominent with water-soluble materials, but well-dispersed lipids display the same phenomenon.

The timing of substrate addition must take into account the physiological state of the microorganism. The position of the cell in its growth cycle determines its enzyme capabilities. Enzymes of interest may occur only at specific times during the growth cycle. Figure 3 illustrates a hypothetical organism that produces high enzyme titers only during the late logarithmic growth phase.

Addition of substrate when the enzyme is not present would be futile. Enzyme production also may be subject to fluctuations in the pH of the medium, the amount and kinds of carbon and nitrogen nutrients in the medium, and the degree of oxygenation of the medium. The optimal time for substrate addition is difficult to predict and is best determined experimentally.

9.7.4 Methods of Addition

Water-soluble solids and liquids

Soluble compounds are added directly to incubation mixtures or cultures without dilution. Additions made early in the growth period must be done aseptically. In most cases, membrane filtration should be used to sterilize ingredients; stable compounds may be autoclaved. Toxic or inhibitory compounds should be added incrementally in small doses as many substances in large doses are inhibitory. If high concentrations of nontoxic substrate are required, incremental addition may help to lessen the inhibition.

Gases, volatile solvents, and other compounds with high vapor pressure

Volatile compounds pose difficulties in shaken-flask incubations, but they are relatively simple to handle in small fermentors and other types of bioreactors, except for the danger of explosions. Gases and other volatile substances can be carried into vessels along with sterile air. Air bubbled over the surface of a volatile solid or through a liquid will carry the compound into the reaction mixture. There may be some difficulty in the estimation of the amount of substrate added by these means, although calibration with a gas chromatograph might be a direct way to do it.

Enclosed incubator-shakers are the best way to handle volatile compounds and gases in shaken-flask reactions. Again, extreme care must be taken to avoid the accumulation of explosive mixtures. For a small number of flasks, individual spargers could be used.

Use of inert supports

A variety of compounds can be adsorbed or dissolved within the lattices of inert materials, such as zeolites (R. L. Raymond, U.S. patent 3,224,946, Chem. Abstr. **64**:7327c, 1965), molecular sieves, and diatomaceous earth. The resulting ultrafine particle sizes and large surface areas promote a high degree of dispersion of lipophilic substrates. These adsorbed substrates are remarkably available to biocatalysts. Liquid paraffins are adsorbed to the supports from solvent solutions. After evaporation of the solvent, bound compounds are added directly to incubation mixtures.

Organic solvent vehicles

The most common method for adding water-insoluble substrates is as solutions in water-miscible organic solvents. Preferably, these solvents should have low toxicity to the biocatalyst. Solvents commonly used are DMF, dimethyl sulfoxide, ethanol, methanol, and acetone. Others, including ethylene glycol, ethyl acetate, chloroform, and benzene, are of limited use.

The organic compound is dissolved in dry DMF (stored over molecular sieves) at a concentration from 0.1 to 1 g/liter. This may be hastened by gentle treatment with a sonifier for a few seconds. The concentrated substrate solution is added to incubation mixtures at the required level by pipette. A milky precipitate forms instantly as the vehicle dissolves in the aqueous medium. To prevent reaggregation of the substrate and to ensure even distribution, each flask or vessel should be shaken immediately upon addition of substrate solution. If necessary, the substrate solution in DMF should be sterilized by filtration before addition to the medium. This technique works with most water-miscible solvents (27, 103).

Since many hygroscopic organic solvents lose solvent power as they take up water, they should be kept dry. A small amount of water in DMF greatly reduces the capacity of this excellent solvent.

PVP as a vehicle

PVP disperses many types of aromatic compounds in aqueous media by formation of coprecipitates. The aromatic substrate and PVP-40,000 are dissolved in chloroform-methanol (9:1), and the mixture is evaporated to dryness in a rotary evaporator. Ratios of aromatic compound to PVP of 1:5 up to 1:120 (27) should be used.

To prepare concentrated solutions or dispersions, suspend compounds in 10 to 30% solutions of PVP-40,000 by grinding them together with a glass mortar and pestle. Add mixtures to incubation mixtures as needed.

Milling and wetting techniques

A milling and wetting method is used to disperse steroids such as progesterone (149), as follows. Grind USP-grade progesterone with a "Jet-O-Mizer" model 202 grinder (Fluid Energy Processing and Equipment Co., Philadelphia, Pa.) to a fine particle size and an apparent density of about one-third of the starting material, as judged by volumes occupied by equal weights. Weigh the ground progesterone into 250-ml Erlenmeyer flasks; wet down with a suitable amount of 0.01% aqueous Tween 80 surfactant, and shake well. Sterilize by exposure to steam at atmospheric pressure for about 30 min; do not autoclave. By this technique, 20 to 50 g of steroid substrate can be dispersed per liter in the aqueous fermentation medium.

9.8 REACTIONS IN SOLVENT MIXTURES

The use of organic solvents as the primary solvent in a biocatalytic reaction is one of the newest developments in biotransformations (9, 21, 133). Previously it was thought that enzyme reactions could be run only in water solutions. As it turns out, a certain amount of water must be present near the enzyme to maintain spatial integrity of the active site and to participate in some of the catalytic reactions. Zaks and Klibanov have shown that vacuum-dried lipase (containing as little as 0.48% water) is stable and catalytically active in organic media at 100°C (158).

Water emulsions have been used successfully for biotransformations with either whole cells or purified enzymes (9). The aqueous phase maintains the activities of the enzymes while the organic phase dissolves lipophilic reactants and products. By keeping a very low aqueous content in emulsions, it is possible to favor catalysis of the reverse of hydrolytic reactions. Thus, for example, esterifications with esterases become possible (78, 95, 155). Amidation and phosphorylation reactions also are possible by the reversal of corresponding hydrolytic reactions.

For catalysis in very low-water organic solvents, whole cells should be used because they retain water as long as the organic solvent does not extract it. For this reason, polar solvents that tend to dehydrate cells should be avoided.

Biocatalytic reactions requiring oxygen can be run advantageously in organic solvent-water emulsions because oxygen usually is more soluble in such solvents than in water. Other advantages of using solvent mixtures, alluded to above, are the improved solubility of lipophilic substrates and the possibility of reversing hydrolytic reactions. It is clear that the use of hydrolytic enzymes in a reverse mode opens many new possibilities in biotransformations.

Examples of the application of biocatalysts in solvent mixtures have been compiled (9, 133). A variety of substrates, including steroids, amino acids, and other types of organic compounds, have been converted to useful products by enzymes and resting cells. Reactions involving oxidoreductases, epoxidases, dehydrogenases, isomerases, and esterases have been described. Ethyl acetate, butyl acetate, carbon tetrachloride, cyclohexane, benzene, heptane, and chloroform have been used as solvents in these reactions.

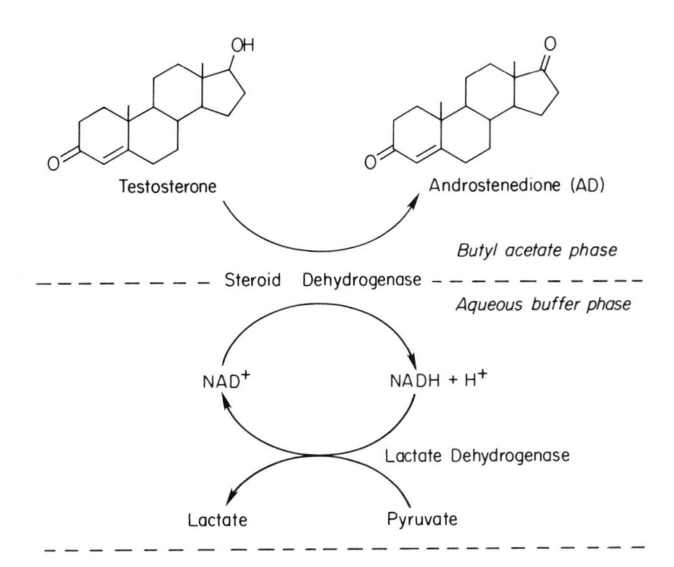

FIG. 4. Enzymatic dehydrogenation of testosterone in a biphasic medium.

The enzymatic dehydrogenation of testosterone is an example of an enzyme reaction in organic solvent-water emulsion (32). The reaction, catalyzed by hydroxysteroid dehydrogenase from *Pseudomonas testosteroni*, is diagrammed in Fig. 4. The steroid dehydrogenation reaction is driven to completion by being coupled with lactate dehydrogenase. The latter reaction regenerates the NAD^+ required by steroid dehydrogenase. Pyruvate serves as the cosubstrate to drive the reaction to completion. Enzymes, NAD^+, and pyruvate are dissolved in the aqueous buffer phase; steroid is dissolved in the organic phase. Butyl acetate is superior to both ethyl acetate and butanol. Yields of androstenedione depend directly on the amount of pyruvate added and the amount of pyruvate reduced to lactate.

In another example of a reaction in biphasic solvents, the coupling of 20β-hydroxysteroid dehydrogenase with alcohol dehydrogenase has been shown to specifically reduce 20-oxo-steroids (33).

9.9 EXAMPLES OF TYPICAL BIOCONVERSION PROCEDURES

The following five procedures serve as models for procedures used for microbial transformation of organic compounds.

9.9.1 Aerobic Screening (58)

A screening method is necessary for checking the activity of large numbers of microorganisms on large numbers of compounds. Culturing is done in conical flasks on shakers, so that the number that can be handled is limited only by the capacity of the shakers and the ability to handle and analyze the samples that will be generated. The method is used to screen activities of both fungi and bacteria.

In our hands, the most reliable procedure involves a two-stage incubation. During the first stage, the culture is grown to late logarithmic phase in a rich medium to provide a heavy inoculum for the second

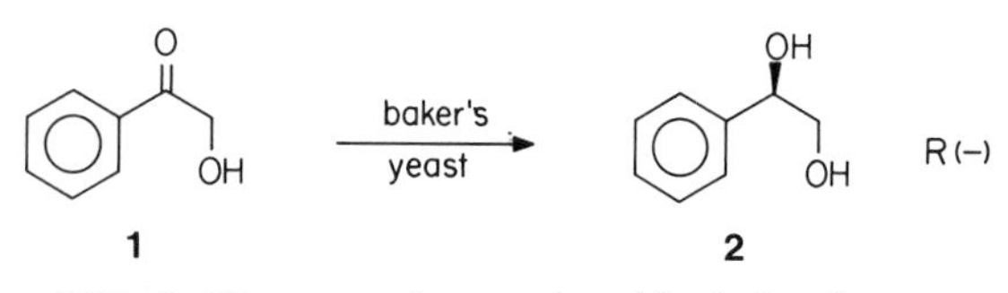

FIG. 5. Bioconversion catalyzed by bakers' yeast.

stage. Compounds to be screened are added to the second-stage culture when it has reached its maximum growth. However, the time of addition of compounds is one variable that should be investigated after preliminary results are obtained. For the first screening experiments, the protocol should be simple so that many samples can be processed. When preliminary leads are obtained, conditions for the later rounds of screening can be refined.

First stage

Prepare an aqueous suspension of the surface growth of freshly grown cells on a rich agar medium. Most bacteria require 18 to 72 h for growth; fungi require 3 to 7 days. Prepare the suspension in 3 to 5 ml of sterile distilled water or physiological saline (0.85% NaCl). Transfer the resulting cell suspension aseptically to 50 ml of a suitable nutrient medium in a 250-ml conical flask. Incubate the flask at 25 to 30°C (or the appropriate temperature) for 1 or 2 days (until the culture reaches maximum growth) on a rotary shaker. Prepare a separate culture for each microorganism to be screened, and use it as a control with no additions to the medium.

Second stage

Inoculate flasks containing fresh nutrient medium with 10% (vol/vol) of the first-stage culture and incubate as before. For the first round of screening, the same nutrient medium is satisfactory. In later studies one might try a less rich medium for the second stage; this sometimes improves yields.

Add compounds to be screened for biotransformation reactions when the cultures have attained maximum growth. This should be within 24 h because of the heavy inoculum used. The best time for compound addition, as discussed above, will have to be determined experimentally.

For the first-round studies, add substrates in the simplest manner, that is, either directly as powders or as concentrated solutions in the appropriate solvent. The chemical nature of the substrate dictates the best solvent for its dispersion. DMF, ethanol, and acetone are relatively nontoxic, water-miscible solvents that have found wide use in screening procedures. Useful surfactants are the Tweens or Spans. Tween 80 surfactant as a 1% water solution is especially recommended. The final concentration of solvent in the experimental culture medium should not exceed 0.5%.

Controls are required for each solvent and surfactant used. Controls without substrate are necessary for each medium and microorganism. Also run controls for each compound without microorganism. For initial studies, take samples for analysis at 24-h intervals for several days. As results are obtained, the sampling time can be adjusted.

With modification of incubation conditions, this general procedure can be used to screen for biotransformation reactions with anaerobic organisms.

9.9.2 Reductions with Yeast (120)

Bioconversions catalyzed by bakers' yeast are some of the easiest reactions to perform because of their simplicity. The asymmetric reduction of carbonyl compounds frequently occurs in good yield. This procedure (Fig. 5) utilizes readily available materials, namely, sucrose and commercial bakers' yeast in a water medium. For reductions, sucrose serves as the ultimate electron donor, operating by the recycling of cofactors, such as the pyridine nucleotides.

A mixture containing substrate 1 of Fig. 5 (10 g), bakers' yeast (100 g), and sucrose (150 g) in 800 ml of water is incubated at room temperature for 1 to 2 days. The mixture is extracted with ethyl acetate, dried, and distilled to yield 4.8 g of compound 2, $[\alpha]_{25}{}^{D} = -45.8°$. The yeast also reduces ketoesters. This general procedure can be used to reduce many carbonyl-containing compounds.

9.9.3 Catalysis with Dried Cells

(See R. C. Frickson, W. E. Brown, and R. W. Thoma, U.S. patent 2,260,439, 1967.)

Bioconversions with dried cells resemble the above procedure for yeast cells. However, dried bacterial and fungal cells usually are not available commercially, so that the preparation of such cells is an additional complication. Preparation of dried cells is discussed in section 9.3.3.

The dehydrogenation of a hydrocortisol derivative is presented in Fig. 6 as a model reaction. 9α-Fluoro-16α-hydrocortisol (compound 3 in Fig. 6) (20 g), acetone-dried cells of *Arthrobacter simplex* (250 g), phosphate buffer (pH 7.0, 0.1 M), and 2-methyl-1,4-naphthoquinone (3 g) in 2.5 liter of water are agitated

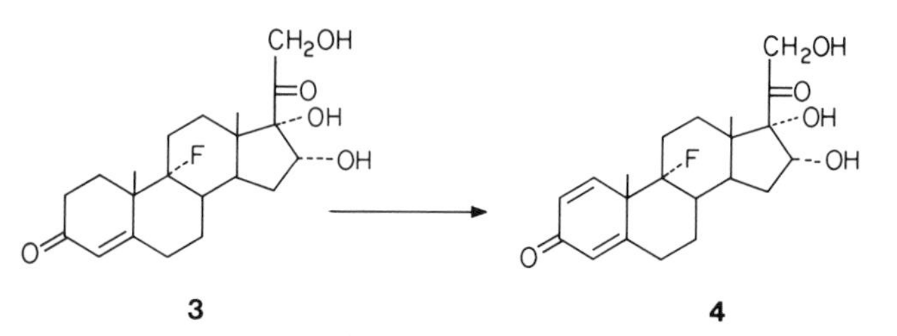

FIG. 6. Model of dehydrogenation of a hydrocortisol derivative.

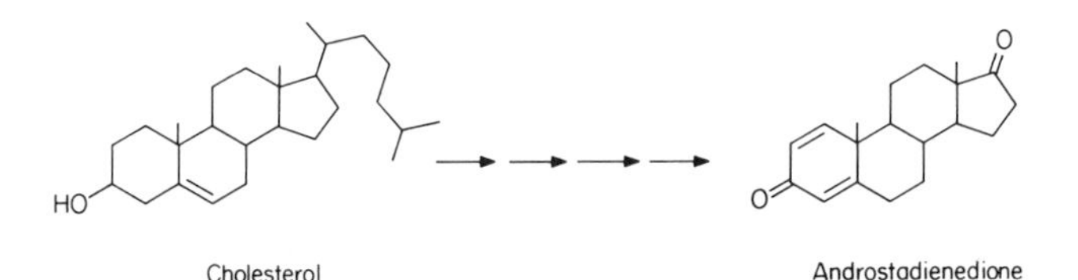

FIG. 7. Accumulation of androstadienedione in degradation of cholesterol.

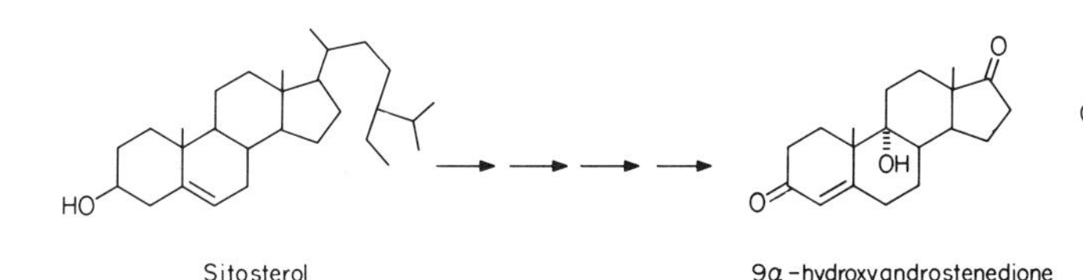

FIG. 8. An example of a useful intermediate of sterol degradation.

and aerated for 2 to 3 h. Triamcinolone (compound 4 in Fig. 6) is obtained in 90% yield. The naphthoquinone serves as a hydrogen acceptor cofactor in the reaction. Acetone-dried cells remain active for relatively long periods when stored in a refrigerator. These cells also may be used for nonaqueous solvent procedures.

9.9.4 Use of Metabolic Inhibitors (104)

During screening experiments, microorganisms frequently destroy compounds without the accumulation of recognizable products. The addition of a metabolic inhibitor will stall utilization of certain intermediates to allow the accumulation of products. An excellent example is the accumulation of a steroid intermediate in the presence of an iron chelator during the degradation of sterols by *Arthrobacter simplex* and *Nocardia corallina*.

The addition of α,α′-dipyridyl (0.001 M), *o*-phenanthroline (0.0001 M), or 8-hydroxyquinoline (0.001 M) to logarithmic-growth-phase cultures with cholesterol as substrate causes the accumulation of androstadienedione by blocking the 9α-hydroxylase reaction that leads to complete destruction of the sterol (Fig. 7). This procedure was first used to elucidate the pathway of sterol degradation in microorganisms, but for economic reasons it is not the method of choice for large-scale androstadienedione production.

9.9.5 Blocked Mutants (35, 89, 154)

Mutants blocked at various stages of a metabolic pathway accumulate isolable intermediates in their culture medium. Such mutants have been used for large-scale production of biotransformation products.

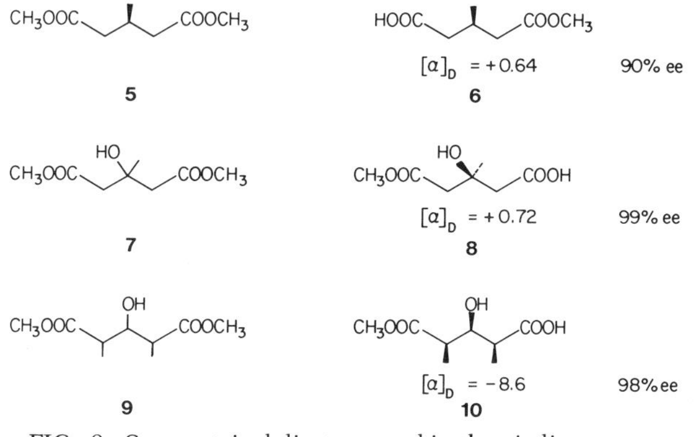

FIG. 9. Symmetrical diesters used in the pig liver esterase procedure.

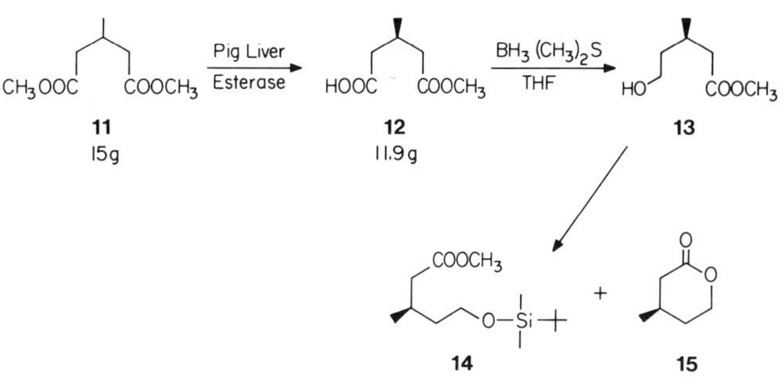

FIG. 10. Products of the pig liver esterase procedure.

The method can be laborious, since it involves the induction and selection of specifically blocked mutants. Induction and selection methods for mutants are detailed elsewhere (51).

Coupled with enrichment culture techniques, the use of mutants is a powerful tool for producing biotransformation products of a wide variety of organic compounds. Thus, if it is possible to induce metabolic pathways for the degradation of a particular compound, it is reasonable to assume that mutants can be obtained which are blocked at points in this pathway, such that transformation products of the starting material will accumulate.

Mutants have been used with great success by scientists at the Upjohn Co. and G. D. Searle Co. Processes were developed for producing useful intermediates of sterol degradation by mutants of *Mycobacterium* species (Fig. 8).

9.9.6 Catalysis with Purified Enzymes (71, 151)

Many purified enzymes available commercially are useful in preparative biocatalytic reactions. Because of their high catalytic activity under mild reaction conditions and their specificity, they are finding increased use in modern synthetic organic chemistry. Enzymes can be used to elicit reactions in the presence of labile functional groups without recourse to troublesome protecting groups. Often, enzyme-catalyzed reactions are highly regioselective, stereoselective, or both.

The use of an enzyme in a simple yet highly stereoselective synthetic reaction is illustrated by the pig liver esterase procedure for producing a chiral monoester by selective hydrolysis of symmetrical diesters (compounds 5, 7, and 9 in Fig. 9) (100).

To 15 g of dimethyl 3-methylglutarate (compound 11, Fig. 10), suspended in 100 ml of 0.01 M phosphate buffer (pH 8.0), are added 1,000 U of pig liver esterase (Sigma Chemical Co., St. Louis, Mo.) with vigorous stirring. The pH is kept constant by the addition of 1 N NaOH. After consumption of 1 mol equivalent of base (overnight), the mixture is homogeneous. The pH is adjusted to 9, and the reaction mixture is extracted with ether. The yield of half-ester 12 (Fig. 10) is 11.9 g (86%). Reduction with borane-methyl sulfide complex of the free carboxyl in compound 12 to the hydroxyester (13) and subsequent protection with *t*-butyldimethylsilyl ether gives compound 14 and the six-membered lactone (15) (Fig. 10). Compound 15 (Fig. 10) was compared with authentic meva-

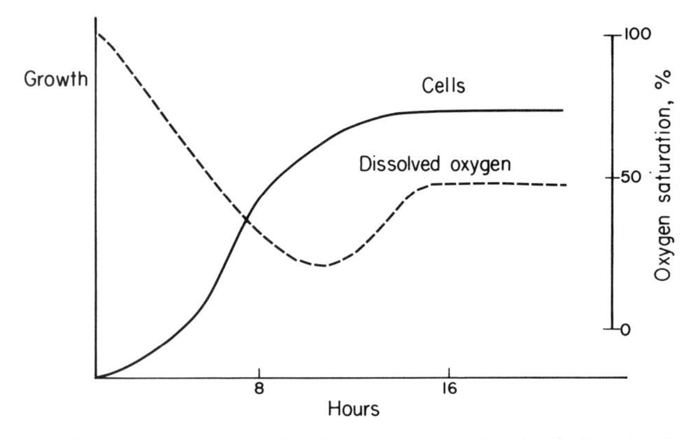

FIG. 11. Influence of cell growth on level of dissolved oxygen.

lonolactone to determine optical purity and was found to be the R enantiomer (90% ee) (61).

9.10 OPTIMIZATION PROCEDURES

To obtain enough product for identification and further testing, preliminary optimization studies are necessary. Yields can be improved substantially by a few systematic studies of nutritional and environmental parameters. Such studies will also be of use in scaling up the process.

9.10.1 Physical Parameters

Temperature

Changes in temperature can drastically affect biocatalytic reactions. The temperature should be varied between 20 and 50°C for mesophilic organisms and enzymes. The variations can start with 5°C increments on either side of the original temperature at which the reaction occurred. If positive effects are observed, the temperature can then be varied in smaller increments in the succeeding experiments. Large differences in yields can occur with a 1 or 2°C difference in temperature, as demonstrated with the bioconversion of isobutyric acid to L-(+)-3-hydroxybutyric acid by stationary-phase cells of *Pseudomonas putida* (53).

The optimum temperature for growth of cells could well be different from the optimum for biocatalysis. Resting cells, stationary-phase cells, and even enzyme preparations frequently perform well at high temperatures. Although no specific example comes to mind, temperatures on the low side are worthy of investigation also.

pH

It is difficult to study the effect of changes in pH in small-scale flask cultures. About the best that one can do is to experiment with growth in media buffered to different pH values. Growth usually is very sensitive to variation in pH. Furthermore, the process of growth changes the pH of the medium. For these reasons, systematic studies of growing cultures should be done in small fermentors where pH can be controlled automatically.

The problem of studying the effect of pH in biocatalytic reactions that do not involve growth (e.g., enzyme reactions or resting-cell reactions) is in the selection of appropriate, noninhibitory buffers. A pH stat is useful to study reactions that produce pH changes.

Aeration

Aerobic organisms require oxygen for growth. It is difficult to achieve optimum growth in shaken flasks because, almost always, as a culture proliferates oxygen becomes the limiting factor. The changes in dissolved-oxygen levels that take place in a growing culture are diagrammed in Fig. 11.

The medium is easily saturated with oxygen at the beginning of the growth cycle. However, rapid growth during the logarithmic phase depletes oxygen faster than it can be supplied to the medium. Then as the growth rate slows, the oxygen level rises. Even in a highly aerated vessel, a rapidly growing culture can bring the dissolved-oxygen concentration to zero (117).

In culture flasks, the efficiency of aeration is determined by the shape of the flask, the volume of liquid it contains, the type of shaking (reciprocal or orbital), the gaseous environment, the culture medium composition, the type of flask closure, and ambient conditions (18, 42, 43, 79, 94, 152).

A few simple rules can be used to obtain the best aeration in shaken flasks.

1. The maximum aeration in a given vessel is attained when the liquid volume is no more than 20% of the total volume.

TABLE 5. Oxygen absorption rates in laboratory culture vessels (131)

Vessel	Medium vol (ml)	Procedure	vvm[a]	OAR
10 by 150-mm tube	10	Stationary		0.03
100-ml conical	20	Stationary		0.32
500-ml conical	100	Stationary		0.10
100-ml conical	20	Shaking, 250 rpm		1.10
500-ml conical	10	Shaking, 250 rpm		2.00
500-ml conical	50	Shaking, 250 rpm		0.60
500-ml conical	100	Shaking, 250 rpm		0.27
20-liter bottle	15,000	8-mm tube	1.0	0.06
20-liter bottle	15,000	Sinter sparger	1.0	0.60
20-liter fermentor	15,000	Sparger, 250 rpm	1.0	2.00

[a] Volume of air per volume of medium per minute.

2. Shaking rates should be high but adjusted so that splashing is not excessive. Rates of 100 to 250 rpm, depending on the flask size, are usually best. For larger flasks, baffles improve aeration efficiency.

The data in Table 5 show how the oxygen adsorption rates (OAR) vary with the type of aeration (131). By far the best aeration is achieved with a stirred and sparged fermentor. Shaking is essential to achieve reasonable aeration with flasks. Shaking a minute amount of medium in a large flask will produce a very good OAR, but such volumes are impractical, owing to evaporation. A 50-ml volume allows sufficient aeration for screening studies. More convenient sizes of flasks for screening studies are 125 and 250 ml. The recommended volumes for these flasks are 25 to 50 ml.

Measurement of dissolved oxygen has been facilitated by the development of the oxygen electrode (41, 67, 87, 116). In scaled-up procedures it is important to match aeration efficiencies, if possible.

OAR method (31) for measurement of aeration

This method is based on the oxidation of sulfite by molecular oxygen.

1. Add the desired volume of sodium sulfite solution (0.1 M) to the vessel.

2. Commence shaking or sparging and stirring.

3. Withdraw 5-ml volumes at appropriate intervals and pipette into test tubes (1 by 10 in.; ca. 2.5 by 25 cm) containing small pellets of dry ice. The resulting cloud of carbon dioxide over the sample prevents further oxidation of the sulfite.

4. Add 2 drops of starch indicator and titrate with standard iodine solution.

5. Calculate OAR according to the equation: OAR = [(ml of titration difference × normality iodine solution)/4] × (1000/5) × min^{-1}.

The importance of oxygen levels to the growth of organisms used in biotransformation has been discussed above. Also, oxygen is a substrate in many important biocatalytic reactions, such as aromatic hydroxylation, *N*-dealkylation, *O*-dealkylation, and sulfur oxidation. These reactions are catalyzed by monooxygenases, dioxygenases, and other enzymes that activate molecular oxygen. The dynamics of medium oxygenation are important to all these types of biotransformations, whether the reactions are performed with growing cells, dried cells, or enzyme preparations (4, 124).

9.10.2 Nutritional Parameters

The nutritional components of a culture medium should be tested for effect one at a time, if possible. Sampling times for monitoring the effect on biocatalytic reactions should be arranged to account for possible changes in growth kinetics. The concentration and type of the carbon source should be evaluated first by testing a number of different sources at the same concentration, then at different concentrations. Carbon sources to compare include glucose, other carbohydrates, and glycerol. Citric acid cycle intermediates are good candidates, as are pyruvate and acetate. Combinations of carbon sources can be very effective for improvement of cell growth and enzyme induction.

Variations in medium components should be checked in the order of their decreasing concentration. After carbon, the next most abundant nutrient is the nitrogen source. The first nitrogen compounds to compare are the simple organic salts, ammonium sulfate, ammonium nitrate, and potassium nitrate. After this, urea, glutamate, asparagine, and glutamine should be tested, then the various complex nitrogen sources such as yeast extracts, peptones, and tryptones. Combinations of inorganic salts may then be tried, followed by vitamins, purines and pyrimidines, amino acids, sulfur and phosphorus sources, and the various required inorganic salts and trace elements.

A discussion of the nutritional improvement of processes is given by Greasham and Inamine in chapter 4 of this manual.

9.11 LITERATURE CITED

1. **Aaronson, S.** 1978. Enrichment culture, p. 759–769. *In* A. I. Laskin and H. A. Lechevalier (ed.), CRC handbook of microbiology, vol. 2. CRC Press, Inc., Boca Raton, Fla.
2. **Abbott, B. J.** 1976. Preparation of pharmaceutical compounds by immobilized enzymes and cells. Adv. Appl. Microbiol. **20:**203–257.
3. **Abe, K., M. Onda, H. Isaka, and S. Okuda.** 1981. Studies on morphine alkaloids. VII. Microbial transformation of 14-β-bromocodeinone. Chem. Pharm. Bull. **18:**2070–2073.
4. **Adlercreutz, P., and B. Mattiasson.** 1982. Oxygen supply to immobilized cells. Eur. J. Appl. Microbiol. Biotechnol. **16:**165–170.
5. **Aiba, S., A. E. Humphrey, and N. F. Millis.** 1973. Biochemical engineering. Academic Press, Inc., New York.
6. **Ainsworth, G. C., P. W. James, and B. B. Hawksworth.** 1971. Dictionary of the fungi. Commonwealth Mycological Institute, New Surrey, England.
7. **Alexopoulos, C. J., and C. W. Mims.** 1979. Introductory mycology, 3rd ed. John Wiley & Sons, Inc., New York.
8. **American Type Culture Collection.** 1982. American Type Culture Collection catalogue, 15th ed. American Type Culture Collection, Rockville, Md.
9. **Antonini, E., G. Carrea, and P. Cremonesi.** 1981. Enzyme catalyzed reactions in water—organic solvent two-phase systems. Enzyme Microbiol. Technol. **3:**291–296.
10. **Ashwood-Smith, M.** 1965. On the genetic stability of bacteria to freezing and thawing. Cryobiology **2:**39–43.
11. **Atkinson, B.** 1974. Biochemical reactors. Pion Limited, London.
12. **Beck, J. V.** 1971. Enrichment culture and isolation techniques particularly for anaerobic bacteria. Methods Enzymol. **22:**57–70.
13. **Beukers, R., A. F. Marx, and M. H. J. Zuidweg.** 1972. Microbial conversion as a tool in the preparation of drugs, p. 1–31. *In* E. J. Ariens (ed.), Drug design, vol. 3. Academic Press, Inc., New York.
14. **Blunt, J. W., I. M. Clark, J. M. Evans, E. R. H. Jones, G. D. Meakins, and J. T. Pinkey.** 1971. Microbiological hydroxylation of steroids. III. A convenient microbiological route to 15-oxygenated 5α-androstanes. J. Chem. Soc. Sect. C **1971:**1136–1138.
15. **Bobbitt, J. M., A. E. Schwarting, and R. J. Gritten.** 1968. Introduction to chromatography. Van Nostrand Reinhold Co., Inc., New York.
16. **Booth, C.** 1971. Fungal culture media, p. 49–94. *In* C. Booth (ed.), Methods in microbiology, vol. 4. Academic Press, Inc., New York.

17. **Bridson, E. Y., and A. Brecker.** 1970. Design and formulation of microbial culture media, p. 229–295. *In* J. R. Norris and D. W. Ribbons (ed.), Methods in microbiology, vol. 3A. Academic Press, Inc., New York.
18. **Brown, D. E.** 1970. Aeration in the submerged culture of microorganisms, p. 125–174. *In* J. R. Norris and D. W. Ribbons (ed.), Methods in microbiology, vol. 2. Academic Press, Inc., New York.
19. **Brown, P. R.** 1973. High-pressure liquid chromatography: biochemical and biomedical applications. Academic Press, Inc., New York.
20. **Buchanan, R. E., and N. E. Gibbons (ed.).** 1974. Bergey's manual of determinative bacteriology, 8th ed. The Williams & Wilkins Co., Baltimore.
21. **Butler, L. G.** 1979. Enzymes in nonaqueous solvents. Enzyme Microb. Technol. **1:**253–259.
22. **Calam, C. T.** 1969. The culture of microorganisms in liquid culture, p. 255–326. *In* J. R. Norris and D. W. Ribbons (ed.), Methods in microbiology, vol. 1. Academic Press, Inc., New York.
23. **Charney, W., and H. L. Herzog.** 1967. Microbial transformations of steroids. Academic Press, Inc., New York.
24. **Chibata, I.** 1978. Immobilized enzymes, research and development. John Wiley & Sons, Inc., New York.
25. **Chibata, I.** 1979. Immobilized microbial cells with polyacrylamide gel and carageenan and their industrial applications, p. 187–202. *In* K. Venkatasubramanian (ed.), Immobilized microbial cells, vol. 2. ACS Symposium Series 106. American Chemical Society, Washington, D.C.
26. **Chibata, I., and T. Tosa.** 1977. Transformation of organic compounds by immobilized microbial cells. Adv. Appl. Microbiol. **22:**1–27.
27. **Chien, M. M., and J. P. Rosazza.** 1980. Microbial transformation of natural antitumor agents: use of solubilizing agents to improve yields of hydroxylated ellipticines. Appl. Environ. Microbiol. **40:**741–745.
28. **Claridge, C. A., and H. Schmitz.** 1978. Microbial and chemical transformations of some 12,13-epoxytrichothec-9,10-enes. Appl. Environ. Microbiol. **36:**63–67.
29. **Codner, R. C.** 1969. Solid and solidified growth media in microbiology, p. 427–454. *In* J. R. Norris and D. W. Ribbons (ed.), Methods in microbiology, vol. 1. Academic Press, Inc., New York.
30. **Collins, C. H., and P. M. Lyne.** 1970. Microbiological methods, 3rd ed. Butterworths, London.
31. **Corman, J., H. M. Tsuchiya, H. J. Koepsell, R. G. Benedict, S. E. Kelley, V. H. Feger, R. G. Dworschack, and R. W. Jackson.** 1957. Oxygen adsorption rates in laboratory and pilot plant equipment. Appl. Microbiol. **5:**313–318.
32. **Cremonesi, P., G. Carrea, L. Ferrara, and E. Antonini.** 1974. Enzymatic dehydrogenation of testosterone coupled to pyruvate reduction in a two-phase system. Eur. J. Biochem. **44:**401–405.
33. **Cremonesi, P., G. Carrea, L. Ferrara, and E. Antonini.** 1975. Enzymatic preparation of 20-β-hydroxysteroids in a two-phase system. Biotechnol. Bioeng. **17:**1101–1108.
34. **Daily, W. A., and C. E. Higgens.** 1973. Preservation and storage of microorganisms in the gas phase of liquid nitrogen. Cryobiology **10:**364–367.
35. **Davis, B. D., and E. S. Mingioli.** 1950. Mutants of *Escherichia coli* requiring methionine or vitamin B_{12}. J. Bacteriol. **60:**17–28.
36. **Davis, P. J., M. E. Gustafson, and J. P. Rosazza.** 1976. Formation of indole-3-carboxylic acid by *Chromobacterium violaceum*. J. Bacteriol. **126:**544–546.
37. **Demain, A. L., and J. Birnbaum.** 1968. Alteration of permeability for the release of metabolites from the microbial cell. Curr. Top. Microbiol. Immunol. **46:**1–25.
38. **Deo, Y. M., and G. M. Gaucher.** 1983. Semicontinuous production of antibiotic patulin by immobilized cells of *Penicillium urticae*. Biotechnol. Lett. **5:**124–130.
39. **Dietz, A.** 1981. Pure culture methods for industrial microorganisms, p. 411–434. *In* H.-J. Rehm and G. Reed (ed.), Biotechnology, vol. 1. Verlag Chemie, Weinheim, Federal Republic of Germany.
40. **Eckenrode, F., W. Peczynska-Czoch, and J. P. Rosazza.** 1982. Microbial transformations of natural antitumor agents. 18. Conversions of vindoline with copper oxidases. J. Pharm. Sci. **71:**1246–1250.
41. **Elsworth, R.** 1972. The value and use of dissolved oxygen measurement in deep culture. Chem. Eng. (London) **258:**63–71.
42. **Feren, C. J., and R. W. Squires.** 1969. The relationship between critical oxygen level and antibiotic synthesis of capreomycin and cephalosporin C. Biotechnol. Bioeng. **11:**583–592.
43. **Finn, R. K.** 1967. Agitation and aeration. Biochem. Biol. Eng. Sci. **1:**69–99.
44. **Fischli, A. E.** 1980. Chiral building blocks in enantiomer synthesis using enzymatic transformations, p. 268–350. *In* R. Scheffold (ed.), Modern synthetic methods. Conference Papers International Seminar. Otto Salle Verlag, Frankfurt/Main.
45. **Freedman, D.** 1969. The shaker in bioengineering. Process Biochem. **4:**35–40.
46. **Fukui, S., K. Sonomoto, N. Itoh, and A. Tanaka.** 1980. Several novel methods for immobilization of enzymes, microbial cells, and organelles. Biochimie **62:**381–386.
47. **Fukui, S., and A. Tanaka.** 1982. Immobilized microbial cells. Annu. Rev. Microbiol. **36:**145–172.
48. **Fukui, S., A. Tanaka, and G. Gell.** 1978. Immobilization of enzymes, microbial cells, and organelles by inclusion with photo-crosslinkable resins. Enzyme Eng. **4:**294–306.
49. **Fukui, S., A. Tanaka, T. Iida, and E. Hasagawa.** 1976. Application of photo-crosslinkable resin to immobilization of an enzyme. FEBS Lett. **66:**179–182.
50. **Fukushima, S., T. Nagai, K. Fujita, A. Tanaka, and S. Fukui.** 1978. Hydrophilic urethane prepolymers: convenient materials for enzyme entrapment. Biotechnol. Bioeng. **20:**1465–1469.
51. **Gerhardt, P. (ed.).** 1982. Manual of methods for general bacteriology. American Society for Microbiology, Washington, D.C.
52. **Ghisalba, O., R. Roos, T. Schupp, and J. Nuesch.** 1982. Transformation of rifamycin S into rifamycins B and L. A revision of the current biosynthetic hypothesis. J. Antibiot. **35:**74–80.
53. **Goodhue, C. T.** 1982. The methodology of microbial transformation of organic compounds, p. 9–44. *In* J. P. Rosazza (ed.), Microbial transformation of bioactive agents. CRC Press, Inc., Boca Raton, Fla.
54. **Goos, R. D., E. E. Davis, and W. Butterfield.** 1967. Effect of warming rates on the viability of frozen fungus spores. Mycologia **59:**58–66.
55. **Gray, T. R. G., and S. T. Williams.** 1975. Soil microorganisms. Longman Inc., New York.
56. **Gunsalus, I. C.** 1955. Extraction of enzymes from microorganisms. Methods Enzymol. **1:**51–62.
57. **Gunsalus, I. C., and R. Y. Stanier (ed.).** 1960. The bacteria. Academic Press, Inc., New York.
58. **Hamilton, P. B., D. Rosi, G. P. Peruzzotti, and E. D. Nielson.** 1969. Microbiological metabolism of naphthyridines. Appl. Microbiol. **17:**237–241.
59. **Hamilton, R. J., and P. A. Sewell.** 1978. Introduction to high-performance liquid chromatography. Chapman & Hall, Ltd., London.
60. **Haynes, W. C., L. J. Wickerham, and C. W. Hesseltine.** 1955. Maintenance of cultures of industrially important microorganisms. Appl. Microbiol. **3:**361–368.
61. **Herold, P., P. Mohr, and C. Tamm.** 1983. Synthesis of optically active verrucarinic acid. Helv. Chim. Acta **66:**744–754.

62. **Hicks, G. P., and S. J. Updike.** 1966. The preparation and characterization of lyophilized polyacrylamide enzyme gels for chemical analysis. Anal. Chem. **38:**726–730.
63. **Holland, H. L.** 1981. Microbial and *in vitro* enzymic transformation of alkaloids, p. 324–400. *In* R. B. A. Rodrigo (ed.), The alkaloids, vol. 18. Academic Press, Inc., New York.
64. **Hufford, C. D., C. C. Collins, and A. M. Clark.** 1979. Microbial transformations and ^{13}C-NMR analysis of colchicine. J. Pharm. Sci. **68:**1239–1243.
65. **Iizuka, H., and A. Naito.** 1967. Microbial transformations of steroids and alkaloids. University Park Press, Baltimore.
66. **Jakoby, W. B. (ed.).** 1971. Enzyme purification and related techniques. Methods Enzymol. **22.**
67. **Johnson, M. J., J. Borkowski, and C. Engblom.** 1964. Steam sterilizable probes for dissolved oxygen measurement. Biotechnol. Bioeng. **6:**457–468.
68. **Johnson, R. A., H. C. Murray, L. M. Reinke, and G. S. Fonken.** 1969. Stereochemistry of microbiological hydroxylation. 11. Oxygen of benzoyl alkylpiperidines. J. Org. Chem. **34:**2279–2284.
69. **Jones, D. F., R. H. Moore, and G. C. Crawley.** 1970. Microbial modification of mycophenolic acid. J. Chem. Soc. Sect. C **1970:**1725–1737.
70. **Jones, J. B., and N. Baskevitch.** 1973. Steroids and steroidases. 20 (1). Aggregation in aqueous solution of steroids with stigmastane type C-17 side chains and its influence on their enzymic transformations.
71. **Jones, J. B., C. J. Sih, and D. Perlman.** 1976. Applications of biochemical systems in organic chemistry, parts 1 and 2. Techniques in organic chemistry. John Wiley & Sons, New York.
72. **Jones, J. B., and K. E. Taylor.** 1976. Nicotinamide coenzyme regeneration. Flavin mononucleotide (riboflavin phosphate) as an efficient, economical, and enzyme-compatible recycling agent. Can. J. Chem. **54:**2969–2973.
73. **Karow, E. O., and D. N. Petsiavas.** 1956. Effect of physical variables on microbial transformation of steroids. Ind. Eng. Chem. **48:**2213–2217.
74. **Kennedy, J. F.** 1979. Facile methods for the immobilization of microbial cells without disruption of their life processes, p. 119–131. *In* K. Venkatasubramanian (ed.), Immobilized microbial cells. ACS Symposium Series 106. American Chemical Society, Washington, D.C.
75. **Kennedy, J. F., S. A. Barker, and J. D. Humphreys.** 1976. Microbial cells living immobilized on metal hydroxides. Nature (London) **261:**242–244.
76. **Kieslich, K.** 1976. Microbial transformation of nonsteroid cyclic compounds. John Wiley & Sons, New York.
77. **Klibanov, A. M.** 1983. Immobilized enzymes and cells as practical catalysts. Science **219:**722–727.
78. **Klibanov, A. M., G. P. Samokhin, K. Martinek, and I. V. Berezin.** 1977. A new approach to preparative enzymatic synthesis. Biotechnol. Bioeng. **19:**1351–1361.
79. **Kobayashi, T., G. van Dedem, and M. Moo-Young.** 1973. Oxygen transfer into mycelial pellets. Biotechnol. Bioeng. **15:**27–45.
80. **Koshcheyenko, K. A., M. V. Turkina, and G. K. Skryabin.** 1983. Immobilization of living microbial cells and their application for steroid transformations. Enzyme Microb. Technol. **5:**14–21.
80a. **Krieg, N. R., and J. R. Holt (ed.).** 1984. Bergey's manual of systematic bacteriology, vol. 1. Williams & Wilkins, Baltimore.
81. **Kurtzman, C. P.** 1980. Preservation of fungi by lyophilization and liquid-nitrogen freezing. A workshop, August 9–10. Society of Industrial Microbiology, Northern Arizona University, Flagstaff.
82. **Lapage, S. P., J. E. Shelton, and T. G. Mitchell.** 1970. Media for the maintenance and preservation of bacteria, p. 1–133. *In* J. R. Norris and D. W. Ribbons (ed.), Methods in microbiology, vol. 3A. Academic Press, Inc., New York.
83. **Laskin, A. I., and H. Lechevalier (ed.).** 1974. CRC handbook of microbiology, vol. 4. CRC Press, Inc., Cleveland, Ohio.
84. **Lee, B. K., W. E. Brown, D. Y. Ryu, H. Jacobson, and R. W. Thoma.** 1970. Influence of mode of steroid substrate addition on conversion of steroid and growth characteristics in a mixed culture fermentation. J. Gen. Microbiol. **61:**97–105.
85. **Lin, Y. Y., and L. L. Smith.** 1970. Microbial hydroxylation. 6. Hydroxylation of rac-17β-hydroxyestra-4,8(14)-dien-3-one by *Curvularia lunata*. Biochim. Biophys. Acta **210:**319–327.
86. **Linko, P., K. Poutanen, L. Weckstrom, and Y.-Y. Linko.** 1980. Preparation and kinetic behavior of immobilized whole cell biocatalysts. Biochimie **62:**387–394.
87. **Liu, M. S., R. M. R. Branion, and D. W. Duncan.** 1973. Determination of the solubility of oxygen in fermentation media. Biotechnol. Bioeng. **15:**213–216.
88. **Marsheck, W. J.** 1971. Current trends in the microbiological transformation of steroids. Prog. Ind. Microbiol. **10:**49–103.
89. **Marsheck, W. J., S. Kraychy, and R. D. Muir.** 1972. Microbial degradation of sterols. Appl. Microbiol. **23:**72–77.
90. **Martin, C. K. A., and D. Perlman.** 1972. Stimulation by organic solvents and detergents of conversion of L-sorbose to L-sorbosone by *Gluconobacter melanogenus* IFO 3293. Biotechnol. Bioeng. **17:**1473–1484.
91. **Martin, S. M., and V. B. D. Skerman.** 1972. World directory of collections of cultures of microorganisms. John Wiley & Sons, New York.
92. **Mattiasson, B. (ed.).** 1983. Immobilized cells and organelles, vol. 1 and 2. CRC Press, Boca Raton, Fla.
93. **Mazur, P.** 1960. Physical factors implicated in the death of microorganisms at subzero temperatures. Ann. N.Y. Acad. Sci. **85:**610–629.
94. **McDaniel, L. E., and E. G. Bailey.** 1969. Effect of shaking speed and type of closure on shake flask cultures. Appl. Microbiol. **17:**286–290.
95. **McDougall, B., P. Dunnill, and M. D. Lilly.** 1982. Enzymic acylation of 6-aminopenicillanic acid. Enzyme Microbiol. Technol. **4:**114–115.
96. **McGregor, W. C., B. Tabenkin, E. Jenkins, and R. Epps.** 1972. Pilot plant conversion of steroid using resting cell suspension as biocatalyst. Biotechnol. Bioeng. **14:**831–841.
97. **McNair, H. M., and E. J. Bonelli.** 1967. Basic gas chromatography. Varian Aerograph Co., Walnut Creek, Calif.
98. **Mitruka, B. M.** 1975. Gas chromatography applications in microbiology and medicine. John Wiley & Sons, New York.
99. **Mizrahi, A., and G. Miller.** 1969. Role of glycols and Tweens in the production of ergot alkaloids by *Claviceps paspali*. J. Bacteriol. **97:**1155–1159.
100. **Mohr, P., N. Waespe-Sarcevic, C. Tamm, K. Gawronska, and J. K. Gawronski.** 1983. A study of stereoselective hydrolysis of symmetrical diester with pig liver esterase. Helv. Chim. Acta **66:**2501–2511.
101. **Mosbach, K. (ed.).** 1976. Immobilized enzymes. Methods Enzymol. **44.**
102. **Mosbach, K., and P. O. Larsson.** 1970. Preparation and application of polymer entrapped enzymes and microorganisms in microbial transformation processes with special reference to steroid 11-β-hydroxylation and δ-1-dehydrogenation. Biotechnol. Bioeng. **12:**19–27.
103. **Nabih, T., P. Davis, J. Caputo, and J. Rosazza.** 1977. Microbial transformations of natural antitumor agents. 3. Conversions of thalicarpine to (+)-hernandalinol by *Streptomyces punipalus*. J. Med. Chem. **20:**914–917.

104. **Nagasawa, M., N. Watanabe, H. Hashiba, M. Murakami, M. Bae, G. Tamura, and K. Arima.** 1970. Microbial transformation of sterols. V. Inhibitors of microbial degradation of cholesterol. Agric. Biol. Chem. **34**:838–844.
105. **Nakamatsu, T., T. Beppu, and K. Arima.** 1983. Microbial production of 3α-*H*-4α-(3′-propionic acid)-5α-hydroxy-7β-methylhexahydro-1-indanone-δ-lactone from soybean sterol. Agric. Biol. Chem. **47**:1449–1454.
106. **Nakayama, K.** 1981. Sources of industrial microorganisms, p. 357–410. *In* H.-J. Rehm and G. Reed (ed.), Biotechnology, vol. 1. Verlag Chemie, Weinheim, Federal Republic of Germany.
107. **Nishida, Y., T. Sato, T. Tosa, and I. Chibata.** 1979. Immobilization of *Escherichia coli* cells having aspartase activity with carrageenan and locust gum. Enzyme Microb. Technol. **1**:95–99.
108. **Ohlson, S., S. Flygare, P. O. Larsson, and K. Mosbach.** 1980. Steroid hydroxylation using immobilized spores of *Curvularia lunata* germinated *in situ*. Eur. J. Appl. Microbiol. Biotechnol. **10**:1–9.
109. **Omura, S., N. Sadakane, and H. Matsubara.** 1982. Bioconversion and biosynthesis of 16-membered macrolide antibiotics. 22. Biosynthesis of tylosin after protylonolide formation. Chem. Pharm. Bull. **30**:223–229.
110. **Onions, A. H. S.** 1971. Culture collections of micro fungi. Biol. J. Linn. Soc. **3**:189–196.
111. **Parris, N. A.** 1976. Instrumental liquid chromatography: a practical method. Elsevier Science Publishing, Inc., New York.
112. **Peppler, H. J., and D. Perlman.** 1979. Microbial technology, vol. 1 and 2. Academic Press, Inc., New York.
113. **Perlman, D.** 1976. Procedures useful in studying microbial transformations of organic compounds, p. 47–68. *In* J. B. Jones, C. J. Sih, and D. Perlman (ed.), Techniques of chemistry, vol. 10, part 1. John Wiley & Sons, New York.
114. **Perry, J. J.** 1979. Microbial cooxidations involving hydrocarbons. Microbiol. Rev. **43**:59–72.
115. **Petroski, R. J., W. Peczynska-Czoch, and J. P. Rosazza.** 1980. Analysis, production, and isolation of an extracellular laccase from *Polyporus anceps*. Appl. Environ. Microbiol. **40**:1003–1006.
116. **Phillips, D. H., and M. J. Johnson.** 1961. Measurement of dissolved oxygen in fermentations. J. Biochem. Microbiol. Technol. Eng. **3**:261–275.
117. **Phillips, D. H., and M. J. Johnson.** 1961. Aeration in fermentations. J. Biochem. Microbiol. Technol. Eng. **3**:277–309.
118. **Reed, G. (ed.).** 1981. Prescott and Dunn's industrial microbiology, 4th ed. Avi Publishing Co., Westport, Conn.
119. **Ribbons, D. W.** 1970. Quantitative relationships between growth media constituents and cellular yields and composition, p. 297–304. *In* J. R. Norris and D. W. Ribbons (ed.), Methods in microbiology, vol. 3A. Academic Press, Inc., New York.
120. **Ridley, D. D., and M. Stralow.** 1975. The stereospecific asymmetric reduction of functionalized ketones. J. Chem. Soc. Chem. Commun. **1975**:400.
121. **Rinfret, A. P., and B. LaSalle.** 1975. Round table conference on the cryogenic preservation of cell cultures. National Academy of Sciences, Washington, D.C.
122. **Rosazza, J. P., and R. V. Smith.** 1979. Microbial models for drug metabolism. Adv. Appl. Microbiol. **25**:169–208.
123. **Rose, A. H.** 1980. History and scientific basis of commercial exploitation of microbial enzymes and bioconversions, p. 1–47. *In* Economic microbiology, vol. 5. Microbial enzymes and bioconversions. Academic Press, Inc., New York.
124. **Sato, K., and K. Toda.** 1983. Oxygen uptake rate of immobilized growing *Candida lipolytica*. J. Ferment. Technol. **61**:239–245.
125. **Sato, T., T. Mori, T. Tosa, I. Chibata, M. Furui, K. Yamashita, and A. Sumi.** 1975. Engineering analysis of continuous production of L-aspartic acid by immobilized *Escherichia coli* cells in fixed beds. Biotechnol. Bioeng. **17**:1797–1804.
126. **Sato, T., Y. Nishida, T. Tosa, and I. Chibata.** 1979. Immobilization of *Escherichia coli* cells containing aspartase activity with κ-carrageenan. Enzyme properties and application for L-aspartic acid production. Biochem. Biophys. Acta **570**:179–186.
127. **Schnaitman, C. A.** 1981. Cell fractionation, p. 52–61. *In* P. Gerhardt (ed.), Manual of methods for general bacteriology. American Society for Microbiology, Washington, D.C.
128. **Shibata, M., M. Uyeda, and S. Mori.** 1976. Microbial transformations of antibiotics. 2. Additional transformation products of maridomycin 3. J. Antibiot. **29**:824–828.
129. **Silman, I. H., and E. Katchalski.** 1966. Water-insoluble derivatives of enzymes, antigens, and antibodies. Annu. Rev. Biochem. **35**:873–908.
130. **Skerman, V. B. D.** 1967. A guide to the identification of the genera of bacteria, 2nd ed. The Williams & Wilkins Co., Baltimore.
131. **Smith, C. G., and M. J. Johnson.** 1954. Aeration requirements for the growth of aerobic microorganisms. J. Bacteriol. **68**:346–350.
132. **Smith, J. E., and D. R. Berry.** 1975. The filamentous fungi. Industrial mycology, vol. 1. Halsted Press, John Wiley & Sons, New York.
133. **Smith, R. V., D. Acosta, Jr., and J. P. Rosazza.** 1977. Microbial and cellular models of mammalian metabolism. Adv. Biochem. Eng. **5**:70–100.
134. **Snyder, L. R., and J. J. Kirkland.** 1979. Introduction to modern liquid chromatography, 2nd ed. John Wiley & Sons, New York.
135. **Solomons, G. L.** 1969. Materials and methods in fermentation. Academic Press, Inc., New York.
136. **Sonomoto, K., M. M. Hoq, A. Tanaka, and S. Fukui.** 1983. 11β-Hydroxylation of cortexolone (Reichstein compound S) to hydrocortisone by *Curvularia lunata* entrapped in photo-cross-linked resin gels. Appl. Environ. Microbiol. **45**:436–443.
137. **Stahl, E. (ed.).** 1969. Thin layer chromatography. A laboratory handbook, 2nd ed. Springer-Verlag, New York.
138. **Stanier, R. Y., E. A. Adelberg, and J. C. Ingraham.** 1976. The microbial world, 4th ed. Prentice Hall, Inc., Englewood Cliffs, N.J.
139. **Stewart, J. B.** 1974. Methods of media preparation for the biological sciences. Charles C Thomas, Springfield, Ill.
140. **Takata, I., T. Tosa, and I. Chibata.** 1977. Screening of matrices suitable for immobilization of microbial cells. J. Solid-Phase Biochem. **2**:225–236.
141. **Tempest, D. W.** 1970. The continuous cultivation of microorganisms. 1. Theory of the chemostat 12, p. 259–276. *In* J. R. Norris and D. W. Ribbons (ed.), Methods in microbiology, vol. 2. Academic Press, Inc., New York.
142. **Tosa, T., T. Sato, T. Mori, K. Yamamoto, I. Takata, Y. Nishida, and I. Chibata.** 1979. Immobilization of enzymes and microbial cells using carrageenan as matrix. Biotechnol. Bioeng. **21**:1697–1709.
143. **Veldkamp, H.** 1970. Enrichment cultures of prokaryotic organisms, p. 305–361. *In* J. R. Norris and D. W. Ribbons (ed.), Methods in microbiology, vol. 3A. Academic Press, Inc., New York.
144. **Vézina, C., and K. Singh.** 1975. Transformation of organic compounds by fungal spores, p. 158–192. *In* J. E. Smith and D. R. Berry (ed.), The filamentous fungi, vol. 1. Edward Arnold, London.

145. **Vieth, W. R., S. S. Wang, and R. Saini.** 1973. Immobilization of whole cells in a membranous form. Biotechnol. Bioeng. **15:**565–570.
146. **Walker, J. C.** 1957. Plant pathology. McGraw-Hill Book Co., New York.
147. **Wallen, L. L., F. H. Stodola, and R. W. Jackson.** 1959. Type reactions in fermentation chemistry. Bull. ARS-71-13. Agriculture Research Service, U.S. Department of Agriculture, Washington, D.C.
148. **Wang, D. I. C., C. L. Cooney, A. L. Demain, P. Dunnill, A. E. Humphrey, and M. D. Lilly.** 1979. Fermentation and enzyme technology. John Wiley & Sons, New York.
149. **Weaver, E. A., H. E. Kenney, and M. E. Wall.** 1960. Effect of concentration on microbiological hydroxylation of progesterone. Appl. Microbiol. **8:**345–348.
150. **Webster, J.** 1970. Introduction to fungi. Cambridge University Press, Cambridge.
151. **Whitesides, G. M., and C. H. Wong.** 1983. Enzymes as catalysts in organic synthesis. Aldrichim. Acta **16:**27–34.
152. **Wimpenny, J. W. T.** 1969. Oxygen and microbial metabolism. Process Biochem. **1969:**19–22.
153. **Wodzinski, R. S., and J. E. Coyle.** 1974. Physical state of phenanthrene for utilization by bacteria. Appl. Microbiol. **27:**1081–1084.
154. **Wovcha, M. G., F. J. Antosz, J. C. Knight, L. A. Kominek, and T. R. Pyke.** 1978. Bioconversion of sitosterol to useful steroidal intermediates by mutants of *Mycobacterium fortuitum*. Biochim. Biophys. Acta **531:**308–321.
155. **Yokozeki, K., S. Yamanaka, K. Takinami, Y. Hirose, A. Tanaka, K. Sonomoto, and S. Fukui.** 1982. Application of immobilized lipase to regiospecific interesterification of triglyceride in organic solvent. Eur. J. Appl. Microbiol. Biotechnol. **14:**1–5.
156. **Zaborsky, O. R. (ed.).** 1973. Immobilized enzymes. CRC Press, Inc., Cleveland, Ohio.
157. **Zajic, J. E., and C. J. Panchal.** 1976. Bio-emulsifiers. Crit. Rev. Microbiol. **5:**39–66.
158. **Zaks, A., and A. M. Klibanov.** 1984. Enzymatic catalysis in organic media at 100°C. Science **224:**1249–1251.
159. **Zhou, B., A. S. Gopalan, F. VanMiddlesworth, W. R. Shieh, and C. J. Sih.** 1983. Stereochemical control of yeast reductions. 1. Asymmetric synthesis of L-carnitine. J. Am. Chem. Soc. **105:**5925–5926.

Chapter 10

Substrates for Large-Scale Fermentations

THOMAS L. MILLER AND BRUCE W. CHURCHILL

Reviews of fermentation media are usually organized in one of two ways: (i) a listing of different media that have been described in the literature for various procedures or (ii) a simple listing of various kinds of medium components, i.e., sources of carbon, nitrogen, minerals, growth factors, etc. The following discussion will take neither of these forms. Rather, an attempt will be made to define the rationale for media designed for various purposes, e.g., production of cells, antibiotics, steroids, solvents, etc. A few examples of media are given only to serve as illustrations and should not be construed as being recommended for any specific use.

The design of media for large-scale fermentations has historically been much more a matter of art than science. The choices of medium ingredients have been based almost solely on factors other than what is best for synthesis of a desired product. These factors include material cost and availability, location of the fermentation plant, factors affecting the prices of commodity products and their by-products, etc. However, it is possible in many cases to design media based on more scientific considerations. A discussion of the development of a chemically defined medium will provide an insight into how media may be designed if some thought is given to microbial metabolism and the nature of the desired product. This will be followed by a discussion of medium ingredients used in the production of several categories of fermentation products.

10.1 CHEMICALLY DEFINED MEDIA

A large number of aerobic and facultative organisms will grow on simple chemically defined media that do not contain growth factors or other complex ingredients. Since the elemental composition of most microorganisms is very similar (23), one may logically design media based upon this analysis. Thus, the medium is composed of the elements found in the "typical" cell in the ratios in which they are present in the cell. Cellular composition depends on the cell type, i.e., bacterium, yeast, or mold, and usually falls within the following ranges (percent on a dry basis): carbon, 45 to 55%; nitrogen, 6 to 14%; potassium, 0.5 to 2%; phosphorus, 1 to 3%; magnesium, 0.1 to 1%; sulfur, 0.02 to 1%; and calcium, trace to 1%. Minor minerals are also present at (dry weight per 100 g): copper, 0.1 to 1 mg; iron, 1 to 10 mg, zinc, ~1mg; and manganese, 0 to 5 mg.

With the above information, a defined medium may be designed that will give any desired cell concentration (within obvious limits). The first requirement is to choose the limiting nutrient. Carbon is usually selected for the purpose since growth will normally stop when the carbon source is depleted, whereas depletion of other nutrients sometimes results in abnormal growth, e.g., lipid production under conditions of limiting nitrogen. Therefore, let it be assumed that the desired cell concentration is 10 g/liter and the carbon source is glucose. Under aerobic conditions, cell yields are about 45 to 50% (dry basis) with glucose as the carbon source. Thus, the medium given in Table 1 was derived using the elemental composition of bacterial cells as a guide. The group A components contain all needed nutrients except nitrogen. The group B ingredients provide nitrogen in the form of the ammonium ion, which is the preferred nitrogen source for almost all microorganisms, and the phosphate anions provide some buffering capacity. The function of group C, a chelator, is to keep the metal ions in solution; this also prevents their toxic effects, especially when small inocula are used. The buffering properties of group D components are necessary to maintain a pH reasonably close to 7.0 (optimal for most bacteria) because of the acidity that is produced as a result of the uptake of the ammonium ion. When higher cell concentrations are desired, the medium components must be adjusted proportionately. However, in this case the pH is usually controlled by the addition of a base, for example ammonium hydroxide, rather than by the use of buffers, because the high

TABLE 1. Example of a chemically defined medium

Constituent	Purpose	Concn (g/liter)
Group A		
Glucose	C, energy	20
KH_2PO_4	K, P	1.0
$MgSO_4 \cdot 7H_2O$	Mg, S	0.4
$CaCl_2$	Ca	0.03
$Fe_2(SO_4)_3$	Fe	12×10^{-4}
$ZnSO_4 \cdot 7H_2O$	Zn	4×10^{-4}
$CuSO_4 \cdot 5H_2O$	Cu	4×10^{-4}
$MnSO_4 \cdot H_2O$	Mn	4×10^{-4}
Group B		
$(NH_4)_2HPO_4$	N	4
$(NH_4)H_2PO_4$	N	3
Group C		
$C_6H_5Na_3O_7 \cdot 2H_2O$	Chelator	4
Group D		
Na_2HPO_4	Buffer	20
KH_2PO_4	Buffer	10

concentrations of buffers needed to obtain high cell densities would probably be toxic to the organisms. The pH must be controlled within the proper ranges depending on the type of organism being grown, namely, pH 4.5 to 7 for molds, pH 4.5 to 6 for yeasts, or 6.0 to 7.5 for bacteria.

When cells are grown in flasks, oxygen, a most important nutrient, is usually supplied by shaking the flask. With 100 ml of medium in 500-ml flasks on a rotary shaker, oxygen uptake rates of 0.3 to 9.5 mmol of oxygen per liter per min may be achieved, depending on the type of flask used, e.g., plain or baffled. Sparged, aerated, and agitated fermentors give oxygen uptake rates of 0.1 to 10 mmol of oxygen per liter per min.

Table 1 and this discussion, taken largely from the unpublished notes of the late Marvin J. Johnson, are intended to illustrate a scientific approach to medium design. The desired product may be either the cells themselves or some cellular component, e.g., a recombinant DNA product. When the product is a metabolite such as an antibiotic, complex media are almost always used. In these media, crude natural materials serve as sources of carbon, nitrogen, cofactors, and energy. Supplementation with additional minerals may or may not be required.

In the production of fermentation products on a large scale, economics play a key role in medium development. With antibiotic fermentation, inexpensive, abundant, and readily available raw materials are usually required because the products are usually low cost and exotic, and expensive raw materials are prohibitive. However, with steroid fermentations, the products are relatively high-cost items and yields become the predominant consideration. Therefore, higher-priced raw materials may be used in the medium for steroid bioconversion, and in some cases chemically defined media are used. In most conventional fermentations, relatively inexpensive raw materials that are by-products of grain, meat, and fiber processing usually serve as key medium ingredients. However, within the newly developing field of biotechnology, very different and sometimes expensive media will be required in some cases, such as for the growth of mammalian and plant cells.

10.2 ANTIBIOTICS

The studies necessary to design an improved medium for the production of an antibiotic are very important since only small to moderate increases in the level of production can result in (i) the annual savings of hundreds of thousands of dollars with major antibiotics, (ii) substantial cost reduction in the selling prices of the antibiotic, or (iii) an actual reduction in the cost of producing an antibiotic so that it can economically be sold in competition with other antibiotics. There is little published literature on the complex substrates that have been developed for the production of the various antibiotics. Most fermentation processes, which include the production of fermentation media, are closely guarded trade secrets. The methods that are used in developing complex media, the choice of nutrients which must be made, and the related fermentation conditions which must be studied will be described.

The fermentation medium designed for the initial production of an antibiotic usually does not have to be developed very skillfully since the potential for antibiotic production is quite low with wild-type strains. However, the fermentation media for ultra-high antibiotic-producing strains which have been developed through repeated genetic manipulations must be formulated with the utmost care. In the past, strain improvement and media development were the responsibilities of different research groups. Today we know that each higher-producing clone, after mutation and screening, requires a medium optimized for its performance. Its true potential can be realized only after another round of media studies in which the nutrients for the new strain have been "fine tuned," balanced, and supplemented. Sometimes the improved culture may have additional requirements (auxotrophic) not present in the parent culture.

The nutrients in an antibiotic production medium are usually different from those used in the primary and secondary seed stages. The seed stages are designed to give rapid and reproducible growth without nutrient depletion, autolysis, or an adverse change in pH. There is less concern with the cost of ingredients in the seed stages since the volume is usually only 5% of the fermentation volume and excellent uniformity of medium ingredients is highly desirable. Some specially prepared dairy products have been used quite extensively in primary and secondary seeds. The complex nitrogen sources which are available and desirable for antibiotic fermentation media fall into a few classes. The most common nitrogen sources are grain and bean commodity products, which include soybean meal and flour, cottonseed flour, corn steep liquor and powder, peanut meal, barley malt meal, corn gluten meals, dried distillers' solubles, linseed meal, rice meal, and wheat meal. The next most commonly used complex nitrogen sources are meat and fish waste products such as blood meal, various fish meals, fish hydrolysates, and meat and bone meal. Some very crude products from this group have occasionally been used. Years ago there was a product called "animal stick liquor," which was readily available in large quantities but was very variable in quality. There is also a relatively large group of dairy by-products, but, other than whey, most are too expensive to be used as a major nitrogen source. It should be stressed that, in most complex fermentation media for antibiotic production, part of the nitrogen is usually supplied as ammonium sulfate, chloride, or occasionally nitrate. A combination nitrogen, vitamin, and growth factor source that is used frequently in many fermentation media is brewers yeast or, occasionally, torula yeast. Yeasts are always added at much lower levels than the major nitrogen source. The composition of amino acids in complex nitrogen sources is seldom very important, but the overall level of protein is extemely important.

In general, two or more complex nitrogen sources can be used together, but medium development studies are much more difficult if the complex nitrogen sources are mixed. The three most important features to consider when choosing nitrogen sources are (i) uniform composition from one lot to another, (ii) stability of formulation as it is affected by moisture, temperature changes, and aging, and (iii) relative cost (a desirable nitrogen source would cost about $0.40/lb with 50% protein). It is sometimes found that unusual complex nitrogen sources are extremely effective, but the reason they are so beneficial for high antibiotic production is seldom known. One anticancer antibiotic was found to be produced in significant levels only with sesame seed meal. Pharmamedia is very effective for the production of lincomycin; the only other effective complex nitrogen source was castor pomace (a residue of castor beans after castor oil removal), which is now unavailable. The use of fish protein has been found to be essential for the production of high levels of some antibiotics, although the fish protein is not the major protein source in those fermentations. An excellent chart of the composition of nitrogen sources, from *Traders Guide to Fermentation Media Formulations* (25), appears in appendix A.

A great variety of carbon source samples are available for testing from the sample stocks maintained by most pharmaceutical companies. Slow-feed carbon sources are increasingly used. Many antibiotic-producing cultures, such as those producing neomycin, do not synthesize high levels of antibiotic while significant levels of glucose remain in the medium. It is not unusual to use mixed carbon sources so that rapid growth initially occurs on glucose supplied in the medium mix. Then, by the action of α- and β-amylases, glucose is slowly released from starch during antibiotic production. Frequently oil, e.g., lard oil or soybean oil, is supplied during the extended fermentation cycle for continued antibiotic production. Cultures that do not utilize triglycerides or fatty acids to produce an antibiotic are often slowly fed glucose continuously during the latter part of the fermentation cycle. The most commonly used carbohydrates are lactose, glucose (usually supplied as the monohydrate), starch, sucrose, maltose, dextrins, and sometimes xylose. Many fermentation media incorporate whey, which contains lactose, corn flour for its starch, molasses for sucrose, and products such as Enzose, which are cheap and contain a mixture of carbohydrates, largely glucose. The production of an antibiotic almost always closely parallels carbon source utilization so that one of the first tasks is to select a carbon source. It should be pointed out that with the cheap carbohydrates, medium development is much more difficult and variations in antibiotic titers are much more erratic than with refined carbohydrates. Fortunately, a complex fermentation medium can usually be designed for the production of an antibiotic for a short cycle, and after conditions are optimized the cycle can be significantly lengthened.

Continuous or intermittent feeding of a soluble inorganic nitrogen source as an ammonium salt or a water-soluble hydrolysate of a nonsoluble protein, in addition to either a carbohydrate or an oil, is common. In some processes acetic, citric, or another organic acid is fed, but the most commonly fed carbon sources are oils. Oils (saturated and unsaturated fatty acids) are widely used as cosubstrates in antibiotic fermentation media. The oils act as slow-feed carbon sources, they are reasonably priced in comparison with their nutrient value, they can be readily fed continuously or in "shots" to a fermentation, they tend to prevent excessive foaming, and they appear in some cases to protect the fragile microbial cells in the fermentor. The use of oils does not cause depression of growth or antibiotic production as is frequently encountered when glucose is added. Oils have found widespread use in the production of penicillins, tetracyclines, streptomycin, erythromycin, cephalosporins, and lincomycin, to name only a few. The common oils are lard, soybean, corn, cottonseed, menhaden or other fish oils, peanut, palm, safflower, and more recently sunflower oil. Sometimes individual saturated or unsaturated fatty acids are fed with pumps to supply carbon source as well as to control pH within the optimum antibiotic production range. A methyl oleate-based fermentation medium, supporting high levels of cephalosporin C production, was reported (17). Methyl oleate was fed continuously, but the feed rate was varied during the stages of the fermentation cycle. All four previously used carbohydrates were replaced with methyl oleate, and the actual residual level needed for optimum production was less than 0.05% (wt/vol).

Supplementing the complex fermentation media with minerals for antibiotic production becomes extremely critical as the level of production increases. Generally, somewhat excessive levels of added minerals do not substantially lower antibiotic production. A nitrogen source such as corn steep liquor can be used; this contains a large variety of minerals as well as amino acids, vitamins, growth factors, purines, and pyrimidines (13). In many complex-medium studies, however, the optimum effect is the result of very carefully balancing the minerals present in water and raw materials and those added as salts. Workers at the Upjohn Co. developed a medium that gave a fourfold increase in antibiotic production, compared with the previously used medium, simply by balancing the minerals in the medium (unpublished data). In such cases the use of a medium ingredient such as corn steep liquor is worthless since the level of any mineral from one lot of corn steep liquor to another may vary as much as 10-fold (13).

Generally the minerals in a medium are separated into major and minor classes by specifying that major minerals are required at levels greater than 0.2 g/liter and minor or trace minerals are needed at levels less than 25 mg/liter. The major minerals most often added are sulfates or chlorides of calcium, potassium, sodium, magnesium, and mono- or dibasic phosphate compounds. The minor minerals are salts of manganese, iron, zinc, cobalt, copper, and occasionally molybdenum. The level of phosphate is usually the most critical when it is added as potassium monobasic phosphate. An example of the critical nature of phosphate levels can be seen in the production of cycloheximide. Small changes in phosphate level can cause drastic changes in cycloheximide production. Optimum titers have been obtained with KH_2PO_4 levels of 0.2 to 0.25 g/liter, whereas 0.4 and 2.0 g of KH_2PO_4 per liter gave respective titers of 80 and 0% of the controls (B. W. Churchill, U.S. patent 2,885,326, 1959).

A great deal of success in developing complex media for antibiotic production can be achieved by using statistically designed factorial 2^5 experiments (21, 22). These experiments are done with 32 shake-flask fermentations so that two levels of five compounds, conditions, or selected groups of variables can be tested in all possible combinations. The initial studies of balancing mineral levels with the protein and carbon sources used are done in distilled water, thus avoiding the influence of varying levels of minerals in tap water. The selection of new levels of the variables can be determined by using a computer program (6). The positive and negative levels of the various ingredients in the media are reflected by antibiotic yields. The results direct whether more of an ingredient should be added in a subsequent 2^5 run. Using this method, it is possible to balance the mineral level precisely for optimum antibiotic production. Subsequent studies with the mineral levels found in tap water from a previous analysis can be run. These mineral additions (high and low level from wells) can be made with consideration given to the mineral levels present in various water supplies. The use of the computer has made statistically designed factorial 2^5 experiments an extremely powerful tool since countless hours are not required to analyze the data. Improvements of from 20 to 400% in antibiotic titers after 25 to 30 cycles of factorial 2^5 experiments have been realized with various developmental programs of complex substrates for antibiotic fermentations. In addition, the factorial 2^5 experimental parameters can include a great many variables other than mineral levels, including time of temperature shift, level of complex nitrogen source, type and level of carbon source, comparison of antifoams, effect of vitamins and other growth factors, seed level, and level of aeration or agitation. The following are a few general observations which might be helpful to those who are developing media for antibiotic production.

1. The seed medium, especially the carbon source, can greatly influence the fermentation. An isolate that produces high levels of amylase should not be grown in starch containing seed medium since starch in the fermentation medium will be used much too rapidly.

2. Fermentation parameters such as temperature, level of aeration, level and type of antifoam used, and even residual carbon dioxide in the medium can greatly influence the selection of complex medium ingredients.

3. Increased medium strength, 10 to 20% stronger than that which gave optimum antibiotic titers, may be used to screen new isolates and can even be used in production to reduce fluctuations in antibiotic titers.

4. The pH of the fermentation medium in relation to the harvest pH of the seed is sometimes very important since large differences may interfere with growth.

5. The order of ingredients added in the medium mix, temperature during mixing, length of time of mix, and pH during the mix cycle can all substantially affect antibiotic production in a complex medium.

6. The solubility of the protein sources (nitrogen soluble index) during mixing has a great influence on the eventual production of antibiotic. It has been reported (7) that interrelations between soy protein, calcium, and phytic acid may occur which could greatly influence the eventual utilization of soy protein.

7. If addition of a crude ingredient, such as molasses, substantially increases antibiotic yields, the next step is to test the ash of this material. If addition of the ash has the same effect, then its mineral composition must be determined, and the optimum levels of the various minerals must be established.

8. The control of pH by the addition of acid or alkali or through addition of oil or soluble protein during a fermentation run is frequently necessary to obtain optimum titers.

9. Levels of 10 to 20 g of residual starch per liter remaining at the end of a fermentation need not be of concern. The amylase produced by most antibiotic-producing cultures either fails to hydrolyze starch if the starch concentration is too low or does not destroy all of the bonds in the starch. Some amylases hydrolyze straight-chain linkages and some hydrolyze branched chains; they rarely hydrolyze both types.

Recent trends in complex-medium development for antibiotic fermentation have largely been dictated by economic conditions such as manpower costs, utility costs, and of course costs of nutrient ingredients. Most fermentors now being installed are at least 150,000 liters (about 40,000-gal operating volume), and the labor costs for preparing and sterilizing media are about the same for a large fermentor as they are for a smaller tank. The goal, once a fermentation is started, is to keep antibiotic production going as long as possible before it slows down dramatically. To lengthen runs, pH control equipment has become more sophisticated, regardless of whether control is by acid or alkali additions or whether it is accomplished by nutrient additions, such as oil or ammonia feeding, which may supply nutrients as well as pH control. The discovery that only certain fatty acids are being used from an oil has resulted in more single fatty acids or esters of fatty acids being used. Fatty acids as carbon source in fermentations may allow fermentations to continue for as long as 3 to 4 weeks. The development of cultures which utilize fatty acids efficiently has resulted in a need for nonmetabolizable antifoams such as Ucon LB625 (brake fluid) and silicones.

Nitrogen and carbon sources must be stable and uniform from lot to lot since production of antibiotics may be run in campaigns. Complex nutrients such as liquid suspensions of yeasts, meat peptones, corn steep liquor, distillers solubles, and molasses are not stable enough for storage for extended periods. Pharmamedia (cottonseed flour) seems to be one of the most uniform nitrogen sources. Other nutrients, such as soy flours, spray-dried corn steep solids, spray-dried lard water, fish protein hydrolysates, and dried yeasts remain stable upon storage and are uniform enough from lot to lot to be used in complex media.

10.3 STEROIDS

The media used in the bioconversion of steroids are generally much less complicated than those used in antibiotic fermentations. Usually it is only necessary to devise a medium that gives good growth of the desired microorganism, although the nature of the steroid substrate may also influence medium choice. For example, if the substrate is quite nonpolar, such as progesterone, then a complex medium high in protein helps to keep the substrate properly suspended. Thus, when progesterone at a high substrate level is converted by *Rhizopus nigricans*, the medium used consists of 41.25 g of soybean meal and 37.5 g of dextrose per liter (F. R. Hansen and W. D. Maxon, U.S. patent 3,201,324, 1965). A surfactant may also be added to the medium to help keep the steroid substrate suspended. Other studies have shown that magnesium and phosphate ions are essential for this bioconversion and that acetate, malate, and riboflavin enhance the bioconversion (8).

When more soluble steroids or steroid derivatives (e.g., steroid 21-hemisuccinates) are bioconverted, a semidefined medium may be used. For example, Kominek accomplished 1-dehydrogenation of 11, 21-dihydroxy-4, 17(20)-pregnadien-3-one-21-hemisuccinate in a medium of the following composition (per liter): 15 g of glucose, 2.75 g of ammonium sulfate, 0.5 g of dibasic potassium phosphate, 1.3 g of calcium chloride, 0.7 g of potassium chloride, 0.05 g of ferrous sulfate, and 1.0 g of corn steep liquor (L. A. Kominek, U.S. patent 3,770,586, 1973). One advantage of such a semidefined medium is that product isolation is easier and more efficient than when dealing with complex media.

With steroid bioconversion, the composition of the medium may have a profound effect on the final product. For example, in the 16-hydroxylation of 9α-fluorohydrocortisone by *Streptomyces roseochromogenes*, iron in the medium leads to the formation of the undesired D-homoannulated steroid (9). The rearrangement of a D-ring is a nonenzymatic reaction and can be blocked by the elimination of iron from the medium. In this example the fermentation medium consisted of (per liter); starch, 40 g; corn steep liquor, 25 g; calcium carbonate, 5 g; and lard oil, 0.2% (vol/vol). The addition of potassium phosphate (5 g of dibasic potassium phosphate per liter) to this medium resulted in the binding of iron and the accumulation of the desired product.

For steroid bioconversions, any of a wide variety of medium ingredients that give good cell growth may be used. However, it also must be recognized that the composition of the medium may determine the nature of the bioconversion, and therefore medium development studies should be undertaken with each steroid bioconversion to obtain the optimum medium. As mentioned above, higher-priced medium ingredients may be more generally used in steroid bioconversions than in antibiotic fermentations since steroids are high-cost products and yields are more important than process costs.

10.4 BEVERAGES

Alcoholic beverages have been produced by fermentation for centuries. No review of substrates for large-scale fermentations would be complete without some discussion of this topic. However, the processes are relatively simple and more a matter of art than science; therefore, a detailed analysis will not be presented.

In the brewing industry, a variety of grains including barley, corn, rice, wheat, sorghum, etc., are used as carbon and nitrogen sources. The starch in these grains is hydrolyzed by amylase in malt (sprouted barley) in a process known as mashing. The mash is then heated to about 75°C and coarse-filtered. The filtrate, containing organic solubles, is called wort. Hops (dried blossom clusters from the hops plant) are added to the wort; the mixture is boiled for several hours and then cooled, and the insolubles are removed. Next the wort is inoculated with a strain of *Saccharomyces cerevisiae* or other brewing yeast and incubated under anaerobic or aerobic conditions. The product is clarified, usually aged, and packaged. Small differences in the brewing process determine whether the product will be beer, ale, stout, etc.

Distilled products such as whiskey, rum, brandy, vodka, and gin are produced by a fermentation process similar to that used in brewing. However, in these processes the insolubles are not removed from the mash. Whiskey, gin, vodka, and spirits are produced from various grains; rum is made from molasses, and brandy is made from fruit. Again an anaerobic fermentation is carried out using strains of *S. cerevisiae* or other yeasts; sometimes lactic acid bacteria are also present. The products are recovered by distillation and have ethanol contents ranging from 40 to 90% (19).

Wines are produced from fruit, usually grapes. Sometimes additional sugar is added to increase the alcohol production and help regulate the dryness of the product. After the anaerobic fermentation period, the insolubles are separated and the clarified product is aged and bottled.

It should be noted that the alcoholic beverage industry also generates ingredients for large-scale fermentation, e.g., distillers' solubles, distillers' dried grain, brewers' yeast, etc.

10.5 FOOD AND FEED YEASTS

A great deal of yeast is produced for use as a leavening agent in baking. Usually strains of *S. cerevisiae* are used. In the United States, bakers' yeast is generally produced in media containing clarified beet or cane molasses as the carbon source; these substrates also contain nitrogen in the form of amino acids as well as numerous growth factors. However, the medium must still be supplemented with nitrogen, as ammonia, ammonium salts, or ammonium hydroxide, and also a source of phosphate. Growth is carried out in highly aerated and agitated fermentors to achieve maximum cell yields and growth rates. If conditions are properly controlled, little or no ethanol is formed.

Although strains of bakers' yeast cannot metabolize some disaccharides or pentoses, torula yeast (*Candida utilis*) can utilize pentoses. It has therefore traditionally been grown on sulfite waste liquor, a waste stream from paper making. In this process sulfur dioxide is steam-stripped from the waste liquor and ammonia and phosphate are added as needed. The composition of the sulfite waste liquor depends mainly on the type of wood being processed; in the northern Midwest the principal trees are tamarack, spruce, and balsam, while the western hemlock is processed in the Northwestern United States. Aeration is accomplished in air-lift Waldhof-type fermentors operated in a continuous manner. The yeast product is separated, spray-dried, and used in food products and as a substrate in large-scale fermentations, where it serves as a source of protein and growth factors.

Other substrates such as whey, ethanol, and hydrocarbons have been explored as substrates for yeast production. However, significant quantities of yeast are not currently produced by these processes.

10.6 SOLVENTS

The production of ethanol for beverages has already been discussed. Earlier in this century ethanol was also produced by fermentation for use as a solvent. In these fermentations the same kinds of substrates as in beverage production were used. However, with the advent of the petrochemical industry, solvent ethanol produced by fermentation could not compete economically with that produced chemically. Recently, when crude oil became more expensive and less available, the production of ethanol by fermentation was reexamined. The principal substrates considered were corn, sorghum (24), other grains, and sugar cane (18). It has been reported that Brazil is producing large amounts of ethanol for fuel use in motor vehicles by fermentation of sugar cane. The economics of making ethanol from grain are marginal at best (10). Much more favorable economics can be obtained by the use of a cheap raw material such as cellulose (24). However, ethanol-producing yeasts do not metabolize cellulose, so it must be enzymatically or chemically hydrolyzed before fermentation can occur.

Early in this century fermentations were developed for the production of *n*-butanol and acetone. The most commonly used organism was the strict anaerobe *Clostridium acetobutylicum*. The substrate was glucose (or fructose) from molasses. As with all anaerobic fermentations, the carbon source, e.g., glucose, must be partially oxidized, since one portion of the molecule is oxidized while another portion is reduced. For acetone-butanol fermentations the medium contains molasses, ammonium salt, and phosphate in water, and the pH is controlled by ammonia addition. Distribution of products is influenced by the fermentation pH, with butyric acid being converted to butanol at lower pH values. In this fermentation butanol and acetone always occur together to maintain the oxidation-reduction balance. Under some conditions the acetone may be further reduced to isopropanol. Fermentations for the production of organic solvents have not been used commercially in Europe and America for several decades, but they could become important again as supplies of crude oil are depleted.

10.7 CELL CULTURE

10.7.1 Mammalian Cells

For more than 30 years mammalian cell lines such as HeLa (human carcinoma) and L cells (mouse fibroblasts) have been cultured in vitro. Such living cells can be used to study the effects of drugs, hormones, etc. on various tissues. However, large-scale production of these cells has not been required for this kind of work. This is fortunate since the media have been exceedingly complex and expensive. Most of the early media contained about 10% serum, e.g., fetal calf serum, as well as a variety of other organic and inorganic components. Later studies resulted in the development of chemically defined serum-free media (1, 11). These media were far from simple, usually containing amino acids, vitamins, coenzymes, nucleic acids, inorganic salts, carbohydrates, hormones, etc. Therefore, they would also be very expensive to use on a large scale.

With the discovery that hybridomas (antibody-producing cells) can be produced by the fusion of lymphocytes and myeloma cells, the prospect for large-scale cell culture has become very attractive. Hybridomas, for producing antibodies of varius kinds, can be grown on complex serum-containing media (16). However, the wide variety of naturally occurring physiological compounds found in serum makes antibody purification difficult.

Therefore, protein-free media have been developed for the growth of some hybridoma cell lines (5). Again, these media are not simple, and the transfer of cells from a serum-containing medium to a serum-free medium must be done in numerous stages, each at reduced serum levels.

In the laboratory, cultured cells are grown as monolayers on the walls of various kinds of glass vessels. For large-scale production of monoclonal antibodies for clinical uses, suspension culture in stirred tank reactor-type vessels would be preferred. However, early attempts at growth in suspension culture have not been successful (5). The addition of microcarriers to the medium for cell attachment has shown some promise for growth of cultured cells in suspension (4).

It is clear that many problems remain to be solved before the production of mammalian cells in pure culture on a commercial scale becomes an economic reality. Currently, the requirements for very pure and expensive medium components, including high-purity water, make large-scale growth of cultured cells a challenge. The future will certainly see changes in the kinds of substrates used for large-scale cell culture, but at this time one can only speculate as to what these substrates will be.

10.7.2 Plant Cells

Plant cell culture on a large scale is under investigation. The synthesis of a variety of medicinal compounds as well as secondary products such as alkaloids, antibiotics, cardenolides, terpenoids, steroids, and furanthromones by plants has been reported (12). However, there are scant data on media that might be employed on the large scale. It is clear that the growth of plant cells would be as complicated as growth of mammalian cells. The media that have been used are usually very complex, containing a wide array of organic and inorganic compounds (2). Growth regulators such as gibberellin(s), kinetin, indoleacetic acid, naphthalenacetic acid, and 2,4-dichlorophenoxyacetic acid are also required. Even under optimum conditions, growth was found to be exceedingly slow (doubling times of 2 days or longer [2]). Though there is clearly a potential for the manufacture of important products by plant cell culture, it is unrealized at this time.

10.8 POLYPEPTIDES FROM RECOMBINANT DNA TECHNOLOGY

Protein products, principally enzymes, have been produced by fermentation for many years. These products, such as amylases, proteases, cholesterol oxidase, etc., are usually produced only as a crude material. Therefore, various complex media that support good growth and product formation are used.

With the advent of genetic engineering techniques for inserting genes from higher organisms into microorganisms came the capacity to produce new classes of protein products. Some of these products, such as insulin, human growth hormone, and tissue plasminogen activator, are proposed for clinical use in humans and will require a high degree of purity. Other products for agricultural uses, such as bovine growth hormone, must also be very pure. The range of products arising from the new biotechnology seems unlimited, but it is clear that the production and recovery of the products will offer special challenges. It is essential to grow cells at very high densities to maximize product yields, and it will be necessary to keep the medium as simple as possible to facilitate product isolation. Much of the work in progress uses strains of *Escherichia coli* for expression of foreign genes. *E. coli* and many other bacteria and yeasts grow very well on the chemically defined media described above. Of course, these media must be tailored to achieve the desired cell densities. This will undoubtedly require feeds of carbon and nitrogen sources as well as external pH control. Although the use of defined media will make the isolation of purified protein products easier, it will not reduce product isolation to a trivial exercise since the microorganisms produce a variety of proteins that will complicate the purification process. The challenge will be to combine chemically defined medium design and new isolation procedures to give protein products at reasonable costs.

10.9 BIOCONTROL PRODUCTS

Biocontrol is the use of microorganisms to control weed and insect pests. The first priority for the production of spores or, rarely, hyphal fragments of biocontrol microorganisms is the selection of the production medium. A wide variety of nitrogen and carbon sources are screened for the mass production of spores of a particular microorganism. The primary objective is to produce the spores economically; this is achieved by using crude agricultural commodities that are abundant, cheap, and of relatively uniform quality as medium ingredients. The production broth is generally similar to the screening media even though the screening may have been done on agar plates. Customarily, the primary and secondary seed stages are used as a means of increasing spore titers rapidly, since the alternative of using massive inocula in the fermentor is not practical. The length of the fermentation cycle can be substantially reduced if primary and secondary seeds are used. In the case of biocontrol agents, it is not necessary to have distinctly different media in seeds and fermentations, as is required with antibiotics, since the production phase is considered to be only an extension of growth in the seed vessels (3).

In the case of production of spores of *Colletotrichum gloeosporioides* f.sp. *aeschynomene*, Daniel et al. (7) reported good sporulation on lima bean agar and also in shake flasks with liquid medium composed of V-8 juice, sucrose, potassium nitrate, and a mineral supplement. It was found that although this organism grew on most media, the optimum spore yields were obtained only when nitrogen, carbon, and the mineral supplement were carefully balanced (14). A similar balancing of nutrients was shown to be necessary to obtain optimum spore production with *Colletotrichum lindemuthianum* (15). In addition, that culture was reported to require vitamins and growth factors for optimum spore production (15). The three biocontrol agents, *C. gloeosporioides* f.sp. *aeschynomene*, *Colletotrichum gloeosporioides* f.sp. *jussiaeae*, and *Colletotrichum malvarum*, did not require as critical a ratio of carbon to nitrogen as was required by *Hirsutella thompsoni* (20). It was found, however, that each required different carbon sources.

The requirement that microbial pesticides compete economically with chemical pesticides imposes some restrictions on the selection of nutrients for spore production. The usual grain and bean commodity products are relatively cheap sources of nitrogen. Such products as soybean flour and meal, cottonseed flour, corn steep liquor or powder, peanut meal, corn protein, dried distillers' solubles, wheat bran, barley malt, linseed meal, and wheat meal are screened for the production of microbial pesticides. Brewers' or, occasionally, torula yeast may frequently be a necessary addition to the fermentation medium as a nitrogen, vitamin, and growth factor source if the biocontrol organism has additional requirements. These yeasts are added at a much lower level than the major nitrogen source. The carbon sources usually tested are corn starch, corn flour, malto dextrins, glucose monohydrate, hydrolyzed corn-derived products, lactose, glycerol, and sucrose. Occasionally microorganisms are found that lack either amylases or sucrase; the selection of a carbon source is then much more restricted.

The optimum medium for growth frequently results in excellent mycelial growth but no submerged sporulation. An "out-of-balance" fermentation medium is one of the ways that sporulation can be induced. Smith and Anderson (20) reported that the reduction or absence of assimilable nitrogen, coupled with available carbon, generally induces sporulation. It has been found that not only nutritional factors but also certain physical factors, such as aeration, temperature, and light requirements, are necessary for some microorganisms to sporulate profusely. The abundant spores of *C. gloeosporioides* f.sp. *aeschynomene* obtained in tank fermentations result from a fermentation medium designed for optimum growth. The optimum growth medium, combined with the adverse physical factor of too high aeration, results in little vegetative growth of this organism but abundant fission spores, as well as a few conidia, blastospores, and arthrospores (14).

Shake-flask fermentations are essential in developing processes for optimum production of asexual spores. However, it is common knowledge that it is extremely difficult to control the pH, temperature, agitation/aeration, and foaming in shake flasks. pH can frequently be successfully controlled with combinations of various levels of potassium nitrate, ammonium nitrate, and other utilizable ammonium salts. Temperature can usually be controlled with carefully timed temperature shifts during the course of the fermentation run. Agitation/aeration is varied by changing the speed and throw of the shaker, using flasks with various neck openings (amplified by using restrictive closures over the caps), using baffled flasks, changing the volume in flasks, and varying the osmotic strength of the medium. The design of a medium for optimum growth, by carefully balancing the selected nitrogen and carbon sources as well as the selected mineral supplement, is the first step to the successful mass production of spores for biocontrol of weeds or insects. The second, equally important step is to create adverse chemical or physical conditions to induce sporulation.

10.10 APPENDIX A

See Table 2.

10.11 APPENDIX B

Carbon and Nitrogen Sources for Fermentation Media Development

The following list of raw materials has been assembled for use by those developing fermentation media. The materials are largely agricultural commodity products, so generally the price is such that they can be used in commerical production. Most of the products are also available in large enough quantities that supply should not be a problem in fermentation production on the large scale. Full names and addresses of suppliers appear at the end of this appendix.

Products

Corn

Archer-Daniels-Midland
- No. 131C Brewers Corn Starch
- No. 655 Dextrin
- No. 700 Canary Dextrin
- 290B Industrial Starch
- CFS Concentrate

American Maize Products Co.
- Corn Starch, Pure Food Powder
- Amioca Starch (Waxy Maize Starch)
- No. 5 Starch, High Amylose Starch
- Amylomaize, High Amylose Starch
- No. 843 Starch, Modified Waxy Maize
- No. 1895 Dextrin, Waxy Maize Dextrin
- No. 1104 Dextrin
- No. 1402 Dextrin
- No. 1752 S Dextrin, Canary Dextrin

CPC International
- Cerelose (glucose monohydrate)
- Starch 3005 modified

Grain Processing Corp.
- A-100 Industrial Corn Syrup
- Maltrin M 040, Malto Dextrin
- Maltrin M 050, Malto Dextrin
- Maltrin M 100, Malto Dextrin
- Maltrin M 150, Malto Dextrin
- Maltrin M 200, Malto Dextrin
- Maltrin M 250, Malto Dextrin
- Maltrin M 365, Malto Dextrin

Illinois Cereal Mills
- Yellow Corn Grits
- Yellow Corn Meal
- Yellow Corn Flour
- High Protein Germ

Krause Milling Co.
- Corn flour, degerminated

Lauhoff Grain Co.
- Defatted Corn Germ Meal
- FCG 200 Corn Meal
- FCM 350 Corn Meal
- CCF 600 Corn Flour
- CCF 610 Corn Flour
- DCF 1000 Corn Flour

National Starch & Chemical Corp.
- Tapon 82
- Tapon 85
- Tapon 88
- Nadex 772
- Melojel Corn Starch
- Kofilm 80, Dextrin

Roquette Frères
- High Maltose Corn Syrup
- Spray Dried Glucose Syrup MD50
- MD 02, Malto Dextrin
- MD 01, Malto Dextrin
- MD 005, Malto Dextrin

J. L. Short Milling Co.
- Sunlite Hominy Feed
- Sunlite Yellow Corn Flour
- Sunlite Yellow Corn Cones
- Sunlite Yellow Dusting Meal
- Sunlite Yellow Brewer's Grits
- Standard Meal

TABLE 2. Appendix A: fermentation media ingredient analysis

Ingredient	Major components						Mineral content					
	Dry matter (%)	Protein (%)	Carbohydrates (%)	Fat (%)	Fiber (%)	Ash (%)	Calcium (%)	Magnesium (%)	Phosphorus (%)	Available phosphorus (%)	Potassium (%)	Sulfur (%)
Alfalfa meal, dehydrated	89.0	14.5	40.3	2.7	74.2	7.3	1.3	0.29	0.23	0.18	2.49	0.22
Barley	90.0	11.5	68.0	1.8	7.0	2.5	0.07	0.13	0.36	0.12	0.49	0.15
Barley malt	96.0	13.0	70.0	2.0	3.5	2.5	—	—	—	—	—	—
Beet pulp	90.0	8.9	59.1	0.6	18.3	3.1	0.56	0.15	0.1	0.03	0.2	0.22
Blood meal	93.0	80	2.5	<1	<1	3	0.3	0.21	0.22	0.22	0.09	0.60
Cerelose (commercial glucose)	91.5	—	91.5	—	—	—	—	—	—	—	—	—
CFS concentrate	95.0	27	50.0	2.0	7.0	8.5	—	—	—	—	—	—
Citrus pulp, dried	90.0	6.0	62.7	3.4	13.0	6.9	2.0	0.16	0.1	0.03	0.62	0.06
Corn	82.0	9.9	69.2	4.4	2.2	1.3	0.02	0.11	0.28	0.1	0.31	0.08
Corn germ meal	93.0	22.6	53.2	1.9	9.5	3.3	0.3	0.16	0.5	0.16	0.34	0.32
Corn gluten feed, 21%	90.0	22.3	57.1	3.8	8.0	2.5	0.44	0.29	0.57	0.19	0.57	0.16
Corn gluten meal, 41%	91.0	42.0	40.2	2.5	4.3	2.0	0.14	0.05	0.46	0.16	0.03	0.50
Corn gluten meal, 60%	90.0	62.0	20.0	2.5	1.6	1.8	0.0	0.06	0.54	0.19	0.05	0.83
Corn steep liquor	50.0	24.0	5.8	1.0	1.0	8.8	—	—	—	—	—	—
Corn steep powder	95.0	48.0	—	0.4	—	17	0.06	1.5	3.3	1.1	4.5	0.58
Cottonseed meal, expeller	94.0	41.0	28.9	3.9	13.5	6.7	0.15	0.49	0.93	0.31	1.25	0.40
Cottonseed meal, solvent	92.5	41.4		1.5	12.4		0.15	0.47	0.98	0.33	1.26	0.21
Dried distillers' solubles	92.0	26.0	45.0	9.0	4.0	8.0	0.30	0.65	1.3	1.2	1.75	0.37
Edamine (protein hydrolysates)	—	7.5	—	—	—	6.0	—	—	—	—	—	—
Enzose (commercial starch hydrolysates)	75.0	—	70.0	—	—	3.5	—	—	—	—	—	—
Fermamine (protein hydrolysates)	94.0	55.0	—	—	—	15.0	—	—	—	—	—	—
Fish meal (anchovy), 65%	92.0	65.0	—	3.8	1.0	21.3	4.0	0.25	2.6	2.60	0.74	0.54
Fish meal (herring), 70%	93.0	72.0	—	7.5	1.0	—	2.0	0.14	1.5	1.50	1.12	0.62
Fish meal (menhaden), 60%	92.0	61.0	—	7.5	1.0	—	5.2	0.14	2.9	2.9	0.73	0.45
Fish solubles	50.0	32.0	7.0	3.0	1.0	10.0	0.05	0.04	0.49	0.49	1.48	0.13
Linseed meal	92.0	36.0	38.0	0.5	9.5	6.5	0.4	0.56	0.9	0.3	1.22	0.39
Meat and bone meal	92.0	50.0	0.0	8.0	3.0	31	8.9	1.13	4.4	4.4	1.46	0.26
Milo	89.0	9.3	73.0	2.8	2.5	1.4	0.02	0.13	0.3	0.1	0.35	0.09
Molasses, beet	77.0	6.7	65.1	0.0	0.0	5.2	0.16	0.23	0.02	0.01	4.71	0.47
Molasses, blackstrap	78.0	3.0	54.0	0.4	—	9.0	0.74	0.35	0.08	—	3.67	—
NZ-Amine B (protein hydrolysates)	96.0	80	—	—	—	5.5	—	—	—	—	—	—
Oats	86.5	12.0	54.0	4.5	12.0	4.0	0.05	0.16	0.35	0.1	0.42	0.21
Peanut meal and hulls	90.5	45.0	23.0	5.0	12.0	5.5	0.15	0.32	0.55	0.2	1.12	0.28
Pharmamedia (cottonseed flour)	99.0	59.2	24.13	4.02	2.55	6.71	0.25	0.74	1.31	0.31	1.72	0.6
Proflo (cottonseed flour)	98.8	61.1	23.2	4.1	3.19	6.73	0.25	0.75	1.31	0.35	1.73	1.56
Rice	89.5	8.0	65.0	2.0	10.0	4.5	0.04	0.06	0.23	—	—	—
Rice bran	91.0	13.0	45.0	13.0	14.0	16.0	—	—	—	—	—	—
Soybean meal, expeller	90.0	42.0	29.9	4.0	6.0	6.5	0.25	0.25	0.63	0.16	1.75	0.32
Soybean meal, solvent	90.0	45.0	32.2	0.8	6.5	5.5	0.25	0.27	0.6	0.15	1.92	0.32
Wheat	90.0	13.2	69.0	1.9	2.6	1.8	0.05	0.17	0.35	0.17	0.45	0.12
Whey, dried	95.0	12.0	68.0	1.0	0.0	9.6	0.9	0.13	0.75	0.75	1.20	1.04
Yeast, brewers	95.0	43.0	39.5	1.5	1.5	7.0	0.1	0.25	1.4	1.4	1.48	0.49
Yeast, hydrolysate	94.5	52.5		0.0	1.5	10.0	—	—	—	—	—	—
Yeast, torula	94.0	50.0	32.0	4.8	0.8	9.0	0.11	0.27	1.5		1.4	0.1

Vitamin content							Amino acids												
Biotin (mg/kg)	Choline (mg/kg)	Niacin (mg/kg)	Pantothenic acid (mg/kg)	Pyridoxine (mg/kg)	Riboflavin (mg/kg)	Thiamine (mg/kg)	Arginine (%)	Cystine (%)	Glycine (%)	Histidine (%)	Isoleucine (%)	Leucine (%)	Lysine (%)	Methionine (%)	Phenylalanine (%)	Threonine (%)	Tryptophan (%)	Tyrosine (%)	Valine (%)
—	1,496	44.0	29.92	—	12.1	—	0.51	0.25	0.52	0.22	1.15	0.62	0.61	0.19	0.59	0.73	0.22	0.26	0.71
0.22	1,100	52.8	6.16	3.52	1.54	5.5	0.5	0.26	0.45	0.3	0.5	0.8	0.42	0.22	0.6	0.4	0.17	0.4	0.6
—		49.74	8.58		2.86	3.74	0.4	—	0.1	0.1	0.1	0.1	0.1	0.1	0.1	0.1	0.1	0.1	0.1
—	814	16.28	1.54	—	0.66	—	0.3	0.0	—	—	—	—	0.6	0.04	—	0.25	0.09	—	—
—	748	31.46	1.1	—	1.43	0.22	3.4	0.2	4.2	4.2	1.2	10.0	6.5	1.1	5.4	4.1	1.3	2.4	6.7
—	—	—	—	—	—	—	—	—	—	—	—	—	—	—	—	—	—	—	—
0.44	—	—	—	14.3		3.74	1.2	0.4	1.1	0.8	0.8	2.5	0.9	0.4	0.9	0.9	0.2	0.8	1.4
—	770	22.0	14.96	—	22.2	—	0.28	0.11	—	—	—	—	0.2	0.11	—	—	0.06	—	—
	528	22.0	5.72	7.6	1.1	—	0.50	0.09	0.43	0.20	0.40	1.10	0.20	0.17	0.50	0.40	0.10	—	0.40
—	1,760	41.8	4.4	—	3.74	—	1.3	0.4	—	—	—	—	0.9	0.57	—	1.1	0.18	—	—
—	2,420	74.8	17.16	—	2.42	—	1.0	0.5	—	—	—	—	0.6	0.2	—	0.4	0.1	—	—
—	1,320	77.0	10.34	—	2.2	—	1.4	0.8	—	—	—	—	0.8	1.2	—	1.4	0.2	—	—
—	220	81.4	2.86	—	2.2	0.1	1.9	1.1	—	—	—	—	1.0	1.9	—	2.0	0.25	—	—
0.88	—	—	—	19.36		0.88	0.4	0.5	1.1	0.3	0.9	0.1	0.2	0.5	0.3	—	—	0.1	0.5
—	5.6	0.16	0.03	0.02	0.01	0.01	3.3	1.9	5.1	2.8	3.6	11.3	2.5	1.9	4.4	4.0	—	3.4	5.8
—	2,794	37.84	7.7	—	4.18	—	4.3	0.59	2.30	1.02	1.52	2.35	1.6	0.45	2.00	1.45	0.5	1.21	1.89
—	2,706	39.16	9.9	—	4.4	—	4.6	0.62	—	—	—	—	1.76	0.46	—	1.45	0.52	—	—
2.86	4,400	110.0	19.8	—	15.4	5.5	1.0	0.6	1.1	0.7	1.6	2.1	0.9	0.6	1.5	1.0	0.2	0.7	1.5
—	—	—	—	—	—	—	3.1	2.2	1.7	1.9	5.4	10.0	10.0	1.8	3.4	3.8	2.1	3.3	4.1
—	—	—	—	—	—	—	—	—	—	—	—	—	—	—	—	—	—	—	—
—	—	—	—	—	—	—	—	—	—	—	—	—	—	—	—	—	—	—	—
—	3,740	93.5	9.68	—	9.46	—	3.6	0.6	6.31	1.25	2.43	4.27	4.7	1.9	2.37	2.80	0.7	1.91	2.83
—	3,960	88.0	8.8	—	9.02	—	4.2	0.7	3.53	1.34	2.86	4.70	5.7	2.0	2.52	2.96	0.8	1.76	3.61
—	2,860	55.0	8.8	—	4.84	—	3.6	0.5	—	—	—	—	4.6	1.7	—	2.96	0.6	—	—
	2,860	165.0	35.2	12.54	1.32	5.5	2.2	1.4	5.3	2.4	1.5	2.2	7.6	2.4	1.3	1.1	0.7	—	1.4
—	1,848	35.2	17.6	—	3.08	8.8	2.5	0.6	0.23	0.5	1.3	2.1	1.0	0.8	1.8	1.4	0.7	1.7	1.8
—	1,914	55.0	8.8	—	4.4	1.1	4.0	1.4	6.6	0.9	1.7	3.1	3.5	0.7	1.8	1.8	0.2	1.22	2.4
—	660	39.6	11.0	—	1.1	3.96	0.29	0.14	0.30	0.23	0.38	1.22	0.19	0.13	0.47	0.3	0.09	0.33	0.55
—	880	39.6	4.62	—	2.2	—	—	—	—	—	—	—	—	—	—	—	—	—	—
—	660	46.86	42.9	44	4.4	0.88	—	—	0.10	—	—	0.01	0.01	—	—	0.06	—	—	0.02
—	—	—	—	—	—	—	3.9	0.5	1.8	2.9	4.6	19.4	7.6	2.4	5.0	4.2	1.6	3.2	6.1
—	935	13.2	12.76	—	1.1	6.6	0.8	0.2	0.2	0.2	0.6	1.0	0.4	0.2	0.7	0.4	0.2	0.6	0.7
—	1,672	167.2	48.4	—	5.28	7.26	4.6	0.7	3.0	1.0	2.0	3.1	1.3	0.6	2.3	1.4	0.5	—	2.2
1.52	3,270	83.30	12.40	16.40	4.82	3.99	12.28	1.52	3.78	2.96	3.29	6.11	4.49	1.52	5.92	3.31	0.95	3.42	4.57
0.792	33.44	83.82	12.47	0.88	5.1	4.36	7.15	1.45	2.75	1.98	2.42	6.46	3.3	1.4	4.17	2.46	0.86	2.53	3.25
—	990	33.0	11.0	—	1.32	3.08	0.5	0.1	—	0.1	0.4	0.6	0.3	0.3	0.4	0.3	0.1	0.7	0.6
—	1,254	297	23.1	—	2.64	22	0.5	0.1	0.9	0.2	0.4	0.6	0.5	0.2	0.4	0.4	0.1	—	0.6
—	2,420	30.36	14.08	—	3.08	—	2.9	0.62	—	—	—	—	2.8	0.59	—	1.72	0.59	—	—
—	2,673	26.4	14.52	—	3.3	—	3.2	0.67	2.92	1.06	2.25	3.42	3.0	0.65	2.14	1.9	0.63	1.71	2.44
—	880	61.6	13.2	—	1.1	5.06	0.8	0.2	—	0.3	0.6	1.0	0.5	0.2	0.7	0.4	0.2	0.5	0.6
—	2,420	11.0	48.4	2.86	19.8	3.96	0.4	0.4	0.7	0.2	0.7	1.2	1.0	0.4	0.5	0.6	0.2	0.5	0.6
—	4,840	498.3	121.44	49.72	35.2	74.8	2.2	0.6	3.4	1.3	2.7	3.3	3.4	1.0	1.8	2.5	0.8	1.9	2.4
—	—	—	—	—	—	—	3.3	1.4	—	1.6	5.5	6.2	6.5	2.1	3.7	3.5	1.2	4.6	4.4
—	290.4	495.0	81.4	28.6	44.0	6.16	2.6	0.6	2.7	1.4	2.9	3.5	3.8	0.8	3.0	2.6	0.5	2.1	2.9

A. E. Staley Manufacturing Co.
- Staclipse Starch
- Star-Dri 1 Malto Dextrin
- Star-Dri 5 Malto Dextrin
- Star-Dri 10 Malto Dextrin
- Star-Dri 24R Corn Syrup Solids
- Star-Dri 42R Corn Syrup Solids
- Corn Starch, Pure Food Powder
- Waxy 7350 no. 1, Waxy Maize Starch
- Stadex 9, Dextrin
- Stadex 15, Dextrin

Vitamins, Inc.
- Defatted Corn Germ Granules
- Defatted Corn Germ Flour

Wheat

Henkel Corp.
- Edigel 100 Wheat Starch
- Provim Vital Gluten
- Aytex P Wheat Starch

Knappen Milling Co.
- Wheat Flour
- Wheat Germ
- Wheat Heavy Bran
- Wheat Gage Bran

Viobin Corp.
- Defatted Wheat Germ no. 1
- Defatted Wheat Germ no. 3
- Defatted Wheat Germ no. 8
- Defatted Wheat Germ no. 9
- Viobran, Mixture Germ and Bran

Vitamins, Inc.
- Vi-Vax Wheat Germ

Oats

National Oats Co.
- Oat Flour
- Oat Bran

Vitamins, Inc.
- Prote Defatted Oat Flour

Cottonseed

Southern Cotton Oil Co.
- Cottonseed Meal

Trader's Protein
- ProFlo, Cottonseed Flour
- ProFlo Oil
- Pharmamedia

Soybean

Archer-Daniels-Midland
- Nutrisoy Flour
- Toasted Nutrisoy Flour
- Baker's Nutrisoy Flour
- Toasted Nutrisoy Flour
- KaySoy Grits 20-80
- KaySoy 200A
- KaySoy 200C
- KaySoy 200D

Cargill Inc.
- Poly Soy, 60 mesh, soy feed
- Poly Soy, PA100, soy feed

Central Soya
- 3040 Soyafluff 200W Soy Flour
- 3470 Soyarich 115W
- Soy Protein Concentrate, Promosoy 4970
- Soy Protein Concentrate, Promosoy 4950
- Cenpro 70 Soy Protein Concentrate with lecithin
- Soy Molasses

DeLamar, Inc.
- Soy peptone

Grain Processing Corp.
- Profam G-902 Isolated Soy Protein
- Profam S-646 Isolated Soy Protein
- Profam S-950 Isolated Soy Protein
- Profam S-972 Isolated Soy Protein

Gunther Products
- NV Protein

Lauhoff Grain Co.
- 20/0 Soy Grits

J. L. Short Milling Co.
- Pro-Fam S-955

Safflower

PVO International
- Safflower meal

Potato

Boise Cascade
- Potato Starch

Roquette Frères
- Potato starch

A. E. Staley Manufacturing Co.
- Potato starch, unmodified

Tapioca

National Starch & Chemical Corp.
- Tapioca flour

A. E. Staley Manufacturing Co.
- Tapioca flour, unmodified

Yeast

Amoco Pure Culture Products
- Toruway 49

Anheuser-Busch, Inc.
- AYE-Light
- Prime Gro

Busch Industrial Products
- Budweiser Autolyzed Yeast
- AYE Light no. 2200 Autolyzed Bakers Yeast

Food Ingredient Development Co.
- Autolyzed yeast, DAY
- Food yeast
- Autolyzed Yeast Extract FG Light

Hercules Inc.
- Zyest-70 Autolyzed Yeast
- Zyest-SLF Autolyzed Yeast
- Zyest-HF Autolyzed Yeast

Lake States Div.
- Torula Yeast

Philadelphia Dry Yeast Co.
- Brewers Yeast
- Washed Brewers Dried Yeast, P40%, F1%

Sheffield Products
- Yeast, Autolysate, Technical

Universal Foods Corp.
- Tastone 50, Bakers Yeast Extract
- Tastone 154, Bakers Yeast Extract
- Tastone 730, Bakers Yeast Extract
- Amberex 500, Brewers Yeast Extract
- Amberex 1003, Brewers Yeast Extract
- Amberex 1400, Brewers Yeast Extract

Amberex 215, Brewers Yeast Extract
Nutrex, 540, Food Yeast
BYF-100, Brewers Yeast Autolyzed
DBY-1, Brewers Yeast
Amberyeast WWBY
Brewers Yeast
Yeast Products, Inc.
Ardamine YEP Yeast Extract
Ardamine pH Yeast Extract
Ardamine Z Yeast Extract
Debittered Brewers Yeast

Dairy products
Amber Laboratories
Amber EHC
Erie Casein
Edible Ecco Blend
Devon Milk Protein
Sodium Caseinate
Rennet Casein
Foremost Foods Co.
Teklac Sweet Dairy Whey
60% Hydrolyzed Whey Syrup, Permeate
Hercules Inc.
Toraway-30, Lactose-Reduced Whey
Kraft
Krafen Sweet Whey
Land O Lakes
Puretein 20
Puretein 29
35% Whey Protein Concentrate
55% Whey Protein Concentrate
Extra Grade Sweet Dairy Whey
Whey Permeate
Low Heat Non-Fat Dry Milk
Mutchler Chemical Co.
HMS Lactose, XIX
Coarse Powder, USP
Sheffield Products
NZ Amine AT
NZ Amine BT
NZ Amine ET
NZ Amine HD
NZ Amine YTT
Peptonized Milk Nutrient
Primatone
Sheftene
Hycase S.F.
Pepticase
Edamin T
Calcium Lactate Powder-USP
Calcium Lactate Granular-NF
Universal Foods
EHC, Enzyme Hydrolyzed Casein
Vitamins, Inc.
Vikase
Western Dairy Products
Savortone 25
Savortone 100
Wisconsin Dairy Cooperative
Whey Protein Concentrate

Animal products
American Laboratories Inc.
Anatone, Peptic Digest of Pork
Beef Peptone
Microbiotone
Microbiotone-L
Microbiotone-T
Pharmatone
Pork Heart Digest
Proteose-Micro
Costec
Peptein 2000
Erie Casein
Animal Protein Extract
Hormel
Peptone no. 5
Poly Pro 5000
Dried Meat Solubles
Peptone PSR-5
Inland Molasses
Bovine Blood Plasma
Blood Meal
Pork Liver Meal
Spray Dried Lard Water
Spray Dried Whole Blood
Inolex Chemical Co.
Lexein F1000 D, Pork Collagen Hydrolysate
Lexein F-159
Lexein F-152D Beef Collagen Hydrolysate
Marcor Development
Meat Peptone SB, Beef Hydrolysate
Meat Peptone PS, Pork Hydrolysate
Spray Dried Products
Lard Water, Connective Tissue Digest
Zapata Haynie Corp.
Atlantic Menhaden Fish Meal
Atlantic Menhaden Condensed Fish Solubles, Spray Dried

Miscellaneous products
Argo Ingredients
Rapeseed Oil, Low Acid
Blommer Chocolate Co.
Cocoa Shell
Henkel Corp.
Super Col. U. (Guar Gum)
Hudson Industries
Hydro no. 1070
Knappen Milling
Badger Meat and Bone Meal
Krause Milling
Adcol
National Starch & Chemical Corp.
Hylon (Amylose Starch)
Hylon VII (Amylose Starch)
Stepan Chemical
Nacconol 90F (Sorghum Starch)
Suffolk Oil Mill
Peanut meal

Suppliers

Amber Laboratories
1101 N. Teutonia Ave.
Milwaukee, WI 53209

American Laboratories Inc.
4410 South 102nd St.
Omaha, NE 68127

American Maize Products Co.
113th and Indianapolis Blvd.
Hammond, IN 46326

Amoco Pure Culture Products
200 East Randolph Drive
Chicago, IL 60601

Anheuser-Busch Inc.
Industrial Products Division
10877 Watson Rd.
St. Louis, MO 63127

Archer-Daniels-Midland
466 Farigs Parkway
P.O. Box 1470
Decatur, IL 62525

Argo Ingredients
International Plaza
Englewood Cliffs, NJ 07632

Argo Ingredients
600 E. Algonquin Rd.
Des Plaines, IL 60016

Blommer Chocolate Company
600 West Kinzie St.
Chicago, IL 60610

Boise Cascade Chemical Operations
1600 S. Fourth Ave.
P.O. Box 1414
Portland, OR 97207

Busch Industrial Products
Industrial Products Division
10877 Watson Rd.
St. Louis, MO 63127

Cargill Inc.
P.O. Box 2817
Cedar Rapids, IA 52402

Cargill Inc.
Attention Soy Special Product Dept.
Cargill Bldg.
Minneapolis, MN 55402

Central Soya
Fort Wayne, IN 46802

Costec Inc.
P.O. Box 693
Palatine, IL 60067

CPC International
International Plaza
Englewood Cliffs, NJ 07632

DeLamar, Inc.
2653 Greenleaf Ave.
Elks Grove Village, IL 60007

Erie Casein Co.
International Proteins Division
P.O. Box 648
Erie, IL 61250

Food Ingredient Development Co.
4 Garnett Dr.
White Plains, NY 10604

Foremost Foods Company
Crocket Plaza
One Post St.
San Francisco, CA 94104

Grain Processing Corp.
1600 Oregon St.
St. Muscatine, IA 52761

Gunther Products
2200 Eldorado St.
Decatur, IL 62525

Henkel Corp.
120 W. Eastman
Suite 205
Arlington Heights, IL 60004

Hercules Inc.
Hercules Marketing Center
Wilmington, DE 19894

Hormel (George A. Hormel)
Austin, MN 55803

Hudson Industries
19 Hutton Ave.
West Orange, NJ 07052

Illinois Cereal Mills
Paris, IL 61944

Inland Molasses
American Trust Bldg.
Dubuque, IA 52001

Inolex Chemical Co.
3 Science Rd.
Glenwood, IL 60425

Knappen Milling Co.
Augusta, MI 49012

Kraft Inc., Dairy Group
P.O. Box 7380
Philadelphia, PA 19101

Krause Milling Co.
P.O. Box 1156
Milwaukee, WI 53201

Lake States Division
St. Regis Paper Co.
Rhinelander, WI 54501

Land O Lakes
Industrial Sales Division
P.O. Box 1087
Eau Clair, WI 54501

Lauhoff Grain Co.
Box 571
Danville, IL 61832

Marcor Development
206 Park St.
Hackensack, NJ 07601

Mutchler Chemical Co.
99 Kinderkamack Rd.
Westwood, NJ 07675

National Oats Co., Inc.
Cedar Rapids, IA 52402

National Starch & Chemical Corp.
3641 S. Washtenaw Ave.
Chicago, IL 60632

Pacific Vegetable Oil (PVO) Corp.
World Trade Center
San Francisco, CA 94111

Philadelphia Dry Yeast Co.
157-167 West Oxford St.
Philadelphia, PA 19122

Roquette Frères
17 Boulevard Vauban
Lille, France

Seacoast Products
P.O. Box D
Port Monmouth, NJ 07758

Sheffield Products
P.O. Box 398
Memphis, TN 38101

J. L. Short Milling Co.
Mt. Vernon Milling Co., Division
Mt. Vernon, IN 47620

Southern Cotton Oil Co.
Buckeye St.
Little Rock, AR 72114

Spray Dried Products
1201 American Trust Building
Dubuque, IA 52001

A. E. Staley Manufacturing Co.
2200 Eldorado St.
Decatur, IL 62525

Stepan Chemical Co.
Northfield, IL 60093

Suffolk Oil Mill Inc.
P.O. Box 1458
Suffolk, VA 23434

Traders Protein Div.
Traders Oil Mill Co.
P.O. Box 1837
Fort Worth, TX 76101

Universal Foods Corp.
433 East Michigan St.
Milwaukee, WI 53201

Viobin Corp.
Monticello, IL 61856

Vitamins, Inc.
200 East Randolph Dr.
Chicago, IL 60601

Western Dairy Products
118 World Trade Center
San Francisco, CA 94111

Wisconsin Dairy Cooperative
P.O. Box 429
Waukon, WI 52172

Yeast Products, Inc.
25 Styertowne Rd.
Clifton, NJ 07012

Zapata Haynie Corp.
P.O. Box 175
Reedville, VA 22539

10.2 LITERATURE CITED

1. **Barnes, D., and G. Sata.** 1980. Methods for growth of cultured cells in serum-free medium. Anal. Biochem. **102:**255–270.
2. **Carew, D. P., and E. J. Staba.** 1965. Plant tissue culture: its fundamentals, applications and relationship to medical plant studies. Lloydia **28:**1–26.
3. **Churchill, B. W.** 1982. Mass production of microorganisms for biological control, p. 139–156. *In* R. Charudalttan and H. L. Walker (ed.), Biological control of weeds with plant pathogens. John Wiley & Sons, New York.
4. **Clark, J. M., and M. D. Hirtenstein.** 1981. Optimizing culture conditions for the production of animal cells in microcarrier culture. Ann. N.Y. Acad. Sci. **369:**33–46.
5. **Cleveland, W. L., I. Wood, and B. F. Erlanger.** 1983. Routine large scale production of monoclonal antibodies on a protein-free culture medium. J. Immunol. Methods **56:**221–234.
6. **Daniel, C.** 1976. Applications of statistics to industrial experimentation, p. 1–11, 175–179. J. Wiley & Sons, New York.
7. **Daniel, J. T., G. E. Templeton, R. J. Smith, Jr., and W. T. Fox.** 1973. Biological control of northern jointvetch in rice with an endemic fungal disease. Weed Sci. **21:**303–307.
8. **El-Rafi, A.-M., L. Sallam, and I. El-Kady.** 1970. The transformation of progesterone by *Rhizopus nigricans* REF 129 as influenced by modification of the fermentation medium. Bull. Chem. Soc. Japan **43:**2878–2884.
9. **Goodman, J. J., and L. S. Smith.** 1960. 16α-Hydroxy steroids. IX. Effect of medium composition on isomerization of 9α-fluoro-16α-hydroxyhydrocortisone and 9α-fluoro-16α-hydroxyprednisolone (Triamcimolone) during microbiological fermentations. Appl. Microbiol. **8:**363–366.
10. **Hawley, M. C., J. R. Black, and E. A. Grulke.** 1981. Ethanol for gasohol: production and economics. Feedstocks **22:**22–25.
11. **Higuchi, K.** 1973. Cultivation of animal cells in chemically defined media, a review. Adv. Appl. Microbiol. **16:**111–136.
12. **Kaul, B., and E. J. Staba.** 1967. Ammi Visnaga (L) Lam. tissue culture. Z. Arzneipflanzenforsch. **2:**146–155.
13. **Ligget, R. W., and H. Koffler.** 1948. Corn steep liquor in microbiology. Bacteriol. Rev. **12:**297–311.
14. **Mathur, R. S., H. L. Barnett, and V. G. Lilly.** 1950. Sporulation of *Colletotrichum lindemuthianum* in culture. Phytopathology **40:**104–114.
15. **McCoy, C. W., T. L. Couch, and R. Weatherwax.** 1978. A simplified medium for the production of *Hirsutella thompsonii*. J. Invertebr. Pathol. **31:**137–139.
16. **McCullough, K. C., R. N. Butcher, and D. Parkenson.** 1983. Hybridoma cell lines secreting monoclonal antibiodies against foot and mouth disease virus. I. Cell culturing requirements. J. Biol. Stand. **11:**171–181.
17. **Pan, C. H., S. V. Speth, E. McKillip, and C. H. Nash III.** 1982. Methyloleate-based fermentation medium for cephalosporin C production. Dev. Ind. Microbiol. **23:**315–325.

18. **Polack, J. A., H. S. Birkett, and M. D. West.** 1981. Sugarcane: positive energy source for alcohol. Chem. Eng. Proc. **81**:62–65.
19. **Prescott, S. C., and C. G. Dunn.** 1959. Industrial microbiology. McGraw-Hill Book Co., Inc., New York.
20. **Smith, J. E., and J. G. Anderson.** 1973. Differentiation in the aspergilli, p. 295–337. *In* J. M. Ashworth and J. E. Smith (ed.), Microbial differentiation: 23rd Symposium of General Microbiology. Cambridge University Press, London.
21. **Snedecor, G. W., and W. G. Cochran.** 1980. Statistical methods, 7th ed. The University of Iowa Press, Ames.
22. **Steel, R. G. D., and J. H. Tosrie.** 1960. Principles and procedures of statistics. McGraw-Hill Book Co., Inc., New York.
23. **Weinshank, D. J., and J. C. Garver.** 1967. Theory and design of aerobic fermentations, p. 417–449. *In* H. J. Peppler (ed.), Microbial technology. Reinhold Publishing Corp., New York.
24. **Weiss, L. H., and C. J. Mikulka.** 1981. Gasohol: a realistic assessment. Chem. Eng. Prog. **81**:35–41.
25. **Zabriskie, D. W., W. B. Armiger, D. H. Phillips, and P. A. Albano.** 1982. Traders guide to fermentation media formulation. Traders Protein, Memphis, Tenn.

Chapter 11

Continuous Fermentation

KEVIN STAFFORD

Continuous cultivation of microorganisms allows one to separate and define parameters which are interdependent in batch fermentation, namely, culture growth rate, nutritional and physical environment, and cell density. For the scientist exploring microbial form, function, or population dynamics, the continuous-culture technique is an invaluable tool. For the engineer attempting to optimize the industrial-scale production of biomolecules, the control of these parameters can be translated into increased productivity over batchwise fermentation. The theory, equipment, and methodology of continuous culture are described in this chapter. There are few such descriptions available in the literature, perhaps due to the diversity of fermentations and the consequent proliferation of equipment and methods. Although the systems presented here can in no way encompass all possible needs, it is hoped that they can be used as a convenient point of departure.

11.1 GENERAL THEORY AND DESCRIPTION

A microbial population can adapt to an environment which is modified by its own metabolism. This adaptation causes environmental modification, which causes further adaptation, in a cycle that ultimately ends in starvation and death, or sporulation and stasis. At times the investigator of microbial physiology requires a sample of cells which is not constantly changing. Although exponentially growing cells in an exponentially changing batch fermentation may suffice for some studies, this is not always satisfactory.

A batch fermentation may be extended by feeding it the nutrients which limit cell growth. This "fed-batch" technique can forestall but never prevent the inevitable accumulation of an unsustainable cell mass. And even though growth may be prolonged, depletion of selected nutrients and accumulation of metabolic by-products change the environment.

Again, the microbe adapts to the environment that it changes.

In the absence of genetic selection, continuous fermentation offers the means to obtain a cell population which grows indefinitely in an unchanging environment. This is accomplished by feeding complete medium to a fermentation and removing whole broth to maintain a fixed volume. The turbidostat and the chemostat are the two most widely used techniques for growing cells continuously. In a turbidostat, cells grow at their maximal rate and the cell density is controlled by washing the cells out of the vessel to maintain a certain turbidity. In a chemostat, cell growth is limited by a selected nutrient. Thus the rate at which the medium is supplied dictates the growth rate of the organism, and the concentration in the nutrient feed determines the cell density which can be attained. Of these operating modes, the chemostat is more generally used and will be treated in detail. The turbidostat and other techniques of continuous fermentation are briefly discussed at the end of this chapter. With only minor exceptions, the equipment and procedures described for the chemostat apply to these other techniques as well.

11.1.1 Material Balance on Biomass

The chemostat is usually started as a batch fermentation. Before nutrient limitation, the nutrient feed is begun at a flow rate, F, expressed in liters per minute. The liquid phase volume, V, expressed in liters, is maintained by controlled removal of whole broth. The cells grow until the chosen nutrient is limiting. After this, cell growth is limited by the rate of medium addition.

By accounting for all the biomass produced, exiting, and accumulating in the fermentor, a material balance on cell mass can be written:

Accumulation = Incoming medium + Cell growth − Washout − Death

$$\frac{dX}{dt} = \frac{F}{V}X_0 + uX - \frac{F}{V}X - aX \quad (1)$$

where dX/dt is rate of accumulation of cell mass per unit of time (grams of dry cell weight [DCW] per hour); X_0 is cell concentration (grams of DCW per liter) in the incoming medium; X is cell concentration (grams of DCW per liter) in the fermentor; u is specific growth rate of the culture ($hour^{-1}$); and a is specific death rate of the culture ($hour^{-1}$). This expression can be simplified if it is assumed that the chemostat is at steady state, that there are no cells in the incoming medium, and that cell death is negligible (4, 6). The assumption of steady state implies that there is no net accumulation of cell mass in the fermentor, so the left-hand side of the equation is equal to zero. Equation 1 reduces to:

$$\frac{F}{V}X = uX \quad (2)$$

and

$$F/V = u \quad (3)$$

Thus, by maintaining a constant volume and changing the nutrient feed rate, control over the specific growth rate up to the maximum specific growth rate can be achieved. The rate at which the culture is replenished by fresh medium is referred to as the dilution rate, D. The dilution rate is expressed in $hour^{-1}$ and is defined as:

$$F/V = D \quad (4)$$

11.1.2 Determination of Maximal Specific Growth Rate

Occasionally it may be helpful to determine the maximal specific growth rate of a culture in the fermentor rather than in shake flasks. To use this in situ method, the dilution rate is increased to approximately 25% above the estimated maximal growth rate of the organism. All nutrients are delivered faster than the culture can utilize them, and the limiting nutrient concentration begins to increase. Unlike the chemostat at steady state, the material balance of this system reflects a nonzero rate of change in biomass concentration:

$$\frac{dX}{dt} = uX - \frac{F}{V}X \quad (5)$$

Substituting equation 3 into equation 4 and rearranging gives specific growth rate = washout rate + dilution rate, or:

$$u = \frac{1}{X}\frac{dX}{dt} + D \quad (6)$$

The cell density declines exponentially as the culture washes out. By plotting the log of the cell density against time, one obtains a straight line with a slope equal to the washout rate. The difference between the washout rate and the dilution rate equals the maximal specific growth rate.

11.1.3 Material Balance on Substrate

The material balance on substrate is expressed as: rate of change = substrate in − substrate out − substrate for cells − substrate for product − substrate for maintenance, or:

$$\frac{dS}{dt} = DS_0 - DS - \frac{uX}{Y_{x/s}} - \frac{q_pX}{Y_{p/s}} - mX \quad (7)$$

where S_0 is the concentration (grams per liter) of limiting nutrient in feed; S is the concentration (grams per liter) of limiting nutrient leaving the vessel; $Y_{x/s}$ is the yield of cells on substrate (grams of DCW per gram of substrate); $Y_{p/s}$ is the yield of product on substrate (grams of product per gram of substrate); q_p is the specific product formation rate (grams of product per gram of DCW · hour); and m is the maintenance requirement (grams of substrate per gram of DCW · hour). Under certain circumstances this expression can be simplified. For example, in the case of a rapidly growing culture much more substrate is required for growth than for maintenance metabolism and the maintenance term is negligible. If little substrate is used for product formation, then at steady state equation 7 can be rearranged and combined with equation 5 to yield:

$$Y_{x/s}(S_0 - S) = X \quad (8)$$

Provided that S is the growth-limiting substrate, the concentration of cells in the vessel can be manipulated independently of the growth rate by increasing or decreasing the substrate concentration in the feed. This allows the researcher or engineer to tailor the demands of the fermentation to the capacity of the equipment by increasing or decreasing the biomass concentration.

11.1.4 Determination of Yield and Maintenance Coefficients

The maintenance and yield coefficients for the limiting nutrient can be estimated from data obtained in a chemostat (4, 6, 70). An acceptable method for obtaining these values derives from a material balance on the substrates. Once again, a number of assumptions can be made to simplify this expression. At steady state the nutrient concentration should not change: $dS/dT = 0$. The inlet substrate concentration, S_0, is much greater than the exit concentration; thus $F/V \times S$ can be ignored. In addition, one must assume that the substrate required for product formation is negligible compared to that for maintenance and cell growth. This final assumption is difficult to justify in cases where product formation rates are high. However, in some cases the expression may simplify to:

$$\frac{FS_0}{V} - mX = \frac{uX}{Y_{x/s}} \quad (9)$$

Because $F/V = D = u$, the equation can be solved for $1/Y_{x/s}$:

$$\frac{1}{Y_{x/s}} \text{(theoretical yield)} + \frac{m}{u} = \frac{1}{Y_{x/s}} \text{(actual yield)} \quad (10)$$

With the above assumptions, the "actual yield" is the cell density at a given dilution rate divided by the substrate concentration in the nutrient feed. This value will decrease with decreasing dilution rate. As growth rate declines, an organism utilizes a greater percentage of the available substrate for "endogenous" or "maintenance" metabolism than for growth. To obtain the maintenance and yield coefficients, the reciprocal of the apparent yield is plotted against the reciprocal of the specific growth rate. A straight line should be obtained having a y intercept equal to the reciprocal of the theoretical yield and a slope equal to the maintenance coefficient (Fig. 1).

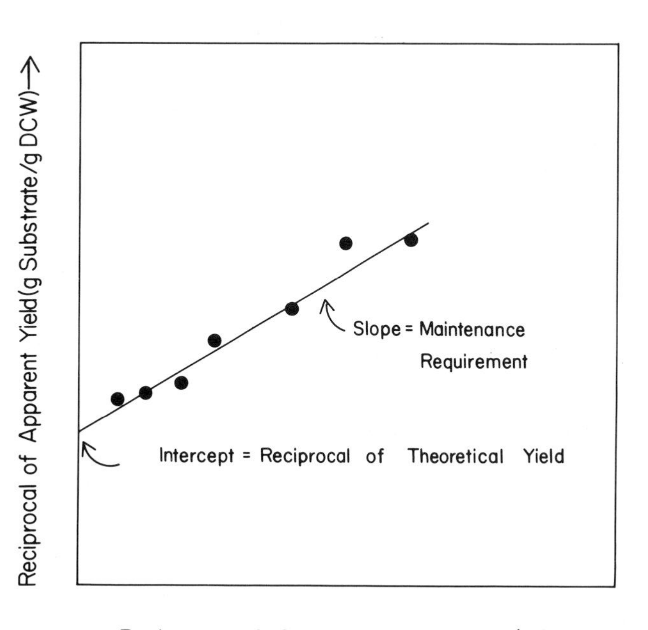

FIG. 1. A reciprocal plot from which the theoretical yield and maintenance requirements of an organism can be estimated. When the reciprocal of the apparent yield of cells on substrate is plotted against the reciprocal of the specific growth rate, the slope of the line is equal to the maintenance requirement. The y intercept is equal to the reciprocal of the theoretical yield.

11.1.5 Monod Model of Growth

The most common model for relating cell growth rate to substrate concentration in chemostats derives from an observation by Monod. He noted that the rate of growth of a microbial culture relative to its maximal growth rate is proportional to the limiting substrate concentration and a constant:

$$\frac{u}{u_{\max}} = \frac{S}{K_s + S} \quad (11)$$

where u is the specific growth rate (hour^{-1}); $u_{\max}$ is the maximal specific growth rate (hour^{-1}); S is the limiting substrate concentration (grams per liter); and K_S is the substrate concentration (grams per liter) at which the maximum specific growth rate is reduced by one-half. This expression, which relates substrate concentration to specific growth rate, resembles the Michaelis-Menten model for enzyme kinetics. The value K_S indicates the efficiency with which an organism can remove the substrate, S, from its environment.

Other expressions have been formulated in an effort to model microbial growth more accurately. A modification of the Monod equation by Edwards et al. includes product inhibition (22), and the model of Droop, first developed for algae, uses intracellular substrate concentration to predict growth rates (18).

11.1.6 Industrial Application

The chemostat may offer a significant increase in productivity over batch or fed-batch operation. Increased productivity is the result of reduced fermentor down time per unit of product manufactured (30). Also, because the growth rate and cell concentration can be controlled independently, critical demands of the fermentation for oxygen supply and heat removal can be adjusted to match the equipment's capabilities. In addition, by controlling the dilution rate of the continuous fermentation, it may

TABLE 1. Medium requirements at different dilution rates

Fermenter working vol	Liters of medium required per day at dilution rate:			
	0.01 h^{-1}	0.05 h^{-1}	0.50 h^{-1}	1.0 h^{-1}
250 ml	0.06	0.3	3	6
1,000 ml	0.24	1.2	12	24
5 liters	1.2	6.0	60	120
10 liters	2.4	12.0	120	240

be possible to maintain an optimal specific growth rate for product expression. The productivity gains which can be achieved by continuous culture allow for smaller manufacturing plants and thus the reduction of the capital costs for new facilities.

However, continuous culture is not widely used as an industrial process. Three disadvantages must be overcome before continuous culture becomes a viable alternative to the batch or fed-batch processes. First, the continuous addition of feed streams over several weeks greatly increases the probability of contamination. This risk can be minimized by careful operation and equipment design. Second, equipment and instrumentation used to operate and control the process must be extremely reliable for long periods of continuous operation. The overall failure rate can be reduced by redundancy of critical equipment and by frequent monitoring of the process.

The most difficult problem to correct is strain degeneration, a gradual and irreversible loss of product expression (21, 27, 29, 59). Strain degeneration occurs when a spontaneous mutation produces a new strain with a selective growth advantage but a reduced capacity for product formation. In this way a population of low producers can become the dominant species. A further discussion of selection may be found in section 11.5.

Most overproducing strains used for the industrial production of biopolymers, amino acids, and antibiotics are the result of mutation and selection programs. These strains are generally unstable and are highly likely to revert to poorer producers and better growers in continuous cultures. Genetic instability is a major obstacle to the introduction of continuous fermentation technology into the biochemical process industry.

11.2 EQUIPMENT DESIGN

Although the theory of continuous fermentation is straightforward, successful translation of the theory into practice and interpretable data is more formidable than one might at first imagine. At the bench scale, the most challenging job is to assemble reliable equipment and develop appropriate techniques. Generally, this is an iterative process in which equipment and procedures are replaced and modified as experience is gained. The type of fermentation will always dictate the choice of equipment. Medium composition (chemically defined or complex) and organism morphology (filamentous or unicellular) set restrictions on the type of equipment that can be used.

11.2.1 Vessel

In selecting equipment one must determine what scale of operation is most suitable. This is the variable that has the greatest influence on the day-to-day operation of the chemostat. The data in Table 1 show medium consumption rate as a function of dilution rate and fermentor working volume. Thus a benchtop-scale fermentor is more convenient to operate. In addition, investment and space requirements are modest, and peristaltic pumps of adequate reliability are available. A benchtop vessel is also easily dismantled for cleaning. This can be especially advantageous after long mycelial fermentations.

Small-scale continuous culture offers some unique problems, however. When operating at volumes of less than 500 ml, sampling can reduce the volume, affecting the dilution rate and thus the metabolism of the organism. Feed rates that are required to obtain low dilution rates can be very difficult to control, and feeding a complex medium at low flow rates can result in settling out of insoluble medium components in feed lines, causing plugging. It is often difficult to get a sample for off-gas analysis due to low back-pressures and low aeration rates. Also, wall growth can represent a larger fraction of the total population in small vessels. Thus, wall growth may become a significant variable in studies of population dynamics (21, 48). Teflon vessels and silicone coating reagents can be used which may reduce or prevent wall growth in particular fermentations.

11.2.2 Nutrient Feed Reservoir

If carboys are used for feed reservoirs, they should be calibrated by volume or placed on a weighing device so that integrated nutrient feed rates can be determined accurately over several hours or days. An addition port on the reservoir should be included so that heat-labile nutrients can be added separately after sterilization. A vent through a sterilizing filter must be attached and a nutrient feed line should be installed. The feed line should be a stainless-steel tube going to the bottom of the vessel. A magnetic stirring bar must be included to suspend any particulates and mix nutrients which are added after sterilization of the bulk medium.

Silicone rubber tubing should be used where tubing is required. It can be sterilized repeatedly, retains its resilience, and has excellent chemical resistance. The method of connecting tubing should be aseptic, fast, and reliable. One method of connecting tubing is shown in Fig. 2 (17). These connectors are made of stainless steel and can be sterilized by flaming immediately before connecting.

A pipette, attached to each reservoir bottle as shown in Fig. 3, will allow measurement of the instantaneous medium flow rate and adjustment of the pump rate. To obtain the medium flow rate, the clamp that isolates the pipette from the system is removed, allowing medium to fill the pipette by gravity to the same height as in the reservoir. The medium reservoir is then clamped off. The time required to pump a known volume of medium from the pipette is used to calculate the liquid flow rate. The pump is adjusted and the procedure is repeated until the desired flow

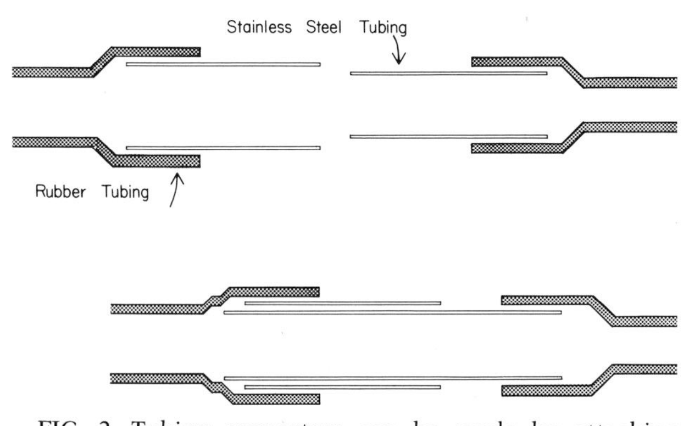

FIG. 2. Tubing connectors can be made by attaching silicone tubing to a stainless-steel tube (2.5 by 0.25 in. OD; ca. 63.5 by 6.4 mm OD). The other silicone tubing should be attached to a stainless-steel tube (2 by 0.3 in. OD; ca. 50.8 by 7.6 mm OD). The ends can be flamed for sterilization and inserted one into the other. The longer piece of stainless tubing seals against the silicone tubing of the shorter piece.

rate is achieved. Over the course of a long fermentation the pipette can clog and may need to be replaced. Therefore it should be connected in such a way that it can be removed if necessary. One should be aware that the flow rate through a peristaltic pump is dependent on the resilience of the tubing. Because silicone rubber tubing loses its resilience with use, a long piece should be used so that the pump can be moved every other day to a new section. For extended operation, replacement of the tubing may be necessary.

The medium inlet should be designed with a drip-tube in the line immediately before the vessel. A drip-tube provides a break in the medium flow path that prevents microorganisms from growing into the feed line. A drip-tube can be purchased from New Brunswick Scientific Co. or constructed as shown in Fig. 4. Some fermentor manufacturers recommend that the medium feed go into the vessel with the air. In certain cases this may suffice, but where insoluble medium components make up even a small percentage of the feed stream, the air inlet can plug after prolonged use.

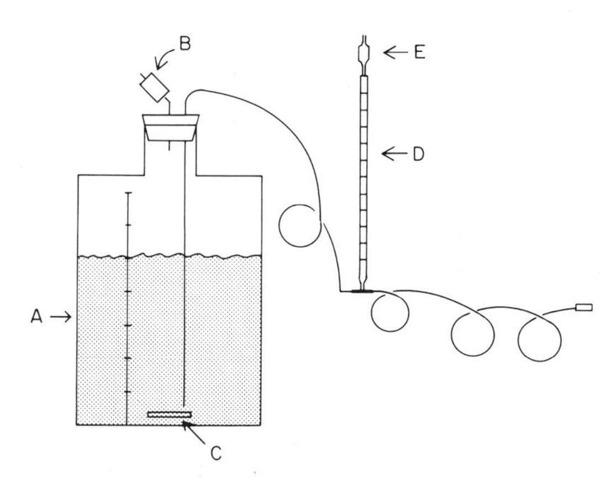

FIG. 3. Nutrient reservoir with pipette for calibrating medium flow rate. A calibrated polycarbonate carboy reservoir (A) contains a sterile air filter (B) and a magnetic stirring bar (C). A pipette for instantaneous feed rate measurement (D) is included and has its own sterile air filter (E).

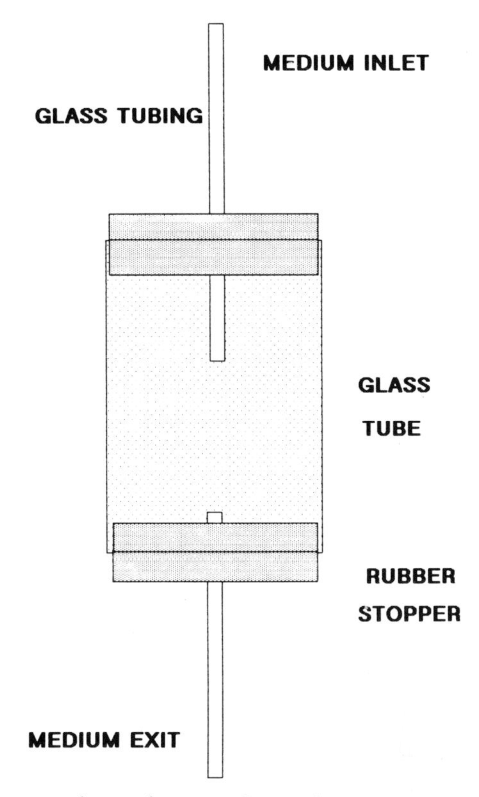

FIG. 4. A drip tube provides a physical barrier to the growth of microorganisms up the nutrient feed line. A drip tube can be constructed from two rubber stoppers and a glass cylinder.

In some cases very low feed rates may be required. It is difficult to find pumps that provide reliable, low flow rates. In such cases, a timer may be used to turn the pump on only a certain percentage of the time. In this way a flow rate can be attained which is a fraction of the minimum flow rate of the pump. This is particularly helpful when nonlimiting substrates are to be fed, such as product precursors or antifoaming agents. The use of this system for feeding the limiting nutrient, however, will probably result in metabolic cycling which may not be desirable.

11.2.3 Broth Removal and Level Control

To maintain accurate control of the dilution rate, it is very important that the ungassed volume be determined and controlled. There are two parameters used to control the ungassed vessel volume: gassed liquid level (assuming gas hold-up is constant), and mass of the vessel contents. Because control of gassed volume is more practicable at the bench scale, it is the most widely used. However, constant gas hold-up is difficult to achieve. Alternatively, mass control can be more accurate if properly installed, but it is much more expensive to implement on the bench scale of operation.

There are two methods of liquid level control. In the technique most often recommended by vendors of continuous-culture vessels, the air outlet is placed at

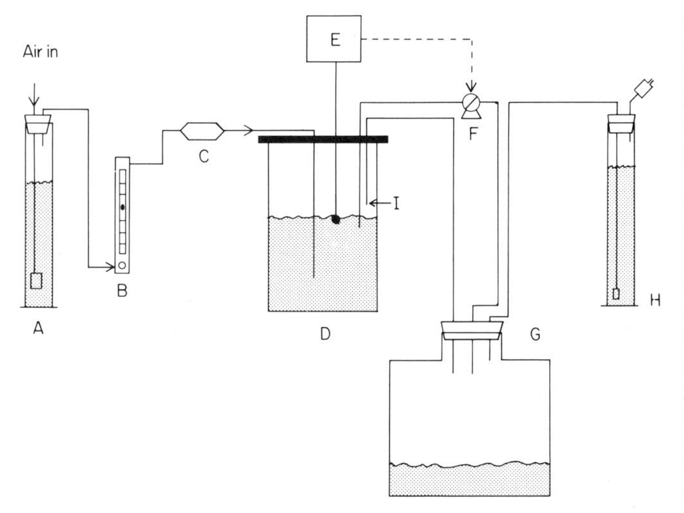

FIG. 5. Airflow and level control in a chemostat. Air enters via a humidifier (A) and passes through a rotometer for control (B) and a sterile filter (C). The liquid level in the vessel is detected by a sensor, and a level controller (E) activates a pump (F) to remove broth from the vessel (D). The product reservoir (G) receives broth from the pump and foam outs from the air exit (I). All off-gas is scrubbed (H) before leaving the system, as described in the text.

the broth surface. The exhaust air is then directed into a reservoir which is vented aseptically to the atmosphere. When the broth volume increases, broth is forced out into the product reservoir with the air. Such a system suffers from the fact that cells can sometimes be at a different concentration at the air/liquid interface than in the bulk liquid. This can cause a nonrepresentative sample to be pumped from the vessel. Such partitioning can be affected by pH, ionic strength, and surface-active agents. In some cases the effect is so dramatic that foam fractionation has been used to harvest bacterial cells (68).

In contrast, a liquid level controller can be used. Although this technique requires more equipment, it allows withdrawal of whole broth from below the liquid surface. A number of liquid level controllers are available. Most are based on a sensor within the vessel that makes physical contact with the culture fluid. Because the sensor is located within the tank, failure of the probe will end the run. Depending on the type of sensor used, inaccuracies in volume can be caused by variations in agitation, aeration, foam generation, and wall growth. One also loses the flexibility of being able to change the level in the vessel without contaminating the culture. Ultrasonic level controllers are available which can be attached to the outside of a glass or steel vessel. These sensors can be attached after autoclaving, and if a sensor malfunctions it can be replaced without violating sterility. Sensitivity can be adjusted so that effects due to foam can be minimized. Problems caused by gas hold-up and foam cannot be fully eliminated, however.

A redundant control scheme that I have found to be quite useful is shown in Fig. 5. In this system an automatic level controller is backed up by the exhaust gas system. Note that the medium exit tube is located just below the liquid level. The exhaust gas exit is located just above the liquid level. The exhaust gas is directed into the product reservoir. If the level controller fails "on," the overflow pump cannot reduce the liquid level below the medium take-off line. If the level controller fails "off" or the pump stops or the tubing becomes clogged, then the liquid level cannot rise above the exhaust gas exit tubing.

The use of mass as a control parameter eliminates the problems of variable gas hold-up and foaming, but may introduce other problems. Because of high equipment weight relative to the weight of the broth being measured, the use of a load cell to control volume is difficult at volumes less than 5 liters. Tubing connections must be carefully supported to avoid inadvertent changes in weight, and on larger vessels piping must be designed with flexible couplings to avoid supporting the vessel. Differential pressure sensors can also be used to measure the mass in a fermentor. These devices measure the hydrostatic pressure between the top and the bottom of the vessel. The pressure is then used to calculate the fermentor volume. The practical lower limit of these sensors is about 500 to 1,000 liters.

A stable continuous culture requires an accurate and reliable pump. The pump should be cleanable and steam sterilizable. Most importantly, it should be capable of long-term aseptic operation. On the benchtop scale, peristaltic pumps meet these criteria. As fermentation scale increases, peristaltic pumps become impractical. Silicone tubing should be replaced by piping. Generally, in situ steam sterilization is required at greater than the 10-liter scale. Diaphragm pumps can be used which provide reliable, adjustable flow rates. These pumps have several advantages over centrifugal and gear pumps. Perhaps the greatest advantage is the lack of a mechanical seal. Over the course of time mechanical seals wear and can be a source of contamination. The diaphragm pump has no such seal and contamination problems are minimized. Second, the pump head and the lines leading to and from the head can be sterilized by steam. Finally, the flow rate can be maintained over a wide range of pressures. This means that changes in the back-pressure of the vessel required to obtain maximal oxygen transfer can be accomplished without substantially changing the pump flow rate.

11.2.4 Sampling

When complex media and mycelial organisms are studied, the culture broth can attain high viscosity, and removing a sample aseptically can be difficult. One technique requires the air inlet to be connected to the sample port as shown in Fig. 6. To purge the sample line, the accessory air line is opened and sterile air forces a portion of the purge volume from the sample line. The accessory air line is then closed. By unplugging a portion of the sample line in this way, the rest of the sample line is more easily cleared. Caution must be exercised because this procedure can generate aerosols. The use of large-diameter tubing in the sample line can also ease plugging problems, but this should be weighed against the problems created by requiring too great a purge volume. In many cases, the total sample and purge volume may become so large that the dilution rate will be significantly af-

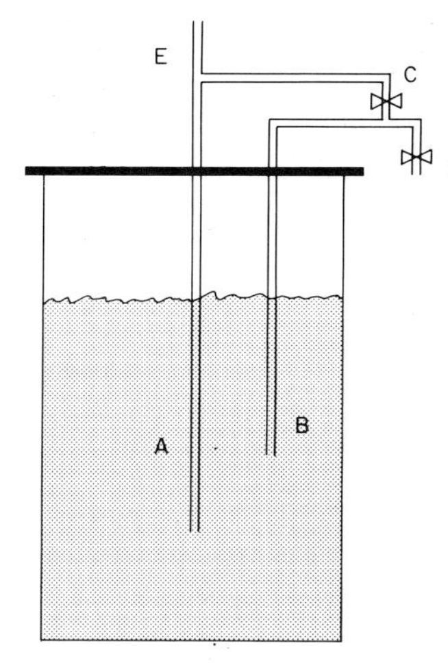

FIG. 6. Sampling device for high-viscosity broth. (A) Air sparger; (B) sample dip-tube; (C) supplementary air to sampler; (D) sample exit; (E) air inlet.

fected. In general, the sample volume should not exceed 10% of the vessel volume.

11.2.5 Environmental Control

Temperature, pH, and dissolved oxygen (or DO_2) measurement and control are similar in batch and continuous operations. Dissolved oxygen measurements may present problems that are not usually encountered in short fermentations. Over the course of time, a biofilm may begin to cover the membrane of some types of polarographic probes. Microorganisms adhere to the steel screen protecting the probe's membrane. A partial solution to the problem is to cover the screen with a 1.5-mil (ca. 2.5×10^{-3} cm) Teflon film. The response of the probe is greatly reduced, however. In small vessels there may not be room in the headplate for a commercially available polarographic probe, necessitating the use of a galvanic probe. If a galvanic probe is used, it must be calibrated with nitrogen and air after autoclaving. In contrast, polarographic probes have excellent stability after autoclaving, and the electronic zero feature can be used to zero them.

11.2.6 Air Inlet and Exit

An example of an air supply and exhaust system is shown in Fig. 5. Humidification of the air prevents evaporation of medium from the vessel and improves the estimation of dilution rate. This is particularly important when operating at temperatures of 30°C or more, when using compressed air which is very dry, or when operating at dilution rates of less than 0.1 h^{-1}. Sterilization of the inlet air can be provided by an absolute membrane filter or a packed glass-wool filter. Two air filters connected by a tee (not shown) are recommended as a precaution, especially when using humidifiers. In the event that one filter becomes plugged, the other filter can be unclamped and used immediately. A regulated air supply is important for safety purposes. In the event that the air exit becomes plugged, the air pressure should be sufficiently low that the vessel does not explode. Wrapping the vessel with tape is an added precaution to prevent flying glass.

The exhaust gas should exit to the overflow reservoir. In this way foam-out can be contained. The overflow reservoir is then connected to a graduated cylinder filled with 0.5% bleach. The exit air is sparged through the liquid and is vented through a sterilizing filter. This additional piece of equipment serves several useful functions. First, it provides a constant back-pressure on the vessel which can be changed by changing the liquid level in the cylinder. This back-pressure is necessary if an exit gas sample is to be sent to an analyzer. It also facilitates sampling by providing enough pressure to force the sample out of the vessel. Second, one has an instantaneous visual verification of airflow through the vessel. Absence of bubbling may indicate loss of containment, a particular concern if pathogenic or recombinant organisms are being cultured. Third, the use of bleach reduces the potential of viable microorganisms escaping to the environment. This includes the microbe under investigation as well as those which may potentially invade the vessel: bacteriophage, sporeformers, and the like. Finally, it acts as a scrubber to reduce unpleasant odors in the laboratory.

11.3 OPERATION

11.3.1 Equipment Sterilization

Proper sterilization of culture vessels and nutrient reservoirs in excess of 1 liter requires more time than one might expect. Table 2 gives some helpful data in this regard, although the times may depend on the autoclave used and the position of the items in the autoclave. Also, sterilization of medium containing insoluble components requires more time than that of synthetic medium, due to the suspended solids. It should be emphasized that glass bottles are very dangerous when hot; a cool draft can easily break a hot bottle. Sterilizable polycarbonate and polypropylene carboys are much safer, but may release low levels of plasticizer and monomer into the medium.

11.3.2 Start-Up

The chemostat is usually started as a batch fermentation. When the culture is in the exponential growth phase, the nutrient feed and level controller are turned on. If the start-up medium has the same

TABLE 2. Effect of medium volume on sterilization time (38)

Liquid vol (liters) per containers	No. of containers per load	Initial liquid temp (°C)	Time (min) for liquid to reach 121°C
2.0	10	27	37
4.0	5	26	52
6.0	4	26	62
10.0[a]	2	25	105

[a] These data were obtained by the author, using a Finn Aqua autoclave to sterilize a minimal salts medium at 120°C with a 5-min purge cycle.

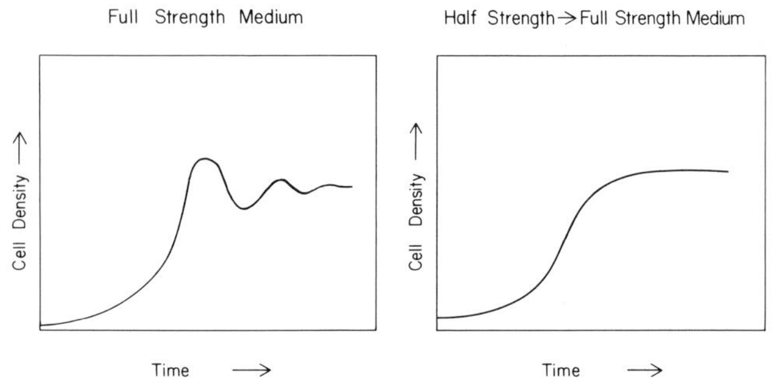

FIG. 7. When a chemostat is started in full-strength medium, the sudden nutrient limitation that cells encounter upon entering continuous mode can result in damped oscillations of cell density and metabolic function. Beginning the fermentation in half-strength medium and feeding full-strength medium can prevent this behavior.

composition as the feed medium, the transition from batch to continuous fermentation may be accompanied by an oscillation in certain fermentation parameters such as cell density or the concentration of a secondary metabolite (Fig. 7). If the substrate is toxic to the cells, the oscillation may become so severe that washout occurs. This problem occurs because the exponentially growing culture abruptly becomes substrate limited. The metabolism of the organism must shift to accommodate the nutritional limitation, and oscillations ensue. An excellent discussion of the transition from batch to chemostat operation is provided by Dunn and Mor (19). An additional discussion is provided in section 11.4 below.

A smooth transition from batch to continuous fermentation can be obtained by using half-strength medium for the initial batch growth. The nutrient feed is started when the cell concentration is about one-half of that expected at steady state, and the culture becomes a nutrient-limited fed batch. The specific growth rate declines as the cell concentration increases until the steady-state concentration is achieved. At this time the level controller is turned on and the continuous fermentation begins.

11.3.3 Determination of Steady State

The assumption of steady state is fundamental to the development of the kinetics of continuous culture. Steady state is achieved when there is no net change in the contents of the reactor: the biomass, substrate, and product concentrations are unchanging, and the metabolic state of the organism is constant. Theoretically, a continuous chemical reactor will reach 95% of steady state after three residence times. A residence time is the time it takes for one reactor volume to pass through the vessel and is equal to the reciprocal of the dilution rate. Although cell concentration generally follows these kinetics, the biological processes which accompany cell growth may not reach steady state so quickly (37). Thus, using cell density as a measure of steady state can often be misleading. The best indication of steady state is a constant value of the cell product being studied for a period of three residence times. Mutation and selection within a chemostat can affect the attainment of steady state. A further discussion is provided in section 11.5 below.

11.3.4 Medium Formulation and Optimization

Before operating a nutrient-limited fermentation it is useful to approximate the yield of cells on substrate. For many fermentations, a chemically defined medium is available. Cell density and substrate concentration are easily determined. However, for some fermentations, most notably the antibiotics, insoluble complex nutrients are used. These present the investigator with potential problems. First, how is biomass to be measured, if at all? Assuming the techniques of biomass estimation are translated from batch fermentations to continuous fermentation, one still has the problem of insoluble nutrients and their slow release. Phosphates, amino acids, sulfate, and carbohydrates can all be slowly liberated by enzymatic or chemical hydrolysis of complex medium components such as soy flour, cottonseed meal, etc. Particle size and residence time in the fermentor then become additional process variables. An alternative which can eliminate this problem is to formulate a synthetic medium.

The choice of a growth-limiting nutrient depends on a basic understanding of the processes one wishes to study. In cases where the overproduction of a particular biomolecule is desired, a rational approach to the selection of a nutrient limitation may be possible (34, 53). Nutrient limitations can cause physiological and morphological changes which can be exploited (64, 65). The most direct of these changes occur in fermentations where nutrients repress the synthesis of certain enzymes (13, 60). Many antibiotics are produced poorly in media containing excess glucose or phosphate (12, 45, 46). In some cases protease production is reduced in the presence of excess nitrogen or sulfur (1, 66). The production of eucaryotic proteins by genetically engineered strains of *Escherichia coli* is frequently designed to be repressed by tryptophan (25). In such systems a nutritional limitation of the repressing compound can be advantageous. Nutrient limitation can sometimes be used to uncouple growth from other metabolic pathways (15). For example, nitrogen limitation can cause excess carbon to be redirected from cell mass into the overproduction of biopolymers such as xanthan gum (36) and polyhydroxybutyrate (16). Simple oxidation products of glucose, such as gluconic acid and citric acid (39, 43), can also be overproduced when a nutrient is limiting. Finally, nutritional deficiencies can alter the structure of a microorganism (65). Phosphate and magnesium limitations are known to alter cell membranes in some bacteria to such an extent that permeability characteristics are changed (62). Such microorganisms may become more susceptible to antibiotics (14, 28), and some lose pools of intracellular metabolites to the surrounding medium (5). The limitation of biotin in biotin auxotrophs of *Corynebacterium glutamicum* has been shown to cause leakage of glutamic acid by changing membrane permeability (49).

The yield of cells on substrate, $Y_{x/s}$, must be determined to calculate the concentration of limiting substrate to be used in the feed medium. The yield can be approximated in shaker-flask culture if cell lysis and sporulation do not occur. One should be aware that growth yields derived from batch experiments can be quite different in continuous culture, and this technique can only give a rough approximation (26). Several flasks are prepared with a medium containing all essential nutrients except the limiting nutrient. This substrate is then added back to several flasks in incremental concentrations. An inoculum is centrifuged and washed twice in the basal medium to prevent carryover of the substrate. The flasks are then inoculated, and cell mass is determined after growth ceases. A linear relationship between cell density and substrate added will be obtained up to a certain substrate concentration. After this point other factors begin to limit growth, such as pH and dissolved oxygen. A linear regression of cell mass against substrate concentration in the linear region gives an approximation of the yield which can be used to formulate the chemostat medium. A subinhibitory excess of all other nutrients should be provided. A fivefold excess is recommended as a starting point.

11.3.5 Experimental Design

When designing an experimental program in which continuous fermentation is to be used, one must be aware of the selective pressure to which the culture is subjected. Spontaneous mutations arising at a rate of about 1 in 10^6 cell divisions may sooner or later generate an organism which has a competitive advantage over the original strain in the vessel. In some cases this selection may be desired. However, if the investigator is interested in studying a homogeneous cell population, a subpopulation of mutants can invalidate the data. Experiments should be designed to separate the metabolic response of the culture to experimental manipulations from genetic changes which may occur in response to the same manipulations.

One way to detect changes in the genetic makeup of the culture is to follow the productivity of the fermentation in satellite shake flasks. Samples are taken periodically from the continuous culture and reinoculated into shake flasks. These parallel batch fermentations can be assayed for growth rate and product formation. Increasing growth rate or decreasing product formation in shake flasks over the course of the fermentation may indicate a changing cell population. This approach suffers from the fact that whole broth is used as an inoculum. A subpopulation of mutants may not be seen until it becomes a significant portion of the culture. An even more rigorous approach involves dilution plating of samples from the culture as the fermentation progresses. Individual colonies can then be reinoculated into shake-flask cultures. In this way a statistically significant number of clones can be evaluated and small changes in the population can be identified.

11.4 TRANSIENT PHENOMENA

Cellular adaptation is the equilibration of the myriad reactions a cell embodies to a new environment. The changes of metabolic activity required to establish the new equilibrium are known as transient phenomena. The study of transients in chemostat culture has led to successful strategies for on-line process control (J. R. Swartz, thesis, Massachusetts Institute of Technology, Cambridge, 1978) and a greater understanding of the regulation of growth (26, 37) and product formation (11, 42, 44, 57, 72). An excellent review of transient phenomena is provided by Cooney et al. (10).

Three different types of environmental changes can induce transient behavior in chemostats. Nutritional transient phenomena will result from a step change or pulse in either dilution rate or limiting-substrate concentration. The response of a cell population to a transient nutrient increase is dependent on the limiting nutrient (5), the magnitude and direction of the change (47), and the presence of other limiting nutrients (12). By suddenly removing a nutrient limitation, one can study the intrinsic metabolic limitation to the rate of growth or product formation. Step changes in temperature (28) and dissolved oxygen (61) have been used to elucidate metabolic regulation of energy production. Finally, inhibitors such as antibiotics or toxic metabolites (8, 13, 42) have been used to inhibit specific functions of microbial metabolism.

The measurement of unsteady-state phenomena can be difficult. Mateles et al. showed that the measurement of specific growth rate during transient growth can be underestimated if cell density measurements are used (47). The relatively small changes in cell density which occur during transient growth can be difficult to determine accurately. A more reliable measurement of specific growth rate can be obtained by determining the concentration of an element in the medium, such as nitrogen, incorporated into cell mass. The concentration of this element will change severalfold as compared to 10 to 50% change in cell mass. Assuming there is no significant product formed containing the element being measured, the expression for specific growth rate can be determined as follows:

$$u = D - \frac{1}{(E_0 - E_c)} \frac{dE}{dt} \tag{12}$$

where E_0 is the concentration of element E in the incoming medium and E_c is the concentration of noncellular element E in the vessel.

The dynamic nature of cellular metabolism can make the measurement of instantaneous rates difficult. Metabolic activities change at varying rates in response to environmental change. Intracellular pools of ATP and ADP change in fractions of a second (31). Synthesis of mRNA (52) and ppGpp (23) can occur in seconds, and de novo synthesis and degradation of proteins (55) can occur in minutes. Therefore, the sampling methods used to investigate transient phenomena should be designed to rapidly quench and stabilize the activity being investigated. Denaturation of enzymes by diluting the sample into cold acid or base is frequently employed to stop further enzymatic activity. Alternatively, the sample may be quickly chilled. All techniques should be validated for complete and reproducible inactivation.

11.5 GENETIC SELECTION

11.5.1 Selective Growth Advantage

The chemostat has long been observed to select different genotypes from initially homogeneous populations. In 1959, Powell (54) detailed the theoretical basis for the takeover of a chemostat culture by an intruding microorganism, either a mutant or a contaminant. The successful predominance of the intruder can only occur under certain conditions of u_{max}, K_S, and dilution rate. The purpose of this section is to discuss techniques for quantifying the selective growth advantage one organism has over another.

If the only interaction between microorganisms in a chemostat is one of simple competition for a single substrate, i.e., no predation, commensalism, antagonism, etc., then:

$$\frac{d \ln (a_t/b_t)}{dt} = \frac{u_a S}{K_a + S} - \frac{u_b S}{K_b + S} \tag{13}$$

where a_t is the concentration of organism a at time t; b_t is the concentration of organism b at time t; u_a is the maximal specific growth rate of organism a; u_b is the maximal specific growth rate of organism b; K_a is the K_S of organism a; K_b is the K_S of organism b; and S is the limiting substrate concentration. It follows from this relationship that:

$$\ln (a_t/b_t) = \ln (a_0/b_0) + st \tag{14}$$

where a_0 and a_t are the concentration of organism a at times 0 and t; b_0 and b_t are the concentration of organism b at times 0 and t; t is time; and s is coefficient of selection (hour^{-1}). Thus, the coefficient of selection, s, can be determined by regressing linearly ln a_t/b_t against time.

11.5.2 Wall Growth

A major obstacle to the study of population dynamics in a chemostat can be wall growth. This is the adhesion of cells to the sides of the vessel in a very thin film which may not be visible to the naked eye. The cells are generally deposited shortly after inoculation and may become a permanent population in the chemostat (67). These cells tend to display different properties of selection and greater genetic stability than the cell population in the bulk fluid (21). Thus, the shedding of cells into the bulk population can maintain a stable cell population at low frequency and the rate of selection will appear to decline with time. The percentage of cells due to wall growth will depend on the surface-area-to-volume ratio of the fermentor. The dilution rate, the number of generations the culture has undergone, and the cell density all influence the degree to which wall growth may affect the experiment. In addition, the affinity of a culture for the glass is species specific. *E. coli* and *Serratia marcescens* have a high affinity while *Bacillus* and *Saccharomyces* species have a low affinity (41).

Wall growth can be detected relatively easily. A new chemostat is started with culture fluid from a chemostat suspected of having wall growth. An increase in the rate of selection indicates that wall growth is present in the initial fermentation. If selection continues at the same rate as in the primary vessel, then some phenomenon other than wall growth may be present (21).

11.5.3 Functional Effects of Mutations

Does the presence of a particular enzyme give an organism a selective growth advantage over an otherwise isogenic strain? Does the metabolic energy investment required to synthesize a protein result in a competitive disadvantage for a cell population under certain physical and nutritional conditions? The chemostat can be used to answer such questions.

In a technique described by Dykhuizen and Hartl (20), three control experiments are required. In the positive control, the parent and mutant are inoculated into a chemostat under a nutrient limitation which will provide selective pressure. For example, a parent strain and its *lac*$^-$ derivative would be inoculated into a lactose-limited chemostat. The purpose of this control is to verify that selection will occur due to the mutation. The negative control is identical to the positive control except that the medium is designed with the selective pressure removed. In this example, the lactose-limited medium would be replaced with a glucose-limited medium. If selection occurs, one may surmise that unidentified genetic changes are present in the strains or that the mutation introduced has pleiotropic effects. The final control involves the strain markers. To easily differentiate between the parent and the mutant, a neutral marker must be transferred into either parent or mutant. The criterion of neutrality implies that no selective growth advantage or disadvantage should result from the marker under the growth conditions of the experiment. Markers such as *tonA* in *E. coli* (40) and concanavalin A resistance in *Saccharomyces cerevisiae* (2) have been reported to be neutral in glucose-limited chemostats. The effect of the marker can be easily tested by inoculating both marked and unmarked strains into a chemostat. The ratio of the two populations should not change significantly. Finally, the two stains are inoculated into a chemostat and the ratio of the marked to the unmarked strain is determined. Dykhuizen and Hartel claim that differences in growth rate of 0.005 h^{-1} can be detected by their methods.

11.5.4 Plasmid Stability

The use of artificially constructed plasmids to express eucaryotic proteins in procaryotes is responsible in large part for the burgeoning interest in biotechnology. The understanding of plasmid stability therefore has value which extends beyond the laboratories of academia.

A plasmid is a self-replicating genetic element that confers traits nonessential for growth on the microorganism. The survival of a plasmid in a population depends on the faithful replication and partitioning of the plasmid into daughter cells. Zund and Lebek studied approximately 100 R factors and showed that about 25% decreased maximal specific growth rate more than 15% (77). Most of those were larger than 80 kilobases in size (77). Perhaps the most widely utilized plasmid vector, pBR322, has been shown to be lost from *E. coli* when grown in a phosphate-limited chemostat (35). Wouters et al. showed that hereditary instability of pBR322 in *E. coli* increased with temperature and decreased with dilution rate (74).

One method of counteracting the loss of plasmids from host cells is to incorporate an antibiotic resistance gene into the plasmid. Plasmid-free segregants are selected against by including the antibiotic in the medium. This type of selective pressure has been used to amplify genes in mammalian cell culture (56). Cells from the murine sarcoma line S-180 were cultivated in increasing concentrations of the folate analog methotrexate. The resulting increase in resistance to methotrexate was shown to be due to gene amplification. *E. coli* mutants which have been shown to overexpress β-lactamase also have increases in gene dosage (51). Another technique of stabilizing a plasmid in a culture is to delete from the host chromosome a gene required for synthesis of a nutrient required for growth. The gene is then incorporated into the plasmid. In this way only cells retaining plasmid can survive (73).

The selective growth disadvantage of cells harboring plasmids is exacerbated by the overexpression of foreign proteins. Metabolic resources normally used for growth may be diverted into the synthesis of a nonfunctional protein. In addition, the high intracellular concentration achieved with some cloned proteins (25) may disrupt normal metabolic functions.

Loss of plasmid copy number is one way in which cells can reduce the burden imposed on cellular metabolism by the expression of foreign proteins. A technique which is commonly employed to detect loss of plasmid copy number is to measure the activity of a protein expressed by the plasmid, such as antibiotic resistance or β-galactosidase activity (9). This technique assumes that the expression of the plasmid is proportional to the number of plasmids per cell. Although such an assumption is sometimes accurate, it is not always justified. For example, if the limitation to protein synthesis is at the level of translation, changes in plasmid copy number will have a negligible effect on protein accumulation.

Using DNA-DNA hybridization to analyze plasmid copy number, Adams and Hatfield have shown that the number of plasmids per cell changes as a function of growth conditions and promoter strength (3). In this technique a ^{32}P probe was synthesized by nick translation of plasmid DNA. The probe was then hybridized to DNA obtained from cell lysate. The amount of the probe bound to cellular DNA was quantified in a scintillation counter. Hybridization of the same probe to known concentrations of plasmid DNA was used to prepare a standard curve. The authors claim that the assay is linear over a 100-fold range. All values were normalized to protein concentration to account for differences in cell disruption.

The loss of the structural integrity of a plasmid can also result in a population of low-producing microorganisms. This can occur by excision of some portion of the plasmid or recombination of plasmid DNA with chromosomal DNA. There have been several reports of loss of unselected markers from plasmids in continuous culture (24, 50). There are two methods by which the loss of plasmid integrity can be demonstrated. The plasmid may be recovered and mapped by an appropriate restriction endonuclease. This technique suffers from the poor resolving power of gel electrophoresis: small changes in the plasmid may not be detected. Alternatively, the recovered plasmid can be retransformed into fresh host. Reduced expression of the trait of interest is indicative of changes in plasmid rather than changes in plasmid copy number or modifications of the host chromosome.

11.5.5 Selection of Mutants with Altered Enzyme Expression

Continuous culture has been used to select organisms that overexpress intracellular enzymes (27). Horiuchi et al. isolated β-galactosidase overproducers of *E. coli* by growing these cultures in lactose-limited chemostats (33). Increases of 5- to 10-fold in specific activity were reported, but many of the strains were unstable and reverted to lower producers when the selection pressure was removed. The addition of a gratuitous inducer could accelerate the selection process. The enzyme produced by the selected strains was shown to be identical to the enzyme produced by the starting cultures. Therefore, the increased specific activity was postulated to be due to mutations in regulation of gene expression or to gene amplification.

A D-serine deaminase constitutive strain of *E. coli* was isolated by Bloom and McFall in a chemostat in which D-serine was the growth-limiting nutrient (7). It was insensitive to catabolite repression by glucose. A mutagen was also added to enhance mutation frequency. Sikyta et al. (58) have used a similar technique to select strains which overproduce ribitol dehydrogenase. The starting culture was a strain of *E. coli* that had been transduced with the genes for ribitol catabolism from *Klebsiella aerogenes*. The transductants were grown in a xylitol-limiting chemostat. The high K_m of ribitol dehydrogenase for xylitol resulted in a strong selective pressure for strains overproducing ribitol dehydrogenase.

Strains which have lost an enzymatic activity may also be selected in continuous culture. The increase in energetic efficiency resulting from the loss of enzyme expression gives the mutant a selective growth advantage over its prototrophic parent. Zamenhof and Eichhorn showed that a histidine-requiring mutant outgrew a prototroph in histidine-sufficient medium (76).

11.6 OTHER CONTINUOUS-CULTURE TECHNIQUES

Other continuous-culture techniques are shown in Table 3.

11.6.1 Turbidostat

The turbidostat is a continuous-culture apparatus in which the cell concentration is not limited by a nutrient. In cases where nutrient limitation is not desirable, such as degradation of toxic wastes, the turbidostat can be a valuable tool. Cells are grown to the desired cell density as measured by turbidity or light scattering. Medium with all nutrients in excess is then used to wash the culture out of the vessel as quickly as it accumulates. This then maintains the culture at its maximal growth rate. The medium inflow and outflow are balanced so that no increase in volume occurs. The turbidity is constantly measured

TABLE 3. Other continuous-culture control techniques

Continuous-culture technique	Control parameter	Reference
Turbidostat	Cell density (turbidity)	71
Nutristat	Residual substrate concentration	22
pH-auxostat	pH	45
CER-stat	Carbon dioxide evolution rate	69
DO_2-stat	Dissolved oxygen	32
OUR-stat	Oxygen uptake rate	75

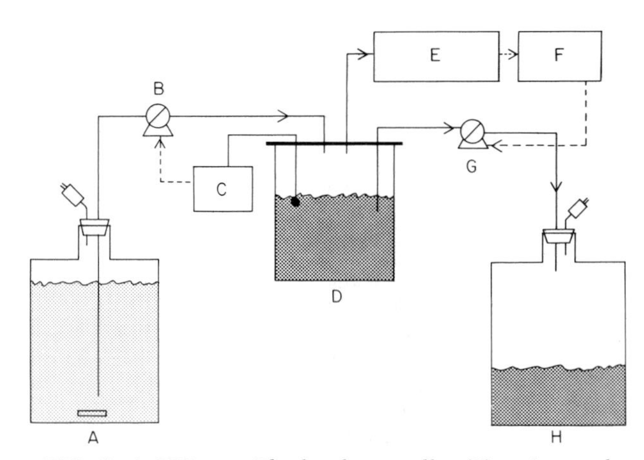

FIG. 8. A CER-stat. The level controller (C) activates the nutrient addition pump (B) to maintain a preset volume. Off-gas is monitored by a mass spectrometer (E), and data are fed to a minicomputer (F), which calculates the CER and compares it with a preprogrammed set point. If the measured value is higher than the set point, the computer increases the pump rate (G), and cells flow out of the vessel at a faster rate. If the CER is lower than the set point, the computer slows the pump and cells accumulate in the vessel. A proportional integral derivative control loop is incorporated into the software to enhance contol.

and used to control the exit pump, and a level controller is used to maintain the liquid volume.

The turbidostat suffers from the use of turbidity as the control parameter. A biofilm usually develops over the optics with time, and special designs must be employed to eliminate bubbles in the sample stream. To date, turbidostats have found their greatest utility in the study of algae, where these problems are not so likely to occur.

11.6.2 CER-Stat

An alternative to the turbidostat is the CER-stat, which was demonstrated by Watson in 1969 (69). The CO_2 evolution rate (CER) is proportional to the cell concentration and the cell growth rate. Because all the cells in a CER-stat are growing at u_{max}, the CO_2 concentration in the off-gas at constant airflow rate will be proportional to the cell concentration. In this case, CO_2 is a sensitive and quickly responding biological indicator of cell density. The advent of mass spectrometers which can measure CO_2, O_2, N_2, and water simultaneously, along with inexpensive minicomputers for convenient data acquisition and control, allows easy implementation of the CER-stat.

A schematic view of a CER-stat that we have built and operated successfully is given in Fig. 8. As in the chemostat, a level controller is required and inflow and outflow pumps are used. The rate of broth removal is controlled to maintain a constant rate of CO_2 evolution. Off-gas is analyzed for CO_2 with a mass spectrometer, and the data are transferred into a minicomputer. The CER values are computed and compared to a CER setpoint. A proportional integral derivative control loop incorporated into the software then activates a variable-speed pump which removes cell broth to maintain the desired CER. The medium inflow is maintained by the level controller. When the liquid level falls too low, it calls for the addition of more medium. (Note that this control is inverted with respect to operation in a chemostat.)

There are several points which should be emphasized with regard to proper control of the CER-stat. First, it is important that delay between a change in CO_2 in the vessel and its subsequent analysis be minimized. This means that the vessel should have a small headspace and the distance from the vessel to the analyzer should be minimized. Second, it is important to arrange the equipment as shown in Fig. 8. The alternative would be to have the broth removal controlled by level and the medium addition controlled by CER. However, this method is unsatisfactory because it substantially increases the response time of the system to changing CER. Finally, the proportional integral derivative control loop is included to allow the control system to be conveniently tuned.

11.7 LIST OF VENDORS

Chemostat vessels

B. Braun Instruments
875 Stanton Rd.
Burlingame, CA 94010

Bioengineers
Sagenrainstrasse 7
CH 8636 Wald, Switzerland

Biolafitte, Inc.
719 Alexander Rd.
Princeton, NJ 08540

Chemapec, Inc.
230-c Crossway Park Dr.
Woodbury, NY 11797

Gallenkamp
P.O. Box 290
Technico House
Christopher Street
London EC2P 2ER, England

L.E. Marubushi, Co., Ltd.
Koki Bldg. 8-7, 2-Chome
Kaji-cho, Chiyoda
Tokyo, Japan

New Brunswick Scientific Co., Inc.
P.O. Box 986
Edison, NJ 08818

Queue Systems
Box 1901
Parkersburg, WV 26102

Setric (Astra Scientific)
410 Martin Ave.
Santa Clara, CA 95050

Nutrient reservoirs

Bellco Glass, Inc.
340 Edrudo Rd.
Vineland, NJ 08360
Nalge Co.
Box 365
Rochester, NY 14602

New Brunswick Scientific Co., Inc.
P.O. Box 986
Edison, NJ 08818

Peristaltic pumps

Cole Parmer Instrument Co.
7425 N. Oak Park Ave.
Chicago, IL 60648

LKB Instruments, Inc.
9319 Gaithers Rd.
Gaithersburg, MD 20877

Diaphragm pumps

Milton Roy
Flow Control Division
201 Ivyland Rd.
Ivyland, PA 18974

ProMinent Fluid Controls, Inc.
Parkway West Industrial Park
503 Parkway View Dr.
Pittsburgh, PA 15205

Pulsafeeder
77 Ridgeland Rd.
Rochester, NY 14623

Level controllers

Xertex, National Sonics Division
250 Marcus Blvd.
Happauge, NY 11787

11.8 LITERATURE CITED

1. **Acevedo, F., and C. L. Cooney.** 1973. Penicillin amidase production by *Bacillus megaterium*. Biotechnol. Bioeng. **15:**493–503.
2. **Adams, C. W., and P. E. Hanasche.** 1974. Population studies in microorganisms. I. Evolution of diploidy in *Saccharomyces cerevisiae*. Genetics **76:**327–338.
3. **Adams, C. W., and G. W. Hatfield.** 1984. Effects of promoter strengths and growth conditions on copy number of transcription-fusion vectors. J. Biol. Chem. **259:**7399–7403.
4. **Aiba, S., A. E. Humphrey, and N. F. Millis.** 1973. Biochemical engineering, 2nd ed., p. 128–162. Academic Press, Inc., New York.
5. **Alton, T. H., and A. L. Koch.** 1974. Unused protein synthetic capacity of *Escherichia coli* grown in phosphate-limited chemostats. J. Mol. Biol. **86:**1–9.
6. **Bailey, J. E., and D. F. Ollis.** 1977. Biochemical engineering fundamentals, p. 500–508. McGraw-Hill Book Co., New York.
7. **Bloom, F. R., and E. McFall.** 1975. Isolation and characterization of D-serine deaminase constitutive mutants by utilization of D-serine as sole carbon or nitrogen source. J. Bacteriol. **121:**1078–1084.
8. **Borzani, W., and M. L. R. Vairo.** 1973. Observations of continuous culture responses to additions of inhibitors. Biotechnol. Bioeng. **15:**299–306.
9. **Clewell, D. B., and D. R. Helinski.** 1969. Supercoiled circular DNA-protein complex in *Escherichia coli*: purification and induced conversion to an open circular DNA form. Proc. Natl. Acad. Sci. U.S.A. **62:**1159–1166.
10. **Cooney, C. L., H. M. Koplove, and M. Haggstrom.** 1981. Transient phenomena in continuous culture, p. 143–168. *In* P. H. Calcott (ed.), Continuous culture of cells, vol. 2. CRC Press, Inc., Boca Raton, Fla.
11. **Cooney, C. L., J. Leung, and A. J. Sinskey.** 1976. Growth and physiology of *Streptococcus mutans* during transients in continuous culture, p. 799–807. *In* Proceedings of the microbial aspects of dental caries. Microbiol. Abstr. **3:**799–808.
12. **Cooney, C. L., and D. I. C. Wang.** 1976. Transient response of *Enterobacter aerogenes* under a dual nutrient limitation in a chemostat. Biotechnol. Bioeng. **86:**1–9.
13. **Dean, A. C. R.** 1972. Influence of environment on the control of enzyme synthesis. J. Appl. Chem. Biotechnol. **22:**245–259.
14. **Dean, A. C. R., D. C. Ellwood, J. Melling, and A. Robinson.** 1976. The action of antibacterial agents on bacteria grown in continuous culture, p. 251–261. *In* A. C. R. Dean, D. C. Ellwood, J. Melling, and A. Robinson (ed.), Continuous culture 6: applications and new fields. Ellis Horwood Ltd., Chichester, U.K.
15. **Demain, A. L.** 1972. Cellular and environmental factors affecting the synthesis and excretion of metabolites. J. Appl. Chem. Biotechnol. **22:**345–362.
16. **Doudoroff, M.** 1966. Metabolism of poly-β-hydroxybutyrate, p. 385–400. *In* N. O. Kaplan and E. P. Kennedy (ed.), Current aspects of biochemical engineering. Academic Press, Inc., New York.
17. **Drew, S. W.** 1981. Liquid culture, p. 151–176. *In* P. Gerhardt (ed.), Manual of methods in general bacteriology. American Society for Microbiology, Washington, D.C.
18. **Droop, M. R.** 1969. Vitamin B-12 and marine ecology. IV. The kinetics of uptake, growth, and inhibition in *Monochrysis lutheri*. J. Mar. Biol. Assoc. U.K. **48:** 689–733.
19. **Dunn, I. J., and J. R. Mor.** 1975. Variable volume continuous cultivation. Biotechnol. Bioeng. **17:**1805–1822.
20. **Dykhuizen, D. E., and D. L. Hartl.** 1980. Selective neutrality of 6-PGD allozymes in *E. coli* and the effects of genetic background. Genetics **96:**801–817.
21. **Dykhuizen, D. E., and D. L. Hartl.** 1983. Selection in chemostats. Microbiol. Rev. **47:**150–168.
22. **Edwards, V. H., R. C. Ko, and S. A. Balogh.** 1972. Dynamics and control of continuous microbial propagators to subject substrate inhibition. Biotechnol. Bioeng. **15:**939–944.
23. **Gallant, J., G. Margason, and B. Finch.** 1972. On the turnover of ppGpp in *Escherichia coli*. J. Biol. Chem. **247:**6055–6061.

24. **Godwin, D., and J. H. Slater.** 1979. The influence of the growth environment on the stability of a drug resistance plasmid in *E. coli* K-12. J. Gen. Microbiol. **111**:210–216.
25. **Goeddel, D. V., E. Yelverton, A. Ullrich, H. L. Heyneker, G. Miozzari, W. Molmes, P. S. Seeburg, T. Dull, L. May, N. Stebbing, R. Crea, S. Maeda, R. McCandliss, A. Sloma, J. M. Tabor, M. Gross, P. C. Familletti, and S. Pestka.** 1980. Human leucocyte interferon produced by *E. coli* is biologically active. Nature (London) **287**:411–416.
26. **Goldberg, I., and Z. Er-el.** 1981. The chemostat—an efficient technique for medium optimization. Process Biochem. (Oct./Nov.), p. 2–8.
27. **Harder, W., J. G. Kuenen, and A. Matin.** 1977. Microbial selection in continuous culture. J. Appl. Bacteriol. **43**:1–24.
28. **Harder, W., and H. Veldkamp.** 1969. Regulation of glucose uptake under nitrogen limitation in continuous culture, p. 59–71. *In* I. Malek, K. Beren, Z. Fencl, V. Munk, J. Ricica, and H. Smrekova (ed.), Continuous cultivation of microorganisms. Academic Press, Inc., New York.
29. **Heineken, F. G., and R. J. O'Connor.** 1972. Continuous culture studies on the biosynthesis of alkaline protease, neutral protease, and α-amylase by *Bacillus subtilis* NRRL-B3411. J. Gen. Microbiol. **73**:35–45.
30. **Herbert, D., R. Elsworth, and R. C. Telling.** 1956. The continuous culture of bacteria; a theoretical and experimental study. J. Gen. Microbiol. **14**:601–622.
31. **Holmes, W. D., I. D. Hamilton, and A. G. Robertson.** 1972. The rate of turnover of the ATP pool of *Escherichia coli* growing aerobically in simple defined media. Arch. Microbiol. **83**:95–101.
32. **Hopkins, T. R.** 1981. Feed-on-demand control of fermentation by cyclic changes in dissolved oxygen tension. Biotechnol. Bioeng. **23**:2137–2143.
33. **Horiuchi, T., J. Tomizawa, and A. Novick.** 1962. Isolation and properties of bacteria capable of high rates of beta-galactosidase synthesis. Biochim. Biophys. Acta **55**:152–163.
34. **Hu, W. S., and A. L. Demain.** 1979. Regulation of antibiotic biosynthesis by utilizable carbon sources. Process Biochem. (Aug./Sept.), p. 2–5.
35. **Jones, S. A., K. Dearnley, P. M. Bennet, and J. Melling.** 1980. The stability of antibiotic resistance plasmids in *Escherichia coli* hosts grown in continuous culture. Soc. Gen. Microbiol. Q. **8**:44–52.
36. **Kang, K. S., and I. W. Cottrell.** 1967. Polysaccharides, p. 418–475. *In* H. J. Peppler (ed.), Microbial technology. Reinhold Publishing Co., New York.
37. **Koplove, H. M., and C. L. Cooney.** 1979. Enzyme production during transient growth. Adv. Biochem. Eng. **12**:1–40.
38. **Korczynski, M. S.** 1981. Sterilization, p. 476–486. *In* P. Gerhardt (ed.), Manual of methods in general bacteriology. American Society for Microbiology, Washington, D.C.
39. **Kristiansen, B., and C. G. Sinclair.** 1979. Production of citric acid in continuous culture. Biotechnol. Bioeng. **21**:315–323.
40. **Kubitschek, H. E.** 1974. Evolution in the microbial world. Symp. Soc. Gen. Microbiol. **24**:105–130.
41. **Larsen, D. H., and R. L. Dimmick.** 1964. Attachment and growth of bacteria on surfaces of continuous culture vessels. J. Bacteriol. **88**:1380–1387.
42. **Lee, I. H., A. G. Fredrickson, and H. M. Tsuchiya.** 1976. Damped oscillations in continuous culture of *Lactobacillus plantarum*. J. Gen. Microbiol. **93**:204–209.
43. **Lockwood, L. B.** 1967. Production of organic acids by fermentation, p. 356–386. *In* H. J. Peppler (ed.), Microbial technology. Reinhold Publishing Co., New York.
44. **Luedeking, R., and E. L. Piret.** 1959. Transient and steady states in continuous fermentation. Theory and experiment. J. Biochem. Microbiol. Tech. Eng. **1**:453–459.
45. **Martin, G. A., and W. P. Hempfling.** 1976. A method for the regulation of microbial population density during continuous culture at high growth rates. Arch. Microbiol. **107**:41–47.
46. **Martin, J. F.** 1977. Control of antibiotic synthesis by phosphate. Adv. Biochem. Eng. **6**:105–127.
47. **Mateles, R. I., D. Y. Ryu, and T. Yasuda.** 1965. Measurement of unsteady-state growth rates of micro-organisms. Nature (London) **208**:263–265.
48. **Munson, R. J., and B. A. Bridges.** 1964. "Take-over"—an unusual selection process in steady-state cultures of *Escherichia coli*. J. Gen. Microbiol. **37**:411–418.
49. **Nakao, Y. M., M. Kikuchi, J. Suzuki, and M. Doe.** 1972. Microbial production of glycerol auxotrophs and production of 1-glutamic acid from n-paraffins. Agric. Biol. Chem. **36**:490–496.
50. **Noack, D., M. Roth, R. Geuther, G. Muller, K. Undisz, C. Hoffmeier, and S. Gaspar.** 1982. Maintenance and genetic stability of vector plasmids pBR322 and pBR325 in *E. coli* K12 strains grown in a chemostat. Mol. Gen. Genet. **184**:121–124.
51. **Normark, S., T. Edlund, T. Grunstrom, S. Bergstrom, and H. Wolf-Watz.** 1977. *Escherichia coli* K-12 mutants hyperproducing chromosomal beta-lactamase by gene repetitions. J. Bacteriol. **132**:912–922.
52. **Norris, T. E., and A. L. Koch.** 1972. Effect of growth rate on the relative rates of synthesis of messenger, ribosomal and transfer RNA in *Escherichia coli*. J. Mol. Biol. **64**:633–649.
53. **Pirt, S. J.** 1975. Principles of microbe and cell propagation, p. 117–136. Blackwell Scientific Publications, Oxford.
54. **Powell, E. O.** 1959. Criteria for the growth of contaminants and mutants in continuous culture. J. Gen. Microbiol. **18**:259–268.
55. **St. John, A. C., and A. L. Goldberg.** 1978. Effects of reduced energy production on protein degradation, guanosine tetraphosphate, and RNA synthesis in *Escherichia coli*. J. Biol. Chem. **253**:2705–2711.
56. **Schimke, R. T., R. J. Kaufman, F. W. Alt, and R. K. Kellems.** 1978. Gene amplification and drug resistance in cultured murine cells. Science **202**:1051–1055.
57. **Senior, P. J., G. A. Beech, G. A. F. Ritchie, and E. A. Dawes.** 1972. The role of oxygen limitation in the formation of poly-β-hydroxybutyrate during batch and continuous culture of *Azotobacter beijerinckii*. Biochem. J. **128**:1193–1198.
58. **Sikyta, B., E. Pavlasova, P. Kyslik, and E. Stejskalova.** 1984. Population changes of *Escherichia coli* strains hyperproducing endoenzymes in continuous cultures, p. 305–312. *In* A. C. R. Dean, D. C. Ellwood, and C. G. T. Evans (ed.), Continuous culture 8: biotechnology, medicine and the environment. Ellis Horwood Limited, Chichester, U.K.
59. **Silman, R. W., and S. P. Rogovin.** 1972. Continuous fermentation to produce xanthan biopolymer: effect of dilution rate. Biotechnol. Bioeng. **14**:23–31.
60. **Smith, R. W., and A. C. R. Dean.** 1972. Beta-galactosidase synthesis in *Klebsiella aerogenes* growing in continuous culture. J. Gen. Microbiol. **72**:37–43.
61. **Sterkin, V. E., I. M. Chirkov, and V. A. Samoylenko.** 1973. Study of transitional stages in continuous culture of microorganisms. Biotechnol. Bioeng. Symp. **4**:53–62.
62. **Strange, R. E., and J. R. Hunter.** 1967. Effect of magnesium on the survival of bacteria in aqueous suspension, p. 102–116. *In* E. O. Powell, C. G. T. Evans, R. E. Strange, and D. W. Tempest (ed.), Microbial physiology and continuous culture, 3rd International Symposium on Continuous Culture. H. M. Stationery Office, London.
63. **Tamaki, S., and M. Matsuhashi.** 1973. Increase in sensitivity to antibiotics and lysozyme on deletion of lipo-

polysaccharides in *Escherichia coli* strains. J. Bacteriol. **114**:453–458.

64. **Tempest, D. W.** 1969. Quantitative relations between inorganic cations and anionic polymers in growing bacteria, p. 87–94. *In* P. Meadow and S. J. Pirt (ed.), Microbial growth, 19th Symposium of the Society for General Microbiology. Cambridge University Press, Cambridge.
65. **Tempest, D. W., and O. M. Neijssel.** 1976. Microbial adaptation to low nutrient environments, p. 283–296. *In* A. C. R. Dean, D. C. Ellwood, J. Telling, and A. Robinson (ed.), Continuous culture 6: applications and new fields. Ellis Horwood Ltd., Chichester, U.K.
66. **Tomonaga, G.** 1966. Preferential synthesis of extracellular protease by *Aspergillus niger* in sulfur deficiency. J. Gen. Appl. Microbiol. **12**:267–276.
67. **Topiwala, H. H., and G. Hamer.** 1971. Effect of wall growth in steady-state continuous cultures. Biotechnol. Bioeng. **18**:919–922.
68. **Wang, D. I. C., and A. J. Sinskey.** 1970. Collection of microbial cells. Adv. Appl. Microbiol. **12**:121–132.
69. **Watson, T. G.** 1969. Steady state operation of a continuous culture at maximum growth rate by control of carbon dioxide production. J. Gen. Microbiol. **59**:83–87.
70. **Watson, T. G.** 1970. Effects of sodium chloride on steady-state growth and metabolism of *Saccharomyces cerevisiae*. J. Gen. Microbiol. **64**:91–99.
71. **Watson, T. G.** 1972. The present status and future prospects of the turbidostat. J. Appl. Chem. Biotechnol. **22**:229–243.
72. **Welles, J. B., and H. W. Blanch.** 1976. Effect of discontinuous feeding on ethanol production by *Saccharomyces cerevisiae*. Biotechnol. Bioeng. **18**:129–133.
73. **Windass, J. D., M. D. Worsey, D. Pioli, P. T. Barth, K. T. Atherton, E. C. Dart, D. Byron, K. Powell, and P. J. Senior.** 1980. Improved conversion of methanol to single-cell protein by *Methylophilus methylotrophus*. Nature (London) **287**:396–401.
74. **Wouters, J. T., F. L. Driehuis, P. J. Polaczek, M. L. H. A. Van Oppenraag, and J. G. Van Andel.** 1980. Persistence of the pBR322 plasmid in *Escherichia coli* grown in chemostat cultures. Antonie van Leeuwenhoek J. Microbiol. Serol. **46**:353–362.
75. **Yamada, S., M. Wada, and I. Chibata.** 1979. Oxygen transfer as a parameter of automatic control of the continuous cultivation for the conversion of sorbitol to sorbose by *Acetobacter suboxydans*. J. Ferment. Technol. **57**:210–225.
76. **Zamenhof, S., and H. H. Eichhorn.** 1967. Study of microbial evolution through loss of biosynthetic functions: establishment of "defective" mutants. Nature (London) **216**:456–458.
77. **Zund, P., and G. Lebek.** 1980. Generation time-prolonging R plasmids: correlation between increases in generation time of *Escherichia coli* caused by R plasmids and their molecular size. Plasmid **3**:65–69.

III. Strain Improvement

Strain Improvement: Introduction

RICHARD H. BALTZ

There are three basic genetic approaches to improving fermentation product titers in industrial microorganisms: mutation, recombination, and gene cloning. Each has distinct advantages, and in some cases all three techniques can be used in concert to maximize production. Mutation has the advantage of simplicity: it requires little knowledge of the genetics and physiology of the important pathways involved in the biosynthesis of the desired product and requires minimal technical manipulation. It predictably leads to rapid improvement in product titer if appropriate procedures can be established to monitor product yields accurately. Its effectiveness is determined by the overall efficiency of the mutagen and the accuracy of the screen.

Genetic recombination becomes an important adjunct to mutagenesis once several lineages of mutants have been established. It provides a means to construct strains with many different combinations of mutations that influence product titers. With the development of protoplast fusion techniques, it is now relatively simple to recombine properties of a wide variety of industrial microorganisms that are relatively uncharacterized genetically or biochemically. For better characterized microorganisms, the use of natural fertility systems may have specific advantages over protoplast fusion, particularly for genetic mapping.

Gene cloning has distinct advantages for highly developed systems where the limiting factors for product titer are relatively well understood. This method can be used to relieve rate-limiting enzyme reactions by increasing the concentration of the enzyme. It can also be used in conjunction with site-directed mutagenesis to alter promoter strength, to alter ribosome binding sites, or to modify enzymes to eliminate undesirable properties such as product inhibition.

This section addresses the procedures used to carry out mutation, recombination, and gene cloning in traditional industrial microorganisms. We have not included chapters specifically devoted to *Escherichia coli*, an important academic microorganism which has recently become important for the industrial production of mammalian polypeptides, nor *Bacillus subtilis*, except for a section on protoplast fusion. Techniques for classical mutation and recombination in *E. coli* have been described in great detail in an earlier ASM manual (4) and in a Cold Spring Harbor manual (1), while procedures for gene cloning are described in detail in another Cold Spring Harbor manual (5). Several other books also cover genetic techniques in *E. coli* (6) and *B. subtilis* (2, 3, 6).

We have instead, for several reasons, emphasized the mutation, recombination, and gene cloning procedures developed for the procaryotic *Streptomyces* spp. and the eucaryotic *Saccharomyces* spp. First, streptomycetes are major industrial producers of antibiotics, antitumor agents, and enzymes, and they are closely related to other actinomycetes such as *Nocardia*, *Micromonospora*, and *Actinoplanes*, which are also important industrial microorganisms. Thus procedures developed for *Streptomyces* spp. should have wide applicability to many important industrial microorganisms. *Saccharomyces* strains are major sources of bakers' yeast and wine, beer, and saki yeasts and are used in other industrial fermentations. Since *Streptomyces* spp. are filamentous, they pose special technical problems for mutagenesis. Thus, procedures developed for *Streptomyces* can be applied to a wide variety of filamentous and nonfilamentous procaryotic microorganisms. Procedures for genetic recombination in *Streptomyces* spp. and gene cloning in *Streptomyces* and *Saccharomyces* spp. employ protoplasts. These techniques are general and can be adapted readily to single-celled eubacteria, actinomycetes, and filamentous fungi of industrial importance. Finally, and most importantly, much more is known about the processes of mutation, recombination, and gene cloning in *Streptomyces* and *Saccharomyces* spp. than in other traditional industrial microorganisms.

LITERATURE CITED

1. **Davis, R. W., D. Botstein, and J. R. Roth (ed.).** 1980. Advanced bacterial genetics. Cold Spring Harbor Laboratory, Cold Spring Harbor, N.Y.
2. **Dubnau, D. A. (ed.).** 1982. The molecular biology of the bacilli, vol. 1. Academic Press, Inc., New York.
3. **Dubnau, D. A. (ed.).** 1985. The molecular biology of the bacilli, vol. 2. Academic Press, Inc., New York.
4. **Gerhardt, P. (ed.).** 1981. Manual of methods for general bacteriology. American Society for Microbiology, Washington, D.C.
5. **Maniatis, T., E. F. Fritsch, and J. Sambrook (ed.).** 1982. Molecular cloning. Cold Spring Harbor Laboratory, Cold Spring Harbor, N.Y.
6. **Scaife, J., D. Leach, and A. Galizzi (ed.).** 1985. Genetics of bacteria. Academic Press, Inc., New York.

Chapter 12

Screening and Selection for Strain Improvement

STEPHEN W. QUEENER AND DAVID H. LIVELY

Success in bringing a fermentation product to market and success in competing in that market are, in part, dependent on continuous improvement of the microorganism that makes the product. Improvements can be, for example, increased yield of desired metabolites, removal of unwanted cometabolites, improved utilization of inexpensive carbon and nitrogen sources, or alteration of cellular morphology to a form better suited for separation of organism from the product.

In our experience (which concurs with the experience reported by others; 43, 52, 70, 81, 90), the mutation/fermentation screening process is an essential, and often the most direct and least expensive, means of improving most industrial microorganisms. The screening method is general and does not require a wealth of biochemical and genetic information on the organism to which it is applied.

For most primary metabolites, such as amino acids, enrichment procedures have been developed and published as part of fundamental studies. Some of these published procedures select for mutants deregulated in the biosynthesis of these primary metabolites. Others enrich for auxotrophs in which the flow of carbon can sometimes be shunted into the biosynthesis of a desired primary metabolite. Application of these procedures can drastically reduce the number of isolates that have to be screened in fermentations to identify an improved strain.

Enrichment procedures can also apply to improving the yield of secondary metabolites derived from primary metabolites. Overproduction of the primary metabolite precursors may improve yields by saturating the enzyme system which converts the precursors to the secondary metabolite(s).

Direct application of enrichment procedures to improve the conversion of the precursors to antibiotic, pigment, or other secondary metabolites is more difficult. Here the selection must be indirect since secondary metabolites are dispensable to the cell. In addition, much less is known about the chemistry of the biosynthetic pathways in secondary metabolism, and even less is known about the specific mechanisms which regulate the biosynthesis of secondary metabolites.

This chapter summarizes experience with successful screens for improved strains, outlines the general characteristics of the automated screen for strain improvement, and surveys some of the enrichment procedures.

12.1 SCREENING FOR STRAIN IMPROVEMENT

12.1.1 History and Background

Initial effort to improve a fermentation by screening microorganisms probably began with the recognition that some yeasts were preferable to others as agents for the production of alcoholic beverages. The choice of the best wild-type organism for a particular fermentation remains as the important first step for the subsequent improvement of the fermentation. However, the present discussion deals only with a process that begins after an organism is chosen for a particular fermentation. The process involves repeated application of three basic processes: (i) induction of genetic variability in cell populations of a chosen organism, (ii) fermentation in small scale of many individuals from those populations, and (iii) assays of the fermentations to allow identification of an improved strain. Each improved strain can then be used as a parent strain in a new cycle of mutation induction, fermentation, and assay. The prototype system for this process, sometimes referred to as "screening for strain improvement," probably originated with the work of Thom and Steinberg (101). The system was developed further at the University of Wisconsin, the Northern Regional Research Laboratory (U.S. Department of Agriculture) in Peoria, Ill., the Carnegie Institution of Washington, D.C., and the University of Minnesota. At these locations, screens were established for improved penicillin production by variants of a *Penicillium chrysogenum* wild-type strain isolated from a rotten cantaloupe. Strains selected in these screens were further improved in separate, competitive programs of mutation and screening at many pharmaceutical companies (1, 34, 73, 89). Articles by Elander (34, 35),

Queener and Swartz (89), and Aharonowitz and Cohen (1) illustrate the dramatic effectiveness of screening for strain improvement and document a detailed example of a lineage of improved strains from an industrial program. By use of a combination of mutation induction, screening, and improvements in fermentation technology, the yield of penicillin has been increased from 0.06 to over 26 mg/ml.

The same general approach of screening many survivors of mutagen treatments to find more productive variants has been employed successfully to improve many other fermentations. For example, Stauffer et al. (100) discussed improvement of fungal strains for the production of cephalosporin antibiotics. Alikhanian et al. (3) and Mandal et al. (70) discussed improvement of *Streptomyces* strains for the production of oxytetracycline and tetracycline fermentations. Dulaney described strain improvement for streptomycin fermentation (33). As a result of the success of this approach to strain improvement, it is now applied as a conventional method throughout the world. Conventional screening for strain improvement may be counted on to deliver small but additive increases in the yield of fermented products via poorly defined genetic changes in large numbers of mutant strains. The methodology can be combined with classical genetic and recombinant DNA programs. The latter programs employ mutant strains characterized by stable, well-defined phenotypic and genotypic properties to construct special improved strains of a more defined character. The genetic and recombinant DNA programs require a greater initial investment of time and resources to develop and apply than does conventional screening, but recently, investments related to strain improvement by these programs have increased because of examples demonstrating that this approach can succeed (see reference 66). Screening for strain improvement, genetics, and recombinant programs are complementary efforts that function most effectively for strain improvement when sufficient resources are available to mount a coordinated effort employing each. However, the scope of the present discussion must remain limited to the first of these.

Other articles related to the topic of screening for strain improvement are recommended for reading: general articles dealing with genetics and population analysis in strain improvement programs (11, 28, 29, 36, 38, 52, 57, 90; R. H. Baltz *in* L. E. Day and S. W. Queener, ed., *The Bacteria*, vol. 9, *Antibiotic-Producing Streptomyces*, in press), articles focusing on selection of improved strains for production of primary metabolites such as amino acids, nucleosides, and nucleotides (e.g., reference 60), and articles reviewing the use of mutants blocked in the biosynthesis of secondary metabolites (e.g., references 12, 13, and 88).

12.1.2 Design of a Screening Program

General concepts and allocation of screening capacity

Publications by Davis (26), Brown and Elander (16), Elander (34), and Dahl et al. (23) are useful in designing a conventional screen for strain improvement. The paper by Davis remains valuable 20 years after its publication. Each of these reports discusses the results of employing the general procedure of a conventional screen. Mutations were induced in a cell population of a chosen microbial strain, and a large number of individual survivors of the mutagen treatment were compared with the untreated parent strain in test fermentations. The concept of a reference or control strain fermented in parallel with the mutagen-treated isolates was introduced by Dahl et al. (23). The difference in titer between the reference strain and each experimental strain was calculated for fermentations harvested on each day. These differences were used for comparison of experimental strains in fermentations harvested on the same or different days. Mutation induction and fermentation comparisons were repeated until a significantly improved strain was indicated by a fermentation test with a product titer that was statistically superior to that observed in the control (fermentation employing the reference strain). To prove that the significantly superior product titer was due to a truly superior strain (and not due to other factors, e.g., error in media makeup, inoculation, or assay), each strain identified as superior in the primary fermentation test was retested using an increased number of replicate fermentations. Repetition of the superior product titer in this secondary test served to rule out random experimental error in the primary test and verified the strain as the cause for the observed increase. A strain with verified improvement was again exposed to a mutagen, and the screening process was repeated using the improved strain as the control in the comparative fermentation tests. Experience derived from these studies showed that testing error, both fermentation and assay, could be extensive and played an important role in the design of an effective screen. A large amount of testing was necessary to find a truly superior strain. An important relationship between frequency and degree of titer increase was observed: small improvements occurred more frequently than large increases. Because of this latter observation, Davis (26) chose to select mutants with small titer improvements from a relatively small number of isolates tested after each mutagen exposure. Screening of a sequential series of strains, each with a small titer increase over that of its parent, was judged more effective than attempting to select a strain that would incorporate the sum of individual improvements in one step. The latter mutants were too rare or nonexistent. The choice of the number of replicates for each fermentation test was important in allocating assay, shaker, and manpower resources, which limit any practical screening effort. If capacity was N, the number of isolates which could be tested with one fermentation test per strain, then only N/r strains could be screened when r tests per strain were employed. However, sufficient replications were required to ensure that true titer differences could be distinguished from random testing error. Therefore, it was of extreme importance to design the screen to limit and monitor testing error. Statistical analysis of the data for parameters such as distribution of the antibiotic biosynthetic capacities in populations of mutants, testing error, number of testing stages, number of replications for each test fermentation, and number of strains selected from each testing stage resulted in the conclusion that a two-stage screen was adequate over a wide range of mutation distributions. A typical protocol for a two-stage screen from this early work is given in Table 1.

Mutagenesis, expression of mutation, segregation

Effective mutagenesis contributes substantially to the success of any screen for strain improvement. The choice of an appropriate mutagen and dose has been reported by many authors to be essential for successful strain improvement. For each mutagen and each organism, there is a combination of mutagen concentration, time of exposure, and conditions of treatment which produces the highest proportion of a particular class of mutants among survivors of the mutagen treatment (51). Brown and Elander (16) used UV light in a screen for strain improvement and discussed proper dose based on their results. More mutants with improved titer were found at intermediate doses that allowed survival of about 10% of the treated cell population. Most of these improved strains exhibited a normal colony morphology. Calam (18) discussed the rationale and general techniques of mutagenesis by UV light, X rays, gamma rays, and chemical agents such as *N*-methyl-*N'*-nitro-*N*-nitrosoguanidine. A detailed description of the conditions necessary for optimal induction of mutations in *Streptomyces* spp. is described by Baltz and Seno (12) and Baltz (see chapter 14, this volume; Baltz, in press). For gain mutations, the appropriate mutagen will, in general, produce a population of cells in which about one-third of the cells are unaffected (as may be judged by their ability to produce control titers) in fermentation tests. The rationale for this dose and the concept of multiplicity (mutations per cell) is discussed in depth by Baltz (see chapter 14, this volume).

After mutagenesis, treated cells should be allowed a period of DNA replication and cell growth to convert damaged DNA into stable, altered DNA encoding reproducibly inheritable mutations. The rationale and molecular basis for allowing genetic segregation before screening is dealt with in detail by Baltz (chapter 14, this volume; Baltz, in press).

The last step in the segregation procedure involves processes (e.g., sporulation) or manipulations (e.g., fragmentation of mycelia) designed to ensure that each fermentation starts with a pure culture so that the genetic potential of each individual is measured. If isolates for screening are derived from multicellular hyphae or another multicellular form, the chances for genetic homogeneity of the isolate are reduced. Reduction of mold or actinomycete hyphae to single cells by the natural process of sporulation, or by mechanical fragmentation before mutagenesis and after postmutagenesis growth, will increase the chance for genetic homogeneity (12). In a screen for better lipase producers, Betz et al. (15) prepared a suspension of *Rhizopus arrhizus* Fischer spores by rinsing slants with a saline-Tween 80 solution and exposing the suspended material to ultrasonic treatment to break up chains and clumps. Filtering through a 50-μm nylon mesh removed residual vegetative mycelium and any aggregates of spores that remained. Queener and Capone (87) reported a similar method employing membrane filters.

TABLE 1. Typical protocol for a small two-stage screen for strain improvement[a]

Stage	Step	No.
1	1. Mutation of parent culture	8 different doses
	2. Plating of mutated cultures	
	3. Selection of colonies	25 from each mutagen dose
	4. Fermentation	200 strains (1 replication)
	5. Select highest titers	40 strains
2	1. Retest selected strains (5 replications)	40 × 5 = 200 strains
	2. Select highest strains	8 strains (=20% of secondary, 4% of primary)
Repeat stages 1 and 2 (1 of 8 strains chosen for mutation)		

[a] Developed from the early studies of Brown and Elander, Dahl et al., Davis, Elander, and others (16, 23, 26, 35).

Fermentation

Special techniques employing surface cultures have been used to increase the numbers of isolates that could be screened with conventional methodology (9, 30, 80, 99). Direct screening of surface cultures growing on solid medium (9, 30, 95, 99) can be accomplished for diffusible products with a biological assay incorporated in an overlay. Alternatively, the product can be assayed after extraction from a standard amount of medium on which the colony has grown. The detection of mutants with 10 to 50% yield improvement by screens employing these techniques has been reported (34, 54), but the techniques are limited to water-soluble metabolites. The conditions of surface fermentation on solid medium differ so greatly from the stirred, highly aerated, liquid fermentation employed in the typical manufacturing environment that the surface culture screening techniques are often unreliable in predicting a strain's performance in the submerged fermentation. A number of fermented products are mixtures of similar metabolites that must be controlled in relative amounts because of the product definition or because the metabolites differ in bioactivity. Usually the relative production of these metabolites is highly sensitive to medium composition and conditions, and a solid medium will rarely give the ratios observed in a stirred fermentor. As a general rule, strain programs have searched for improved strains among random survivors of mutagenesis under conditions approximating the conditions in production fermentors, e.g., medium composition, fermentation temperature, and cycle time.

Sample preparation, assay, and data analysis

For certain products, mechanical homogenization and extraction of the substance into an organic solvent may be required before assay. For example, this may be required if the product is not water soluble, is not excreted, or has a high binding affinity for the surface of the producing organism. For nonexcreted proteins, extraction into highly concentrated solutions of certain salts may be required. For excreted water-soluble products, dilution may be necessary to minimize the effect of a substance that interferes with the assay of the product or to minimize the effect of biomass solids on the apparent titer of the product. Adjustment of pH may be necessary if the product is unstable at the pH of the fermentation at harvest or if the pH at harvest is not compatible with the assay conditions. Refrigeration or freezing may be required if fermentation broths cannot be assayed immediately. Proper sample preparation requires an understanding of the chemistry of both the product and the assay of the product. A detailed discussion of this subtopic is outside the scope of this chapter.

Assay development is an extremely important part of the screen for strain improvement. An inadequate assay can nullify or drastically limit the effectiveness of the screening effort.

Direct assay of metabolites on surface culture is rapid and simple and consistent with the numerical objective of surface culture screens (see above), but as pointed out by Rowlands (90), any screen based on a biozonal assay is limited by exponentially decreasing assay response as titers of the measured metabolite increase; i.e., the method is less effective for detecting small increases in titer than a screen employing a linearly responsive assay.

Desired features for an assay in a screen for strain improvement are product specificity, low cost, short duration, simplicity, reliability, and precision (i.e., detection of 5% titer increases). Biological assays (bioassays) do not meet many of these requirements and, in general, should be avoided if possible. In practice, bioassays are sometimes used to assay fermentations from primary screens when the assay is less expensive than an alternative chemical assay or when a product has been defined by a bioassay early in its development. Choice of a cheaper bioassay over a superior chemical assay can be very costly in the long run if the statistical effectiveness of the screen is reduced substantially. Particular caution should be exercised when the desired fermentation product is multifactored (56); bioassays often respond differently to the various factors because of different specific activities. For multifactor products (13, 59), high-pressure liquid chromatography is probably the assay of choice, but cost, assay time, and limited ability to handle large numbers of samples represent obstacles that can preclude the use of this assay in screens. To overcome these obstacles, high-pressure liquid chromatography can be reserved for the secondary or tertiary testing stages.

Data analysis can be handled manually or with off-line computer assistance. Accepted statistical practices should be followed but should be simple so that data analysis does not hinder the screening process.

Selection of candidate strains

Selection of strains for further testing in the secondary stage of a conventional screen is based on proof of superiority by standard statistical procedures. In general, only a relatively small titer increase is demanded, and the probability of superiority demanded is less than in an automated screen (see section 12.1.3).

Secondary stage of screening

The secondary stage of the screen begins with the retrieval of the strains chosen from the primary stage. In general, steps of the primary fermentation stage (see outline) are repeated using more replications of candidate strains to improve the statistics. In conventional (nonautomated) screens, the secondary test may be more complicated than the primary and employ additional vegetative stages that more closely mimic inoculum buildup stages used in the actual manufacturing procedure at the plant site. The secondary stage of the screen ends with the preservation of strains chosen from the secondary fermentation test.

Tertiary evaluation of candidate strains

After selection and preservation of candidate strains, the evaluation of the cultures becomes increasingly complex with the introduction of another unit of operation, the stirred fermentor. The factors associated with scale-up from laboratory operations to pilot plant equipment to industrial production are well documented in the fields of biochemical and chemical engineering. Such items as shear force, mixing, oxygen transfer, cell mass, feed strategy, temperature control, and media composition are a few of the variables that influence the successful testing of a new strain. Chapters by Trilli (chapter 22) and by Hamilton et al. (chapter 24) which appear elsewhere in this volume should be consulted for additional information.

12.1.3 Automation of Screens for Strain Improvement

Background and general concepts

Nonautomated screens involving liquid fermentation are slow and labor intensive; however, many programs have been successful. Strains for penicillin production and for several other commercial antibiotic fermentations were developed by this methodology. Many of these programs produced 20 or more stepwise improvements in product yield in a period spanning several decades. In the early 1970s, the techniques and procedures that were developed during the early penicillin programs were still being used with minor modifications. Lack of automation was dictated in part by the lack of mechanical and electrical engineers assigned to strain improvement programs and by the complacency generated by the continued, albeit less frequent, isolation of improved strains by the manual systems. Automation of screens for strain improvement was markedly stimulated by the work of Donald Glaser and associates at the University of California at Berkeley (43, 44). Glaser (honored as a Noble Laureate in Physics for his contribution of the bubble chamber for studying high-energy particles) became interested in studies that required manipulation and optical analysis of massive microbial populations. He assembled a large group of experts with diverse backgrounds to automate standard microbiological techniques. The group developed a massive and sophisticated technology which employed complex setups for medium preparation, plate pouring, and incubation. A single-droplet dispenser and X-Y stages driven by step motors and controlled by computers were employed. Computers also controlled and recorded information on colony and morphology obtained from cameras and optical scanning equipment. Summaries of Glaser's work, such as an article appearing in *Fortune* magazine (17) which reported the existence of automated equipment able to screen 100×10^6 microbial strains at once, stimulated the development of automated screening programs in the fermentation industry. Typically, many individuals with a variety of backgrounds (microbiology, analytical chemistry, engineering, statistics, and computer science) collaborated to establish these automated programs.

The design and allocation of resources for automated screens could be based on the same concepts originally proven for the conventional screen (see section 12.1.2). However, provision for the high capacity of an automated system, e.g., 100,000 tests per week, and provision for the standardization required by an automated system suggest that a greater percentage of total screening capacity would be assigned to the initial or primary screen in a typical automated program relative to the percentage assigned to the primary stage in a conventional screen. Only a very small percentage of the strains from the primary test might be selected for secondary testing in the automated screen. Standardization required by automated systems might dictate the use of the same equipment for the primary and secondary stages of the automated screen. High capacity would allow for a significant increase in the number of replicate test fermentations employed in the various stages of an automated screen relative to that employed in a conventional screen.

An important feature in the design of an automated screen involves establishing the amount of mechanical effort needed, i.e., whether to utilize operators and machines or only machines. For example, the development of a system for total machine transfer of colonies is difficult and expensive. The system would involve a sophisticated optical system, computer techniques, and mechanical handling equipment; however, this transfer of colonies is easily performed by mechanically aided human effort. The next 10 years will see great development in this area as the rapidly expanding field of robotics finds uses in microbiology. The astute manager of an automated strain program will carefully evaluate the reliability and cost of new equipment that becomes available against those of alternative mechanically aided human methods, so that increased automation will steadily increase productivity and lower the cost of a screen, rather than making it complicated and prone to mechanical failures.

Key features of an automated screen

In the following discussion, key features of an automated screen for strain improvement are dealt with individually. Steps in which conventional technology have been replaced with technology more suited for automation are emphasized. The strategy and tactics of strain selection in an automated program will differ from system to system because the biological variation inherent in the organism and the fermentation conditions employed in each screen will be different. Each system will differ in precision and variation of the assay used to analyze fermentation yields. The efficiency of mutation induction and the capacity of the screens may also vary. In addition, each automated strain program may have unique features that remain as trade secrets. However, similarities in operation and common equipment in several documented systems suggest proven utility for a generally accepted technology (15; G. Nesemann, W. Sittig, and G. Stanke, European patent application 70,466, 1982). Our focus is on that technology.

Strain and culture conditions in the automated screen

The procedure and equipment for an automated system should be designed to place no constraints on the strain chosen for a fermentation. For example, equipment for dispensing and aeration should not require use of fragmenting variants of a filamentous organism, or efficiently sporulating mutants, merely to comply with limitations of the system. Likewise, dispensing equipment should allow the use of complex, partially insoluble media. The factors governing choice of strain and choice of fermentation medium should be based on the ability of the microorganism to produce the desired product under conditions similar to a production environment. In general, these factors are the same for conventional and automated screens.

Mutation induction and expression of mutation

The concepts and procedures for mutation induction and expression of mutation are the same for conventional and automated screens. These are described by Baltz (see chapter 14).

Genetic segregation in the automated screen

The need for pure cultures derived from genetically segregated isolates is the same for conventional and automated screens. However, the last procedure in the genetic segregation step of the conventional screen is not readily compatible with automation. In the conventional screen, small aliquots of a suspension of the mutagen-treated, segregated population of single cells are manually pipetted and spread on the surface of agar medium (see section 12.1.2). This operation is labor intensive and produces a randomly oriented population of colonies. A random orientation means that a very sophisticated and expensive system for locating nonoverlapping colonies is required for automation of the transfer of segments of each colony to storage and to fermentation vessels. The labor-saving and automation-compatible features of special dispensing equipment make this equipment and associated procedures attractive for automated screens. The current technology allows automatic dispensing of massive numbers of single cells by a single operator. For the most sophisticated (and more expensive) systems, cells of a particular morphology or size can be selected if desired.

Articles by Horan and Wheeless (53) and Cram and Salzman (21) describe the features of a flow cytometer used for cell dispensing. Improvements in flow cytometry in recent years have included the use of various types of lasers to scan drops containing a cell. Photosensors, coupled with photomultipliers and computers, detect and convert these interactions into digital signals. Software has been developed to analyze these signals so that cells can be classified and sorted according to size and shape.

To collect appropriate cells, the computer sends a signal to energize a deflection plate to direct the drop containing the cell to a collection vessel or to the surface of an agar medium. Betz et al. (15) and Nesemann et al. (European patent application 70,466, 1982) used a Cytoflorogram Systems 50H flow cytometer sorter (Ortho Diagnostics Systems) to analyze and sort spores and vegetative cells in an automated screen for improvement of lipase production in *R. arrhizus* Fischer. The instrument utilized a He-Ne laser (633 nm) to scan cells and induce red forward-scattered light and axial light loss signals. The criterion was sufficient to discriminate subpopulations within a suspension and mold spores based on size and shape of cells. The sorter was adapted to permit single-cell selection and to accept commands from software to control an X-Y scanning stage. Petri dishes or microtiter plates which were placed on the stage received the drops from the dispensor. Dispensing and X-Y movement were coordinated so that an ordered array of drops containing a single cell was achieved, and single cells were delivered to each well of the microtiter plate. The cytometer was sterilized by rinsing the connecting tubing from the cell reservoir with 70% ethanol. The sheath water was autoclaved and filtered through a 0.05-μm cartridge. In working with bioparticles in the range of 0.5-μm diameter, it is essential to use particle-free solutions.

Colony selection and transfer to fermentation vessels

The inoculum for fermentation cultures in a screen for improved antibiotic variants should, in general, consist of the minimum cell mass that will allow each inoculum to grow reproducibly to a mature culture without undue lag. If single cells of certain antibiotic-producing organisms are dispensed directly to a liquid fermentation or vegetative medium, a large percentage of the cultures fail to grow out in the time available for this step. However, the same cells plated on special agar medium usually produce colonies at high frequency, and a portion of the colony will serve as a reliable inoculum. An excessive growth lag in the fermentation step of a screen must be avoided to prevent the need for excessive numbers of expensive rotary shakers.

Preparation of inocula as colonies on agar surfaces is convenient and relatively inexpensive. As indicated in the previous section, if the viability of individual cells of a particular organism is high and a very small fermentation vessel is used, direct inoculation of fermentation vessels with single cells can sometimes be employed economically for the fermentation step of a screen with satisfactory results. However, this eliminates a possibly useful aspect of employing colonies as inocula. Correlations of colony morphology and product yield may exist. Colonies with certain morphologies may always be associated with poor yield and therefore can be eliminated before the liquid fermentation test. The technology for automation of colony selection on the basis of morphology now exists (2, 24, 43, 44, 62, 63), but this technology is very expensive and may not be a good investment since elimination of colony types with poor production is easily accomplished by human effort. To our knowledge, measurements of morphological parameters have never been successfully used to replace direct bioassay or chemical assay of a fermentation product. However, in the future, automated screening of colony morphology may be extended to microscopic examinations that might be more predictive than the current macroscopic examinations.

The current technology for automated screening of colony morphology can be illustrated with a system developed by Glaser and colleagues (43, 44). In this system (at the University of California at Berkeley), the optical scanning was interfaced with a mainframe computer for data analysis and storage of colony profiles. The original system was designed to scan a plate using five colors and four camera heights. Photographic film was exposed, and profiles were obtained using a complex optical scanner. The scanner employed a mercury arc lamp, focused by a complex mirror system to ensure uniform light intensity across the plate. Each frame of film had a color standard placed next to the subject image and a code to identify individual plates and colonies. A frame of film was scanned by flying spot scanner, and information related to the colony's morphology was transmitted to a computer. The center and diameter of each colony were measured in eight vectors and stored as Fourier coefficients. Colony profile data and colony location could be recalled and displayed as a colony profile using a high-speed printer or monitor.

Transfer of segments of chosen colonies to storage or fermentation vessels theoretically might be accomplished by robotics; however, high-speed robots are very expensive, and mechanically aided human effort may be competitive unless very large capacity screens (e.g., millions of strains per week) are desired. A detailed discussion of robotics is outside the scope of this chapter.

Fermentation in automated screen

Automation of surface culture screens to improve strains may not be advisable for the reasons previously outlined (see section 12.1.2) and is not considered here. The rationale for choice of medium is the same for conventional and automated screens for strain improvement (see sections 12.1.2 and 12.1.3). In an automated screen employing liquid medium, equipment and quality control for the dispensing of media and design of the fermentation vessel require special attention. Requirements for automation related to vessel size and dispensing can be illustrated through the following example (15).

In operating a screen employing thousands of fermentation vessels, space and material problems are encountered and miniaturization becomes essential. Betz et al. (15) employed microtiter plates in place of shake flasks to miniaturize an automated screen for improved lipase production. These investigators used commercially available automated pipettors to dis-

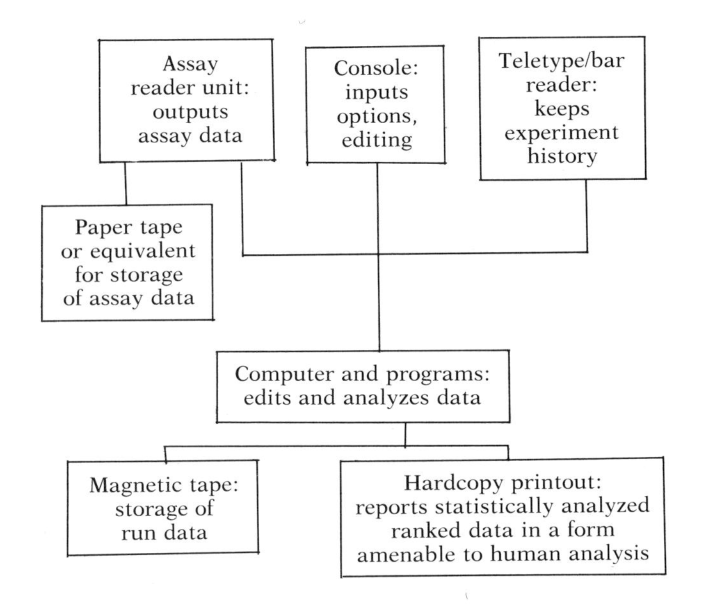

FIG. 1. Flow diagram of typical information transfer for an automated strain screen.

pense small volumes of culture medium into wells of microtiter plates. Operators placed the filled microtiter plates on an X-Y stage, and the automatic dispensor inoculated each well, one well at a time, with a single cell. Operators placed inoculated plates in special containers which were incubated for 3 days on rotary shakers at 95% relative humidity at 28°C. After the culture growth step, the plates were removed and placed in special buckets for centrifugation. The organism was separated (pellet) from the culture broth (supernatant) at approximately 5,000 × *g*.

Commercial instruments for filling are available. The Pro/Pette (Cetus Corp., Emeryville, Calif.) operates like a computer-controlled machine tool with a stepping motor carrying pipettes in the X direction while the dispensing head is driven up and down by a second stepping motor. The level 2 Gibson 212 Liquid Handler (Gibson International, Middleton, Wis.) is a filling device that employs a computer-controlled peristaltic pump to accurately deliver liquids at preset volumes. The second device may be more suitable for viscous fermentation media and fermentation media containing solid particles.

As the scale of the automated operation increases, the need for quality control of the dispensed media also increases. In a small conventional screen, routine assays to check for correct make-up are probably not cost effective, and careful training of the operators who prepare media will usually suffice.

Assay

The huge increase in numbers of fermentations made possible by automation dictates a corresponding increase in assay operation. Depending on the particular fermentation product, various techniques and procedures may be utilized. At the Hoechst laboratories (15), the assay for lipase involved removal of 25 μl of supernatant fluid from each of the 96 centrifuged wells in the titer plate with a 96-channel pipettor (Dynatech Deutschland GmbH) and transfer to a test plate. An eight-channel pipettor (Flow Laboratories) added 125 μl of substrate suspension to each broth location. An enzyme-linked immunosorbent assay reader determined the turbidity of the plate every 90 s and transferred the data to a computer, which calculated the slope of the turbidity/time function for each culture and compared that slope with dilutions of a lipase standard.

For measuring antibiotic concentrations, a number of machines may be adapted for large numbers of assays. Two examples are the zone reader for disk diffusion assay (Geigert et al. [42]) and the autoturbidometric reader employed in an assay for tylosin (74) described by Baltz and Seno (12). The "autoturb" system is available commercially from Eli Lilly and Co.

If a biozone assay is deemed adequate for the particular screen involved (see discussion in section 12.1.2), automated systems for this type of assay do exist. Geigert et al. (42) suggested an analytical procedure for measuring secondary metabolites in large numbers of fermentation broths. An automatic zone reader, consisting of a console, scanner, microprocessor, and computer, permitted automation of the zone-reading process in an agar diffusion assay. The mechanism allowed one technician to read 100,000 zones per week.

Data processing and strain selection

The final operation in the primary stage of an automated strain screening program begins with the compilation and statistical analysis of assay data and ends with the choice of a small minority of strains, designated candidate strains, for further testing in the secondary stage of the screen.

With the large number of samples to be assayed each day, it is essential to develop a computer-based data analysis system as an integral part of the automated screen. All information generated by various parts of the operation may be transmitted to a computer for compiling, editing, statistical analysis, and ranking of strains (see flow diagram, Fig. 1).

Choice of strains for further testing in the secondary stage is characterized by strict criteria which demand high statistical significance in results such that very few strains are passed on to the secondary stage (see discussion at the beginning of this section). Human judgment still plays a major role in strain choice. Strains ranking near the top in the primary screen but just outside the normal cut-off are sometimes passed if the strain has special history (e.g., a deregulated mutant previously isolated from a special biochemical selection or the product of a genetic cross of two different candidate strains).

Outline of an automated screen

The outline below is a useful adjunct to the above description of the automated screen's primary stage. The outline divides the primary stage as adopted from Betz et al. (15) into the six major steps that make up the screen in a daily operational sense.

Automated screen: primary stage

1. Preparation and storage of mutagenized single cells
   a. Mutation
   b. Expression/genetic segregation
   c. Fragmentation or sporulation
   d. Storage in liquid nitrogen
   e. Viable count on stored single cells
2. Preparation of isolated, genetically pure colonies: single-cell dispensing
   a. Dilution of single-cell suspension
   b. Operation of dispenser unit
3. Preparation and dispensing of fermentation media
4. Fermentation culture
   a. Inoculation
   b. Incubation in appropriate conditions and vessel
5. Assay
6. Data processing
   a. Compilation
   b. Statistical analysis, ranking
   c. Monitoring coefficients of variation
   d. Choice of strains for test in secondary stage

Automated screen: after the primary stage

The parameters governing the secondary test or any subsequent stages of the automated screen for strain improvement are essentially the same as for conventional screen. Because standardization is important in an automated screen, there will be a preference for using the same type vessels and equipment in the primary and secondary stages of the screen, whereas larger vessels would often be used in the secondary stage of a conventional screen. Selection criteria will probably tend to be more stringent in the automated screen because so many more isolates are available from which to choose. The final selection of strains for scale-up is made using the same criteria regardless of origin (conventional or automated screen). In the final analysis the automation of a screen is judged on its ability to produce results. A well-designed automated system can rapidly improve the ability of microorganisms to produce a fermentation product. For example, in their automated screen, Betz et al. (15) reported a rapid five- to sixfold improvement in lipase production from a series of selected *R. arrhizus* strains. From approximately 8,000 clones, 10 were isolated that were superior to the original strain.

Additional reading

Review articles by Aharonowitz and Cohen (1) and Rowlands (90) outline recent modifications of strain programs that have greatly expanded operational capacity.

12.2 ENRICHMENT PROCEDURES APPLIED TO STRAIN IMPROVEMENT

Microbiologists frequently apply special environmental conditions, toxic to a majority cell type but less or nontoxic to a desired minority of cells, to enrich a cell population for the desired mutants. Such selection techniques are referred to as enrichment procedures. The term "strain selection," as used by microbiologists in industry, is erroneously used both for enrichments as defined above and for screens as just described in section 12.1.

As described above (section 12.1), a single industrial screen can examine annually hundreds of thousands of individual microbial isolates for an economically desirable phenotype, usually increased product formation. The great power of this screening method is that it is general and does not require sophisticated understanding of the molecular biology and physiology of the microorganism being manipulated. Enrichment procedures, on the other hand, are developed out of an understanding of cellular metabolism and biosynthesis of the product of interest and are usually adapted from procedures involving more intensively studied microorganisms such as *Escherichia coli* and *Saccharomyces cerevisiae*. This extrapolation from intensively studied systems to less intensively studied ones can pay significant economic dividends as judged by the many current reports on the application of enrichment procedures to organisms of interest to the fermentation industry.

12.2.1 Enrichments for Deregulated Mutants

Many enrichment procedures take advantage of the evolutionarily conserved natural regulatory mechanisms that control primary metabolism in microorganisms. By supplying analogs of amino acids and vitamins which regulate their own synthesis, mutants can be selected which lack feedback regulation and overproduce an amino acid or vitamin. In the wild-type parental cell, the analog (acting through the normal regulatory mechanism) prevents synthesis of an adequate amount of the primary metabolite. Since this metabolite is required for growth and maintenance, most normal cells die and the remaining population is thereby enriched for deregulated mutants.

Deregulated mutants in fermentative production of primary metabolites

Selection of deregulated mutants has been applied extensively to the fermentative production of amino acids, vitamins, and nucleic acid precursors. Recent patents include claims for mutants resistant to proline analogs, which overproduce proline (I. Chibata, M. Kisumi, M. Sugiura, and T. Takagi, European patent application 76,516, 1983); mutants of *Bacillus* sp. resistant to 5-(2-thienyl)valeric acid and actithiazic acid, which produce increased amounts of biotin (Nippon Zeon Co., Ltd., Japan KoKai 58,152,495, 1983); and mutants of *Bacillus* strains resistant to tryptophan analogs and to azaserine and 6-diazo-5-oxo-L-norleucine (analogs of glutamine, which is used in tryptophan biosynthesis), which overproduce L-tryptophan (Ajinomoto Co., Inc., Japan KoKai 58,107,190, 1983; Ajinomoto Co., Inc., Japan KoKai 58,107,194, 1983; O. Kurahashi, M. Kamada, and H. Enei, European patent application 81,107, 1983; T. Tsuchida, N. Kawashima, S. Nakamori, H. Enei, and O. Kurahashi, European patent application 80,378, 1983). The use of enrichment procedures in the selection of strains with high capacity for production of lysine has recently appeared in patents (e.g., Ajinomoto Co., Inc., Japan KoKai 58,149,689, 1983) and in a review of lysine fermentation (104). Improved production of 5′-inosinic acid was reported for mutants of *Corynebacterium* sp. resistant to 2-deoxyribose (Ajinomoto Co., Inc., Japan KoKai 58,111,696, 1983) and resistant to glutamine analogs (Ajinomoto Co., Inc., Japan KoKai 58,111,695, 1983). Improved production of inosine was obtained from 8-azaguanine-resistant mutants of *Bacillus subtilis* (Ajinomoto Co., Inc., Japan KoKai 58,158,197, 1983). Purine analog-resistant strains of *Bacillus* sp. were patented for their overproduction of guanosine monophosphate (Ajinomoto Co., Inc., Japan KoKai 58,175,492, 1983). With the advent of recombinant DNA techniques, mutant alleles coding for feedback-resistant enzymes can be moved into a desirable strain and amplified. Thus a DNA fragment from a *B. subtilis* mutant resistant to the histidine analog 1,2,4-triazolealanine was inserted into plasmid pVB110 and used to transform histidine auxotrophs of a strain otherwise suited for histidine production. A resulting transformant was capable of producing 23 g of histidine per liter (T. Tsuchida, N. Kawashima, S. Nakamori, H. Enei, and A. Kurahashi, European patent application 82,637, 1983). Not all enrichment procedures can be correlated with documented deregulation of specific metabolic steps. Thus, anthracycline-resistant mutants of *Corynebacterium* sp. were isolated which overproduced L-glutamic acid (Ajinomoto Co., Inc., Japan KoKai 58,141,789, 1983).

Some microorganisms are resistant to certain amino acid, purine, or pyrimidine analogs which cause growth inhibition of *E. coli*. This probably is sometimes due to high intracellular pools of the natural metabolite corresponding to the analog. In such cases, this problem can be overcome by carrying out the selections on a poor carbon source (19) or by using auxotrophs and controlling the size of the key metabolic pool by feeding the metabolite corresponding to the analog.

Deregulated mutants in the fermentative production of enzymes

Many enzymes which break down complex polymers are produced commercially or are being developed for such use. Biosynthesis of these enzymes is often controlled by carbon catabolite repression. Selection for mutants resistant to analogs of glucose has been used to isolate mutants which are deficient in carbon catabolite repression of these enzymes. For example, resistance to 2-deoxyglucose was recently used to isolate deregulated mutants of *Trichoderma* sp. which overproduce cellulase (64). Carbon catabolite repression-resistant mutants of *B. subtilis* which overproduce amylase have been isolated (94).

Mutants of *B. subtilis* selected for resistance to the nucleoside-containing antibiotic tunicamycin were found to produce five times the amount of amylase as the sensitive parent but normal amounts of RNase and protease (55). Tunicamycin is known to inhibit the biosynthesis of cell wall carbohydrates in gram-positive bacteria, but the mechanism by which tunicamycin resistance affects amylase production is unknown and is probably pleiotropic rather than direct.

Deregulated mutants in the fermentative production of secondary metabolites

Enrichment procedures have also been used to isolate improved strains for the production of secondary metabolites, which are not required by cells for growth and maintenance. For antibiotic-producing *Streptomyces* spp., there are several reports of the utility of mutants selected for increased resistance to the toxic effects of the antibiotic produced, e.g., chlorotetracycline-resistant strains of *Streptomyces aureofaciens* (58), aurodox-resistant strains of *Streptomyces goldiniensis* (105), and kanamycin-resistant strains of *Streptomyces kanamyceticus* (32).

Selection procedures which enrich for mutants overproducing primary metabolite precursors of antibiotics have been used to improve antibiotic fermentations. Thus, tryptophan analog-resistant mutants of *Pseudomonas fluorescens* overproduce pyrrolnitrin (37). Among a small number of *Penicillium stoloniferum* mutants resistant to polyene antibiotics, a high producer of ergosterol was isolated which also overproduced the broad-spectrum antibiotic mycophenolic acid (S. W. Queener and C. H. Nash III, United States patent 4,115,197, 1978). Farnesylpyrophosphate is a common intermediate in the biosynthesis of ergosterol and mycophenolic acid.

More studies of the regulation of antibiotic synthesis are being undertaken today, and often selection procedures germane to these studies also exist. In such cases, it is logical to attempt a practical application. For example, ammonium ion suppresses the synthesis of some secondary metabolites (e.g., reference 84). Selection for resistance to methylammonium ion can be used to enrich for mutants derepressed with regard to ammonium repression (5, 6). Cesium ion can mimic the effects of ammonium repression (50). Hence, for ammonia-suppressed antibiotics, it may be useful to screen a number of methylammonium-resistant and cesium-resistant strains for superior antibiotic production.

Poor sporulation is a characteristic which sometimes can be detrimental for a production strain or parent in a mutation/screening program. Poorly sporulating strains often arise after many rounds of mutation and screening. Treatment of poorly sporulating yeast cultures with diethyl ether has been reported as an effective means of enrichment for rare spores in a population of vegetative cells (27).

In some cases the useful selective agent is biological rather than chemical. Resistance to actinophage has been used to isolate superior vancomycin-producing strains of *Streptomyces orientalis* (Soviet Union patent 235-244-A, 1969).

Recognition of undesirable enrichments which occur in nature can be exploited. For example, a decarboxylase present in the brewing yeast *Saccharomyces diastaticus* breaks down toxic cinnamic acids present in nature. This enzyme is governed by the *POF1* nuclear gene and is responsible for conversion of ferulic acid in wort to 4-vinylguaiacol, a compound associated with a medicinal off-flavor. *POF1*-negative strains were genetically crossed by protoplast fusion to glucamylase-positive strains to produce hybrids which could utilize wort dextrins and produced palatable beers (91).

12.2.2 Enrichments for Auxotrophs

The utility of auxotrophs in fermentation processes is not surprising, as blockage of a known metabolic pathway represents a simple and direct method of shunting metabolism in predictable patterns. The biosynthetic pathways for amino acids, vitamins, and other primary metabolites have been elucidated in many microorganisms and have been found, for the most part, to be conserved. Thus, procedures for the enrichment of auxotrophs can be valuable tools for strain improvement.

Examples of auxotrophs in fermentative processes

The titers of many economically important primary metabolites have been improved by isolating auxotrophs. A guanine-requiring mutant of *Streptomyces coelicolor* was patented as an overproducer of xanthine (V. Derkos-Sojak and V. Delic, Federal Republic of Germany patent 3,212,380, 1983). Increased L-ornithine production was reported for an arginine auxotroph of *Acinetobacter lwoffi* on hydrocarbon as sole carbon source (4). A homoserine auxotroph of *Brevibacterium lactofermentum* was patented for its overproduction of L-lysine (I. Nitelea, C. Jonescu, E. Matica, and D. Dobrescu, Romania patent 72,660, 1981). An adenine-requiring mutant of *Corynebacterium* was patented for its overproduction of 5′-inosinic acid (Ajinomoto Co., Inc., Japan KoKai 58,111,696, 1983). An adenine/xanthine-requiring strain of *Brevibacterium ammogenes* was patented for its production of guanosine 5′-diphosphate (13 g/liter) and guanosine 3′-phosphosphate (14 g/liter) (R. Saruno, Japan KoKai 58,098,094, 1983). A tyrosine auxotroph of *B. lactofermentum* has been patented for the production of L-phenylalanine (Ajinomoto Co., Inc., Japan KoKai 58,116,693, 1983).

Auxotrophs are also sometimes found to be useful in the production of secondary metabolites. For example, a mutant of *Cephalosporium acremonium* blocked in the conversion of sulfate to cysteine was a superior producer of cephalosporin C on methionine-based medium (82).

Development of auxotroph enrichment procedures

Usually modifications of existing methods developed with organisms studied intensively in basic research serve as adequate starting points for the development of procedures applicable for less studied microorganisms being developed for an industrial fermentation. One of the earliest techniques used to enrich for auxotrophs relied on physical separation of spores from filamentous mycelia after germination of conidia in minimal medium (39). This technique has been used with several modifications (20, 65, 93, 110) to isolate auxotrophs of both procaryotes and eucaryotes.

Auxotroph enrichment procedures for procaryotes

Penicillin enrichment. The most widely used, best understood, and generally most effective method for isolating auxotrophs of procaryotic microorganisms is the penicillin enrichment technique (25). Penicillin inhibits a transpeptidation reaction required to cross-link the peptidoglycan component of the cell wall of most procaryotes. Penicillin reacts covalently with a small number of different proteins (penicillin-binding proteins) located in the membranes of these microorganisms (108). Some of these proteins exhibit transpeptidase activity. The interactions of penicillin with these proteins result in the death and lysis of actively growing cells when they are exposed to the antibiotic. If viable cells are prevented from growing and synthesizing cell wall by environmental conditions, such cells are protected from the toxic effects of penicillin. These cells can be rescued from a solution of penicillin by first destroying the penicillin and then establishing environmental conditions suitable for growth. In the case of the penicillin enrichment for auxotrophs, cells are grown in a minimal medium containing penicillin so that prototrophs initiate active growth and as a result are killed by the penicillin. Auxotrophs present in the population cannot initiate growth because of the absence of required nutrients, and as a result these nongrowing cells survive. Penicillin is then destroyed enzymatically. Finally, nutrients required by the auxotrophs are added and the cells are incubated so that a dense population of cells, now containing a greater percentage of auxotrophs, can be reestablished. In practice, the killing of prototrophic cells during exposure to penicillin will not be complete. Therefore, for populations of cells containing very few auxotrophs, several cycles of selection and rescue may be required before the auxotrophs represent the dominant portion of the population.

A typical application of the penicillin enrichment technique to a *Streptomyces* species to be used in the production of a new antibiotic would begin with the fragmentation of mutagen-treated mycelia. After washing, the fragmented mycelia would be "starved" by incubation in a suitable minimal medium for 4 to 8 h at 30 to 37°C. (If the parent culture is an amino acid auxotroph, it is generally sufficient to include 200 to 300 μg of the appropriate amino acid supplement per ml of the minimal medium.) Freshly dissolved penicillin G is then added to the starved mycelia to achieve a penicillin concentration of 1,000 to 2,000 U/ml. Incubation is then continued for 4 to 24 h. (If the strain has undergone many rounds of mutation and selection, it may be slower growing and require longer exposure to penicillin.) At the end of penicillin exposure, sufficient penicillinase is added to inactivate the penicillin in 1 h. After the additional hour of incubation, the treated mycelia are washed in water to destroy any unlysed spheroplasts and cells with damaged walls. The wash also serves to remove any residual penicillin. Washed mycelia are inoculated into a complete nutrient medium, grown overnight, and fragmented. The fragmented mycelia can be used to begin a new cycle of penicillin enrichment, or divided into aliquots and frozen at −80°C until the investigator is ready to screen the population for number and type of auxotrophs present in the enriched population. A modification of the basic procedure which may prove helpful for some streptomycetes is the simultaneous use of penicillin and D-cycloserine (750 μg/ml) in the procedure (61).

Other auxotroph enrichments applied to procaryotic cells. Auxotrophic enrichment procedures employing agents other than penicillin to select against growing procaryotic cells have been reported, e.g., nalidixic acid (109), D-cycloserine used without penicillin (22), and 5-fluorouracil (106). In general, these methods have not gained the broad acceptance attained by the penicillin procedure.

Auxotroph enrichment procedures for eucaryotes

For eucaryotes, the techniques for auxotroph enrichment are in general less effective than the penicillin enrichment method employed for procaryotes. Nevertheless, there are a variety of techniques to choose from, and the correct choice and application of technique can save time when auxotrophs of a yeast or mold strain are required. The highly polyploid strains of yeast employed in the brewing industry pose a special problem. Special steps to reduce ploidy are required before auxotrophs can be used in strain improvement programs for these strains (e.g., see reference 97). Historically, except for a few special cases (see above), most auxotrophic mutants of *Penicillium* and *Cephalosporium* were counterproductive for the synthesis of beta-lactam antibiotics (e.g., see reference 69). However, the utility of protoplast fusion for strain improvement (45) in these systems makes isolation of marked strains of more interest. The ability to rapidly isolate auxotrophs which have equivalent titers of their parents is something to be desired, and effective enrichment procedures for auxotrophs of these genera take on more significance. In general, the impact of protoplast fusion (e.g., references 10 and 11) on breeding should broadly enhance the utility of auxotroph enrichment procedures.

Some of the earliest enrichment procedures for obtaining auxotrophs of lower eucaryotes involved genetic crosses (65). Such methods were inconvenient and were not widely adopted. In another early procedure, germination of *Neurospora crassa* spores in minimal medium and filtration were employed to separate nutritionally deficient spores from prototrophic ones capable of germinating and forming mycelia that could not pass the filters employed (20, 110). This method was simple and direct, and a 29-fold enrichment of auxotrophs was reported.

Several techniques applicable to molds or yeasts are based on a phenomenon observed by Fries many years ago, that double auxotrophs of *Ophiostoma* survived longer in minimal medium than their parents with single nutritional requirements (39, 40). This phenomenon was exploited in the development of auxotroph enrichment procedures for *Aspergillus* (86), *Allomyces* (83), and *Ustilago* (102) species. When inositol auxotrophs are available, inositol-less death provides significant enrichment to double auxotrophs. This method has been applied broadly, for example, to *N. crassa* (67), *C. acremonium* (82), and *S. cerevisiae* (47). Fatty acid auxotrophs and thymine auxotrophs have similarly been used in enrichment procedures to elicit fatty acid-less death (48) and thymineless death (14).

Several methods of enrichment for eucaryote auxotrophs are based on the greater physiological resistance of nongrowing, stationary-phase cells over rapidly growing cells under certain conditions. For example, polyene antibiotics preferentially act on rapidly growing eucaryote cells. Nystatin can be used to enrich for yeast and *Penicillium* auxotrophs (98). Another polyene antibiotic, *N*-glycosyl-polyfungin, was used successfully to enrich for auxotrophs of *Aspergillus* (8). This polyene antibiotic is convenient because of its water solubility. Significantly, application of the polyfungin method yielded all types of auxotrophs, with 50% of the colonies tested after enrichment representing nutritional mutants. Fungizone, a more soluble form of amphotericin B, when used in carefully defined physiological conditions produced a significant enrichment in auxotrophs of *S. cerevisiae* without the development of respiration-deficient mutants (77).

Echinocandin B is a member of a group of fatty acid-containing oligopeptide antibiotics which appear to inhibit glucan synthesis required for formation of the fungal cell wall (76). The compound has been used to develop an enrichment procedure for auxotrophs of *S. cerevisiae*. The method bears some analogy to the penicillin method for procaryotes in that both procedures rely on inhibition of cell wall synthesis in actively growing cells (72). A number of other methods have been reported for auxotroph enrichment in eucaryotic microorganisms. Sodium pentachlorophenate was reported to preferentially kill prototrophic cells over auxotrophic ones (41). This observation was verified and applied to *Penicillium chrysogenum* (71). Certain lytic enzymes which act on glucans in the yeast and fungal cell wall also act preferentially on rapidly growing cells and can be used to enrich for auxotrophs in minimal media (49, 85, 96). High temperature also appears to affect growing cells more than stationary cells and has been reported to enrich for auxotrophs of yeasts (46, 107). A fivefold enrichment for auxotrophs of *Rhodotorula* using griseofulvin has been reported (79).

Methods for selecting specific classes of eucaryote auxotrophs have been reported. Cell death can be caused by the decay of tritium incorporated into their macromolecules. Auxotrophs of *S. cerevisiae*, *Aspergillus*, and *Neurospora* have been enriched for by feeding isotopically labeled precursors which failed to be incorporated into the macromolecules of the nutritionally deficient strains (31, 68, 92).

Enrichment techniques have been developed to obtain mutants with lesions in certain loci governing enzymes involved in pyrimidine biosynthesis in *Coprinus* and certain strains of *S. cerevisiae* (78). Ureidosuccinic acid, an intermediate in pyrimidine biosynthesis, accumulates in dihydroorotaseless mutants of *Coprinus* and is toxic to the fungus. This observation was used to develop an enrichment technique for mutants lacking either carbamoyl phosphate synthetase or aspartic transcarbamylase. Lesions which inactivated either of these enzymes prevented formation of ureidosuccinic acid (78). The method is not general, in that one must have a characterized dihydroorotaseless mutant to employ the procedure. In addition, in systems with strong feedback regulation by uracil, the uracil supplement to the dihydroorotaseless mutants may be sufficient to lower formation of ureidosuccinic acid to below the toxic concentration. This problem was encountered and overcome in *S. cerevisiae* by starting with a mutant defective in uracil uptake as well as in dihydroorotase. Thus, the level of intracellular uracil could be controlled by varying the external concentration such that the potent feedback mechanism was not activated (7). Bach and Lacroute (7) modified the method to allow enrichment for pyrimidine auxotrophs deficient in orotidylic acid decarboxylase or orotidylic acid pyrophosphorylase. Again, however, the method is not general, in that it employs a mutant lacking dihydroorotase and requires that the strain possess a permease capable of accumulating ureidosuccinic acid which is supplied exogenously. A completely general and highly effective enrichment procedure for uracil auxotrophs lacking orotidylic acid decarboxylase might be very useful for the development of vector systems in poorly characterized yeast species since the corresponding gene has been cloned and characterized in *S. cerevisiae*.

12.2.3 Enrichment Procedure for Application of Recombinant DNA Technology to Secondary Metabolism

Application of recombinant DNA technology to fermentation processes has, to date, focused on the development of systems that allow the fermentative production of mammalian hormones, cytokines, and enzymes. The application of recombinant DNA techniques to fermentation pathways requiring a series of genes for the production of a primary metabolite is beginning to receive serious attention, and some practical contributions to the yields of amino acids have been reported (75; J. C. Patte, L. M. Lebeault, and A. Deschamps, France patent 2,511,032, 1983; Tsuchida et al., European patent applications 80,378 and 82,637, 1983). In some cases where amino acid precursors are rate limiting in the production of secondary metabolites, this recombinant DNA methodology has been claimed to improve the titer of a secondary metabolite (Kaken Industry Co., Ltd., Japan KoKai 58,162,290, 1983).

Application of recombinant DNA methods to improvement of the conversion of primary metabolite precursors into antibiotics, pigments, and other secondary metabolites is more difficult. Much more is understood about biosynthesis, regulation, and gene organization in primary metabolism. However, a methodology referred to as "cascade hybridization" by Timberlake (103) may decrease the time required to clone the large numbers of genes that control the conversion of primary to secondary metabolites. The basic method is to isolate mRNAs from both vegetative and differentiated cell states and use the vegetative mRNAs to discard those mRNAs in the differentiated state which are not specific to the differentiated state (111). The mRNAs specific to the differentiated state (e.g., antibiotic production phase of a streptomycete) would then be used to make corresponding complementary DNAs for cloning. In the case of an antibiotic-producing strain, complementary DNA related to antibiotic yield might be identified by transformation into a low-titer strain where the effect of the gene would be more easily recognized. Thus, genes could be screened for effect on antibiotic titer in this manner. This approach would augment but not replace the more classical approaches represented by a step-by-step pathway elucidation and complementation of well-characterized blocked mutants by DNA from a genomic library.

12.3 FINAL COMMENTS

Through the use of automated equipment and data processing by computers, a strain selection program may be developed which dramatically increases the number of strains tested. Such a system allows a reduction in the manpower required to operate screens and increases the potential of processing many variables. The system can be operated under high microbiological and statistical standards. In the operation of an automated strain program, a high number of variables influencing a program must be controlled, and all aspects of the system must function efficiently, e.g., assays, proper media makeup, correct agitation and growth conditions. The key to a successful program is constant monitoring and evaluation.

Special biochemical conditions can sometimes be applied to enrich a cell population for improved mutants, but successful choice and application of these selection techniques often require information not yet available for the organism of interest.

12.4 LITERATURE CITED

1. **Aharonowitz, Y., and G. Cohen.** 1981. The microbiological production of pharmaceuticals. Sci. Am. **245:** 141–152.
2. **Aidells, B. D., M. W. Konrad, and D. A. Glaser.** 1979. Growth and morphology of colonies of Chinese hamster ovary cells growing on agar is affected by insulin. Proc. Natl. Acad. Sci. U.S.A. **76:**1863–1867.
3. **Alikhanian, S. I., S. Z. Mindlin, S. U. Goldat, and A. V. Vladimizov.** 1959. Genetics of organisms producing tetracyclines. Ann. N.Y. Acad. Sci. **81:**914–949.
4. **Amund, O. O., G. Mackinnon, and I. J. Higgins.** 1983. Increased L-ornithine production by an arg mutant of *Actinetobacter lwoffi*. Eur. J. Appl. Microbiol. Biotechnol. **17:**252–253.
5. **Arst, H. N., Jr., and D. J. Cove.** 1969. Methylammonium resistance in *Aspergillus nidulans*. J. Bacteriol. **98:**1284–1293.
6. **Arst, H. N., Jr., and M. M. Page.** 1973. Mutants of *Aspergillus nidulans* altered in the transport of methylammonium and ammonium. Mol. Gen. Genet. **121:** 239–245.
7. **Bach, M., and F. Lacroute.** 1972. Direct selective techniques for the isolation of pyrimidine auxotrophs in yeast. Mol. Gen. Genet. **115:**126–130.
8. **Bal, J., E. Balbin, and N. J. Pienazek.** 1974. Method for isolating auxotrophic mutants in *Aspergillus nidulans* using N-glycosyl-polyfungin. J. Gen. Microbiol. **84:**111–116.
9. **Ball, C., and M. P. McGonagle.** 1978. Development and evaluation of a potency index screen for detecting mutants of *Penicillium chrysogenum* having increased penicillin yield. J. Appl. Bacteriol. **45:**67–74.
10. **Baltz, R. H.** 1978. Genetic recombination in *Streptomyces fradiae* by protoplast fusion and cell regeneration. J. Gen. Microbiol. **107:**93–102.
11. **Baltz, R. H.** 1980. Genetic recombination by protoplast

fusion in *Streptomyces*. Dev. Ind. Microbiol. **21**:43–54.

12. **Baltz, R. H., and E. T. Seno.** 1981. Properties of *Streptomyces fradiae* mutants blocked in biosynthesis of the macrolide antibiotic tylosin. Antimicrob. Agents Chemother. **20**:214–225.
13. **Baltz, R. H., E. T. Seno, J. Stonesifer, and G. M. Wild.** 1983. Biosynthesis of the macrolide antibiotic tylosin. A preferred pathway from tylactone to tylosin. J. Antibiot. **36**:131–141.
14. **Barclay, B. J., and J. G. Little.** 1977. Selection of yeast auxotrophs by thymidylate starvation. J. Bacteriol. **132**:1036–1037.
15. **Betz, J. W., W. Aretz, and W. Hartel.** 1984. Use of flow cytometry in industrial microbiology for strain improvement programs. Cytometry **5**:145–150.
16. **Brown, W. F., and R. P. Elander.** 1966. Some biometric considerations in an applied antibiotic AD-464 strain development program. Dev. Ind. Microbiol. **7**:114–123.
17. **Bylinsky, G.** 1974. Industry is finding more jobs for microbes. Fortune **89**:96–102.
18. **Calam, C.** 1970. Improvement of microorganisms by mutation, hybridization and selection, p. 435–459. *In* J. R. Norris and D. W. Ribbons (ed.), Methods in microbiology, vol. 3A. Academic Press, Inc., New York.
19. **Calhoun, D. H., and R. A. Jensen.** 1972. Significance of altered carbon flow in aromatic amino acid synthesis: an approach to the isolation of regulatory mutants in *Pseudomonas aeruginosa*. J. Bacteriol. **109**:365–372.
20. **Catcheside, D. G.** 1954. Isolation of nutritional mutants of *Neurospora crassa* by filtration enrichment. J. Gen. Microbiol. **11**:34–36.
21. **Cram, L. S., and G. C. Salzman.** 1976. Developments in rapid cell analysis and sorting techniques applicable to biomedical problems. Dev. Ind. Microbiol. **17**:141–151.
22. **Curtiss, R., III., L. J. Charamella, C. M. Berg, and P. E. Harris.** 1965. Kinetic and genetic analysis of D-cycloserine inhibition and resistance in *Escherichia coli*. J. Bacteriol. **90**:1238–1250.
23. **Dahl, S. G., T. Waaler, S. Thomassen, B. Oystese, and T. Hoyland.** 1972. Testing error in the screening of bacitracin producing strains for high yielding mutants. Pharm. Acta Helv. **47**:424–432.
24. **Dairkee, S. H., and D. A. Glaser.** 1982. Dimethyl sulfoxide affects colony morphology on agar and alters distribution of glycosaminoglycans and fibronectin. Proc. Natl. Acad. Sci. U.S.A. **79**:6927–6931.
25. **Davis, B. D.** 1949. The isolation of biochemically deficient mutants of bacteria by means of penicillin. Proc. Natl. Acad. Sci. U.S.A. **35**:1–10.
26. **Davis, O. L.** 1964. Screening for improved mutants in antibiotic research. Biometrics **20**:576–591.
27. **Dawes, I. W., and I. D. Hardie.** 1974. Selective killing of vegetative cells in sporulated yeast cells by exposure to diethyl ether. Mol. Gen. Genet. **131**:281–289.
28. **Demain, A.** 1973. Mutation and the production of secondary metabolites. Adv. Appl. Microbiol. **16**:177–202.
29. **Demain, A. L.** 1971. Overproduction of microbial metabolites and enzymes due to alteration of regulation. Adv. Biochem. Eng. **1**:113–142.
30. **Ditchburn, P., B. Giddings, and K. D. MacDonald.** 1974. Rapid screening for the isolation of mutants of *Aspergillus nidulans* with increased penicillin yields. J. Appl. Bacteriol. **37**:515–523.
31. **Donkersloot, J. A., and R. I. Mateles.** 1968. Enrichment of auxotrophic mutants of *Aspergillus flavus* by tritium suicide. J. Bacteriol. **96**:1551–1555.
32. **Du, R., and J. Zhao.** 1983. Selection of high yielding strains of *Streptomyces kanamyceticus* by a kanamycin-resistant mutant. Kangshengsu **8**:24–28.
33. **Dulaney, E. L.** 1953. Observations on *Streptomyces griseus*. VI. Further studies on strain selection for improved streptomycin production. Mycologia **45**:481–487.
34. **Elander, R. P.** 1966. Two decades of strain development in antibiotic-producing microorganisms. Dev. Ind. Microbiol. **7**:61–73.
35. **Elander, R. P.** 1967. Enhanced penicillin biosynthesis in mutant and recombinant strains of *Penicillium chrysogenum*, p. 403–423. *In* K. Grober, F. Scholz, and M. Zacharias (ed.), Induzierte Mutationen und ihre Nutzung. Abhandlungen der Deutschen Akademie der Wissenschaften zu Berlin. Akademie Verlag, Berlin.
36. **Elander, R. P., L. T. Chang, and R. W. Vaughn.** 1977. Genetics of industrial microorganisms, p. 1–40. *In* D. Perlman (ed.), Annual report on fermentation processes, vol. 1. Academic Press, Inc., New York.
37. **Elander, R. P., J. A. Mabe, R. L. Hamill, and M. Gorman.** 1971. Biosynthesis of pyrrolnitrins by analogue-resistant mutants of *Pseudomonas fluorescens*. Folia Microbiol. **16**:157–165.
38. **Fantini, A. A.** 1975. Strain development. Methods Enzymol. **43**:24–41.
39. **Fries, N.** 1947. Experiments with different methods of isolating physiological mutations of filamentous fungi. Nature (London) **159**:199.
40. **Fries, N.** 1948. Viability and resistance of spontaneous mutations in *Ophiostoma* representing different degrees of heterotrophy. Physiol. Plant. **1**:330–341.
41. **Ganju, P. L., and M. R. S. Iyengar.** 1968. An enrichment technique for isolation of auxotrophic mutants using sodium pentachlorophenate. Hind. Antibiot. Bull. **11**:12–21.
42. **Geigert, J., D. Hansen, C. McDowell, R. Merrill, and C. Ward.** 1976. Microbiological assay utilizing an automatic zone screen. Dev. Ind. Microbiol. **17**:153–156.
43. **Glaser, D. A.** 1976. Symposium: automated microbiology. Dev. Ind. Microbiol. **17**:139–156.
44. **Glaser, D. A., and W. H. Wattenberg.** 1966. An automated system for the growth and analysis of large numbers of bacterial colonies using an environmental chamber and computer controlled flying-spot scanner. Ann. N.Y. Acad. Sci. **139**:243–257.
45. **Gore, A.** 1984. Commercial biotechnology: an international analysis. Office of Technology Assessment, Washington, D.C.
46. **Hardie, I. D., and I. W. Dawes.** 1977. Optimal conditions for selecting specific auxotrophs of *Saccharomyces cerevisiae* using temperature-sensitive suicide mutants. Mutat. Res. **42**:215–222.
47. **Henry, S. A., T. F. Donahue, and M. R. Culbertson.** 1975. Selection of spontaneous mutants by inositol starvation in yeast. Mol. Gen. Genet. **143**:5–11.
48. **Henry, S. A., and B. Horowitz.** 1975. A new method for mutant selection in *Saccharomyces cerevisiae*. Genetics **79**:175–186.
49. **Herrera, L.** 1976. Use of snail enzyme for the selection of glucose- and mannose-negative mutants in *Saccharomyces cerevisiae*. Mutat. Res. **34**:123–130.
50. **Holloman, W. K., and C. A. Dekker.** 1971. Control by cesium and intermediates of the citric acid cycle of extracellular ribonuclease and other enzymes involved in the assimilation of nitrogen. Proc. Natl. Acad. Sci. U.S.A. **68**:2241–2245.
51. **Hopwood, D. A.** 1970. The isolation of mutants, p. 363–433. *In* J. R. Norris and D. W. Ribbons (ed.), Methods in microbiology, vol. 3A. Academic Press, Inc., New York.
52. **Hopwood, D. A., and M. J. Merrick.** 1977. Genetics of antibiotic production. Bacteriol. Rev. **41**:595–635.
53. **Horan, P. K., and L. L. Wheeless, Jr.** 1977. Quantitative single cell analysis and sorting. Science **198**:149–157.
54. **Ichikawa, T., M. Date, T. Ishikura, and A. Ozaki.** 1971. Improvement of kasugamycin-producing strain by the agar piece method and the prototroph method. Folia Microbiol. **16**:218–224.
55. **Ingle, M. B., and R. J. Erickson.** 1978. Bacterial alpha-

amylases. Adv. Appl. Microbiol. **24**:257–278.

56. **Jensen, A. L., M. A. Darken, J. S. Schultz, and A. J. Shay.** 1964. Relomycin: flask and tank fermentation studies, p. 49–53. Antimicrob. Agents Chemother. 1963.
57. **Johnston, J. R.** 1975. Strain improvement and strain stability in filamentous fungi, p. 59–78. *In* J. Smith and D. Berry (ed.), The filamentous fungi, vol. 1. Wiley, New York.
58. **Katagiri, K.** 1954. Study on the chlortetracycline: improvement of chlortetracycline-producing strain by several kinds of methods. J. Antibiot. **7**:45–52.
59. **Kennedy, J. H.** 1978. High performance liquid chromatography analysis of fermentation broths: cephalosporin C and tylosin. J. Chromatogr. Sci. **16**:492–495.
60. **Kikuchi, M.** 1980. Application of genetics for strain improvement in industrial microorganisms. Biotechnol. Bioeng. **22**(Suppl. 1):195–208.
61. **Kirkpatrick, J. R., and O. W. Godfrey.** 1973. The isolation and characterization of auxotrophs of the aspartic acid family from *Streptomyces lipmanii*. Folia Microbiol. **18**:90–101.
62. **Konrad, M. W., J. L. Couch, and D. A. Glaser.** 1980. A grey level frequency histogram representation of animal cell colonies seen by scattered light. Computers Biomed. Res. **13**:333–349.
63. **Konrad, M. W., B. Storrie, D. A. Glaser, and L. H. Thompson.** 1977. Clonal variation in colony morphology and growth of CHO cells cultured on agar. Cell **10**:305–312.
64. **Labudova, I., and V. Farkas.** 1983. Enrichment technique for the selection of catabolite repression-resistant mutants of *Trichoderma* as producers of cellulase. FEMS Microbiol. Lett. **20**:211–215.
65. **Lein, J., H. K. Mitchell, and M. B. Houlahan.** 1948. A method for selection of biochemical mutants of *Neurospora*. Proc. Natl. Acad. Sci. U.S.A. **34**:435–442.
66. **Leive, L. (ed.).** 1985. Microbiology—1985. American Society for Microbiology, Washington, D.C.
67. **Lester, H. E., and S. R. Gross.** 1958. Efficient method for selection of auxotrophic mutants of *Neurospora*. Science **129**:572.
68. **Littlewood, B. S., and J. E. Davies.** 1973. Enrichment for temperature-sensitive and auxotrophic mutants in *Saccharomyces cerevisiae* by tritium suicide. Mutat. Res. **17**:315–322.
69. **MacDonald, K. D.** 1968. The selection of auxotrophs of *Penicillin chrysogenum* with nystatin. Genet. Res. **11**:327–330.
70. **Mandal, S. K., P. K. Dey, M. P. Singh, and D. K. Roy.** 1983. Oxytetracycline fermentation-strain selection of *Streptomyces rimosus* for improved production of oxytetracycline. J. Food Sci. Technol. **20**:30–32.
71. **Masurekar, P. S., M. P. Kahgan, and A. L. Demain.** 1972. Mutagenesis and enrichment of auxotrophs in *Penicillium chrysogenum*. Appl. Microbiol. **24**:995–996.
72. **McCammon, M. T., and L. W. Parks.** 1981. Enrichment for auxotrophic mutants in *Saccharomyces cerevisiae* using the cell wall inhibitor, echinocandin B. Mol. Gen. Genet. **186**:295–297.
73. **McGuire, J. M.** 1961. The antibiotics—past, present and future. Proc. Indiana Acad. Sci. **71**:248–257.
74. **McGuire, J. M., W. S. Boniece, C. E. Higgens, M. M. Hoehn, W. M. Stark, J. Westhead, and R. Wolfe.** 1961. Tylosin, a new antibiotic. I. Microbiological studies. Antibiot. Chemother. **11**:320–327.
75. **Miwa, K., T. Tsuchida, O. Kurahashi, S. Nakamori, K. Sano, and H. Momose.** 1983. Construction of L-threonine overproducing strains of *Escherichia coli* K-12 using recombinant DNA techniques. Agric. Biol. Chem. **47**:2329–2334.
76. **Mizoguchi, J., T. Saito, K. Mizuno, and K. Hayano.** 1977. On the mode of action of a new antifungal antibiotic, aculeacin A: inhibition of cell wall synthesis in *Saccharomyces cerevisiae*. J. Antibiot. **30**:308–313.
77. **Moat, A. G., I. J. Barnes, and E. H. McCurley.** 1966. Factors affecting the survival of auxotrophs and prototrophs of *Saccharomyces cerevisiae* in mixed populations. J. Bacteriol. **92**:297–301.
78. **Motta, R.** 1967. Method de selection de mutants uracile exigeants au locus ur l de *Coprinus radiatus*. C. R. Hebd. Seances Acad. Sci., Ser. D Sci. Nat. **264**:654–657.
79. **Nand, K., R. Joseph, and T. N. R. Rao.** 1973. Use of griseofulvin for the isolation of auxotrophic mutants of *Rhodotorula* sp. Experientia **29**:237–239.
80. **Nkanga, E. J., and C. Hagedorn.** 1978. Detection of antibiotic-producing *Streptomyces* inhibiting forest soils. Antimicrob. Agents Chemother. **14**:51–59.
81. **Normansell, I. D.** 1982. Strain improvement in antibiotic producing microorganisms. J. Chem. Technol. Biotechnol. **32**:296–303.
82. **Nuesch, J., H. J. Treichler, and M. Liersch.** 1973. The biosynthesis of cephalosporin C, p. 309–334. *In* Z. Vanek, L. Hostalek, and J. Cudlin (ed.), Genetics of industrial microorganisms, vol. 2. Academia, Prague.
83. **Olson, L. W., and T. A. B. Nielsen.** 1981. Isolation of auxotrophic mutants of *Allomyces macrogynus*. Mycologia **73**:493–499.
84. **Omura, S., Y. Tanaka, H. Mamada, and R. Masuma.** 1983. Ammonium ion suppresses the biosynthesis of tylosin aglycone by interference with valine catabolism. J. Antibiot. **36**:1792–1794.
85. **Piedra, D., and L. Herrera.** 1976. Selection of auxotrophic mutants in *Saccharomyces cerevisiae* by snail enzyme digestion method. Folia Microbiol. **21**:337–341.
86. **Pontecorvo, G., J. A. Roper, L. M. Hemmons, K. D. MacDonald, and A. W. Bufton.** 1953. The genetics of *Aspergillus nidulans*. Adv. Genet. **5**:141–238.
87. **Queener, S. W., and J. J. Capone.** 1974. Simple method for preparation of homogenous spore suspensions useful in industrial strain selection. Appl. Microbiol. **28**:498–500.
88. **Queener, S. W., O. K. Sebek, and C. Vézina.** 1978. Mutants blocked in antibiotic synthesis. Annu. Rev. Microbiol. **32**:593–636.
89. **Queener, S. W., and R. W. Swartz.** 1979. Penicillins: biosynthetic and semisynthetic, p. 35–122. *In* A. H. Rose (ed.), Economic microbiology, vol. 3. Academic Press, Inc., New York.
90. **Rowlands, R. T.** 1984. Industrial strain improvement: mutagenesis and random screening procedures. Enzyme Microb. Technol. **6**:3–10.
91. **Russell, I., I. F. Hancock, and G. G. Stewart.** 1983. Construction of dextrin fermentative yeast strains that do not produce phenolic off-flavors in beer. J. Am. Soc. Brew. Chem. **41**:45–51.
92. **Russell, P. J., and M. P. Cohen.** 1976. Enrichment for auxotrophic and heat-sensitive mutants of *Neurospora crassa* by tritium suicide. Mutat. Res. **34**:359–366.
93. **Saito, H., and Y. Ikeda.** 1957. An improved method of isolating biochemical mutants of *Streptomyces griseoflavus*. J. Gen. Appl. Microbiol. **3**:240–249.
94. **Saito, N., and K. Yamamoto.** 1975. Regulatory factors affecting α-amylase production in *Bacillus licheniformis*. J. Bacteriol. **121**:848–856.
95. **Shomura, T., J. Yoshida, S. Amano, M. Kojima, S. Inouye, and T. Niida.** 1979. Studies on *Actinomycetales* producing antibiotics only on agar culture. I. Screening taxonomy and morphology-productivity relationship of *Streptomyces halstedii*, strain SF-1993. J. Antibiot. **32**:427–435.
96. **Sipiczki, M., and L. Ferenczy.** 1978. Enzyme methods for enrichment of fungal mutants. I. Enrichment of *Schizosaccharomyces pombe* mutants. Mutat. Res. **50**:163–173.
97. **Skatrud, P. L., D. M. Jaeck, E. J. Kot, and J. R. Helbert.**

1980. Fusion of *Saccharomyces uvarum* with *Saccharomyces cerevisiae*: genetic manipulation and reconstruction of a brewer's yeast. J. Am. Soc. Brew. Chem. **38**:49–53.

98. **Snow, R.** 1966. An enrichment method for auxotrophic yeast mutants using the antibiotic 'nystatin.' Nature (London) **211**:206–207.
99. **Spagnoli, R., and L. Cappelletti.** 1981. New screening method for the rapid identification of blocked mutants and their production: application to erythromycin and oleandomycin producing strains. Agric. Biol. Chem. **45**:761–763.
100. **Stauffer, J. F., L. J. Schwartz, and C. W. Brady.** 1966. Problems and progress in a strain selection program with cephalosporin producing fungi. Dev. Ind. Microbiol. **7**:104–113.
101. **Thom, C., and R. Steinberg.** 1939. The chemical induction of genetic change in fungi. Proc. Natl. Acad. Sci. U.S.A. **25**:329–335.
102. **Thomas, P. L.** 1972. Increased frequency of auxotrophic mutants of *Ustilago hordei* after combined UV irradiation and inositol starvation. Can. J. Genet. Cytol. **14**:785–788.
103. **Timberlake, W. E.** 1980. Developmental gene regulation in *Aspergillus nidulans*. Dev. Biol. **78**:497–510.
104. **Tosaka, O., H. Enei, and Y. Hirose.** 1983. The production of L-lysine by fermentation. Trends Biotechnol. **1**:70–74.
105. **Unowsky, J., and D. C. Hoppe.** 1978. Increased production of the antibiotic aurodox (x-5108) by aurodox-resistant mutants. J. Antibiot. **31**:662–666.
106. **Wachsman, J. T., and L. Hogg.** 1964. Use of 5-fluorouracil for the isolation of auxotrophic mutants of *Bacillus megaterium*. J. Bacteriol. **87**:1137–1139.
107. **Walton, E. F., B. L. A. Carter, and J. R. Pringle.** 1979. An enrichment method for temperature-sensitive and auxotrophic mutants of yeast. Mol. Gen. Genet. **171**:111–114.
108. **Waxman, D. T., and J. L. Strominger.** 1982. Beta-lactam antibiotics: biochemical modes of action, p. 210–286. *In* R. B. Morin and M. Gorman (ed.), Chemistry and biology of beta-lactam antibiotics, vol. 3. Academic Press, Inc., New York.
109. **Weiner, R. M., M. J. Voll, and T. M. Cook.** 1974. Nalidixic acid for enrichment of auxotrophs in cultures of *Salmonella typhimurium*. Appl. Microbiol. **28**:579–581.
110. **Woodward, V. W., J. R. DeZeeuw, and A. M. Srb.** 1954. The separation and isolation of particular biochemical mutants of *Neurospora* by differential germination of conidia followed by filtration and selective plating. Proc. Natl. Acad. Sci. U.S.A. **40**:192–200.
111. **Zimmerman, C. R., W. C. Orr, R. F. Leclerc, E. C. Barnard, and W. E. Timberlake.** 1980. Molecular cloning and selection of genes regulated in *Aspergillus nidulans* development. Cell **21**:709–715.

Chapter 13

Protoplast Fusion

PATTI MATSUSHIMA AND RICHARD H. BALTZ

Protoplast fusion is a versatile general technique to induce genetic recombination in a variety of procaryotic and eucaryotic microorganisms. Protoplast fusion is particularly useful for industrial microorganisms, such as the procaryotic actinomycetes which have not been subjected to extensive genetic analysis, since it does not require transducing phages, plasmid sex factors, or competency development. It does, however, require identification of procedures to form stable protoplasts, to fuse protoplasts, and to regenerate viable cells from fused protoplasts.

Protoplasts are prepared by removing the cell wall by treating cells with lytic enzymes in the presence of osmotic stabilizers. In the presence of a fusogenic agent, such as polyethylene glycol (PEG), protoplasts are induced to fuse and form transient hybrids or diploids. During this hybrid state, the genomes (or chromosomes) may reassort and genetic recombination can occur. For the procaryotic bacilli and streptomycetes, recombination occurs at very high frequencies since their chromosomes are free in the cytoplasm. For the eucaryotic fungi, protoplast fusion must be followed by nuclear fusion to obtain true recombinants. The final crucial step in the protoplast fusion technique is the regeneration of viable cells from the fused protoplasts, without which no viable recombinants can be obtained.

PEG-induced protoplast or cell fusion was first observed in plants (36) and has since been used successfully to fuse animal (34, 47), filamentous fungal (8,

10, 15–18, 37, 46), yeast (51, 52, 55), *Bacillus* sp. (19, 50), *Streptomyces* sp. (3, 11, 20, 23, 40, 41), *Brevibacterium* sp. (35), *Streptosporangium* sp. (43), and *Staphylococcus* sp. (24) protoplasts.

In the gram-negative bacteria, removal of the outer membrane from spheroplasts may present special problems. Without the formation of true protoplasts, membrane fusion may be hindered. Two cases of true or partial (incomplete removal of the outer membrane) protoplast fusion have been reported for gram-negative bacteria (14, 54).

Protoplast fusion can promote high frequencies of genetic recombination between organisms which have previously demonstrated poor or no genetic exchange or which are uncharacterized for genetic exchange. The fusions may be intraspecific, interspecific, or even intergeneric, involving two or more complete parental genomes. Thus, a number of desirable genes from divergent strains can potentially be introduced into a single strain by this procedure. Mating type or incompatibility factors do not hinder hybrid formation (7), and sex factor mediation is not required to obtain high frequencies of recombination (6). There is no need for introducing selectable genetic markers into the parental strains for intraspecific recombination, at least for gram-positive procaryotes, because of the high frequency of recombinants obtained (11, 27, 30).

The potential of protoplast fusion to generate high frequencies of genetic recombinants, particularly in procaryotes, has practical applications for industrial strain development. The fungi, bacilli, and streptomycetes produce many economically important products (e.g., antibiotics, enzymes, and antitumor agents). Thus the protoplast fusion technique may be applied to improvement of product yield and to the development of hybrid biosynthetic pathways to produce novel chemical structures.

The reader is referred to references 1 through 7 for a general review of protoplast fusion in bacteria and fungi.

13.1 GENERAL CONSIDERATIONS

The ability to form, fuse, and regenerate protoplasts may vary greatly from organism to organism. Efficient formation of protoplasts does not ensure that efficient regeneration of protoplasts will occur. Subtle differences in experimental conditions and techniques before protoplast formation and during cell regeneration can have profound effects on protoplast regeneration and genetic recombination. The techniques described in this chapter for forming, fusing, and regenerating protoplasts of *Streptomyces*, *Micromonospora*, *Bacillus*, and fungal species have been found to be generally applicable for industrial strain development.

In selecting strains to assess protoplast fusion and genetic recombination frequencies, mutants with easily scorable genetic markers (e.g., nutrition, antibiotic resistance, morphology, etc.) are useful. Crossing over between the appropriate markers is scored to assess the efficiency of protoplast fusion and genetic recombination.

To simplify the discussion of the various procedures in this chapter, it is assumed that two strains auxotrophic for different markers are used to prepare protoplasts: the protoplasts are fused and plated to obtain regeneration of viable cells. Prototrophic recombinant frequencies among the viable cells are used to assess protoplast fusion and genetic recombination.

13.2 *STREPTOMYCES*

The procedures for protoplast fusion in *Streptomyces* sp. unless otherwise noted, are those in Baltz (11), Baltz and Matsushima (12, 13), and Matsushima (unpublished data). The procedures described have been used successfully with several species of *Streptomyces*.

13.2.1 Conditions for Cell Growth before Protoplast Formation

The strains are grown in Trypticase soy (TS) broth (see section 13.7.1) on a gyratory water bath shaker at ~240 rpm. Lyophilized or frozen vegetative mycelia or spores are used as inoculum.

Growth temperature

The temperature for cell growth before protoplast formation is very important to obtain efficient protoplast regeneration with some species (12). The correct growth temperature can be determined by doing a preliminary experiment in which the cells are grown at various temperatures, and protoplasts are formed and then plated on modified R2 medium for regeneration at 29 to 37°C. The temperature for cell growth that gives the best regeneration frequency should be used in subsequent experiments. For *Streptomyces* spp. the optimum temperature for cell growth before protoplast formation to maximize cell regeneration does not always correspond to the optimum temperature for the regeneration of cells from protoplasts (12). However, the best temperature for regeneration is usually about 30°C.

Homogenization

Grow the cells in TS broth to an absorbancy at 600 nm (A_{600}) of 2 to 9 (24 to 48 h). Homogenize the culture with a Teflon pestle tissue grinder (A. H. Thomas Co., Philadelphia, Pa.). Homogenization breaks the large mycelial masses into smaller units which are readily fractured by ultrasound.

Ultrasonic treatment

Dilute the homogenized mycelia twofold in TS broth and add about 20 ml to a 50-ml sterile plastic vessel. Submerge an ultrasonic probe (Branson, model W185) about 1 to 2 cm in the liquid, and sonicate the sample for 5 to 10 s at 76 W. Sonication breaks the mycelial masses into mostly single cell units (fracturing approximately every other cell [11]), which facilitates rapid exponential growth upon subculturing and helps to eliminate the formation of large mycelial clumps which tend to be more resistant to protoplast formation.

Glycine treatment

Growth of cells in partially inhibitory concentrations of glycine renders the cells more susceptible to lysozyme digestion, resulting in more efficient protoplast formation (44, 49). The concentration of glycine needed to render *Streptomyces* sp. more susceptible to lysozyme treatment is usually between 0.1 and 2.0%. To determine the appropriate concentration of glycine, set up a series of flasks, each containing TS broth, sonicated cells (diluted 1:20), and glycine ranging from 0 to 2.0%. Grow the cultures for 16 to 24 h on a gyratory water bath shaker, homogenize the culture, and take the absorbance readings (A_{600}). The concentration of glycine that reduces the absorbance reading to approximately one-half that of the control culture with no glycine added is the concentration that should be used for protoplast formation.

The homogenized mycelia grown in TS broth with the appropriate concentration of glycine are sonicated for 1 to 3 s rather than the normal 5 to 10 s since growth in the presence of glycine usually increases the sensitivity of the cells to ultrasonic fracture. Dilute the cells 1:20 and grow them in TS broth plus glycine to an A_{600} between 2.0 and 7.0. Homogenize and sonicate the cells and plate serial dilutions of the cells on a complete medium such as TS agar to determine CFU. When the homogenization and sonication techniques described above are used, approximately one-half of the potential viable cells initiate the formation of colonies. When the efficiency of protoplast regeneration is calculated, the CFU should be doubled (see section 13.2.5) to approximate the initial viable cells in the mycelia before sonication or protoplast formation.

Potential problems

(i) Lack of growth. If the culture does not grow in TS broth prepared as stated, a more alkaline pH may be required. The pH should be adjusted accordingly. Also, although TS broth is considered a complete medium, many auxotrophic strains require additional supplementation for good growth.

(ii) Mycelial clumping. Some strains tend to grow in large mycelial clumps even after vigorous homogenization and sonication. This tendency to aggregate greatly reduces the efficiency of protoplast formation. The addition of 10% sucrose to the TS broth may help alleviate this problem.

13.2.2 Formation of Protoplasts

1. Wash the homogenized cells three times by centrifugation (1,060 × g) and resuspension in equal volumes of medium P (see section 13.7.2), a hypertonic solution which will provide an osmotically stable environment for the protoplasts.
2. Resuspend the washed cells in an equal volume of medium P containing 1 mg of lysozyme per ml. Incubate the suspension at 37°C. A rapid clearing of the suspension is generally observed and is indicative of protoplast formation. The conversion of cells to stable protoplasts may be complete in 15 min or may take as long as an hour. Confirm the formation of individual spherical protoplasts by phase-contrast microscopy. Very few, if any, remaining mycelial fragments should be observed.
3. Wash the protoplasts two times in equal volumes of medium P by centrifugation and gentle resuspension.
4. Determine the number of viable protoplasts by plating on a regeneration medium (section 13.2.5). Make serial 10-fold dilutions of the protoplasts in medium P and add 0.1-ml volumes to R2 modified soft agar overlays (section 13.7.1) at 46°C and mix (section 13.2.4). Pour the overlays onto modified R2 bottom agar (section 13.7.1) in standard petri dishes and allow them to solidify. The modified R2 plates are supplemented with the appropriate amino acids for the growth of the parental protoplasts. Determine the frequency of regeneration (section 13.2.5) from the viable protoplast counts.
5. Determine the reversion frequency of the parental strains to prototrophy by plating dilutions of the protoplasts on unsupplemented modified R2 plates.
6. Determine the frequency of nonprotoplasted, viable cells remaining by plating several dilutions of the protoplasts on a nonhypertonic medium such as TS agar. Since the protoplasts will be osmotically lysed, only the nonprotoplasted cells will initiate colony formation.

Potential problems

(i) Lack of protoplast formation. If protoplast formation does not occur, it is often because the cells are in the incorrect stage of growth. Cells in rapid exponential ($A_{600} < 2.0$) or stationary ($A_{600} > 7.0$) growth sometimes form protoplasts less efficiently than those in the transition phase between exponential and stationary phases (A_{600} of 2.0 to 7.0).

Sometimes concentrations of lysozyme greater than 1 mg/ml are needed to cause protoplast formation. If there is no evidence of protoplast formation after 1 h, pellet the cells by centrifugation and suspend them in a higher concentration of lysozyme. Concentrations of lysozyme of up to 20 mg/ml can be used. Some strains may still resist protoplast formation. The reasons for this recalcitrance are not known. Different growth conditions or enzymes may be needed to induce protoplast formation in these strains. It has recently been reported (42) that the use of achromopeptidase in conjunction with lysozyme can result in better protoplast formation and regeneration than is obtained with lysozyme treatment alone.

(ii) Protoplast clumping. Often protoplasts of strains that tend to clump during cell growth also tend to clump during protoplast formation. If the clumps cannot be dispersed by vortexing, the suspension may be very gently homogenized.

A common cause of protoplast clumping in strains that do not tend to clump during growth is overcompaction of the protoplasts during centrifugation. The use of round-bottomed centrifuge tubes rather than conical tubes can help to avoid this problem.

(iii) "Ghosts." Occasionally, large numbers of protoplast "ghosts" and irregularly shaped protoplasts are observed by phase-contrast microscopy. These protoplast ghosts generally are not viable, and their presence in high numbers may suggest some detrimental condition during the protoplasting technique. A temperature that is too high during cell growth is a common cause. Many *Streptomyces* spp. are also sensitive to elevated temperatures during protoplast formation and may need to be treated with lysozyme at temperatures below 37°C. Slight adjustments in the concentration of divalent cations or in the pH of medium P may also be needed with some strains to ensure stable formation of protoplasts.

13.2.3 Fusion of Protoplasts

1. Mix 0.5 ml of protoplasts of each of the two strains to be fused in a round-bottomed sterile centrifuge tube.
2. Centrifuge (1,060 × *g*) the mixture for 8 to 10 min.
3. Decant the supernatant and suspend the protoplast pellet in 0.1 ml of medium P.
4. Add 0.9 ml of 50% (wt/vol) PEG 1000.
5. Gently mix the protoplast-PEG suspension and then let the suspension stand at room temperature for about 30 s to 1 min.

Conditions affecting recombination

(i) Temperature. The incubation temperature for cell growth before protoplast formation (section 13.2.1) and for protoplast regeneration (section 13.2.4) can influence greatly the frequency of genetic recombinants obtained from a protoplast fusion (12).

(ii) PEG. The concentration of PEG used to induce protoplast fusion affects the frequency of recombinants obtained. PEG preparations of molecular weights other than 1,000 may be used successfully. However, the concentration of PEG is very important. With PEG 1000 and 6000, concentrations in the range of 40 to 60% give the maximum frequencies of genetic recombinants (12). PEG 4000 at 65% has been used to induce maximum recombination in *Streptomyces parvulus* (40).

13.2.4 Regeneration of Protoplasts

The serial dilution method and the confluent lawn method may be used to recover recombinants from a protoplast fusion.

Serial dilution method

The serial dilution method employs direct selection for recombinants on the regeneration medium. Serial dilutions of the fusion mixture are plated on modified R2 agar with and without the necessary supplements of amino acids required for growth of the parental strains and the recombinants. Protoplasts diluted in medium P are added to modified R2 soft agar overlays maintained at 46°C in a water bath, mixed, and poured onto modified R2 agar. Modified R2 is an osmotically stable medium that allows the protoplasts to grow and revert to the cellular state. The use of soft agar overlays minimizes physical manipulations (such as spreading with a glass rod) which may rupture many of the protoplasts.

Recombinants and nonrecombinant regenerated colonies may appear as early as 1 day or as late as 14 days, depending on the species and strain of a specific species.

Confluent lawn method

The confluent lawn method is nonselective, and all protoplasts are allowed to regenerate on the surface of fully supplemented modified R2 plates. The resulting colonies are then replica plated to CDA, a nonhypertonic minimal medium, and AS-1, a complete medium (section 13.7.4), for the detection of recombinants and parental clones. This method is suitable if the strains used sporulate efficiently on the regeneration medium. If adequate sporulation does not occur, substrate mycelia must be analyzed, thus making this technique less desirable (12).

Conditions affecting regeneration

(i) Growth temperature of the cells before protoplasting. See section 13.2.1.

(ii) Incubation temperature during regeneration. The temperature at which the plates are incubated for regeneration can have a large effect on the efficiency of regeneration. Incubation at high temperatures (37 to 42°C) greatly decreases the number of protoplasts which regenerate viable colonies with some species. Lower incubation temperatures (29 to 30°C) are generally optimum for the regeneration of *Streptomyces* protoplasts (12).

(iii) Autoinhibition. With many *Streptomyces* spp., autoinhibition may occur during the regeneration of the protoplasts; that is, protoplasts which regenerate the fastest inhibit the regeneration of surrounding protoplasts (11, 12, 30). Partial dehydration of the modified R2 agar (15 to 22% loss of weight) can essentially eliminate autoinhibition, shorten regeneration times, and result in increased regeneration frequencies (12).

13.2.5 Calculation of the Frequency of Regeneration

The frequency of regeneration is calculated by dividing the number of viable protoplasts by two times the initial CFU. (Protoplast numbers by hemacytometer counts can be used to calculate the frequency of regeneration, but this method is less accurate since not all of the protoplasts observed in the hemacytometer contain DNA.) Regeneration frequencies approaching 100% may be obtained with many species under optimal conditions (12).

13.2.6 Calculation of the Frequency of Recombination

For both the serial dilution and the confluent lawn methods the total viable cells obtained on supplemented modified R2 or AS-1, a complete medium, and recombinants obtained on unsupplemented modified R2 or CDA, a minimal medium, after protoplast fusion are scored. The recombination frequency is determined by calculating one of the following ratios: (i) the number of prototrophic colonies arising on unsupplemented or minimal medium to the number of colonies arising on fully supplemented or complete medium, (ii) the number of prototrophic colonies arising on unsupplemented or minimal medium to the sum of the colonies arising on plates supplemented for growth of both parental types; or (iii) the number of prototrophic colonies arising on unsupplemented or minimal medium to the number of colonies arising on plates supplemented for the growth of only one parental type (major or minor parent).

Using the first method, recombinant frequencies of greater than 10% of the total progeny are routinely obtained (12, 28, 32, 40). Higher recombinant frequencies have been observed with the other two methods.

13.3 *MICROMONOSPORA*

The procedures for forming, fusing, and regenerating protoplasts of *Micromonospora* spp. are very similar to the procedures for *Streptomyces*. Two recent papers (38, 48) outline the conditions for optimizing protoplast fusion and regeneration in *Micromonospora rosaria*. These procedures are summarized below.

13.3.1 Conditions for Cell Growth before Protoplast Formation

1. Grow the cultures in germination medium (GER) (see section 13.7.2) at 32°C using vial or slant stock cultures as the inoculum.
2. Harvest the cells during late logarithmic growth phase (a cell concentration of 5.0 to 5.5 g [dry weight]/liter).
3. Sonicate the mycelial fragments for 1 min at 100 W (7 kilocycles; Ultrasonic Dismembrator, model 300, Artex Corp.).
4. Use the sonicated mycelial fragments to inoculate (5% [vol/vol]) fresh GER containing 0.075% (wt/vol) glycine. (*Micromonospora* is much more sensitive to glycine than *Streptomyces* [53].) Concentrations of glycine less than 0.2% are usually sufficient.
5. Harvest the cells during the late logarithmic growth phase.
6. Determine the number of CFU per milliliter by plating serial dilutions of cells on a nonhypertonic complete medium such as GER agar (GERA; section 13.7.2).
7. Wash the cells by centrifugation at 3,000 × *g* for 10 min in medium P.

13.3.2 Formation of Protoplasts

1. Resuspend the cells in an equal volume of medium P containing 2 mg of lysozyme per ml.
2. Incubate the suspension at 32°C. Protoplast formation should normally be complete in 1 h.
3. Wash the protoplasts by centrifugation and resuspension in medium P.
4. Determine the number of total protoplasts by counting in a hemacytometer. Determine the number of viable protoplasts by plating serial dilutions of the protoplasts in medium P on regeneration medium (RM) (section 13.7.2) supplemented as necessary for the growth of the parental strains. Incubate the plates at 32°C.
5. Determine the frequency of nonprotoplasted cells remaining after protoplast formation by diluting the protoplasts in sterile distilled water and plating on RM, supplemented as necessary.

Factors affecting protoplast formation

Growth phase of the cells. The growth phase of the cells used to prepare protoplasts influences the efficiency of protoplast formation and regeneration. Cells in early stationary phase are much more susceptible to lysozyme treatment and protoplast formation than cells from late stationary phase.

13.3.3 Fusion of Protoplasts

1. Mix 0.5 ml of each protoplast suspension to be fused.
2. Pellet the protoplasts by centrifugation at 3,000 × *g* for 10 min.
3. Suspend the pellet in 0.1 ml of medium P.
4. Add 0.9 ml of 55.6% (wt/vol) PEG 1000. Mix gently.
5. Let the suspension incubate at 32°C for 3 min with periodic gentle shaking.

13.3.4 Regeneration of Protoplasts

Either the confluent lawn method or the serial dilution method may be used to obtain the recombinants (see section 13.2.4). The serial dilution method is preferred.

Prepare dilutions of the fusion mixture in medium P and plate on RM supplemented and unsupplemented with appropriate amino acids. Incubate the plates at 32°C. Protoplasts of *M. rosaria* regenerate slowly (7 to 12 days).

Factors affecting regeneration of protoplasts

(i) Growth phase of cells. Regeneration is most efficient when protoplasts are prepared from cells in early exponential phase.

(ii) pH. A pH of 7.6 to 7.7 (rather than the standard pH 7.2 used for *Streptomyces*) is optimum for regeneration of *Micromonospora* protoplasts.

13.3.5 Calculation of the Frequency of Regeneration

The frequency of regeneration of protoplasts is calculated by dividing the number of protoplasts resulting in colony formation by the number of protoplasts obtained by hemacytometer counting. Thirty percent regeneration has been obtained with *M. rosaria*.

13.3.6 Calculation of the Frequency of Recombination

The frequency of recombination is calculated as the ratio of the number of prototrophic colonies arising on unsupplemented RM to the total number of colonies arising on supplemented RM. Recombinant frequencies of up to 10% have been obtained with *M. rosaria*.

13.4 *BACILLUS*

Protoplast fusion has been demonstrated with *Bacillus subtilis* (22, 50) and *Bacillus megaterium* (19–21, 25). The protocols for growing the cells, forming and fusing protoplasts, and regenerating viable cells of *B. subtilis* have been optimized by Gabor and Hotchkiss (22) and are summarized below.

13.4.1 Conditions for Cell Growth before Protoplast Formation

The cultures are grown overnight in hypertonic nutrient broth (HNB) (section 13.7.3) at 30°C on a gyratory shaker. The cultures are harvested before stationary phase and used to inoculate 20 ml of fresh HNB. Incubate the culture with shaking at 30°C until an A_{650} of 0.3 to 0.4 is reached. Plate the cells for viable counts on a complete medium such as HNA (section 13.7.3).

13.4.2 Formation of Protoplasts

1. Centrifuge the cells at 7,000 rpm for 10 min at about 18°C.
2. Suspend the cell pellets in SMMAD (section 13.7.3) buffer to an A_{650} of 2.0.
3. Add lysozyme in SMMD (section 13.7.3) buffer to a final concentration of 100 μg/ml.
4. Incubate at 42°C for 30 min.
5. Confirm protoplast formation by phase-contrast microscopy.
6. Centrifuge the protoplasts and suspend the pellet in 1/5 to 1/10 volume of SMMAD buffer.
7. Viable protoplasts are determined by plating dilutions of the protoplasts on regeneration medium RD (section 13.7.3) supplemented as necessary for the growth of the parental strains.
8. Nonprotoplasted viable cells are determined by plating dilutions of the protoplasts on modified NA (section 13.7.3).

13.4.3 Fusion of Protoplasts

1. Mix equal volumes of the two protoplast suspensions to be fused.
2. Add 0.1 ml of the mixture to 0.9 ml of 40% PEG 6000.
3. Shake vigorously and then let the suspension stand for 2 min at room temperature.

13.4.4 Regeneration of Protoplasts

As with *Streptomyces* spp., the serial dilution or confluent lawn methods may be used to score recombination frequencies after *Bacillus* protoplast fusion. However, in *B. subtilis* it has been observed that plating of the fusion mixture directly on a complete or rich medium before scoring recombinants results primarily in the retrieval of parental types. Gabor and Hotchkiss (22) have proposed that the majority of postfusion pairs segregate parental cells very rapidly during regeneration. Under conditions favoring early and efficient regeneration, such as regenerating on a complete or rich medium, the diploid protoplast is induced to form a cell wall and divide before the two chromosomes have been replicated, thus reducing the opportunity for genetic recombination to occur. Therefore, Gabor and Hotchkiss (22) suggest that selective adjustments of the postfusion environment, such as regenerating on a minimal medium rather than on a complete medium, may stabilize the formation of diploids and increase the frequency of subsequent genetic recombinations. For these reasons, the confluent lawn method of scoring recombinants on minimal medium is recommended.

1. Dilute the fusion mixture in SMMAD and plate on nonselective RD plates, allowing all protoplasts to regenerate.
2. Incubate the plates at 37°C for 2 days.
3. Replica plate the colonies, using sterile velveteen, to modified S, a nonhypertonic selective medium containing appropriate supplements to allow growth of the individual parental strains or recombinants (22). A third class of colonies, termed biparentals, may constitute 1 to 10% of the fusion products (31). Biparentals are semistable colonies that exhibit the exact phenotype of one parent but are able to segregate the phenotype of the other parent and are able to produce true recombinants.
4. Incubate the plates at 37°C for 1 to 2 days and then score for parental and recombinant colonies.

Factors affecting regeneration

(i) Temperature. The temperature during bacterial growth and lysozyme treatment can affect the regeneration frequency. It was found that a differential between the temperature used for the growth of the cells before protoplast formation and the temperature of the lysozyme treatment increases the number of regenerating bacteria (22). For example, if the cells are grown at 37°C, followed by lysozyme treatment at 42°C, improved frequencies of regeneration and recombination are obtained compared to growth of the cells at 37°C and lysozyme treatment also at 37°C. Likewise, growth at 30°C and lysozyme treatment at 37 or 42°C is better than growth and lysozyme treatment at 37°C.

(ii) PEG. Exposure of the protoplasts to PEG for more than 5 min may decrease the efficiency of regeneration (22).

13.4.5 Calculation of the Frequency of Regeneration

The frequency of regeneration is calculated by dividing the number of protoplasts resulting in colony formation by the number of CFU before protoplast formation. Regeneration frequencies of as high as 100% may be obtained.

13.4.6 Calculation of the Frequency of Recombination

The frequency of recombination is determined by dividing the number of prototrophic recombinants by the number of regenerated cells of the minority parent. Recombinant frequencies of 1×10^{-4} to 9×10^{-4} may be expected.

13.5 FUNGI

The recovery of genetic recombinants from fungal protoplast fusions is more difficult than with bacterial systems. Not only must the fusion of two or more protoplasts occur, but nuclear fusion and subsequent haploidization must also occur if genetic recombinants are to be obtained. With both intraspecific and interspecific fungal protoplast fusions, the plating of the fusion mixture on a minimal medium results in the appearance of colonies which appear to be nutritionally complemented (i.e., prototrophic). These nutritionally complemented colonies are generally the result of heterokaryon formation. In intraspecific protoplast fusions, complementation frequencies of 40 to 60% may be obtained, whereas the frequencies obtained from interspecific fusions are much lower, on the order of 10^{-5} (18). The fusion products must then be cytologically, biochemically, and genetically analyzed to determine ploidy and stability.

This chapter will deal only with the steps through the regeneration of the fusion products and will not discuss the further analysis of the fusion products. The reader is referred to several review articles for more information and references regarding the analysis of intraspecific and interspecific fusion products (1, 2, 5, 6).

13.5.1 Conditions for Cell Growth before Protoplast Formation

Penicillium

Stock cultures of *Penicillium* spp. are maintained on malt extract agar (see section 13.7.4). Cultures are grown in Vogel medium (section 13.7.4) (33) on a rotary shaker at 180 rpm at 28°C (26). A spore suspension is generally used as the inoculum.

Aspergillus

Stock cultures of *Aspergillus* spp. should be maintained on malt extract agar. Cultures are inoculated with a spore suspension and grown in a modified minimal medium (MMP; see section 13.7.4) (26) for 19 h on a rotary shaker at 180 rpm and 28°C.

Cephalosporium acremonium

Stock cultures of *C. acremonium* are maintained on a complete medium (CM) (section 13.7.4). Cultures are inoculated with a spore suspension and grown in a modified medium (MMD; section 13.7.4) of Demain et al. (26) on a rotary shaker at 180 rpm and 28°C.

13.5.2 Formation of Protoplasts

Harvest the mycelium from *Penicillium*, *Aspergillus*, and *Cephalosporium* cultures by aseptic vacuum filtration.

Penicillium

1. Wash the filtered mycelia two times with 0.7 M NaCl, the osmotic stabilizer.
2. Resuspend the mycelia (50 mg/ml, wet weight) in 0.7 M NaCl containing 5 mg of Novozyme 234 (26).
3. Incubate the suspension at 28°C with gentle shaking on a reciprocal shaker at 120 strokes per min for 3 h.
4. Separate the protoplasts from the mycelial debris by filtration through a sintered glass filter, porosity 1 (45).
5. Recover the protoplasts from the filtrate by centrifugation at 500 × *g* for 10 min (45).
6. Wash the protoplasts two times with 0.7 M NaCl by centrifugation and resuspension.
7. Suspend the protoplasts in a small volume of 0.7 M NaCl.
8. Estimate the number of protoplasts by counting in a hemacytometer. Dilute the protoplasts in 0.7 M NaCl to a final concentration of 5×10^6 to 1×10^7 protoplasts per ml for fusion.

Aspergillus

The procedure for forming protoplasts of *Aspergillus* spp. is essentially the same as for *Penicillum* spp. except for the following two differences.

1. Throughout the procedure, the osmotic stabilizer is 0.6 M NaCl in a 0.2 M phosphate buffer (pH 5.8) rather than 0.7 M NaCl.
2. For the cell wall digestion process a combination of enzymes is used; Novozyme 234 and Cellulase CP, both at 5 mg/ml (26).

Cephalosporium

1. Wash the filtered mycelia two times using 0.7 M NaCl in a citrate-phosphate buffer (26) as the osmotic stabilizer (OSS).
2. For *C. acremonium* it is necessary to pretreat the mycelium with a thiol reagent to help reduce disulfide formation before treatment by lytic enzymes. Pretreat the mycelium (0.2 g, wet weight) for 1 h in citrate-phosphate buffer (pH 7.3) containing 0.01 M dithiothreitol (26).
3. Filter and wash the mycelia by centrifugation and resuspension three times with OSS.
4. Suspend the mycelia (50 mg/ml, wet weight) in OSS containing Novozyme 234 at 5 mg/ml.
5. Incubate the suspension at 28°C with gentle shaking on a reciprocal shaker at 120 strokes per min for 3 h.
6. Separate the protoplasts from the mycelial debris by filtration through a sintered glass filter, porosity 1.
7. Recover the protoplasts from the filtrate by centrifugation at 500 × *g* for 10 min.
8. Wash the protoplasts two times in OSS by centrifugation.
9. Suspend the protoplasts in a small volume of OSS.
10. Estimate the number of protoplasts in the suspension by counting in a hemacytometer. Dilute the protoplasts in OSS to a final concentration of 5×10^6 to 1×10^7 protoplasts per ml.

Alternative method

For many species of filamentous fungi, the cultivation of cells in a liquid medium is unfavorable for protoplast formation (16). Instead the strains should be grown on cellophane sheets or disks (16, 46). In this procedure the cellophane is placed on the surface of a complete medium in petri dishes. The cellophane is inoculated with a dense spore suspension, and the plates are incubated for abundant growth of the cells. For the formation of protoplasts the cellophane is removed to a clean, sterile petri dish and a small amount (2 to 5 ml) of the appropriate osmotic stabilizer containing the lytic enzyme(s) is added. The protoplasts are recovered by washing the cellophane with the osmotic stabilizer. The protoplasts are separated from any mycelial debris by filtration through a sintered glass filter, porosity 1, and recovered as outlined above.

Factors affecting protoplast formation

Enzymes. The ability of different commercial enzymes to produce protoplasts from different fungi may vary greatly. There are other enzymes besides those used in the above procedures that may be tried, singly or in combination, to induce protoplast formation (26). The other available enzymes are Cellulase CT (John and E. Sturge Ltd., Selby, U.K.), Cereflo 200 L (Novo Enzyme Products, Ltd., Windsor, Ontario, Canada), *Oxyporus* cellulase (Merck, Darmstadt, Federal Republic of Germany), Helicase (L'Industrie Biologique, Grennevilliers, France), β-D-glucuronidase (EC 3.2.1.31) and chitinase (Sigma London Chemical Company, Ltd., Poole, U.K.), and *Cytophaga* enzyme L and β-D-glucanase (BDH Chemicals Ltd., Poole, U.K.) With some fungi, pretreatment with a thiol compound may be necessary for the production of protoplasts (see *C. acremonium*).

13.5.3 Fusion of Protoplasts

The procedure of Anne and Peberdy (10) may be used to fuse protoplasts from *Penicillium*, *Aspergillus*, and *Cephalosporium* spp.

1. Mix 1 ml of each of the parental protoplasts to be fused.
2. Centrifuge the mixture at 700 × *g* for 10 min.
3. Suspend the protoplast pellet in 1 ml of a prewarmed (30°C) solution of 30% (wt/vol) PEG (Koch-Light) in 0.01 M $CaCl_2$ and 0.05 M glycine, adjusted to pH 7.5 with 0.01 M NaOH.
4. Incubate the suspension for 10 min at 30°C.
5. Dilute the suspension with 6 ml of minimal medium (MM; section 13.7.4) adjusted to 0.6 M NaCl for osmotic stability.
6. Centrifuge the suspension at 700 × *g* for 5 min.
7. Wash the protoplasts twice with 8-ml portions of the appropriate osmotic stabilizer (see 13.4.2).
8. Resuspend the protoplasts in 5 ml of the appropriate osmotic stabilizer.

Factors affecting fusion

(i) Cations and pH. There is an absolute requirement for Ca^{2+} in the form of $CaCl_2$ during the protoplast fusion process. For both *Penicillium* and *Aspergillus* a concentration of 0.01 M $CaCl_2$ is optimum for pH values of 7.2 to 7.5 (9, 17). In *Penicillium*, if the pH of the fusion mixture is increased, the optimum molarity of $CaCl_2$ for fusion is increased from 0.01 to a maximum value of 0.6 M at pH 9.0. The addition of Na^+ or K^+ to the fusion mixture decreases the fusion frequency (9).

(ii) PEG. For *Penicillium* and *Aspergillus* spp., PEG 4000 and PEG 6000 are equally efficient in promoting protoplast fusion. Concentrations of 25 to 30% (wt/vol) are optimal (9, 17). At concentrations of PEG less than 20% (wt/vol), the stabilizing effect of PEG is lost and protoplasts lyse. At concentrations of 30% (wt/vol) PEG or greater, the medium becomes hypertonic and causes shrinking of the protoplasts and reduction in viability (9). Incubation of protoplasts in PEG for 5 min is sufficient to maximize fusion frequencies and to minimize loss of viability of protoplasts (9).

(iii) Temperature. The temperature during fusion generally affects the fusion frequency. With *Penicillium* spp., protoplast fusion occurs at temperatures as low as 4°C, but the frequency increases as the temperature increases up to a maximum of 37.5°C (9).

13.5.4 Regeneration of Protoplasts

Almost invariably, methods employing positive selection for recombinants are used in fungal protoplast fusions (10, 15) to maintain the heterokaryotic state. Relaxation of positive selection for heterokaryons (i.e., plating on a complete medium) usually leads to the segregation of the parental types.

Penicillium

1. Prepare dilutions of the protoplasts in 0.7 M NaCl.
2. Plate the dilutions of the protoplasts on minimal and complete regeneration media (MMA and PMA, respectively; see section 13.7.4).
3. Incubate the plates at 28°C. Colonies should be visible after 5 days.

Aspergillus

1. Prepare dilutions of the protoplasts in 0.6 M KCl in 0.2 M phosphate buffer.
2. Plate the diluted protoplasts on minimal and complete regeneration media (RMMPA and RCMPA, respectively; see section 13.7.4).
3. Incubate the plates at 28°C. Colonies are normally visible after 3 to 5 days.

Cephalosporium

1. Prepare dilutions of the protoplasts in OSS.
2. Plate the diluted protoplasts on minimal and complete regeneration media (HMMD and HCM, respectively; see section 13.7.4).
3. Incubate the plates at 28°C.

The majority of the fusion products arising on the minimal regeneration plates from fusions involving *Penicillium*, *Aspergillus*, or *Cephalosporium* will be heterokaryons. The heterokaryons must be subcultured on complete or supplemented minimal medium for further analysis of parental segregants and recombinants.

Factors affecting regeneration

(i) Temperature. With *Penicillium* spp. the number of regenerating protoplasts steadily decreases as the regeneration temperature is increased from 4 to 37.5°C. A dramatic drop in regeneration occurs at temperatures greater than 37.5°C (9).

(ii) Miscellaneous. The reversion frequency of protoplasts to viable cells containing cell walls in fungi is highly variable. The reasons are unknown, but speculation centers on the absence of a nucleus in some regenerating protoplasts and on the inability of protoplasts arising from distal areas of the hyphae to regenerate (5).

13.5.5 Calculation of the Frequency of Fusion

The frequency of protoplast fusion is determined by dividing the number of colonies arising on minimal regeneration plates by the number of colonies arising on complete regeneration plates. It must be stressed that the frequency of protoplast fusion does not reflect the frequency of true genetic recombinants, since the majority of the colonies arising on the minimal regeneration plates are heterokaryons which will segregate different genotypes upon further subculturing.

13.6 SPECIAL PROCEDURES

13.6.1 Heat Inactivation of the Donor

Often it is desirable to inactivate one of the parents of a fusion cross. For example, one parental strain may produce an enzyme or antibiotic that is lethal to the recipient protoplasts of the other parental strain. This problem may be circumvented by using heat inactivation of the protoplasts of the strain that is to serve as the donor of genetic information. The heat-inactivated protoplasts are unable to revert to cellular form but may donate their genetic material to the viable recipient protoplasts during the fusion process. Successful heat inactivation, fusion, and recovery of genetic recombinants has been demonstrated in *Bacillus* (21), *Micromonospora* (53), and *Streptomyces* (13, 39) species.

The temperature at which the heat inactivation is performed is critical. The temperature should be high enough to cause a substantial decrease in viability of the strain, yet it should have very little effect on the integrity of the cytoplasmic membrane and the genome. A too-high temperature for heat inactivation can be deleterious to the protoplasts and results in the decreased recovery of recombinants (13). For *Bacillus*, 50°C for 2.0 h was used for heat inactivation. In *Streptomyces*, Ochi (39) used 60°C for 5 min whereas Baltz and Matsushima (13) used 50°C for 2.5 hours. The procedure of Ochi resulted in a much poorer recovery of recombinants than that of Baltz and Matsushima (13). Poor recovery of recombinants after exposure to 60°C may be due to disruption of the cytoplasmic membrane (13).

13.6.2 UV Inactivation of the Donor

UV irradiation may be used to inactivate one or both parental protoplast populations to be used in a protoplast fusion experiment. UV irradiation may enhance crossing over but also can be used to reduce the length of linkage groups. If both parents are irradiated, the recombinants obtained are the result of selective cross-overs between the two genomes which eliminate the sites of potentially lethal UV hits. Since one can regulate the number of lethal UV hits per genome, the use of UV irradiation is helpful when a limited region of one parental genome is to be introduced into the genome of another strain. The technique has been used successfully in *Streptomyces* protoplast fusions (29).

13.6.3 Mitochondrial Transfer

In eucaryotic protoplast fusions, genetic elements other than nuclei (such as mitochondria, plasmids, or viruses) may be transferred from one protoplast to another. The term "genetic transfusion" (1) is often used to denote the transfer of these components between protoplasts. Unlike a true protoplast fusion, in which two complete genomes may interact with no directional transfer of material, genetic transfusion is always directional (2).

There are two types of mitochondrial transfer: nonselective, in which the mitochondria are transferred in conjunction with the nuclear material, and selective, in which the mitochondria are transferred without the nuclear genome. Intraspecific and interspecific nonselective and selective mitochondrial transfers have been carried out (2).

Before the development of protoplast fusion techniques, cell matings were confined to cells of the opposite nuclear mating type. Protoplast fusion now allows nonselective mitochondrial transfers to occur between protoplasts of the same nuclear mating type since the transfer of mitochondria does not appear to be dependent on the nuclear mating type of the fusion partners. This helps facilitate the isolation of mitochondrial recombinants even though conditions satisfying nuclear marker requirements may be lacking.

In selective mitochondrial transfers, isolated mitochondria are used. The process is more like a transformation than a fusion. Increased mitochondrial transfers may be achieved by a process termed mitochondrial transfusion (2). To do so, anucleate protoplasts of the donor strain carrying a mitochondrial marker are separated from the much larger, nucleated protoplasts by selective centrifugation. The anucleated protoplasts are induced to fuse with protoplasts of the mitochondrial negative recipient. Under selective conditions the majority of the fusion products should contain the nucleus of the recipient strain and the mitochondrial genome of the donor strain.

The technique of mitochondrial transfer has aided in the genetic analysis and strain improvement of several fungal strains (2).

13.7 MEDIA AND REAGENTS

13.7.1 *Streptomyces*

Nutritional requirements for auxotrophs should be added at 100 μg/ml.

TS broth

Complex medium for growth of Streptomyces

Trypticase soy broth (BBL) .... 30 g
Distilled water .... 1 liter

Autoclave for 30 min at 121°C.

TS agar

TS broth containing 15 g of agar per liter. Autoclave for 30 min at 121°C.

Glycine

Glycine, 10% (wt/vol) in distilled water. Filter sterilize.

Medium P

Base solution

Sucrose	103 g
K_2SO_4	0.25 g
$MgCl_2 \cdot 6H_2O$	2.03 g
Distilled water	to 700 ml

Autoclave for 30 min at 121°C.
Sterilize separately stocks of the following:

KH_2PO_4	0.5 g/liter
$CaCl_2 \cdot 2H_2O$	27.8 g/liter
0.25 M *N*-tris(hydroxymethyl)methyl-2-aminoethanesulfonic acid (TES) (pH 7.2)	

At the time of use, add 100 ml each of the KH_2PO_4, $CaCl_2$, and TES solutions to the 700 ml of base solution.

Lysozyme

Lysozyme (Calbiochem) dissolved in medium P and filter sterilized.

Modified R2 agar

Base solution

Sucrose	103 g
K_2SO_4	0.25 g
$MgCl_2 \cdot 6H_2O$	10.12 g
Glucose	10.0 g
L-Asparagine · $1H_2O$	2.0 g
Casamino Acids	0.1 g
Trace elements solution (see below)	2 ml
Agar	22.0 g
Distilled water	to 700 ml

Autoclave for 25 min at 121°C.
Autoclave for 30 min at 121°C, separately, the following solutions:

$CaCl_2 \cdot 2H_2O$	22.2 g/liter
KH_2PO_4	0.5 g/liter
0.25 M TES (pH 7.2)	

Before pouring plates, aseptically add 100 ml each of the $CaCl_2$, KH_2PO_4, and TES solutions to the 700 ml of modified R2 agar base solution.

Trace elements solution

$ZnCl_2$	40 mg
$FeCl_3 \cdot 6H_2O$	200 mg
$CuCl_2 \cdot 2H_2O$	10 mg
$MnCl_2 \cdot 4H_2O$	10 mg
$Na_2B_4O_7 \cdot 10H_2O$	10 mg
$(NH_4)_6Mo_{24} \cdot 4H_2O$	10 mg
Distilled water	1 liter

Filter sterilize.

R2 modified soft agar overlays

Sucrose	103 g
$MgCl_2 \cdot 6H_2O$	10.12 g
$CaCl \cdot 2H_2O$ solution (see above)	100 ml
0.25 M TES (pH 7.2)	100 ml
Agar	4.10 g
Distilled water	to 1 liter

Dispense in 3-ml volumes in capped culture tubes (13 by 100 mm). Autoclave for 25 min at 121°C.

PEG 1000

PEG, 50% (wt/vol), dissolved in medium P. Filter sterilize.

CDA

Per liter:

Czapek Dox broth (Difco)	35 g
L-Asparagine	2 g
Agar (Difco)	15 g

Autoclave for 30 min at 121°C.

AS-1

Per liter:

Yeast extract	1.0 g
L-Alanine	0.2 g
L-Arginine	0.2 g
L-Asparagine	0.5 g
Soluble starch	5.0 g
NaCl	2.5 g
Na_2SO_4	10.0 g
Agar	20.0 g

Adjust the pH to 7.5. Autoclave for 30 min at 121°C.

13.7.2 *Micromonospora*

Nutritional requirements for auxotrophs should be supplemented at 30 μg/ml.

Germination medium (GER)

Per liter:

Beef extract	3 g
Tryptone	5 g
Dextrose	1 g
Soluble starch	24 g
Yeast extract	5 g
$CaCO_3$	2 g

Adjust pH to 7.6.

GERA

GER plus 20 g of agar per liter.

Medium P (MP)

Sucrose	103 g
K_2SO_4	0.25 g
$MgCl_2 \cdot 6H_2O$	5.09 g
Trace element solution (section 13.7.1)	2 ml
Distilled water	to 700 ml

Adjust the pH to 7.6 and autoclave for 15 min at 121°C.

Sterilize separately stocks of the following:

KH_2PO_4 .... 0.5 g/liter
$CaCl_2 \cdot 2H_2O$ .... 73.7 g/liter
0.25 M TES (pH 7.6)

At the time of use, add 100 ml each of KH_2PO_4, $CaCl_2$, and TES to the 700 ml of base solution.

Regeneration medium (RM)

Sucrose .... 125 g
K_2SO_4 .... 0.25 g
Trace element solution (section 13.7.1) .... 1.0 ml
$MgCl_2 \cdot 6H_2O$ .... 5.09 g
Glucose .... 10 g
L-Asparagine .... 4 g
Casamino Acids .... 0.1 g
Agar .... 20 g
Distilled water to .... 700 ml

Autoclave for 15 min at 121°C.

Sterilize separately the following solutions:

$CaCl_2 \cdot 2H_2O$ .... 73.7 g/liter
KH_2PO_4 .... 0.5 g/liter
0.25 M TES (pH 7.6)

Before pouring the plates, aseptically add 100 ml each of the $CaCl_2$, KH_2PO_4, and TES solutions to the 700 ml of RM.

PEG 1000

PEG 1000, 55.6% (wt/vol) in medium P. Filter sterilize.

13.7.3 *Bacillus*

Nutritional requirements of auxotrophs should be supplemented at 25 μg/ml.

DNase has been added to several media as a precautionary measure to eliminate the possibility that DNA-mediated transformation or transfection might occur.

Modified nutrient agar (NA)

Per liter:
Nutrient broth (Difco) .... 8.0 g
KCl .... 1.0 g
$MgSO_4 \cdot 7H_2O$ .... 0.25 g
$MnCl_2 \cdot 4H_2O$ .... 0.002 g
Agar .... 17.0 g

Adjust the pH to 7.0; autoclave. Add, separately, sterile solutions of $CaCl_2 \cdot 2\ H_2O$ and $FeSO_4 \cdot 7H_20$ to 5×10^{-4} M and 1×10^{-6} M, respectively.

Hypertonic nutrient agar (HNA)

Modified NA containing 0.5 M sucrose.

Hypertonic nutrient broth (HNB)

HNA minus the agar.

Sucrose maleate magnesium DNase (SMMD)

0.5 M sucrose
0.02 M maleate buffer (pH 6.5)
20 mM $MgCl_2$
5 μg of DNase I (Worthington) per ml

SMMAD

SMMD containing 1% bovine serum albumin (Sigma).

PEG 6000

PEG 6000, 40% (wt/vol) in SMMAD. Filter sterilize.

Regeneration medium (RD)

Per liter:
NH_4NO_3 .... 1.0 g
K_2HPO_4 .... 3.5 g
KH_2PO_4 .... 1.5 g
Agar 20.0%
Sodium succinate (0.33 M, pH 7.3) .... 81.0 g
Gelatin .... 5.0 g
$MgCl_2 \cdot 6H_2O$ .... 4.07 g
Glucose .... 5 g

Autoclave; at time of use, aseptically add calf serum (GIBCO) diluted 1:200 and DNase I (crude grade, Sigma) at a 5-μg/ml final concentration.

Modified S

Per liter:
NH_4NO_3 .... 2.0 g
K_2HPO_4 .... 14.0 g
KH_2PO_4 .... 6.0 g
Sodium citrate .... 1.0 g
$MgSO_4 \cdot 7H_2O$ .... 0.2 g
L-Glutamate .... 0.0025 g
L-Lysine .... 0.0050 g
L-Asparagine .... 0.0125 g
L-Valine .... 0.0025 g
$MnCl_2 \cdot 4H_2O$ .... 0.005 g
$MgCl_2$ .... 0.024 g
$CaCl_2$ .... 0.017 g
$FeSO_4$ .... 0.004 g
Agar (Difco) .... 15.0 g

DNase (5 μg/ml, final concentration) and glucose (0.5%) are added aseptically at the time of use.

13.7.4 Fungi

Nutritional requirements of auxotrophs should be supplemented at 50 μg/ml.

Citrate-phosphate buffer
For OSS for Cephalosporium, Aspergillus

0.2 M, pH 5.8.

Complete medium (CM)
Cephalosporium

Per liter:	
Maltose (BDH)	40 g
Peptone (Oxoid)	10 g
Malt extract (Oxoid)	24 g
Oxoid no. 3 agar	20 g

Adjust pH to 7.5 and autoclave for 20 min at 121°C.

HCM
Cephalosporium

CM containing 0.8 M NaCl and 2% sucrose for osmotic stability.

HMMD
Cephalosporium

MMD containing 0.8 M NaCl and 2% sucrose for osmotic stability.

Malt extract agar
Aspergillus, Penicillium

Use at 2% (wt/vol).

Minimal medium (MM)
Penicillium, Aspergillus, Cephalosporium

Per liter:	
$NaNO_3$	3 g
KCl	0.5 g
$MgSO_4 \cdot 7H_2O$	0.5 g
$FeSO_4 \cdot 7H_2O$	0.01 g
KH_2PO_4	1 g
Glucose	40 g

pH 6.0.

MMA
Penicillium

MM containing 18 g of agar per liter; 0.7 M NaCl added for osmotic stability.

MMD
Cephalosporium

Per liter:	
Sucrose	36 g
L-Asparagine	7.5 g
K_2HPO_4	21 g
KH_2PO_4	15 g
Na_2SO_4	0.75 g
$MgSO_4 \cdot 7H_2O$	0.18 g
$CaCl_2$	0.06 g
$Fe(NH_4)_2(SO_4)_2 \cdot 6H_2O$	0.15 g
$MnSO_4$	0.03 g
$ZnSO_4 \cdot H_2O$	0.03 g
$CuSO_4 \cdot 5H_2O$	0.008 g

MMP
Aspergillus

Per liter:	
$NaNO_3$	6.0 g
$MgSO_4 \cdot 7H_2O$	0.52 g
KCl	0.52 g
$FeSO_4 \cdot 7H_2O$	trace*
$ZnSO_4 \cdot 7H_2O$	trace

*The original reference cited in Peberdy and Kevei (46) does not define "trace."

Adjust pH to 6.5 and autoclave for 20 min. Add glucose aseptically to a final concentration of 10 g/liter, yeast extract (Difco) to 0.5%, and Casamino Acids (Difco) to 0.2%.

PMA
Penicillium

MMA plus 2 g of yeast extract per liter.

RCMPA
Aspergillus

Per liter:	
$NaNO_3$	6.0 g
Yeast extract (Difco)	1.0 g
Peptone (Difco)	1.0 g
Casamino Acids (Difco)	1.0 g
Adenine	0.15 g
Vitamin solution (below)	10 ml
Agar	15.0 g
$MgSO_4 \cdot 7H_2O$	0.52 g
KCl	0.52 g
$FeSO_4 \cdot 7H_2O$	trace*
$ZnSO_4 \cdot 7H_2O$	trace

* "Trace" not defined in original reference (46).

Adjust pH to 6.0 and autoclave. Add aseptically, at the time of use, glucose to a final concentration of 10 g/liter and KCl to 0.6 M for osmotic stability.

Vitamin solution

Per 100 ml:	
Biotin	10 mg
Pyridoxine HCl	10 mg
Thiamine HCl	10 mg
Riboflavin	10 mg
p-Aminobenzoic acid	10 mg
Nicotinic acid	10 mg

Filter sterilize and store in the dark.

RMMPA
Aspergillus

MMP containing 20 g of agar (Difco) per liter and 0.6 M KCl, added aseptically at the time of use.

Vogel medium
Penicillium

Per liter:	
Trisodium citrate	3.0 g
KH_2PO_4	5.0 g
NH_4NO_3	2.0 g
$MgSO_4 \cdot 7H_2O$	0.2 g
$CaCl_2 \cdot 2H_2O$	0.1 g
Glucose	10.0 g

13.8. LITERATURE CITED

13.8.1 General References

1. **Ferenczy, L.** 1981. Microbial protoplast fusion. Symp. Soc. Gen. Microbiol. **31**:1–34.
2. **Ferenczy, L.** 1984. Fungal protoplast fusion: basic and applied aspects, p. 145–169. *In* R. F. Beers, Jr., and E. G. Bassett (ed.), Cell fusion; gene transfer and transformation. Raven Press, New York.
3. **Hopwood, D. A.** 1981. Genetic studies with bacterial protoplasts. Annu. Rev. Microbiol. **35**:237–272.
4. **Hopwood, D. A., and J. Merrick.** 1977. Genetics of antibiotic production. Bacteriol. Rev. **41**:595–635.
5. **Peberdy, J. F.** 1979. Fungal protoplasts: isolation, reversion and fusion. Annu. Rev. Microbiol. **33**:21–39.
6. **Peberdy, J. F.** 1980. Protoplast fusion—a tool for genetic manipulation and breeding in industrial microorganisms. Enzyme Microb. Technol. **2**:23–29.
7. **Queener, S. W., and R. H. Baltz.** 1980. Genetics of industrial microorganisms. Annu. Rep. Ferment. Processes **3**:5–45.

13.8.2 Specific References

8. **Anne, J., H. Eyssen, and P. DeSomer.** 1976. Somatic hybridization of *Penicillium roquefortii* with *P. chrysogenum* after protoplast fusion. Nature (London) **262**:719–721.
9. **Anne, J., and J. F. Peberdy.** 1975. Conditions for induced fusion of fungal protoplasts in polyethylene glycol solutions. Arch. Microbiol. **105**:201–205.
10. **Anne, J., and J. F. Peberdy.** 1976. Induced fusion of fungal protoplasts following treatment with polyethylene glycol. J. Gen. Microbiol. **92**:413–417.
11. **Baltz, R. H.** 1978. Genetic recombination in *Streptomyces fradiae* by protoplast fusion and cell regeneration. J. Gen. Microbiol. **107**:93–102.
12. **Baltz, R. H., and P. Matsushima.** 1981. Protoplast fusion in *Streptomyces*: conditions for efficient genetic recombination and cell regeneration. J. Gen. Microbiol. **127**:137–146.
13. **Baltz, R. H., and P. Matsushima.** 1983. Advances in protoplast fusion in *Streptomyces*. Exper. Suppl. (Basel) **46**:143–148.
14. **Coetzee, J. N., F. A. Sirgel, and G. Lecatsas.** 1979. Genetic recombination in fused spheroplasts of *Providence alcalifaciens*. J. Gen. Microbiol. **114**:313–322.
15. **Ferenczy, L., F. Kevei, and M. Szegedi.** 1975. High frequency fusion of fungal protoplasts. Experientia (Basel) **31**:1028–1030.
16. **Ferenczy, L., F. Kevei, and M. Szegedi.** 1975. Increased fusion frequency of *Aspergillus nidulans* protoplasts. Experientia (Basel) **31**:50–52.
17. **Ferenczy, L., F. Kevei, M. Szegedi, A. Franko, and I. Rohik.** 1976. Factors affecting high-frequency fungal protoplast fusion. Experientia (Basel) **32**:1156–1158.
18. **Ferenczy, L., M. Szegedi, and F. Kevei.** 1977. Interspecific protoplast fusion and complementation in *Aspergilli*. Experientia (Basel) **33**:184–186.
19. **Fodor, K., and L. Alfoldi.** 1976. Fusion of protoplasts of *Bacillus megaterium*. Proc. Natl. Acad. Sci. U.S.A. **73**:2147–2150.
20. **Fodor, K., and L. Alfoldi.** 1979. Polyethylene-glycol induced fusion of bacterial protoplasts. Mol. Gen. Genet. **168**:55–59.
21. **Fodor, K., E. Demiri, and L. Alfoldi.** 1978. Polyethylene glycol-induced fusion of heat-inactivated and living protoplasts of *Bacillus megaterium*. J. Bacteriol. **135**:68–70.
22. **Gabor, M. H., and R. D. Hotchkiss.** 1979. Parameters governing bacterial regeneration and genetic recombination after fusion of *Bacillus subtilis* protoplasts. J. Bacteriol. **137**:1346–1353.
23. **Godfrey, O., L. Ford, and M. L. B. Huber.** 1978. Interspecies matings of *Streptomyces fradiae* with *Streptomyces bikiniensis* mediated by conventional and protoplast fusion techniques. Can. J. Microbiol. **24**:994–997.
24. **Gotz, F., S. Ahrne, and M. Lindberg.** 1981. Plasmid transfer and genetic recombination by protoplast fusion in staphylococci. J. Bacteriol. **145**:74–81.
25. **Hadlaczky, G., K. Fodor, and L. Alfoldi.** 1976. Morphological study of the reversion to bacillary form of *Bacillus megaterium* protoplasts. J. Bacteriol. **125**:1172–1179.
26. **Hamlyn, P. F., R. E. Bradshaw, F. M. Mellon, C. M. Santiago, J. M. Wilson, and J. F. Peberdy.** 1981. Efficient protoplast isolation from fungi using commercial enzymes. Enzyme Microb. Technol. **3**:321–325.
27. **Hopwood, D. A., and H. M. Wright.** 1978. Bacterial protoplast fusion: recombination in fused protoplasts of *Streptomyces coelicolor*. Mol. Gen. Genet. **162**:307–317.
28. **Hopwood, D. A., and H. M. Wright.** 1979. Factors affecting recombinant frequency in protoplast fusions of *Streptomyces coelicolor*. J. Gen. Microbiol. **111**:137–143.
29. **Hopwood, D. A., and H. M. Wright.** 1981. Protoplast fusion in *Streptomyces*: fusions involving ultraviolet-irradiated protoplasts. J. Gen. Microbiol. **126**:21–27.
30. **Hopwood, D. A., H. M. Wright, M. J. Bibb, and S. N. Cohen.** 1977. Genetic recombination through protoplast fusion in *Streptomyces*. Nature (London) **268**:171–174.
31. **Hotchkiss, R. D., and M. H. Gabor.** 1980. Biparental products of bacterial protoplast fusion showing unequal parental chromosome expression. Proc. Natl. Acad. Sci. U.S.A. **77**:3553–3557.
32. **Hranueli, D., J. Pigac, T. Smokvina, and M. Alacevic.** 1983. Genetic interactions in *Streptomyces rimosus* mediated by conjugation and protoplast fusion. J. Gen. Microbiol. **129**:1415–1422.
33. **Isaac, S.** 1979. Isolation and ultrastructure of protoplasts from fungi, p. 17–22. *In* J. F. Peberdy (ed.), Protoplasts—applications in microbial genetics. University of Nottingham, Nottingham, U.K.
34. **Jones, C. W., I. A. Mastrangelo, H. H. Smith, and H. Z. Liu.** 1976. Interkingdom fusion between human (HeLa) cells and tobacco hybrid (GGLL) protoplasts. Science **193**:401–403.
35. **Kaneko, H., and K. Sakaguchi.** 1979. Fusion of protoplasts and genetic recombination of *Brevibacterium flavum*. Agric. Biol. Chem. **43**:1007–1013.
36. **Kao, K. N., and M. R. Michayluk.** 1974. A method for high-frequency intergeneric fusion of plant protoplasts. Planta **115**:355–367.
37. **Kevei, F., and J. F. Peberdy.** 1977. Interspecific hybridization between *Aspergillus nidulans* and *Aspergillus rugulosus* by fusion of somatic protoplasts. J. Gen. Microbiol. **102**:255–262.
38. **Kim, K. S., D. D. Y. Ryu, and S. Y. Lee.** 1983. Application of protoplast fusion technique to genetic recombination of *Micromonospora rosaria*. Enzyme Microb. Technol. **5**:273–280.
39. **Ochi, K.** 1982. Protoplast fusion permits high-frequency transfer of a *Streptomyces* determinant which mediates actinomycin synthesis. J. Bacteriol. **150**:592–597.
40. **Ochi, K., J. M. Hitchcock, and E. Katz.** 1979. High frequency fusion of *Streptomyces antibioticus* protoplasts induced by polyethylene glycol. J. Bacteriol. **139**:984–992.
41. **Ochi, K., and E. Katz.** 1980. Genetic analysis of the actinomycin-producing determinants (plasmid) in *Streptomyces parvulus* using the protoplast fusion technique. Can. J. Microbiol. **26**:1460–1464.
42. **Ogawa, H., S. Imai, A. Satoh, and M. Kojima.** 1983. An improved method for the preparation of streptomycetes and *Micromonospora* protoplasts. J. Antibiot. **36**:184–186.
43. **Oh, Y. K., J. L. Speth, and C. H. Nash.** 1980. Protoplast fusion with *Streptosporangium viridogriseum*. Dev. Ind. Microbiol. **21**:219–226.

44. **Okanishi, M., K. Suzuki, and H. Umezawa.** 1974. Formation and reversion of Streptomycete protoplasts: cultural condition and morphological study. J. Gen. Microbiol. **80:**389–400.
45. **Peberdy, J. F., and R. K. Gibson.** 1971. Regeneration of *Aspergillus nidulans* protoplasts. J. Gen. Microbiol. **69:**325–330.
46. **Peberdy, J. F., and F. Kevei.** 1979. Interspecific hybridization in fungi by protoplast fusion, p. 23–26. *In* J. F. Peberdy (ed.), Protoplasts—applications in microbial genetics. University of Nottingham, Nottingham, U.K.
47. **Pontecorvo, G., P. N. Riddle, and A. Hales.** 1977. Time and mode of fusion of human fibroblasts treated with polyethylene glycol (PEG). Nature (London) **265:** 257–258.
48. **Ryu, D. D. Y., K. S. Kim, N. Y. Cho, and H. S. Pai.** 1983. Genetic recombination in *Micromonospora rosaria* by protoplast fusion. Appl. Environ. Microbiol. **45:**1854–1858.
49. **Sagara, Y., K. Kukui, F. Ota, N. Yoshida, T. Kashiyama, and M. Fujimoto.** 1971. Rapid formation of protoplasts of *Streptomyces griseoflavus* and their fine structure. Jpn. J. Microbiol. **15:**73–84.
50. **Schaeffer, P., B. Cami, and R. D. Hotchkiss.** 1976. Fusion of bacterial protoplasts. Proc. Natl. Acad. Sci. U.S.A. **73:**2151–2155.
51. **Sipiczki, M., and L. Ferenczy.** 1977. Protoplast fusion of *Schizosaccharomyces pombe* auxotrophic mutants. Mol. Gen. Genet. **151:**77–81.
52. **Svoboda, A.** 1978. Fusion of yeast protoplasts induced by polyethylene glycol. J. Gen. Microbiol. **109:**169–175.
53. **Szvoboda, G., T. Lang, I. Gado, G. Ambrus, C. Kari, K. Fodor, and L. Alfoldi.** 1980. Fusion of *Micromonospora* protoplasts, p. 235–240. *In* L. Ferenczy and G. L. Farkas (ed.), Advances in protoplast research. Budapest Akademiai Kiado, Oxford. Pergamon Press, New York.
54. **Tsenin, A. N., G. A. Karimov, and V. N. Ribchin.** 1978. Recombinaciya prisliyanii protoplastov *Escherichia coli* K12. Dokl. Akad. Nauk SSSR **243:**1066–1068.
55. **Van Solingen, P., and J. B. vander Plaat.** 1977. Fusion of yeast spheroplasts. J. Bacteriol. **130:**946–947.

Chapter 14

Mutagenesis in *Streptomyces* spp.

RICHARD H. BALTZ

In the past several years, important advances have been made in the development and exploitation of protoplast fusion (7, 9, 10, 17) and protoplast transformation for gene cloning (10, 13, 14, 16, 18, 20a, 22) in the economically important *Streptomyces* spp. While these techniques hold great promise for solving specific problems in antibiotic yield improvement and in the formation of novel (hybrid) antibiotics (8), there is still an important need for efficient random mutagenesis for strain improvement. The main reasons for this continued need are that antibiotic biosynthesis is under multigenic control and that in most cases little is known about the structure, function, and regulation of antibiotic biosynthetic genes, the biosynthesis and regulation of primary metabolic pathway enzymes involved in supplying precursors and cofactors for antibiotic biosynthesis, the rate-limiting steps in the overall metabolic scheme, the subtle regulatory interactions between different biosynthetic and catabolic pathways which may control levels of important small-molecule effectors, and other important factors. In the absence of comprehensive knowledge of all of these factors, a prudent approach is to develop very efficient mutagenesis procedures and, when feasible, to develop screening procedures to identify particular mutants of interest. As information on rate-limiting steps is generated and mutant strains which overproduce antibiotics are identified, then gene recombination and cloning can be used effectively in conjunction with mutagenesis. It is difficult, however, to conceive how recombination and gene cloning might be used effectively for strain improvement without mutagenesis. Thus efficient mutagenesis procedures should provide the foundation for any strain improvement program. To develop such procedures, it is important to understand what efficient mutagenesis means and to be able to measure mutation induction accurately.

For recent reviews and background information, the reader is referred to references 1–6.

14.1 EFFICIENT MUTAGENESIS IN *STREPTOMYCES* spp.

14.1.1 The Poisson Model

To understand and define efficient mutagenesis, it is important to understand the target for mutagenesis, the *Streptomyces* genome. *Streptomyces* chromosomes are composed of circular double-stranded DNA (19, 20) of about 6.9×10^9 daltons (12) or approximately 10^7 base pairs. This is about three times the size of the *Escherichia coli* genome, which codes for about 2,240 transcribable genes (15). Thus, by analogy, *Streptomyces* genomes could contain about 6,720 transcribable genes. We estimated that in *Streptomyces fradiae* as many as 1,000 to 2,000 genes may encode functions which influence the antibiotic yield either directly or indirectly (R. H. Baltz, E. T. Seno, and J. Stonesifer, unpublished data). For instance, virtually all auxotrophic, sporulation-defective, and antibiotic resistance mutations alter antibiotic production, to name just a few. Also, mutations in antibiotic biosynthetic genes and in genes encoding enzymes in metabolic pathways which control the availability of precursors, cofactors, energy source, etc., also influence the yields of antibiotic. With this many genes involved in controlling antibiotic yield, the potential problems associated with isolating mutants containing only mutations affecting antibiotic yields in positive ways can be readily appreciated.

Since the productivity of an antibiotic-producing streptomycete can be influenced by many genes, the major objective of random mutagenesis strategies to improve antibiotic yields is to maximize the frequen-

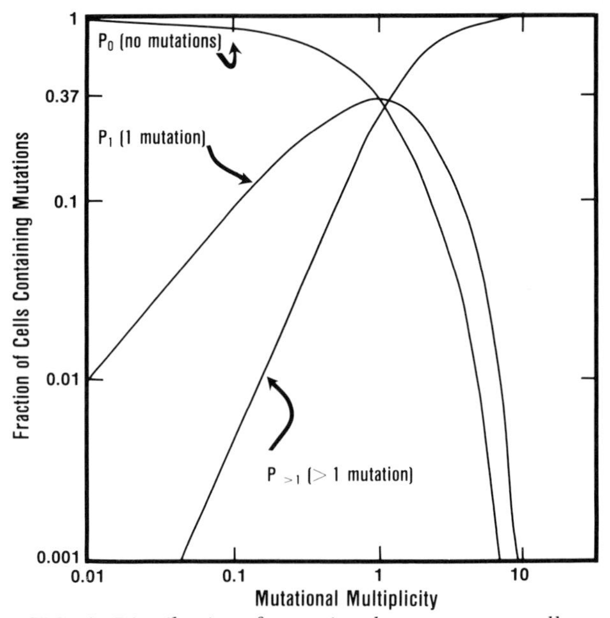

FIG. 1. Distribution of mutational events among cells as a function of mutational multiplicity.

cy of desired mutations in a population of cells, while minimizing the frequency of nonlethal second-site mutations which might offset the positive effects of the desired mutations. This problem is of course simplified if selection procedures for desired mutants are available and heavy mutagenic treatments can be avoided. However, if no known selectable or easily identifiable traits are associated with the desired mutations, then the solution is to maximize the frequency of survivors of mutagenesis which contain on the average only one mutation affecting antibiotic yield, since most mutations which affect antibiotic productivity do so in a negative way, particularly in highly developed strains. If we assume that mutations that affect antibiotic yield are accumulated randomly within a population after mutagenesis, then such mutations should be distributed within a population of cells according to the Poisson equation ($P = m^x e^{-m}/x!$), where P is the fraction of cells containing x mutations at a multiplicity of m mutations per cell. Thus optimum mutagenesis can be achieved when the mutational multiplicity (m) = 1, and the fraction of cells containing no mutations (and producing control antibiotic yields) equals 37% (i.e., $P_0 = 1^0 \times e^{-1}/0! = e^{-1} = 0.37$). At this multiplicity, the fraction of cells which contain only one mutation is also 37% (or $P_1 = 1^1 \times e^{-1}/1! = e^{-1} = 0.37$) (Fig. 1).

It is clear from Fig. 1 that overmutagenesis can dramatically reduce the frequency of cells containing single mutations. For instance, at $m = 4$, $P = 0.018$. Thus a 3-fold excess of mutational hits can reduce the frequency of desired mutants among survivors by 20-fold.

On the other hand, reduction of the mutational multiplicity from 1.0 to 0.25 should cause only a twofold decrease in the fraction of cells containing only one mutation affecting antibiotic yields. Therefore, it is much better to err on the low side than on the high side of the optimum multiplicity.

A practical way to measure mutational multiplicity directly derives from the observation that most mutations which influence antibiotic yields do so in a negative way. For example, at a mutational multiplicity (m) of 1, only 37% of the survivors of mutagenesis should produce control levels of antibiotic (see shaded area, Fig. 2), while 63% should produce less than control levels. Since the fraction of cells producing control yields (the P_0 of the Poisson distribution) equals e^{-m}, the mutational multiplicity can be simply calculated as $m = -\ln(P_0)$.

Since it may not be practical to routinely measure the dose by calculating the P_0 of the Poisson distribution by scanning an antibiotic productivity histogram, particularly at $m < 0.3$, it is convenient to establish a relationship between the mutant frequency at easily measured loci (e.g., antibiotic resistance or reversion of auxotrophy) and the mutational multiplicity as defined by the Poisson equation. Once this is established, routine determinations of mutational multiplicity can be made by simply determining mutant frequency at the appropriate easily quantified loci.

The foregoing discussion of course does not apply to situations where specific rare mutations of particular phenotype or genotype are desired and when antibiotic productivity is not an important consideration. In these cases, heavy mutagenesis yielding multiplicities greater than 1 may be desirable. Second- and third-site mutations can be back-crossed out by conjugation or protoplast fusion if needed.

14.2 CURRENT PROCEDURES FOR RANDOM MUTAGENESIS

14.2.1 Preparation of Cells

Since *Streptomyces* spp. are filamentous microorganisms, they are not as easy to manipulate by mutation as the single-celled eubacteria. Streptomycetes form spores (which contain single genomes) under certain environmental conditions, but general procedures to efficiently induce sporulation of a wide variety of *Streptomyces* species are not available. Thus general mutational procedures which do not rely on spore formation are more readily applied to a wide variety of species and mutants of particular species.

Since *Streptomyces* mycelia contain cross walls which define individual cells when growing in liquid media, appropriate levels of ultrasonic vibration can be employed to rupture, on the average, every other cell, thus leaving about 50% of the cells as viable CFU (7). The general procedure for ultrasonic fragmentation of mycelia is as follows. First, spores or mycelial fragments (either from scraping a slant culture or from homogenizing a liquid culture with a Teflon pestle tissue grinder [A. H. Thomas Co.]) are used to inoculate Trypticase soy broth (TS broth), and cells are grown for about 24 h or until late-exponential or stationary phase is achieved (i.e., A_{600} of about 2.0 to 9.0). The resulting mycelia are homogenized in a tissue grinder and diluted 1:1 in TS broth. About 20 ml of diluted mycelia is added to a sterile plastic vessel (usually about 50 ml of total volume), and the tip of an ultrasonic probe (Branson Sonifier, model

W185) is submerged about 1 to 2 cm into the liquid. Ultrasonic fracture of mycelia is achieved by exposing the medium to about 75 W output for 3 to 10 s. For most *Streptomyces* spp., this treatment will result in about 50% inactivation of potential viable cells, while the remaining cells exist primarily as single-cell CFU. Thus individual mycelial aggregates containing about 1,000 viable cells can yield about 500 CFU (7).

If exponentially growing cells are desired for mutagenesis, the homogenized and sonicated cells are used to inoculate fresh TS broth (usually at 1 to 10% inoculum), and cells are grown for about 2 to 3 h at 34°C at about 270 rpm in a New Brunswick water bath shaker to initiate exponential cell growth. Cells prepared in this way are used routinely for *N*-methyl-*N'*-nitro-*N*-nitrosoguanidine (MNNG) mutagenesis. Alternatively, cells can be grown to stationary phase for other types of mutagenesis. In this case, the resulting mycelial aggregates must be homogenized and sonicated again before mutagenesis. In the folowing sections, I refer to homogenized and sonicated mycelia as cells.

In some cases it may be desirable to carry out mutagenesis of spores, as has been done with *Streptomyces coelicolor* (19). Since the following procedures have been developed for mutagenesis of mycelial fragments rather than spores, it would be prudent to expect that spores may yield different results in some cases, particularly since little is known about the levels of expression of the various error-free and error-prone repair systems in spores, spore germlings, and well-developed mycelia.

14.2.2 Mutant Segregation

After mutagenesis, cells are washed free of the mutagen or the mutagen is inactivated as described below. Cells are diluted 10- to 20-fold in TS broth and are grown for 18 to 24 h at 34°C as described above to allow for segregation before mutant frequencies are determined. Segregated cells (mycelia) are homogenized and sonicated, diluted, and plated for viable cells on AS-1 agar and for spectinomycin-resistant (Spc^r) mutants on AS-1 plus 50 μg of spectinomycin per ml.

Mutagen-treated and untreated cells are also diluted and plated directly (no segregation) on AS-1 agar to determine surviving fractions.

14.2.3 Precautions to Avoid Contamination of Experiment or Experimenter

Since *Streptomyces* spp. grow relatively slowly (doubling times range from about 1.5 to 3.0 h in TS broth at 34°C), and since the complete procedure for mutagenesis requires several transfers and manipulations (homogenization and sonication), it is not unusual for the novice to experience contamination problems. Much of this can be avoided by carrying out all manipulations and transfers in a biosafety cabinet, preferably one which can exhaust to the outside. If the mutagenic agent is volatile (e.g., ethyl methanesulfonate [EMS] and methyl methanesulfonate [MMS]), or if it gives off toxic gas (e.g., MNNG liberates diazomethane), then incubations should also be carried out in a hood which exhausts to the outside of the building. Alternatively, special devices can be fabricated to place over New Brunswick water bath shakers to permit volatile substances to exhaust to the outside.

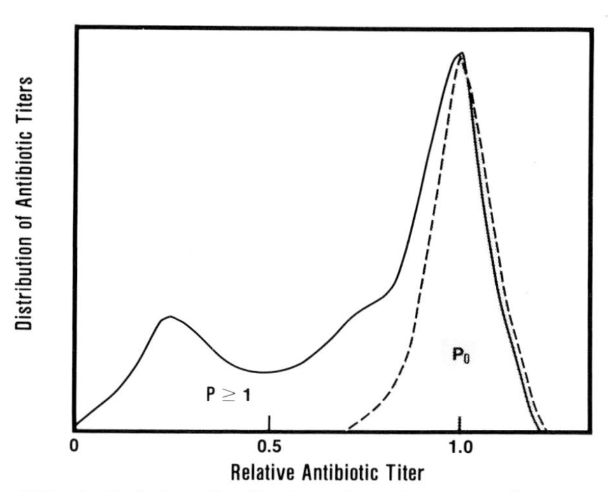

FIG. 2. Relative distribution of antibiotic production by mutagenized (solid line) and unmutagenized (dashed line) cells. Since the mutations which result in increased antibiotic production are relatively rare (1 in 10^3 to 1 in 10^4), they are not readily visualized on such a plot, but are quite apparent upon computer analysis of antibiotic titers relative to control titers, particularly when standard deviations are low.

It is important to avoid any contact with mutagens since all should be considered as potential carcinogens. Thus all weighing and transfers of powder or flakes of mutagens should be done in a biosafety cabinet, and the experimenter should wear a disposable gown, gloves, and mask. All exposed surfaces should be immediately decontaminated before proceeding, and gloves, gowns, and masks should be discarded for incineration or other suitable decontamination.

Also, all contaminated glassware, etc., should be decontaminated before washing or disposal.

14.2.4 UV Light

A 35-ml volume of washed, late-exponential- or stationary-phase cells is adjusted to an A_{600} of 1.0 in 0.01 M 2-tris(hydroxymethyl)methyl-aminoethanesulfonic acid (TES) buffer (pH 7.2) and gently swirled with a magnetic stir bar in a petri dish. Cells are then irradiated with UV light from a germicidal lamp placed about 30 to 40 cm above the surface of the liquid. Samples of the irradiated cells are removed at appropriate times to achieve doses of UV light ranging from about 20 to 160 J/m^2. Treatment times to obtain the appropriate doses are determined by measuring the fluence of UV at the appropriate distance from the source with a standard UV monitor. All manipulations are carried out in the dark (or semidark) to avoid photoreactivation (photo reversal of pyrimidine dimers).

While UV light is a fairly potent mutagen for *E. coli* and other microorganisms, it is only poorly mutagenic in *S. fradiae* (Fig. 3). This was somewhat surprising since *S. fradiae* has an error-prone DNA repair pathway which in many ways is similar to that in *E. coli* (1, 21). Also UV light is apparently much more mutagenic

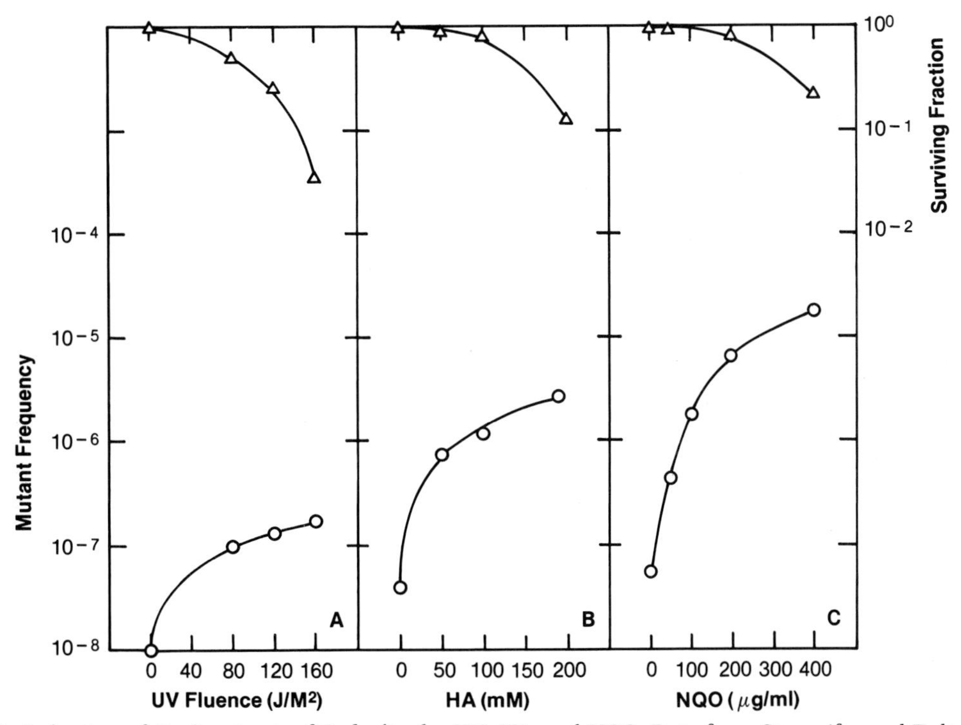

FIG. 3. Induction of Spc^r mutants of *S. fradiae* by UV, HA, and NQO. Data from Stonesifer and Baltz (21).

in *S. coelicolor* and *Streptomyces clavuligerus* than in *S. fradiae* (1). The reason(s) for the apparent differences in mutability within these three species is not yet apparent, but several possibilities have been suggested (1).

14.2.5 Hydroxylamine

Cells from the late exponential growth phase are suspended in 0.1 M phosphate buffer (pH 7.0) and incubated for 2 h at 37°C in 100 to 400 mM hydroxylamine (HA). Reactions are terminated by diluting the cells 10-fold in chilled nutrient broth plus 10% acetone. Cells are further diluted to determine surviving fractions and segregated before determining mutant frequencies as described in sections 14.2.1 and 14.2.2.

While HA is a more potent mutagen for *S. fradiae* than UV light, it is still a relatively weak mutagen (Fig. 3). HA apparently causes mutations by direct mispairing and error-prone repair in *S. fradiae* (1, 21). Mutations accumulate proportionally with the first power of HA dose, rather than with a higher power of dose as is the case with EMS, MMS, 4-nitroquinoline-1-oxide (NQO), and MNNG (at low doses). The relatively weak mutagenic properties of HA in *S. fradiae* suggest that it may be of marginal utility for routine strain development programs.

14.2.6 4-Nitroquinoline-1-Oxide

Cells from late exponential phase are suspended in 0.05 M phosphate buffer (pH 7.0) and are incubated for 2 h at 37°C with 100 to 400 μg of freshly prepared NQO. (To prepare a 0.5-mg/ml stock solution, 5.0 mg of NQO is dissolved in 5.0 ml of acetone, mixed with 5.0 ml of 0.05 M phosphate buffer, pH 7.0, and filter sterilized.) The reaction is terminated by diluting the cells 10-fold in 0.16 M sodium thiosulfate to inactivate the mutagen.

NQO induces mutations in *S. fradiae* primarily by an error-prone DNA repair mechanism (1, 21). It is a more efficient mutagen than UV or HA (Fig. 3), but less efficient than MNNG. NQO induces mutations proportionally with the square of the dose, so it is important to identify accurately the appropriate dose to obtain the desired mutational multiplicity. NQO may be a suitable alternative to MNNG since it produces both transitions and transversions, whereas MNNG produces mostly GC → AT transitions (6).

14.2.7 Ethyl Methanesulfonate

Late-exponential-phase cells are washed by centrifugation, suspended in 0.2 M potassium phosphate buffer (pH 7.0), and incubated for 10 min at 37°C with 0.5 to 3.0% (wt/vol) EMS. The reaction is terminated by diluting the cells 10-fold in 0.16 M sodium thiosulfate to inactivate the mutagen. The mutagenized cells are plated to determine surviving fractions and segregated before determining mutant frequencies (see above).

EMS induces mutations primarily by an inducible error-prone DNA repair pathway in *Streptomyces* (1, 21), and the frequency of mutations increases proportionally with the square of dose (Fig. 4). Therefore, defining the optimum dose accurately is important to ensure efficient mutagenesis.

14.2.8 Methyl Methanesulfonate

Late-exponential-phase cells are washed by centrifugation, suspended in 0.2 M potassium phosphate buffer (pH 7.0), and incubated for 30 min at 37°C with

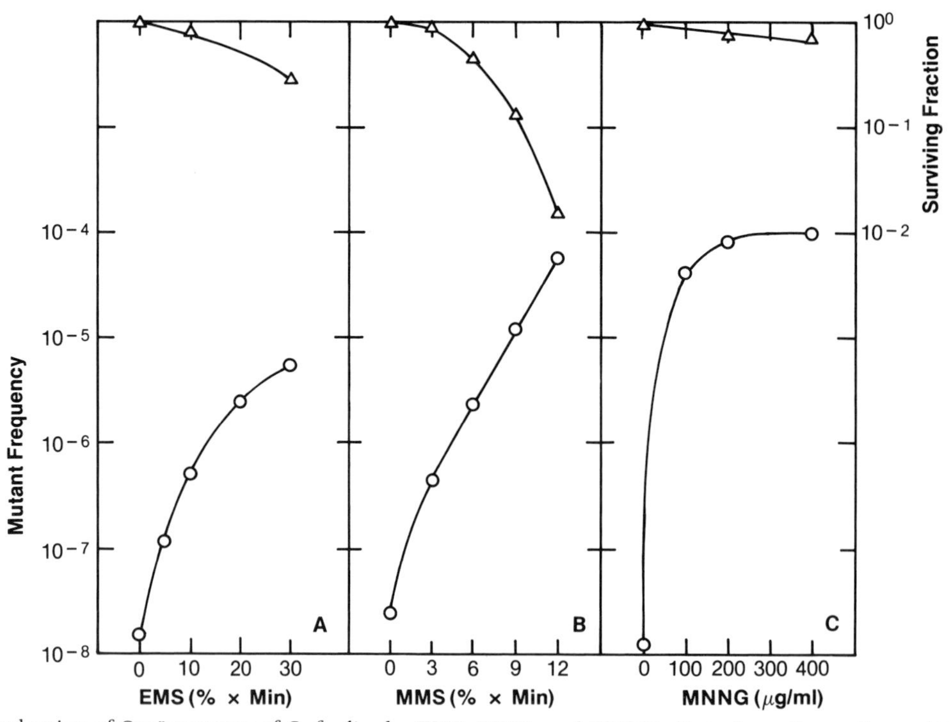

FIG. 4. Induction of Spc^{r} mutants of *S. fradiae* by EMS, MMS, and MNNG. Data from Stonesifer and Baltz (21).

0.05 to 0.4% (vol/vol) MMS. Reactions are terminated by diluting the cells 10-fold in 0.16 M sodium thiosulfate to inactivate the mutagen. The mutagenized cells are plated to determine surviving fractions and segregated before determining mutant frequencies (see above).

MMS also causes mutations by error-prone repair (1, 21), but appears to differ from EMS and NQO in that mutations accumulate proportionally with the third or fourth power of dose. While this complex dose response is not yet fully understood (1), it is clear that great care must be taken to adjust the MMS dose to achieve the desired mutational multiplicity. In practice, it is best to mutagenize at several doses, to measure mutation induction at one or two loci (e.g., *spc*, *rif*, etc.), and then to retain only those mutagenized cells which contain the desired frequency of mutants (the appropriate mutational multiplicity).

14.2.9 *N*-Methyl-*N'*-Nitro-*N*-Nitrosoguanidine

A culture grown exponentially for 3 h in TS broth (see above) is adjusted to pH 8.5 with 2% sodium hydroxide. MNNG is added to a final concentration of 100 to 400 μg/ml, and the culture is incubated at 37°C in a New Brunswick water bath shaker at about 250 to 300 rpm. After 20 min, 1 ml of culture is diluted into 19 ml of TS broth (chilled to 0°C) to determine the viable cell count (surviving fraction). The remaining cells are washed free of MNNG by centrifugation in a clinical centrifuge for 5 min. The supernatant is decanted, and the cell pellet is resuspended in 50 ml of TS broth. Cultures are grown at 32°C overnight to allow segregation and then homogenized and sonicated. Cells are diluted and plated to determine mutant frequencies (see above).

Like UV, NQO, MMS, and EMS, MNNG also induces mutations primarily by an error-prone DNA repair pathway (1, 21). However, MNNG is more potent than any of the other mutagens and induces high frequencies of mutants at doses which result in little or no lethality (Fig. 4). Further, while MNNG-induced mutations accumulate proportionally with the square of dose at low doses, the frequency of mutants plateaus at higher doses (Fig. 4). Thus it is much easier to obtain reproducibly high frequencies of mutants with MNNG since the mutant frequency is relatively insensitive to dose changes in the optimum range. This differs dramatically from the situation noted for MMS, for instance. Even higher frequencies of mutants are obtained if the MNNG treatment is carried out in the presence of chloramphenicol (1). The reason for this response is not yet fully understood.

14.3 COMPARATIVE MUTAGENESIS

The maximum frequency of mutations induced by the various mutagens discussed in this chapter ranged from about 2×10^{-7} with UV light to about 1×10^{-4} with MNNG. UV light gave only a 10-fold increase in mutant frequency over the spontaneous background frequency, whereas MNNG gave about a 10,000-fold increase. Thus, induced mutation frequencies can range over about 1,000-fold. The question thus arises as to what level of induced mutagenesis is most appropriate for strain development. The answer is the treatment that results in a mutational multiplicity of about 1 mutation per cell which affects antibiotic yield. This dose cannot be predicted accurately a priori, but needs to be determined experimentally. In practice, however, it appears that only the most efficient mutagens can approach a mutational multiplicity of 1. Thus a relatively safe approach is simply to use MNNG, which has a relatively insensitive dose-response profile (Fig. 4). However, MNNG may cause mostly GC → AT transition mutations (6). NQO may also be advantageous in that it produces both transitions and transversions (6).

14.4 MEDIA

The following media are quite suitable for *S. fradiae* and for many other *Streptomyces* species. Some species, however, may not grow particularly well on these media; in these cases other more suitable media should be substituted.

TS broth

(Complex medium for growth of Streptomyces)

Trypticase soy broth (BBL)	30 g
Distilled water	1 liter

Autoclave for 15 min at 121°C.

AS-1 agar

(Rich plating medium for Streptomyces)

Yeast extract	1 g
L-Alanine	0.2 g
L-Arginine	0.2 g
L-Asparagine	0.5 g
Soluble starch	5 g
NaCl	25 g
Na_2SO_4	10 g
Agar	20 g
Distilled water	1 liter

Adjust to pH 7.5. Autoclave for 15 min at 121°C. Pour 30 to 35 ml into sterile plastic petri plates (100 by 15 mm).

AS-1 plus spectinomycin

Prepare AS-1 agar. After autoclaving and cooling to about 45°C, add 5 ml of spectinomycin solution from a filter-sterilized stock at 10 mg/ml to 1 liter of AS-1 agar. Mix and pour into petri plates.

Soft agar overlays

Nutrient broth	8.0 g
Agar	6.5 g
Distilled water	1 liter

Autoclave for 15 min at 121°C. Aseptically distribute 3 ml to sterile, capped culture tubes (13 by 100 mm). Melt the soft agar in a steam bath or other suitable device just before use. Cool the melted soft agar to 45 to 46°C in a water bath before adding cells. After adding cells to the soft agar, mix well, pour over AS-1 or AS-1–plus–spectinomycin agar plates, and allow several minutes for the agar to solidify before placing the plates in an incubator.

14.5 SOURCES OF STRAINS AND MATERIALS

Strains

Streptomyces strains can be obtained through the American Type Culture Collection or the Northern Regional Research Laboratories.

Media and chemicals

TS broth: BBL Microbiology Systems, Cockeysville, MD 21030.

Agar and nutrient broth: Difco Laboratories, Detroit, MI 48232.

HA, MMS, EMS: Eastman Organic Chemicals, Rochester, NY 14650.

MNNG, NQO: Aldrich Chemical Co., 940 W. St. Paul Ave., Milwaukee, WI 53233.

Mitomycin C: Calbiochem-Behring, San Diego, CA 92112.

14.6 LITERATURE CITED

14.6.1 General References

1. **Baltz, R. H.** 1986. Mutation in *Streptomyces*. *In* The bacteria, vol. 9. Antibiotic-producing *Streptomyces*. Academic Press, Inc., New York.
A summary of current knowledge on mutagenic mechanisms in Streptomyces.
2. **Carlton, B. C., and B. J. Brown.** 1981. Gene mutation, p. 222–242. *In* P. Gerhardt (ed.), Manual of methods for general bacteriology. American Society for Microbiology, Washington, D.C.
An important reference from an ASM book with similar but more general scope than Manual of Industrial Microbiology and Biotechnology. This chapter should be read as background for the present chapter.
3. **Drake, J. W.** 1970. The molecular basis of mutation. Holden-Day, San Francisco.
A good reference on mechanisms of mutation. Also provides detailed background on statistical considerations and the use of the Poisson equation.
4. **Drake, J. W., and R. H. Baltz.** 1976. The biochemistry of mutagenesis. Annu. Rev. Biochem. **45:**11–37.
A review of mutagenic mechanisms relevant to the present chapter.
5. **Lindahl, T.** 1982. DNA repair enzymes. Annu. Rev. Biochem. **51:**61–87.
A review of enzyme systems which interact with DNA treated with mutagenic agents.
6. **Miller, J. H.** 1983. Mutation specificity in bacteria. Annu. Rev. Genet. **17:**215–238.
A review of the specificity of mutagenic agents in bacteria, primarily E. coli.

14.6.2 Specific References

7. **Baltz, R. H.** 1978. Genetic recombination in *Streptomyces fradiae* by protoplast fusion and cell regeneration. J. Gen. Microbiol. **107:**93–102.
8. **Baltz, R. H.** 1982. Genetics and biochemistry of tylosin production: a model for genetic engineering in antibiotic-producing *Streptomyces*, p. 431–444. *In* A. Hollaender (ed.), Genetic engineering of microorganisms for chemicals. Plenum Publishing Corp., New York.
9. **Baltz, R. H., and P. Matsushima.** 1981. Protoplast fusion in *Streptomyces*: conditions for efficient genetic recombination and cell regeneration. J. Gen. Microbiol. **127:**137–146.
10. **Baltz, R. H., and P. Matsushima.** 1983. Advances in protoplast fusion and transformation in *Streptomyces*. Exper. Suppl. **46:**143–148.
11. **Baltz, R. H., and J. Stonesifer.** 1985. Mutagenic and error-free DNA repair in *Streptomyces*. Mol. Gen. Genet. **200:**35–355.
12. **Benigni, R., P. A. Petrov, and A. Carere.** 1975. Estimate of the genome size by renaturation studies in *Streptomyces*. Appl. Microbiol. **30:**324–326.
13. **Bibb, M. J., K. F. Chater, and D. A. Hopwood.** 1983. Developments in *Streptomyces* cloning, p. 53–82. *In* M.

Inouye (ed.), Experimental manipulation of gene expression. Academic Press, Inc., New York.

14. **Chater, K. F., D. A. Hopwood, T. Kieser, and C. J. Thompson.** 1982. Gene cloning in *Streptomyces*. Curr. Top. Microbiol. Immunol. **96:**69–95.
15. **Hahn, W. E., D. E. Pettijohn, and J. Van Ness.** 1977. One strand equivalent of the *Escherichia coli* genome is transcribed: complexity and abundance classes of mRNA. Science **197:**582–585.
16. **Hershberger, C. L., J. L. Larson, and S. E. Fishman.** 1983. Uses of recombinant DNA for analysis of *Streptomyces* species. Annu. N.Y. Acad. Sci. **413:**31–46.
17. **Hopwood, D. A.** 1981. Genetic studies with bacterial protoplasts. Annu. Rev. Microbiol. **35:**237–272.
18. **Hopwood, D. A., and K. F. Chater.** 1982. Cloning in *Streptomyces*: systems and strategies, p. 119–145. *In* J. K. Setlow and A. Hollaender (ed.), Genetic engineering, vol. 4. Plenum Publishing Corp., New York.
19. **Hopwood, D. A., K. F. Chater, J. E. Dowding, and A. Vivian.** 1973. Advances in *Streptomyces coelicolor* genetics. Bacteriol. Rev. **37:**371–405.
20. **Hopwood, D. A., and M. J. Merrick.** 1977. Genetics of antibiotic production. Bacteriol. Rev. **41:**595–635.

20a. **Matsushima, P., and R. H. Baltz.** 1985. Efficient plasmid transformation of *Streptomyces ambofaciens* and *Streptomyces fradiae* protoplasts. J. Bacteriol. **163:**180–185.

21. **Stonesifer, J., and R. H. Baltz.** 1985. Mutagenic DNA repair in *Streptomyces*. Proc. Natl. Acad. Sci. U.S.A. **82:**1180–1183.
22. **Thompson, C. J., J. M. Ward, and D. A. Hopwood.** 1982. Cloning of antibiotic resistance and nutritional genes in *Streptomyces*. J. Bacteriol. **151:**668–677.

Chapter 15

Genetic Mapping by Conjugation in *Streptomyces* spp.

DAVID A. HOPWOOD

The material in this chapter is an abridged version of a section of a detailed and comprehensive manual of techniques for the genetic manipulation of *Streptomyces* spp. (10). That manual should be consulted for further practical details and caveats.

Conjugation in *Streptomyces* spp. occurs in growing cultures of two strains on the surface of an agar medium. There is no equivalent of the familiar interrupted mating of *Escherichia coli* genetics, and meaningful kinetic analysis of the mating process in *Streptomyces* spp. has not proved possible. One therefore analyzes the end results of a series of events taking place over a period of days, during which many rounds of conjugation and recombination may have taken place. Nevertheless, much useful information is obtainable: for example, gross chromosomal linkage mapping can be a very simple operation (in those streptomycetes which readily yield recombinants). Because the linkage map is relatively short in terms of the probability of crossing over in a mating, a single cross with as few as four markers can yield an unambiguous circular linkage map by a method of analysis called the "four-on-four" procedure (because it involves the plating of a four-factor cross in parallel on four selective media). The practical techniques and the mapping logic for linkage analysis were originally worked out for *Streptomyces coelicolor* (9) but have been found to be easily applicable to several other strains, such as *Streptomyces glaucescens* (2), *Streptomyces rimosus* (6), *Streptomyces lividans* (11), and *Nocardia mediterranei* (16).

The production of genetic recombinants in *Streptomyces* matings probably always depends on the activity of plasmid sex factors, and many strains carry such plasmids (not always in a physically isolable form). Thus, genetically marked strains derived from the same starting strain by mutation will often generate recombinants when they are mated together. However, the frequency of recombination for some strains is very low. In certain cases, this has been related to the fact that the particular plasmid present in the strain shows the property of "entry disadvantage." This term describes the situation when recombinants are very infrequent in crosses with the same plasmid present in both partners in the mating, but are much more frequent when only one partner carries it. This is the case in *S. lividans* 66; good recombination frequencies were obtained only when it was possible to eliminate the SLP2 plasmid from certain marked derivatives of strain 66 so that SLP2$^+$ × SLP2$^-$ matings could be made (11). In *S. coelicolor* A3(2), a so-far unique situation is found: the SCP1 plasmid interacts with the chromosome to form various donor strains which yield very high recombinant frequencies in matings with SCP1$^-$ strains (12); the NF type of strain is an example. Use of such strains allows specific mapping procedures for certain regions of the chromosome. Some strains that lack a convenient sex plasmid may be rendered fertile by introducing a broad-host-range sex plasmid into them (14).

In *S. coelicolor* A3(2) and a few other strains, such as *S. rimosus* (1), heteroclones have been used in mapping. Heteroclones are colonies arising from partially heterozygous plating units which give rise to spores representing a population of recombinant and parental genotypes whose frequencies can be used to generate linkage data. They are recognized as small colonies on a medium selecting for recombinants between two closely linked genes, and by the fact that they fail to grow on a medium of the same composition as that on which they were selected. The analysis of heteroclones is laborious and specialized and can lead to inaccuracies of mapping caused by the selection of fast-growing recombinant sectors within the colony. Heteroclone analysis is not a generally convenient mapping tool, but has found a use in certain strains, such as *Streptomyces achromogenes* subsp. *rubradirans* (5), in which more conventional mapping procedures are unsuitable.

Apart from linkage mapping, crosses are useful for strain construction, when new combinations of chromosomal mutations or plasmids originally carried by different strains are combined into a single strain. Protoplast fusion, although it can be used for mapping, is particularly adapted to strain construction since the very high frequencies of recombinants produced allow combinations of nonselectable chromosomal markers to be built up. Because there is much more recombination per unit map length in a fusion than in a mating (leading to a higher frequency of multiple crossover recombinants), mapping of distant markers may be rendered ambiguous, but close markers are easier to resolve (11, 13). Protoplast fusion is described by Matsushima and Baltz elsewhere in this manual.

15.1 MAKING, HARVESTING, AND PLATING CROSSES

15.1.1 Making the Cross

Crosses are prepared by mixing the two strains together on the surface of an agar slant (complete medium [CM] or R2YE) with added growth factors. Fairly fresh slants are best. They usually have a drop of liquid at the bottom of the slant; if not, add a drop of sterile water. To this, add a drop of spore suspension of each strain from a pipette, or a loopful of spore suspension, or a loopful of dry spores, mycelium, or both from a growing culture. Then spread the liquid uniformly over the surface of the slant with circular movements of the loop, or with the tip of the pipette. Incubate the slant with the agar surface horizontal so that the liquid soaks into the medium, until sporulation occurs.

15.1.2 Harvesting the Cross

1. Add ca. 9 ml of sterile water to the sporulation slant (for some strains a wetting agent is needed, e.g., 0.1% Tween 80 or 0.001% Triton X-100) and scrape the surface of the culture with an inoculating loop, first with gentle pressure and then gradually more vigorously, to suspend the spores.
2. Pour the crude suspension back into the container that held the sterile water and agitate the liquid as violently as possible on a Vortex mixer for a minute or so.
3. Filter the suspension through nonabsorbent cotton wool, pour the filtered suspension into a centrifuge tube, and spin for 5 to 10 min at ca. 3,000 rpm to pellet the spores.
4. Pour off the supernatant, and agitate the tube on the Vortex mixer for a few seconds to disperse the pellet in the drop of water remaining in the tube.
5. Add sterile 20% glycerol (usually 1 to 2 ml for the spores from a well-sporulating slant culture) and briefly agitate again. The suspension can be plated immediately or can be kept frozen at −20°C for very long periods.

15.1.3 Plating the Cross

Streptomyces spores (and even mycelial fragments) are quite resistant to osmotic damage, and so dilutions can be made in distilled water, even when diluting directly from 20% glycerol. For plating, it is usual to spread 0.1 ml of spore suspension on each standard petri plate of suitably supplemented minimal medium (MM), using a glass spreader. There is usually no need to dry the plates; just incubate them with the agar surface facing upwards for the first day or so. If necessary, dry the plates in a laminar-flow cabinet after spreading (drying them beforehand, so that the suspension soaks in rapidly, may lead to patchy growth).

With most strains, several hundred colonies can be counted accurately on each plate. Countable plates carrying unselected progeny, or progeny with the genotype of one or the other parent, will normally be observed with 10^{-4}, 10^{-5}, or 10^{-6} dilutions (regarding the original suspension as 10^{0}). The right dilutions for selective platings will depend on the expected frequency of recombination; sometimes even 0.1 ml of the 10^{0} suspension will yield only a few recombinants.

15.2 CHARACTERIZING RECOMBINANT GENOTYPES

Usually, colonies growing on a particular selective medium need to be classified with respect to nonselected markers. It is best to pick off a sample of colonies to a "master plate" of medium with the same composition as the original plate and replicate the ordered array of (usually 50) patches to a series of diagnostic media, each one testing for one of the markers. For example, a cross between two strains of genotypes *his leu phe str* (parent 1) and *arg* (parent 2) might be plated on a selective medium containing leucine (L), phenylalanine (P), arginine (A), and streptomycin (S), thereby selecting for recombinants inheriting *str* (streptomycin resistance) from parent 1 and his^{+}(histidine nonrequirement) from parent 2. A sample of the recombinants is then classified for their genotypes with respect to the three nonselected markers (*leu, phe, arg*); there are eight possible combinations of these markers. Recombinant colonies are picked to master plates of the same selective medium as that on which they were growing. Then, when the patches have grown, each master plate is replicated to a set of four diagnostic media: (i) PAS (minus leucine); (ii) LAS (minus phenylalanine); (iii) LPS (minus arginine); and (iv) LAPS (control plate). After incubation of the replicas, each recombinant is classified for growth or no growth on each replica plate and its genotype is thus determined; e.g., growth on PAS and LAPS but not LAS and LPS indicates a *phe arg str* genotype.

Master plates are made by placing a fresh plate of medium over a grid and inoculating a colony into each square. (Colonies can be picked, even when they are not sporulating, by digging out part of the mycelium with a flattened needle or toothpick.)

Master plates can usually be replica plated (using sterile velvet) after a few days of incubation, even if they are not sporulating copiously, especially if the strain has a tendency to give "friable" growth (that is, soft as opposed to tough and leathery). Usually, the same velvet can be used for making replicas on several diagnostic plates in succession.

TABLE 1. Analysis of a four-factor cross (*ade-2 his-1* × *arg-1 lys-2*)

Class pair no.	Genotypes of progeny (in complementary pairs)	No. of each genotype per unit vol on selective medium[a]: Adenine and arginine	Adenine and lysine	Arginine and histidine	Histidine and lysine	Avg frequency of each pair of genotypes
1	*ade* + *his* +	Parental: do not grow on selective media				
	+ *arg* + *lys*					
2	+ + + +	(14) 24	(45) 36	(14) 33	(6) 26	30
	ade arg his lys	—	—	—	—	
3	*ade* + + +	(17) 29	(30) 24	—	—	27
	+ *arg his lys*	—	—	—	—	
4	+ *arg* + +	(42) 72	—	(48) 112	—	92
	ade + *his lys*	—	—	—	—	
5	+ + *his* +	—	—	(86) 202	(92) 393	298
	ade arg + *lys*	—	—	—	—	
6	+ + + *lys*	—	(57) 41	—	(3) 13	27
	ade arg his +	—	—	—	—	
7	*ade arg* + +	(76) 130	—	—	—	100
	+ + *his lys*	—	—	—	(16) 70	
8	*ade* + + *lys*	—	(0) 0	—	—	3
	+ *arg his* +	—	—	(2) 5	—	
	Heterokaryons[b]	(1) 2	(24) 19	(0) 0	(33) 142	
	Total (colony count on medium)	257	120	352	644	

[a] Dashes indicate recombinant classes which do not grow because the selective medium does not contain all their growth requirements; the analysis assumes that such recombinants are present in numbers equal to the corresponding reciprocal classes, and this assumption is justified if the frequencies of those complementary classes that can be recovered (class pairs 7 and 8) are not grossly dissimilar (as here). Numbers of colonies actually scored are given in parentheses for use in Table 2.

[b] Heterokaryons are defined as prototrophic colonies generating spores of parental genotypes only. (They grow on the master plates because mycelial fragments are transferred when the master plate patches are inoculated, but not on the replicas because only spores are carried on the velvet pads.) They are disregarded in the analysis.

15.3 CALCULATING RECOMBINATION FREQUENCY

Normally, dilutions of the spore suspension derived from a mating are plated on the two media each selective for one of the parental genotypes, as well as on one or more media selective for recombinants. Growing colonies are then counted. A convenient way to express the results is as a fraction obtained by dividing the average colony count on the media that select for the recombinants by the sum of the counts on the two media that select for the parents. (If the cross is unbalanced, with a much higher count on one of the two "parental" media, one might use the "minority" parent as a basis.) In using a single average figure for the frequency of recombinants, the total frequency of recombination is underestimated, but usually an order of magnitude only is needed for comparative purposes. (In the four-on-four mapping procedure [see below], accurate counts on each of the four media selecting recombinant classes are needed for the calculations.)

TABLE 2. 2 × 2 tabulations of segregation of pairs of unselected markers, using numbers in parentheses from Table 1

	arg^+	arg^-	lys^+	lys^-
ade^+	14	42	45	51
ade^-	17	76	30	0
Probability of independence (from χ^2 test)	0.30		<0.001	
Selected alleles	his^+/lys^+		arg^+/his^+	

	arg^+	arg^-	lys^+	lys^-
his^+	14	48	6	3
his^-	86	2	92	16
Probability of independence (from χ^2 test)	<0.001		0.15	
Selected alleles	ade^+/lys^+		arg^+/ade^+	

15.4 GENETIC MAPPING

The best way to start building up a chromosomal linkage map is to use four-factor crosses, with two selectable markers in each parent, analyzed by the four-on-four procedure. Once some markers have been located on the map, new markers can be mapped by making one or two selections and considering the

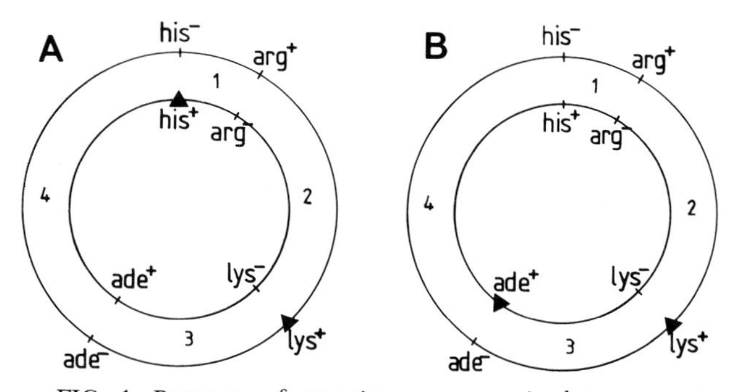

FIG. 1. Patterns of crossing over required to generate various genotypes of recombinants recovered on two of the selective media from Table 1. (A) Selection of recombinants on arginine + adenine (*his*$^+$ and *lys*$^+$ selected). (B) Selection of recombinants on arginine + histidine (*ade*$^+$ and *lys*$^+$ selected). Intervals in which crossing over is required to form detectable recombinants:

For (A),

	arg$^+$	*arg*$^-$
ade$^+$	1.3	2.3
ade$^-$	1.4	2.4

For (B),

	arg$^+$	*arg*$^-$
his$^+$	1.3	2.3
his$^-$	3.4	1,2,3,4

segregation of the new marker in relation to that of nonselected markers of known map locations.

15.4.1 Establishing a Genetic Map in a New Streptomycete by the "Four-on-Four" Procedure

From a starting strain, two strains are derived that differ by four characters (commonly by virtue of two auxotrophic mutations in each strain, or two auxotrophs plus antibiotic resistance in one strain and one auxotroph in the other). The strains are grown together, and spores are harvested from the mixed culture. Recombinants are selected from unit volumes of the same spore suspension by plating on four selective media, each with a different nonparental combination of two growth factors. All, or a sample, of the colonies from each medium are classified with respect to the two nonselected markers on that medium, giving four possible genotypes on each medium. The number of each genotype per unit volume is thus obtained. The data (from reference 6) are handled as in Table 1 to give an estimate of each of the seven pairs of complementary recombinant genotypes.

Next, we ask whether the two pairs of unselected alleles on each medium behave independently of each other. Thus, on the medium containing arginine and adenine, all progeny must be *his*$^+$ and *lys*$^+$ to grow (the selected alleles), but they may be *arg*$^+$ or *arg*$^-$ and *ade*$^+$ or *ade*$^-$ (the unselected alleles). We can therefore draw up 2 × 2 tables (Table 2) of the frequency of each of the four possible combinations of unselected alleles for each of the four media, and then ask whether the ratio of, for example, *arg*$^+$ to *arg*$^-$ is the same among *ade*$^+$ as among *ade*$^-$ colonies. (A simple statistical test [χ^2] may be applied here to estimate the probability that this is not true. This test requires that the actual numbers scored in the samples be used, rather than the frequencies of genotypes per unit volume, if a valid estimate of the probability of independence is to be obtained.) We see in this example that the ratio is the same. The *arg* and *ade* genes are therefore said to be segregating independently, and this implies that they are separated on the linkage map by the *his* and *lys* genes. (For further explanation, see below.) However, on the medium containing arginine and histidine, *arg* and *his* do not segregate independently, and are therefore adjacent on the linkage map. The rarest class (*arg*$^-$ *his*$^-$) is the one that requires a quadruple crossover for its formation, whereas the other two arise by double crossovers. By combining the information from all four 2 × 2 tables about whether or not genes are adjacent, their order on the linkage map can be deduced.

The remaining step is to use the average frequencies of pairs of complementary genotypes (from the rightmost column in Table 1) to give the relative recombination frequencies between pairs of markers (Table 3). For example, recombination between *ade-2* and *arg-1* is required to form any recombinant in which *arg*$^+$ is combined with *ade*$^+$ or *arg*$^-$ with *ade*$^-$, i.e., the class pairs 2, 5, 6, and 7 from Table 1. In this example, the interval *arg-1–his-1* is clearly shorter than the others, since recombination frequency is related to map distance. The fact that the intervals *arg-1–lys-2* and *his-1–lys-2* are shorter than the remaining three intervals which all involve *ade-2* suggests that *ade-2* may be farther from the other three markers than they are from each other, but this conclusion would require further confirmation.

Note that the new double mutant combinations (class pairs 7 and 8) arising in the cross may be used in two further crosses with each other to strengthen the conclusions on linkage derived from the first cross; a good example of this is found in a paper on *S. glaucescens* (2).

Test of independent segregation

How are dependence and independence of two unselected pairs of alleles related to their locations on the linkage map? The important point is that, since the *Streptomyces* chromosome is circular, an even number of crossovers is required to generate a viable haploid recombinant chromosome by recombination between a chromosome (or a fragment of it) from the

TABLE 3. Calculation of the relative recombination frequencies in each map interval[a]

	Recombination frequencies in map interval:					
	ade/arg	*ade/his*	*ade/lys*	*arg/his*	*arg/lys*	*his/lys*
Components	30	27	30	30	92	30
	298	298	92	27	27	27
	27	100	298	27	100	92
	100	3	3	3	3	100
Total	455	428	423	87	222	249

[a] The four components, which are added to give the total recombination frequency in each map interval, come from Table 1, column 7.

donor strain and the chromosome of the recipient. Assuming the linkage arrangement shown in Fig. 1, the observed dependent and independent segregations can easily be explained.

Triangles in Fig. 1 indicate selected alleles. Circles represent the chromosomes of the two strains (zygotes usually contain a complete chromosome from only one strain and a fragment from the other, but this does not affect the argument). The numbers identify intervals in which crossing over can occur to generate recombinants. In Fig. 1A, all four recombinant classes can be generated by a single crossover in each of the two arcs separating the selected markers, that is, in interval 1 or 2 and in interval 3 or 4. Therefore the pairs of alleles should segregate independently because the ratio of probabilities 1,3 to 2,3 should equal the ratio of 1,4 to 2,4. In Fig. 1B the situation is different, because *arg*⁻ *his*⁻ recombinants require multiple crossing over in their formation and so the pairs of alleles will not segregate independently. Hence, departure from independent segregation indicates that the two genes are not separated by the selection points; consideration of which recombinant class is the rarest allows the deduction of the order of these genes in relation to the selection points.

Note that, although it may be useful to calculate χ^2 values as an aid to recognizing independent segregation or lack of it, these values should not be followed too slavishly. Often, viability disadvantages of certain genotypes, or a tendency for the cross to be polarized (with the complete chromosome coming preferentially from one parent), causes a segregation to depart from independence even when the markers are not adjacent. One should be prepared to make allowances for such imperfections in the data. However, if

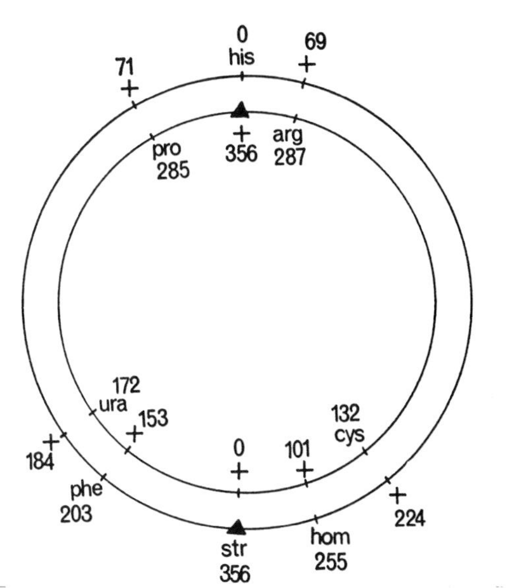

FIG. 2. Arrangements of markers, and allele ratios at each locus, in an eight-factor cross. The experiment consisted of selecting recombinants inheriting the two alleles indicated by triangles and classifying them with respect to alleles at the other six loci. Numbers indicate the frequency of each allele in the classified sample. The numbers of recombinants with each genotype are in Table 4.

TABLE 4. Analysis of the data of Fig. 2 to choose between two alternative positions for the locus *hom* (either between *str* and *cys* or between *str* and *phe*)[a]

Genotype in respect of left-hand arc	No. of progeny having genotype in respect of right-hand arc:					
	hom	*arg hom*	*arg cys hom*	*arg cys*	*arg*	*cys*
ura pro	13	49	12	70	4*	0
phe ura pro	3	7	6	6†	0	0
phe pro	16	68	14	14†	0	1‡
phe	34	28	3	3†	0	0
pro	0	0	0	2‡	0	0
ura	1‡	0	0	0	0	1‡
—	0	1‡	0	0	0	0

[a] All progeny arise by the minimum number of two crossovers, one in each arc between the selected markers, except for the following: * these 4 progeny require multiple crossovers if *hom* lies between *str* and *cys*; † these 23 progeny require multiple crossovers if *hom* lies between *str* and *phe*; ‡ these 6 progeny require multiple crossovers on either hypothesis. The order *str-hom-cys* results in a minimum number of crossover progeny (10 instead of 29) and is therefore chosen.

marked distortions of the data are apparent, revealed by gross differences between the frequency of the same genotype recovered on different media or between the frequencies of members of pairs of complementary genotypes (class pairs 7 and 8), it will probably be impossible to deduce a meaningful linkage arrangement of the markers.

15.4.2 Mapping a New Marker by Single Selection

Once a number of markers have been located on the circular linkage map by the four-on-four procedure, further markers can be assigned to a particular interval by a simple procedure involving plating the products of a cross on a single selective medium (8). The parents are chosen to differ in several markers in addition to a new one. Selection is best made for two approximately diametrically situated markers, and the resulting recombinants are classified in respect to the nonselected markers, including the new gene. From the resulting genotype frequencies, the position of the new marker is deduced by a two-stage process.

The first stage consists of considering the percentage frequencies of the two alleles (the "allele ratio") at each locus. The allele ratios at the selected loci must be 100:0 and 0:100; the allele ratios at nonselected loci take intermediate values, falling on a continuous gradient in each of the two arcs. It follows that any new locus can be assigned two alternative positions by virtue of its allele ratio, one in each arc. Thus, in Fig. 2, the marker *hom* could lie between *str* and *cys* (as shown) or between *str* and *phe* (Fig. 2 and Table 4).

The second stage in assignment is a choice between these two alternative map positions. This choice derives from the frequencies of particular genotypes of recombinants, and it depends, as in all genetic mapping arguments, on minimizing the total number of crossovers required to explain the observed data, bearing in mind the particular constraints of circular linkage: that the simplest pattern of recombinant formation requires two crossovers, one in each arc.

The consequences, in terms of crossovers, of these two alternatives for the *hom* locus are analyzed in Table 4. It is apparent that the position between *str* and *cys* is very much more economical in terms of crossovers (Table 4), and it is therefore chosen.

15.4.3 Plate Crosses

The "plate-crossing" technique is useful when there are several or many strains to be crossed with a common parent to assay the frequency of chromosomal recombination. It is usually a two-stage procedure. First, a master plate carrying the set of strains to be tested is replicated to a lawn of spores of the common parent on CM or R2YE medium to form the crosses. When the crossing plate has sporulated well, it is replicated to one or more selective media to recover recombinants. This procedure may be extended through further rounds of replication, since *Streptomyces* replication gives very accurate "prints." This opens the possibility of rapid mapping by plate crossing, such that the recombinants initially selected may be crudely assessed for the frequency among them of a second marker. This procedure is particularly effective when the second marker is very closely linked to one of the initially selected markers. Thus, rifampin-resistant mutations were classified into those which were or were not closely linked to *strA* (4), and glycerol nonutilization mutations closely linked to *argA* were identified (17).

15.4.4 Mapping Closely Linked Markers

Mapping of closely linked markers in *Streptomyces* spp. is not very efficient. A good system of generalized transduction would be very useful for this purpose and is being developed in *Streptomyces venezuelae* (18). Reliance has to be placed on matings made in the normal way, with selection usually for recombination between selectable markers as closely linked as possible to, and on either side of, the mutations to be mapped (see, for example, references 3, 7, and 15).

15.5 MEDIA

Minimal medium (MM) (8)

Agar	10	g
L-Asparagine	0.5	g
K_2HPO_4	0.5	g
$MgSO_4 \cdot 7H_2O$	0.2	g
$FeSO_4 \cdot 7H_2O$	0.01	g
Glucose (added after autoclaving)	10	g
Distilled water	1,000	ml

Place 2 g of agar powder each in 250-ml Erlenmeyer flasks. Dissolve the other ingredients (except glucose) in the distilled water, adjust to pH 7.0 to 7.2 with NaOH, and pour 200 ml into each flask. Close the flasks and autoclave (121°C, 15 min). At the time of use, remelt the medium and add 4 ml of a 50% solution of glucose to each flask.

Supplements. See Table 5. Stock solutions (or suspensions in the case of poorly soluble compounds) are made in water and sterilized by autoclaving (or filtration in the case of the antibiotics).

TABLE 5. Supplements for minimal medium

Compound	Stock soln (mg/ml)	Working concn (μg/ml)
Histidine	10	50
Other amino acids	7.5	37
Adenine, guanine, thymine, uracil	1.5	7.5
Vitamins	0.1	0.5
Streptomycin (sulfate)	10	10[a]
Spectinomycin	50	50[b]

[a] 50 μg/ml in CM or R2YE.
[b] 100 μg/ml in CM or R2YE.

Complete medium (CM) (8)

Agar	10	g
K_2HPO_4	5	g
NaCl	0.5	g
$MgSO_4 \cdot 7H_2O$	0.5	g
Bacto-Peptone (Difco)	2	g
Yeast extract (Difco)	1	g
Casamino Acids (Difco)	1.5	g
*L-Histidine	50	mg
*L-Proline	50	mg
**Yeast nucleic acid hydrolysate	5	ml
***Vitamin solution	1	ml
Glucose	25	g
Distilled water	1,000	ml

* If histidine or proline auxotrophs are not in use, these amino acids need not be added; if tryptophan or tyrosine auxotrophs are in use, add 50 mg of L-tryptophan or L-tyrosine.

** Boil 2 g of yeast nucleic acid (e.g., "ribonucleic acid, from yeast," BDH) in 15 ml of 1 N HCl for 10 min. Boil 2 g of yeast nucleic acid in 15 ml of 1 N NaOH for 10 min. Mix the two solutions, adjust to pH 6, filter while hot, and make up to 40 ml with water. Do not sterilize.

*** Riboflavin, 100 mg; nicotinamide, 100 mg; *p*-aminobenzoic acid, 10 mg; pyridoxine hydrochloride, 50 mg; thiamine hydrochloride, 50 mg; biotin, 20 mg; water, 100 ml. Autoclave (115°C, 10 min).

Place 2 g of agar powder each in 250-ml Erlenmeyer flasks. Dissolve all the other ingredients (except vitamin solution, but including glucose) in the distilled water; adjust to pH 7.2 with 1 N HCl. Add vitamin solution. Pour 200 ml into each flask, close the flasks, and autoclave (121°C, 15 min).

R2YE medium (17)

Make up the following basic solution:

Sucrose	103	g
K_2SO_4	0.25	g
$MgCl_2 \cdot 6H_2O$	10.12	g
Glucose	10	g
Casamino Acids (Difco)	0.1	g
*Trace element solution	2	ml
Yeast extract (Difco)	5	g
**TES buffer	5.73	g
Distilled water	to 1,000	ml

* Trace element solution (per liter):

$ZnCl_2$	40 mg
$FeCl_3 \cdot 6H_2O$	200 mg
$CuCl_2 \cdot 2H_2O$	10 mg
$MnCl_2 \cdot 4H_2O$	10 mg
$Na_2B_4O_7 \cdot 10H_2O$	10 mg
$(NH_4)_6Mo_7O_{24} \cdot 4H_2O$	10 mg

** *N*-Tris(hydroxymethyl)methyl-2-aminoethanesulfonic acid.

Place 2.2 g of agar (Difco) each in 250-ml Erlenmeyer flasks and add 100 ml of the basic solution. Close the flasks and autoclave (121°C, 15 min).

At the time of use, remelt the medium and add to each flask the following sterile solutions:

KH_2PO_4 (0.5%)	1 ml
$CaCl_2 \cdot 2H_2O$ (5 M)	0.4 ml
L-Proline (20%)	1.5 ml
NaOH (1 N) (does not need sterilization)	0.7 ml

15.6 STRAINS

Two strains of *S. lividans* 66 can be used for a model cross to illustrate the four-on-four procedure, namely, 3078 (*his-2 leu-2 spc-1* pIJ303) (pIJ303 is a derivative of the high-copy-number, broad-host-range plasmid pIJ101 [14]) and TK64 (*pro-2*). Recombination can be assessed by plating on the four selective media (histidine, proline, and spectinomycin; leucine, proline, and spectinomycin; histidine; or leucine) and plasmid transfer by comparing colony counts on proline medium with and without thiostrepton (pIJ303 carries a cloned gene, *tsr*, for thiostrepton resistance [19]). (Thiostrepton, obtainable from E. R. Squibb and Sons, New Brunswick, N.J., is dissolved at 50 mg/ml in dimethyl sulfoxide to give a stock solution and is used at a final concentration of 50 μg/ml.) The analysis of crosses involving these four markers is described in reference 11.

For a simple six-factor cross, the following pairs of *S. coelicolor* A3(2) strains are suitable: M110 (*hisA1 uraA1 strA1* SCP2*) × M124 (*proA1 argA1 cysD18*), or M130 (*hisA1 uraA1 strA1*) × M107 (*proA1 argA1 cysD18* SCP2*). (SCP2* makes a good sex plasmid for general linkage mapping in *S. coelicolor*.) The most suitable primary selection for recombinants is on a medium containing arginine, cystine, proline, uracil, and streptomycin (selecting $hisA^+$ and *strA1*). Recombinants are then picked and classified with respect to the four nonselected markers (*argA1*, *cysD18*, *proA1*, and *uraA1*), giving a total of 16 possible genotypes of recombinants. The map positions of the six markers are shown in Fig. 2.

15.7 LITERATURE CITED

1. **Alačević, M.** 1976. Recent advances in *Streptomyces rimosus* genetics, p. 513–519. *In* K. D. Macdonald (ed.), Second International Symposium on the Genetics of Industrial Microorganisms. Academic Press, London.
2. **Baumann, R., R. Hütter, and D. A. Hopwood.** 1974. Genetic analysis in a melanin-producing streptomycete, *Streptomyces glaucescens*. J. Gen. Microbiol. **81**:463–474.
3. **Chater, K. F.** 1972. A morphological and genetic mapping study of white colony mutants of *Streptomyces coelicolor*. J. Gen. Microbiol. **72**:9–28.
4. **Chater, K. F.** 1974. Rifampicin-resistant mutants of *Streptomyces coelicolor* A3(2). J. Gen. Microbiol. **80:** 277–290.
5. **Coats, J. H.** 1976. Genetic recombination in *Streptomyces achromogenes* var. *rubradirans*, p. 521–530. *In* K. D. Macdonald (ed.), Second International Symposium on the Genetics of Industrial Microorganisms. Academic Press, London.
6. **Friend, E. J., and D. A. Hopwood.** 1971. The linkage map of *Streptomyces rimosus*. J. Gen. Microbiol. **68**:187–197.
7. **Harold, R. J., and D. A. Hopwood.** 1970. Ultraviolet-sensitive mutants of *Streptomyces coelicolor*. II. Genetics. Mutat. Res. **10**:439–448.
8. **Hopwood, D. A.** 1967. Genetic analysis and genome structure in *Streptomyces coelicolor*. Bacteriol. Rev. **31**:373–403.
9. **Hopwood, D. A.** 1972. Genetic analysis in microorganisms, p. 29–158. *In* J. R. Norris and B. W. Ribbons (ed.), Methods in microbiology, vol. 7B. Academic Press, London.
10. **Hopwood, D. A., M. J. Bibb, K. F. Chater, T. Kieser, C. J. Bruton, H. M. Kieser, D. J. Lydiate, C. P. Smith, J. M. Ward, and H. Schrempf.** 1985. Genetic manipulation in Streptomyces: a laboratory manual. John Innes Foundation, Norwich, U.K.
11. **Hopwood, D. A., T. Kieser, H. M. Wright, and M. J. Bibb.** 1983. Plasmids, recombination and chromosome mapping in *Streptomyces lividans* 66. J. Gen. Microbiol. **129**:2257–2269.
12. **Hopwood, D. A., D. J. Lydiate, F. Malpartida, and H. M. Wright.** 1984. Conjugative plasmids in *Streptomyces*, p. 615–634. *In* D. Helinski, S. N. Cohen, D. B. Clewell, D. A. Jackson, A. Hollaender, L. Hager, S. Kaplan, J. Konisky, and C. M. Wilson (ed.), Plasmids in bacteria. Plenum Press, New York.
13. **Hranueli, D., J. Pigac, T. Smokvina, and M. Alačević.** 1983. Genetic interactions in *Streptomyces rimosus* mediated by conjugation and by protoplast fusion. J. Gen. Microbiol. **129**:1415–1422.
14. **Kieser, T., D. A. Hopwood, H. M. Wright, and C. J. Thompson.** 1982. pIJ101, a multi-copy broad host-range *Streptomyces* plasmid: functional analysis and development of DNA cloning vectors. Mol. Gen. Genet. **185**:223–238.
15. **Rudd, B. A. M., and D. A. Hopwood.** 1979. Genetics of actinorhodin biosynthesis by *Streptomyces coelicolor* A3(2). J. Gen. Microbiol. **114**:35–43.
16. **Schupp, T., R. Hütter, and D. A. Hopwood.** 1975. Genetic recombination in *Nocardia mediterranei*. J. Bacteriol. **121**:128–136.
17. **Seno, E. T., C. J. Bruton, and K. F. Chater.** 1983. The glycerol utilization operon of *Streptomyces coelicolor*: genetic mapping of *gyl* mutations and the analysis of cloned *gyl* DNA. Mol. Gen. Genet. **193**:119–128.
18. **Stuttard, C.** 1979. Transduction of auxotrophic markers in a chloramphenicol-producing strain of *Streptomyces*. J. Gen. Microbiol. **110**:479–482.
19. **Thompson, C. J., J. M. Ward, and D. A. Hopwood.** 1980. DNA cloning in *Streptomyces*: resistance genes from antibiotic-producing species. Nature (London) **286:** 525–527.

Chapter 16

Gene Cloning in *Streptomyces* spp.

DAVID A. HOPWOOD

Since the first successful DNA cloning experiments with *Streptomyces* spp. were reported in 1980 (3, 26, 28), in vitro genetic manipulation with streptomycetes has developed rapidly. Some members of this genus (notably *Streptomyces lividans*) are now among the handful of microorganisms, which also includes *Escherichia coli*, *Bacillus subtilis*, and *Saccharomyces cerevisiae*, in which gene cloning is developed to the point that the direct isolation and analysis of most genes have become routine. In 1982, it was possible to provide a fairly comprehensive description of the techniques then available for gene cloning in *Streptomyces* spp. in a chapter of reasonable length (6). A much lengthier treatment is now needed, such as that in the new practical laboratory manual *Genetic Manipulation of Streptomyces*, which runs to some 350 pages (12). The book describes in experimental detail a wide range of in vitro and in vivo procedures, including the use of plasmid and bacteriophage cloning vectors for the isolation and analysis of DNA and RNA from *Streptomyces* spp. Thus, it would be impossible to abridge all of the protocols required for gene cloning in *Streptomyces* spp. so as to fit them into the present chapter without losing much essential information; the detailed manual should be consulted instead. Here I attempt to provide a few pointers on the available host strains and vectors and their use for different cloning strategies, followed by an outline of a simple shotgun cloning experiment using a high-copy-number plasmid vector.

16.1 HOST STRAINS

In dealing with streptomycetes, the genetic engineer confronts a problem different from that facing the cloner working with *E. coli*, *B. subtilis*, or *S. cerevisiae*. Because of the need to work with a wide range of antibiotics and other industrial streptomycete products, gene cloning in *Streptomyces* spp. potentially concerns dozens of diverse species and strains. Not surprisingly, techniques for the introduction of DNA, as well as the behavior of vectors, vary from strain to strain. However, many of the vectors already developed have a comparatively broad host range and can be used without change in a variety of hosts, and techniques for handling the hosts themselves can often be devised without too much difficulty by modifying procedures already worked out for "standard" strains.

Cloning into a specific host is required for the isolation of genes by complementation of host mutations or by mutational cloning using bacteriophage vectors (2, 7, 11, 15); it will in any case usually be required eventually if the objective is to manipulate a strain to generate a new product. However, for isolating genes by other strategies—such as by the detection of individual gene products by genetic, biochemical, or immunological tests, or of genes by nucleic acid probe techniques—as well as for much analysis of gene structure and function, the cloning and further manipulation of the DNA can be done in a standard host in which techniques and vectors have already been optimized. Currently, the most developed freely available host for this purpose is *S. lividans* 66. Particularly useful are strains of this organism that have been cured of its two known plasmids, SLP2 and SLP3; two such strains are TK64 and 3104 (14), which each carry useful chromosomal markers (*pro-2* and either *str-6* or *spc-1*, for proline auxotrophy and resistance to streptomycin and spectinomycin, respectively). When using such a strain we have a host that (i) grows and sporulates well on solid media; (ii) grows well in liquid media (without sporulation) and is then easily converted to protoplasts which regenerate readily and can be transformed very effectively by all of a wide variety of plasmids (13) and transfected by

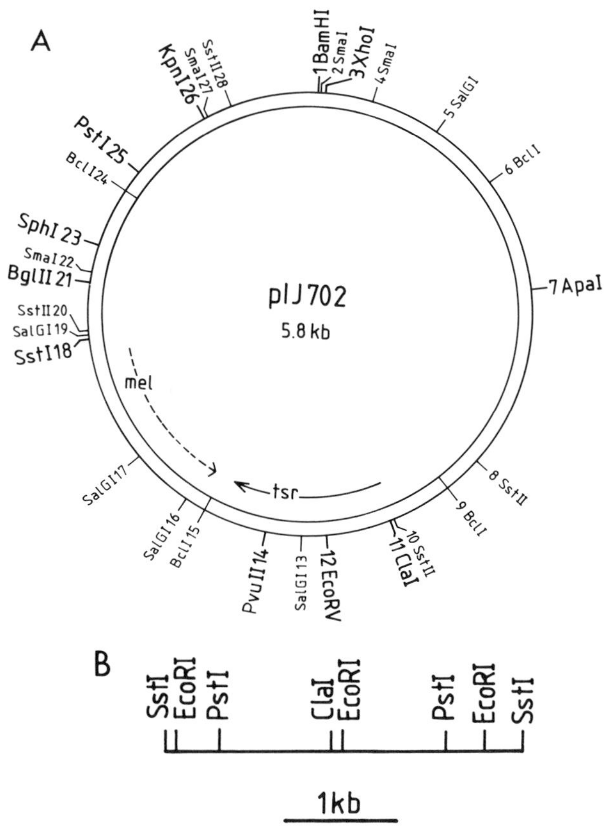

FIG. 1. (A) Restriction map of pIJ702 (from reference 17 and unpublished results from the John Innes Institute). (B) Restriction map of the *sph* fragment from *S. glaucescens* ETH22794 (from reference 10).

bacteriophage ϕC31 vectors (4); (iii) lacks any known replicons that might interfere with replication or analysis of plasmids or phages used as cloning vectors; (iv) shows no obvious restriction of DNA introduced into it from a wide range of donor organisms, including many other streptomycetes (it is of course possible that restriction of certain foreign DNAs by *S. lividans* will be found in the future, but there appears to be no good evidence for restriction to date); and (v) carries two of a set of chromosomal markers that have been incorporated into a linkage map of 10 loci (13), providing a useful background for the study of extrachromosomal genes.

16.2 DNA UPTAKE

The only practicable route for the introduction of isolated DNA into streptomycetes is by the treatment of protoplasts with the DNA in the presence of calcium and polyethylene glycol; no efficient system of natural competence or artificial uptake of DNA by whole cells has been described. For plasmid DNA, high frequencies of transformation can be achieved: up to 10^7/μg for plasmids of 10 to 20 kilobases (kb), falling only in proportion to the decrease in the number of molecules per microgram for larger plasmids, at least for *S. lividans* 66 and *Streptomyces coelicolor* A3(2). Over 10^7 transfectants per μg of the DNA of ϕC31-derived phages can be obtained by adding positively charged DNA-free liposomes to the transfection mixture (25) (without liposomes the frequency is 2 orders of magnitude lower). Transformation of protoplasts by chromosomal DNA occurs at very low frequencies unless the DNA is incorporated into liposomes which are then fused with protoplasts of the recipient (21); however, no use of this finding in relation to gene cloning has yet been reported.

16.3 VECTORS

16.3.1 Plasmid Vectors

High-copy-number vectors

Many naturally occurring *Streptomyces* plasmids are now known that have been, or could easily be, developed into cloning vectors (13). Most are of high copy number and broad host range within the genus *Streptomyces*. Of these high-copy-number plasmids, derivatives of pIJ101 (19) have been used far more than the others for gene cloning, a particularly popular vector being pIJ702 (Fig. 1A; 17). This vector is nonconjugative and non-pock forming, thus avoiding superinfection of clones by unaltered vector plasmids, and carries both the *tsr* gene (for thiostrepton resistance) of *Streptomyces azureus*, for use in selecting transformants, and the *mel* sequences (for melanin synthesis via tyrosinase production) of *Streptomyces antibioticus*. The *mel* region contains three convenient cloning sites (for *Bgl*II, *Sph*I, and *Sst*I) into which insertion of DNA results in a Mel^- phenotype (white instead of black colonies on tyrosine-containing indicator plates). Most, but not all, *Streptomyces* strains are capable of melanin production when transformed by pIJ702. Several genes have been isolated or analyzed by cloning into pIJ702 (13).

Low-copy-number vectors

Other things being equal, a high copy number of cloned DNA fragments will lead to a higher specific activity of enzymes coded by them, making recognition of the clones and study of the gene products, as well as of the DNA itself, simpler; however, attempts to clone genes on high-copy-number vectors do not always work. As in other bacteria, there are some documented cases, and other probable examples, of failure to isolate desired clones, caused by deleterious effects on the host of high-copy-number cloning: either the clones do not appear at all, or they appear in deleted or rearranged form. In these circumstances (as well as for physiological studies of the cloned DNA), low-copy-number vectors are desirable. Vectors derived from SLP1.2 have a narrow host range but can be used very effectively in *S. lividans* 66 and some other strains for cloning at a copy number of around 5. pIJ61 (27), the most widely used SLP1 derivative (13), is a conjugative, pock-forming plasmid with two markers, *tsr* for selection and *aph* (for neomycin resistance, from *Streptomyces fradiae*) for insertional inactivation by cloning at unique *Bam*HI or *Pst*I sites. Several other unique sites on the plasmid can also be used, without the benefit of insertional inactivation. For cloning at a copy number of only one or a little

TABLE 1. A selection of ϕC31-based phage cloning vectors[a]

Cloning mode	Phage	Markers	Host and cloning enzymes
Screening by plaque hybridization	KC518	c^- *att*$^-$ *vph tsr*	Inserts of up to 6 kb in *Pst*I or *Pvu*II sites, 6.5 kb in the *Sst*I sites, or 2 to 8 kb (positively selected) in the *Bam*HI sites
Screening by complementation of mutants or acquisition of new capabilities	KC121	c^+ *att*$^+$	Host need not be a lysogen or carry DNA homologous with the insert; inserts up to 4.5 kb in the *Bam*HI site
	KC518	c^- *att*$^-$ *vph tsr*	Host must be a lysogen; see above for inserts
	KC515, KC516	c^+ *att*$^-$ *vph tsr*	Host need not be a lysogen but then must carry DNA homologous with the insert; inserts up to 4 kb (4.5 kb with *Sst*I) in *Pst*I, *Pvu*II, *Sst*I, *Bam*HI, *Bgl*II, or *Xho*I sites
Mutational cloning	KC515, KC516	c^+ *att*$^-$ *vph tsr*	Host must be a nonlysogen and must carry DNA homologous with the insert; see above for inserts

[a] See references 7 and 24 for further details.

higher, vectors derived from SCP2* have proved very useful (20a) and have the advantage of a fairly broad host range. Several are available, all carrying *tsr* for selection and some carrying a second marker (*mel*, or *hyg* for hygromycin resistance, from *Streptomyces hygroscopicus*). All the vectors (like pIJ702 and pIJ61) have a unique site for either *Bam*HI or *Bgl*II, thereby allowing the cloning of fragments of DNA generated by partial digestion with *Mbo*I or *Sau*3A, as well as by digestion with *Bam*HI, *Bgl*II, *Bcl*I, or *Xho*II. pIJ922 has been well authenticated as a vector for the stable cloning of large DNA fragments, such as one of 32.5 kb in a clone (pIJ2303) carrying the entire set of *S. coelicolor* genes for the biosynthesis of the antibiotic actinorhodin (22). The extended host range of these vectors was invaluable in the transfer of the actinorhodin genes stably to other species in the production of hybrid antibiotics (14a). Other SCP2* derivatives, yet to be used extensively, are pIJ940 and pIJ941, carrying *hyg*, and pIJ943, carrying *mel*. These vectors are 19 to 25 kb in size. Much smaller SCP2*-derived vectors have been made (20), but they lack a function that controls stable inheritance of the plasmid (3) and are lost at high frequency.

Shuttle vectors

Shuttle plasmids for *E. coli* and *Streptomyces* spp. are quite easy to construct (e.g., references 19, 20, 20a, 23, and 28). They have advantages for certain types of studies, for example, mutagenesis of cloned DNA by Tn*5* (9) or the isolation of suppressor or suppressible mutations. However, in view of the high efficiency of direct transformation of *Streptomyces* strains and the possibility of constructing good libraries carrying large DNA fragments via this route, shuttle vectors are not fundamental to gene cloning in streptomycetes as they are in some other organisms. Shuttle cosmids could be useful if they allowed libraries of large DNA fragments to be constructed in *E. coli* by lambda packaging, followed by transfer back to *Streptomyces* for functional testing. Attempts have been made in several laboratories to construct such vectors. In most cases extensive deletions of the vector or insert DNA have been observed in *Streptomyces* spp. after transfer from *E. coli*, but promising results are now being talked about and useful vectors of this class will probably be developed soon.

16.3.2 Phage Vectors

The *Streptomyces* bacteriophage that has been developed into the most versatile and extensively used series of vectors is the broad-host-range temperate phage ϕC31. The phage resembles lambda in many respects, with cohesive genome ends, a *c* region that controls the lysogenic response, and an attachment region (*attP*) that allows crossing over with a chromosomal site (*attC*) to produce lysogens. Through the extensive work of K. F. Chater and his collaborators, ϕC31 has been developed into a whole series of vectors suitable for cloning DNA in a variety of modes (4). They include: (i) c^- *att*$^-$ vectors (such as KC518) for making libraries to be screened by DNA probing by the Grunstein-Hogness technique; (ii) c^+ *att*$^+$ phages (such as KC121) for use when the clones are to be recognized by conferring an "added" phenotype on the host or by the complementation of specific mutations in nonlysogenic hosts (c^- *att*$^-$ phages, because of their greater cloning capacity, are often used instead to clone into a lysogen, or c^+ *att*$^-$ phages may be used provided the recipient strain contains DNA homologous with the cloned DNA to provide an artificial attachment site); and (iii) c^+ *att*$^-$ vectors (such as KC515 and KC516) for the recognition or analysis of clones by mutational cloning (5). Most of the vectors carry *tsr*, *vph* (for viomycin resistance, from *Streptomyces vinaceus*), or both, so that lysogens (and therefore potential clones) can be selected directly by virtue of the thiostrepton or viomycin resistance conferred on the host by the vector. Several of the vectors can shuttle between *E. coli* and *Streptomyces* spp. because they carry an intact pBR322 replicon. The properties of a few of the most used vectors are listed in Table 1.

16.4 CLONE RECOGNITION

Transformants produced in a cloning experiment with a plasmid vector are usually recovered by delayed selection. The mixture containing transformed protoplasts is plated on the surface of regeneration plates and incubated for a period to allow antibiotic resistance to be expressed. The plates are then overlaid with soft agar containing the antibiotic (usually thiostrepton) and reincubated to allow transformant colonies to grow through the agar and sporulate. Tyrosine is incorporated in the overlay if insertion into the *mel* region of the vector has been attempted so that the Mel^+ versus Mel^- phenotype is diagnosed directly; other insertional inactivation markers can be tested by replica plating of the transformant colonies. The required clones are then recognized by some suitable test applied to the colonies. An alternative procedure, recommended when using pock-forming plasmid vectors, is to allow complete regeneration to occur nonselectively and then to recognize transformants (or clones directly) by replica plating to media containing thiostrepton (or additional plates containing other diagnostic agents); this procedure works well for pock-forming vectors because they give rise to patches of plasmid-containing mycelium (the pocks) on the regeneration plates, but it is inefficient for non-pock-forming vectors because many of the plasmid-containing colonies are overgrown on the regeneration plates by the untransformed background growth. Sometimes, individual transformed colonies can be screened for clones by a color, biochemical, or other test; otherwise, populations of transformants can be pooled and clones can be isolated by rounds of sib-selection (e.g., reference 16).

In phage cloning, the primary step is usually to transfect *S. lividans* protoplasts and allow them to regenerate in a lawn of *S. lividans* spores to yield a population of plaques. Unless screening for a DNA sequence by probing is intended, the phages from the plaques are then transferred by replica plating to a lawn of spores of an *S. lividans* or other strain, leading to infection and lysogenization of that strain (note that the second strain need not be one in which protoplast technology has been optimized, or even attempted). Subsequent phenotypic tests are used to detect the desired clones.

16.5 FURTHER ANALYSIS

One of the first tests to apply to a presumptive clone is to reintroduce the purified plasmid or phage DNA (either naked or as virions) into a recipient strain to prove that the desired phenotype is actually determined by the cloned DNA. A second important test is to prove the origin, in unaltered form, of the clone from the presumptive donor by Southern hybridization. Further stages in analysis will involve restriction mapping of the cloned DNA, subcloning to define the limits of particular genes, studies of the transcribed RNA, etc.

16.6 A MODEL SHOTGUN CLONING EXPERIMENT

A straightforward model shotgun cloning experiment suitable to test one's ability to handle *Streptomyces* spp. would involve the use of a high-copy-number plasmid vector, such as pIJ702, to isolate an easily selected gene, such as one coding for resistance to an antibiotic. Of several antibiotic resistance genes that have been isolated (13), only a few come from freely available strains and are known to be isolable on unique restriction fragments suitable for simple insertion into one of the three main cloning sites (for *Bgl*II, *Sph*I, or *Sst*I) of pIJ702. Of these, the *tsr* gene of *S. azureus*, isolable on a 1,055-base-pair *Bcl*I fragment (1), is already carried by pIJ702 and most other developed *Streptomyces* vectors. The *vph* gene of *S. vinaceus*, which lies on a 5.8-kb *Bam*HI fragment (29), is not ideal because viomycin is difficult to obtain. The *cat* gene of *Streptomyces acrimycini* can be cloned on a 1.7-kb *Bcl*I fragment (J. A. Gil, H. M. Kieser, and D. A. Hopwood, submitted for publication), but isolation of good-quality chromosomal DNA from this strain is somewhat difficult because of unusually high nuclease activity which can lead to extensive DNA degradation. A suitable gene for a model experiment is *sph* (for streptomycin resistance, via a streptomycin phosphotransferase) from *Streptomyces glaucescens* ETH22794, which lies on a 3.2-kb *Sst*I fragment (Fig. 1b, from reference 10). The steps needed to clone such a gene are outlined in this section; the experimental details for all the steps, as well as for the further analysis of the cloned DNA, can be found in the John Innes manual (12).

16.6.1 Isolation of Donor DNA

A variety of procedures are suitable for the isolation of a sample of genomic DNA from a streptomycete. They all involve the lysis of liquid-grown mycelium (which may be frozen indefinitely as a paste after growth and before lysis) by treatment with lysozyme followed by a detergent (e.g., Sarkosyl, sodium dodecyl sulfate, or "Kirby mix"). The DNA is then recovered in pure form by one of several alternative series of manipulations, usually starting with a deproteinization with phenol-chloroform. Several of the procedures do not require the use of an ultracentrifuge or other large equipment.

16.6.2 Isolation of Vector DNA

Again, several alternative procedures are available for isolating plasmid DNA from streptomycetes, in this case *S. lividans* TK150 carrying pIJ702. Especially since pIJ702 is a high-copy-number plasmid, large amounts of the DNA can be isolated by a rapid alkaline lysis procedure (18) which requires no large equipment such as an ultra-centrifuge. However, one of the other procedures may yield DNA of greater purity such as is required for good results in cloning experiments.

16.6.3 Restriction Endonuclease Digestion of Donor and Vector DNA

The *sph* gene of *S. glaucescens* ETH22794 was originally cloned on a 10-kb *Bam*HI fragment (10) in a low-copy-number SLP1.2-based vector (pIJ41). It was later subcloned on a 3.2-kb *Sst*I fragment into a high-copy-number pIJ101 derivative (pIJ364). The simplest way to clone the gene is therefore to cleave a sample of *S. glaucescens* genomic DNA to completion with *Sst*I (*Sac*I). A sample of pIJ702 is cut with the same enzyme to convert it to full-length linear molecules. Small aliquots of the two DNA samples, before and after endonuclease digestion, are then analyzed by electrophoresis on an agarose gel (usually a minigel) to check for the completeness of digestion (this is particularly important for the vector since one wants to avoid, as far as possible, the presence of any intact vector molecules which will contribute to the population of transformants without DNA inserts and may be preferentially taken up by protoplasts).

16.6.4 Treatment of Vector DNA with Alkaline Phosphatase

Calf intestine alkaline phosphatase may optionally be used to remove the 5′-terminal phosphate groups from the vector DNA to prevent its recircularization (or concatenation of more than one vector molecule) without the insertion of a segment of donor DNA.

16.6.5 Ligation

Theoretical consideration (8) of the conditions for ligation needed to maximize the yield of intermolecular recombinants (i.e., desired clones) in a ligation mixture must take into account the absolute DNA concentration and the relative concentrations of vector and donor DNA ends, allowing for the sizes of the molecules. For the present experiment, one would normally ligate 1 μg of pIJ702 DNA with 5 μg of *S. glaucescens* DNA at a concentration of 15 μg/ml. After ligation, a small sample of the DNA mixture is monitored by agarose gel electrophoresis, in comparison with cut and uncut starting DNAs, for the success of the ligation.

16.6.6 Preparation of Recipient Protoplasts

The recipient strain in this experiment should, of course, be *S. lividans* strain 3104 rather than TK64, which is already streptomycin resistant. Protoplasts of strain 3104 are prepared, using lysozyme in an appropriate buffer, from mycelium grown with shaking in baffled flasks for 36 to 40 h in liquid YEME medium. After preparation of the protoplasts, they may be stored for long periods by freezing in P buffer at −70°C. (It is convenient to freeze the protoplasts in samples suitable for individual experiments: ca. 4×10^9 protoplasts based on a hemacytometer count.) Since protoplasts of different batches may vary in their ability to regenerate and be transformed, it is often useful to test a sample of the protoplasts for transformation proficiency, using uncut plasmid alone, before using protoplasts from the same batch for a cloning experiment.

16.6.7 Transformation, Regeneration, and Clone Recognition

The ligation mixture described above is normally used to transform about 4×10^9 protoplasts, which are then spread on the surfaces of 10 plates of predried R2YE regeneration medium. After incubation for about 16 h (or longer, until a thin film of growth is visible on the plates), soft nutrient agar containing thiostrepton and tyrosine is poured over each plate, and the plates are incubated for a further few days. Up to a few thousand colonies should appear on each plate. The colonies are then examined for white or black phenotype. If alkaline phosphatase has been used under optimal conditions, almost all should be white; otherwise, the proportion will probably be in the 10 to 20% range. The colonies are then replica plated to plates containing streptomycin, whereupon several colonies should grow as putative clones carrying the *sph* gene.

16.6.8 Further Analysis

The putative *sph*-containing clones are now picked and purified by streaking for single colonies on thiostrepton-containing plates and replicated again to streptomycin to test for resistance. Cultures are prepared from the purified colonies, and plasmid DNA is extracted and used to retransform protoplasts of strain 3104. Assuming that streptomycin-resistant transformants are obtained (there should now be 100% correlation between resistance to thiostrepton and streptomycin), the DNA can then be analyzed by restriction endonuclease cleavage. Clones carrying a simple insertion of the *sph* gene into pIJ702 should yield an *Sst*I fragment of 3.2 kb in addition to the 5.8-kb vector band. Digestion with other enzymes or combinations of enzymes will reveal the orientation of the insert in the vector. Restriction maps of pIJ702 and the *sph* fragment of *S. glaucescens* are in Fig. 1.

16.7 LITERATURE CITED

1. **Bibb, M. J., M. J. Bibb, J. M. Ward, and S. N. Cohen.** 1985. Nucleotide sequences encoding and promoting expression of three antibiotic resistance genes indigenous to *Streptomyces*. Mol. Gen. Genet. **199:**26–36.
2. **Bibb, M. J., K. F. Chater, and D. A. Hopwood.** 1983. Developments in *Streptomyces* cloning, p. 53–82. *In* M. Inouye (ed.), Experimental manipulation of gene expression. Academic Press, Inc., New York.
3. **Bibb, M. J., J. L. Schottel, and S. N. Cohen.** 1980. A DNA cloning system for interspecies gene transfer in antibiotic-producing *Streptomyces*. Nature (London) **284:**526–531.
4. **Chater, K. F.** 1985. *Streptomyces* phages and their application to *Streptomyces* genetics. *In* S. W. Queener and L. E. Day (ed.), Antibiotic-producing *Streptomyces*. Academic Press, Inc., New York.
5. **Chater, K. F., and C. J. Bruton.** 1983. Mutational cloning in *Streptomyces* and the isolation of antibiotic production genes. Gene **26:**67–78.
6. **Chater, K. F., D. A. Hopwood, T. Kieser, and C. J. Thompson.** 1982. Gene cloning in *Streptomyces*. Curr. Top. Microbiol. Immunol. **96:**69–95.
7. **Chater, K. F., A. A. King, M. R. Rodicio, C. J. Bruton, S. H. Fisher, J. M. Piret, C. P. Smith, and S. G. Foster.** 1985. Cloning and analysis of *Streptomyces* DNA in ϕC31-derived vectors, p. 421–426. *In* L. Leive (ed.), Microbiology—1985. American Society for Microbiology, Washington, D.C.
8. **Dugaiczyk, A., H. W. Boyer, and H. M. Goodman.** 1975. Ligation of *Eco*RI endonuclease-generated DNA fragments into linear and circular structures. J. Mol. Biol. **96:**171–184.
9. **Gil, J. A., and D. A. Hopwood.** 1983. Cloning and expression of a *p*-aminobenzoic acid synthetase gene of the candicidin-producing *Streptomyces griseus*. Gene **25:**119–132.

10. **Hintermann, G., R. Crameri, M. Vögtli, and R. Hütter.** 1984. Streptomycin-sensitivity in *Streptomyces glaucescens* is due to deletions comprising the structural gene coding for a specific phosphotransferase. Mol. Gen. Genet. **196:**513–520.
11. **Hopwood, D. A., M. J. Bibb, C. J. Bruton, K. F. Chater, J. S. Feitelson, and J. A. Gil.** 1983. Cloning of *Streptomyces* genes for antibiotic production. Trends Biotechnol. **1:**42–48.
12. **Hopwood, D. A., M. J. Bibb, K. F. Chater, T. Kieser, C. J. Bruton, H. M. Kieser, D. J. Lydiate, C. P. Smith, J. M. Ward, and H. Schrempf.** 1985. Genetic manipulation in Streptomyces: a laboratory manual. John Innes Foundation, Norwich, U.K.
13. **Hopwood, D. A., T. Kieser, D. J. Lydiate, and M. J. Bibb.** 1985. *Streptomyces* plasmids: their biology and use as cloning vectors. *In* S. W. Queener and L. E. Day (ed.), Antibiotic-producing *Streptomyces*. Academic Press, Inc., New York.
14. **Hopwood, D. A., T. Kieser, H. M. Wright, and M. J. Bibb.** 1983. Plasmids, recombination and chromosome mapping in *Streptomyces lividans* 66. J. Gen. Microbiol. **129:**2257–2269.
14a. **Hopwood, D. A., F. Malpartida, H. M. Kieser, H. Ikeda, J. Duncan, I. Fujii, B. A. M. Rudd, H. G. Floss, and S. Ōmura.** 1985. Production of "hybrid" antibiotics by genetic engineering. Nature (London) **314:**642–644.
15. **Hopwood, D. A., F. Malpartida, H. M. Kieser, H. Ikeda, and S. Ōmura.** 1985. Cloning genes for antibiotic biosynthesis in *Streptomyces* spp: production of a hybrid antibiotic, p. 409–413. *In* L. Leive (ed.), Microbiology—1985. American Society for Microbiology, Washington, D.C.
16. **Jones, G. H., and D. A. Hopwood.** 1984. Molecular cloning and expression of the phenoxazinone synthase gene from *Streptomyces antibioticus*. J. Biol. Chem. **259:**14151–14157.
17. **Katz, E., C. J. Thompson, and D. A. Hopwood.** 1983. Cloning and expression of the tyrosinase gene from *Streptomyces antibioticus* in *Streptomyces lividans*. J. Gen. Microbiol. **129:**2703–2714.
18. **Kieser, T.** 1984. Factors affecting the isolation of CCC DNA from *Streptomyces lividans* and *Escherichia coli*. Plasmid **12:**19–36.
19. **Kieser, T., D. A. Hopwood, H. M. Wright, and C. J. Thompson.** 1982. pIJ101, a multi-copy broad host-range *Streptomyces* plasmid: functional analysis and development of DNA cloning vectors. Mol. Gen. Genet. **185:** 223–238.
20. **Larson, J. L., and C. L. Hershberger.** 1984. Shuttle vectors for cloning recombinant DNA in *Escherichia coli* and *Streptomyces griseofuscus* C581. J. Bacteriol. **157:** 314–317.
20a. **Lydiate, D. J., F. Malpartida, and D. A. Hopwood.** 1985. The *Streptomyces* plasmid SCP2*: its functional analysis and development into useful cloning vectors. Gene **35:**223–235.
21. **Makins, J. F., and G. Holt.** 1981. Liposome-mediated transformation of streptomycetes by chromosomal DNA. Nature (London) **293:**671–673.
22. **Malpartida, F., and D. A. Hopwood.** 1984. Molecular cloning of the whole biosynthetic pathway of a *Streptomyces* antibiotic and its expression in a heterologous host. Nature (London) **309:**462–464.
23. **Richardson, M. A., J. A. Mabe, N. E. Beerman, W. M. Nakatsukasa, and J. T. Fayermann.** 1982. Development of cloning vectors from the *Streptomyces* plasmid pFJ103. Gene **20:**451–457.
24. **Rodicio, M. R., C. J. Bruton, and K. F. Chater.** 1985. New derivatives of the temperate phage ϕC31 useful for the cloning and functional analysis of *Streptomyces* DNA. Gene **34:**283–292.
25. **Rodicio, M. R., and K. F. Chater.** 1982. Small DNA-free liposomes stimulate transfection of *Streptomyces* protoplasts. J. Bacteriol. **151:**1078–1085.
26. **Suarez, J. E., and K. F. Chater.** 1980. DNA cloning in *Streptomyces*: a bifunctional replicon comprising pBR322 inserted into a *Streptomyces* phage. Nature (London) **286:**527–529.
27. **Thompson, C. J., T. Kieser, J. M. Ward, and D. A. Hopwood.** 1982. Physical analysis of antibiotic-resistance genes from *Streptomyces* and their use in vector construction. Gene **20:**51–62.
28. **Thompson, C. J., J. M. Ward, and D. A. Hopwood.** 1980. DNA cloning in *Streptomyces*: resistance genes from antibiotic-producing species. Nature (London) **286:** 525–527.
29. **Thompson, C. J., J. M. Ward, and D. A. Hopwood.** 1982. Cloning of antibiotic resistance and nutritional genes in streptomycetes. J. Bacteriol. **151:**668–677.

Chapter 17

Genetic Manipulation of *Saccharomyces cerevisiae*

THOMAS D. INGOLIA AND JOHN S. WOOD

Saccharomyces cerevisiae (bakers' yeast) can be genetically manipulated in many ways. Classical genetic techniques include the isolation of spontaneous and chemically induced mutations, generation of diploids from haploids by fusion of opposite mating types, and generation of haploids from diploids by sporulation followed by separation and germination of the spores. These techniques have been utilized in the past half century to identify, map, and characterize hundreds of genetic loci (1, 20). More recently, techniques have been developed which allow introduction of foreign DNA into the cell, either as autonomously replicating DNA or integrated into chromosomal DNA. These techniques, coupled with the ability to isolate, clone, and amplify small segments of DNA by recombinant methods, have dramatically increased the genetic diversity which can be introduced into yeast cells (2). In this chapter several of the most useful techniques will be described and exemplified. For discussion purposes, the techniques have been divided into two groups: those which alter the genome and those which recombine or rearrange the genome. The first group includes techniques that alter only limited regions of the genome, such as mutation or transformation, while the second group includes those with a larger or more generalized effect, such as meiotic and mitotic recombination. This division is somewhat arbitrary since a given technique can have a large or small effect on the genotype of a cell depending on how it is applied.

The source of the *S. cerevisiae* strains and plasmids and the recipes for the media used in the examples are given in the appendix. The genotypes of all strains and plasmids should be checked before they are used.

17.1 ALTERING THE GENOTYPE

The ability to induce mutations is basic to any genetic manipulations. *S. cerevisiae* is useful because, unlike most other eucaryotes, it has both stable haploid and diploid stages in its life cycle. This allows recessive mutations to be picked up in the haploid phase and tested for allelism with other mutations in the diploid phase. Mutants can be isolated from a population of cells either by direct selection of a phenotype or by screening for a visible phenotype among many individual colonies. The average spontaneous mutation frequency for *S. cerevisiae* at any particular locus is about 10^{-6} per generation. For many experiments, a higher mutation rate is required, so mutagenic agents are used. Most mutagenic agents active against bacteria are also active against *S. cerevisiae*, including UV radiation, nitrous acid, ethyl methanesulfonate (EMS), and *N*-methyl-*N'*-nitro-*N*-nitrosoguanidine (10, 15, 16, 19, 21). In the following example, EMS is used to mutagenize yeast cells, and the target genes include the arginine permease gene *CAN1* and the adenine biosynthetic genes *ADE1* and *ADE2*.

17.1.1 EMS Mutagenesis

Grow 20 ml of strain DBY746 (*MATα ura3-52 leu2-3 leu2-112 his3-Δ1 trp1-289*) in YPD to an A_{600} of about 0.7. Collect the cells by centrifugation, and suspend in 2.0 ml of sterile 0.05 M KPO_4 buffer (pH 7.0). Remove 0.1 ml, add this to 5 ml of sterile 5% sodium thiosulfate, and set aside as a control; to the remaining cells add 60 μl of EMS (Sigma Chemical Co.) and shake at 30°C for 1 h. Inactivate the EMS and stop the mutagenesis by transferring 0.1 ml of the treated cells to 5 ml of sterile 5% sodium thiosulfate. Determine the survival rate of the mutagenized cells by plating 0.1 ml of 10^{-3}, 10^{-4}, and 10^{-5} dilutions of the mutagenized and control cells on YPD plates. A survival rate of about 10% is expected under these conditions (15). Collect the remaining cells by centrifugation, suspend in 5 ml of liquid YPD, and shake at 30°C for 90 min. The stock cultures of mutagenized and control cells can be stored for several weeks at 4°C.

17.1.2 Canavanine-Resistant Mutants

Wild-type yeast strains are sensitive to the arginine analog canavanine, but canavanine-resistant mutants can be obtained. Mutations to canavanine resistance are recessive and occur at a single locus, *CAN1*, the arginine permease gene (9, 31). Resistant mutants can be easily selected by growing cells in the presence of canavanine and the absence of arginine. Since most complex media contain arginine, a defined medium without arginine must be used for selecting canavanine resistance.

Spread about 10^4 cells from the mutagenized and control cultures on separate plates containing C-arginine agar plus 60 mg of canavanine sulfate per liter. After several days of incubation at 30°C, canavanine-resistant colonies will be visible and will be more frequent on the plates with mutagenized cells than on those with nonmutagenized cells.

17.1.3 Adenine Biosynthetic Mutants

In medium which is limiting for adenine, mutations in the adenine biosynthetic genes *ADE1* or *ADE2* cause accumulation of a red pigment (7, 27). Mutant cells form red colonies which can be detected easily in a background of normal, wild-type, white colonies.

To isolate *ade1* or *ade2* mutations from the stock mutagenized culture, repeat the plating experiment described above except use YPD plates and reduce the plating density to about 500 colonies per plate. After several days of growth at 30°C, red colonies should be visible in a background of normal white colonies. Since the red pigment is made only in response to adenine starvation, the plates may have to be incubated for longer than usual before red colonies become apparent. These red colonies should contain mutations at the *ade1* or *ade2* locus.

17.1.4 Complementation Test

Once a series of mutants with a particular phenotype is isolated, one needs to determine whether their mutations are at the same or different genetic loci. This is done by complementation testing, i.e., determining whether the defect caused by a mutation in one strain can complement, or correct, the defect caused by a mutation in another strain. Because *S. cerevisiae* has a stable diploid phase, complementation testing is easily done by crossing two haploid mutant strains and assaying the phenotype of the resulting diploid. If both mutations are recessive, then a diploid carrying both will have a mutant phenotype if the mutations are at the same locus (no complementation), or a nonmutant, wild-type phenotype if the mutations are at different loci (positive complementation). This is illustrated in the following example, which demonstrates how the red, adenine-requiring mutations isolated in the previous section can be assigned to either the *ade1* or *ade2* locus.

Grow each of the red, adenine-requiring mutants of strain DBY746 isolated previously in a narrow strip or line of cells on a YPD agar plate (six to eight mutants can be put on a single plate) at 30°C. Also grow *ade1* and *ade2* tester strains N248-1A (*MAT***a** *ade1 gal1 leu1 his2 ura3 trp1 met14*) and N442-4A (*MAT***a** *ade2 his6 lys9 ura1 trp5 met2 arg4* Mal^- Suc^-) and ADE^+ strain N435-1A (*MAT***a** *his7 lys7 met6 arg1 gal4 MAL2* Suc^+) on a YPD plate, again as narrow strips, with one strip of each strain on the plate. The tester strains must be able to mate with the unknown strains for the complementation test to work, so the tester strains and unknown strains must have opposite mating types (see section 17.2). In this case, the new adenine-requiring mutants were obtained from a *MATα* strain so the tester strains must be *MAT***a**.

Replica plate the tester strains to several YPD plates (as many as there are plates containing red mutants to be tested). Replica plate one of the plates containing the adenine-requiring mutants to one of the tester strain replica plates. Rotate the tester strain plate 90° when doing this, so that the strips of the tester strains and the mutant strains are at right angles to each other (i.e., they are "cross-stamped"). This allows multiple crosses (in this case 18 to 24) to be done on a single petri dish.

Incubate the cross-stamped plates at 30°C overnight to allow mating and diploid formation, then replica plate them to C-histidine (C-his) plates. Only diploids formed between one of the mutants and one of the tester strains will grow on these plates. The diploids will be heterozygous for two different histidine mutations and therefore will be able to grow on media lacking histidine. After the C-his plates have grown at 30°C overnight, replica plate each to another C-his plate and incubate this for 24 to 48 h. This eliminates any possibility of carryover of haploid cells which did not mate from the cross-stamp plates.

Replica plate the second set of C-his plates to C-adenine (C-ade) plates and incubate at 30°C for 24 to 48 h. Scoring is as follows:

1. All mutants crossed with strain N435-1A should be growing and form white colonies. Since N435-1A is Ade$^+$, this indicates that the adenine-requiring mutations are recessive.
2. Diploids formed between the mutants and strain N248-1A which can grow and are white on C-ade plates must NOT contain a mutation in the *ade1* gene.
3. Diploids formed between the mutants and strain N248-1A which cannot grow and are red on C-ade plates must have a mutation in the *ade1* gene.
4. Diploids formed between the mutants and strain N442-4A which can grow and are white on C-ade plates must NOT contain a mutation in the *ade2* gene.
5. Diploids formed between the mutants and strain N442-4A which can grow and are red on C-ade plates must have a mutation in the *ade2* gene.

17.1.5 Complementation of Genomic Mutations with Cloned DNA

With the advent of recombinant DNA and transformation techniques it has become possible to alter the genotype with DNA sequences introduced into a cell on plasmids. Many genomic mutations can be complemented in *trans* by providing DNA containing the wild-type allele of the mutated gene to a haploid strain (22). The wild-type allele can be maintained extrachromosomally, creating a partial diploid, or it can be integrated into the genome.

Several types of vectors are available which facilitate introduction of genetic information into a yeast cell. YRp and YEp vectors (yeast replicating plasmid and yeast episomal plasmid) are maintained extrachromosomally, whereas YIp vectors (yeast integrating plasmid) must integrate into the genome to be stably maintained. Typically these vectors contain both yeast and bacterial selection and maintenance elements, for ease of handling and manipulation. A complete description of the available systems is beyond the scope of this chapter, but a number of excellent reviews can be found in *Methods in Enzymology*, vol. 101 (1983).

Some of the most useful *S. cerevisiae* vectors are the YEp vectors, which contain replication and maintenance functions from 2μm, an endogenous yeast plasmid (3). One example is YEp24 (Fig. 1). This plasmid contains the beta-lactamase gene and origin of replication from pBR322 (to facilitate growth in *Escherichia coli*), the *S. cerevisiae URA3* gene, and a segment of 2μm DNA containing the origin of replication and stabilizer functions (to allow growth in *S. cerevisiae*) (14). The *URA3* gene on this plasmid will complement the *ura3-52* mutant allele of strain DBY746. Thus, DBY746 cells transformed with YEp24 can be selected on the basis of their ability to grow in minimal medium in the absence of uracil. This selection is the basis of the transformation experiment described below.

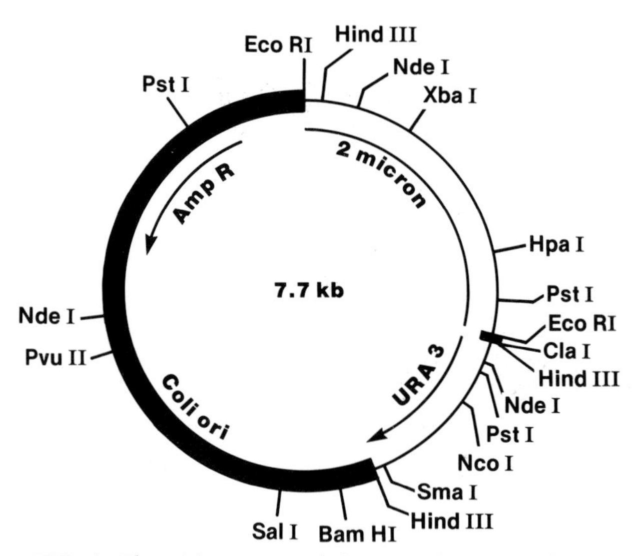

FIG. 1. Plasmid YEp24, a bifunctional plasmid able to replicate autonomously in *S. cerevisiae* and *E. coli* (3).

17.1.6 Transformation of *S. cerevisiae* with YEp24 DNA

This technique is an adaptation of the method described by Hinnen et al. (13). Yeast cells are rendered competent to take up foreign DNA by enzymatic digestion of most of the outer cell wall. Competent cells are mixed with DNA, and polyethylene glycol is added to promote uptake of DNA. Cells are then plated in a medium that permits regeneration of the cell wall. The cells can be selected for uptake of foreign DNA by using an auxotrophic mutant yeast cell as the host and putting a wild-type allele of the auxotrophic marker on the foreign DNA. In the case described below, *S. cerevisiae* DBY746 contains a mutated *ura3* gene and YEp24 contains a wild-type *URA3* gene.

Grow *S. cerevisiae* DBY746 (*MAT*α *his3-Δ1 leu2-3 leu2-112 ura3-52 trp1-289*) in 100 ml of YPD broth to an A_{600} of 1.0. A convenient way to obtain a culture at about the right density is to inoculate 10^{-5}, 10^{-4}, and 10^{-3} dilutions of a stationary culture into YPD and grow overnight.

Using sterile solutions and operations, wash cells once in 15 ml of 1.2 M sorbitol and then resuspend in 15 ml of 1.2 M sorbitol. Add 100 ml of 2.5-mg/ml Zymolyase 100T (Seikagaku Kogyo Co., Tokyo) suspended in 5 mM KPO_4 (pH 7.6)–1.2 M sorbitol. (The Zymolyase solution is stable for several months at −20°C if it is also made 50% in glycerol.) Incubate at 30°C, and monitor protoplasting by removing 20-μl samples and adding to 180 μl of 10% sodium dodecyl sulfate. Under a phase-contrast microscope, protoplasts appear black in the sodium dodecyl sulfate whereas whole cells are much more refractile. In a typical digestion, a culture contains about 50% protoplasts after a 20-min incubation with Zymolyase and about 90% protoplasts after 40 min. When

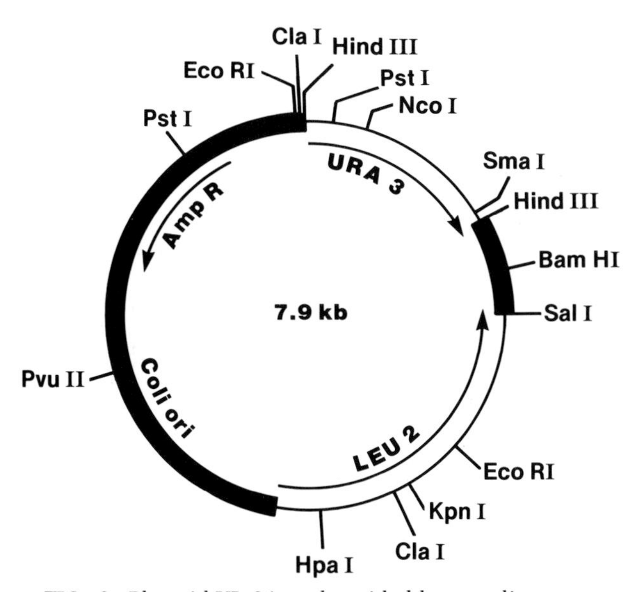

FIG. 2. Plasmid YIp26, a plasmid able to replicate autonomously in *E. coli* but not in *S. cerevisiae*. This plasmid must integrate into *S. cerevisiae* DNA for stable maintenance of YIp26-carried information.

protoplasting is about 90% complete, the cells are collected by centrifugation at 1,000 × *g* for 3 to 5 min and washed twice with 15 ml of 1.2 M sorbitol. (The protoplasts should be treated gently—no vortexing or high *g* forces—and kept away from detergents, using sterile disposable plasticware whenever possible.) The cells are then suspended in 600 μl of YPD containing 1.2 M sorbitol, 10 mM $CaCl_2$, and 10 mM Tris hydrochloride (pH 7.5).

One microgram of YEp24 DNA is then mixed with an equal volume of sterile 2.4 M sorbitol. (The amount of DNA required for yeast transformation will depend on the experiment, the yeast strain, and the transforming plasmid. Using YEp24 and DBY746 under the conditions described, transformation frequencies of about 10^4 to 10^5 transformants per μg are typical.) Add 200 μl of the competent yeast cells both to the DNA and to a control tube containing 1.2 M sorbitol without DNA. Gently mix, and incubate at room temperature for 10 min. Then add 1 ml of 20% polyethylene glycol (Baker's; 3,350, average molecular weight)–10 mM $CaCl_2$–10 mM Tris hydrochloride (pH 7.5) and incubate at room temperature for 60 min.

Add 10, 50, 200, and 800 μl of the YEp24-containing mixture to separate tubes containing 30 ml of melted regeneration agar without uracil held at 47°C. Invert to mix, and rapidly pour into a sterile petri dish. Plate 50 and 200 μl of the control cells in the same way. After the agar solidifies, incubate the plates at 30°C. Transformants should be visible after 3 to 5 days as disk-shaped colonies embedded in the agar. The control cells should produce few or no colonies, since the *ura3-52* allele reverts at a low frequency. The regeneration efficiency can be checked by plating several 100-fold serial dilutions of the cells before and after the protoplasting treatment in 1.2 M sorbitol on complete (nonselective) regeneration agar.

17.1.7 Cloning a Gene

The methods described in the previous two sections can be used to isolate a gene when a selection or screen is available for the gene of interest. For example, a strain unable to grow on minimal medium because of an auxotrophic mutation can be used to select plasmids containing the wild-type allele of that gene. The mutant strain is transformed with a clone bank or gene bank, which is a large collection of individual plasmid clones like YEp24, each containing a random fragment of yeast DNA. Detailed methods for preparing gene banks can be found in references 17 and 29. Plasmids containing the wild-type allele will permit the mutant strain to grow without the particular auxotrophic requirement because the extrachromosomal wild-type allele complements the defective genomic allele. DNA from the transformed yeast cells that can grow on minimal medium can then be used to transform *E. coli* (recall that YEp24 also contains sequences allowing plasmid growth and selection in *E. coli*) so that the plasmid can be purified and characterized. This technique has been successfully employed for scores of yeast genes, as well as genes from other organisms which can be expressed in *S. cerevisiae* and complement the yeast mutation.

17.1.8 DNA Insertion into the Genome

The yeast vector YEp24, used to transform *S. cerevisiae* in the past several sections above, contains origin of replication and stabilizer functions from *S. cerevisiae* 2μm DNA and therefore can be maintained extrachromosomally. Plasmids that do not contain an endogenous origin of replication, called YIp vectors, can be stably maintained only when they integrate into the genome. Integration of the YIp vectors occurs by recombination between homologous regions on the plasmid and the genome (13). Linearizing the YIp vector increases the frequency of integration and targets the plasmid to the region of the genome homologous to the linear ends (25, 26). That is, if the vector is cut within the *URA3* gene, integration will be more efficient than for the circular vector and integration will occur predominantly at the genomic *URA3* locus.

The following procedure illustrates integrative transformation of *S. cerevisiae*. The integrating vector used will be YIp26 (Fig. 2) (3). YIp26 is similar to YEp24 except that it contains the *S. cerevisiae LEU2* gene as well as the *URA3* gene and lacks the 2μm DNA sequences which facilitate autonomous replication.

Prepare *S. cerevisiae* DBY746 for transformation as described in section 17.2.5 and split into two aliquots. One aliquot should be mixed with 5 μg of uncut YIp26, and the second aliquot should be mixed with 5 μg of YIp26 which has been previously digested with restriction enzyme *Sma*I, which cuts once within the *URA3* locus. Two classes of uracil prototrophs should be observed in the regeneration plates. Cells that received but did not integrate the DNA will appear as tiny colonies since the YIp vector is unstable when it is not integrated. A smaller number of larger colonies should also be present, and these cells should have an integrated copy (or copies; see reference 24) of the vector. Cells from these large colonies should be stable uracil prototrophs. To test the stability of the uracil prototrophy, grow transformed cells for about 30 generations in nonselective rich medium (YPD). Grow cells transformed with YEp24 (from section 17.1.6) under the same conditions. Then measure the proportion of uracil prototrophs by diluting the cells and plating on minimal plates with or without uracil. Less than half of the YEp24-transformed cells should have retained uracil prototrophy, whereas greater than 90% of the YIp26-transformed cells should still be uracil prototrophs.

The stable transformants arising from cells mixed with linearized YIp26 should contain vector sequences predominantly at the *URA3* locus, since the YIp26 was cut within DNA sequences homologous to the *URA3* locus and the linear ends target the integration site (25, 26). The transformants arising from cells mixed with uncut YIp26 should have vector sequences at either the *LEU2* or *URA3* loci, since YIp26 contains DNA homologous to both regions (13). The location of the integrated sequences can be determined by mapping the wild-type *URA3* gene in the transformed cells, using techniques described later in this chapter.

17.1.9 Insertional Inactivation of Genes

Insertion or deletion mutations can be simply constructed when the cloned gene is available. The DNA sequence encoding the wild-type allele of a convenient auxotrophic marker (usually *TRP1*, *HIS3*, *URA3*, or *LEU2*) is cloned into the coding region of the cloned gene of interest, thereby disrupting the cloned gene (for a review, see reference 28). In the example shown in Fig. 3, the *URA3* gene is inserted into the gene of interest. If desired, DNA sequences from the coding region of the gene of interest can also be deleted (Fig. 3). A linear fragment containing the mutated gene is purified and used to transform cells according to the techniques described in sections 17.1.6 and 17.1.8. The ends of the linear fragment must be homologous to the region of the genome containing the gene of interest to obtain efficient gene replacement. In a high proportion of the uracil prototrophs recovered from the hypothetical experiment shown in Fig. 3, the wild-type gene of interest is replaced, presumably by a double crossover event, with the in vitro-altered gene. The result is a strain containing a deletion mutation, insertion mutation, or both in a particular gene. Possible lethality of the desired mutation can be checked by transforming an appropriate diploid strain, selecting for the wild-type allele on the fragment (*URA3* in our example), and then sporulating the diploid to assess the viability of haploid cells containing the mutated gene of interest (see section 17.2).

17.2 RECOMBINING THE GENOME

Section 17.1 describes several useful methods for mutating and altering small segments of the *S. cerevisiae* genome. It is often useful to be able to rearrange, manipulate, delete, or add large segments of genetic information. In this section we will explore

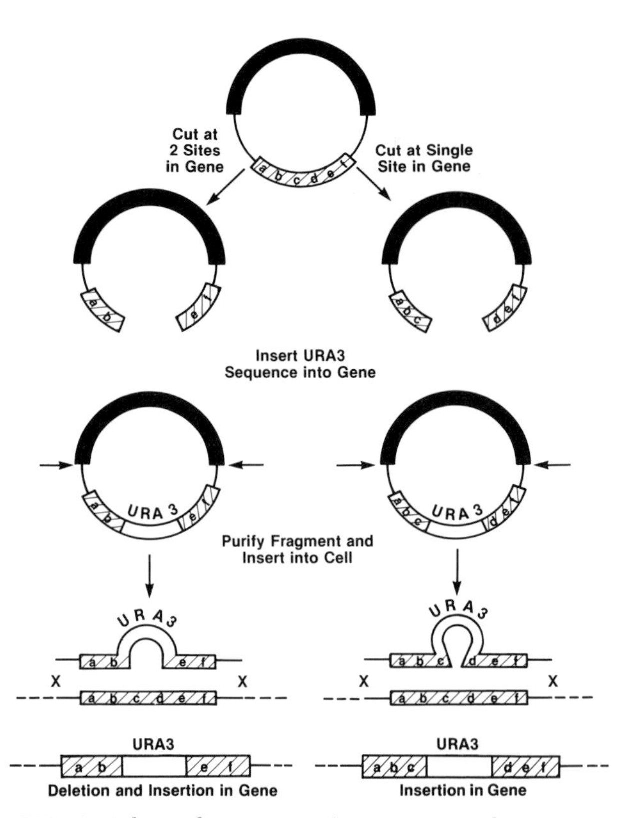

FIG. 3. Scheme for insertional inactivation of an *S. cerevisiae* gene, adapted from Rothstein (28). Details are described in the text (section 17.1.9).

some of the most useful "macro" genetic techniques applicable to this yeast.

17.2.1 Mating Type and Ploidy

S. cerevisiae has two sexes or mating types: *MAT***a** and *MAT*α. Cells of opposite mating type can "mate," fusing both cytoplasms and nuclei to form a stable cell containing the entire chromosome complement of the two mating cells. The only requirements for mating are that each cell be homozygous for mating type (its ploidy level is unimportant as long as it is homozygous for mating type) and that the two mating cells be of opposite mating type.

Although *S. cerevisiae* can exist in either a haploid or diploid state, most wild-type strains are homothallic. This means that cells frequently switch mating type (12, 30). In a population of cells, rapid mating-type switching causes a large percentage of nonmating (mating type heterozygous) cells to accumulate as individual cells switch mating type and mate with a cell of the opposite mating type. The preponderance of nonmating cells in a population interferes with the ability to do matings between different strains in the laboratory; therefore, most laboratory strains contain a mutation which greatly reduces the frequency of mating type switches (from 0.5 switch per cell division to 10^{-6} switch per cell division). These mutated cells are called heterothallic (designated "*ho*," where "*HO*" is the wild-type allele which confers homothallism). Heterothallic strains can be grown and maintained with a single, stable mating type.

17.2.2 Protocol for Mating

Strains can be easily mated if they have complementing markers; the only requirement is that cells must be growing (undergoing cell division) to mate. One easy procedure is to inoculate 0.1 ml each of two stationary stocks (one a *MAT***a** strain and the other *MAT*α) into 5 ml of YPD liquid medium and incubate WITHOUT SHAKING OR AGITATION for 12 to 24 h. An incubation temperature of 34°C is optimal for mating wild-type cells, but mating will occur over a wide range of temperatures (20 to 35°C). The mating mix is then streaked out on solid medium that does not permit either of the two haploid strains to grow but does allow the diploid to grow. For strains that do not have complementing markers (i.e., it is not possible to select against the two haploid parents and for the diploid at the same time), the mating can be done in the same way, and the diploid can be physically isolated by micromanipulating zygotes (the large, misshapen cells which are the immediate result of the fusion between two haploid cells during mating) away from parental, haploid cells which did not mate.

The following experiment details the mating of haploid *S. cerevisiae* strains of opposite mating type. The haploid strains chosen will produce diploids with genotypes and phenotypes useful for further experiments. Almost all haploid strains of opposite mating type will mate under the conditions described.

Grow *S. cerevisiae* DBY746 (*MAT*α *his3-Δ1 ura3-52 leu2-3 leu2-112 trp1-289*) and STXZF-1C-2D (*MAT*α *ade7 ura3 met8 his7 trp1 gal7 can1*) on YPD plates. Mix together a loopful of each strain in liquid YPD and incubate overnight at 30°C. Cells from this mating are then streaked onto a C-his plate to select diploids from the background of unmated haploid cells. After incubating the minimal plates at 30°C for 2 or 3 days, pick several colonies and grow up liquid stocks. These His^+, presumed diploid clones can be tested for their ploidy level with the following test.

17.2.3 Protocol for Assaying Haploid/Diploid

The recessive mutation for resistance to canavanine, *can1*, described in section 17.1.2, is the basis of a simple, useful test to distinguish haploid and diploid cells. The assay is based on the observation that mitotic recombination events (discussed below in section 17.2.6) which will make the *CAN* locus homozygous spontaneously occur at a significant frequency. When a diploid heterozygous for canavanine resistance (*can1/CAN1* or resistant/sensitive [R/S]) is replica plated to a canavanine-containing plate, frequent resistant individual colonies, or papillations, growing in a patch of sensitive (nongrowing) cells are seen. If these papillations are retested they are found to be resistant to canavanine and still diploid (i.e., they are now R/R).

The test is simply to cross the strain in question, which must be canavanine sensitive, with a haploid *can1* (R) strain. The resulting diploid (if the strain being tested is haploid) or triploid (if the strain is already a diploid) is replica plated to a plate containing canavanine and, after incubation, scored for the frequency of canavanine-resistant papillations. If the original strain was haploid, then the "crossed" strain will be diploid and heterozygous for *can1* (R/S) and will show a high frequency of canavanine-resistant papillations. If the original strain was diploid, then the crossed strain will be triploid and again heterozygous for *can1*. However, in this case the cells contain two canavanine-sensitive alleles and one resistant allele (S/S/R). Because two mitotic recombination events must occur to yield a cell resistant to canavanine, the frequency of canavanine-resistant papillations will be greatly reduced from that observed for the heterozygous diploid (under most conditions none will be seen).

17.2.4 Meiotic Recombination

Cells that are heterozygous for mating type can undergo meiosis and sporulate. Starting with a diploid strain, this results in the formation of four haploid spores. However, sporulation will occur under the proper conditions in any strain that is heterozygous for mating type; e.g., a tetraploid strain will give four diploid spores. Strains that have an odd-number ploidy (3*N*, 5*N*) will usually sporulate but will have very low spore viability.

To recover haploid cells which have undergone meiosis and therefore are genotypic hybrids of the parent haploid cells, it is first necessary to sporulate the diploid cells. The following method is adopted from Fowell (8). Grow 10 ml of diploid cells in YPD medium to late log phase (A_{600} of a 1:10 dilution of about 1.0). Wash cells twice in sterile water, and then suspend the cells in 10 ml of sterile 0.3 M potassium acetate. Put 2 ml of these cells into a sterile 20-ml tube, and also set up 2-ml cultures of 10-fold and 100-fold dilutions in 0.3 M potassium acetate. Incubate at 30°C with shaking, and monitor spore formation microscopically. Four spores are contained in a tetrahedral structure called an ascus, which is approximately one-third to one-half the size of a yeast cell. The efficiency and rapidity of spore formation is strain dependent; typically 5 to 50% of the cells will be sporulated in 2 to 5 days. Spores are stable in this liquid culture for months at 4°C.

17.2.5 Isolation of Individual Spores

There are two common methods used for isolating spores, i.e., tetrad dissection and random spore analysis. Tetrad dissection has the advantage that all four products of meiosis in *S. cerevisiae* are recoverable. This permits a sensitive assay of recombination distances between genetic loci, as well as allowing extensive reassortment of the genome. However, it is time consuming and requires specialized equipment and techniques. Therefore, we will concentrate on random spore analysis.

Random spore analysis is used when tetrad dissection is not feasible or necessary, i.e., when a relatively uncomplicated genotype reassortment is desired from meiosis. It is a rapid technique and requires no special equipment or skills.

For random spore analysis, the spores must be separated from one another and spread out on a selective plate. The method we use to separate the spores is sonication of enzymatically digested asci, followed by plating the cells on a medium that will allow only haploid cells to grow. This method is possible when the diploid strain is heterozygous for a recessive resistance gene such as *can1* (resistance to canavanine) or *cyh2* (resistance to cycloheximide).

The diploid constructed as described in section 17.2.2, which is heterozygous for canavanine resistance, can be used to isolate haploid strains. Remove 1 ml of the sporulated culture described in section 17.2.4 and incubate with 0.1 ml of Glusulase (DuPont Pharmaceuticals) at 30°C for about 30 min. (Glusulase is a crude enzyme mixture which digests the ascus wall without affecting the viability of the spores.) Monitor the extent of Glusulase digestion of the asci walls by microscopically following breakup of the tetrahedral structure. When most of the asci are disrupted, sonicate the mixture until the spores are separated (absence of asci under microscopic examination). After sonication, spread aliquots onto minimal plates containing canavanine and all the auxotrophic requirements of both parent haploid strains. In our example, the minimal plates should be supplemented with histidine, uracil, leucine, tryptophan, adenine, and methionine. Incubate the plates at 30°C. Colonies that grow on these plates should be haploid cells containing the canavanine resistance gene *can1*. The phenotypes of individual spore clones can then be checked by monitoring growth on minimal plates containing all but one of the nutritional requirements of the parent haploid strains.

17.2.6 Mitotic Recombination

Mitotic recombination is useful because it allows more limited genotypic changes than meiosis, which necessarily reassorts the entire genome. Mitotic recombination is usually observed by the appearance of a homozygous sector within a colony which is otherwise heterozygous at a particular genetic locus and occurs as a result of DNA repair activities. In *S. cerevisiae*, these repair activities frequently are nonreciprocal; homozygosity at a locus occurs when a nonreciprocal crossover takes place within, or centromere proximal to, the gene. As a genetic technique, mitotic recombination can be used to make genes homozygous in a strain and to determine the genetic distance between a locus and the centromere or between hetero-alleles within a single gene. The spontaneous frequency of mitotic recombination events which lead to homozygous alleles at a locus is relatively low in *S. cerevisiae* and is proportional to the distance between a locus and its centromere; the average frequency is less than 1 in 10^5 divisions. Therefore, to use mitotic recombination as a genetic technique, cells are treated with DNA-damaging agents such as UV or X rays to increase the frequency 100 to 1,000-fold. Since *S. cerevisiae* is relatively resistant to UV, X rays are typically used. The following example will demonstrate this with the *ade1/ade6* and *ade2/ade6* heterozygous diploids made previously for complementation testing (section 17.1.4).

From the crosses made for complementation testing (red, adenine-requiring DBY746 mutants × *ade1,2,6* tester strains) grow one diploid heterozygous for *ade1* and *ade6* and one diploid heterozygous for *ade2* and *ade6* in 5 ml of liquid YPD medium.

Pipette 2 ml of these exponentially growing cultures (approximately 5×10^6 cells per ml) into separate 90-mm glass petri plates. (This gives a thin layer of cells in the bottom of the plate, avoiding any problems with uneven radiation dose due to settling of cells in the culture liquid.) Treat each plate with 5×10^3 to 1×10^5 rads of X rays. (A typical setup used a Picker X-ray machine with a beryllium window delivering 106 rads/s 10 cm from the window for 1 min to give 6,400 rads total.) Plate for single colonies on YPD agar from both the treated and untreated cells. After irradiation, the viability of the treated cells should be approximately 50% of the pretreatment value. Incubate until red colonies can be seen; score the total number of colonies and the fraction which are red. The untreated cells and the treated *ade1/ade6* cells should have a very low frequency of red colonies (*ade1* is tightly linked to the centromere of chromosome 1); the treated *ade2/ade6* cells should have 0.1 to 1% red cells (*ade2* is greater than 70 centimorgans from the centromere of chromosome 15).

17.3 OTHER TECHNIQUES

There are a number of other genetic techniques which have been developed in *S. cerevisiae*. Most of these are specialized and have only limited application compared with the relatively general techniques discussed so far. Because of this they will not be discussed in as much detail as the other techniques.

17.3.1 Mitochondrial Mutants

S. cerevisiae is a facultative anaerobe; this makes it possible to obtain viable mitochondrial mutants, which are then obligate aerobes. These mutations can be either in nuclear genes coding for mitochondrial functions or within the mitochondrial genome itself. Nuclear mutations are inherited in Mendelian fashion and can be manipulated in the same manner as any other nuclear mutation. Most respiratory-deficient mutants, however, are inherited in a complex, cytoplasmic manner, indicating that they are within the mitochondrial genome (reviewed in reference 5). Although extensive work has been done characterizing cytoplasmic mutants defective in mitochondrial function (called "petites" because they form smaller colonies than mitochondria$^+$ cells, called "grandes"), petites are most often used simply as a cytoplasmic marker with no concern about the actual molecular defect in mitochondrial function. Cytoplasmic petite mutants occur frequently; they are found at a frequency of 1 to 3% in most strains of *S. cerevisiae*. Petite mutations can also be easily induced by treating cells (exponentially growing or stationary) with 10 μg of ethidium bromide per ml for 24 h in the dark. Ethidium bromide interferes with replication of the mitochondrial chromosome, causing large deletions which abolish mitochondrial function.

17.3.2 Protoplast Fusion

S. cerevisiae cells that have had their cell wall partly removed or "damaged" can be induced to fuse if they are mixed in a polyethylene glycol solution. This is useful for making crosses between strains which have the same mating type, strains which are defective for mating (sterile), or strains which are heterozygous for mating type.

17.3.3 Heterokaryons

When two cells of the opposite mating type conjugate, cytoplasmic fusion (cytogamy) is closely followed by nuclear fusion (karyogamy). Conde and Fink (4) isolated a mutant, *kar1*, defective in karyogamy. In this mutant, after conjugation with a wild-type (*KAR1*) strain and cytoplasmic fusion, an unstable heterokaryon cell with two nuclei is formed. The heterokaryon can form daughter cells containing either one or both of the parental nuclei types or, at a low frequency, a diploid nucleus. Although the daughter cells containing one or the other nucleus types have not undergone nuclear fusion, they contain a mixed cytoplasm, hence their designation "cytoductants." These can be utilized genetically to cross cytoplasmic or mitochondrial markers without forming a diploid and reassorting the nuclear genome. An additional use for the *kar1* mutation stems from the observations of Nilsson-Tillgren et al. (23) and Dutcher (6) that at a low frequency (1 in 10^6 to 1 in 10^3 cytoductants) there is a transfer of entire chromosomes between the nuclei in a heterokaryon. With the proper selective markers, this allows changing the genotype of a strain a single chromosome at a time.

17.3.4 Chromosome Loss

S. cerevisiae cells are very tolerant of aneuploidy, making it possible to select cells which have lost one or more chromosomes. The spontaneous frequency of loss of any single chromosome is approximately 1 in 5×10^4 divisions (11), which is too low to be useful as a genetic tool. However, there are now a number of techniques for inducing chromosome loss in *S. cerevisiae*. These include the temperature-sensitive cell cycle mutation *cdc6* (G. Kawasaki, Ph.D. Thesis, University of Washington, Seattle, 1979), the mitotic inhibitor methyl-benzimidazole carbamate (32), or *rad52*, a radiation sensitivity mutation (18).

17.4 APPENDIX

17.4.1 Sources

All *S. cerevisiae* strains used in this chapter can be obtained from the Yeast Genetic Stock Center (Rebecca Contopoulou), Department of Biophysics and Medical Physics, University of California, Berkeley, CA 94720. A catalog containing descriptions of these strains and many other available strains is available. The *S. cerevisiae* plasmid vectors used can be obtained from the American Type Culture Collection, c/o Professional Services Department, 12301 Parklawn Drive, Rockville, MD 20852.

17.4.2 Recipes

YPD medium (rich medium)

Ingredient	Amount
Yeast extract (Difco)	10 g
Bacto-Peptone (Difco)	20 g
Agar (for YPD plates)	20 g
Water	900 ml
*Glucose (20% solution, wt/vol)	100 ml

*The glucose is added from a sterile concentrated stock after autoclaving to prevent the caramelization which occurs when solutions containing yeast extract and glucose are autoclaved.

Minimal medium (B medium)

Ingredient	Amount
Yeast Nitrogen Base without amino acids or ammonium sulfate (Difco)	1.6 g
Succinic acid	10 g
Sodium hydroxide	6 g
Ammonium sulfate	1 g
Agar (for minimal medium plates)	20 g
*Nutritional supplements, each	4 mg
Water	900 ml
**Glucose (20% solution, wt/vol)	100 ml

*Amino acid, purine, or pyrimidine supplements are added at 4 mg/liter as necessary for growth of particular strains.
**Glucose is added from a sterile concentrated stock after autoclaving to prevent caramelization.

Synthetic complete medium (C medium)

Ingredient	Amount
Yeast Nitrogen Base without amino acids or ammonium sulfate (Difco)	1.6 g
Succinic acid	10 g
Sodium hydroxide	6 g
Ammonium sulfate	1 g
*Amino acids, purines, and pyrimidines, each	4 mg
Agar (for plates)	20 g
Water	900 ml
**Glucose (20% solution, wt/vol)	100 ml

*Adenine, arginine, asparagine, cysteine, histidine, isoleucine, leucine, lysine, methionine, phenylalanine, serine, threonine, tryptophan, tyrosine, uracil, and valine were added at 4 mg/liter.
**Glucose is added from a sterile concentrated stock after autoclaving to prevent caramelization.

Auxotrophic mutations were assayed on synthetic complete medium plates made without one or more of the amino acid or nucleotide supplements (e.g., C-adenine, C-arginine, etc.).

Canavanine plates

C-arginine plates (see synthetic complete medium, above) plus 60 mg of canavanine sulfate per liter.

Regeneration medium

Yeast nitrogen base without amino acids or ammonium sulfate (Difco)	1.6	g
Succinic acid	10	g
Sodium hyrdroxide	6	g
Ammonium sulfate	1	g
*Amino acids, purines, and pyrimidines, each	4	mg
L-Sorbitol	218	g
Agar	30	g
Water	900	ml
**Glucose (20% solution, wt/vol)	100	ml

*Nutritional requirements are added, as for synthetic complete medium (above), except that the supplement used to select for uptake of foreign DNA is omitted. In the case of selection of uptake of YEp24, which contains the *URA3* allele, selective regeneration medium does not contain uracil.
**Glucose is added from sterile stock after autoclaving to prevent caramelization.

Regeneration medium should be made fresh each time and maintained at about 48°C to prevent premature solidification.

17.5 ACKNOWLEDGMENTS

We thank Eli Lilly and Co. for support and Cheryl Alexander for typing the manuscript.

17.6 LITERATURE CITED

17.6.1 General References

1. **Strathern, J. N., E. W. Jones, and J. R. Broach.** 1981. The molecular biology of the yeast *Saccharomyces*: life cycle and inheritance. Cold Spring Harbor Laboratory, Cold Spring Harbor, N.Y.
2. **Strathern, J. N., E. W. Jones, and J. R. Broach.** 1982. The molecular biology of the yeast *Saccharomyces*: metabolism and gene expression. Cold Spring Harbor Laboratory, Cold Spring Harbor, N.Y.

References 1 and 2 provide an excellent overview of the genetics and molecular biology of Saccharomyces.

17.6.2 Specific References

3. **Botstein, D., S. C. Falco, S. E. Stewart, M. Brennan, S. Scherer, D. T. Stinchcomb, K. Struhl, and R. W. Davis.** 1979. Sterile host yeasts (SHY): a eukaryotic system of biological containment for recombinant DNA experiments. Gene **8:**17–24.
4. **Conde, J., and G. Fink.** 1976. A mutant of *Saccharomyces cerevisiae* defective for nuclear fusion. Proc. Natl. Acad. Sci. USA **73:**3651–3655.
5. **Dujon, B.** 1981. Mitochondrial genetics and functions, p. 505–635. *In* J. N. Strathern, E. W. Jones, and J. R. Broach (ed.), The molecular biology of the yeast *Saccharomyces*: life cycle and inheritance. Cold Spring Harbor Laboratory, Cold Spring Harbor, N.Y.
6. **Dutcher, S. K.** 1981. Internuclear transfer of genetic information in *kar1-1/KAR1* heterokaryons in *Saccharomyces cerevisiae*. Mol. Cell. Biol. **1:**245–253.
7. **Ephrussi, B., H. Hottinguer, and J. Taulitski.** 1949. Action de l'Acroflavine sur levures. II. Etude génétique du mutant "petite colonie." Ann. Inst. Pasteur **76:**419–450.
8. **Fowell, R. R.** 1969. Sporulation and hybridization of yeasts, p. 303–383. *In* A. H. Rose and J. S. Harrison (ed.), The yeasts. Academic Press, Inc., New York.
9. **Grenson, M., M. Mousset, J. M. Wiome, and J. Bechet.** 1966. Multiplicity of the amino acid permeases in *Saccharomyces cerevisiae*. I. Evidence for a specific arginine transporting system. Biochim Biophys. Acta **127:** 325–338.
10. **Guglielminetti, R., S. Bonatti, N. Loprieno, and A. Abbondandolo.** 1967. Analysis of the mosaicism induced by hydroxylamine and nitrous acid in *Schizosaccharomyces pombe*. Mutat. Res. **4:**441–447.
11. **Hartwell, L. H., S. K. Dutcher, J. S. Wood, and B. Garvik.** 1982. The fidelity of mitotic chromosome reproduction in *Saccharomyces cerevisiae*. Rec. Adv. Yeast Mol. Biol. **1:**28–38.
12. **Herskowitz, I., and Y. Oshima.** 1981. Control of cell type in *Saccharomyces cerevisiae*. Mating type and mating-type interconversion, p. 181–209. *In* J. N. Strathern, E. W. Jones, and J. R. Broach (ed.), The molecular biology of the yeast *Saccharomyces*: life cycle and inheritance. Cold Spring Harbor Laboratory, Cold Spring Harbor, N.Y.
13. **Hinnen, A., J. B. Hicks, and G. R. Fink.** 1978. Transformation of yeast. Proc. Natl. Acad. Sci. USA **75:**1929–1933.
14. **Kikuchi, Y.** 1983. Yeast plasmid requires a cis-acting locus and two plasmid proteins for its stable maintenance. Cell **35:**487–493.
15. **Lindegren, G., L. Y. Hwang, Y. Oshima, and C. Lindegren.** 1965. Genetical mutants induced by ethyl methanesulfonate in *Saccharomyces*. Can. J. Genet. Cytol. **7:**491–499.
16. **Loprieno, N., and C. H. Clarke.** 1965. Investigation on reversions to methionine independence induced by mutagens in *Schizosaccharomyces pombe*. Mutat. Res. **2:**312–319.
17. **Maniatis, T., E. F. Fritsch, and J. Sambrook.** 1982. Molecular cloning: a laboratory manual. Cold Spring Harbor Laboratory, Cold Spring Harbor, N.Y.
18. **Mortimer, R. K., R. Contopoulou, and D. Schild.** 1981. Mitotic chromosome loss in a radiation sensitive strain of the yeast *Saccharomyces cerevisiae*. Proc. Natl. Acad. Sci. USA **78:**5778–5782.
19. **Mortimer, R. K., and D. C. Hawthorne.** 1969. Genetic mapping in yeast, p. 385–460. *In* A. H. Rose and J. S. Harrison (ed.), The yeasts. Academic Press, Inc., New York.
20. **Mortimer, R. K., and D. Schild.** 1985. Genetic map of *Saccharomyces cerevisiae*, edition 9. Microbiol. Rev. **49:**181–213.
21. **Nasim, A., and C. Auerbach.** 1967. The origin of complete and mosaic mutants from mutagenic treatment of single cells. Mutat. Res. **4:**1-14.
22. **Nasmyth, K. A., and S. I. Reed.** 1980. Isolation of genes by complementation in yeast: molecular cloning of a cell-cycle gene. Proc. Natl. Acad. Sci. USA **77:**2119–2123.
23. **Nilsson-Tillgren, T., J. G. Litske, S. Holmberg, and M. C. Kielland-Brandt.** 1980. Transfer of chromosome III during kar mediated cytoduction in yeast. Carlsberg Res. Commun. **45:**113–117.
24. **Orr-Weaver, T. L., and J. W. Szostak.** 1983. Multiple tandem plasmid integration in *Saccharomyces cerevisiae*. Mol. Cell. Biol. **3:**747–749.
25. **Orr-Weaver, T. L., J. Szostak, and R. Rothstein.** 1981. Yeast transformation: a model system for the study of recombination. Proc. Natl. Acad. Sci. USA **78:**6354–6358.
26. **Orr-Weaver, T. L., J. Szostak, and R. Rothstein.** 1983. Genetic applications of yeast transformation with linear and gapped plasmids. Methods Enzymol. **101:**228–245.
27. **Reaume, S. E., and E. L. Tatum.** 1949. Spontaneous and nitrogen mustard-induced nutritional deficiencies in *Saccharomyces cerevisiae*. Arch. Biochem. **22:**331–338.
28. **Rothstein, R. J.** 1983. One step gene disruption in yeast. Methods Enzymol. **101:**202–211.

29. **Schlief, R. F., and P. C. Wensink.** 1981. Practical methods in molecular biology. Springer-Verlag, New York.
30. **Thorner, J.** 1981. Pheromonal regulation of development in *Saccharomyces cerevisiae*, p. 143–180. *In* J. N. Strathern, E. W. Jones, and J. R. Broach (ed.), The molecular biology of the yeast *Saccharomyces*: life cycle and inheritance. Cold Spring Harbor Laboratory, Cold Spring Harbor, N.Y.
31. **Wehlan, W., E. Gocke, and T. Manney.** 1979. The *Can*1 locus of *Saccharomyces cerevisiae*: fine structure analysis and forward mutation rates. Genetics **91**:35–51.
32. **Wood, J. S.** 1982. Genetic effects of methyl benzimidazole-2-yl-carbamate on *Saccharomyces cerevisiae*. Mol. Cell. Biol. **2**:1064–1079.

IV. Immobilization and Cell Culture

Immobilization and Cell Culture: Introduction

BERNARD J. ABBOTT

Techniques of enzyme and cell immobilization continue to evolve and improve, although at a diminished rate since the 1970s. Sustaining, and even rekindling, interest in this area are the improvements in the production of microbial enzymes afforded by genetic engineering. Now it is possible to engineer a microorganism to produce a substantial portion of its total cell protein as a specific enzyme. This capability, together with methods to economically reuse and stabilize enzymes, will lead to more examples of the commercial exploitation of immobilization technology. The two chapters by Chibata et al. and by Bernath and Venkatasubramanian will enable the reader to rapidly explore the many processing advantages that can be realized from enzyme and cell immobilization.

A dramatic resurgence of interest in mammalian cell culture can be directly tied to attempts to clone and express mammalian proteins in *Escherichia coli*. These studies have shown that proteins with numerous disulfide bonds cannot be produced in bacterial cells because of the intracellular reducing environment. Forming the correct disulfide bonds after protein purification can be a formidable task. With mammalian cells, proteins can be produced with the correct disulfide structures. Also, mammalian cells, unlike bacteria, are capable of posttranslationally modifying proteins by glycosylation, γ-carboxylation of glutamate, etc. Some of these modifications may be needed for biological activity or to prevent an immune response when used clinically. Thus, to make some proteins by recombinant DNA technology, mammalian cells must be engineered to produce them. An equally important use for mammalian cell culture is the production of monoclonal antibodies. In addition to the potential therapeutic use, these antibodies are finding widespread commercial applications as diagnostic tools. The chapter by Boder and Hull provides numerous tips and techniques for cell culture garnered through years of hands-on experience.

Finally, genetic engineering also has stimulated renewed interest in plant cell culture. The cultivation of plant tissues in culture is presently required for the successful introduction of heterologous genes into plants. Indeed, the inability to grow appropriate cells in culture and the failure to regenerate the entire plant are currently major obstacles to transformation of agronomically important crops. The methods for culturing plant cells differ from species to species and will vary with different tissues from the same plant. Methods and guidelines for successful plant tissue culture and protoplast regeneration are concisely outlined in the chapter by Gamborg.

Chapter 18

Methods of Cell Immobilization

ICHIRO CHIBATA, TETSUYA TOSA, AND TADASHI SATO

Over the past two decades there have been rapid developments in the use of enzymes as catalysts for industrial, analytical, and medical purposes, and a new field of research known as enzyme technology has appeared. In this field of study, enzymes are immobilized to make their use more convenient in such a way that they resemble ordinary solid-phase catalysts used for synthetic chemical reactions.

Although enzymes are produced by all living organisms, enzymes from microbial sources are the most suitable for industrial purposes for the following reasons. (i) The production cost is low. (ii) The conditions for production are not restricted by location and season. (iii) The time required for production is short. (iv) Mass production is possible.

Microbial enzymes can be classified into two groups: extracellular, if the enzyme is excreted from the cells into the growth medium, and intracellular, if the enzyme is retained in or on the cells during cultivation.

To utilize intracellular enzymes, it is necessary to extract them from the cells. However, such extracted enzymes may be unstable in some cases and not suitable for practical use as immobilized enzymes. In addition, the process of extraction is generally expensive. To avoid extraction and purification by procedures that are difficult to scale up, and to avoid the problems of enzyme instability, whole cells have been immobilized and used as solid catalysts. Many industrial applications using continuous enzyme reactions are now carried out using immobilized microbial cells.

With immobilized microbial cells, there may be a limitation in the permeability of the substrate and product through the cell membrane. In addition, the presence of many enzymes in the cells may lead to side reactions. These problems can be solved, however, and the enzyme systems within the microorganism can be efficiently utilized.

Reactions by immobilized microbial cells are advantageous when: (i) enzymes are intracellular; (ii) enzymes extracted from cells are unstable during and after immobilization; (iii) the microorganism contains no interfering enzymes or when interfering enzymes are readily inactivated or removed; or (iv) the substrates and products are low-molecular-weight compounds. Under the conditions described above, the following advantages of immobilized microbial cells may be expected. (i) The processes for extraction or purification, or both, of enzymes are not necessary. (ii) The yield of enzyme activity on immobilization is high. (iii) The operational stability is generally high. (iv) The cost of the enzyme is low. (v) Applications for a multistep enzyme reaction may be possible.

Another aspect that should be considered is the volume of liquid to be processed. For the unit production of a desired compound, the required volume of fermentation broth is much smaller in the case of continuous reactions using immobilized cells as compared to conventional batch fermentation. Thus, the continuous process may reduce plant pollution problems.

Many papers and reviews on immobilization of microbial cells have been published (7, 10, 12, 51). However, no ideal general methods applicable to immobilization of all types of microbial cells have been developed. In practice, it is necessary to choose suitable methods and conditions for immobilization of each type of cell.

In this chapter, we summarize cell immobilization methods and describe in detail the most commonly used techniques and matrices.

18.1 IMMOBILIZATION OF CELLS

A large number of immobilization methods have been published. These techniques can be classified into three categories: carrier-binding, cross-linking, and entrapping methods.

18.1.1 Carrier-Binding Method

The carrier-binding method is based on direct binding of cells to water-insoluble carriers by physical adsorption, ionic bonds, or covalent bonds. As carriers, water-insoluble polysaccharides (cellulose, dextran, and agarose derivatives), proteins (gelatin and albumin), synthetic polymers (ion-exchange resins and polyvinylchloride), and inorganic materials (brick, sand, and porous glass) are used.

One of the first attempts to immobilize cells (27, 28) employed *Escherichia coli* and *Azotobacter agile* cells adsorbed to Dowex-1 for the oxidation of succinic acid. Adsorption of cells to carriers is dependent on the characteristics of the environment because the adsorption phenomenon is based primarily on electrostatic interactions (van der Waals forces) and on ionic and hydrogen bonds between the cell surface and the carrier material.

The covalent-binding method is one of the most frequently used techniques for the immobilization of enzymes because enzyme preparations with high operational stabilities are obtained. However, for cells this process of immobilization is not widely used because of the toxicity of the coupling agents involved, sometimes causing the loss of both enzyme activity and cell viability.

Covalent coupling is the direct linkage of cells to an activated support. Any reactive component of the cell surface, e.g., the amino, carboxyl, thiol, hydroxyl, imidazole, or phenol groups of proteins, can be used for linking. To introduce the covalent linkage, chemical modification of the carrier is usually necessary. In the few reports describing this immobilization method, inorganic and organic carriers have been used. Silica and ceramics are the major inorganic carriers that have been used. Navarro and Durand (64) immobilized *Saccharomyces carlbergensis* to porous silica beads which had been treated with aminopropyltriethoxysilane and activated with glutaraldehyde. Messing et al. (53) coupled cells of *Serratia marcescens*, *Saccharomyces cerevisiae*, and *Saccharomyces amurcae*, by using an isocyanate coupling agent, to borosilicate glass and zirconia ceramics.

In the case of organic carriers, Jack and Zajic (31) immobilized *Micrococcus luteus* cells on the carboxyl groups of agarose beads in a two-step process that avoided exposure of the cells to carbodiimide; thus, the urocanic acid-forming activity of these cells was retained even though cell viability was lost. This method has the disadvantage that leakage of the enzymes may occur due to autolysis during the enzyme reaction.

18.1.2 Cross-Linking Method

Cells can be immobilized by cross-linking with bi- or multifunctional reagents such as glutaraldehyde, toluene diisocyanate, and others. We used this method to immobilize *E. coli* cells having high aspartase activity by cross-linking cell walls or cell membranes with the bifunctional reagents glutaraldehyde and toluene diisocyanate (8). A 34% retention of aspartase activity was obtained by using glutaraldehyde, but inactivated and immobilized cell preparations were obtained with toluene diisocyanate and toluene diisocyanate plus 1,6-diaminohexane. Since very few studies on this method have appeared, there remains a possibility that better cross-linking reagents for immobilization of cells will be found in the future.

18.1.3 Entrapping Method

The method of directly entrapping cells into polymer matrices has been extensively investigated. The following matrices have been employed: collagen, gelatin, agar, alginate, carrageenan, cellulose triacetate, polyacrylamide, epoxy resin, photo-cross-linkable resin, polyester, polystyrene, and polyurethane. Of these matrices, polyacrylamide gel, alginate gel, carrageenan, and photo-cross-linkable resin have been used most extensively.

Polyacrylamide gel

The polyacrylamide gel method was used for the immobilization of enzymes by Bernfeld et al. in 1963. The technique was subsequently applied to the cells of the lichen *Umbilicaria pustulata* by Mosbach and Mosbach in 1966.

The procedure for the preparation of insoluble-gel networks is identical to that employed for the preparation of gels commonly used for electrophoresis. The method is based on the free-radical polymerization of acrylamide monomer in aqueous solution. The free-radical polymerization of acrylamide occurs in solution containing the cells and a cross-linking agent such as *N,N'*-methylenebisacrylamide. Polymerization is commonly carried out at a low temperature to avoid cell damage. The polymerizing reaction is initiated by potassium persulfate or riboflavin and accelerated by β-dimethylaminopropionitrile or *N,N,N',N'*-tetramethylenediamine. The resulting gel can be mechanically dispersed into particles of any desired size. However, the gels are structurally weak. This problem can be overcome by optimizing the degree of cross-linking. We extensively investigated the conditions for immobilization of cells (8) (e.g., *E. coli* with high aspartase activity) and achieved highly active, stable, and mechanically strong immobilized cells. Table 1 lists examples of the use of polyacrylamide for the immobilization of cells.

A major disadvantage of polyacrylamide is the toxicity of the acrylamide monomer, the cross-linking agent, the initiator, and the accelerator. In some cases, free-radical polymerization results in a decrease not only in enzyme activity but also in cell viability. Therefore, microorganisms should be exposed to the reagents for the shortest time possible.

The optimum conditions for the immobilization of *E. coli* cells in polyacrylamide are as follows (9).

Cell suspension. *E. coli* cells (1 kg, wet weight) are suspended in 2 liters of physiological saline or broth, and the suspension is cooled to 8°C.

Monomer solution. Acrylamide monomer (750 g) and *N,N'*-methylenebisacrylamide (40 g) are dissolved in 2.4 liters of water, and the monomer solution is cooled to 8°C.

Immobilization. The cell suspension and monomer solution are mixed at 8°C. Portions of 25% (vol/vol) β-dimethylaminopropionitrile (100 ml) and 1% potassium persulfate (500 ml) are added to the reaction mixture, which is then maintained at 20 to 25°C. The polymerization starts in about 5 min, and the temperature increases as the reaction proceeds. When the temperature increases to 30°C, the resulting stiff gel is cooled rapidly with ice-cold water and held for 15 to

TABLE 1. Microbial cells immobilized with polyacrylamide and their products

Cells	Products	References
Acetobacter xylinum	Dihydroxyacetone	63
Achromobacter aceris	NADP	99
Achromobacter butyri	Glucose 6-phosphate	62
Achromobacter guttatus	ε-Aminocarpoic acid	34
Achromobacter liquidum	Urocanic acid	109
Achromobacter sp.	*d*-Tartaric acid	Patent[a]
Arthrobacter globiformis	Prednisolone	38
Aspergillus tereus	Itaconic acid	30
Bacillus sp.	Bacitracin	55, 57
	D-α-Phenylglycine	107
	D-α-Hydroxyphenylglycine	107
Bacillus subtilis	α-Amylase	36
Brevibacterium ammoniagenes	L-Malic acid	112, 113
	Coenzyme A	81
	NADP	60
Corynebacterium dismutans	L-Alanine	74
Corynebacterium glutamicum	L-Glutamic acid	83
Curvularia lunata and *Corynebacterium simplex*	Prednisolone	43, 59
Escherichia coli	L-Aspartic acid	8, 75, 96
	L-Tryptophan	Patent[b]; 3, 47
	5-Hydroxytryptophan	Patent[c]
	6-APA	78
Escherichia freundii	Glucose 6-phosphate	72
	Glucose 1-phosphate	72
Gluconobacter melanogenus and *Pseudomonas syringus*	2-Ketogulonic acid	49, 50
Kluyvera citrophila	Ampicillin	56
Microbacterium ammoniaphilum	L-Lysine	Patent[d]
Penicillium chrysogenum	Penicillin G	54
Pseudomonas putida	L-Citrulline	110
Saccharomyces cerevisiae and *Escherichia coli*	Glutathione	61
Streptomyces clavuligerus	Cephalosporin C	20
Streptomyces fradiae	Protease	37
Xanthomonas citri	Cephalexin	33

[a] Y. Kawabata and S. Ichikura, Japanese patent 77-102496, August 1977.
[b] I. Chibata, T. Kakimoto, and K. Nabe, Japanese patent 74-81591, August 1974.
[c] I. Chibata, T. Kakimoto, and K. Nabe, Japanese patent 74-81590, August 1974.
[d] O. Kanemitsu, Japanese patent 75-132181, October 1975.

20 min to complete the polymerization. In this process it is very important to keep the temperature of the gel below 50°C. The gel is then cut into 3- to 4-mm cubes.

κ-Carrageenan

κ-Carrageenan is a readily available nontoxic polysaccharide isolated from seaweed. It is widely used in the food and cosmetic industries as a gelling, thickening, and stabilizing agent. It is composed of a unit structure of β-D-galactose sulfate and 3,6-anhydro-α-D-galactose.

κ-Carrageenan becomes a gel by cooling or by contact with a solution containing a gel-inducing agent such as K^+, NH_4^+, Ca^{2+}, Cu^{2+}, Mg^{2+}, Fe^{3+}, amines, or water-miscible organic solvents (97). If the proper gel-inducing agent is used, a preparation with high enzyme activity can be obtained.

An advantage of using κ-carrageenan is that immobilization can be performed under very mild conditions without the use of chemicals that might destroy the desired enzyme activity. Generally, the enzyme activities and yields of immobilized microbial cells using K^+ ion as a gel-inducing reagent are relatively high.

Another advantage of this method is that various shapes of immobilized cells can be tailor-made for particular applications. In addition, when immobilized cells prepared with gel-inducing reagents are suspended in physiological saline, the κ-carrageenan gel rapidly dissolves and a free-cell suspension is obtained. This property is useful for investigating characteristics of cells after immobilization, e.g., viability.

A disadvantage of κ-carrageenan is that when a gel-inducing reagent is not present in the reaction mixture, cells are released by dissolution of the gel. Also, the gel-inducing reagent may inhibit a desirable enzyme activity. Table 2 lists some examples in which κ-carrageenan was used for the immobilization of cells.

Two basic procedures are available for immobilization of cells with κ-carrageenan. The first is a one-step procedure which directly entraps a large amount of cells in the gel. The second is a two-step procedure in which a small amount of cells is entrapped in carrageenan gel and allowed to grow in the gel by incubation in a nutrient medium.

(i) The one-step procedure

In this procedure, a large number of cells are homogeneously immobilized in carrageenan gel. The cells may be growing, resting, or in an autolyzed state. In many cases, the immobilized cells are dead, although the desired enzyme remains in an active and stable state. The pore size of the gel matrix is small enough to prevent high-molecular-weight compounds, such as enzymes, from leaking out from the gel lattice, while the low-molecular-weight substrates and products easily pass through. Practical examples for immobilizing *E. coli* with aspartase activity are described below (97). Typical procedures for preparation of various shapes of immobilized cells such as cube, bead, and membrane are schematically illustrated in

TABLE 2. Microbial cells immobilized with κ-carrageenan and their products

Cells	Products	References
Brevibacterium flavum	L-Malic acid	90, 91
Brevibacterium fuscum	12-Ketochenodeoxycholic acid	79
Candida tropicalis	α,ω-Dodecanedioic acid	114
	α,ω-Tridecanedioic acid	114
Corynebacterium dismutans	L-Alanine	74
Enterobacter aerogenes	2,3-Butanediol	14
Erwinia carotovora	*N*-Acetyl-L-methionine	65
Escherichia coli	L-Aspartic acid	66, 76
Escherichia coli and *Pseudomonas dacunhae*	L-Alanine	77, 88
Penicillium urticae	Patulin	17
Propionibacterium sp.	Vitamin B_{12}	117
Pseudomonas dacunhae	L-Alanine	111
Saccharomyces bayanus	Ethanol	105
Saccharomyces cerevisiae	Glycerol	4
Serratia marcescens	L-Arginine	22
	L-Isoleucine	106
Tolypocladium inflatum	Cyclosporin A	19
Trichoderma reesei	Cellulase	21
Zymomonas mobilis	Ethanol	2, 26

Fig. 1, and a scanning electron micrograph of immobilized cells is shown in Fig. 2.

Cube type. Whole cells of *E. coli* (16 g, wet weight) are suspended in 16 ml of physiological saline at 40°C, and 3.4 g of κ-carrageenan are dissolved in 60 ml of the saline at 45°C. The solutions are mixed, cooled to 10°C, and held for 30 min. To increase gel strength, the gel is soaked in a cold 0.3 M KCl solution. After this treatment, the resulting stiff gel is cut into cubes (3 mm^3) with a knife.

Bead type. To form beads, 50 ml of the mixture of κ-carrageenan and *E. coli* cells described above is dropped at a constant speed into 0.3 M KCl solution through an injector with an orifice 1 mm in diameter. The solution should be gently stirred at 20°C, and gel beads 3 to 4 mm in diameter are then obtained.

Membrane type. Membranes are formed by spreading 50 ml of the mixture described above on nylon net in a vessel (3 by 250 by 200 mm). The film is soaked in a cold 0.3 M KCl solution, and the resulting gel membrane is cut to the desired size.

(ii) The two-step procedure

In the first step, cells are cultured and the whole fermentation broth is mixed with a κ-carrageenan solution under sterile conditions. The mixture is added dropwise with gentle stirring into an aqueous solution containing a gel-inducing agent. Gel beads with a homogeneous distribution of a small number of cells are obtained. The gel beads are then incubated in a nutrient medium. An example experiment utilizing live *S. cerevisiae* for ethanol production is described in detail below (105).

***S. cerevisiae* IFO 2367.** The microorganism is grown under aerobic conditions for 18 h at 30°C in a medium containing 0.5% glucose, 1.25% yeast extract, 1.0% peptone, and 0.5% NaCl (pH 7.0). The whole broth (1 ml) is mixed with 50 ml of 4% carrageenan in physiological saline at 37°C under aseptic conditions. The mixture is dropped through an injector (with an orifice 1 mm in diameter) at a constant speed into a 2% KCl solution with gentle stirring at 20°C. Gel beads (3 to 4 mm in diameter) are obtained that contain a small number of cells (3.5×10^6 cells per ml of gel). Cell beads (10 ml) are incubated on a rotary shaker at 30°C in 100 ml of a medium (pH 5.0) containing 10% glucose, 0.15% yeast extract, 0.25% NH_4Cl, 0.55% K_2HPO_4, 0.025% $MgSO_4 \cdot 7H_2O$, 0.1% NaCl, 0.001% $CaCl_2$, and 0.3% citric acid. After 60 h of incubation, the number of cells in the gel increases by 1,000-fold (5.4×10^9 cells per ml of gel). The number of cells in a unit volume of gel is about 10 times higher than that of free cells in ordinary liquid culture. In addition, the cells form a thin layer of concentrated cells near the surface of the gel beads (Fig. 3). These immobilized living cells efficiently produce ethanol from glucose.

Estimation of number of living cells in the gel beads. The beads (two particles) isolated from the medium described above are dissolved in 5 ml of sterilized physiological saline by gentle shaking for 15 min at 37°C and then converted to a cell suspension. The cell suspension is serially diluted, and the number of viable cells is determined by colony counting on agar plates.

Calcium alginate

Alginate, which is extracted from seaweed, is a linear copolymer of D-mannuronic and L-guluronic acid. It can be gelled by multivalent ions such as calcium and aluminum. For immobilization of cells, the cross-linking of alginate with Ca^{2+} is the preferred method. This method has been used to immobilize very sensitive cells such as plant cells (5, 101) and protoplasts (80). The flexibility of this method is illustrated by the following parameters. (i) Alginates of different molecular weight and chemical composition (ratio of mannuronic and guluronic acid residues) can be used. (ii) Alginate solutions of concentrations between 0.5 and 10% can be used. (iii) The $CaCl_2$ concentration in the precipitation bath can be varied between 0.05 and 2%. (iv) The working temperature can be between 0 and 80°C. (v) Beads ranging from 0.1 to 5 mm in diameter can be prepared with uniform size distribution. (vi) Cell loading by direct entrapment can be high (up to 30 g [wet weight] of cells per ml of catalyst).

When a medium containing calcium-chelating agents (such as phosphates) or cations (such as Mg^{2+} or K^+) is used during an enzyme reaction, disruption of the gel occurs by solubilization of the bound Ca^{2+}. Cell leakage from the matrix may also occur when cell division within the gel beads takes place and the beads are used in stirred vessels. Table 3 lists some applications of calcium alginate for the immobilization of microbial cells. Some examples of immobilization in alginates are described below in more detail.

***Arthrobacter simplex* (67).** Wet cells of *A. simplex* (0.25 to 0.5 g) are suspended in 0.5 ml of water. The cell suspension is mixed with 9 ml of an autoclaved

3% (wt/vol) sodium alginate solution. The alginate-cell mixture is added to 0.1 M $CaCl_2$ as droplets with the aid of a syringe equipped with a hypodermic needle (0.1 to 1 mm in diameter). The speed of the extrusion of the alginate solution through the needle in combination with the choice of the needle gauge determines the size of the beads. The beads are allowed to "cure" at 20 to 22°C for about 3 h. The gel beads are then washed briefly with water. The immobilized *A. simplex* cells can be used for the production of prednisolone from cortisol.

***S. cerevisiae* (102).** Sodium alginate (50 ml; 4%, wt/vol) is mixed with 50 ml of cell suspension (containing 25 g [wet weight] of cells) until homogeneous. The mixture is passed through a narrow tube and dropped into 0.05 M $CaCl_2$ solution (pH 6 to 8) at 37°C, forming beads 2.8 to 3.0 mm in diameter. The beads are allowed to "cure" at 20 to 22°C for 1 h, rinsed with water, and equilibrated overnight in a 0.05 M $CaCl_2$ solution at 4°C. The immobilized *S. cerevisiae* cells can produce ethanol from glucose.

Plant cells (*Daucus carota*) (101). A 5- to 6-day-old culture of *D. carota* is harvested by low-speed centrifugation, and the pelleted cells are adjusted to a 2.5-ml packed-cell volume. Culture medium (2.5 ml) is added to the cells, and the resulting suspension is mixed with 15 ml of 3.2% sodium alginate. The alginate-cell suspension is dropped under gravity through a 1-mm orifice into 500 ml of 0.05 M $CaCl_2$ solution. The beads (3 to 4 mm in diameter) of immobilized cells are annealed in the $CaCl_2$ solution with stirring for 30 min. The immobilized plant cells catalyze the 5β-hydroxylation of digiotoxigenin.

Photo-cross-linkable resin prepolymers

Polyglycol oligomers are functionalized with polymerizable vinyl end groups. After the addition of a photosensitizer and under illumination with UV light for several minutes, flat network sheets are obtained. The oligomer chain length can be controlled to determine the network porosity, and the chemical composition of the polyglycol precursor determines the hydrophilicity of the matrix.

Fukui and co-workers (68, 71) have developed a method to entrap enzymes and whole cells with photo-cross-linkable resin prepolymers. The structures of typical photo-cross-linkable resin prepolymers are shown in Fig. 4. A typical procedure for the preparation of the prepolymers consists of reacting equimolar amounts of hydroxyethyl acrylate and isophorone diisocyanate at 70°C in the presence of a suitable catalyst. After 2 h a half-molar ratio of the polyethylene glycol is added. The reaction proceeds for 5 h at 70°C. The resulting product is the photo-cross-linkable resin prepolymer. In this process, 1 part of the water-soluble prepolymer is mixed with 0.01 part of a photosensitizer, usually benzoin ethyl ether, and the mixture is melted by warming at 60°C. To this molten mixture is added a whole-cell suspension, and the resultant mixture is illuminated at 370 nm for 3 min. The resin gel formed is cut into small pieces and used. The immobilized cells obtained are primarily used in organic solvents for transformations of steroids because of the low solubility of both the substrate and the product.

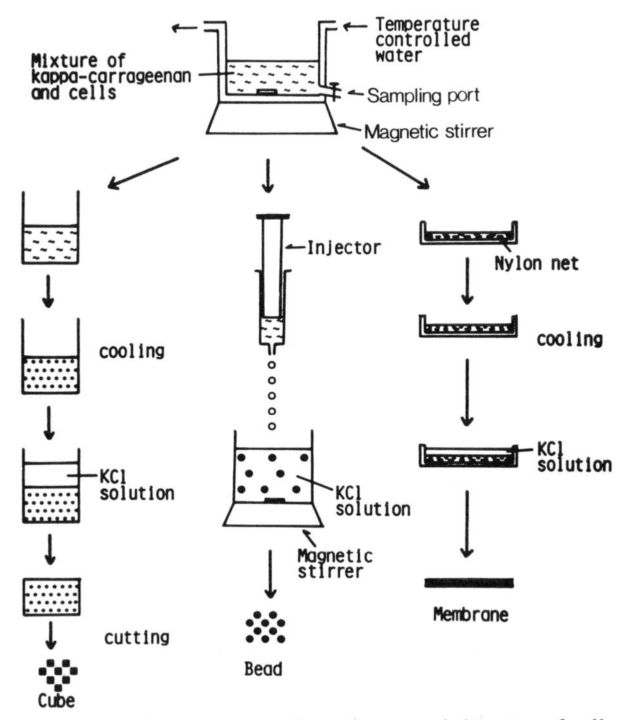

FIG. 1. Schematic procedures for immobilization of cells using κ-carrageenan.

Table 4 lists some examples of the use of photo-cross-linkable resin prepolymers for the immobilization of microbial cells. More detailed descriptions on the use of this method are presented below.

***A. simplex* (71).** *A. simplex* acetone-dried cells (200 mg) are suspended in 2 ml of a water-acetone mixture (1:19, vol/vol) containing 1 g of maleic polybutadiene derivative. The mixture is layered on a sheet of transparent polyester and radiated from an appropriately placed UV light source for 3 min. The resin film which forms is cut into small pieces (approximately 5 by 5 mm) and used for transforming hydrocortisone.

***Nocardia rhodocrous* (68).** Thawed cells of *N. rhodocrous* (1 g of wet cells) suspended in 2 ml of a water-saturated mixture of benzene and *n*-heptane (1:1, vol/vol) by mild sonication are mixed with 1 ml of the same solvent containing 1 g of hydrophobic polypropylene derivative and 10 mg of benzoin ethyl ether as a photosensitizer. The mixture is placed on a sheet of transparent polyester and illuminated with near-UV light for 3 min. The resin gel that forms is cut into small pieces (approximately 1.5 by 1.5 mm) and used for transformation of 3β-hydroxy-Δ^5-steroid.

***Rhodotorula minuta* (69).** Thawed cells of *R. minuta* (1 g [wet weight] of cells) are suspended in 3 ml of water and immobilized with 1 g of the hydrophilic polyethylene glycol derivative described above. The immobilized cells are used for hydrolysis of methyl ester in an organic solvent.

Other entrapping methods

Immobilization of microbial cells using collagen can be carried out by the technique developed by Vieth et al. (104) for immobilization of enzymes. This method is suitable for the preparation of cells immo-

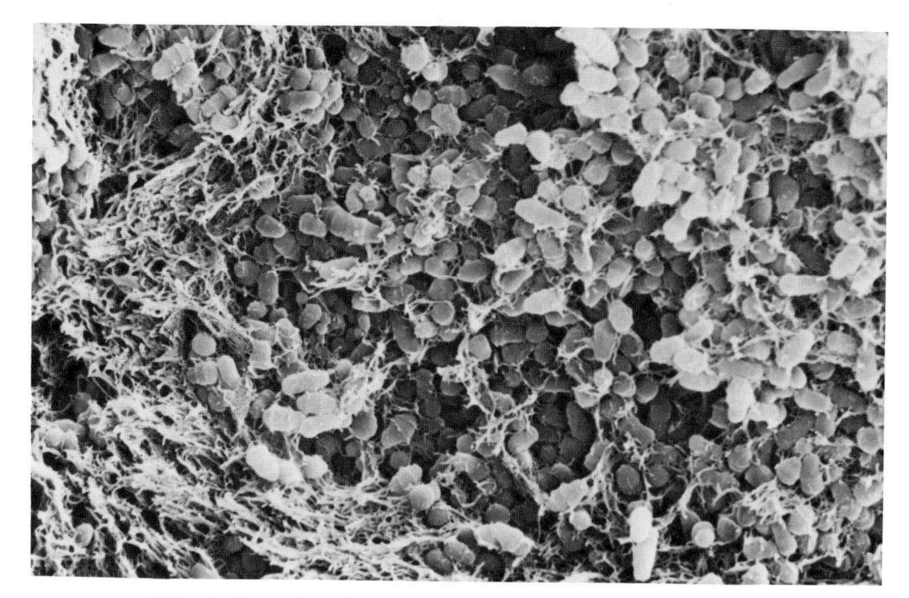

FIG. 2. Scanning electron micrograph of immobilized cells.

bilized in membranes. Since the collagen membrane is weak and enzymes are readily lost from the matrix, various hardening treatments are carried out after immobilization. Formaldehyde, glutaraldehyde, and dialdehyde starch are used as hardening agents.

For example, *Corynebacterium simplex* cells can be immobilized with collagen as follows (15). *C. simplex* cells (20 g, dry weight) are suspended in 100 ml of water, and the suspension is gently added to a 2% collagen solution. The mixture is adjusted to pH 8.5 with 0.1 N NaOH. To this mixture 3 ml of 50% glutaraldehyde is added, and the mixture is allowed to incubate for 5 to 10 min before being cast as a membrane on a Mylar sheet. The membrane is dried at room temperature. The dried membrane is hardened by dipping in alkaline glutaraldehyde solution, and the hardened cell-collagen membrane is cut into small pieces.

Entrapment of cells in an agar gel is an obvious method of cell immobilization. This method is not widely used, however, because of the poor mechanical strength of the gel. Cells of the yeast *Saccharomyces pastorianus* have been immobilized by direct injection of a warm (50°C) 2.5% (wt/vol) agar suspension containing the whole cells into toluene or tetrachloroethylene (93).

Immobilization of microbial cells using cellulose fibers was developed by Dinelli (18). The enzyme activity of immobilized microbial cells prepared by this method is very stable, and the technique can be applied to many enzymes or microbial cells. For example, cellulose triacetate is dissolved in methylene chloride, and a microbial cell suspension containing glycerol is added dropwise with gentle stirring. The mixture is emulsified, and the emulsion is spun through a spinneret into a coagulating bath containing toluene. The resulting fibers are dried in vacuo to give immobilized cell fibers. This method has several advantages. The fibers are resistant to both weak acid (or weak alkali) solutions of high ionic strength and some organic solvents. However, their use is limited to substrates and products of low molecular weight because of restricted diffusion into and out of the fibers. In addition, the method is limited by the necessity of using water-immiscible liquids as polymer solvents and coagulants, which may cause damage to the cells.

18.1.4 Miscellaneous Immobilization Methods

Fixation of enzymes within microbial cells can be carried out by heat treatment at a temperature that does not inactivate the desired enzyme. For example, whole cells of a *Streptomyces* sp. with glucose isomerase activity are heated at 60 to 85°C for 10 min (89). This procedure fixes the enzyme inside the cells and prevents leakage when the cells are incubated under reaction conditions. Such immobilized cells have been used for the industrial production of high-fructose syrup from D-glucose. The glucose isomerase of *Streptomyces phaeochromogenes* has also been fixed inside the cells by β-irradiation (N. Tsumura and T. Kasumi, Japanese patent 77-44285, April 1977). A method for immobilization of *S. phaeochromogenes* cells having glucose isomerase activity and *E. coli* cells having L-tryptopan synthetase activity by complex formation with chitosan has also been reported (98, 103). When cells are added to a chitosan solution in acetic acid and the mixture is adjusted to pH 6.0 with 0.1 N NaOH, a complex of cells and chitosan is formed which can be precipitated by aggregation.

18.2 APPLICATION OF IMMOBILIZED CELLS

As shown in the tables for the section above, microorganisms can be immobilized by various methods. In this section, some examples of the use of immobilized cells for production of useful chemicals are briefly described.

18.2.1 Production of Amino Acids

Amino acids are widely used in the food, feed, medicine, and cosmetic industries and also as starting materials for synthetic chemicals. For food and nutritional applications, only the L-isomers of the amino acids are active.

As an alternative to our immobilized aminoacylase process for continuous production of L-amino acids from acyl-DL-amino acids (11, 94, 95), *Aspergillus ochraceus* cells having aminoacylase activity were immobilized by cross-linking with egg albumin and glutaraldehyde. These cells catalyzed the continuous optical resolution of acetyl-DL-methionine (29).

L-Aspartic acid has been industrially produced from ammonium fumarate by fermentation or enzymatic batch processes that utilize aspartase. To improve the productivity of this system, we studied the continuous production of L-aspartic acid using immobilized enzymes and immobilized cells. We developed an efficient process using *E. coli* cells immobilized with polyacrylamide gel (8). By using a column packed with these cells, we commercialized a continuous process for L-aspartic acid production (75, 96). This process is very efficient and is superior to the conventional fermentative and enzymatic batch methods. It represents the first industrial use of immobilized cells.

In addition, we developed a novel method for immobilization of cells using κ-carrageenan as the gel matrix (97). The method is simple and does not harm cells. The new immobilization technique replaced the polyacrylamide gel method for industrial production of L-aspartic acid in 1978 (76).

L-Alanine was initially produced by an enzymatic batch process from L-aspartic acid using the enzyme L-aspartate 4-decarboxylase from *Pseudomonas dacunhae* cells. To provide a more efficient method, we developed a continuous process using *P. dacunhae* cells immobilized with κ-carrageenan (111). To improve the process further, we developed a system for the continuous production of L-alanine from ammonium fumarate using a mixture of immobilized *E. coli* cells with aspartase activity and *P. dacunhae* cells with L-aspartate 4-decarboxylase activity (77, 88). In 1982, we commercialized this mixed immobilized cell system. A flow diagram for continuous production of L-alanine by the mixed immobilized cells is shown in Fig. 5.

Many other microbial cells have been immobilized and investigated for production of various L-amino acids as shown in Tables 1 and 2. For example, living cells of *Brevibacterium flavum* immobilized in collagen have been used for the continuous production of L-glutamic acid in a recycle reactor system (16).

In addition, we have studied the continuous production of L-isoleucine and L-arginine using immobilized cell systems (22, 106).

18.2.2 Production of Antibiotics

Only a few studies have been published on the utilization of immobilized cells for the production of antibiotics. Mycelium and protoplasts of *Penicillium chrysogenum* were immobilized with polyacrylamide (54) and calcium alginate (41) to produce penicillin G from glucose. Cephalosporin C was synthesized by whole cells of *Streptomyces clavuligerus* immobilized with polyacrylamide (20). Acetone-dried cells of an *Achromobacter* sp. adsorbed on DEAE-cellulose or hydroxylapatite were used for the production of cephalexin (T. Fujii, K. Matsumoto, Y. Shibuya, K. Hanamitsu, T. Yamaguchi, T. Watanabe, and S. Abe, Japanese patent 73-75792, October 1973).

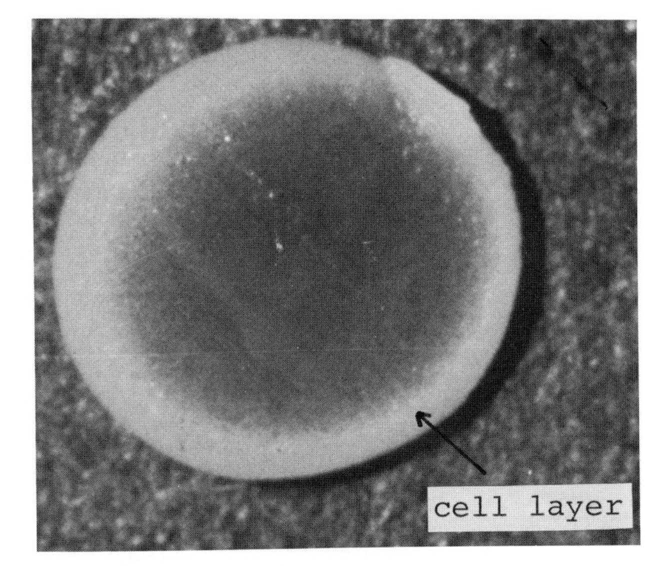

FIG. 3. Photograph of condensed cells in immobilized preparation.

Whole cells of a *Bacillus* sp. were immobilized in a polyacrylamide gel lattice and used for the production of bacitracin in batch and continuous processes (55, 57). The macrolide antibiotic tylosin and the nucleoside peptide antibiotic nikkomycin were produced by cells of a *Streptomyces* sp. and *Streptomyces tendae*, respectively, immobilized with calcium alginate (100).

Conidia of *Penicillium urticae* immobilized in κ-carrageenan beads were germinated in situ in a growth-supporting medium to form a patulin-producing cell mass, and the germinated cells produced patulin from glucose (17).

In another process, immobilized cells were used to synthesize ampicillin from 6-aminopenicillanic acid (6-APA) and D-phenylglycine methyl ester. This process used an *Achromobacter* sp. or *Bacillus megaterium* immobilized by ionic bindng on DEAE-cellulose (T. Fujii, K. Matsumoto, Y. Shibuya, K. Hanamitsu, T. Yamaguchi, T. Watanabe, and S. Abe, Belgian patent 803832, December 1973).

6-APA is an important intermediate for the production of semisynthetic penicillins. We studied a continuous method for the production of 6-APA from penicillin G using *E. coli* cells immobilized with polyacrylamide (78). By cloning the penicillin amidase gene of *E. coli* ATCC 11105 with multicopy plasmids, a hybrid strain of *E. coli* (5K [PHM 12]) with high enzyme activity was obtained (52). The cells of the new *E. coli* strain were immobilized in calcium alginate or in an epoxy matrix (obtained by polycondensation of water-soluble epoxy precursor and polyfunctional amine), and the immobilized cells were used for production of 6-APA from penicillin G (35).

TABLE 3. Microbial cells immobilized with calcium alginate and their products

Cells	Products	References
Arthrobacter simplex	Prednisolone	67
Aspergillus niger	Citric acid	45
	Gluconic acid	45
Clostridium butyricum	Isopropanol	39, 40
	n-Butanol	39, 40
Lactobacillus delbruekii	Lactic acid	92
Lactobacillus lactis	Lactic acid	45
Pachysolen tannophilus	Ethanol	82
Penicillium chrysogenum	Penicillin G	41
Propionibacterium sp.	Vitamin B_{12}	116, 117
Saccharomyces cerevisiae	Ethanol	13, 32, 42, 44, 46, 102
Saccharomyces formosensis	Ethanol	23
Streptomyces sp.	Tylosin	100
Streptomyces tendae	Nikkomycin	100
Trigonopsis variabilis	α-Keto acid	6
Zymomonas mobilis	Ethanol	26, 48

18.2.3 Production of Organic Acids

Organic acids are widely used in food and medicine, and some are produced by conventional fermentation. Immobilized microbial cells are occasionally utilized to improve productivities.

In 1974 we developed a continuous commercial process for making L-malic acid from fumaric acid using the enzyme fumarase in *Brevibacterium ammoniagenes* cells immobilized with polyacrylamide gel (112, 113). We subsequently used the carrageenan method to improve L-malic acid productivity (90). By screening various microorganisms for higher fumarase activity, *B. flavum* was found to exhibit higher enzyme activity, after immobilization with κ-carrageenan, than did *B. ammoniagenes* (90, 91). Therefore, the *B. ammoniagenes* method was replaced by the *B. flavum* method for commercial production in 1977.

We also immobilized cells of *Achromobacter liquidum* having high L-histidine ammonia lyase activity into a polyacrylamide gel lattice. These cells were used for the production of urocanic acid from L-histidine (109).

Immobilized cells have also been used for aerobic multistep reactions such as the production of acetic acid and citric acid. Cells of *Acetobacter aceti* were immobilized with porous ceramic and produced acetic acid in a simple medium containing glucose (24, 25).

Citric acid was produced from glucose using cells of *Aspergillus niger* immobilized with calcium alginate (45).

18.2.4 Production of Alcohols

More recently, the production of alcohols, primarily ethanol, using immobilized cells has been the subject of considerable interest. These systems seem promising for the commercial production of ethanol. We found that ethanol can be continuously and efficiently produced by using yeast cells immobilized with κ-carrageenan (105).

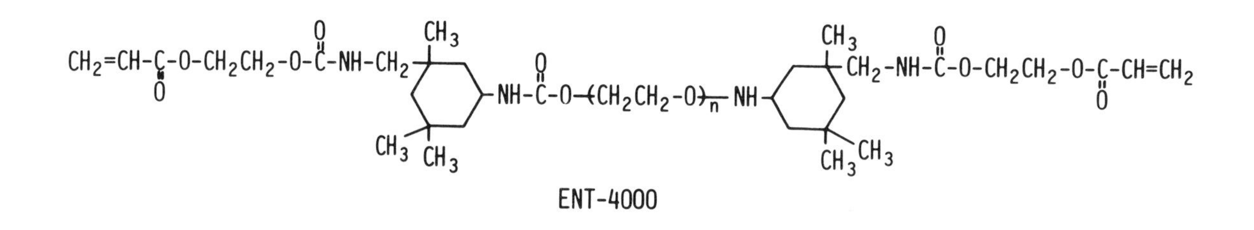

ENT-4000

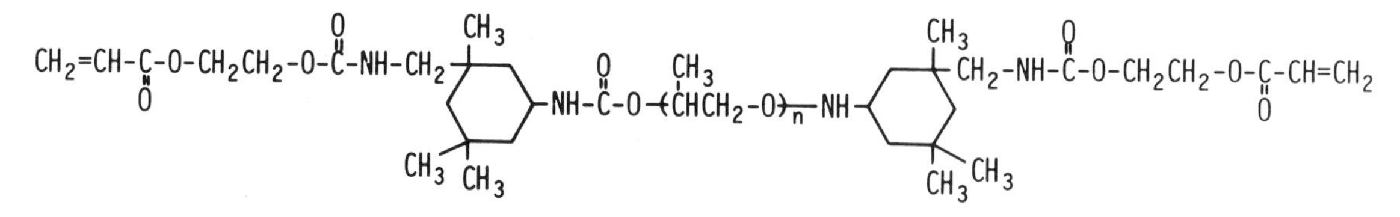

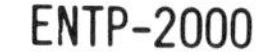

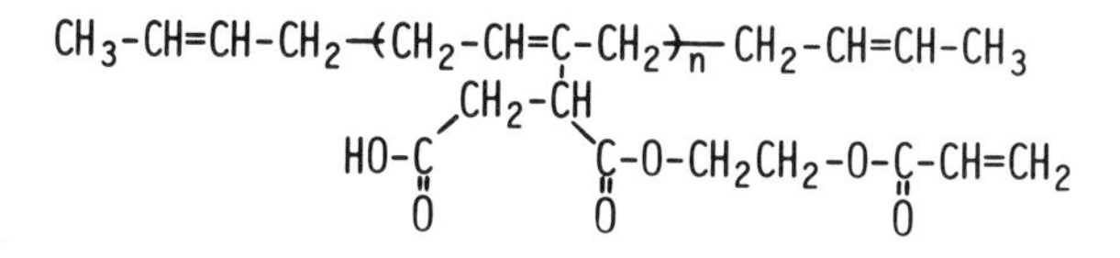

PBM-2000

FIG. 4. Structures of photo-cross-linkable resin prepolymers. ENT-4000, a derivative of polyethylene glycol (molecular weight, ≈4,000), water soluble; ENTP-2000, a derivative of polypropylene glycol (molecular weight, ≈2,000), water insoluble; PBM-2000, maleic polybutadiene (molecular weight, ≈2,000), water insoluble.

Other investigators have entrapped large amounts of *S. cerevisiae* cells in calcium alginate gel and then used these for the production of ethanol from glucose (32). However, good results were not obtained due to low enzyme activity and stability. *S. cerevisiae* cells have also been immobilized in calcium alginate gel for the continuous production of ethanol from cane molasses (46).

The continuous production of *n*-butanol and isopropanol using *Clostridium butyricum* entrapped in calcium alginate gel has been demonstrated (39, 40).

Glycerol can be produced using living cells of *S. cerevisiae* immobilized with κ-carrageenan by adding sodium sulfite to an ethanol production medium (4). In this case the sulfite caused excess formation of NADH, which reduces dihydroxyacetone phosphate to glycerol phosphate.

18.2.5 Other Useful Organic Compounds

Steroid transformation reactions have been carried out in organic solvents using immobilized microbial cells. *Arthrobacter simplex* cells immobilized with urethane polymer and *N. rhodochrous* cells immobilized with photo-cross-linkable resin were used in a nonaqueous solvent for the conversion of hydrocortisone and testosterone (70, 84, 85).

We immobilized *Achromobacter aceris* cells having a high NAD kinase activity into polyacrylamide gel for the continuous production of NADP from NAD and ATP (99). In this process, the utilization of ATP was not economically efficient. *S. cerevisiae* cells with high ATP regenerating activity and *B. ammoniagenes* cells with high NAD kinase activity were coimmobilized by entrapment in polyacrylamide gel (61) or by microencapsulation with cellulose acetate-butyrate (1). The coimmobilized cells were used for the production of NADP.

TABLE 4. Microbial cells immobilized with photo-cross-linkable resins and their products

Cells	Products	References
Arthrobacter simplex	Prednisolone	71
Corynebacterium sp.	9α-Hydroxy-4-androstene-3,17-dione	87
Curvularia lunata	Hydrocortisone	84, 87
Enterobacter aerogenes	Adenine arabinoside	84, 115
Nocardia rhodochrous	4-Androstene-3,17-dione	68, 115
Rhizopus stolonifer	11α-Hydroxyprogesterone	86
Rhodotorula minuta	*l*-Menthol	69

Flavin adenine dinucleotide was produced from flavin mononucleotide and ATP by whole cells of *Arthrobacter oxydans*. These cells possessed high flavin adenine dinucleotide pyrophosphorylase activity and were immobilized in a film of polyvinyl alcohol cross-linked with ethyl silicate (108). By the same immobilization procedure, whole cells of *Pseudomonas fluorescens* with high pyridoxine 5′-phosphate oxidase activity were immobilized and used for the production of pyridoxal 5′-phosphate from pyridoxine 5′-phosphate (1).

Continuous production of coenzyme A from pantothenic acid, L-cysteine, and ATP was investigated using *B. ammoniagenes* cells immobilized with

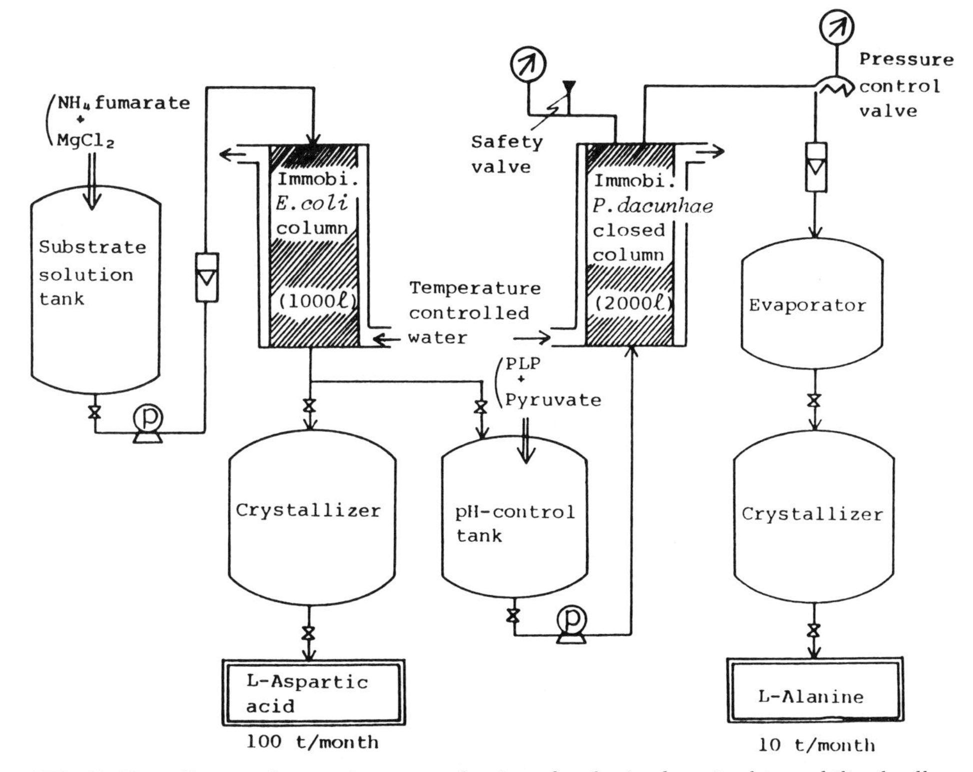

FIG. 5. Flow diagram for continuous production of L-alanine by mixed immobilized cells.

polyacrylamide gel (81). Another system for producing coenzyme A was developed using *S. cerevisiae* cells and *B. ammoniagenes* cells coimmobilized by microencapsulation with ethyl cellulose (73).

Glucose 6-phosphate and glucose 1-phosphate were produced from glucose and *p*-nitrophenylphosphate by cells of *Escherichia freundii* with acid phosphatase activity. The cells were immobilized in a polyacrylamide gel lattice (72).

Cells of *Enterobacter aerogenes* with transglycosylation activity were immobilized with a hydrophilic photo-cross-linkable resin and used to produce adenine arabinoside from uracil arabinoside and adenine (115). By a similar immobilization procedure, cells of a *Propionibacterium* sp. were immobilized and used for the production of vitamin B_{12} (117).

The production of dihydroxyacetone, which is used as a pharmaceutical intermediate and as a suntanning agent, was accomplished by the oxidation of glycerol using *Acetobacter xylinum* cells immobilized in polyacrylamide gel (63).

Glutathione has been produced from glutamic acid, cysteine, glycine, and ATP with immobilized cells. When *S. cerevisiae* cells with high ATP regeneration activity and *E. coli* cells with high glutathione synthetase activity were coimmobilized in a polyacrylamide gel lattice, the immobilized cells showed high productivity (61).

Attempts to produce extracellular enzymes by using immobilized microbial cells have been reported. The production of α-amylase was attempted in a batch system using cells of *Bacillus subtilis* immobilized in a polyacrylamide gel lattice (36). In a similar system, immobilized *Streptomyces fradiae* cells were used for protease production (37). Cellulase production by *Trichoderma reesei* cells immobilized with κ-carrageenan was demonstrated in continuous culture (21).

Recently, *Bacillus subtilis* cells carrying plasmids encoded for rat proinsulin were immobilized in agarose beads. The immobilized cells were used for continuous production of proinsulin in a small, continuously stirred tank reactor (58). This technology may be applicable to a variety of eucaryotic proteins produced by recombinant DNA techniques.

18.3 CONCLUSION

Many immobilization techniques have been applied to cells, but an ideal general method has not yet been developed. Immobilization of cells by chemical methods may cause cellular damage because of the toxic character of the chemicals used. To reduce this disadvantage, these processes are carried out under conditions which are as mild as possible. Immobilized cells can be prepared by adsorption under mild conditions, although with physical adsorption the forces between the cells and the carrier are generally weak, and leakage of the cells from the matrix can readily occur. One advantage of the adsorbed preparations is the ability to regenerate the biomass by regrowth.

With entrapment methods, no binding should occur between the cells and the carrier, and a high retention of activity is expected. However, this method is limited to substrates and products of low molecular weight because of diffusional restrictions. For retaining cell viability, nonchemical methods are preferred, especially when nontoxic materials such as κ-carrageenan and calcium alginate are used.

For practical applications, it is necessary to find an immobilization procedure that is simple and inexpensive and that achieves good retention of activity and good operational stability. The use of immobilized cells for the production of useful organic compounds is expanding and should be an important adjunct to recombinant DNA technology.

18.4 LITERATURE CITED

1. **Ado, Y., K. Kimura, and H. Samejima.** 1980. Production of useful nucleotides with immobilized microbial cells. Enzyme Eng. **5:**295–304.
2. **Amin, G., and H. Verachtert.** 1982. Comparative study of ethanol production by immobilized-cell systems using *Zymomonas mobilis* or *Saccharomyces bayanus*. Eur. J. Appl. Microbiol. Biotechnol. **14:**59–63.
3. **Bang, W.-G., U. Behrendt, S. Lang, and F. Wagner.** 1983. Continuous production of L-tryptophan from indole and L-serine by immobilized *Escherichia coli* cells. Biotechnol. Bioeng. **25:**1013–1025.
4. **Bisping, B., and H. J. Rehm.** 1982. Glycerol production by immobilized cells of *Saccharomyces cerevisiae*. Eur. J. Appl. Microbiol. Biotechnol. **14:**136–139.
5. **Brodelius, P., B. Deus, K. Mosbach, and M. H. Zenk.** 1979. Immobilized plant cells for the production and transformation of natural products. FEBS Lett. **103:** 93–97.
6. **Brodelius, P., B. Hägerdal, and K. Mosbach.** 1980. Immobilized whole cells of the yeast *Trigonopsis variabilis* containing D-amino acid oxidase for the production of α-keto acids. Enzyme Eng. **5:**383–387.
7. **Chibata, I.** 1978. Immobilized enzymes—research and development. John Wiley & Sons, Inc., New York.
8. **Chibata, I., T. Tosa, and T. Sato.** 1974. Immobilized aspartase-containing microbial cells: preparation and enzymatic properties. Appl. Microbiol. **27:**878–885.
9. **Chibata, I., T. Tosa, and T. Sato.** 1976. Production of L-aspartic acid by immobilized cells. Methods Enzymol. **44:**739–746.
10. **Chibata, I., T. Tosa, and T. Sato.** 1983. Immobilized cells in the preparation of fine chemicals. Adv. Biotechnol. Processes **10:**203–222.
11. **Chibata, I., T. Tosa, T. Sato, T. Mori, and Y. Matsuo.** 1972. Preparation and industrial application of immobilized aminoacylases, p. 383–389. Proceedings of the Fourth IFS. Fermentation Technology Today.
12. **Chibata, I., and L. B. Wingard, Jr.** 1983. Immobilized microbial cells. Appl. Biochem. Bioeng. **4:**1–349.
13. **Cho, G. H., C. Y. Choi, Y. D. Choi, and M. H. Han.** 1982. Ethanol production by immobilized yeast and its CO_2 gas effects in a packed bed reactor. J. Chem. Tech. Biotechnol. **32:**959–967.
14. **Chua, J. W., S. Erarslan, S. Kinoshita, and H. Taguchi.** 1980. 2,3-Butanediol production by immobilized *Enterobacter aerogenes* IAM 1133 with κ-carrageenan. J. Ferment. Technol. **58:**123–127.
15. **Constantinides, A.** 1980. Steroid transformation at high substrate concentrations using immobilized *Corynebacterium simplex* cells. Biotechnol. Bioeng. **22:**119–136.
16. **Constantinides, A., D. Bhatia, and W. R. Vieth.** 1981. Immobilization of *Brevibacterium flavum* cells on collagen for the production of glutamic acid in a recycle reactor. Biotechnol. Bioeng. **23:**899–916.
17. **Deo, Y. M., and G. M. Gaucher.** 1983. Semi-continuous production on the antibiotic patulin by immobilized cells of *Penicillium urticae*. Biotechnol. Lett. **5:**125–130.
18. **Dinelli, D.** 1972. Fiber-entrapped enzymes. Process Biochem. **7:**9–12.

19. **Foster, B. C., R. T. Coutts, F. M. Pasutto, and J. B. Dossetor.** 1983. Production of cyclosporin A by carrageenan-immobilized *Tolypocladium inflatum* in an airlift reactor with external loop. Biotechnol. Lett. **5**:693–696.
20. **Freeman, A., and Y. Aharonowitz.** 1981. Immobilization of microbial cells in crosslinked, prepolymerized, linear polyacrylamide gels: antibiotic production by immobilized *Streptomyces clavuligerus* cells. Biotechnol. Bioeng. **23**:2747–2759.
21. **Frein, E. M., B. S. Montenecourt, and D. E. Eveleigh.** 1982. Cellulase production by *Trichoderma reesei* immobilized on κ-carrageenan. Biotechnol. Lett. **4**:287–292.
22. **Fujimura, M., J. Kato, T. Tosa, and I. Chibata.** 1984. Continuous production of L-arginine using immobilized growing *Serratia marcescens* cells: effectiveness of supply of oxygen gas. Appl. Microbiol. Biotechnol. **19**: 136–139.
23. **Fukushima, S., and S. Hanai.** 1982. Pilot operation for continuous alcohol fermentation of molasses in an immobilized bioreactor. Enzyme Eng. **6**:347–348.
24. **Ghommidh, C., and J. M. Navarro.** 1982. A study of acetic acid production by immobilized *Acetobacter* cells: product inhibition effects. Biotechnol. Bioeng. **24**: 1991–1999.
25. **Ghommidh, C., J. M. Navarro, and G. Durand.** 1981. Acetic acid production by immobilized *Acetobacter* cells. Biotechnol. Lett. **3**:93–98.
26. **Grote, W., K. J. Lee, and P. L. Rogers.** 1980. Continuous ethanol production by immobilized cells of *Zymomonas mobilis*. Biotechnol. Lett. **2**:481–486.
27. **Hattori, T., and C. Furusaka.** 1960. Chemical activities of *Escherichia coli* adsorbed on a resin. J. Biochem. **48**:831–837.
28. **Hattori, T., and C. Furusaka.** 1961. Chemical activities of *Azotobacter agile* adsorbed on a resin. J. Biochem. **50**:312–315.
29. **Hirano, K., I. Karube, and S. Suzuki.** 1977. Aminoacylase pellets. Biotechnol. Bioeng. **19**:311–321.
30. **Horitsu, H., Y. Takahashi, J. Tsuda, K. Kawai, and Y. Kawano.** 1983. Production of itaconic acid by *Aspergillus terreus* immobilized in polyacrylamide gels. Eur. J. Appl. Microbiol. Biotechnol. **18**:358–360.
31. **Jack, T. R., and J. E. Zajic.** 1977. The enzymatic conversion of L-histidine to urocanic acid by whole cells of *Micrococcus luteus* immobilized on carbodiimide activated carboxymethyl cellulose. Biotechnol. Bioeng. **19**:631–648.
32. **Kierstan, M., and C. Bucke.** 1977. The immobilization of microbial cells, subcellular organelles, and enzymes in calcium alginate gels. Biotechnol. Bioeng. **19**: 387–397.
33. **Kim, I. H., D. H. Nam, and D. D. Y. Ryu.** 1983. Cephalexin synthesis using immobilized *Xanthomonas citri* cells. Appl. Bioeng. Biotechnol. **8**:195–202.
34. **Kinoshita, S., M. Muranaka, and H. Okada.** 1975. Hydrolysis of ε-aminocaproic acid cyclic dimer by cells entrapped in acrylamide gel. J. Ferment. Technol. **53**:223–229.
35. **Klein, J., and F. Wagner.** 1980. Immobilization of whole microbial cells for the production of 6-amino penicillanic acid. Enzyme Eng. **5**:335–345.
36. **Kokubu, T., I. Karube, and S. Suzuki.** 1978. α-Amylase production by immobilized whole cells of *Bacillus subtilis*. Eur. J. Appl. Microbiol. Biotechnol. **5**:233–240.
37. **Kokubu, T., I. Karube, and S. Suzuki.** 1981. Protease production by immobilized mycelia of *Streptomyces fradiae*. Biotechnol. Bioeng. **23**:29–39.
38. **Koshcheenko, K. A., G. V. Sukhodolskaya, V. S. Tyurin, and G. K. Skryabin.** 1981. Physiological, biochemical and morphological changes in immobilized cells during repeated periodical hydrocortisone transformations. Eur. J. Appl. Microbiol. Biotechnol. **12**:161–169.
39. **Krouwel, P. G., W. J. Groot, N. W. F. Kossen, and C. G. van der Lean.** 1983. Continuous isopropanol-butanol-ethanol fermentation by immobilized *Clostridium beijerinckii* cells in a packed bed fermenter. Enzyme Microb. Technol. **5**:46–54.
40. **Krouwel, P. G., W. F. M. van der Lean, and N. W. F. Kossen.** 1980. Continuous production of *n*-butanol and isopropanol by immobilized, growing *Clostridium butylicum* cells. Biotechnol. Lett. **2**:253–258.
41. **Kurzatkowski, W., W. Kurytowicz, and A. Paszkiewicz.** 1982. Penicillin G production by immobilized fungal vesicles. Eur. J. Appl. Microbiol. Biotechnol. **15**: 211–213.
42. **Larsson, P.-O., and K. Mosbach.** 1979. Alcohol production by magnetic immobilized yeast. Biotechnol. Lett. **1**:501–506.
43. **Larsson, P.-O., S. Ohlson, and K. Mosbach.** 1976. New approach to steroid conversion using activated immobilized microorganisms. Nature (London) **263**:796.
44. **Lee, T. H., J. C. Ahn, and D. D. Y. Ryu.** 1983. Performance of an immobilized yeast reactor system for ethanol production. Enzyme Microb. Technol. **5**:41–45.
45. **Linko, P.** 1981. Immobilized live cells, p. 711–716. *In* M. Moo-Young, C. W. Robinson, and C. Vezina (ed.), Advances in biotechnology. Pergamon Press, Inc., Elmsford, New York.
46. **Linko, Y. Y., and P. Linko.** 1981. Continuous ethanol production by immobilized yeast reactor. Biotechnol. Lett. **3**:21–26.
47. **Marechal, P. D.-L., R. Calderon-Seguin, J. P. Vandecasteele, and R. Azerad.** 1979. Synthesis of L-tryptophan by immobilized *Escherichia coli* cells. Eur. J. Appl. Microbiol. Biotechnol. **7**:33–44.
48. **Margaritis, A., and J. B. Wallace.** 1982. The use of immobilized cells of *Zymomonas mobilis* in a novel fluidized bioreactor to produce ethanol. Biotechnol. Bioeng. Symp. **12**:147–159.
49. **Martin, C. K. A., and D. Perlman.** 1976. Conversion of L-sorbose to L-sorbosone by immobilized cells of *Gluconobacter melanogenus* IFO 3293. Biotechnol. Bioeng. **18**:217–237.
50. **Martin, C. K. A., and D. Perlman.** 1976. Conversion of L-sorbose to 2-keto-L-gulonic acid by mixtures of immobilized cells of *Gluconobacter melanogenus* IFO 3293 and *Pseudomonas* species. Eur. J. Appl. Microbiol. Biotechnol. **3**:91–95.
51. **Mattiason, B.** 1983. Immobilized cells and organelles. Chemical Rubber Co., Cleveland, Ohio.
52. **Mayer, H., J. Collins, and F. Wagner.** 1980. Cloning of the penicillin G-acylase gene of *Escherichia coli* ATCC 11105 on multicopy plasmids. Enzyme Eng. **5**:61–69.
53. **Messing, R., R. A. Oppermann, and F. B. Kolot.** 1979. Pore dimensions for accumulating biomass, in immobilized microbial cells. ACS (Am. Chem. Soc.) Symp. Ser. **106**:13–28.
54. **Morikawa, Y., I. Karube, and S. Suzuki.** 1979. Penicillin G production by immobilized whole cells of *Penicillium chrysogenum*. Biotechnol. Bioeng. **21**:261–270.
55. **Morikawa, Y., I. Karube, and S. Suzuki.** 1980. Continuous production of bacitracin by immobilized living whole cells of *Bacillus* sp. Biotechnol. Bioeng. **22**: 1015–1023.
56. **Morikawa, Y., I. Karube, and S. Suzuki.** 1980. Enhancement of penicillin acylase activity by cultivating immobilized *Kluyvera citrophila*. Eur. J. Appl. Microbiol. Biotechnol. **10**:23–30.
57. **Morikawa, Y., K. Ochiai, I. Karube, and S. Suzuki.** 1979. Bacitracin production by whole cells immobilized in polyacrylamide gel. Antimicrob. Agents Chemother. **15**:126–130.
58. **Mosbach, K., S. Birnbaum, K. Hardy, J. Davies, and L. Bulow.** 1983. Formation of proinsulin by immobilized

Bacillus subtilis. Nature (London) **302:**543–545.

59. **Mosbach, K., and P.-O. Larsson.** 1970. Preparation and application of polymer-entrapped enzymes and microorganisms in microbial transformation process with special reference to steroid 11β-hydroxylation and Δ^1-dehydrogenation. Biotechnol. Bioeng. **12:**19–27.
60. **Murata, K., J. Kato, and I. Chibata.** 1979. Continuous production of NADP by immobilized *Brevibacterium ammoniagenes* cells. Biotechnol. Bioeng. **21:**887–895.
61. **Murata, K., K. Tani, J. Kato, and I. Chibata.** 1981. Glycolytic pathway as an ATP generation system and its application to the production of glutathione and NADP. Enzyme Microb. Technol. **3:**233–242.
62. **Murata, K., T. Uchida, K. Tani, J. Kato, and I. Chibata.** 1979. Continuous production of glucose-6-phosphate by immobilized *Achromobacter butyri* cells. Eur. J. Appl. Microbiol. Biotechnol. **7:**45–52.
63. **Nabe, K., N. Izuo, S. Yamada, and I. Chibata.** 1979. Conversion of glycerol to dihydroxyacetone by immobilized whole cells of *Acetobacter xylinum*. Appl. Env. Microbiol. **38:**1056–1060.
64. **Navarro, J. M., and G. Durand.** 1977. Modification of yeast metabolism by immobilization onto porous glass. Eur. J. Appl. Microbiol. **4:**243–254.
65. **Nishida, Y., K. Nabe, S. Yamada, and I. Chibata.** 1984. Enzymatic continuous production of *N*-acetyl-L-methionine from *N*-acetyl-DL-methionineamide with *Erwinia carotovora* containing a new amidase activity. Enzyme Microb. Technol. **6:**85–90.
66. **Nishida, Y., T. Sato, T. Tosa, and I. Chibata.** 1979. Immobilization of *Escherichia coli* cells having aspartase activity with carrageenan and locust bean gum. Enzyme Microb. Technol. **6:**85–90.
67. **Ohlson, S., P.-O. Larsson, and K. Mosbach.** 1979. Steroid transformation by living cells immobilized in calcium alginate. Eur. J. Appl. Microbiol. Biotechnol. **7:**103–110.
68. **Omata, T., T. Iida, A. Tanaka, and S. Fukui.** 1979. Transformation of steroids by gel-entrapped *Nocardia rhodocrous* cells in organic solvent. Eur. J. Appl. Microbiol. Biotechnol. **8:**143–155.
69. **Omata, T., N. Iwamoto, T. Kimura, A. Tanaka, and S. Fukui.** 1981. Stereoselective hydrolysis of DL-methyl succinate by gel-entrapped *Rhodotorula minuta* var. *texensis* cells in organic solvent. Eur. J. Appl. Microbiol. Biotechnol. **11:**199–204.
70. **Omata, T., A. Tanaka, and S. Fukui.** 1980. Bioconversions under hydrophobic conditions: effect of solvent polarity on steroid transformations by gel entrapped *Nocardia rhodocrous* cells. J. Ferment. Technol. **58:**339–343.
71. **Omata, T., A. Tanaka, T. Yamane, and S. Fukui.** 1979. Immobilization of microbial cells and enzymes with hydrophobic photo-crosslinkable resin prepolymers. Eur. J. Appl. Microbiol. Biotechnol. **6:**207–215.
72. **Saif, S. R., Y. Tani, and K. Ogata.** 1975. Preparation of glucose phosphate through the transphosphorylation with immobilized cells. J. Ferment. Technol. **53:** 380–385.
73. **Samejima, H., K. Kimura, Y. Ado, Y. Suzuki, and T. Tadokoro.** 1978. Regeneration of ATP by immobilized microbial cells and its utilization for the synthesis of nucleotides. Enzyme Eng. **4:**237–244.
74. **Sarkar, J. M., and J. Mayaudon.** 1983. Alanine synthesis by immobilized *Corynebacterium dismutans* cells. Biotechnol. Lett. **5:**201–206.
75. **Sato, T., T. Mori, T. Tosa, I. Chibata, M. Furui, K. Yamashita, and A. Sumi.** 1975. Engineering analysis of continuous production of L-aspartic acid by immobilized *Escherichia coli* cells in fixed beds. Biotechnol. Bioeng. **17:**1797–1804.
76. **Sato, T., Y. Nishida, T. Tosa, and I. Chibata.** 1979. Immobilization of *Escherichia coli* cells containing aspartase activity with κ-carrageenan. Biochim. Biophys. Acta **570:**179–186.
77. **Sato, T., S. Takamatsu, K. Yamamoto, I. Umemura, T. Tosa, and I. Chibata.** 1982. Production of L-alanine from ammonium fumarate using two types of immobilized microbial cells. Enzyme Eng. **6:**271–272.
78. **Sato, T., T. Tosa, and I. Chibata.** 1976. Continuous production of 6-aminopenicillanic acid from penicillin by immobilized microbial cells. Eur. J. Appl. Microbiol. **2:**153–160.
79. **Sawada, H., S. Kinoshita, T. Yoshida, and H. Taguchi.** 1981. Continuous production of 12-ketochenodeoxycholic acid in a column reactor containing immobilized living cells of *Brevibacterium fuscum*. J. Ferment. Technol. **59:**111–114.
80. **Scheurich, P., H. Schnabel, U. Zimmerman, and J. Klein.** 1980. Immobilization and mechanical support of individual protoplasts. Biochim. Biophys. Acta **598:** 645–651.
81. **Shimizu, S., H. Morioka, Y. Tani, and K. Ogata.** 1975. Synthesis of coenzyme A by immobilized microbial cells. J. Ferment. Technol. **53:**77–83.
82. **Slininger, P. J., R. J. Bothast, L. T. Black, and J. E. McChee.** 1982. Continuous conversion of D-xylose to ethanol by immobilized *Pachysolen tannophilus*. Biotechnol. Bioeng. **24:**2241–2251.
83. **Slowinski, W., and S. E. Charm.** 1973. Glutamic acid production with gel-entrapped *Corynebacterium glutamicum*. Biotechnol. Bioeng. **15:**973–979.
84. **Sonomoto, K., M. M. Hoq, A. Tanaka, and S. Fukui.** 1983. 11β-Hydroxylation of cortexolone (Reichstein compound S) to hydrocortisone by *Curvularia lunata* entrapped in photo-cross-linked resin gels. Appl. Environ. Microbiol. **45:**436–443.
85. **Sonomoto, K., I. N. Jin, A. Tanaka, and S. Fukui.** 1980. Application of urethane prepolymers to immobilization of biocatalysts: Δ^1-dehydrogenation of hydrocortisone by *Arthrobacter simplex* cells entrapped with urethane prepolymers. Agric. Biol. Chem. **44:**119–1126.
86. **Sonomoto, K., K. Nomura, A. Tanaka, and S. Fukui.** 1982. 11α-Hydroxylation of progesterone by gel-entrapped living *Rhizopus stolonifer* mycelia. Eur. J. Appl. Microbiol. Biotechnol. **16:**57–62.
87. **Sonomoto, K., N. Osui, A. Tanaka, and S. Fukui.** 1983. 9α-Hydroxylation of 4-androstene-3,17-dione by gel-entrapped *Corynebacterium* sp. cells. Eur. J. Appl. Microbiol. Biotechnol. **17:**203–210.
88. **Takamatsu, S., I. Umemura, K. Yamamoto, T. Sato, T. Tosa, and I. Chibata.** 1982. Production of L-alanine from ammonium fumarate using two immobilized microorganisms. Eur. J. Appl. Microbiol. Biotechnol. **15:**147–152.
89. **Takasaki, Y., and A. Kanbayashi.** 1969. Studies on sugar-isomerizing enzyme. 5. Fixation of glucoseisomerase by heat-treatment of cells of *Streptomyces* sp. and its application. Kogyo Gijutsu-in Biseibutsu Kogyo Gijutsu Kenkyusho Kenkyu Hokoku **37:**31–37.
90. **Takata, I., K. Yamamoto, T. Tosa, and I. Chibata.** 1979. Screening of microorganisms having high fumarase activity and their immobilization with carrageenan. Eur. J. Appl. Microbiol. Biotechnol. **7:**161–172.
91. **Takata, I., K. Yamamoto, T. Tosa, and I. Chibata.** 1980. Immobilization of *Brevibacterium flavum* with carrageenan and its application for continuous production of L-malic acid. Enzyme Microb. Technol. **60:**431–437.
92. **Tipayang, P., and M. Kozaki.** 1982. Lactic acid production by a new *Lactobacillus* sp nov immobilized in calcium alginate. J. Ferment. Technol. **60:**595–598.
93. **Toda, K., and H. Shoda.** 1975. Sucrose inversion by immobilized yeast cells in a complete mixing reactor. Biotechnol. Bioeng. **17:**481–497.
94. **Tosa, T., T. Mori, and I. Chibata.** 1971. Studies on continuous enzyme reactions. VIII. Kinetics and pres-

sure drop of aminoacylase column. J. Ferment. Technol. **49**:522–528.
95. **Tosa, T., T. Mori, N. Fuse, and I. Chibata.** 1966. Studies on continuous enzyme reactions. I. Screening of carriers for preparation of water-insoluble aminoacylase. Enzymologia **31**:214–224.
96. **Tosa, T., T. Sato, T. Mori, and I. Chibata.** 1974. Basic studies for continuous production of L-aspartic acid by immobilized *Escherichia coli* cells. Appl. Microbiol. **27**:886–889.
97. **Tosa, T., T. Sato, T. Mori, K. Yamamoto, I. Takata, Y. Nishida, and I. Chibata.** 1979. Immobilization of enzymes and microbial cells using carrageenan as matrix. Biotechnol. Bioeng. **21**:1697–1709.
98. **Tsumura, N., T. Kasumi, and M. Ishikawa.** 1976. Processing methods for microbial cells and their effects on enzyme retention. Rep. Natl. Food Res. Inst. **31**:71–75.
99. **Uchida, T., T. Watanabe, J. Kato, and I. Chibata.** 1978. Continuous production of NAD by immobilized *Achromobacter aceris* cells. Biotechnol. Bioeng. **20**: 255–266.
100. **Veelken, M., and H. Pape.** 1982. Production of tylosin and nikkomycin by immobilized *Streptomyces* cells. Eur. J. Appl. Microbiol. Biotechnol. **15**:206–210.
101. **Veliky, I. A., and A. Jones.** 1981. Bioconversion of gitoxigenin by immobilized plant cells in a column bioreactor. Biotechnol. Lett. **3**:551–554.
102. **Veliky, I. A., and R. E. Williams.** 1981. The production of ethanol by *Saccharomyces cerevisiae* immobilized in polycation-stabilized calcium alginate gels. Biotechnol. Lett. **3**:275–280.
103. **Verlop, K.-D., and J. Klein.** 1981. Formation of spherical chitosan biocatalysts by ionotropic gelation. Biotechnol. Lett. **3**:9–14.
104. **Vieth, W. R., S. S. Wang, and R. Saini.** 1973. Immobilization of whole cells in a membraneous form. Biotechnol. Bioeng. **15**:565–569.
105. **Wada, M., J. Kato, and I. Chibata.** 1979. A new immobilization of microbial cells: immobilized growing cells using carrageenan gel and their properties. Eur. J. Appl. Microbiol. Biotechnol. **8**:241–247.
106. **Wada, M., T. Uchida, J. Kato, and I. Chibata.** 1980. Continuous production of L-isoleucine using immobilized growing *Serratia marcescens* cells. Biotechnol. Bioeng. **22**:1175–1188.
107. **Yamada, H., S. Shimizu, H. Shimada, Y. Tani, S. Takahashi, and T. Ohashi.** 1980. Production of D-phenylglycine related amino acids by immobilized microbial cells. Biochimie **62**:395.
108. **Yamada, H., H. Shimizu, Y. Tani, and T. Hino.** 1980. Synthesis of coenzymes by immobilized cell system. Enzyme Eng. **5**:405–411.
109. **Yamamoto, K., T. Sato, T. Tosa, and I. Chibata.** 1974. Continuous production of urocanic acid by immobilized *Achromobacter liquidum* cells. Biotechnol. Bioeng. **16**:1601–1610.
110. **Yamamoto, K., T. Sato, T. Tosa, and I. Chibata.** 1974. Continuous production of L-citrulline by immobilized *Pseudomonas putida* cells. Biotechnol. Bioeng. **16**: 1589–1599.
111. **Yamamoto, K., T. Tosa, and I. Chibata.** 1980. Continuous production of L-alanine using *Pseudomonas dacunhae* immobilized with carrageenan. Biotechnol. Bioeng. **22**:2045–2054.
112. **Yamamoto, K., T. Tosa, K. Yamashita, and I. Chibata.** 1976. Continuous production of L-malic acid by immobilized *Brevibacterium ammoniagenes* cells. Eur. J. Appl. Microbiol. **3**:169–183.
113. **Yamamoto, K., T. Tosa, K. Yamashita, and I. Chibata.** 1977. Kinetics and decay of fumarase activity of immobilized *Brevibacterium ammoniagenes* cells for continuous production of L-malic acid. Biotechnol. Bioeng. **19**:1101–1114.
114. **Yi, Z.-H., and H. J. Rehm.** 1982. Formation of α,ω-tridecanedioic acid from different substrates by immobilized cells of a mutant of *Candida tropicalis*. Eur. J. Appl. Microbiol. Biotechnol. **16**:1–4.
115. **Yokozeki, K., S. Yamanaka, T. Utagawa, K. Takanami, Y. Hirose, A. Tanaka, K. Sonomoto, and S. Fukui.** 1982. Production of adenine arabinoside by gel-entrapped cells of *Enterobacter aerogenes* in water-organic cosolvent system. Eur. J. Appl. Microbiol. Biotechnol. **14**: 225–231.
116. **Yongsmith, B., and K. Chutima.** 1983. Production of vitamin B_{12} by living bacterial cells immobilized in calcium alginate gels. J. Ferment. Technol. **61**:593–598.
117. **Yongsmith, B., K. Sonomoto, A. Tanaka, and S. Fukui.** 1982. Production of vitamin B_{12} by immobilized cells of a propionic acid bacterium. Eur. J. Appl. Microbiol. Biotechnol. **16**:70–74.

Chapter 19

Methods of Enzyme Immobilization

F. R. BERNATH AND K. VENKATASUBRAMANIAN

Enzyme immobilization has been the subject of considerable research during the past two decades, with the result that it has been transformed from a novelty to a fully accepted and useful technology. Some commercial applications are now firmly established, and several reliable methods are available for binding enzymes to insoluble and soluble carriers for routine use in the laboratory and in industrial processes. Further potential remains for widespread use of immobilized enzymes in a variety of industrial, environmental, analytical, and medical applications. Researchers have identified several factors which have slowed the realization of that potential, and work is continuing in an effort to find solutions to these problems.

In this chapter we discuss current methods for immobilizing enzymes and some recently proposed techniques which appear promising. We shall attempt to present some general guidelines from the great variety of combinations of carrier materials, methods, and enzymes that have been reported in the literature. In the process, we will provide specific examples of several types of immobilized enzymes, including details regarding their preparation and their performance. We begin with an overview of immobilized enzyme technology to gain some perspective on the current status of the field.

19.1 OVERVIEW OF IMMOBILIZED ENZYME TECHNOLOGY

Immobilized biocatalysts are enzymes, cells, and organelles, bound individually or in combination, which are in a state permitting their reuse (80). This rather broad definition enables us to consider a variety of different types, all of which have the common goal of allowing continuous or intermittent use of enzymes over an extended period of time.

Immobilized enzymes are usually classified by the method of binding (or localization) which allows reuse. Some of the major classes are: (i) adsorbed enzymes (including ionic, metal bridge, biospecific and hydrophobic binding); (ii) entrapped enzymes, including those in a matrix (polymeric), in microcapsules, liposomes, and erythrocytes, and in hollow fibers (ultrafiltration); and (iii) covalently bound enzymes. Classification in this manner should be viewed only as a general guideline. In fact, there continues to be considerable overlap, especially with the use of bifunctional cross-linking agents in conjunction with adsorption and entrapment methods.

Immobilized whole cells and organelles are receiving considerable attention among researchers today (14, 47, 56, 94). This approach appears to offer significant advantages over enzyme systems, such as removing the need for expensive purification steps and stabilizing the enzyme in adverse environments. Fixed-cell preparations are also probably superior for the synthesis of products involving several enzymes and requiring coenzyme regeneration. On the other hand, the application of these preparations may be limited by restricted substrate access due to cell wall or membrane resistances, maintenance of viability or coenzyme activity, and inhibition effects from products or side products. Future developments in biotechnology will require a wide range of choices for methods of enzyme immobilization, including those involving enzymes as well as whole cells or organelles.

In the literature are literally hundreds of articles on immobilized enzymes describing a wide range of immobilization methods, carriers, reactor types, and methods of data analysis for a large number of enzymes. Fortunately, some excellent books have been published which have attempted to organize and evaluate this massive amount of work (5, 13, 48, 72, 74, 81, 91, 96, 98).

This quantity of effort, however, has not been sufficient to bring about the major technological impact that had been predicted for immobilized enzymes. One reason for this, according to The Working Party on Immobilized Biocatalysts within the European Federation of Biotechnology (80), is that researchers have often not sufficiently characterized their immobilized biocatalysts or reported the details of preparation and performance. To remedy this situation, this committee has provided detailed guidelines for characterizing immobilized biocatalysts (1). Its recommendations are based on the observation that most current work in the field stems from an explicit or implicit practical motivation, and that research should be conducted and reported in such a way as to contribute to the assessment of each system's potential. In short, the guidelines are aimed at encouraging researchers to provide sufficient details to allow others to reproduce the experiments and evaluate the systems.

According to a report by L. Hepner and Associates (61), current worldwide sales of enzymes total about \$390 million, with amylase, glucose isomerase, amyloglucosidase, and detergent proteolytic enzymes accounting for 70% of the market. With the exception of glucose isomerase (85), immobilized enzymes represent only a small fraction. The future role of biotechnology in the enzyme industry is still, after all these years, uncertain. It appears, however, that it is beginning to develop a specific direction. Rather than the replacement of existing soluble-enzyme processes with immobilized enzymes, the future will see an emphasis on the development of specially tailored or unexploited enzymes which can produce new products or dramatically improve existing processes. It is predicted that recombinant DNA techniques will contribute significantly to this strategy.

Although there are still few large-scale industrial applications of immobilized enzymes (26, 85), many preparations are being used routinely in laboratory procedures, especially in analytical applications (33, 69). In addition, a number of companies now market immobilized enzymes and the associated technology related to process design. It is also possible to purchase derivatized carriers for direct binding with enzymes. These developments are all signs of the growing importance of immobilized enzymes as an accepted class of active, stable biocatalysts.

19.2 IMMOBILIZATION METHODS (GENERAL COMMENTS)

Immobilized biocatalysts can be prepared in almost an unlimited range of compositions and morphologies. Furthermore, there are no a priori requirements on the values of individual characteristics such as specific activity, storage and operational stability, compression or abrasion properties, enzyme leaching, ease and cost of preparation, etc. (80). Rather, it is the combination of these characteristics in relation to the overall objective, i.e., quality of performance of the enzyme in its actual application, that is important. When we add to this the facts that each combination of enzyme, carrier, and binding technique produces different properties and that hundreds of different combinations have been studied, we are faced with an extremely complex situation.

Although there is no universal carrier or binding technique, several methods and carriers have been proven reliable over the years and are now practically recognized as standards. Also, some variations of these and a few newer methods appear to be promising additions. In this chapter we discuss both types of preparations, chiefly by classifying methods according to the mode of binding and discussing variations associated with specific carriers under each heading.

19.3 ADSORPTION

The adsorption method has traditionally referred to the binding of an enzyme by weak attractive forces to an inert carrier which has not been functionalized for covalent binding (p. 79 of reference 72). It is generally regarded as the simplest and most economical procedure for binding an enzyme to a solid support. Normally, immobilization is achieved by simply incubating the carrier material with a given amount of enzyme in solution at a specific temperature, pH, and ionic strength. Usually the support requires minimal pretreatment such as washing or presoaking in a buffered solution.

A variety of inorganic supports have been used to adsorb enzymes, including clay, alumina, titania, sand, glass, and silica. These carriers generally result in rather low amounts of bound protein (1 mg/g of support) and continuous elution of the enzyme (p. 25 of reference 98). Even for these so-called inert carriers, the binding mechanism is not simple and probably involves a combination of van der Waal's ionic and hydrogen bonds. With some supports, such as titania (a transition metal oxide), the bond may even have a partial covalent character which seems to produce a more active and stable complex (p. 153 of reference 74).

The more successful carriers have chemical compositions which increase the strength and stability of the bonds discussed above. Thus, ion-exchange resins have been used to obtain stronger electrostatic interactions, while protein carriers result in cooperative multifunctional binding. Some materials designed with a controlled number and type of nonpolar groups may produce stable hydrophobic interactions, and others may contain a certain ligand or be of a chemical nature which offers a specific affinity for the bound enzyme. Each of these methods has been studied extensively and has been shown to result in improved activity and stability characteristics. In most instances, elution of the enzyme will still occur, but it can often be reduced by altering pH, temperature, and ionic strength.

The first commercially successful immobilized enzyme was prepared by adsorption, a method which still offers many advantages for large-scale industrial applications (p. 329 of reference 98). Some have suggested that adsorption methods, specifically those involving hydrophobic bonding (89) or transition metal mediation (78), may also be useful in developing applications in organic chemistry. The relative attributes of this method are summarized in Table 1.

Perhaps the most important advantage of adsorption from a commercial viewpoint is the capability of regenerating the immobilized enzyme activity in place. This usually involves a brief shutdown of an operating reactor and incubation of the spent carrier with a fresh solution of enzyme for a specific amount of time. As long as the process can tolerate the elution of some enzyme into the product stream, and if the rate of elution can be controlled to produce economical operational half-lives, adsorption remains a very attractive technique for large-scale, continuous industrial processes.

To reduce elution and improve operational stability, the carrier and adsorbed enzyme are sometimes exposed to a bifunctional cross-linking agent, usually glutaraldehyde. The nature of the resulting complex is uncertain, but if the carrier is inert, the cross-linking agent probably ties individual enzyme molecules together, thereby slowing their release from the carrier. Some carriers, such as collagen or chitin, have amino functionality and may be able to react with glutaraldehyde so that a covalent carrier-enzyme bond is formed (83; p. 243 of reference 74). Irrespective of the mechanism, cross-linking adsorbed enzymes usually improves initial activity levels and operational half-lives; however, it may interfere with the desirable property of regeneration by reducing binding sites for subsequently applied enzymes.

Activating a carrier such as chitin with glutaraldehyde before it comes into contact with the enzyme also generally improves the initial activity of the immobilized enzyme (83), but often at a cost of reduced coupling yield (expressed activity of immobilized enzyme divided by loss of soluble enzyme activity from the supernatant). We will discuss some aspects of the glutaraldehyde reaction in a later section.

19.3.1 Adsorption of Enzymes on Chitin

Chitin is a complex amino-polysaccharide which is the principal component of the exoskeleton of crustaceans. As such, it is a major waste product of the crab, shrimp, and krill processing industries. Chitin can be isolated from ground crustacean shells by demineralization in strong acid (6 N HCl at a ratio of 10 ml/g of chitin at room temperature for 2 h), followed by deproteinization in strong base (5 N KOH at 10 ml/g of chitin at 95°C for 2 h). It can be further purified by washing in 1% NaCl followed by 1% acetic acid (both at 50°C) to remove pigment, or by contact with methanol (5 ml/g of chitin) at room temperature. After a thorough wash with distilled water, the chitin can be dried, milled, and screened to a desired particle size (usually 20 to 50 mesh). Purified chitin is also available commercially.

Purified chitin has a coarse, porous structure with a capacity to bind large amounts of enzyme by adsorption. As a waste product, it should be relatively inexpensive, and it has a very good potential for immobilized enzyme applications (12, 28, 83, 86, 87).

TABLE 1. Enzyme immobilization by adsorption

Advantages	Disadvantages
1. Simple	1. Elution
2. Inexpensive	2. Nonspecific binding
3. General	3. Controlled environment (pH, temperature, ionic strength)
4. Regeneration in situ	

Binding method

1. Add distilled water or buffer to dry chitin to prepare moist chitin (50% solids).
2. Add desired amount of enzyme in distilled water or buffer at a ratio of 2 ml/g of moist chitin.
3. Allow mixture to stand at room temperature for 1 h with intermittent mixing and then store at 4°C for 12 h.
4. Wash thoroughly with distilled water until wash solution shows no enzyme activity.
5. Suspend in desired buffer and store moist at 4°C.

Stanley et al. (83) used the binding method to bind a crude liquid preparation of glucoamylase to crab chitin. One milliliter of the enzyme was diluted to 20 ml with distilled water and added to 10 g of moist chitin. The method was slightly modified by adding a 3 M NaCl wash to remove loosely bound enzyme before the final wash. Despite this relatively harsh treatment, the resulting conjugate still expressed about 25% of the activity of the applied soluble enzyme, as assayed with a soluble starch. No stability data were given for the uncross-linked preparation, but Flor and Hayashida (28) found that glucoamylase bound to chitin by a similar procedure (except 0.02 M acetate buffer, pH 3.8, and a 2-h binding period were used) was stable for 80 days when stored moist at 4°C. More importantly, it showed no significant decline in activity after 20 days of continuous operation in a column reactor operating on solubilized potato starch (45% solids) as substrate at pH 3.8 (0.02 M acetate) and 50°C. When the activity decreased by about 20% after 30 days, Flor and Hayashida regenerated the column in situ to its original activity by first washing it in buffer to remove sugar and then flowing a solution of 200 mg of enzyme in 100 ml of acetate buffer through the column at a rate of 5 ml/h.

As with any adsorption procedure, the amount of stably bound enzyme depends on pH and ionic strength, whose optimal values must be determined experimentally. The optimum pH is often near the isoelectric point of the enzyme, and optimum ionic strength levels are quite low (0.01 to 0.05 M). For example, when invertase is adsorbed on krill chitin (86), the optimum values for pH and ionic strength are 5.0 and 0.01 M, respectively. The amount of bound enzyme decreases by 84% as ionic strength increases from 0.01 to 0.25 M. This characteristic has led some researchers to state that simple adsorption of enzymes to chitin might not be adequate for commercial applications (83, 87) and that a cross-linking agent such as glutaraldehyde should be employed to improve enzyme loading and stability. However, it appears that after contact with high-ionic-strength solutions, the residual enzyme is quite stably bound and activity levels may be acceptable in some cases, depending on the overall economics of the process.

Alternative method with glutaraldehyde cross-linking

1. Prepare moist chitin as described above.
2. Add glutaraldehyde to the enzyme solution to a final concentration of 0.1 to 1% and add to moist chitin as described above. Determine optimal glutaraldehyde concentration for each specific enzyme.
3. If the enzyme cannot tolerate direct contact with a glutaraldehyde solution, preactivate the carrier with glutaraldehyde (1 to 10% solution), wash, and treat with enzyme solution in ratios given in the adsorption method.

When Stanley et al. (83) used this alternative technique to bind glucoamylase to chitin, they were able to improve the final activity of their preparation, after a 3 M NaCl wash, to about 70% of the applied soluble enzyme activity (compared to 25% for the adsorption method). Their optimal conditions for cross-linking were 1% glutaraldehyde in the 20 ml of enzyme-distilled water solution added to 10 g of moist chitin. They obtained virtually identical results when the carrier was preactivated with 1% glutaraldehyde before contact with the enzyme solution (20 ml/10 g of moist, preactivated chitin). This preparation retained 90% of its activity after 17 days of continuous operation in a column reactor using Maltrin 150 (30% solids) as substrate at 45°C and pH 4.5 (0.1 M acetate).

The use of a cross-linking agent in conjunction with adsorption generally improves operational stabilities, but unlike the example given above, it may not always improve the initial coupling yield. For example, bound invertase activities were reduced by up to 40% when glutaraldehyde was present in the binding solution (86). This effect, of course, will vary for each enzyme. A more serious limitation of this method, however, may be that it reduces the capability of carrier regeneration. There are few data in the literature to quantify these effects, but it is clear that coupling yield, operational stability, and regenerability are important properties that should be measured and compared for both the cross-linked and noncross-linked preparations.

19.3.2 Hydrophobic Bonding

Supports containing nonpolar groups, especially aromatics, are capable of binding enzymes by hydrophobic interaction with the protein's tryptophan, tyrosine, and isoleucine residues. Many researchers have ignored nonpolar carriers because it is generally recognized that enzyme stability is related to the hydrophilicity of the support. This is a general guideline, however, not an absolute truth, and several enzymes have been successfully bound to hydrophobic materials (p. 43 of reference 74). Enzymes which are bound to membranes in their natural environment are especially good candidates for this method.

Aliphatic and aromatic ether derivatives of polysaccharides appear to be especially promising; among these, trityl (triphenylmethyl) agarose has demonstrated excellent potential as a general carrier material (9, 10). Trityl agarose is available commercially or may be prepared by the following procedure (9).

Method for preparing trityl agarose

1. Prepare a 100-ml bed of beaded, cross-linked agarose (Sepharose CL-4B) and make anhydrous by suction filtration on a sintered-glass funnel with a few bed volumes of 95% ethanol, followed by a few bed volumes of anhydrous dimethylformamide.
2. Transfer resin to a 500-ml round-bottom flask, from which three bed volumes of dimethylformamide are then azeotropically flash evaporated.
3. To a total volume (agarose and dimethylformamide) of 200 ml, add 8 g of trityl chloride and allow reaction to occur at 30°C with gentle agitation. Care must be taken to keep the beads wetted so they do not dry and crack.
4. The reaction is stopped after the desired level of activation (usually 4 h) by pouring the suspension into a sintered-glass funnel and washing with many bed volumes of ethanol. Ethanol-washed beads may be stored in ethanol at 5 to 10°C or under aqueous conditions at pH 5 to 10. The number of trityl groups on the carrier can be controlled by varying the concentration and time of the reaction. Saturation appears to occur at 100 μmol of trityl per ml of beads, which represents one trityl group per disaccharide repeat unit (9).

Enzyme binding method

1. Dissolve or dilute enzyme in appropriate buffer and apply to a trityl agarose column.
2. Wash the applied enzyme through the column using two bed volumes of buffer. Recycle eluate until all enzyme is bound.

Cashion et al. (9, 10) have immobilized more than 20 different enzymes by this method, including enzymes of interest in gene cloning such as RNA ligases, polynucleotide kinases, and restriction endonucleases. Their results indicate that some enzymes are bound at levels of up to 30 mg of protein per ml of carrier with 50 to 100% recovery of the applied soluble enzyme activity. Some enzymes, however, are not successfully immobilized by this method. Little pancreatic RNase is bound, probably because it lacks tryptophan and contains low levels of tyrosine and leucine. Chymotrypsin and carboxypeptidase A are bound but express low activities due to inhibition by phenyl groups on the carrier.

This appears to be a mild, simple method, which produces a tightly bound enzyme complex with coupling yields and activity levels that are competitive with or superior to other adsorption techniques. Unlike ionic adsorption, this method produces an immobilized enzyme complex which may be used over a wide range of pH (4 to 11) and ionic strength (5 mM to 2 M) with no elution. One disadvantage is the expense of the carrier, which probably limits the method to small-scale analytical or preparative applications. The concept, however, may be applicable to other, less expensive hydrophobic carriers.

One requirement of this method is that chaotropic salts (NaSCN, NaI) and detergents must be absent from the environment of the immobilized enzyme because these agents will cause elution. This is not a serious limitation, however, and in fact provides a method for reversible immobilization. Cashion et al. (10) have completely eluted bound enzymes from trityl agarose columns with 0.1% sodium dodecyl sulfate or Triton X-100 and found that the support retains its original capacity for binding fresh enzyme through several cycles. Furthermore, if elution is conducted by a mild organic solution such as 25% glycerol, most of the eluted enzyme retains its activity and can subsequently be completely reimmobilized. Thus, this method provides the unique property of reversibility, allowing reuse of both carrier and soluble enzyme.

19.3.3 Biospecific Binding

The biospecific binding method is based on biospecific interactions between enzymes and other molecular species, such as antibodies and inhibitors, which have been exploited by affinity separation methods. Antibodies and inhibitors are not good species for enzyme immobilization, however, because the bound enzyme is usually inactive. A more promising interaction is that which occurs between lectins and the carbohydrate moiety of enzymes (84). Lectins are glycoproteins, chiefly of plant origin, that bind tightly to specific carbohydrate residues. Perhaps the best known lectin is concanavalin A, which is derived from jack bean. The interaction between certain lectins and glycoenzymes during immobilization apparently results in improved activity and stability of the bound enzyme. Since it is recognized that many enzymes are glycoproteins, the method could be widely used.

Binding method for L-ascorbate oxidase

1. Perfuse a column containing 0.5 ml of concanavalin A-Sepharose (obtained commercially) with several volumes of 0.1 M sodium acetate buffer (pH 5.5).
2. Add 4.5 U of L-ascorbate oxidase as a short pulse into the buffer flow, and wash through the column (flow rate, 0.85 ml/min).
3. Wash with 1 M NaCl to remove unspecifically bound enzyme and then with 0.1 M sodium acetate buffer (pH 5.5).

Mattiasson and Danielsson (70) immobilized L-ascorbate oxidase by this method, as well as D-glucose oxidase (in 0.1 M sodium acetate, pH 5.0) and invertase (in 0.1 M sodium citrate-phosphate, pH 5.5). Utilization of a small amount of enzyme (4.5 U/0.5 ml of gel) was apparently sufficient for their analytical technique and produced coupling yields approaching 100%. No detailed stability data were given, but the researchers followed a procedure in which the bound enzyme was used for 1 day, all enzyme was eluted by perfusing with 0.1 M glycine hydrochloride buffer (pH 2.2), and the column was reequilibrated with buffer and reimmobilized with enzyme at the beginning of the following day. By this method, the ascorbate oxidase column was eluted and recharged daily with fresh enzyme for 10 consecutive days with no significant variation in assay performance. The authors claim this method is particularly advantageous for analytical applications using labile or expensive enzymes, or for the analysis of irreversible inhibitors.

Iqbal and Saleemuddin (40) have recently used this method to immobilize invertase for the continuous conversion of sucrose. Their technique involved washing the concanavalin A-Sepharose matrix in 0.1 M sodium acetate buffer (pH 5.6), adding 0.45 mg of invertase (1,161 IU) to 1 g of carrier in a total volume of 5.0 ml, incubating for 12 h, and washing in buffer. A preparation with 1,089 U/g of matrix (71% coupling yield) resulted. To improve the operational stability of the bound enzyme, a 10% suspension was subjected to cross-linking with 0.2% glutaraldehyde in phosphate buffer (pH 6.1) at 30°C. After 2 h of cross-linking, only about 10% of the original activity was lost. The cross-linking reaction was stopped by adding ethanolamine to a final concentration of 0.01%, holding for 1 h, and then washing in acetate buffer. When utilized in a column reactor to convert 1 M sucrose at 30°C, the cross-linked complex lost only 18% of its activity after 60 days of continuous operation at a flow rate of 100 ml/h.

19.3.4 Transition Metal Activation

The method of transition metal activation was first described by Emery et al. in the early 1970s (27). It has also been studied by Messing, who refers to the technique as inorganic bridge formation (p. 166–169 of reference 74). Over the years, it has been shown to improve the activity and stability of immobilized enzymes as compared to simple adsorption, when used in conjunction with a variety of materials including cellulose, glass, ceramics, nylon, chitin, sand, and some organic polymers (6). The principle involves chelation by a transition metal between the protein and support material. This method retains the advantage of regeneration (27).

Binding method (glucoamylase on CPG)

1. Prepare a mixture of controlled-pore glass (CPG; e.g., CPG 1000-80, 177 to 250 μm, Sigma Chemical Co.) in a 15% (wt/vol) solution of titanium IV ($TiCl_4$) in 15% (wt/vol) HCl (available commercially) at a ratio of 10 ml/g of support.
2. Dry the resulting slurry in an oven at 45°C for 30 h.
3. Wash the activated support thoroughly in 0.02 M sodium acetate buffer (pH 4.5).
4. Mix the activated carrier with enzyme solution in the buffer described in step 3 (200 ml/g of support) and incubate for ca. 16 h at 4°C. The glucoamylase preparation used in the example (7) had an activity of 340 U/ml or 68 U/mg of protein (assay based on liberation of 1 μmol of glucose per min at 45°C for 1% soluble starch in 0.02 M sodium acetate buffer, pH 4.5).
5. Wash thoroughly in sodium acetate buffer (0.02 M, pH 4.5).

This method produced a CPG-glucoamylase complex with an initial activity of approximately 72 U/g of matrix. Since no information was given concerning residual activity in the enzyme solution, it is not possible to estimate the coupling yield (6–8). The amount of bound protein was estimated at 3.5 mg/g of support, which gives a specific activity for the immo-

TABLE 2. Entrapment methods[a]

1. Gel polymerization (single or comonomer)
   a. Solution
   b. Suspension
2. Protein copolymerization
3. Gel formation by cross-linked polymers
4. Coarctation: polyelectrolyte gel by polyvalent ions
5. Interpenetrating networks (gel formation within a porous support)
6. Physical localization
   a. Microcapsules, liposomes, erythrocytes
   b. Hollow fibers
   c. Ultrafiltration

[a] From reference 74.

bilized enzyme of 21 U/mg of bound protein (7). During continuous contact with 1% soluble starch (45°C, 0.02 M sodium acetate, pH 4.5) the complex lost half of its activity within approximately 40 h, due to elution and denaturation, but then remained relatively stable over the next 80 h. No attempt was made to estimate or calculate a half-life after the stabilization period was reached.

Titanium chloride has been used recently with promising results to activate other carriers including alumina (1, 77) and Spheron, a macroporous hydrophilic gel prepared by suspension polymerization of hydroxyethyl methacrylate and ethylene dimethacrylate. When glucoamylase was bound by Spheron P 100,000 (available commercially) by the above method with some modifications, a coupling yield of 71% and a loading of 25 mg of protein per g of carrier were obtained (32). No stability data were given.

In an attempt to improve operational stabilities for glucoamylase on CPG, an alternative method was developed involving amination and carbonylation of the transition metal-activated carrier.

Transition metal activation with amination and carbonylation

1. Using the same carrier and $TiCl_4$ solution as above, mix in a ratio of 2.5 ml of solution per g of carrier.
2. Dry as in the previous method at 45°C for 30 h.
3. Wash with chloroform.
4. Aminate at 45°C for 30 min in 1% (wt/vol) 1,6-diaminohexane in chloroform (5 ml/g of matrix).
5. Wash the aminated support in methanol and water and contact with 5% (wt/vol) glutaraldehyde (5 ml/g of matrix) in 0.05 M phosphate buffer (pH 7.0) for 1 h at 25°C.
6. React the aldehyde derivative with an enzyme solution in 0.02 M sodium acetate buffer (pH 4.5) at 4°C for 2 h. The characteristics of the enzyme solution in the example were: 200 ml/g of matrix, 495 U/ml, 8.25 mg of protein per ml, and 60 U/mg of protein (8).

This method produced a CPG-glucoamylase complex with an initial activity of 67 U/g of carrier and an estimated operational half-life at 45°C and pH 4.5 (1% soluble starch in 0.02 M sodium acetate buffer) of 60 days. When the matrix was activated with $ZrCl_4$, an initial activity of 63 U/g and a half-life of 71 days resulted.

19.4 ENTRAPMENT

Several methods have evolved over the years for entrapping enzymes in a wide variety of carrier matrices. While entrapment techniques have generally not been among the most popular for enzyme immobilization, they have recently become the method of choice for binding whole cells and organelles. According to O'Driscoll (p. 171 of reference 74), one reason for the lack of popularity of enzyme work, especially in industrial processes, is the misconception that all resulting complexes suffer from elution and poor mechanical properties. Table 2 lists the various types of entrapment methods that have been applied, and Table 3 presents a list of advantages and disadvantages generally associated with this technique.

The most popular monomer by far for gel polymerization has been acrylamide, although recently more work has been done with 2-hydroxyethyl methacrylate (HEMA), as well as some hydrophobic gels (59). Some polymers that have been used for gel formation are gelatin, chitosan, and polyvinyl alcohol. Alginate, agar, and carrageenan gels have been most popular for whole-cell immobilization. Some specific examples of entrapment methods are discussed below.

19.4.1 Polyacrylamide Gel Formation

In the polyacrylamide gel procedure, the enzyme is mixed in a buffered aqueous solution of acrylamide monomer and a cross-linking agent, often *N,N'*-methylenebisacrylamide. Polymerization may be initiated by a redox system, such as potassium persulfate and tetramethylene diamine, or by a photochemical system, such as riboflavin in the presence of light. The reaction may be conducted in aqueous solution or by rapidly mixing the aqueous reagents into an organic solution (such as 3:1 toluene-chloroform) containing a

TABLE 3. Characteristics of entrapment methods

Advantages	Disadvantages
1. General to enzyme, cell, multienzyme, coenzyme systems	1. Elution from gels
2. Minimum constraints on enzyme	2. Poor mechanical properties of some gels (abrasion, compaction)
3. Can incorporate particles to increase density or impart magnetic properties	3. Inactivation by monomer or cross-linker (low coupling yield)
	4. Diffusional restrictions for large substrates[a]

[a] Note: Ultrafiltration systems offer the special advantage of having no diffusional restrictions for large-molecular-weight or insoluble substrate, but are adversely affected by shear or concentration polarization effects.

small amount of ionic surfactant as a suspension stabilizer. The former method produces a rather weak gel which can be broken up by mixing, whereas in the latter suspension polymerization results in compact, spherical particles. The resulting complex may be freeze-dried (p. 169–183 of reference 74). The polymerization reaction, which is exothermic, is usually conducted at low temperatures (0 to 25°C) to protect the enzyme against thermal denaturation (97). The system also should be deoxygenated by sparging with N_2 gas before polymerization.

The expressed activity of the complex and the activity yield of the immobilization process are affected greatly by monomer concentration and cross-linker/monomer ratio. Total monomer (acrylamide + methylenebisacrylamide) per 100 ml of solution usually varies from 20 to 50%, while cross-linker/monomer ratios generally fall in the range of 1 to 10%. It has been shown that acrylamide monomer is an enzyme denaturant, probably acting in a manner similar to urea (21, 63). The presence of substrate in the mixture may protect somewhat against this effect (63). Since increasing the monomer concentration increases bound protein, as well as the rate of denaturation, an optimum concentration exists which should be determined experimentally. Likewise, the optimum cross-linker/monomer ratio should be determined.

Although it appears that the free radicals formed during the reaction do not usually affect enzyme activity (p. 172 of reference 74), it has been shown recently (63) that the initiator potassium persulfate may reduce enzyme activity by oxidation of sulfhydryl groups. The presence of reducing agents such as glutathione, dithiothreitol, and cysteine, at a concentration of 0.125% (wt/vol) for a typical persulfate concentration of 0.625%, seems to protect against this effect (63). The surfactant used in suspension polymerization has also been identified as a potential denaturant, which may decrease the level of bound enzyme activity (p. 177 of reference 74).

19.4.2 Radiation Polymerization

Acrylamide polymerization may also be initiated by gamma-ray irradiation. This method, which was developed by Kawashima and Umeda (46), is growing in popularity, especially among Japanese researchers. The polymerization, which may be conducted in either solution or suspension, is carried out at very low temperatures (less than −70°C, e.g., in a petroleum ether-dry ice bath). In this way, aqueous solutions of acrylamide, methylenebisacrylamide, and the enzyme (at concentrations similar to those given for the previous method) are quick-frozen into slabs, membranes, or beads and irradiated by gamma rays from ^{60}Co for 5 to 30 min at a dose rate of about 800 krad/h. Sparging with N_2 is unnecessary. The resulting material, which is thawed in ice water overnight, is highly porous due to the action of ice crystals and has good activity and stability characteristics. Since this porous nature is destroyed at high buffer concentrations (e.g., 0.5 M phosphate), the immobilized enzyme should be restricted to buffer levels in the 10 to 30 mM range (76). Mechanical and swelling properties can be varied by copolymerization with other monomers, such as calcium, sodium, and magnesium acrylate (76).

The radiation copolymerization method has been extended to the use of several different types of hydrophilic and hydrophobic glass-forming monomers for immobilizing whole cells, antibodies, and enzymes. In this method, higher radiation doses and no cross-linker are utilized, and the buffer restrictions listed above are removed. Radiation damage to the enzyme is minimal at doses of less than 10^6 rads under vacuum and at temperatures of −24 to −78°C (44). Enzyme activity yield (and expressed activity) varies as a function of monomer concentration for both hydrophilic monomers such as HEMA and hydrophobic monomers such as diethylene glycol diacrylate (46). Activity yields of 30 to 70% of applied enzyme have been observed for monomer concentrations in the 30 to 50% range for several enzymes. Monomer concentrations of greater than 50%, however, are necessary to eliminate enzyme elution (44, 58). Activity yields may be improved by copolymerization of hydrophilic and hydrophobic monomers (44). For example, when comonomers HEMA and tetraethylene glycol diacrylate are used to entrap glucose oxidase, a hydrophil/hydrophobe ratio of 1:1 produces an optimum coupling yield of approximately 50% (for a total monomer concentration of 50%). Bifunctional monomers, such as nonatetradecaethylene glycol diacrylate, produce superior mechanical properties (58).

Radiation copolymerization method (glucoamylase and HEMA)

1. Place 500 U of glucoamylase and 2.5 g of HEMA in a test tube with 5 ml of 0.1 M acetate buffer at pH 4.5 and shake rapidly.
2. Immediately after shaking, immerse the tube in a Dewar flask filled with a dry ice-methanol mixture and irradiate using a ^{60}Co source at 1.0×10^6 rads/h for 1 h. Maintain temperature at −78°C throughout the radiation.
3. After irradiation, cut the polymerized enzyme composite into particle form; e.g., 1 to 2 mm, 8/12 mesh.

In the example (58), soluble and immobilized glucoamylase were assayed at 40°C and pH 4.5 using a 20% (wt/vol) sweet potato slurry which had been prepared by crushing sun-dried potatoes in a high-speed planetary mill and mixing with 0.1 M acetate buffer. Activity was related to the amount of D-glucose formed in 1 h. Under these conditions the immobilized enzyme expressed the equivalent of 300 U of soluble enzyme (coupling yield of 60%) and was stable through 15 consecutive batch assays.

When hydrophobic monomers are used, small nonporous spheres are formed in aqueous solution. The diameter (100 to 300 μm) varies with monomer concentration, and enzyme molecules are trapped near or covalently bound to the external surface (44). This is a desirable condition for applications involving larger-molecular-weight substrates because internal diffusional resistances are eliminated. In a variation of the radiation polymerization method which produces a similar result with hydrophilic monomers (57), sodium acetate (1 to 4 M) is added to an aqueous solution of monomer and enzyme, resulting in the salting out of fine, monomer-rich particles. After formation of the particles, the system is frozen at −78°C and irradiated with 10^6 rads by gamma ray from ^{60}Co. The spherical particles have an average diameter in the range of 50 to 900 μm, depending on salt and monomer concentrations. Stable, 500-μm particles were produced by 50% HEMA and 3 M acetate.

19.4.3 Protein Copolymerization

Protein copolymerization is actually a covalent technique in which the enzyme is vinylated with acylating or alkylating monomers and copolymerized with other monomers (p. 195–201 of reference 74). It will be discussed here, however, because it can be viewed as an important variation of the entrapment methods described previously. The procedure, which was first presented by Jaworek (42), has recently been used by Mozhaev et al. (75). Enzymes are modified by incubation with a specified amount of acryloyl chloride or acrolein to acylate or alkylate protein amino groups as follows.

Copolymerization method using acylated enzymes (α-chymotrypsin and acrylamide)

1. Prepare a 300 mM solution of bovine α-chymotrypsin in 0.2 M KH_2PO_2 buffer (pH 8). Cool to 0°C.
2. Add a 1,000-fold excess of acryloyl chloride (based on 17 amino groups per enzyme molecule) over a 5-min period with extensive stirring. Maintain temperature at 0°C. Control pH of solution by addition of KOH.
3. Incubate the modified enzyme at pH 8 and 25°C for ca. 1 h to deacylate functional amino groups at the active site of the enzyme and restore optimal activity.
4. Combine the modified enzyme solution with 30% (wt/vol) acrylamide monomer, 1.5% (wt/wt) *N,N'*-methylenebisacrylamide (cross-linker), and 30 mg of riboflavin (initiator) per liter.
5. Polymerize the mixture in a narrow test tube at 0°C under a 300 W tungsten lamp for 1 h.
6. Grind the polymerized conjugate to a powder (particle size less than 0.1 mm) and wash exhaustively in water.
7. Assay the immobilized preparation by adding 0.1 g of powder to 5 ml of a 5 mM *N*-acetyl-L-tyrosine ethyl ester solution at pH 8.0 and 25°C and titrating for liberated acid. The bound enzyme retains 25% of the activity of the applied native enzyme (67).

Enzymes immobilized in this way have demonstrated outstanding storage and operational stability characteristics and dramatically elevated maximum operating temperatures. Furthermore, these copolymerized enzymes are capable of regeneration from an apparently denatured state, and several cycles of this regeneration process are possible (68). The copolymerization method may also be applied to the immobilization of coenzymes (101).

19.4.4 Enzyme Graft Copolymerization

The enzyme graft copolymerization method is similar to the protein copolymerization technique discussed above. In this case, however, copolymerization takes place on a preexisting carrier matrix, and the enzyme copolymer is grafted to the support. The potential advantages of this technique over the previous one are improved mechanical properties, depending on the support material, and a preparation involving surface-bound enzymes which may reduce inefficient use of the catalyst due to internal diffusion effects observed with large-molecular-weight substrates.

As in the previous method, copolymerization can be obtained by use of a vinylated enzyme, vinyl monomer, and redox initiator (20). Although this technique produces complexes with kinetic properties similar to those of corresponding soluble enzymes, coupling efficiencies and activity levels are not outstanding. However, modifications involving the use of photochemical initiation and a nonvinylated enzyme provide significantly improved results (18, 19). This method uses a polysaccharide carrier, such as Sepharose, and $FeCl_3$ as the photochemical initiator. Separose is soaked in a solution of $FeCl_3$, and the ferric catalyst is strongly adsorbed to the support. This is followed by the addition of the enzyme and monomer, activation by UV radiation (low-pressure mercury vapor source), and polymerization for a specific time. After the reaction, the beads are washed in an NaCl solution (1 M) to remove unbound enzyme.

Immobilized enzyme activities and coupling efficiencies resulting from the above method are strongly influenced by $FeCl_3$ concentration, reaction time, and monomer/support and enzyme/support weight ratios and should be optimized with respect to these variables for each monomer and enzyme. Typical values are 0.15 mM $FeCl_3$ per g of Sepharose, 30-min reaction time, 40 mg of enzyme per g of Sepharose, and 50 to 200 mg of monomer per g of Sepharose. The monomers which have been tested so far include glycidylmethacrylate, bisacryloylpiperazine, and 1,3,5-hexhydrotriacryloyl-*s*-triazine.

When glucose oxidase is immobilized with glycidylmethacrylate, activities vary from about 15 to 50 U/g of support while corresponding coupling efficiencies vary from 50 to 15%. The thermal stability at 60°C is superior to those of the free enzyme and of glucose oxidase covalently bound to Sepharose by the cyanogen bromide technique, which will be described later. For horseradish peroxidase (40 mg of enzyme and 100 mg of monomer per g of support), 1,3,5-hexhydrotriacryloyl-*s*-triazine appeared to give the best properties: 125 U/g of support, a coupling efficiency of approximately 15%, and no loss of activity over 20 h of continuous use.

19.4.5 Gel Formation by Cross-Linked Polymers

Gel formation by cross-linked polymers usually involves mixing enzymes in an aqueous solution with a soluble polymer, pouring the solution onto a nonsticking surface, drying to form a solid membrane, and cross-linking to improve enzyme binding and mechanical strength. Cross-linking may be accomplished by use of a bifunctional reagent or by UV radiation. In some cases, the cross-linker may be included in the polymer enzyme solution before drying occurs, which may allow casting into other than membraneous forms. This is not a pure entrapment method because at least some of the trapped enzyme molecules are covalently bound to the polymer by the cross-linking agent, while others are stabilized by multiple noncovalent interactions.

Gelatin

One of the more popular materials used with this method is gelatin, usually derived from calf skin, tendon, or bone. Because of its proteinaceous nature, gelatin is capable of initiating strong cooperative bonds with enzymes and usually provides a stable environment which protects the enzyme against denaturation and inactivation during cross-linking (90).

Enzyme complexing with cross-linked gelatin

1. Prepare a solution of 6% (wt/vol) gelatin (food grade) in 0.1 M acetate buffer (pH 4.5) at 50°C.
2. Add glucoamylase (lyophilized commercial preparation) in the amount of 40 mg/ml of solution.
3. Inject with a syringe or pump at a constant flow (2 ml/min) through a finely drawn capillary tube into 400 ml of dry, freshly distilled *n*-butyl acetate kept in a salt-ice bath. Stir magnetically for 5 min after formation of the gel particles has reached a bed volume of ca. 30 ml. Particle size is on the order of 0.25 to 0.50 mm.
4. Remove *n*-butyl acetate and add 50 ml of 10% glutaraldehyde in aqueous solution to 30 ml of the glucoamylase-gelatin particles. Stir for 30 min at room temperature.
5. Wash the cross-linked enzyme conjugate with 0.1 M acetate buffer (pH 4.5) and assay using 1% (wt/vol) soluble starch in the same buffer at 55°C.

This method produces glucoamylase activities of 16.3 U (μmol of glucose per min) per g of support (49). No stability or coupling yield data were given in the cited reference, but general experience indicates that results should be competitive with other methods. In fact, results were favorable enough to establish the method as a commercial process for the preparation of immobilized glucoamylase (49).

Activities of the gelatin-entrapped enzyme particles can be improved by stirring freshly cross-linked particles in a fresh enzyme solution. When 5 ml of enzyme-entrapped, cross-linked particles are contacted with 25 ml of acetate buffer (0.1 M, pH 4.5) containing 125 mg of glucoamylase and the mixture is stirred for 18 h at 4°C, the total enzyme activity is increased to 20.9 U/g of carrier. The excess binding is due to free aldehyde groups that exist on the bead surface for several hours after cross-linking.

Gelatin has also been used in the form of membranes or thin films (90). Several variations of the immobilization procedure have been employed, but generally the method involves pouring a thin film of gelatin-enzyme solution, drying, and cross-linking in glutaraldehyde. Enzyme activity, coupling yield, stability, and mechanical strength must be optimized with respect to glutaraldehyde concentration and cross-linking time for each enzyme.

As an alternative to gelatin, acid or alkaline suspensions of reconstituted collagen may be utilized to form a more durable carrier (p. 243–263 of reference 74). Reconstituted collagen and gelatin are also excellent matrices for whole-cell immobilization (49).

Polyvinyl alcohol

Enzymes may also be immobilized within a polyvinyl alcohol (PVA) membrane formed by exposure to UV radiation (39). Imai et al. (39) claim that PVA offers a superior alternative to polyacrylamide for applications in the food industry. In this method, an aqueous solution of 5% PVA (degree of polymerization, 2,400), containing a mixture of benzoic acid and sodium benzoate (1:2.96) as sensitizer is combined with a solution of enzyme (crude invertase, 0.1 M acetate buffer, pH 4.6) at a weight ratio of 1:2:100 (enzyme-sensitizer-PVA). Gelation occurs by casting the mixture on a polyvinyl chloride plate over phosphorus pentoxide in a desiccator at room temperature and exposing to UV light at 253.7 nm from a distance of 3 cm. The gel content of the 30-μm-thick membrane (determined by weight of insoluble fraction after 7 h in boiling water) varies with sensitizer concentration and irradiation time. For the weight ratios given above, a gel content of approximately 90% is obtained after 150 min of irradiation.

Chitosan

Chitosan [(1→4) 2-amino-2-deoxy-β-D-glucan] is a support obtained by N-deacetylation of chitin by 40% NaOH in the presence of 0.01% $NaBH_4$ (36). Among the advantages of this material are its derivation from the waste material chitin and its versatile chemical nature, which provides a variety of active groups for enzyme immobilization.

Chitosan, dissolved in 2% acetic acid by standing at room temperature overnight, is mixed with the enzyme and then gelled by the addition of 25% aqueous glutaraldehyde at 5°C. In a typical preparation, 100 mg of chitosan is dissolved in 2 ml of acetic acid, mixed with 100 mg of enzyme, and cross-linked by the addition of 0.80 ml of glutaraldehyde. After gelation, which occurs in several minutes, the complex is stored overnight at 5°C and washed several times in distilled water. As with any method involving glutaraldehyde, a compromise must be made between optimizing for gel strength or for enzyme activity. In this method, a relatively high glutaraldehyde concentration is required, which may inactivate some enzymes (100).

19.4.6 Coarctation: Polyelectrolyte Gel Formation by Polyvalent Ions

Coarctation has been applied by Horvath and Sovak (37) and is discussed briefly by O'Driscoll (p. 179 of reference 74). The former authors were able to form a gel from a trypsin-polycarboxylic acid conjugate by the addition of Ca^{2+} to the polyelectrolyte solution. The cation serves as a cross-linker, and, as such, the character of the gel is affected by its concentration. As calcium ion levels vary from 10^{-4} to 10^{-2} M, gel strength increases and water content decreases. This is potentially a very mild form of immobilization, but it has not been widely applied, probably because of the requirement for covalent binding of the enzyme to the polymer to reduce elution. Tanaka et al. (88) have shown that elution of glucoamylase entrapped in calcium alginate gel beads is significantly reduced by stirring the gel particles in a 1% polyethyleneimine solution at pH 5 for 5 min after enzyme immobilization.

19.4.7 Physical Localization

Fiber-entrapped enzymes

A method utilizing fiber-entrapped enzymes was first described by Dinelli (23) in 1972. It was patented and developed by Snamprogetti (Italy), and today is available commercially through that company. Many enzymes have been immobilized by this procedure, and some have been used commercially (penicillin amidase, lactase). Unique reactors have been designed to accomodate the spun-fiber configuration of the catalyst (64).

Entrapment by this method is achieved by forming an emulsion between an aqueous enzyme phase and an immiscible organic solvent containing the fiber-forming polymer, extruding the emulsion through a small-diameter opening into a liquid precipitant, and continuously collecting the resultant fibers on a wet spinning apparatus similar to that used in the textile industry (p. 227–243 of reference 74). The fibers are then vacuum dried to remove organic solvents. The enzyme is trapped in microcavities within the fiber structure. Cellulose triacetate (dissolved in methylene chloride) has been the most popular polymer used in conjunction with this method, but any fiber-forming polymer can be used. Precipitants have included toluene and petroleum ether.

Retention of enzyme activity is surprisingly high, probably due to the fact that proteins reside within the aqueous environment of the emulsion and, also, to the short contact time involved. Outstanding characteristics of enzymes immobilized by this technique are their high level of activity and their exceptional operational stabilities. For example, preparations of invertase entrapped in cellulose triacetate fibers express up to 65% of the activity of the applied soluble enzyme and are capable of continuously hydrolyzing a 20% (wt/vol) sucrose solution (0.1 M sodium phosphate, pH 4.5) for over 5 years with only a 10% decline in activity at 25°C. When 50% sucrose solutions (deionized water, pH 4.5 with HCl) are used, the entrapped invertase retains its activity for approximately 5 months at temperatures up to 35°C. Specific activities of up to 1,800 U of invertase per g of carrier have been achieved by this method (1 U corresponds to hydrolysis of 1 mg of sucrose per min at 25°C and pH 4.5 when 20% sucrose is used as substrate).

A similar procedure has been employed by Pitt et al. (79) to immobilize enzymes within the walls of hollow fibers. In this method, an aqueous solution of the enzyme (carboxypeptidase G_1) and stabilizer (bovine serum albumin) is mixed with polysulfone in *N,N*-dimethylformamide solution. The mixture is then either cast into membranes, using water as a precipitant, or spun into hollow fibers with a typical internal diameter of 190 μm and a wall thickness of 45 μm. Unlike the fibers discussed previously, these can be used as microtubular reactors, which usually consist of a bundle of fibers within a common housing.

Microcapsules, liposomes, erythrocytes

Microcapsules, liposomes, and erythrocytes have been developed as carriers essentially for use in medical applications in which they may be injected into the body for the administration of enzyme therapy. Solid, semipermeable microcapsules may be injected subcutaneously or intraperitoneally and held in a fixed location, where the enzyme acts on diffusing substrate from surrounding fluid. Liposomes and erythrocytes, however, can be injected intravenously for delivery of the enzyme to targeted areas, such as the liver.

The preparation of solid, semipermeable microcapsules, developed by Chang (p. 201–218 of reference 74), is quite similar to the method of fiber entrapment described above. In a typical procedure, an emulsion is created from an organic phase, containing a surfactant, and an aqueous enzyme phase. This is followed by the addition of the membrane-forming polymer in a solvent solution, e.g., cellulose nitrate in ether. On standing, the polymer precipitates at the microdroplet interface, forming stable microcapsules which are decanted and washed. The encapsulation of several enzymes has been reported in the literature. Recently, a special lipid-polyamide membrane microcapsule has been developed for the encapsulation of unmodified coenzymes and multienzyme systems (93).

Liposomes and erythrocytes are different from solid, semipermeable microcapsules in that enzyme action occurs only upon the disruption of the carrier. For this reason, these are actually delivery devices rather than true immobilized enzyme preparations. Liposomes are lipid vesicles which contain the enzyme in an aqueous environment surrounded by one or more concentric lipid bilayers. They are formed by evaporation of solvent from a lipid solution, dispersion of the resulting dry lipid film with aqueous enzyme solution, size reduction by sonication, and chromatographic separation from residual free enzyme. Lecithin is a typical phospholipid used in this procedure, but other lipids and phospholipids may be utilized. Chloroform is a common solvent. This system has been developed and described by Gregoriadis (p. 218–227 of reference 74). Yagi et al. (99) have described a novel preparation which may be incorporated into the brain.

Entrapment of enzymes in erythrocytes was first demonstrated by Ihler et al., who suggested using the technique to administer enzyme replacement therapy (38). Although subsequent studies have shown that this is not a useful method for disease management, there is still a strong interest in using erythrocytes as a general drug carrier (22). Tsang et al. have demonstrated the ease and efficiency of entrapping enzymes, viruses, bacteria, and DNA in erythrocytes by endocytosis, which occurs on exposure to membrane-active drugs (primaquine, hydrocortisone), or simply by dialysis against a hypo-osmotic buffer (92).

Ultrafiltration

Localization by ultrafiltration may be preferred when the enzyme is to operate on an insoluble substrate, since the soluble enzyme would have greater access to the substrate than one which was bound to a solid carrier. Such a system may also be useful for multienzyme- or coenzyme-requiring reactions. Any traditional ultrafiltration system may be utilized, including stirred cells, thin channels, or hollow fibers (p. 291–317 of reference 74). A commercial process based on this approach is in operation in West Germany.

The most serious limitations associated with this method are membrane adsorption of enzyme and fouling of the membrane, both of which are the result of concentration polarization. The former effect results in the loss of catalytic activity; the latter reduces membrane flux and, therefore, reactor productivity. These problems can sometimes be reduced by increasing shear parallel to the membrane, immobilizing a degradative enzyme on the membrane to break down the fouling layer, precoating the membrane with an inert protein, and operating with high-molecular-weight cut-off membranes. The last solution may require increasing the effective molecular weight of enzymes or coenzymes by covalent coupling to a soluble macromolecular carrier. In some cases, an insoluble substrate serves as an adequate support for retaining the enzyme within the reactor. Various applications of ultrafiltration and hollow-fiber systems have been described recently in the literature (11, 29, 43, 45, 51).

19.5 COVALENT METHODS

Methods resulting in enzyme immobilization by covalent bond formation have received a great deal of study over the years. As a result, there is a considerable amount of information in the literature detailing procedures for a wide variety of carrier materials and reactive groups (5, 13, 48, 72, 74, 81, 91, 96, 98). There seems to be an increasing trend toward the study of "noncovalent" methods, especially for industrial applications, but even these often involve covalent bonding. For example, we have already discussed cross-linking and copolymerization used in conjunction with adsorption and entrapment methods. In this section we will limit our discussion to covalent techniques based on bonding between specific, identifiable groups of both the enzyme and the carrier.

One of the key advantages of covalent binding is usually thought to be the stability of the enzyme-carrier bond, which prevents elution of protein into the product stream. Although this is generally true, it has been demonstrated that some covalent bonds are reversible or unstable (69). Disintegration or dissolution of the carrier will also cause a loss of enzyme. It is important, therefore, that care be taken in choosing both the carrier and binding techniques for specific operating conditions if elution is to be completely avoided or at least minimized.

Perhaps the greatest advantage of a covalent approach is the wide range of choices one has in selecting carrier materials and binding methods. This allows a great deal of flexibility in designing an immobilized enzyme with specific physical and chemical properties, such as charge distribution, hydrophobe/hydrophile ratio, spacer arm separation, partitioning capabilities, etc. Acrylic copolymers are especially versatile as a family of carrier materials which may be prepared with a wide variety of properties.

TABLE 4. Characteristics of covalent methods

Advantages	Disadvantages[a]
1. Specific binding	1. Expensive
2. Little or no elution	2. Complicated procedures
3. Wide choice of carriers and methods	3. Harsh environment
	4. Regeneration impossible or difficult

[a] As mentioned in the text, these are relative and not necessarily serious problems for all methods.

Several of these are available commercially, including Bio-Gel P (polyacrylamide), Bio-Gel CM2 (with carboxyl groups), aminoethyl and hydrazide Bio-Gel P, Enzacryl AA and AH (polyacrylamide containing aromatic amines and hydrazide residues), and Spheron (hydroxyethylmethacrylate-ethylenedimethacrylate copolymer).

Some of the possible disadvantages of covalent methods are the relative expense and complication of procedures which are involved. Also, activity yields may be low due to exposure of the enzyme to harsh environments or toxic reagents. These adverse effects vary considerably for each combination of enzyme and carrier. A method which does not provide good results with one enzyme may perform well with others. Likewise, several different carriers and methods must be screened for the immobilization of a specific enzyme before satisfactory results are achieved. As with all binding methods, a certain amount of trial and error is in order.

The relative advantages and disadvantages of covalent methods are summarized in Table 4, and Table 5 lists some of the popular techniques that have been applied (see references 5, 13, 48, 72, 74, 81, 91, 96, and 98 for further detail). Some of the many different carrier materials that have been used in conjunction with these methods are: polysaccharides, including Sephadex, Sepharose, cellulose, and derivatives (*p*-aminobenzyl, carboxymethyl, aminoethyl, etc.); acrylic polymers and copolymers, such as polyacrylamides (Bio Gel, Enzacryl), hydroxyalkyl metacrylates (Spherons), maleic andydrides, and styrene derivatives; polyamides, such as nylon; proteins, e.g., collagen; activated carbon; and CPG, aminoalkylated, succinylated, etc.

19.5.1 Cyanogen Bromide

The cyanogen bromide (CNBr) method was first presented by Axén et al. in 1967 (2) and has been used extensively since that time to couple enzymes to polysaccharide carriers via hydroxyl groups on the support. It is probably the most popular covalent technique used today, especially for laboratory applications.

Traditionally, activation of polysaccharide carriers by CNBr is conducted in the presence of a strong base to improve the nucleophilicity of the resin. Unfortunately, the strongly basic medium also causes hydrolysis of CNBr and the reactive cyanate esters which bind to the carrier. This results in the requirement for a large excess of the highly toxic reagent and a low activation yield (0.02 mol of ligand coupled per mol of CNBr applied). In addition to its toxicity, concentrat-

TABLE 5. Summary of typical covalent methods[a]

Method	Reactive groups	
	Carrier	Enzyme
CNBr	-OH	$-NH_2$ (lysine)
Carbodiimide	-COOH	$-NH_2$
	$-NH_2$	-COOH
Woodwards reagent K	-COOH	$-NH_2$
EEDQ[b]	-COOH	$-NH_2$
Diazotization	Aromatic amine	Lysine, tyrosine, histidine
Acyl azide	$-CH_2COOH$ to $-CH_2CON_3$	Lysine, tyrosine, cysteine, serine
Acid anhydride	*N*-Carboxyanhydrides	$-NH_2$ (terminal lysine)
Arylation	-OH modified by aromatic halide (e.g., cyanuric chloride)	$-NH_2$
Thiomide (amidination)	$-NH_2$ to -NCS (by thiophosgene)	$-NH_2$
Aldehyde condensation	-CHO (polymeric or by action of glutaraldehyde on amino groups)	$-NH_2$

[a] References 5, 13, 48, 72, 74, 81, 91, 96, and 98.
[b] EEDQ, *N*-Ethoxycarbonyl-2-ethoxyl-1,2-dihydroquinone.

ed solutions of CNBr in the required organic solvents are capable of decomposing explosively under certain conditions. It is, therefore, necessary to exercise extreme caution when utilizing the traditional or conventional CNBr procedure (54, 55).

According to Kohn and Wilchek (55) there are two versions of the conventional procedure: the "titration technique," in which pH is kept constant by continuous addition of base, and the "buffer technique," which employs a strongly buffered medium. The buffer technique is easier to perform, but produces resins with a slightly lower coupling capacity.

Titration method of CNBr activation (55)

1. Prepare a packed cake of Sepharose 4B by washing the resin with distilled water on a course, sintered-glass funnel and applying mild vacuum suction.
2. Transfer 10 g of packed resin to a reaction vessel and add distilled water to produce a stirrable slurry with a volume of 1.2 times the settled bed volume.
3. Cool slurry to 10°C.
4. Dissolve 1 g of CNBr in 2 ml of *N*-methylpyrrolidone, dimethylformamide, or acetonitrile and add to slurry with vigorous stirring (some CNBr will precipitate).
5. Maintain pH at 11.0 by addition of 2 to 4 N aqueous NaOH. Stir vigorously to facilitate the mixing of base and resolubilization of CNBr.
6. After 15 min, or when base consumption slows significantly, transfer the activated resin to a coarse, sintered-glass funnel and wash with ice-cold distilled water (500 ml).
7. Suspend the washed, activated carrier in a buffered solution of the enzyme, preferably at alkaline pH to reduce the amount of protonized amino groups on the protein. Incubate at 4°C for 12 to 24 h.

This method produces a binding capacity of 20 μmol/g of drained Sepharose 4B when tested with ε-aminocaproic acid. The activated resin can be stored at −20°C in an aqueous acetone solution under airtight conditions with a loss of 10% or less per month of its original binding capacity, but it is recommended that the resin be used for coupling immediately after activation. Dried, stabilized, preactivated Sepharose 4B is also available commercially from Pharmacia Fine Chemicals, Sweden.

The buffer technique is similar except that the suction-dried resin is suspended in 2 M sodium carbonate buffer and a reagent solution of 1 g of CNBr in 1 ml of dimethylformamide is used. Also, the reaction proceeds at 0°C for exactly 2 min. The resultant binding capacity is approximately one-half of that produced by the titration technique.

The principal active group on the activated carrier which is responsible for enzyme binding apparently depends on the type of support used (54). Polyglucose carriers, such as Sephadex, cellulose, or dextran, upon CNBr activation first form unstable cyanate esters which are then rapidly converted in solution to carbamates and imidocarbonates (linear and cyclic). Binding of the enzyme occurs predominantly through the cyclic imidocarbonate to form N-substituted imidocarbonate bonds or isourea derivatives. Since imidocarbonates are most stable in basic media, it is suggested that polyglucose carriers be washed in cold 0.1 M sodium bicarbonate (pH 8.5) after activation.

Agarose-based resins, such as Sepharose, apparently stabilize the cyanate ester formed after contact with CNBr, and enzyme binding occurs directly with this group to form the isourea derivative. Since cyanate esters are stable in acid medium, a wash of 0.001 M HCl is suggested to preserve its coupling capacity. Washing CNBr-activated Sepharose in alkaline medium removes the active cyanate esters and leaves only small amounts of less active imidocarbonates. Kohn and Wilchek have developed analytical methods for determining cyanate ester and imidocarbonate content on CNBr-activated polysaccharide resins, which they recommend for estimating the maximum binding capacity of such carriers (53, 55).

Triethyleneamine (TEA) may be used in place of strong bases as a cyanotransfer agent which increases the electrophilicity of CNBr. This method requires a lower pH (8.0) than the conventional approach, which reduces hydrolysis of active cyanate esters and, therefore, the amount of CNBr required. Also, the activated resin is composed of virtually pure cyanate ester rather than the mixture of cyanate ester, imidocarbonate, and carbamate produced by the conventional method. A linear relationship is observed between active cyanate esters on the resin and added CNBr, up to values greater than 75 μmol of cyanate ester per g of drained resin (55).

CNBr method with TEA

1. Dissolve 10 g of CNBr in 100 ml of acetone.
2. Dissolve 15.2 g of analytical-grade TEA in 100 ml of acetone.

3. Place Sepharose 4B in a coarse, sintered-glass funnel and wash, in order, with water, acetone-water (3:7, vol/vol), and finally acetone-water (6:4).

4. Drain resin. Avoid drying to packed cake.

5. Transfer 10 g of drained resin to a reaction vessel and add 10 ml of acetone-water (6:4) to obtain a dense but stirrable slurry.

6. Cool reaction mixture to −15°C.

7. Add the required volume of CNBr solution for the desired level of activation. While stirring vigorously, add an identical volume of TEA solution dropwise over a 2-min period. A procedure using 40 mg of CNBr per g of drained Sepharose produces 80 μmol of cyanate ester per g of drained resin.

8. Transfer the resin rapidly into a large volume of ice-cold water (500 ml/10 g of resin) and filter on a coarse, sintered-glass funnel. Wash extensively with ice-cold water.

9. The activated resin should be used immediately for coupling as described earlier. It may be stored, however, for several hours in 0.05 to 0.1 N HCl (5 ml/g of resin) at 0°C with little decline in coupling capacity. In this case the resin should be washed with ice-cold water followed by ice-cold solutions of the buffer used for immobilization.

Since the acetone solution may cause slight changes in agarose resins which result in the formation of aggregates during activation and coupling, it is recommended that cross-linked preparations, such as Sepharose CL-4B, be used. Also, because acetone is highly volatile and absorbs light at the same wavelength as proteins (280 nm), other organic solvents such as dimethylformamide may be utilized.

It is possible to use much safer, nontoxic, nonvolatile reagents to produce cyanate esters on Sepharose (52, 55). *N*-Cyanotriethylammonium tetrafluoroborate, 1-cyano-4-(dimethylamino)-pyridinium tetrafluoroborate, and *p*-nitrophenyl cyanate have been used successfully to activate agarose resins, and each offers specific advantages (55). These compounds are available from Sigma. The procedure is similar to the CNBr-TEA method with some minor modifications. *N*-Cyanotriethylammonium tetrafluoroborate is added as a solid to the Sepharose-acetone–water (6:4) slurry, whereas the other two compounds are added in solution with dry acetonitrile. Activation is conducted at 0°C, and 0.2 M aqueous TEA solutions are utilized. Coupling capacities are similar to those produced by the CNBr-TEA method.

The conventional CNBr method and its several variations may also be used to produce soluble enzyme-carrier conjugates.

Method for binding enzymes to soluble dextran

1. Prepare a solution of 185 mg of Dextran T-70 (average molecular weight 70,000) in 3.5 ml of water and adjust pH to 10.7 with 0.2 M NaOH at room temperature.

2. Add a CNBr solution (6 mg of CNBr in 1 ml of water) dropwise over a period of 5 min with vigorous stirring. Maintain the pH at 10.7.

3. After 20 min, adjust pH to 8.4 with 0.2 M HCl. Add the desired enzyme solution (5.5 mg of α-D-galactosidase in 2.5 ml of 0.1 M sodium acetate, pH 6.5, with 16 mg of D-galactose for stability). Adjust final pH to 8.4.

4. Incubate at 4°C for 16 h with end-over-end mixing.

5. Add 0.5 ml of 1 M ethanolamine at pH 8.0 and mix for an additional 2 h to block remaining active groups on the carrier.

6. Separate the conjugated enzyme by gel filtration at 4°C on a Sephadex G-200 column.

A low level of CNBr is used to minimize cross-linking of the dextran molecules. This method results in 24.3 mg of enzyme bound per g of dextran, which represents 80% of the applied enzyme. The bound enzyme expresses 55% of the original specific activity but demonstrates an improved thermal stability (60).

19.5.2 Carbodiimide

Perhaps the second most popular method for covalent coupling is a procedure based on the use of water-soluble carbodiimides. These reagents react preferentially with carrier carboxyl groups at slightly acidic pH values (4.75 to 5) to give highly reactive *O*-acylisourea derivatives which condense primarily with amines to give stable amide linkages. Coupling occurs at pH 7 and 4°C for a period of 4 to 16 h. Alternatively, but less commonly, the reagent is mixed with the enzyme to react with the protein's carboxyl groups, and the resultant derivative is bound to carrier amino groups. The most widely used water-soluble carbodiimides are 1-ethyl-3-(3-dimethylamino propyl)carbodiimide (EDC) and 1-cyclohexyl-3-(2-morpholino-ethyl)-carbodiimide metho-*p*-toluene-sulfonate (CMC). Both are available commercially.

Enzyme immobilization may be achieved by one of two different procedures: the sequential or the simultaneous method. In the sequential method, the carrier is preactivated by carbodiimide, washed to remove excess reagent, and then contacted with enzyme. The simultaneous method involves contacting the carrier with carbodiimide and enzyme at the same time.

Sequential carbodiimide binding method (sulfhydryl oxidase and CPG)

1. Add 0.2 M sodium phosphate buffer (pH 4.75) to 500 mg of succinylated glass beads (120/200 mesh, 200-nm mean pore diameter) to a total volume of 17 ml.

2. Add 380 mg of solid EDC and stir the reaction mixture with N_2 gas at room temperature for 20 min.

3. Wash beads immediately with 250 ml of ice-cold 47 mM sodium phosphate at pH 7 for 2 min.

4. In a column reactor with activated beads supported on a coarse, fritted glass disk, circulate 15 ml of enzyme solution (6 mg/ml in 47 mM phosphate buffer, pH 7.0) for 16 to 24 h at 4°C and a flow rate of 70 ml/h.

The amount of EDC (1 M) represents a 100-fold molar excess compared to estimated carboxyl groups. Key variables to be optimized for each enzyme and carrier combination include diimide concentration and activation time, wash time and volume, and coupling time. The wash step is critical because the use of acidic conditions or long times will result in the hydrolysis of *O*-acylisourea groups and inactivation of the carrier (41).

The simultaneous method requires much less carbodiimide. For sulfhydryl oxidase on succinylated CPG, 0.38 mg of EDC is added to 500 mg of beads in 17 ml of 47 mM phosphate at pH 7.0. The enzyme is added simultaneously, and coupling occurs at 4°C for 60 min. For this system, both methods give similar results, i.e., ca. 30 mg of protein per g of beads and a specific activity of 0.95 U/g of beads (41). Some enzymes may be inactivated by the interaction between diimide and protein carboxyls, however, which would necessitate using the sequential method.

Water-soluble carbodiimides are used with a wide variety of carriers. For example, CMC and EDC are used to immobilize acid phosphatase to polyacrylic acid-polyethylene copolymers by the simultaneous procedure (4). Also, activated carbon is used with EDC in the sequential method to bind glucoamylase and glucose oxidase (3). In the latter case, 150 mg of CMC in 5 ml of 0.1 M acetate buffer (pH 5.5) is added per g (dry weight) of coal-based granular activated carbon which has been ground in a ball mill (to 425 to 686 μm), refluxed with 6 N HCl at 80°C, and washed well with water. The mixture is stirred gently for 15 min at room temperature, washed three times each with 30 ml of buffer, and mixed with 50 to 100 mg of enzyme. Coupling occurs for 24 h at room temperature with periodic readjustment of pH to 5.5 by dilute NaOH. The resultant complex is washed three times each with 30 ml of 0.1 M acetate buffer, and stored wet at 5°C. Loading levels for glucoamylase and glucose oxidase are as high as 90 and 65 mg of protein per g of carbon, respectively (3).

Penicillium funiculosum cellulase can be coupled to soluble PVA by using EDC in the simultaneous procedure at pH 4.8 and 0°C for 18 h (73). The resulting soluble conjugate has been used in an Amico ultrafiltration unit with an XM-300 membrane to hydrolyze alkali-treated bagasse at 37°C. The coupled enzyme system demonstrates an increased rate of hydrolysis compared to the soluble form (59% compared to 30% at 48 h) and retains 50% of its original cellulase activity after 5 days of continuous use.

Finally, carbodiimides are often used in methods involving spacer arms, to bind the spacers either to the carrier or to the enzyme. Commonly used spacer molecules are 1,6-diaminohexane hydrochloride and ε-aminocaproic acid. Mazid and Laidler (71) used the method of Lindberg et al. (62) to couple the former compound to N^6-carboxymethyl-NAD, using CMC, and then attached the spacer-modified coenzyme to carboxyl groups of partially hydrolyzed nylon by using the same carbodiimide (simultaneous method; pH 5, 500 mg of CMC per 100 mg of NAD, 24 h at 21°C). Mazid and Laidler next coimmobilized yeast alcohol dehydrogenase by sequential activation of the carrier with glutaraldehyde and contact with the enzyme. Kitano et al. (50) used EDC to bind the C_6 spacer ε-aminocaproate to very small latex particles (styrene-acrylic acid-divinyl benzene) and then used the same reagent to bind alkaline phosphatase to the spacers. The simultaneous method was used for each step. Clark and Bailey (15) bound the same spacer to CNBr-activated Sepharose and then coupled chymotrypsin using EDC by the sequential method (activation and coupling at pH 4.75).

19.5.3 Novel Carriers

Cross-linked PEI

Polyethyleneimine [PEI; $(C_2H_5N)_n$] is a highly branched synthetic polyamine which is available commercially (e.g., 33% aqueous solution, 50,000 to 100,000 molecular weight). It has long been recognized as a stabilizing agent for soluble enzymes, and recently it has demonstrated potential as a carrier matrix for immobilized enzymes. In this latter application, it is used to coat or otherwise be included with another support material (95) or it is cross-linked into a macroporous form and used alone (102).

Coating glass beads with PEI dramatically improves immobilized glucose oxidase and catalase activities, activity yields, and stability characteristics (95). To prepare the modified carrier material, 100 g of nonporous microbeads (13 to 44 μm) are cleaned extensively by washing in distilled water at room temperature and at 90°C; this is followed by sonication at 50 to 55 kHz. The cleansed beads are then added to 1 liter of sodium aluminate solution and incubated for 1 h at 70°C. The aluminate solution is prepared by taking 10 ml of supernatant from a centrifuged saturated aqueous suspension, which has stood for 1 h at room temperature, and dissolving this in 1 liter of distilled water. The treated beads are washed in water and then combined with 150 ml of aqueous PEI (100 mg/ml) which has been adjusted to pH 10.0 with 6 N HCl, suspended by manual stirring for 5 min at room temperature, centrifuged, resuspended in PEI solution, and either quick-frozen and lyophilized or washed with buffer. The former procedure produces beads with a rough surface whereas the latter results in smooth beads. After a wash in 10 mM phosphate buffer (pH 7.4), the beads are suspended by a mechanical stirring bar in 2.5 ml of 1 M glutaraldehyde for 1.5 h, after which 1.2 g of $NaBH_3CN$ is added (under a fume hood). The beads are then rinsed with phosphate buffer, mixed with 125 ml of aqueous PEI (50 mg/ml, pH 7.0 by 6 N HCl), combined with 0.2 g of $NaBH_3CN$, stirred overnight, washed extensively in phosphate buffer and distilled water, lyophilized, and stored in a desiccator at −20°C until used for immobilization. Wasserman et al. (95) postulate that alumina intercalates between the silanol groups of hydrated glass beads, producing a net negative surface that binds at pH 10 to the cationic PEI, which is then cross-linked in place. Immobilization is performed at 25°C in 50 mM citrate-phosphate buffer (pH 7.5) by soaking the modified beads in enzyme solution for 30 to 60 min followed by the addition of glutaraldehyde at a final concentration of 0.91% and incubation for an additional 30 to 60 min.

PEI may also be cross-linked with epichlorhydrin to form macroporous spherical particles (0.1 to 0.2 mm in diameter) (102). The concentration of amino groups and the length of noncross-linked outer chains which serve as natural spacers can be varied by the degree of cross-linking. The preparation involves mixing 100 g of 50% PEI with epichlorhydrin to form an emulsion, holding for 1 h at 40 to 50°C, washing, drying, and sieving to the desired particle size. A range of PEI/epichlorhydrin ratios from 1:1 to 1:0.05 are used. Isothiocyanate, carboxyl, and hydrazide derivatives of the PEI carrier have been prepared, and several different enzymes have been immobilized to each type of modified polymer. The carbodiimide method using CMC was utilized to activate the carboxyl derivatives. In most cases, the activities of the bound enzymes increase as the length of the non-cross-linked outer chains increases, i.e., as the amount of epichlorhydrin is reduced. The isothiocyanate derivative is far superior to the others for the immobilization of glucose oxidase, glucoamylase, acetylcholinesterase, and butyrylcholinesterase.

Reactive polymers for direct binding

Most carriers require some sort of activation before enzyme immobilization. However, some support materials include active coupling groups within their structures. Polymers containing aldehyde and oxirane (highly strained three-membered electrophilic ring) appear to be the most promising due to their high reactivities. These materials are quite useful for enzyme immobilization, immunoadsorbent preparation, and other biological applications.

The immobilization of β-galactosidase has been achieved by direct contact with a commercially available epoxy-activated polyacrylic carrier (oxirane C; Röhm-Pharma GmbH, Darmstadt, Federal Republic of Germany) at pH 7.6 and room temperature. Optimum incubation times are 12 to 72 h, depending on the applied enzyme/carrier ratio. Kinetic studies were conducted in stirred-batch, continuous stirred-tank, and packed-bed reactors (34).

Drobnik et al. (24) have synthesized a glycidyl metacrylate (70%)-ethylene dimethacrylate copolymer by suspension radical polymerization and have demonstrated its versatility as an enzyme support material. The highly reactive oxirane ring allows direct coupling or easy modification to include spacer arms or other active groups. Free amino groups can be added to the carrier by contacting with ammonium hydroxide (25%) or by direct addition of the C_6 spacer 1,6-diaminohexane (under reflux on a steam bath for 5 h). Immobilization can then be obtained by standard methods using carrier amino functionality. Based on studies with penicillin amidase, Drobnik et al. recommend direct immobilization for stable, inexpensive enzymes. This method requires at least 100 mg of enzyme per g of carrier, incubation for at least 72 h, and a temperature as high as tolerable to the enzyme. Azide is used to inhibit microbial growth. For more expensive enzymes, ammonia modification and glutaraldehyde activation is the method of choice. For enzymes inactivated by glutaraldehyde, other more expensive methods are used.

As a final example of the reactive-polymer carrier, Margel (66) has synthesized microspheres of polyacrolein by cobalt radiation polymerization. In a typical procedure, an aqueous solution containing 12% (wt/vol) acrolein and 0.5% (wt/vol) polyethylene oxide is deaerated with argon and then irradiated with a cobalt source at a dose of 1 Mrad. The resulting microspheres (average diameter, 0.2 μm) are then washed extensively in water and encapsulated in agarose. A special device has been developed which can produce monodisperse agarose-polyacrolein microsphere beads of any diameter in the range of 200 μm to 1 cm (65). Magnetic microspheres can be prepared by adding ferrofluid, a suspension of very small magnetic particles, to the agarose gel solution during the encapsulation process (66).

Enzymes or various affinity ligands can be bound directly to aldehyde groups on the carrier by incubating with the support. Typical conditions are 400 mg of ligand and 20 ml of agarose-polyacrolein microsphere beads in 200 ml of buffer for 24 h at 4°C. Remaining aldehyde groups may be blocked by further incubation of the beads for 12 h in the presence of ethanolamine. In a variation of the procedure, polylysine has been bound directly to the beads as a spacer for subsequent ligand binding. In this case, the polylysine-modified beads are exposed to glutaraldehyde, washed in distilled water, and then incubated with the desired ligand (66). In a similar manner polyacrolein-grafted poly(methyl methacrylate) microspheres may be used for direct enzyme or ligand binding (92).

19.5.4 Glutaraldehyde

As mentioned previously, glutaraldehyde has been used in conjunction with enzyme immobilization to preactivate amino groups on carriers, to cross-link enzymes to carriers after prior adsorption, to form cross-linked gels of soluble polymers, and to cross-link enzymes into soluble oligomers or insoluble membranes. It is by far the most widely used bifunctional cross-linking agent used for these purposes. It has been extremely effective despite the fact that major questions still exist regarding its mechanism of action.

Glutaraldehyde, 1,5-pentanedial, is available commercially, usually in 25% acidic aqueous solution. In its pure form, it has a single absorbance maximum at 280 nm, but a secondary maximum at 235 nm occurs in the presence of polymeric impurities. Polymerization increases with temperature and pH under alkaline conditions, producing a yellow color, but at room temperature and neutral pH this process occurs on the order of weeks. Under acidic storage conditions the monomeric form is preserved. Although the biocidal activity of glutaraldehyde is adversely affected by polymerization, there appears to be no evidence that its cross-linking or tanning ability with proteins is modified (31, 35).

There is still some doubt concerning the cross-linking mechanism with proteins. The reaction is pH dependent, with the rate increasing over the range of pH 4 to 9. The α-amino group of lysine is the principal reactant on the protein, and the reaction products give a chromophore maximum at 265 nm. The resultant bonds are stable to acids and semicarbazide, indicating that the cross-links are probably not due to simple Schiff's base formation. There is a possibility that the reaction involves several different types of products. Some possible mechanisms have been discussed (30, 82).

A complex relationship exists between glutaraldehyde concentration, temperature, pH, and reaction time. To date, however, this has not been studied in detail, and few generalizations can be made. Cross-linking is generally conducted at or near neutral pH. Carrier preactivation is often carried out at room temperature, using relatively high concentrations (up to 10%) followed by extensive washing. When enzymes are present, much lower concentrations (0.01 to 1.0%) are used, with contact times varying from 1 to 60 min depending on the amount of cross-linking agent. Optimal conditions should be determined experimentally for each enzyme-carrier pair. It has been found that enzymes adsorbed or entrapped by polymeric carriers can often tolerate much higher glutaraldehyde concentrations than when in free solution (95).

After immobilization by glutaraldehyde, the enzyme-carrier conjugate is often contacted with dilute solutions of lysine, glycine, Tris, or ethanolamine to react with any free aldehyde groups remaining on the support (p. 263 of reference 74). D'Souza has recommended that 0.5 M hydrazine (pH 7.4) is a more effective quenching agent because the reaction is faster and more complete (25). As a final step, the complex may be subjected to borohydride treatment, presumably to stabilize covalent attachment of the enzyme by reduction of the Schiff's base double bond. For example, for creatinine analysis Colliss et al. used 1% (wt/vol) potassium borohydride in borate buffer (pH 8.5) for 60 min at room temperature to stabilize a nylon tube carrying four different enzymes which had been immobilized by the action of glutaraldehyde (17).

19.5.5 Final Comments

Covalent techniques have probably received the most attention from researchers over the years, with the result that there are now a great number of specific procedures to choose from. By choosing to review a few methods in detail we have attempted to demonstrate examples of the covalent approach to immobilizing enzymes. We are by no means implying that these methods are superior to others in the literature. Table 5 outlines several other techniques which are discussed in some detail in references 5, 13, 48, 72, 74, 81, 91, 96, and 98.

19.6 LITERATURE CITED

1. **Alberti, B. N., and A. M. Klibanoy.** 1982. Preparative production of hydroquinone from benzoquinone catalyzed by immobilized D-glucose oxidase. Enzyme Microb. Technol. **4**:47–49.
2. **Axén, R., J. Porath, and S. Ernback.** 1967. Chemical coupling of peptides and proteins to polysaccharides by means of cyanogen halides. Nature (London) **214**: 1302–1304.
3. **Bailey, J. E., and Y. K. Cho.** 1983. Immobilization of glucoamylase and glucose oxidase in activated carbon: effects of particle size and immobilization conditions on enzyme activity and effectiveness. Biotechnol. Bioeng. **25**:1923–1935.
4. **Beddows, C. G., M. H. Gil, and J. T. Guthrie.** 1982. The immobilization of enzymes, bovine serum albumin, and phenylpropylamine to poly(acrylic acid)-polyethylene-based copolymers. Biotechnol. Bioeng. **24**:1371–1387.
5. **Broun, G. B., G. Manecke, and L. B. Wingard, Jr. (ed.).** 1978. Enzyme engineering, vol. 4. Plenum Press, New York.
6. **Cabral, J. M. S., J. F. Kennedy, and J. M. Novais.** 1982. Investigation of the binding mechanism of glucoamylase to alkylamine derivatives of titanium (IV)-activated porous inorganic supports. Enzyme Microb. Technol. **4**:337–342.
7. **Cabral, J. M. S., J. F. Kennedy, J. M. Novais, and J. P. Cardoso.** 1984. Compositions and compositional-behavioural relationships of enzymes immobilized on porous inorganic supports via titanium (IV) species. Enzyme Microb. Technol. **6**:228–232.
8. **Cabral, J. M. S., J. M. Novais, J. F. Kennedy, and J. P. Cardoso.** 1983. Immobilization of biocatalysts on new route transition metal-activated inorganic supports. Enzyme Microb. Technol. **5**:30–32.
9. **Cashion, P., A. Javed, D. Harrison, J. Seeley, V. Lentini, and G. Sathe.** 1982. Enzyme immobilization on tritylagarose. Biotechnol. Bioeng. **24**:403–423.
10. **Cashion, P., V. Lentini, D. Harrison, and A. Javed.** 1982. Enzyme immobilization on tritylagarose: reusability of both matrix and enzyme. Biotechnol. Bioeng. **24**:1221–1224.
11. **Charm, S. E., and B. L. Wong.** 1981. Shear effects on enzymes. Enzyme Microb. Technol. **3**:111–118.
12. **Chen, F. S., H. S. Weng, and C. L. Lai.** 1983. The performance of immobilized glucose isomerase supported by shrimp chitin in various types of reactors. Biotechnol. Bioeng. **25**:725–733.
13. **Chibata, I., S. Fukui, and L. B. Wingard, Jr. (ed.).** 1982. Enzyme engineering, vol. 6. Plenum Press, New York.
14. **Chibata, I., and L. B. Wingard, Jr. (ed.).** 1983. Applied biochemistry and bioengineering, vol. 4. Immobilized cells. Academic Press, Inc., New York.
15. **Clark, D. S., and J. E. Bailey.** 1983. Structure-function relationship in immobilized chymotrypsin catalysis. Biotechnol. Bioeng. **25**:1027–1047.
16. **Clark, D. S., J. E. Bailey, R. Yen, and A. Rembaum.** 1984. Enzyme immobilization on grafted polymeric microspheres. Enzyme Microb. Technol. **6**:317–320.
17. **Colliss, J., J. M. Knox, and R. Ginman.** 1983. The use of an immobilized enzyme nylon tube reactor incorporating a four enzyme system for creatinine analysis. Appl. Biochem. Biotechnol. **8**:213–226.
18. **Cremonesi, P., and L. D'Angiuro.** 1983. Kinetic and thermal characteristics of enzyme-graft copolymers. Biotechnol. Bioeng. **25**:735–744.
19. **D'Angiuro, L., and P. Cremonesi.** 1982. Immobilization of glucose oxidase on sepharose by UV-initiated graft copolymerization. Biotechnol. Bioeng. **24**:207–216.
20. **D'Angiuro, L., P. Cremonesi, G. Mazzola, B. Focher, and G. Vecchio.** 1980. Preparation and characterization of enzymes immobilized by graft copolymerization to different polysaccharides. Biotechnol. Bioeng. **22**: 2251–2272.
21. **Degani, Y., and T. Miron.** 1970. Immobilization of cholinesterase in cross-linked polyacrylamide. Biochim. Biophys. Acta **212**:362–364.
22. **DeLoach, J. R.** 1982. Comparative encapsulation of cytosine arabinoside monophosphate in human and canine erythrocytes with *in vitro* drug efflux. J. Appl. Biochem. **4**:533–541.
23. **Dinelli, D.** 1972. Fiber-entrapped enzymes. Process Biochem. **7**:9–12.
24. **Drobnik, J., V. Saudek, F. Svec, J. Kalal, V. Vojtisek, and M. Barta.** 1979. Enzyme immobilization techniques on poly(glycidylmethacrylate)-co-ethylene dimethacrylate carrier with penicillin amidase as model. Biotechnol. Bioeng. **21**:1317–1332.
25. **D'Souza, S. F.** 1983. Osmotic stabilization of mitochondria using chemical crosslinkers. Biotechnol. Bioeng. **25**:1661–1664.
26. **Dunnill, P.** 1980. The current status of enzyme technology, p. 28–53. *In* P. Dunnill, A. Wiseman, and N.

Blakebrough (ed.), Enzymic and nonenzymic catalysis. Ellis Horwood Ltd., Chichester, England.

27. **Emery, A. N., J. S. Hough, J. M. Novais, and T. D. Lyons.** 1972. Applications of solid phase enzymes in biological engineering. Chem. Eng. (London) **258:**71–76.
28. **Flor, P. Q., and S. Hayashida.** 1983. Continuous production of high-glucose syrup by chitin-immobilized amylase. Biotechnol. Bioeng. **25:**1973–1980.
29. **Gacesa, P., R. Eisenthal, and R. England.** 1983. Immobilization of urease within a thin channel ultrafiltration cell. Enzyme Microb. Technol. **5:**191–195.
30. **Goldstein, L., and G. Manecke.** 1976. The chemistry of enzyme immobilization. Appl. Biochem. Bioeng. **1:**23–126.
31. **Gorman, S. P., E. M. Scott, and A. D. Russell.** 1980. Antimicrobial activity, uses and mechanism of action of glutaraldehyde. J. Appl. Bacteriol. **48:**161–190.
32. **Gray, C. J., C. M. Lee, and S. A. Barker.** 1982. Immobilization of enzymes on Spheron. Enzyme Microb. Technol. **4:**143–152.
33. **Guilbault, G. G.** 1980. Enzyme electrode probes. Enzyme Microb. Technol. **2:**258–264.
34. **Hannibal-Friedrich, O., M. Chun, and M. Sernetz.** 1980. Immobilization of β-galactosidase, albumin, and γ-globulin on epoxy-activated acrylic beads. Biotechnol. Bioeng. **22:**157–175.
35. **Hardy, P. M., A. C. Nicholls, and H. N. Rydon.** 1976. The nature of the crosslinking of proteins by glutaraldehyde. Part I. Interaction of glutaraldehyde with the amino groups of 6-aminohexanoic acid and of α-N-acetyllysine. J. Chem. Soc. Perkin Trans. **1:**958–962.
36. **Hirano, S., and O. Miura.** 1979. Alkaline phosphatase and pepsin immobilized in gels. Biotechnol. Bioeng. **21:**711–714.
37. **Horvath, C., and M. Sovak.** 1973. Membrane coarctation by calcium as a regulator for bound enzymes. Biochim. Biophys. Acta **298:**850–860.
38. **Ihler, G., R. Glen, and R. Schnure.** 1973. Enzyme loading of erythrocytes. Proc. Natl. Acad. Sci. USA **70:**2663–2666.
39. **Imai, K., T. Shiomi, K. Sata, and A. Fujishima.** 1983. Preparation of immobilized invertase using poly(vinyl alcohol) membrane. Biotechnol. Bioeng. **25:**613–617.
40. **Iqbal, J., and M. Saleemuddin.** 1985. Sucrose hydrolysis using invertase immobilized on concanavalin A-Sepharose. Enzyme Microb. Technol. **7:**175–178.
41. **Janolino, V. G., and H. E. Swaisgood.** 1982. Analysis and optimization of methods using water-soluble carbodiimide for immobilization of biochemicals to porous glass. Biotechnol. Bioeng. **24:**1069–1080.
42. **Jaworek, D.** 1974. New immobilization techniques and supports, p. 105–114. *In* E. K. Pye and L. B. Wingard, Jr. (ed.), Enzyme engineering, vol. 2. Plenum Press, New York.
43. **Jeng, C. Y., S. S. Wang, and B. Davidson.** 1980. Ultrafiltration of raw sewage using an immobilized enzyme membrane. Enzyme Microb. Technol. **2:**145–149.
44. **Kaetsu, I., M. Kumakura, and M. Yoshida.** 1979. Enzyme immobilization by radiation-induced polymerization of 2-hydroxyethyl methacrylate at low temperatures. Biotechnol. Bioeng. **21:**847–873.
45. **Kawakami, K., T. Hamada, and K. Kusunoki.** 1980. Performance of a hollow fibre beaker device for continuous enzymic reactions. Enzyme Microb. Technol. **2:**295–298.
46. **Kawashima, K., and K. Umeda.** 1974. Immobilization of enzymes by radiopolymerization of acrylamide. Biotechnol. Bioeng. **16:**609–621.
47. **Kennedy, J. F.** 1982. A future for immobilized cell technology. Nature (London) **299:**777–778.
48. **Kennedy, J. F., and J. M. S. Cabral.** 1983. Immobilized enzymes, p. 253–391. *In* W. H. Scouten (ed.), Solid phase biochemistry. John Wiley & Sons, New York.
49. **Kennedy, J. F., B. Kalogerakis, and J. M. S. Cabral.** 1984. Immobilization of enzymes on crosslinked gelatin particles activated with various forms and complexes of titanium (IV) species. Enzyme Microb. Technol. **6:**68–72.
50. **Kitano, H., K. Nakamura, and N. Ise.** 1982. Kinetic studies of enzyme immobilized on anionic polymer lattices: alkaline phosphatase, α-chymo-trypsin and β-galactosidase. J. Appl. Biochem. **4:**34–40.
51. **Kohlwey, D. E., and M. Cheryan.** 1981. Performance of a β-D-galactosidase hollow fibre reactor. Enzyme Microb. Technol. **3:**64–68.
52. **Kohn, J., R. Lenger, and M. Wilchek.** 1983. P-nitrophenylcyanate—an efficient, convenient and nonhazardous substitute for cyanogenbromide as an activating agent for sepharose. Appl. Biochem. Biotechnol. **8:**227–235.
53. **Kohn, J., and M. Wilchek.** 1981. Procedures for the analysis of cyanogenbromide-activated sepharose or sephadex by quantitative determination of cyanate esters and imidocarbonates. Anal. Biochem. **115:**375–382.
54. **Kohn, J., and M. Wilchek.** 1982. A new approach (cyano-transfer) for cyanogen bromide activation of sepharose at neutral pH, which yields activated resins, free of interfering nitrogen derivatives. Biochem. Biophys. Res. Commun. **107:**878–884.
55. **Kohn, J., and M. Wilchek.** 1984. Cyanylating agents for polysaccharide activation. Appl. Biochem. Biotechnol. **9:**285–305.
56. **Kolot, F. B.** 1981. Microbial carriers—strategy for selection. Process Biochem. **16:**2–9.
57. **Kumakura, M., and I. Kaetsu.** 1982. New preparation of immobilized enzyme particles by the salting out technique and by radiation polymerization. J. Appl. Biochem. **4:**441–448.
58. **Kumakura, M., and I. Kaetsu.** 1983. Pretreatment of starch raw materials and their enzymatic hydrolysis by immobilized glucoamylase. Enzyme Microb. Technol. **5:**199–203.
59. **Kumakura, M., I. Kaetsu, M. Suzuki, and S. Adachi.** 1983. Immobilization of antibodies and enzyme-labeled antibodies by radiation polymerization. Appl. Biochem. Biotechnol. **8:**87–96.
60. **Kuo, J. Y., and J. Goldstein.** 1983. α-D-Galactosidase immobilized on a soluble polymer. Enzyme Microb. Technol. **5:**285–290.
61. **Layman, P. L.** 1983. Enzymes business attracting new products, technology. Chem. Eng. News **61:**11–13.
62. **Lindberg, M., P. O. Larsson, and K. Mosbach.** 1973. A new immobilized NAD^+ analogue, its application in affinity chromatography as a functioning coenzyme. Eur. J. Biochem. **40:**187–193.
63. **Makkar, H. P. S., O. P. Sharma, and R. K. Dawra.** 1983. Effect of reagents for polyacrylamide gel formation on β-D-galactosidase. Biotechnol. Bioeng. **25:**867–868.
64. **Marconi, W., and F. Morisi.** 1979. Industrial applications of fiber-entrapped enzymes. Appl. Biochem. Bioeng. **2:**219–258.
65. **Marcus, L., M. Offarim, and S. Margel.** 1982. A new immunoadsorbent for hemoperfusion: agarose-polyacrolein microsphere beads. I. *In vitro* studies. Biomater. Med. Devices Artif. Organs **10:**157–171.
66. **Margel, S.** 1982. Agarose polyacrolein microsphere beads. New effective immunoabsorbents. FEBS Lett. **145:**341–344.
67. **Martinek, K., A. M. Klibanov, V. S. Goldmacher, and I. V. Berezin.** 1977. The principles of enzyme stabilization. I. Increase in thermostability of enzymes covalently bound to a complementary surface of a polymer support in a multipoint fashion. Biochim. Biophys. Acta **485:**1–12.

68. **Martinek, K., V. V. Mozhaev, and I. V. Berezin.** 1980. Reactivation of enzymes irreversibly denatured at elevated temperature. Biochim. Biophys. Acta **615:** 426–435.
69. **Mattiasson, B.** 1981. Reversible immobilization of enzymes with special reference to analytical applications. J. Appl. Biochem. **3:**183–194.
70. **Mattiasson, B., and B. Danielsson.** 1982. Calorimetric analysis of sugars and sugar derivatives with aid of an enzyme thermistor. Carbohydr. Res. **102:**273–282.
71. **Mazid, M. A., and K. J. Laidler.** 1982. Kinetics of yeast alcohol dehydrogenase and its coenzyme coimmobilized in a tubular flow reactor. Biotechnol. Bioeng. **24:**2087–2097.
72. **Messing, R. A. (ed.).** 1975. Immobilized enzymes for industrial reactors. Academic Press, Inc., New York.
73. **Mishra, C., V. Deshpande, and M. Rao.** 1983. Immobilization of *Penicillium funiculosum* cellulase on a soluble polymer. Enzyme Microb. Technol. **5:**342–344.
74. **Mosbach, K. (ed.).** 1976. Immobilized enzymes. Methods Enzymol. **44.**
75. **Mozhaev, V. V., U. A. Siksnis, V. P. Torchilin, and K. Martinek.** 1983. Operational stability of copolymerized enzymes at elevated temperatures. Biotechnol. Bioeng. **25:**1937–1945.
76. **Nakakuki, T., T. Hayashi, M. Monma, K. Kawashima, and K. Kainuma.** 1983. Immobilization of the exo-maltohexaohydrolase by the irradiation method. Biotechnol. Bioeng. **25:**1095–1107.
77. **Oguntimein, G., and P. J. Reilly.** 1980. Purification and immobilization of *Aspergillus niger* β-xylosidase. Biotechnol. Bioeng. **22:**1127–1142.
78. **Pelsy, G., and A. M. Klibanov.** 1983. Preparative separation of α- and β-naphthols catalyzed by immobilized sulfatase. Biotechnol. Bioeng. **25:**919–928.
79. **Pitt, A. M., S. M. Cramer, A. B. Czernicki, K. Kalghati, C. Horvath, and B. A. Solomon.** 1983. Anisotropic membranes with carboxypeptidase G_1. Biochem. Biotechnol. **8:**55–68.
80. **Poulsen, P. B.** 1983. Guidelines for the characterization of immobilized biocatalysts. Enzyme Microb. Technol. **5:**304–307.
81. **Pye, E. K., and H. H. Weetall (ed.).** 1978. Enzyme engineering, vol. 3. Plenum Press, New York.
82. **Reichlin, M.** 1980. Use of glutaraldehyde as a coupling agent for proteins and peptides. Methods Enzymol. **70:**159–165.
83. **Stanley, W. L., G. G. Watters, S. H. Kelly, and A. C. Olson.** 1978. Glucoamylase immobilized on chitin with glutaraldehyde. Biotechnol. Bioeng. **20:**135–140.
84. **Sulkowski, E., and M. Laskowski.** 1974. Venom exonuclease (phosphodiesterase) immobilized on concanavalin-A-sepharose. Biochem. Biophys. Res. Commun. **57:** 463–468.
85. **Sweigart, R. D.** 1979. Industrial applications of immobilized enzymes: a commercial overview, p. 209–218. *In* L. B. Wingard, Jr., E. Katchalski-Katzir, and L. Goldstein (ed.), Applied biochemistry and bioengineering, vol. 2. Enzyme technology. Academic Press, Inc., New York.
86. **Synowiecki, J., Z. E. Sikorski, and M. Naczk.** 1981. Immobilization of invertase on krill chitin. Biotechnol. Bioeng. **23:**231–233.
87. **Synowiecki, J., Z. E. Sikorski, M. Naczk, and H. Piotrzkowska.** 1982. Immobilization of enzymes on krill chitin activated by formaldehyde. Biotechnol. Bioeng. **24:**1871–1876.
88. **Tanaka, H., H. Kurosawa, E. Kokufuta, and I. A. Veliky.** 1984. Preparation of immobilized glucoamylase using Ca-alginate gel coated with partially quaternized poly(ethyleneimine). Biotechnol. Bioeng. **26:**1393–1394.
89. **Tramper, J., W. E. Hennink, and H. C. van Der Plas.** 1982. Oxidation of 7-alkylpteridin-4-ones to 7-alkyllumazines by free and immobilized xanthineoxidase. J. Appl. Biochem. **4:**263–270.
90. **Tran, N. D., J. L. Romette, and D. Thomas.** 1983. An enzyme electrode for specific determination of l-lysine: a real-time control sensor. Biotechnol. Bioeng. **25:**329–340.
91. **Trevan, M. S.** 1980. Immobilized enzymes. Wiley-Interscience, New York.
92. **Tsang, H., H. Mollenhauer, and G. Ihler.** 1982. Entrapment of proteins, viruses, bacteria and DNA in erythrocytes during endocytosis. J. Appl. Biochem. **4:**418–435.
93. **Tu, Y. T., and T. M. S. Chang.** 1982. Immobilization of multi-enzymes and cofactors within lipid-polyamide membrane microcapsules for the multistep conversion of lipophilic and lipophobic substrates. Enzyme Microb. Technol. **4:**327–331.
94. **Venkatasubramanian, K. (ed.).** 1979. Immobilized microbial cells. Am. Chem. Soc. Symp. Ser. **106.**
95. **Wasserman, B. P., H. O. Hultin, and B. S. Jacobson.** 1980. High-yield method for immobilization of enzymes. Biotechnol. Bioeng. **22:**271–287.
96. **Weetall, H. H., and G. P. Royer (ed.).** 1980. Enzyme engineering, vol. 5. Plenum Press, New York.
97. **Wheatley, M. A., and C. R. Phillips.** 1983. Temperature effects during polymerization of polyacrylamide gels used for bacterial cell immobilization. Biotechnol. Bioeng. **25:**623–626.
98. **Wingard, L. B., Jr., E. Katchalski-Katzir, and L. Goldstein. (ed.).** 1976. Applied biochemistry and bioengineering, vol. 1. Immobilized enzyme principles. Academic Press, Inc., New York.
99. **Yagi, K., M. Naoi, H. Sakai, H. Abe, H. Konishi, and S. Arichi.** 1982. Incorporation of enzyme into the brain by means of liposomes of novel composition. J. Appl. Biochem. **4:**121–125.
100. **Yamaguchi, R., Y. Arai, T. Kaneko, and T. Itoh.** 1982. Utilization of partially N-succinylated derivatives of chitosan and glycolchitosan as supports for the immobilization of enzymes. Biotechnol. Bioeng. **24:**1081–1091.
101. **Yamazaki, Y., H. Maeda, and A. Kamibayashi.** 1982. The long-term production of L-malate by the coimmobilized NAD and dehydrogenases. Biotechnol. Bioeng. **24:**1915–1918.
102. **Zemek, J., P. Kuniak, P. Gemeiner, J. Zamocky, and S. Kucar.** 1982. Crosslinked polyethylenimine: an enzyme carrier with spacers of various lengths introduced in crosslinking reaction. Enzyme Microb. Technol. **4:**233–238.

Chapter 20

Introduction to Techniques of Mammalian Cell Culture

GEORGE B. BODER AND ROBERT N. HULL

The manner in which mammalian cells are cultivated varies widely, depending not only on the requirements of the cells but also upon the applications in culture. For these reasons it is not practical to cover all aspects of mammalian cell culture in this chapter. Instead, we include only the most basic and useful procedures required to begin mammalian cell culture. In some areas, the methods commonly used are so extensive and variable that we have chosen to present them in terms of general principles. It is recommended that the novice make use of the included general references before and during initial attempts at cell culture.

20.1 COMPARISON OF TISSUE CULTURE METHODOLOGY WITH MICROBIOLOGY

Mammalian cell culture techniques were modeled after the techniques of the microbiologist. Mammalian cells, however, are more sensitive than most microorganisms to their environment. The mammalian cell is larger and is not protected by a hardy cell wall, and its nutritional requirements are more complex. The mammalian cell in vivo is not free living as are microorganisms. The mammalian cell is dependent upon a circulatory system that has developed to ensure a precisely regulated homeostatic environment. There is an abundance of different cell types that appear to have special requirements for growth and function in vitro. Most normal animal cells will not grow (multiply) in suspension culture and require specially treated and charged surfaces for selective attachment as a monolayer and for growth. In recent years a certain amount of mystery has been dispelled about the unique requirements for consistent, controlled mammalian cell growth. Nevertheless, much remains to be learned about cell growth and differentiation.

One of the greatest difficulties in cell culture is the avoidance of bacterial contamination. The use of aseptic techniques is essential to both mammalian and bacterial cell growth, but even more stringent requirements are necessary for successful cell culture since susceptibility is greater. The introduction of a microbial contaminant into a cell culture is more devastating and more readily apparent than in a bacterial culture since animal cells grow much less rapidly than some of the common contaminants such as bacteria, yeasts, and molds. Further, freedom from chemical contamination is generally more important in cell culture than it is in microbial culture and requires that glassware be immaculately clean and that reagents be of high purity. Other physical and chemical conditions such as temperature, light, pH, and osmotic pressure must be closely monitored to provide the optimal conditions for the cells undergoing cultivation.

Early microbiologists were concerned with the problems of obtaining pure cultures, organism identification, and quantitative procedures. These obstacles were overcome with the development of solid media and differential and selective media for both the isolation and the identification of organisms. These problems also existed in the early days of cell culture, and even today they have not been solved to the same extent as they have been for microorganisms. The approaches to these problems, the success obtained, and the current limitations are presented here along with details of cell culture technology.

20.1.1 Primary Culture

Vertebrate, invertebrate, and plant cells can be grown in culture, but only mammalian cell culture is considered here. Primary culture involves the growth of cells taken directly from an animal host, put into culture, and grown for a limited time without further subpassage in culture.

Specimens

The procurement of tissue for primary cultures depends on the ultimate use of the cells. For virological studies, for example, the cells must be susceptible to the viruses. This consideration influences the selection of both the animal species and the tissue to be cultured. If a particular cell function is to be studied, then the selection of tissue may be more critical than the selection of the animal species. The fact that host cells from some species are more readily cultivated than those from other species is well known. A review of the literature generally provides information to assist the investigator in the proper selection of host and tissue for a particular area of study.

Sterility

There are a variety of techniques used to initiate cell cultures, but of primary concern is the need for sterility. Internal organs, glands, and tissues of the host generally are sterile in vivo, but they must be removed and processed aseptically to provide usable cultures. The use of antibiotics in the culture medium may add assurance, but strict aseptic technique is still essential. At times the tissue to be cultivated may come from a host tissue that is not normally sterile, such as the skin, oral cavity, or gut. In this case the area from which the tissue sample is taken should be cleaned as thoroughly as possible and well rinsed with sterile balanced salt solution. A solution of 70% ethanol may be used on the skin provided it is completely rinsed away before dissection. Mercurials should not be used since they are extremely toxic to cells in culture and cannot be adequately removed by rinsing. In some cases it may be advantageous to treat the host with antibiotics for several days before tissue excision; if so, antibiotics must be included in the culture medium during the period of cultivation. Penicillin and streptomycin at 100 U and 100 μg per ml of medium, respectively, are commonly used. Gentamicin may be used alone or in combination with penicillin. A 50-μg/ml dose of gentamicin is generally adequate, but the concentration can be increased severalfold if necessary. In addition to these antibacterial antibiotics, it may be necessary to include antifungal compounds such as amphotericin B or nystatin. These are much more toxic than the other reagents and must be used with care. If germfree animals are available, such procedures can be eliminated. There is always the possibility that the tissue, whether or not it comes from normally sterile tissue, will contain latent viruses or be infected with mycoplasma. Little can be done to control these types of contaminants. Latent viruses can present a safety hazard because they may grow to a high titer in the cultured cells (21). All primary cell cultures, as well as the materials and reagents coming into contact with them, should be handled as if they were infectious.

20.1.2 Cell Strains

The primary cultures discussed above will often contain multiple cell types. These cultures generally are used as primary cultures and then discarded. There are, however, many established cell strains or lines, some of which have been purified by cloning to provide cultures of a single cell type. This type of culture is maintained by serial passage much like cultures of a microorganism. There are two general varieties: (i) the established or continuous line, which can be grown indefinitely in serial passage, and (ii) the finite life-span strain, which undergoes a limited number of population doublings in vitro before growth stops and the cells become senescent. The latter is usually diploid (2*N* normal chromosome numbers), whereas most established cell lines vary in their ploidy. The finite life-span strains are generally limited in their split ratios to low numbers, while many of the established cell lines can be split at ratios of 1:10, 1:20, or greater. A split ratio of 1:10 is produced by subculturing the cells from 1 flask into 10 flasks. Details of the preparation and handling of both primary and cell line cultures are presented below.

20.2 BASIC LABORATORY REQUIREMENTS

20.2.1 Sterile Work Area

Maintenance of sterility in the work area is essential for both short-term and long-term cultures. A sterile room environment is achieved by using high-efficiency particulate air filters under positive pressure so that all of the air is flowing out of the room. When cost and space are of concern, laminar-flow hoods can be used to provide a sterile work area. In some instances, Plexiglas boxes or inflatable germfree isolation chambers are suitable. The use of stainless-steel tables and work surfaces is recommended since they can be more readily disinfected and kept clean. Various arrangements can be made for storing needed supplies within convenient reach as in the microbiological laboratory. Equipment should never be stored or stocked inside the laminar-flow hoods since this interrupts the laminar flow.

In addition to providing a sterile environment for the protection of the cultures, consideration must be given to safeguarding laboratory workers when there is a risk of exposure to known or potential pathogenic agents. Therefore, precautions must be taken, especially when working with human materials. The work area should be thoroughly cleaned and disinfected before and after each use (see reference 3).

20.2.2 Equipment Washing and Preparation, Basic Equipment, and Storage Facilities

In the preparation area, separate dedicated sinks should be available for cleaning glassware and other tissue culture equipment. There are a number of glassware-washing machines currently available that minimize the handling of glassware and so reduce breakage. Sonication baths are also available that appear to be satisfactory in removing even the most stubborn material. Since all glassware and equipment should be scrupulously clean, a source of highly purified water should be available. Some laboratories routinely use triple glass-distilled water, while others depend on ultrafiltration systems. Ultrafiltration systems have been identified as sources of microbial and chemical contamination of water due to growth on filters and resins. Arrangements should be made for periodically checking the purity of the water by measuring both inorganic and organic content.

Proper care of cell culture glassware begins in the laboratory. All equipment should be rinsed with water as soon as possible after use to prevent drying of materials on the glassware. Equipment to be autoclaved should be filled with or submerged in water before being placed in the autoclave.

A glassware cleaning procedure that has worked well for us is described below.

Cleaning glass culture vessels

1. Remove all markings (grease pen or ink) on glassware before the glass is soaked or simmered.
2. Do not allow the glassware to dry between use and the cleaning procedure. (Glassware should be submerged or filled with water immediately after use as previously noted.)
3. Place the glassware in a stainless-steel bucket containing distilled water (12 liters) and 56 g of Calgolac (Calgon Corp., St. Louis, Mo.).
4. Simmer for 1 to 2 h. Be sure no air bubbles are trapped in the glassware. (The use of large sonicators eliminates this requirement.)
5. Remove glassware from the cleaning solution, and rinse in tap water five times and then in glass-distilled water five times.
6. Rinse three times in triple glass-distilled water.
7. Invert, allow to drain dry, plug with gauze-wrapped cotton, and wrap with Patapar paper (James River Corp., Edgly, Pa.). Do not use aluminum foil or brown paper. The foil is coated with lubricant that volatizes during sterilization and condenses on the glassware. Brown paper contains formaldehyde and other chemicals that are toxic.
8. Sterilize glassware in an oven at 135 to 150°C for 2.5 to 3 h.

Cleaning glass pipettes

1. Let pipettes stand in cleaning solution (distilled water and Calgolac as described above) overnight.
2. Rinse in tap water 10 times.
3. Rinse in triple-distilled water 10 times.
4. Drain dry, plug, and wrap with Patapar paper.
5. Sterilize in oven as described above.

It is desirable to dry all equipment thoroughly before sterilization; therefore large drying racks in a clean area are required.

Sterilization of equipment requires a dry-heat oven. The oven may require a mechanism for circulating air to avoid hot and cold spots in the oven, depending upon the load and the capacity of the oven. Items that cannot be sterilized by dry heat may have to be autoclaved. This can be a source of difficulty since the steam used in many systems contains volatile compounds that may deposit on the equipment.

Storage facilities

Media and sera require storage at either −20 or 4°C. There seems to be a wide variation in the storage lives of different media and sera. The length of storage life may be related to the sensitivity of the cells and to the selection of cells permissive to certain environments. In general, several recommendations can be made. Serum should be stored at the lowest available temperature, frozen and thawed as infrequently as possible, and shielded from visible light. Samples of serum sufficient for weekly use can be thawed and subsequently stored at 4°C until used.

Some media can be stored at −20°C, although the solubility of some of the ingredients is minimal at this

TABLE 1. Basal Eagle medium-Earle salts

Component	Concn (mg/liter)	Component	Concn (mg/liter)
Amino acids		Inorganic salts	
L-Arginine	17.40	$CaCl_2$ (anhydrous)	200
L-Cysteine	12.00	KCl 400.00	400
L-Glutamine	292.00	$MgSO_4 \cdot 7H_2O$	200
L-Histidine	8.00	NaCl	6,800
L-Isoleucine	26.00	$NaHCO_3$	2,200
L-Leucine	26.00	$NaH_2PO_4 \cdot H_2O$	140
L-Lysine	29.20		
L-Methionine	7.50	Vitamins	
L-Phenylalanine	16.50	Biotin	1.00
L-Threonine	24.00	Calcium-D-pantothenate	1.00
L-Tryptophan	4.00	Choline chloride	1.00
L-Tyrosine	18.00	Folic acid	1.00
L-Valine	23.50	*i*-Inositol	2.00
		Nicotinamide	1.00
Other components		Pyridoxal · HCl	1.00
D-Glucose	1,000.00	Riboflavin	0.10
Phenol red	10.00	Thiamine · HCl	1.00

temperature. If the precipitate does not redissolve upon thawing, sonication in a water bath may resolve this difficulty. Storage at 4°C results in the depletion of some of the nutritional ingredients due to chemical instability. Glutamine, for example, has a half-life of approximately 30 days at 4°C in solution.

Incubators

Different conditions of incubation may be required for different cells. Thus, it may be necessary to have more than one incubator of the same type adjusted to the appropriate temperature and CO_2 concentration. Incubators include both standard incubators used in closed culture without provision for gassing and CO_2 incubators for cultures that are open to the atmosphere. There are many commercial suppliers of the various types of incubators.

For large-scale cultures a warm room can be constructed, although modular units can be purchased commercially. It is important to have circulating air in these incubators to avoid hot and cold spots.

20.2.3 Glassware and Plasticware

Most items commonly found in a microbiology laboratory are standard items in the cell culture laboratory. Many of these items are now made of disposable plastic. Plastic petri dishes commonly used in microbiology laboratories are not suitable for attachment-dependent cell culture. Special treatment of the plastic surface is required to promote cell attachment and growth. The specific items needed vary depending on the research being conducted and the preferences of the investigator. Polystyrene has been accepted as the standard substrate for disposable culture vessels. The surface must be treated to provide a wettable or charged surface. This is usually done by the manufacturer. Other plastics such as polyvinyl chloride, polycarbonate, polytetrafluoroethylene, and others have been used in special situations. Polystyrene is probably the least expensive, and it is optically clear. One pitfall of plastic labware is that plasticizers can leach out during storage or use, sometimes resulting in cell toxicity and variation in cell growth and function. Plastic vessels ordinarily should not be reused.

Glass culture vessels. Reusable glass culture vessels have been used since the early days of cell culture. The fact that glass can be reused with good, consistent growth-supporting properties if properly cleaned and sterilized is a distinct advantage. The reuse of glassware often requires additional support people depending on the glassware volumes required.

20.3 CULTURE MEDIA

The medium is probably the most important factor in the maintenance and growth of cells. There are no universal guidelines for the selection of medium, serum, or growth factors for a given cell type. For most continuous cell lines it is recommended that the cells be kept in the medium to which they are accustomed. Cells may be adapted to other media by proportionally reducing the concentration of the original medium while inreasing the concentration of the new medium at each passage. Growth curves can be done for each set of variables, determining lag time, generation time, and cell density at the stationary growth phase. Since there are numerous media used in mammalian cell culture, it is recommended that the novice refer to more complete descriptions before a selection is made. The ingredients are chosen to provide maintenance or growth of the cells. An example of a typical minimal medium is shown in Table 1.

Osmotic pressure

The medium environment must satisfy more than just the nutritional requirements of cells. Proper osmotic pressure is one parameter that must be met. The osmotic pressure is maintained by the appropriate concentrations of salts and glucose. Cells from different species vary in their ionic requirement. Sodium/potassium ratios in sera from various species are known to differ, and this should be considered when selecting culture media.

Hydrogen ion concentration

The pH of most biological fluids is maintained near neutrality. Deviations of more than several tenths of a pH unit from normal are usually fatal. Most cells survive at a pH range between 6.8 and 7.6; however, many functions of cells vary with small changes in pH. Cell attachment to substrate is generally enhanced by a pH of ~7.0 or less. pH in most media is controlled by a combination of dissolved gases (sodium bicarbonate buffer system) and products of metabolism by the cell (especially lactic acid). The bicarbonate system is the natural buffer in blood.

The chemical reaction of the sodium bicarbonate buffer system is as follows: $NaHCO_3 + H_2O \rightarrow Na^+ + HCO_3^- + H_2O \rightarrow Na^+ + H_2CO_3 + OH^- \rightarrow Na^+ + OH^- + H_2O + CO_2$. Thus, one of the final products of the reaction is gaseous CO_2. The solubility of CO_2 decreases as temperature rises, causing the CO_2 to be released from the medium. This results in an increase in pH. For this reason cultures must be tightly stoppered and in some cases charged with CO_2 gas to prevent the loss of CO_2 and the rise in pH. (This point is critical during membrane filtration of media under a vacuum.) An alternative procedure is to incubate cultures in an incubator with a CO_2-enhanced atmosphere with the caps loose. The amount of CO_2 gas required to maintain stable pH depends on the concentration of $NaHCO_3$ in the medium. CO_2 incubators must be used for the incubation of cells in petri dishes unless special media without the bicarbonate system are used (24). Another consideration is the pK_a of sodium bicarbonate (6.3 at 37°C), which results in less than optimum buffering at the normal culture pH of 7.2 to 7.4.

Other buffers have been developed to substitute for the bicarbonate buffer. One of these (2) utilizes the buffering capacity of the free bases of the amino acids and the higher concentration of phosphate buffers. Another buffer used extensively for incubation in open systems is HEPES (*N*-2-hydroxyethylpiperazine-*N'*-2-ethanesulfonic acid), which acts as a dipolar ion. A concentration of 25 mM or less is commonly used. For some applications a combination of the sodium bicarbonate system and HEPES can be used. Additional details of pH control and measurement can be found elsewhere (3). It has been demonstrated that some cells have a requirement for CO_2 (30).

Carbohydrates

A source of carbohydrate is necessary in cell culture medium to supply an energy source. The most common source of carbohydrate is glucose, but other monosaccharides (e.g., galactose) (24) can also be used. (Disaccharides, however, cannot be used.)

Amino acids

Amino acids are required for cell growth and function. Most animal cells have a specific requirement for 13 amino acids: arginine, cysteine, glutamine, histidine, isoleucine, leucine, lysine, methionine, phenylalanine, threonine, tryptophan, tyrosine, and valine. Only the L forms are utilized, but some older medium formulations contain DL-amino acids.

Vitamins

Vitamins are used as cofactors in cell metabolic functions. In certain cell types, such as bone cells, ascorbic acid is important for growth and function (11). Vitamin E has recently been shown to protect against the damaging effects of light (14). Vitamins can be provided by the serum, but with a reduced serum concentration vitamins must be provided by suplementation.

Trace elements (minerals)

Most minerals required by cells in culture are provided by the serum. In low serum concentrations (25) or in the complete absence of serum, requirements for iron, copper, zinc, selenium, molybdenum, and other minerals have been demonstrated. The levels of minerals in serum can be influenced by mineral levels in the soil, which affect the animal through its food chain. Thus, selenium levels vary in animal serum depending on the geographic location. Since high levels of these minerals are toxic to cells (14), knowing their concentration in serum and making appropriate adjustments in the medium are important.

20.4 SERA, SERUM-FREE MEDIA, AND GROWTH FACTORS

20.4.1 Sera

Serum contains growth factors and growth inhibitors. Most cell lines require the supplementation of medium with serum, but there are many cell lines that have been adapted to serum-free growth (e.g., HeLa, LLC PK_1, and L929). Although some cell selection may have been involved in addition to adaptation, recent experiments by Peehl and Ham (26) demonstrated that epidermal keratinocytes can be selectively grown from an inoculum that contains fibroblasts by the addition of selective growth factors in serum-free medium.

The sera most frequently used in tissue culture are fetal bovine serum, bovine (calf) serum, equine serum, and human serum. Proteins are a major component of serum. The functions of proteins are not well defined, but they may serve as carriers for minerals, fatty acids, and hormones or as growth factors or hormones themselves. A number of peptides in serum have been identified as growth factors and have been purified for use as supplements in serum-free medium (9). As with media, there are no universal guidelines for the selection of sera for a specific cell type. Generally, trial and error methods are employed to select the serum best suited for the growth and function of cells. Consideration must also be given to lot-to-lot variation. Suppliers will often send samples for testing with specific cell systems and will reserve these lots until they receive the results of stringent testing. Tests on sera should include sterility testing (including mycoplasma), growth curves, cloning efficiency, and microscopic observation of the morphological characteris-

TABLE 2. Growth factors for optimization of serum-free media

Growth factors
Epithelial cell growth factor
Fibroblast growth factor
Nerve cell growth factor
Platelet-derived growth factor
Transforming growth factor
T-cell growth factor
Colony stimulating factor
Triiodothyronine
Hydrocortisone
Selenium
Estradiol
Prostaglandins
Insulin
Prolactin
Growth hormone
Somatomedin
Transferrin
Hemin

tics of the cells. All or none of these tests may be applied as needed. The methodology for growth curves and cloning efficiency is described below.

Problems encountered with serum

Endotoxins. The effects of bacterial endotoxins on cells in culture are numerous. For instance, endotoxins are known to affect macrophage function and migration, stimulate hormone secretion, affect hormone sensitivity, receptor affinity, and hemopoietic cell proliferation, etc. High levels of endotoxins are usually introduced by bacterial contamination during serum collection and processing; therefore, sera should be obtained from a supplier that performs quality control assays routinely.

Hemoglobin. The hemoglobin content of sera can be used as an indicator of the quality of the procedures used in collecting serum. Hemin affects cell differentiation, induces globin mRNA, promotes neurite outgrowth in nerve cells, is a macrophage-dependent T-cell growth factor, and is mitogenic for human peripheral blood mononuclear cells. It is therefore advisable to use serum that has a low hemoglobin content unless these features are desired.

Lipids and fatty acids. Certain lipids and fatty acids are known to affect cell growth and function (10). Fatty acids, in particular, influence cell function because they are an essential and major component of cell membranes and they are determinants of membrane fluidity. It is important to recognize these features since many cell characteristics, including cellular enzyme activity, growth rates, virus survival, and infectivity, can be altered by varying the concentrations of fatty acids in serum.

The lipid requirement for the nutrition of cells remains controversial and has not been systematically studied. Recently, Bettger and Ham (10) demonstrated the preferential delivery of lipids to cells via liposomes in contrast to the established perference for serum lipoproteins. Many cell lines show a growth response to supplemented lipid when serum-supplied lipids are limiting or absent. Requirements for linoleic acid, other fatty acids, cholesterol or cholesterol precursors, the fat-soluble vitamins, and phospholipids have been demonstrated in cell culture.

Hormones. Cell culture plays a major role in delineating hormone activities (9). It is therefore important to recognize the value of identifying and controlling the levels of hormones in serum before setting up experiments to study the effects of particular hormones on cell function.

20.4.2 Serum-Free Media

Optimization of serum-free media is a time-consuming process with no general rules to follow and so must be carried out by trial and error (9). Some supplements, such as insulin, transferrin, and selenium, are commonly used. Other supplements are more cell specific, including estrogens, androgens, triiodothyronine, and others (Table 2).

During the development of serum-free media for well-established cell lines, optimization can be accomplished one step at a time. This may not be practical when establishing primary cultures, since the variability of primary cultures is inherent. The work of Barnes and Sato (9) has clearly demonstrated the ability of complex mixtures of hormones, growth factors, and carrier proteins to replace the functions of serum. In culturing human keratinocytes, Peehl and Ham (26) demonstrated that ethanolamine can be substituted for pituitary extract. By increasing the concentration of iron and decreasing the concentration of zinc, the need for transferrin in keratinocyte culture was eliminated. These examples illustrate the complexity of growth requirements.

20.4.3 Growth Factors

The most commonly used growth factors are listed in Table 2. Obviously, this list is not complete since growth factors for specific cell types are the attention of current research.

Substrate

A proper substratum is essential for the growth of some differentiated cells, especially in defined medium. Studies have demonstrated that plastic is not an optimal substratum. Coating plastic with substances from the extracellular matrix such as fibronectin, chondronectin, laminin, and collagen has been shown to have clear advantages in the attachment and growth of cells.

Defined media

Defined media are important for the discovery of novel substances with growth-promoting activities (9). High-density lipoproteins are essential for the growth of some kinds of cells. Cholera toxins, glucagon, thrombin, vitamin E, prolactin, and small proteins (growth factors) are also required by cells. Studies with defined media have helped to identify growth factors and growth inhibitors in serum.

20.5 METHODS FOR PRIMARY CULTURE

20.5.1 Explants

The oldest procedure for setting up primary cultures is the explant method. In this method, the tissue is washed free of blood and cut into small fragments of about 2 mm^2. The best cutting method is the use of two scalpels held with the blades opposing each other and manipulated like a pair of scissors. The fragments are rinsed well with balanced salt solution or culture medium and held in a small volume of either. The fragments must be placed in a culture vessel and anchored to the growth surface. Chicken plasma (rooster) prepared with heparin is commonly used to embed and to anchor the explants. After the explants are in place, the plasma can be clotted by the addition of a few drops of chicken embryo extract. Some tissues such as testicular explants cause coagulation of the plasma without any additive. When the plasma is firmly clotted, an appropriate volume of liquid culture medium is added as an overlay, and the cultures are incubated at 35 to 37°C. Within a few days new areas of cell growth or migration are seen (microscopically) around the edge of the explant. The cultures should be refed two to three times per week during the growth period.

20.5.2 Enzyme-Dispersed Cells

The preparation of cultures by enzymatic digestion of minced tissue was introduced in the early 1950s as a means of readily preparing large numbers of primary cultures for viral propagation and assay. This method was originally applied to kidney tissue, on which it perhaps works best, but has been found useful for other tissues as well. Trypsin (0.25%) is commonly used, but for some tissues collagenase is more effective. The tissue is minced (barber's shears work well for this), washed to remove blood and debris, and suspended in the enzyme solution. The general procedure involves some form of agitation over time, with harvest of the loosened or free single cells by filtration through a gauze filter or similar device. The coarser, larger fragments retained by the filter are subjected to additional cycles of enzymatic digestion followed by similar harvests. The harvested cells are pooled and held in ice water until the process is completed. The cells are counted with a hemacytometer, and a suspension of a desired density is prepared for planting into culture. (There are a variety of more specific procedures for the enzymatic digestion of tissue which are described elsewhere [2].) The suspended cells (after planting in a stationary vessel) settle, adhere to the culture surface, and replicate. Over a period of time with intermittent medium renewals, the cells form a monolayer of growth over the entire available area and appear to be much like the cultures of cell lines described in the following section. It is important to realize that trypsin is most active at an alkaline pH and that the enzyme solution should be kept at pH 7.6 to 8.0. A typical procedure for preparing rabbit kidney cells is given below.

Preparation of primary rabbit kidney cells

1. Use 21-day-old rabbits.
2. Aseptically remove kidneys and place them in chilled growth medium (medium 199) containing 1.5% $NaHCO_3$ and 5% equine serum or fetal bovine serum.
3. Place kidneys in a petri dish, remove capsule and medulla, and rinse well with medium.
4. Transfer kidneys to mincing tube and mince to fine particles (~0.5 to 1.0 mm).
5. Rinse with 125 ml of warm 0.25% trypsin per kidney. Shake gently until tissue is no longer clumped, and then decant the trypsin.
6. Add 125 ml (per kidney) of warm trypsin and a magnetic stirring bar, and place over magnetic stirring motor for 15 min at speeds below those producing foam.
7. Filter the cell suspension through three layers of sterile gauze into a 250-ml centrifuge tube in an ice bath.
8. Repeat step 7 until no large pieces remain.
9. Centrifuge the filtered cell suspension at 250 × g for 10 min.
10. Remove the trypsin supernatant fluid by aspiration.
11. Resuspend the cells in 200 ml of the above medium. Centrifuge as described above.
12. Filter through gauze and repeat step 11.
13. Resuspend cells in 20 ml of medium and transfer into two 15-ml graduated tubes. Centrifuge at 250 × g for 5 min.
14. Dilute the cells to 1:400 in the above medium on the basis of a volume of packed cells.
15. Disperse in culture flasks (4 to 5 ml/25-cm^2 flask).

20.5.3 Blood and Ascites Cells

In a few instances, cells can be obtained directly from the host as a suspension of cells. A pleural effusion or ascites will provide such cells, as will the peripheral bloodstream. These ascites may be induced in an animal by the inoculation of an irritant into the peritoneal cavity; the ascites can then be used to collect macrophages for culture. Cells can be removed from their suspending medium by centrifugation or by selective adsorption to and elution from appropriate materials before being placed in a culture medium and planted in culture.

Primary cultivation of leukocytes from peripheral blood has become a common practice for immunologists. Of the cellular elements in blood, generally only the lymphocytes and monocytes can be induced to proliferate in culture. Methods for separating these cells into reasonably pure suspension have been developed (27). Monocytes and lymphocytes may be separated from the other elements by density gradient centrifugation. The lymphocytes and monocytes collect at the interface, while erythrocytes and polymorphonuclear leukocytes settle below the interface. Monocytes may be further separated by adherence to filters of glass beads or nylon columns. If the monocytes are to be retained for cultivation, adherence to petri dishes or other plastic culture flasks can be utilized. The remaining lymphocyte population is composed of T cells (about 75 to 80%), B cells (1 to 15%), and a minor but important population of null cells. The lymphocytes collected after one of the above procedures can be further subdivided by adsorption to sheep erythrocytes, layered over Ficoll-Hypaque, and centrifuged. The B cells collect at the interface, whereas the T cells attached to the erythrocytes are found in the lower level of the tube. Most null cells are contained in the B-cell fraction but some are also with the T cells. The T cells are freed from the erythrocytes by lysing the latter with NH_4Cl-Tris or by osmotic shock with distilled water. Further purification of the B cells-null cells can be accomplished by taking advantage of the immunoglobulin on the surface of B cells. An antibody prepared against the $F(ab')_2$ portion of the immunoglobulin of the species from which the cells are obtained will react with the surface immunoglobulin. This antibody may be adsorbed to a Sephadex G-200 column for this purpose. The B cells when added to the column are retained, while the null cells are collected in the effluent. B cells can be recovered by treating the column with immunoglobulin of the species under study (27).

Purified suspensions of lymphocytes do not grow in culture unless induced by mitogens, cytokines, transforming agents (such as viruses), or other cellular elements which provide the necessary growth factors. Macrophages or monocytes are commonly used as helper cells. T-cell growth factor, leukocyte activating factor, and B-cell growth factor are examples of cytokines that are used. Two common mitogens that can be used are phytohemagglutinin and concanavalin A. Epstein-Barr virus may be used for human cells if a transformed population is desired.

Cell suspensions for cultivation can be obtained by mechanical methods from some soft tissues like thymus or spleen. This is done by trituration, stirring, or even forceful pipetting. With such methods the tissue cells must be freed of all blood elements by washing and centrifugation. It may be necessary in some instances to treat the cell suspension with NH_4Cl or by osmotic shock to remove contaminating erythrocytes. The resulting cell suspension is planted in culture where, depending on cell type(s), the cells will either settle, adhere to the growth surface of the flask and proliferate, or else remain in suspension during their growth period.

20.6 CELL STRAINS

20.6.1 Monolayers

Most cell strains grow attached to glass or plastic surfaces. Numerous types of culture vessels are available. The adherent cells can be subcultured by resuspending the cells in fresh medium and dividing into new vessels. Several methods available for detaching adherent cells are as follows. (i) Proteolytic enzymes, most commonly trypsin (crude) and pancreatin (also crude), can be used. Other enzymes that are used are collagenase, elastase, hyaluronidase, pepsin, pronase, etc. The choice of the enzyme depends on cell sensitivity and on the processes for stopping the action of the enzyme. (ii) Chelating agents such as EDTA are used to resuspend some attached cells. (iii) Combinations of enzyme and EDTA are frequently employed. (iv) Physical methods, e.g., scraping from the glass with a rubber policeman and shaking, can sometimes be used to loosen cells. These techniques are not recommended for cells that adhere tenaciously to the culture vessel surface. Some of these cells can be suspended as single cells by gentle pipetting.

A stepwise procedure for suspending a monolayer culture of cells is described below.

1. Prepare a sterile solution of trypsin (1:250; Difco Laboratories, Detroit, Mich.) at a concentration of 0.1 to 0.5% in medium or balanced salt solution at pH 7.6 to 8.0. (More dilute solutions are used for cultures in serum-free medium since natural trypsin inhibitors are absent.)
2. Decant or aspirate medium from the culture. Since serum inhibits trypsin activity, it may be necessary to rinse the cell sheet with serum-free medium.
3. Add enough trypsin solution to cover the cell sheet.
4. The time needed for trypsin treatment varies with cell type, but 1 to 5 min is usually required. Incubate at either room temperature or 37°C. The process is complete when the cell sheet appears to be loosened from the surface.
5. Add sufficient medium to make a uniform suspension. Then aspirate the cells in small-bore pipettes to finally disaggregate the cells.
6. Centrifuge the suspension as slowly as possible to pellet cells and minimize gravitational forces.
7. Resuspend the cells in an appropriate volume of complete medium. If cell dilution is required for subculturing, make appropriate dilutions as needed.

Cold trypsinization results in less damage to cells. Again, the sensitivity of cells to trypsin is the factor to be considered in deciding whether to use 4, 25, or 37°C for treatment. The cells can be suspended in trypsin solution or in complete medium with serum. In serum-free medium, a trypsin inhibitor can be added. However, since crude trypsin contains other enzymes, inhibition of enzyme activity is not complete.

Alternatively, cells can be resuspended as follows. After enzyme solution is added to the flask, incubation is carried out for 1 to 3 min (longer for more firmly attached cells), and the flask is inverted, allowing most of the enzyme solution to drain away from the cells. Digestion is allowed to proceed to the point at which small constrictions can be seen in the cell sheet upon microscopic observation. At this time, the trypsin is removed, and culture medium is added. The cells can be resuspended by gentle pipetting in complete medium and then inoculated into new culture flasks.

Note: Low temperatures are known to cause disaggregation of microtubules, rendering the cells more susceptible to damage from shearing forces produced by pipetting or other physical treatments.

Seeding density

The cell concentration for inoculating fresh vessels is usually about 100,000 cells per ml of medium or 20,000/cm^2 of growth area. The cell concentration is usually determined by the growth rate of the cells, which in turn is a function of the cell type, the medium, and the environment. Most cells have a finite density for both seeding and maximum density.

20.6.2 Suspension Culture, Stationary Suspension Culture, Soft Agar Culture, and Agitated Microcarrier Suspension Culture

As with bacteria, the easiest and simplest way to grow mammalian cells is in suspension culture. Most cells of leukemic or lymphoid origin grow as single cells in either stationary or agitated suspension cultures.

There are many apparatus for growing cells in suspension culture. Cherry and Hull (16) devised suitable vessels from round-bottom flasks which were agitated by a Teflon-coated magnetic stirring bar suspended from stainless-steel swivels by stainless-steel wire. The flasks were then mounted at a suitable distance from a motor-driven magnet to reduce the effect of heat generated by the motor.

Large-scale suspension cultures of mammalian cells have recently been described by Feder and Tolbert (17). Spin filter techniques and flexible plastic sheets designed to rotate at slow speeds are used to retain cells in suspension without developing excessive shear forces that are disruptive of cell membranes. Many variations of these techniques have been designed, including continuous perfusion.

The gas phase requirements of suspension cultures are variable. Some investigators simply sparge with air; others use mixtures of air and CO_2 or O_2. The generation of oxygen bubbles in the medium should be avoided.

Agitated microcarrier suspension cultures

Many anchorage-dependent cells will not grow in suspension culture unless special procedures are used. One of these procedures involves the use of Sephadex beads to which the cells attach. These beads can then be kept in suspension by adequate agitation techniques. Various additives such as 1% methylcellulose or polyethylene glycol can be added to the medium to prevent serum precipitation on the sides of the vessel. Other modifications of the medium can be made such as the reduction of calcium or phosphate to reduce cell clumping. There is general disagreement concerning the benefits of these modifications that is probably related to particular cells and culture conditions. For these reasons, many variations of the method have been devised. Microcarrier beads are made of natural or synthetic polymers of various sizes, such as Sephadex, dextran, etc. A problem encountered with the beads involves collisions which damage the attached cells. If cells need to be harvested, it is often difficult to remove them from the beads by conventional techniques. A variety of bead types is available from commercial suppliers.

Modified perfusion systems have also been used with the microcarrier system. Of particular significance is the microcarrier-perfusion reactor described by Feder and Tolbert (17). Through better control of the environment, cell densities have been obtained that are 5 to 10 times greater than those previously attained. An additional feature of this procedure is an efficient scale-up system that minimizes the lag times usually encountered in such systems.

Soft agar cultures

Semisolid and soft agar techniques have been developed since some malignant or virus-transformed mammalian cells can grow into isolated colonies in highly purified soft agar culture. Double-strength medium is used to suspend cells, and the cell concentration is adjusted to twice that of the desired final concentration. Highly purified agar (agarose) is made up in triple-distilled water at concentrations ranging from 0.6 to 1.0%. The agar is sterilized by autoclaving, brought to 41 to 44°C, and held in a water bath. The medium containing the cells is warmed to 35°C and added to an equal volume of warm agar. The medium and agar are quickly mixed, transferred into petri dishes, and incubated in CO_2 incubators. Colonies can be counted or isolated for cloning purposes. A modification of this technique is the human tumor stem cell assay developed by Hamburger and Salmon (20). In this method a feeder layer of nutrients is provided beneath the agar layer containing the cells.

20.6.3 Hollow-Fiber Systems and Other Substrata

During the last 10 years, hollow-fiber culture equipment has been the subject of much developmental effort, especially for mass cell culture techniques. Typically, in this system a cluster of porous hollow fibers is enclosed in a vessel. The ends of the fibers are arranged so that the medium or the gaseous environment, or both, can be continually changed. The cells are implanted on the outside of the fibers, creating a large surface area for cell growth. Cells can grow to high tissuelike densities and can be maintained for months.

20.7 MAINTENANCE OF CELL CULTURES

20.7.1 Subculturing

Once a culture is established, whether it is a cell line or a primary culture, it requires periodic medium change or subculturing. Intervals between medium change and subculturing are determined by rate of cell growth, depletion of nutrients from the medium, accumulation of metabolites in the medium, and pH change.

Conditions to consider when deciding to subculture or refeed the cultures are an increase in cell concentration, a drop in medium pH, and the microscopic appearance of the cells. Generally, normal cells stop dividing when they reach a stage of confluency, when essential ingredients are depleted from the medium, or when a toxic metabolite builds up in the medium. Cells that are not dividing may still deplete some nutrients from the medium. Careful and timely assessment of the cultures is crucial to the establishment of proliferation and function.

20.7.2 Clones and Cloning

In mixed cell cultures, the fastest growing cell type will eventually dominate. To select and propagate specific cell types, a number of different cloning procedures have been developed. These include limited dilution cloning in a variety of conditions and media designed to select a single cell, or the progeny of a single cell, for further growth. Since single-cell growth is a function of plating efficiency and medium volume, optimization of the medium plays a critical role in successful cloning (23). A number of procedures have been devised for achieving this goal such as cloning in soft agar, the use of cloning rings in monolayer culture, selective irradiation, selective media, interaction with substrate, cell migration, and single-cell sorting by flow cytometry. (See references 2 and 7 for detailed procedures.)

20.8 TYPES OF CELLS CULTIVATED

The use by earlier workers of the plasma clot and media with high concentrations of serum and embryo extract favored the growth of fibroblasts or the connective tissue elements from the tissues put into culture. Most tissues or organs contain connective tissue in the form of capsules and internal structural or binding materials which generally have a greater propensity for growth in vitro than do the more differentiated parenchymal cells. Subcutaneous or skin muscle tissue, especially from embryos, provides a good source of fibroblasts for culture preparation. Other, more specialized, connective tissues such as cardiac or skeletal muscle, bone, cartilage, and adipose tissue and the supporting elements of the central nervous system, such as neuroglial cells, can be grown in culture.

20.8.1 Cardiac and Skeletal Muscle Cells

Embryonic or neonatal hearts provide the best opportunity for culture of myocardial cells (13). These can be cultured as explants or as trypsinized cell suspensions, and, in the proper environment, contracting cells develop and can be maintained in a beating state for fairly long periods of time. When enzyme-dispersed cells are used, contaminating endothelial cells can be eliminated by taking advantage of their quickness to settle and adhere to glass or plastic surfaces. With skeletal muscle, only the mononuclear myoblasts proliferate in culture; but as the cell density increases, cell fusion occurs to produce multinucleated muscle fibers which no longer divide. After fusion, contraction of the multinucleated fibers may be seen, and the typical cross-striation of skeletal muscle cells develops (31).

20.8.2 Bone Cells

Cultures of bone cells (osteoblasts and osteocytes) are usually initiated from embryonic bone or primordia, but one report of the long-term cultivation of adult bone cells has appeared (28). Embryonic calvaria provide a good source of cells, which can be obtained from both the periosteum (endocranial surface) and the bone itself (11). The tissues are separated mechanically, washed, minced, and trypsinized to disperse the tissue into a suspension of single cells for planting into culture. A medium referred to as BGJb or its modification, BGJb/FJ, plus fetal bovine serum (10 to 20%) is recommended for the growth of these cells. The cells settle, attach to the floor of the culture vessel, and proliferate, becoming confluent in about 7 to 14 days. As growth continues the cultures become multilayered, and dense areas of cell growth are observed. Mineral (Ca^{+}) deposition occurs and is most evident in the areas of high cell density. Cultures can be initiated from explants of embryonic bone in which cells migrate out from the explant and proliferate. Other bones such as ribs or femurs have been used as a cell source. In addition to the cell culture methods described above, some truly remarkable studies have been done with organ cultures of rudimentary bone in which development of the organ was demonstrated.

20.8.3 Chondrocytes

Chondrocytes from elastic, hyaline, and articular cartilage (obtained from the ear, sternum, and long bone articular surfaces) can be grown in cell culture (19). Following excision of the tissues, all adhering noncartilaginous tissue is dissected away, and the clear cartilage is cut into small fragments (1 to 2 mm^3). These fragments are dissociated by incubation in a medium of 0.5% collagenase–10% fetal bovine serum in salt solution with mild or no agitation. The suspended cells are collected, washed by centrifugation, and suspended in growth medium. Ham F12 medium (2) with 5% fetal bovine serum is satisfactory. After being planted in culture, the cells settle, attach to the growth surface of the culture vessel, and proliferate to form colonies of epithelial-like cells. Some elongated fibroblastlike cells may be seen which represent contaminating cells from connective tissue not completely removed during the preparation of the cartilage for cultivation. Chondrocytes can be subcultured from the primary cultures by using trypsin, but long-term or continuous cultures of these cells are not available.

20.8.4 Adipocytes (Fat Cells)

Adult fat cells do not proliferate, but cultures of adipocytes can be prepared from precursor cells obtained from adult fatty tissue. The tissue is treated with collagenase in a buffer solution and agitated over time at 37°C. The time of digestion can be determined by microscopic observation of the tissue and cells. The use of siliconized glassware is recommended. After incubation the tissue remnants are removed by filtration of the cell suspension through a nylon screen into a centrifuge tube. The fat cells float to the surface and can be harvested for planting into culture. Medium 199 (12, 29) with a serum supplement, plus such additives as insulin, methylcellulose, increased glucose, and linoleic acid, has been used to cultivate the cells. The cells settle, adhere to the culture vessel, spread out, and become elongated like fibroblasts. As growth continues a monolayer is formed, and at this time granules of lipid-staining material appear in the cytoplasm. With continued incubation the granules or lipid droplets coalesce into a single large lipid deposit, and the cells begin to appear as adult fat cells. These will loosen from the growth surface of the flask and float freely in the medium. After harvest, the floating lipid-laden cells can again be treated with collagenase and returned to culture where the process will be repeated (12, 29).

20.8.5 Nerve Cells

Tissue culture had its origin in the experiments of Harrison (2) with nerve tissue maintained in vitro. In the ensuing years, many investigators have studied cells and tissues of both the central and peripheral nervous systems in culture systems. Neurons from brain and spinal cord ganglia can be maintained in culture where, with optimal conditions, maturation with growth of fibers and axons can be observed. Cell replication does not occur unless neuroblasts from very early embryos are used. Cells of the supporting neuroglial tissue, however, do replicate, and continuous cell lines of glial cells have been developed. Pieces of brain tissue are carefully dissected to remove all membranes and extraneous material and then cut into small pieces, put into medium, and dissociated by vigorous pipetting. The resulting suspension of cells and small aggregates is plated into culture vessels where some cells will slowly settle, adhere, and replicate to form monolayers of glial cells. Several weeks may be required for this to occur. Cells can be subcultured by the trypsin method and serially passed for a limited number of cell generations. Eagle medium or Ham F-10 (2) with fetal bovine serum is satisfactory.

20.8.6 Glandular Epithelial and Vascular Endothelial Cells

Epithelial cells, once thought to be difficult to grow in culture, are readily cultivated from a number of tissues by using media and culture procedures now in common use. The use of either lower concentrations of serum in defined media or serum-free media with added growth factors has made growth of these cells possible. The most readily cultivated epithelial cells are those lining organs or tubules or those from the skin. After careful dissection to remove connective tissue elements, the tissue is minced and dissociated with trypsin or collagenase to prepare cell suspensions for planting into culture. In some instances an organ or tubule might be filled or perfused with the enzyme solution to loosen or free the epithelial lining cells. Kidney tubules may provide the best source of epithelial cells for culture by these methods, but other organs and tissues, such as the alimentary canal, respiratory tract, uterus, prostate gland, etc., are also used. These cells provide monolayer cultures of usually large, polygonal, closely knit cells with round nuclei. Subculture can be accomplished by the trypsin method, although the number of possible serial passages varies with the tissue source and animal species used as donor.

Cultivation of the more specialized glandular epithelia with maintenance of functional activities is more difficult, but the success rate is steadily improving with the introduction of new methods and improved culture media. In most cases a gland contains multiple cell types, and problems arise in trying to isolate and identify the particular cells desired for culture. For many years this made the culture of pancreatic beta cells very difficult; however, methods have now been developed for the harvest of islets essentially free of other cellular elements (15). In the case of the pituitary gland, the types of cells appearing in the cultures can be partially controlled by careful dissection of the gland into its component parts. Continuous cell lines have been produced from pituitary tumors that continue to produce prolactin and growth hormone.

For many years liver cells were difficult to cultivate, but cultures of these cells are now easier to perform. In situ perfusion of the organ with enzymes has enhanced the preparation of viable cell suspensions for cultivation, and cells more prone to divide have been obtained from regenerating liver after partial hepatectomy. Improved culture media with more specific requirements for liver cells have been formulated. Although there are numerous studies on primary or short-term cultivation of hepatocytes, there are few established cell lines.

Cells from other glands (e.g., thyroid, parathyroid, adrenal, ovary, and testicle) have been cultured at least for limited periods of time with the production of specific hormones. This has been accomplished by using explant or organ cultures and by preparation of enzyme-dispersed cell suspensions. The fact that a cell produces a specific product aids greatly in the identification of the cells growing in the cultures.

Culture of endothelial cells lining the blood vessels and the heart has been greatly expanded in recent years, and these cells can now be cultured from arteries, veins, capillaries, and the heart. Further, several continuous lines of endothelial cells have become available. The cells from large arteries or veins are harvested by selecting a section of vessel, ligating one end, and filling it with collagenase. With repeated treatment the harvested collagenase solution will contain loosened endothelial cells which can be planted in culture (18, 22). A similar procedure involves everting the blood vessel onto a rod, thus exposing the endothelial lining, and repeatedly dipping or stirring the rod in a solution of collagenase. For the preparation of capillary endothelium, a tissue rich in microvascular material is selected, minced into small fragments, and homogenized or treated with collagenase. The resulting small fragments of capillary tissue are collected by filtration through nylon mesh. Further cell dispersion may be done with collagenase, or the fragments and aggregates may be planted into culture. Endothelial cells settle and attach to the culture vessel rapidly, and this can be used as a means of reducing the number of contaminating cells (fibroblasts, smooth-muscle cells) by early rinsing and refeeding of the cultures. Several growth factors when added to the medium have been reported to enhance the growth of endothelial cells. These include tumor cell-conditioned medium, thrombin, fibroblast growth factor, and endothelial cell growth supplement.

The culture of cells related to the immune system is essential to the new discipline of cellular immunology. The methods for obtaining and cultivating these cells were covered in Methods for Primary Culture (section 20.5). Numerous continuous cell lines of lymphocytes and macrophages are also available.

20.9 SOME SPECIFIC USES OF CELLS IN CULTURE

The uses and applications of cells in culture are increasing as more scientific information on growth regulation, differentiation, drug effects, function, and hormonal regulation is accumulated.

20.9.1 Effects of Chemical Compounds on Cell Proliferation

There are a variety of procedures for measuring the effects of compounds on cell proliferation. These include measurements of cell numbers by microscopic observation, use of automated equipment such as volume counters and flow cytometers, incorporation of isotopically labeled precursors into macromolecules (such as DNA, RNA, and proteins), and cell plating efficiencies. Quantitation methods similar to those used with microorganisms can be used (2, 3).

20.9.2 Genetic Studies

The use of mammalian cell cultures in genetic studies is expanding rapidly, prompted in part by the developments in molecular biology and by new techniques for measuring genetic changes in cells. Cell culture is playing a key role by permitting direct observations of the cells in question. Skin fibroblasts, peripheral blood leukocytes, and amniocentesis samples have all been used as sources of cells for these studies in humans. Cell banks with specific genetic markers have been established in the American Type Culture Collection and the Human Genetic Mutant Cell Respository. Methods for the growth of cells and specific procedures for identifying chromosomal abnormalities are described in publications from both institutions and in various tissue culture methodology handbooks. Genetic studies are being enhanced by rapid developments in hybridoma technology.

Other specific areas of research in which cell culture plays a major role are host-parasite relationships, virus studies, recombinant DNA studies, and growth regulation and differentiation. Differentiation has received considerable attention during the last several years due to advances in the development of hormonally defined media. The use of mammalian cell cultures in basic research and applied developmental situations is rapidly expanding.

20.10 CELL PRESERVATION AND RECOVERY

To ensure that cells are available for future use and to safeguard against accidental loss due to contamination or inadequate or improper culture conditions, cultures may be stored in liquid nitrogen. The viability of cells in liquid nitrogen is preserved for many years.

Preservation

1. Trypsinize a subconfluent population of rapidly growing cells, or, in the case of suspension cultures, centrifuge cells that are in log phase. Resuspend the pellet in cool complete growth medium (including serum) at a concentration of 2.5×10^6 to 4.0×10^6 cells per ml. Slowly add reagent-grade dimethyl sulfoxide or glycerol to this solution to achieve a final concentration of 5 to 10%.
2. Use a syringe fitted with an 18-gauge needle to transfer 1 ml of the cell suspension to a sterile glass ampoule, which is then flame sealed (plastic vials are also commercially available).
3. Place the ampoule in the vapor phase of a liquid nitrogen freezer for 30 min. Alternatively, place the ampoule in a Styrofoam box with sides and lid that are 5 to 10 cm thick. Place this box in a −70°C freezer for 2 h. Either of these procedures gives a slow cooling rate and eliminates the expense of controlled freezing rate apparatus.
4. Place the ampoule in commercially available racks, and store in a liquid nitrogen freezer.
5. After 24 h, check viability and sterility.

Recovery

Generally, the simplest method involving the fewest operations and the least handling is recommended for cell recovery from storage.

1. Remove an ampoule from the liquid nitrogen freezer. Gloves, face mask, and long-sleeved garments should be worn. There is a danger of ampoule explosion if liquid nitrogen has leaked into it.
2. Quickly thaw the contents of the ampoule by placing it in a 37°C water bath. Agitation increases the rate of thawing.
3. Cleanse the outside of the ampoule with 70% ethanol and allow it to dry.
4. Break off the end of the ampoule in a sterile towel, and carefully remove the contents with an 18-gauge needle and syringe.
5. Add sufficient complete growth medium to dilute the dimethyl sulfoxide to a concentration of less than 0.1%. If glycerol is used, the growth medium should be slowly added to the cell suspension and immediately centrifuged at $200 \times g$ for 5 min. Cell culture medium is used to resuspend the cells at the optimal cell density. Modification of this procedure can be made, but the object is to reduce the glycerol concentration in a stepwise fashion.
6. Incubate the cells at 37°C. After 24 h, replace the medium with fresh growth medium.

20.11 DETECTION OF CONTAMINATION

Check all cultures and each individual flask periodically for contamination by using a phase-contrast microscope. Periodic fluorescent staining and culturing for mycoplasma are also recommended.

Where antibiotics are used, maintain separate cultures in the absence of antibiotics. A quarantine of all new cell lines, media, serum, and enzymes before general use in the laboratory is recommended. Do not share the same bottles of medium, serum, or enzymes among different cell lines or technicians.

It is important to check the morphological characteristics of the cells periodically for contamination with other cell lines. Kits for isozyme determination are now commercially available to determine the presence or absence of enzymes characteristic to various species.

Mycoplasma

Mycoplasma contamination is a concern for every investigator who uses cultured cells. Data derived from cell cultures contaminated with mycoplasma should be considered invalid. Mycoplasma have been shown to have the following effects on cells: (i) interference with growth rates; (ii) cytopathic effects; (iii) depletion of essential nutrients from media; (iv) alteration of carbohydrate content of media; (v) inhibition of nucleic acid and protein synthesis; (vi) changes in enzyme patterns; (vii) chromosomal aberrations; (viii) modification and redistribution of cell antigens; (ix) modification of viral infectivity; (x) interference with the purification of cell organelles (mitochondria, membranes); and (xi) reduction of the oncogenicity of malignant cells.

The best way to avoid the effects of mycoplasma is to prevent cell contamination. Sources of infection are human, cattle, or swine sera. Therefore, only serum from high-quality suppliers should be used. Serum should be checked for the presence of mycoplasma by several of the assays currently available. Mouth pipetting must be avoided, and heat inactivation of serum should be considered. Several antibiotics have been reported to suppress the growth of mycoplasma on cells, but other studies suggest recurrence of the infections upon removal of the antibiotics. Other reports have recommended passage of the malignant cells in animals; upon recovery of the cells, mycoplasma could not be detected.

Detection of infections

Microbiological and viral contamination of cells can be difficult to detect and assess. This is especially true of viral contamination since some viruses are species specific. While detection of viral and mycoplasma infections of mammalian cells is complex and requires special diagnostic methods, detection of ordinary bacterial and yeast infections usually requires only a good phase-contrast microscope. Some bacteria, however, adhere to cell surfaces or grow intracellularly, and special procedures must be used to isolate and detect these organisms. This is especially true for anaerobic and facultative organisms (see reference 3 for procedures for isolating, growing, and identifying these organisms).

Since mycoplasma infections of cells in culture affect many vital cell processes and properties and can remain unrecognized and undetected indefinitely, special procedures must be used periodically to determine their presence. The presence of mycoplasma usually is confirmed by demonstrating their characteristic growth on agar or by fluorescence microscopy.

Culture techniques for detection of mycoplasma infection

In our experience, broth and agar cultures have been the most sensitive means of detecting mycoplasma infections because such cultures have the advantage of rare-event detection with growth amplification. Since most mycoplasmas are bound to the cell surface, it is imperative that a cell suspension be prepared. Some mycoplasmas are shed and can be cultured from medium. Frequently, fragments of cells can be mistaken for mycoplasmas upon microscopic observation.

Mycoplasma broth

1. Combine 4.2 g of Difco PPLO broth and 20 ml of yeast extract (see below).
2. Suspend mixture in 140 ml of distilled water.
3. Autoclave at 15 lb/in^2 for 15 min.
4. Cool to 40°C and add 40 ml of mycoplasma-free horse serum.

Mycoplasma agar

1. Prepare broth as described above.
2. Add 2.4 g of Difco purified agar.
3. Autoclave.
4. Cool to 40°C and add 40 ml of mycoplasma-free horse serum.
5. Pipette 7.0 ml of the solution into 55-mm plates.
6. Invert and cool at 4°C.

Yeast extract preparation

1. Add 250 mg of dry Fleischmann's bakers' yeast to 1 liter of water.
2. Bring mixture to a boil with frequent stirring, and keep at a low boil for 10 min.
3. Remove from heat and let stand until yeast cells settle.
4. Centrifuge at 1,000 × *g* to remove most of the yeast cells.
5. Pipette 20 ml of the supernatant into separate vessels and freeze.

Mycoplasma growth and detection

1. Approximately 0.15 ml of sample is streaked onto the surface of duplicate agar plates and allowed to dry at room temperature.
2. One plate is incubated at 37°C in an anaerobic jar with a BBL GasPak (3). The other plate is incubated in an aerobic, moist atmosphere at 37°C.
3. At the same time 0.5 ml of sample is added to each of two 7-ml tubes of broth. One tube is incubated with the anaerobic plate, and the other tube is incubated with the aerobic plate.
4. Platings and passages are made at 5- and 10-day intervals with 0.5 ml transferred to fresh broth.
5. If no mycoplasma colonies develop on the original plates or after the subsequent two passages, a third passage is recommended.
6. Agar plates are examined by inverting them on the microscope stage and focusing through the agar. Low power (2.5 to 6.0×) is useful for locating the colonies. Mycoplasma colonies range in size (5 to 1,000 μm) and generally exhibit a characteristic "fried-egg" appearance. Higher-power objectives (10 to 20×) should be used to obtain greater detail.
7. Mycoplasma grow to high titer in broth culture without exhibiting turbidity. Microscope observation is necessary.

Nonculture techniques for detection of mycoplasma infection

Some strains of mycoplasma do not grow on standard agar plates or have become so adapted to growth in cell culture that their growth in other media is poor. A direct fluorescence stain may therefore be necessary. Cells should be grown on cover slips in Leighton tubes (2) or petri dishes, or on slides in petri dishes. The conditions for culture should be optimum for that particular cell, and the population should be grown to about 50% confluency.

Reagents

Hoechst stain 33258 (500 μg/ml) in Hanks balanced salt solution without phenol red or sodium bicarbonate.

Mounting medium (citric acid monohydrate, 4,662 mg; Na_2HPO_4, 8,247 mg; glycerol, 500 ml; water to 1,000 ml).

Procedure

1. Dilute the stain to 0.5 μg/ml at room temperature.
2. Aspirate all of the medium from the culture.
3. Fix in Carnoys solution (glacial acetic acid-absolute methanol, 1:3) for 5 min. Remove the fixative and replace with fresh fixative. Let stand for 10 min.
4. Air dry the slide, cover it with stain solution, and hold for 30 min.
5. Remove the stain, and wash the slide three times in deionized water.
6. Air dry the slide and preserve in the mounting medium.
7. Examine the slide under a suitable fluorescent microscope (excitation wave length, 360 nm; maximum emission wavelength, 490 to 500 nm). A negative culture is indicated by only nuclear staining. A culture positive for mycoplasma will be seen as a positive nuclear stain with the cytoplasm or cell membrane covered with small fluorescent material relatively uniform in size. Control cultures with and without deliberate mycoplasma infections should be used as controls.
8. Bacteria, yeast cells, and other organisms can be identified by their typical morphological characteristics.

20.12 LITERATURE CITED

20.12.1 General References

1. **Evans, V. J., V. P. Perry, and M. M. Vincent (ed.).** 1975–1979. Tissue Culture Association manual, vol. 1 through 5. Tissue Culture Association, Inc., Rockville, Md.
2. **Freshney, R. I.** 1983. Culture of animal cells: a manual of basic technique. Alan R. Liss, New York.
3. **Gerhardt, P., R. G. E. Murray, R. N. Costilow, E. W. Nester, W. A. Wood, N. R. Krieg, and G. B. Phillips (ed.).** 1981. Manual of methods for general microbiology. American Society for Microbiology, Washington, D.C.
4. **Jacoby, W., and I. Pasten (ed.).** 1979. Methods in enzymology, vol. 58. Academic Press, Inc., New York.
5. **Kahn, R. H., D. J. Merchant, and W. H. Murphy (ed.).** 1964. Handbook of cell and organ culture, 2nd ed. Burgiss Publishing Co., Minneapolis.
6. **Kruse, P. F., Jr., and M. K. Paterson, Jr. (ed.).** 1973. Tissue culture, methods and applications. Academic Press, Inc., New York.
7. **Maramorosch, Y. (ed.).** 1983. Advances in cell culture, vol. 1 and 2. Academic Press, Inc., New York.
8. **Murrell, L. R. (ed.).** 1979–1983. Journal of tissue culture methods. Tissue Culture Association, Inc. Rockville, Md.

20.12.2 Specific References

9. **Barnes, D., and G. Sato.** 1980. Methods for growth of cultured cells in serum-free medium. Anal. Biochem. **102:**255–270.
10. **Bettger, W. J., and R. G. Ham.** 1982. The critical role of lipids in supporting clonal growth of human diploid fibroblasts in a defined medium. Cold Spring Harbor Conf. Cell Proliferation **9:**61–64.
11. **Binderman, I., D. Diksin, A. Harell, E. Katzir, and L. Sacks.** 1974. Formation of bone tissue in culture from isolated bone cells. J. Cell Biol. **61:**427–439.
12. **Bjorntorp, P., M. Karlsson, H. Pertoft, P. Pettersson, L. Sjostrom, and U. Smith.** 1978. Isolation and characterization of cells from rat adipose tissue developing into adipocytes. J. Lipid Res. **19:**316–323.
13. **Boder, G. B., and I. S. Johnson.** 1972. Comparative effects of some cardioactive agents on contractility of cultured heart cells. J. Mol. Cell. Cardiol. **4:**453–463.
14. **Boder, G. B., W. J. Kleinschmidt, R. J. Harley, and D. C. Williams.** 1983. Visible light inhibits growth of Chinese hamster ovary cells. Eur. J. Cell Biol. **31:**132–136.
15. **Boder, G. B., M. A. Root, R. E. Chance, and I. S. Johnson.** 1969. Extended production of insulin by isolated pancreatic islets in suspension culture. Proc. Soc. Exp. Biol. Med. **131:**507–513.
16. **Cherry, W. R., and R. N. Hull.** 1960. Growth of mammalian cells in agitated fluid medium. J. Biochem. Microbiol. Tech. Eng. **2:**267–287.
17. **Feder, J., and W. R. Tolbert.** 1983. The large-scale cultivation of mammalian cells. Sci. Am. **248:**36–43.
18. **Gimbrone, M. A., Jr., R. S. Cotran, and J. Folkman.** 1974. Human vascular endothelial cells in culture. J. Cell Biol. **60:**673–684.
19. **Ham, R. G., and G. L. Sattler.** 1972. Clonal growth of differentiated rabbit cartilage cells. J. Cell. Sci. **72:** 109–114.
20. **Hamburger, A. W., and S. E. Salmon.** 1977. Primary bioassay of human tumor stem cells. Science **197:**461–463.
21. **Hull, R. N.** 1968. The simian viruses. Virol. Monogr. **2:**1–66.
22. **Jaffe, E. A., R. L. Nachman, C. G. Becker, and C. R. Minick.** 1973. Culture of human endothelial cells derived from umbilical veins: identification by morphologic and immunologic criteria. J. Clin. Invest. **52:**2745–2756.
23. **Lechner, J. F., A. Haugen, A. Autrup, I. A. McClendon, B. F. Trump, and C. C. Harris.** 1981. Clonal growth of epithelial cells from normal adult human bronchus. Cancer Res. **41:**2294–2304.
24. **Leibowitz, A.** 1963. The growth and maintenance of tissue-cell cultures in free gas exchange with the atmosphere. Am. J. Epidemiol. **78:**173–183.
25. **McKeehan, W. L., K. A. McKeehan, and D. Calkins.** 1981. Extracellular regulation of fibroblast multiplication. J. Biol. Chem. **256:**2973.
26. **Peehl, D. M., and R. G. Ham.** 1980. Growth and differentiation of human keratinocytes without a feeder layer or conditioned medium. In Vitro **16:**516–525.
27. **Rose, N. R., and H. Friedman (ed.).** 1976. Manual of clinical immunology. American Society for Microbiology, Washington, D.C.
28. **Williams, D. C., G. B. Boder, R. E. Toomey, D. C. Paul, C. H. Hillman, Jr., K. L. King, R. M. Van Frank, and C. C. Johnston, Jr.** 1980. Mineralization in cultured adult rat bone cells: evaluation by video time lapse, scanning

electron microscopy and energy dispersive X-ray analyses. Calcif. Tissue Int. **30:**233–246.

29. **Van, R. L. R., and D. A. K. Roncari.** 1978. Complete differentiation of adipocyte precursors: culture system for studying the cellular nature of adipose tissue. Cell Tissue Res. **195:**317–329.
30. **Willmer, E. N.** 1965. Morphological problems of cell type, shape and identification, p. 143–174. *In* E. N. Willmer (ed.), Cells and tissue in culture, vol. 1. Academic Press, Inc., New York.
31. **Yaffe, D.** 1973. Rat skeletal muscle cells, p. 106–114. *In* P. F. Kruse and M. K. Patterson, Jr. (ed.), Tissue culture: methods and application. Academic Press, Inc., New York.

Chapter 21

Cells, Protoplasts, and Plant Regeneration in Culture

OLUF L. GAMBORG

The aseptic culture of tissues and cells of higher plants is an established technology for research and practical application. Cells of most plant species can be grown and produced in any quantity on defined nutrient media. In the initial stages, the cells are usually grown on a nutrient agar. Such cells can then be transferred into liquid medium and grown as shake cultures, where they are generally more uniform and multiply at a higher rate. Plant cells in suspension culture consist of a mixture of single cells and cell aggregates. They are normally grown in flasks but can be produced in any quantity in fermentors of different designs and sizes.

The nutrients for aseptic growth of plant cells consist of defined constituents of inorganic salts, vitamins, growth hormones, and sucrose. The last is required since cultured cells normally do not exhibit photosynthesis.

The technology also permits the culture of tissues, cells, or organs for the purpose of regenerating complete plants via a morphogenesis process. This has resulted in the establishment of procedures for plant multiplication as an efficient process for commercial plant propagation. The methods of cell culture and morphogenesis also are becoming integrated into genetic improvement procedures for plants of economic importance.

The maintenance of cells in culture can lead to the occurrence of variations which may be expressed in regenerated plants. Lines cloned and selected in culture are referred to as somaclonal variants (10, 35).

The more sophisticated genetic modification approaches include cell fusion and the transfer of genes to produce new hybrids and transformants. The cell genetic modification methods include an extensive technology for the isolation and manipulation of plant protoplasts. The term protoplast refers to plant cells which have been deprived of their cell walls by enzyme digestion. In many species, protoplasts can be grown back to normal cells and regenerated into mature plants.

The present account is concerned primarily with the basic materials and methods for the manipulation of plant cells and protoplasts in culture. Several book chapters and entire volumes are available to provide more details; these are referred to in the appropriate sections.

21.1 CELL CULTURE METHODS AND NUTRIENTS

Plant cells and tissues can be grown in culture aseptically on nutrient media. The different types of culture can be classified into five categories, as follows.

(i) Organ culture. The aseptic culture of embryos, anthers, ovaries, buds, roots, flowers, or other plant organs to study morphogenesis, to grow complete plants, or both.

(ii) Meristem and tissue culture. The culture of shoot meristems, leaf, stem, or other explant tissue for regeneration and multiplication of complete plants.

(iii) Callus culture. The initiation and culture of undifferentiated cell masses (callus) on agar media; such cultures are initiated from an explant of a seedling or other plant tissue source.

(iv) Cell culture. The culture of cells in liquid medium in vessels and their maintenance as shake cultures or as mass cultures in fermentors.

TABLE 1. Inorganic salt and vitamin compositions of plant tissue culture media

Constituent	Concn (mg/liter) in medium[a]:					
	MS	B5	E1	N6	NN	L2
Macronutrients						
$MgSO_4 \cdot 7H_2O$	370	250	400	185	185	435
KH_2PO_4	170		250	400	68	325
$NaH_2PO_4 \cdot H_2O$		150				85
KNO_3	1,900	2,500	2,100	2,830	950	2,100
NH_4NO_3	1,650		600		720	1,000
$CaCl_2 \cdot 2H_2O$	440	150	450	166	166	600
$(NH_4)_2 \cdot SO_4$		134		463		
Micronutrients						
H_3BO_3	6.2	3	3	1.6	10	5.0
$MnSO_4 \cdot H_2O$	15.6	10	10	3.3	19.0	15.0
$ZnSO_4 \cdot 7H_2O$	8.6	2	2	1.5	10.0	5.0
$NaMoO_4 \cdot 2H_2O$	0.25	0.25	0.25	0.25	0.25	0.4
$CuSO_4 \cdot 5H_2O$	0.025	0.025	0.025	0.025	0.025	0.1
$CoCl_2 \cdot 6H_2O$	0.025	0.025	0.025		0.025	0.1
KI	0.83	0.75	0.8	0.8		1.0
$FeSO_4 \cdot 7H_2O$	27.8			27.8		
Disodium EDTA	37.3			37.3		
EDTA sodium ferric salt		40	40		100	25
Glycine				40	5	
Sucrose	30×10^3	20×10^3	25×10^3	50×10^3	20×10^3	25×10^3
Vitamins						
Thiamine hydrochloride	0.5	10	10	1	0.5	2.0
Pyridoxine hydrochloride	0.5	1	1	0.5	0.5	0.5
Nicotinic acid	0.5	1	1	0.5	5.0	
myo-Inositol	100	100	250		100	250
pH	5.8	5.5	5.5	5.8	5.5	5.8

[a] Source: MS, Murashige and Skoog (39); B5, Gamborg et al. (16); E1, Gamborg et al. (14); N6, Chu (7); NN, Nitsch and Nitsch (42); L2, Phillips and Collins (8, 44).

(v) Protoplast culture. The culture of protoplasts isolated from plant tissues or cultured cells.

Each of these categories is used in research and in a variety of practical applications. The choice of materials and type of method is dictated by the purpose of each particular investigation. Detailed information on specific culture procedures for each category are available. The present discussion is concerned with basic requirements which apply to most plants. In these and in other procedures with biological materials, there may be specific conditions and unique requirements for particular species or tissue sources. The predominant factors that determine the success of any of the culture methods are the explant origin, the composition of nutrient medium, and the culture environment.

21.1.1 Nutritional Requirements

Plant cells and tissues have the same basic requirements for inorganic salts as any soil-grown plant, but additional nutrients are required for satisfactory growth (19). A nutrient medium consists of inorganic salts including inorganic nitrogen, a carbon source, vitamins, and growth hormones. Some species require other components, which may be metabolites, for optimal growth. These include organic nitrogen compounds and tricarboxylic acid cycle acids.

Inorganic salts

The inorganic salts normally required and included in nutrient media are shown in Table 1 (see also references 10, 46, and 49). The optimum concentration of each nutrient varies with the culture type. For most purposes, a nutrient medium should contain 25 mM or more inorganic nitrogen. Nitrate is commonly used at 25 to 40 mM. Nitrate alone can be sufficient, but frequently there is a beneficial effect and occasionally an essential role for an ammonium salt. Cells can be grown with ammonium salts as the sole nitrogen source if succinate or another tricarboxylic acid cycle acid is provided. Potassium must be supplied at concentrations of 20 mM or higher. The optimum concentrations of phosphorus, magnesium, calcium, and sulfur vary from 1 to 3 mM. Sodium as NaCl and chloride as NaCl or KCl are not essential and have not been shown to have deleterious effects at concentrations of up to 60 mM. The essential micronutrients are shown in Table 1.

Carbon sources

Sucrose or glucose is the standard carbon source. Fructose is much less effective, and other carbohydrates or organic carbon compounds are unsuitable as sole carbon sources. Mutants utilizing compounds

TABLE 2. Preparation of growth hormone stock solutions[a]

Compound	Common abbreviation	Mol wt	Quantity (mg) for 50 ml of a 1 mM solution	Prepn
Cytokinins				
Benzyladenine	BA	225.2	11.25	Dissolve in 2 to 5 ml of 0.5 N HCl; heat slightly; make to volume; adjust to pH 5
Isopentenyl adenine	2iP	203.2	10.15	
Kinetin	K	215	10.25	
Zeatin	Z	219	10.95	
Auxins				
2,4-Dichlorophenoxyacetic acid	2,4-D	221	11.05	Dissolve in 2 to 5 ml of ethanol; gradually add water; heat slightly; make to volume; adjust to pH 5
2,4,5-Trichlorophenoxyacetic acid	2,4,5-T	255.5	12.75	
Naphthaleneacetic acid	NAA	186.2	9.3	
Indolebutyric acid[b]	IBA	203.2	10.65	
Indoleacetic acid[b]	IAA	175.2	8.25	
Miscellaneous				
Gibberellic acid[b]	GA	346.4	17.3	Dissolve in 2 to 5 ml of 0.2 M KOH; adjust to pH 5.0
Picloram (9)	P	241.2	12.0	
Abscisic acid[b]	ABA	264	13.2	

[a] Store in refrigerator.
[b] Filter sterilize.

other than sucrose or glucose have been produced. Most media contain *m*-inositol, which is beneficial.

Vitamins

Plants normally synthesize all vitamins. However, cells in culture require thiamine, and other vitamins are frequently added to optimize growth and development (Table 1).

Plant hormones

Both natural and synthetic plant hormones (Table 2) are required for most types of plant cell culture (41). A group of compounds called auxins induce cells to divide. The most effective are indoleacetic acid, 2,4-dichlorophenoxyacetic acid, and naphthaleneacetic acid. Others are listed in Table 2. The auxins also promote root initiation in cell cultures and plant tissues. The compounds most effective for root induction are indolebutyric acid, indoleacetic acid, and naphthaleneacetic acid.

The cytokinins are adenine derivatives. These compounds occur in cells of most organisms, but their growth regulatory roles have been observed only in plants. The compounds most frequently used are benzyladenine, kinetin, and zeatin. Cytokinins are often used jointly with auxins to initiate and promote cell division. In many plant tissues, they specifically induce the production of shoot primordia for plant regeneration.

The gibberellins are used for plant development from cultured meristems. The role of abscisic acid in cultured cells is not well defined, but it is known to have a beneficial effect in the production of embryos from somatic cells (4).

Organic nitrogen sources

Commonly used sources of organic nitrogen include protein hydrolysates (including Casamino Acids), glutamine, or amino acid mixtures. The inclusion of organic nitrogen is often beneficial in the early stages of culture initiation. L-Glutamine may be effective in replacing or supplementing a protein digest (Table 3).

Organic acids

The addition of pyruvate or tricarboxylic acid cycle intermediates such as citrate, succinate, or malate permits growth of plant cells on ammonium as the sole nitrogen source (13, 18). The cells tolerate up to at least 10 mM potassium salts. The tricarboxylic acid cycle intermediates also enhance growth of cells and protoplasts inoculated at low density (29).

Complex substances

A variety of extracts have been used, including yeast extract, malt extract, coconut milk, and fruit juices. These preparations are often replaceable by defined

TABLE 3. Concentrations of some complex additives, amino acids, and growth regulators for culture media

Ingredient	Concn range
Casamino Acids, vitamin free	150–500 mg/liter
Casein hydrolysate (NZ Amine A)[a]	0.2–2.0 g/liter
Coconut milk	1.0–15.0 ml/liter
L-Glutamine	150–1,000 mg/liter
Amino acids	0.1–1.0 mM each
Auxins	0.1–15 μM each
Cytokinins	1.0–50 μM each
Abscisic acid	0.1–5.0 μM
Gibberellic acid	0.1–5.0 μM

[a] Kraft Inc., Norwich, N.Y.

nutrients. In some instances the extracts have deleterious effects. Coconut milk is commonly used at 1 to 15 ml/liter (Table 3).

21.1.2 Environmental Factors

Several factors influence the growth of callus and cell suspension cultures. pH is an important variable. Plant cells in culture require an acidic pH, and an initial level of 5.5 to 5.8 is optimum. The pH varies during the growing cycle of a cell suspension culture. Initially, there is a decrease to below pH 5, which is followed by an increase that may reach 6 or higher.

The growth rate is closely related to the incubation temperature. Growth may occur at temperatures below 20°C, but temperatures between 26 and 28°C are considered optimum for achieving maximum growth, although some cell lines can grow at 32 to 33°C.

Plant cells in culture are aerobic, and aeration is essential for suspension cultures. Shakers are operated at rates which vary from 100 to 150 rpm.

Light is generally not essential for growing of callus or cell cultures. Although the cultures may grow equally well in the dark and light, light may have a profound effect on the metabolism of the cells. A "dark" room is frequently an area which is kept dark but lights are turned on for inspection of the cultures at the time of subculturing. In general, light is provided by white fluorescent lamps at intensities of ca. 300 to 10,000 lx (ca. 6 to 100 μM m^{-2} s^{-1}) at the culture level. The light is given continuously or in photoperiods of 12 to 18 h.

21.1.3 Facilities and Equipment

The facilities for culture of plant cells fall into two categories. One is a laboratory area for the aseptic handling of the cell and tissue materials. The principal features of this area are a space for preparation and sterilization of media, storage space for materials and reagents, and facilities for washing and cleaning of glassware. The equipment and apparatus are listed below. An essential feature of the laboratory is a facility for aseptic transfer and handling of the cell materials. This is usually accomplished in a laminar-flow sterile transfer cabinet. The second category of facility includes incubators, culture cabinets, or rooms. These can be reach-in or walk-in spaces with light, temperature, and humidity control. The latter feature is not essential. Detailed descriptions of facilities for plant tissue culture have been presented by Thorpe (52).

21.2 CALLUS AND CELL SUSPENSION CULTURE

The standard methods for obtaining a plant cell line begin with the production of a callus on nutrient agar. The cells of a sterile explant tissue placed on nutrient agar with hormones begin to proliferate. The induction time and rate of proliferation vary considerably, but visible growth normally occurs in 1 week. This cell mass is termed a callus because of analogy with the growth (callusing) occurring in wound healing of plants.

To obtain a cell suspension culture, callus is transferred to liquid medium and placed on a shaker. A proportion of the cells adapt to liquid shake culture and proliferate. The transition from agar to suspension culture and the establishment of a rapidly growing cell line requires 2 to 4 months, depending upon species and the cell line itself.

The cell line is usually maintained as a stock that is subcultured monthly on nutrient agar. Cell inoculum from the liquid suspension is removed and transferred to agar. The growth rate after the cell line is established as a culture is generally more rapid than that of the original callus. This is due to adaptation and inherent selection of cells with the shortest generation time.

The sections below describe some methods for production of plant cell cultures. Further details can be found in references 23, 43, 51, 52, and 54.

21.2.1 Facilities and Equipment

Cell manipulation and medium preparation

1. A laboratory area with a laminar-flow sterile cabinet. The cabinet often is operated continuously. Prefilters should be cleaned or replaced monthly.
2. Autoclave; oven for dry sterilizing; filter sterilizing equipment; microwave oven for melting agar
3. Provision for distilled or high-purity demineralized water
4. Flasks, jars with caps or stoppers, petri dishes
5. Pipettes with regular size and large bores
6. Clinical knives and forceps

Cell culturing

1. A walk-in room(s) and reach-in cabinets with controlled environment are standard facilities. They should be equipped with temperature control and have controlled artificial light consisting of fluorescent and some incandescent lamps. Humidity control is desirable but not essential.
2. Shelving should be constructed to allow maximum air movement.
3. Platform shakers of the gyratory models are most frequently employed.

In addition to what is specified above, the use of cell and tissue culture requires access to the usual medium preparation equipment and facilities for washing, drying, and storing glassware and apparatus. Also essential is access to balances, light microscopes, and small centrifuges for evaluation of cell growth and development.

21.2.2 Materials for Culture Media

1. Mineral salts, carbon sources, vitamins, and growth regulators should be the highest grade available. Growth regulators such as 2,4-dichlorophenoxyacetic acid and naphthaleneacetic acid may require purification by recrystallization and decolorization by adsorbent treatment in water-ethanol mixtures. Heat-labile compounds should be filter sterilized.
2. Only the L-isomers of amino acids should be used. The most effective compound is L-glutamine, which can also be used as the sole nitrogen source. L-Glutamine should be filter sterilized.

3. Protein hydrolysates are available as acid and enzymatic hydrolysates. The latter are usually used because no amino acids are destroyed during enzymatic hydrolysis.

4. Coconut milk from ripe or immature coconuts can be used. A hole is drilled through a germination pore. The liquid from several nuts is collected, heated to 80°C with stirring, filtered, and stored frozen.

5. Unpurified agar or purified (e.g., Noble) agar is generally used. Newer types such as SeaKem LE from the FMC Corp., Rockland, Maine, have proven effective (1).

6. Water should be distilled or demineralized to high purity. Glass-distilled water is most desirable.

21.2.3 Medium Preparation

Stock solutions are prepared as described below. The chemicals are dissolved in distilled or high-purity demineralized water. When the salts and other ingredients have been added and dissolved, the pH is adjusted by using 0.5 N HCl or 0.2 N NaOH. For convenience, a stock solution can be prepared of 5- or 10-fold-concentrated medium comprising, for example, the inorganic salts, vitamins, and sucrose. After the ingredients are dissolved, the solution is distributed in Whirl-Pak bags (100 ml into 6-oz. [ca. 178-ml] bags or 400 ml into 18-oz. [ca. 535-ml] bags). The bags are stored frozen. One 100-ml bag is used when needed to prepare 500 ml or 1 liter of complete medium.

The medium is distributed in containers before autoclaving, or the containers and medium are autoclaved separately. The medium is autoclaved at 120°C for 15 to 20 min and subsequently removed and cooled as soon as possible. Agar media are usually sterilized in lots of 500 ml and subsequently poured into sterile containers. Media should be stored at about 10°C.

Below is a procedure for medium preparation which can generally be applied to most media. Examples are given for B5 (see Gamborg and Shyluk in reference 52) and MS media. The composition of the media is listed in Table 1. A list of culture medium ingredients and their respective molecular weights is presented in Appendix A.

Stock solutions

B5-micronutrients (store frozen); in 100 ml

$MnSO_4 \cdot H_2O$ . . . . 1,000 mg
H_3BO_3 . . . . 300 mg
$ZnSO_4 \cdot 7H_2O$ . . . . 200 mg
$Na_2MoO_4 \cdot 2H_2O$ . . . . 25 mg
$CuSO_4 \cdot 5H_2O$ . . . . 2.5 mg
$CoCl_2 \cdot 6H_2O$ . . . . 2.5 mg

Vitamins for B5 medium (store frozen); in 100 ml

Nicotinic acid . . . . 100 mg
Thiamine hydrochloride . . . . 1,000 mg
Pyridoxine hydrochloride . . . . 100 mg
myo-Inositol . . . . 10,000 mg

Calcium chloride (B5); in 100 ml

$CaCl_2 \cdot 2H_2O$ . . . . 15 g

Using a solution of the calcium salt minimizes the danger of producing a precipitate with phosphate, which can occur when the salt is added directly to the medium.

Potassium iodide; in 100 ml

KI. . . . . 75 mg

Store in an amber bottle in a refrigerator.

EDTA sodium ferric salt. The sodium ferric salt of EDTA is available commercially. A stock solution may also be prepared by dissolving 7.45 g of disodium EDTA and 5.57 g of $FeSO_4 \cdot 7H_2O$ in 1 liter; 5 ml is then added per liter of medium.

MS-micronutrients (store frozen); per 100 ml

H_3BO_3 . . . . 620 mg
$MnSO_4 \cdot 4H_2O$ . . . . 2,230 mg
$ZnSO_4 \cdot 7H_2O$ . . . . 860 mg
$Na_2MoO_4 \cdot 2H_2O$ . . . . 25 mg
$CuSO_4 \cdot 5H_2O$ . . . . 2.5 mg
$CoCl_2 \cdot 6H_2O$ . . . . 2.5 mg

Growth hormones. See Table 2 for preparation.

Preparation of B5 medium

The ingredients should be added and dissolved in the order listed; per liter:

$NaH_2PO_4 \cdot H_2O$. . . . . 150 mg
KNO_3 . . . . 2,500 mg
$(NH_4)_2SO_4$. . . . . 134 mg
$MgSO_4 \cdot 7H_2O$ . . . . 250 mg
$CaCl_2 \cdot 2H_2O$ (stock solution) . . . . 1.0 ml
Micronutrients (B5 stock solution) . . . . 1.0 ml
Potassium iodide (stock solution) . . . . 1.0 ml
EDTA sodium ferric salt . . . . 40 mg
Vitamins (stock solution). . . . . 1.0 ml
Sucrose . . . . 20 g

Adjust pH to 5.5 with 0.2 N NaOH or 0.5 N HCl.

Preparation of MS medium (per liter)

NH_4NO_3 . . . . 1,650 mg
KNO_3 . . . . 1,900 mg
$MgSO_4 \cdot 7H_2O$ . . . . 370 mg
KH_2PO_4. . . . . 170 mg
$CaCl_2 \cdot 2H_2O$ (B5 stock solution). . . . . 2.9 ml
MS-micronutrients (MS stock solution) . . . . 1.0 ml
KI (B5 stock solution) . . . . 1.0 ml
EDTA sodium ferric salt . . . . 40 mg
Vitamins (B5 stock solution). . . . . 1.0 ml
Sucrose . . . . 30 g

Adjust pH to 5.8.

The heat-stable growth hormones are added, the pH is adjusted, and then the medium is autoclaved. The solution of heat-labile compounds is adjusted to the appropriate pH before filter sterilizing (see Table 2) and added to the medium after autoclaving. For agar media, the agar is dissolved in a microwave oven before autoclaving.

21.2.4 Explant Materials

The starting material for tissue and cell cultures can be taken from a seedling or mature tissue. Although most tissues can be induced to produce a callus and a cell suspension culture, juvenile tissues are most likely to succeed. Specific sources are hypocotyls, cotyledons, immature embryos, swelling buds, root tips, stem disks, leaf sections, and basal sections of bulb scales.

Sterile seedling production

Seed sterilization. Surface sterilization can be achieved by exposing the seeds to chlorine gas. The seeds in a container are put into a desiccator placed in a fume hood. Chlorine gas is produced in the desiccator by adding 3 ml of concentrated HCl to a beaker containing 100 ml of a commercial bleach solution. After 6 to 8 h the seeds are transferred aseptically to sterile water for 2 h, washed, and placed on nutrient agar. The duration of treatment must be determined for each seed source.

An alternative method consists of placing the seeds in a flask, covering them with 70% ethanol, and agitating the flask on a shaker for 2 to 5 min. The alcohol is replaced with 20% commercial bleach (Clorox or Javex) or 5% sodium hypochlorite solution, and agitation on the shaker is continued for 15 to 25 min. The seeds are rinsed three to four times in sterile, distilled water and placed on nutrient agar.

Germination. Seeds can be germinated aseptically on moist cotton or cheesecloth in glass jars with screw caps or sealed with Parafilm. A recommended procedure is to germinate the seeds on a nutrient agar in jars. Germination on nutrient agar provides better uniformity of conditions and ready detection of contamination. It is advantageous to use only a few seeds per container since contamination could spread to all the seeds.

The seeds are usually germinated in the dark at 25 to 28°C. A suitable nutrient agar for germination consists of 0.65% agar with the following nutrients (mg/liter): NH_4NO_3, 100; KNO_3, 420; KH_2PO_4, 60; $MgSO_4 \cdot 7H_2O$, 90; $CaCl_2 \cdot 2H_2O$, 50; EDTA sodium ferric salt, 10; B5-micronutrients, 1 ml; sucrose, 10 g; agar, 6.5 g. The pH is adjusted to 5.8 before autoclaving.

Sterilization of plant materials. The 70% ethanol and bleach procedure described above is used with a shorter time exposure. Adding a few drops of wetting agent such as Tween 80 to the bleach solution is beneficial.

21.2.5 Starting Callus and Cell Cultures

1. Place three to five explant sections of 5 to 8 mm each from sterile tissues onto nutrient agar (0.6 to 1.0%, wt/vol). The medium should contain 2,4-dichlorophenoxyacetic acid or indoleacetic acid at 0.5 to 1.0 mg/liter and a cytokinin at 0.1 to 0.25 mg/liter. Several basic media should be tested unless previous research has established the optimum composition. Suitable containers are glass or plastic jars and petri dishes.
2. Incubate the sections in the darkness or in low light at 26 to 28°C.
3. After 3 to 4 weeks, or when the callus has reached a size 10 times that of the explant, the callus can be subcultured.
4. A cell culture is started by transferring callus into liquid medium in a flask. The ratio of callus mass to liquid should be about 1:10 or less; using a 50-ml flask with 5 to 10 ml of medium may be convenient. The callus preferably should be friable, thus separating easily. The separation may be aided mechanically by using a sterile spatula at the time of transfer. The flasks are placed on a shaker at 150 rpm at 26°C.

In the initial stages, it may be necessary to remove a portion of the spent medium at 2-week intervals by pipetting and replacing the volume with fresh medium. A cell suspension should form within 4 to 6 weeks. Initially, it is advisable to use a dilution ratio of 1:1 to 1:4 of inoculum to fresh medium. Eventually, the cell culture should attain a doubling time of 24 to 36 h and consist of a mixture of single cells and small multicellular clusters.

21.3 PLANT PROTOPLASTS: ISOLATION AND CULTURE

Procedures have been devised for the isolation of protoplasts from many plant species. This generally involves the use of enzymes which degrade the cell walls under conditions which preserve the viability of the protoplasts. Commercial preparations of cellulases and pectinases for protoplast isolation are available with different degrees of purity and effectiveness. These procedures have not become standardized. The methods described below have been used successfully, but each tissue may have particular requirements for achieving maximum yields of viable protoplasts (19, 22, 33). The aim is often to produce viable, high-quality protoplasts which can be regenerated into cells and will undergo sustained division at the highest possible frequency.

Leaf tissues are frequently used as sources of protoplasts. The yields, viability, and quality are influenced significantly by the age of the leaf as well as the environmental conditions (i.e., light, temperature, soil fertility, and humidity) under which the plants are grown. Since the physiological state of the tissue is critical, the plants are usually grown in environmental growth chambers or greenhouses where light and temperature can be carefully controlled. With respect to light intensity, a range of 0.3 to 5 $\mu M\ m^{-2}\ s^{-2}$ has proven suitable, although in special cases more intense illumination can be beneficial. A photoperiod of 18:6 or 16:8 h of light/dark is the most common. In some cases, the yield of protoplasts has been greatly improved by placing the tissue in the dark for 24 to 72 h before isolation, during which time starch degradation occurs. The disappearance of starch can improve the yield of protoplasts. The choice of temperature and humidity is generally dictated by the requirements to ensure vigorous plant growth. Maintaining the plants at a high level of fertility, particularly by using nitrogen fertilizers, can be beneficial. In addition to leaf tissue, other plant materials can be used, including shoot tips, cotyledons, flower petals, and microspores. Moreover, in vitro-grown shoots are a convenient and effective source of leaf tissue for protoplasts. Cell suspension cultures are frequently used as sources of protoplasts. For best production of protoplasts, cell cultures should be subcultured at 3- to 4-day intervals to maintain maximum growth rate and uniformity of the cell population. Cells taken at the early log phase are generally the most suitable. In established cultures, this corresponds to ca. 2 days after subculturing. Cultures which have been newly established in liquid suspension often consist of mixtures of cell sizes and ages as well as a proportion of dead cells and thus yield a low percentage of protoplasts.

21.3.1 Equipment and Materials

Most of the facilities and equipment required for protoplasts are identical to those for plant cell culture. The operations are performed under aseptic conditions.

The resource materials needed specifically for protoplasts include the following:

1. Conical test tubes, graduated, 15 ml and 40 ml with caps or capped with aluminum foil
2. Miller disposable filter units and 0.45- and 0.20-μm membranes
3. Plastipak or other disposable plastic syringes
4. Pasteur pipettes with rubber bulbs, of different bore sizes
5. Stainless-steel mesh of 44, 65, and 88 μm
6. Percoll (Pharmacia)
7. Prep agarose (Gelrite; FMC Corp.)
8. Enzymes:
   - ONOZUKA R10 (Yakult Honsha Co., Japan)
   - Cellulysin (Calbiochem-Behring, San Diego, Calif.)
   - Pectolyase Y23 (Seishin Pharmaceutical Co., Japan)
   - Macerase (Yakult Honsha)
   - Pectinase (Sigma Chemical Co., St. Louis, Mo.)

21.3.2 Solutions

Isolation solution	*(per liter)*
Sorbitol (0.15 M)	27.3 g
Mannitol (0.15 M)	27.3 g
Glucose (0.10 M)	18.0 g
$CaCl_2 \cdot 2H_2O$ (6.0 mM)	441 mg
KH_2PO_4 (0.7 mM)	95 mg
MES buffer (3.0 mM)	650 mg
Enzymes (0.1 to 2.0%)	

Adjust to pH 5.6 and filter sterilize.

Enzyme solutions. Enzymes obtained commercially are stable for a year or more when stored dry below 0°C. In the past, the preparations often contained salts and diatomaceous earth and desalting was necessary. The preparations now available are generally of high purity and can be used directly. The enzymes may be dissolved in water and stored frozen for up to 3 months. A convenient procedure is to make concentrated solutions (10×) and store frozen in smaller lots to limit the frequency of thawing and refreezing.

When needed, the enzyme solution is mixed with an appropriate volume of isolation medium and filter sterilized immediately before incubation with the plant materials.

Protoplast culture medium. See Table 4. After the pH is adjusted, the media are filter sterilized and stored in a refrigerator.

21.3.3 Isolation and Culture Procedures

IMPORTANT: All operations should be performed aseptically.

From leaf materials

1. If leaf material is not already sterile, immerse in 70% ethanol for ca. 0.5 to 2.0 min, transfer to a petri dish (15 by 100 mm) containing 10 to 20% Clorox bleach and Tween 80 (1 drop per 25 ml), and soak for 5 min.
2. Rinse leaves with distilled water (three times) or with a sorbitol solution at the same osmolality as the enzyme solution to be used.
3. Remove the lower epidermis by peeling it off with a curved pair of fine surgical forceps, cut off epidermis-free sections of leaf (ca. 1 cm^2), and incubate in a 1:1 mixture of isolation enzyme solution and protoplast culture medium in a petri dish (100 by 15 mm). Alternatively, cut leaves into fine strips, parallel to veins. Use basal portion of leaf.
4. Seal the dish with Parafilm and incubate for 4 to 16 h at 22 to 24°C with slow shaking (50 rpm) on a gyratory shaker, or agitate gently by hand at hourly intervals.
5. Observe the release of protoplasts with an inverted microscope.
6. Remove protoplast suspension from the petri dish with a Pasteur pipette and pass through a 60- to 80-μm-mesh stainless-steel sieve.
7. Collect filtrate in a centrifuge tube and cover with Parafilm or aluminum foil.
8. Centrifuge at 100 × *g* for 5 min.
9. Remove the supernatant fluid with a Pasteur pipette.
10. Resuspend the protoplasts in 5 ml of isolation solution (with no enzyme).
11. Pipette 5 ml of a 20 to 40% Percoll solution to the bottom of a 40-ml centrifuge tube (40 ml graduated).
12. Centrifuge at 1,000 rpm for 10 min. The protoplasts appear as a band at the solution interphase.
13. Repeat washing if necessary.
14. Resuspend the protoplasts in 1 ml of protoplast culture medium at ca. 10^5 protoplasts per ml as determined by a hemacytometer.
15. For culturing, distribute the protoplast suspension in ca. 100-μl droplets in petri dishes (60 by 15 mm), six to eight droplets per dish.
16. Seal petri dishes with Parafilm.
17. Incubate in diffused light at 25°C using a covered plastic box humidified by moist blotting paper in a beaker with 1% $CuSO_4$ solution.

From cultured suspension cells

1. Mix equal volumes of cell suspension and isolation enzyme solution in a petri dish (60 by 15 mm).
2. Seal the dish with Parafilm and incubate at 25°C for 6 to 16 h with slow shaking (50 rpm) on a gyratory shaker.
3. Follow the same procedure (step 6 to step 17) as for leaf protoplasts (filtering, washing, and culturing).

After 1 to 2 days of culturing, the cell walls begin to form, and after 4 to 5 days, division should occur. Add new culture medium after about 2 weeks. Alternative-

TABLE 4. Culture media for plant protoplasts[a]

Ingredient	Concn		
	(6)[b]	(15)[b]	(40)[b]
KNO_3	2,500	2,100	2,500
NH_4NO_3		600	250
$(NH_4)_2SO_4$	134		134
$MgSO_4 \cdot 7H_2O$	250	400	250
KH_2PO_4		250	
$NaH_2PO_4 \cdot H_2O$	150		150
$CaCl_2 \cdot 2H_2O$	885	600	900
Micronutrients	—[c]	—[c]	—[c]
Disodium EDTA			27.8
$FeSO_4 \cdot 7H_2O$			37.3
EDTA sodium ferric salt	40	40	
Sucrose		150	137,000
Glucose	81,000	36,000	
Sorbitol		27,300	
Ribose	500	125	
Xylose	500		250
Other carbohydrates[d]			
myo-Inositol	100	1,250	100
Thiamine hydrochloride	10	10	10
Pyridoxine hydrochloride	1	1	1
Nicotinic acid	1	1	1
Ascorbic acid[e]	2.0	1.0	
Biotin	0.01	0.1	
Calcium panthothenate	1.0	0.5	
Choline chloride	1.0	0.5	
Folic acid	0.4	0.01	
Riboflavin	0.2	0.2	
p-Aminobenzoic acid	0.02	0.01	
Casamino Acids (vitamin free)	250	125	
L-Glutamine		150	
L-Asparagine	300		
L-Arginine	210		
Sodium pyruvate		5	
Sodium citrate		10	
Coconut milk	1%		
2,4-Dichlorophenoxyacetic acid	1.0	0.22	0.1
2,4,5-Trichlorophenoxyacetic acid	0.5		
Naphthaleneacetic acid			1.0
Picloram		0.06	
Benzyladenine		0.1	0.2
pH	5.8	5.7	5.6

[a] Media for culturing protoplasts to regenerate cells. As cells reform and divide, dilute the medium with cell culture medium. When plant regeneration is desired, an MS medium is frequently used (see section 21.4 and Table 5). Concentrations are milligrams per liter unless otherwise indicated.

[b] References.

[c] See B5 medium, section 21.2.3.

[d] See Kao (29).

[e] For a convenient stock solution of multivitamins, dissolve the following in 100 ml: 50 mg of calcium pantothenate, 10 mg of ascorbic acid, 5 mg of choline chloride, 1 mg of *p*-aminobenzoic acid, 20 mg of folic acid, 10 mg of riboflavin, and 1.0 mg of biotin. Use 1 ml/liter.

ly, protoplasts can be cultured by embedding in agar either immediately after isolation or after cell regeneration when some division has occurred within 72 to 96 h. The procedures for the further cultivation and growth of the cells for plant regeneration vary with the species (Table 5; 5, 21, 56).

21.4 PLANT REGENERATION

The procedures for plant regeneration from a tissue explant or cells or from protoplasts in culture vary somewhat with the species and the process by which morphogenesis occurs. Some plants exhibit organogenesis, in which shoot primordia and, subsequently, a shoot are formed. Shoot formation is followed by root development at the base of the shoot. The root initiation may require the use of an auxin hormone. In other species, the cells grow into complete plants via embryogenesis. The cells of such plants grow in culture, and as they divide they form embryos which are almost identical to the development of zygotic embryos in the seed. These embryos grow into complete plants. The procedures described below can be applied to many tissues. Table 5 provides a list of references describing methods for particular species (see also references 12, 32, and 45).

The required facilities and the materials and nutrients are nearly the same as described for cell and tissue culture. Media and their preparation have also been described (see section 21.2.3).

21.4.1 Organogenesis

Media

1. MS salts with B5 vitamins and 3% sucrose should be the first choice. See Table 5 for specific plant species.

2. Use growth regulator concentrations known to induce differentiation. If these are not known, prepare and evaluate media with combinations of auxins and cytokinins in different ratios in the concentration ranges of 0, 0.1, 1.0, 2.5, 5.0, and 10.0 μM. The compounds to test first are benzyladenine with naphthaleneacetic acid or indoleacetic acid (30, 31).

Environmental conditions

The incubation area requires controlled light intensity, temperature, and light/dark periods. The following are guidelines: (i) a temperature of 15 to 26°C, adjusted to be 4 to 8°C higher in the light than in the dark; (ii) a light/dark cycle of 14:10, 16:8, or 18:6 h; (iii) a light intensity of 2,000 to 10,000 lx (40 to 200 μM m^{-2} s^{-1}).

Method

1. Prepare sterile tissue for explants as described for tissue culture (section 21.2.4).

2. Dissect the tissue in 0.5-mm^2 sections and place on agar medium.

3. Dissect regenerated shoots and place on agar medium for rooting.

4. Transfer plantlets to small pots or Jiffy peat pellets. Care should be taken to maintain a high humidity by using a plastic cover or a mist bed. The period of cover is gradually reduced over the first week. Water the plantlets with Hoagland nutrient solution (Table 6).

5. Grow the plants to maturity in a greenhouse or a growth room.

TABLE 5. References to methods for plant regeneration in culture

Plant	Tissue or cells	Process[a]	Basic medium[b]	Reference
Nicotiana sp. (tobacco)	Stem	O	MS	39
Petunia spp.	Leaf	O	MS	5
Solanum tuberosum (potato)	Leaf	O	MS	47
S. tuberosum (potato)	Protoplasts	O	MS	50, 51
Lycopersicon esculentum (tomato)	Protoplasts	O	MS	34
L. esculentum (tomato)	Leaf	O	MS	31
Solanum melongena (eggplant)	Cell culture	E	MS	25
Pimpinella anisum	Cell culture	E	B5	28
Daucus carota (carrot)	Cell culture	E	B5	24
Trifolium pratense (red clover)	Callus	E	SL	44
T. repens (clover)	Leaf	O	B5	27
Medicago sativa (alfalfa)	Callus	E	SH	55
Stylosanthes guianensis	Callus	O	MS	38
Glycine canescens	Callus	O	B5	57
Brassica napus (rapeseed)	Stem	O	MS	33
B. napus	Anther (pollen)	E	B5/MS	32
Lactuca sativa (lettuce)	Protoplasts	O	MS	11
Citrus spp.	Callus	E	MS	53
Zea mays (corn)	Immature embryos	O	MS	26
Z. mays	Callus	E	MS	37
Sorghum bicolor (sorghum)	Callus	O	MS	20
Triticum aestivum (wheat)	Callus	O	MS	2
Saccharum officinalis (sugarcane)	Callus	E	MS	3

[a] O, Organogenesis; E, embryogenesis.
[b] MS, Murashige and Skoog (39); B5, Gamborg et al. (16); SH, Schenk and Hildebrandt (48); SL, Collins and Phillips (8).

21.4.2 Embryogenesis

The initiation of a callus or cell culture follows the same procedure and conditions as described for tissue culture (section 21.2.5). There are no specific medium compositions, growth regulators, or set of environmental conditions which will ensure the induction of embryogenesis in cultured cells. Table 5 makes reference to specific examples.

Carrot embryogenesis

1. Place several 0.5-mm sections of seedlings (sterile) on solid B5 medium containing 0.1 mg of 2,4-dichlorophenoxyacetic acid per liter and 0.6% agar.
2. Incubate the sections at 26 to 28°C.
3. After 4 to 5 weeks, transfer the friable callus to 5 to 10 ml of liquid medium of the same composition in a 50-ml flask.
4. Incubate on a shaker at 26°C in light. When the cell mass has increased, subculture on the same medium.
5. For embryogenesis, collect and wash cells from 5 to 10 ml of a culture in hormone-free medium.
6. Plate the cells in hormone-free agar medium and incubate at 25°C in a 16:8-h photoperiod chamber.
7. Transfer the small plantlets to pots and proceed as described for organogenesis.

TABLE 6. Hoagland nutrient solution[a]

Ingredients	Concn g/liter	Final molar concn
Macronutrients		
$Ca(NO_3)_2 \cdot 4H_2O$	0.94	4.0 mM
$MgSO_4 \cdot 7H_2O$	0.52	2.0 mM
KNO_3	0.66	6.0 mM
$NH_4H_2PO_4$	0.12	1.0 mM
Sequestrene 330 Fe[b]	0.07	
Micronutrients		
H_3BO_3	28	45 μM
$MnSO_4 \cdot H_2O$	34	20 μM
$CuSO_4 \cdot 5H_2O$	1.0	0.4 μM
$ZnSO_4 \cdot 7H_2O$	2.2	0.7 μM
$(NH_4)_6Mo_7O_{24} \cdot 4H_2O$	1.0	0.2 μM
H_2SO_4 (concentrated)	5 ml	

[a] A 0.1-ml volume of the micronutrient solution is mixed with 1 liter of the macronutrients, and the pH is adjusted to 6.7.
[b] Geigy Agricultural Chemical Corp., Ardsley, N.Y.

21.5 APPENDIX A: MOLECULAR WEIGHT

Mineral salts

Salt	Mol wt
H_3BO_3	61.84
$CaCl_2 \cdot 2H_2O$	147.02
$Ca(NO_3)_2 \cdot 4H_2O$	236.15
$CoCl_2 \cdot 6H_2O$	237.93
$CuSO_4 \cdot 5H_2O$	249.68
EDTA sodium ferric salt (13% Fe)	366.85
KCl	74.56
KH_2PO_4	136.09
KI	166.01
KNO_3	101.1
KOH	56.1
K_2SO_4	174.1
$MgSO_4 \cdot 7H_2O$	246.5
$MnSO_4 \cdot H_2O$	169.01
$MnSO_4 \cdot 4H_2O$	223.09
Conversion $H_2O/4H_2O$ = 0.76; $4H_2O/H_2O$ = 1.32	
NaCl	58.44
$NaH_2PO_4 \cdot H_2O$	137.98
NaOH	40.01
$Na_2MoO_4 \cdot 2H_2O$	241.95
Na_2SO_4	142.06
NH_4Cl	53.49
NH_4NO_3	80.09
$(NH_4)H_2PO_4$	115.03
$(NH_4)_2SO_4$	132.14
$ZnSO_4 \cdot 7H_2O$	287.55

Amino acids

Ala	89.09
Arg	174.2
Asn	132.12
Asp	133.1
Cys	121.16
Gln	146.2
Glu	147.13
Gly	75.1
His	155.16
Ile	131.17
Leu	131.17
Lys	146.19
Met	149.21
Phe	165.19
Pro	115.13
Ser	105.09
Thr	119.12
Trp	204.22
Tyr	181.19
Val	117.15

Sugars

Galactose	180.16
Glucose	180.16
Fructose	180.16
Mannitol	182.17
Sorbitol	182.17
Sucrose	342.30
Xylose	150.13
Ribose	150.13

Miscellaneous

Disodium citrate	258.08
Disodium succinate	162.2
Sodium malate	156.0
Sodium pyruvate	110.0
Morpholineethanesulfonic acid	213.0
Adenine sulfate	184.2
Urea	60.1

Vitamins

myo-Inositol	180.16
Nicotinic acid (niacin)	123.11
Pyridoxine hydrochloride	205.64
Thiamine hydrochloride	337.28
Calcium pantothenate	476.53
Ascorbic acid	176.12
Nicotinamide (niacinamide)	122.12
Choline chloride	139.63
p-Aminobenzoic acid	137.13
Folic acid	441.4
Riboflavin	576.4
Biotin	244.3

21.6 ACKNOWLEDGMENT

I thank Wilfred A. Keller for his critical review of this manuscript.

21.7 LITERATURE CITED

1. **Adams, T. L., and J. A. Townsend.** 1983. A new procedure for increasing efficiency of protoplast plating and clone selection. Plant Cell Rep. **2:**165–168.
2. **Ahloowalia, B. S.** 1982. Plant regeneration from callus culture in wheat. Crop Sci. **22:**405–410.
3. **Ahloowalia, B. S., and A. Maretzki.** 1983. Plant regeneration via somatic embryogenesis in sugarcane. Plant Cell Rep. **2:**21–25.
4. **Ammirato, P. V.** 1974. The effects of abscisic acid on the development of somatic embryos from cells of caraway (*Carum carvi* L). Bot. Gaz. **135:**328–337.
5. **Ausubel, F. M., K. Behnsen, M. Hanson, A. Mitchell, and H. J. Smith.** 1980. Cell and tissue culture of haploid and diploid *Petunia*. Plant Mol. Biol. Newsl. **1:**26–32.
6. **Brar, D. S., S. Rambold, F. Constabel, and O. L. Gamborg.** 1980. Isolation, fusion and culture of *Sorghum* and corn protoplasts. Z. Pflanzenphysiol. **96:**269–275.
7. **Chu, C.-C.** 1978. The N6 medium and its applications to anther culture of cereal crops, p. 43–50. *In* Proceedings of the Symposium on Plant Tissue Culture. Science Press, Beijing, China.
8. **Collins, G. B., and G. C. Phillips.** 1982. *In vitro* tissue culture and plant regeneration in *Trifolium pratense* L, p. 22–34. *In* E. D. Earle and Y. Demarly (ed.), Variability in plants regenerated from tissue culture. Praeger Publishers, New York.
9. **Collins, G. B., W. E. Vian, and G. C. Phillips.** 1978. Use of 4-amino-3,5,6-trichloropicolinic acid as an auxin source in plant tissue cultures. Crop Sci. **18:**286–288.
10. **Conger, B. V. (ed.).** 1981. Cloning agricultural plants via *in vitro* techniques. CRC Press, Boca Raton, Fla.
11. **Engler, D. E., and R. G. Grogan.** 1982. Isolation, culture and regeneration of lettuce leaf mesophyll protoplasts. Plant Sci. Lett. **28:**223–229.
12. **Evans, D. A., W. R. Sharp, and C. E. Flick.** 1981. Plant regeneration from cell cultures. Hort. Rev. **3:**214–314.
13. **Gamborg, O. L.** 1970. The effects of amino acids and ammonium on the growth of plant cells in suspension culture. Plant Physiol. **45:**372–375.
14. **Gamborg, O. L., B. D. Davis, and R. W. Stahlhut.** 1983. Somatic embryogenesis in cell cultures of *Glycine* species. Plant Cell Rep. **2:**209–212.
15. **Gamborg, O. L., B. D. Davis, and R. W. Stahlhut.** 1983. Cell division and differentiation in protoplasts from cell cultures of *Glycine* species and leaf tissue of soybean. Plant Cell Rep. **2:**213–215.
16. **Gamborg, O. L., R. A. Miller, and K. Ojima.** 1968. Nutrient requirements of suspension cultures of soybean root cells. Exp. Cell Res. **50:**151–158.
17. **Gamborg, O. L., T. Murashige, T. A. Thorpe, and I. K. Vasil.** 1976. Plant tissue culture media. In Vitro **12:**473–478.
18. **Gamborg, O. L., and J. P. Shyluk.** 1970. The culture of plant cells using ammonium salts as the sole nitrogen source. Plant Physiol. **45:**598–600.
19. **Gamborg, O. L., and J. P. Shyluk.** 1976. Tissue culture, protoplasts and morphogenesis in flax. Bot. Gaz. **137:**301–306.
20. **Gamborg, O. L., J. P. Shyluk, D. S. Brar, and F. Constabel.** 1977. Morphogenesis and plant regeneration from callus of immature embryos of sorghum. Plant Sci. Lett. **10:**67–74.
21. **Gamborg, O. L., J. P. Shyluk, L. C. Fowke, L. R. Wetter, and D. Evans.** 1979. Plant regeneration from protoplasts and cell cultures of *Nicotiana tabacum* sulfur mutant (su/su). **Z. Pflanzenphysiol. 95**(Suppl.):255–264.
22. **Gamborg, O. L., J. P. Shyluk, and E. A. Shahin.** 1981. Isolation, fusion and culture of plant-protoplasts, p. 115–153. *In* T. Thorpe (ed.), Plant tissue culture methods and application in agriculture. Academic Press, Inc., New York.

23. **Gamborg, O. L., and L. R. Wetter.** 1975. Plant tissue culture methods. National Research Council of Canada, Saskatoon, Saskatchewan, Canada.
24. **Giuliano, G., D. Rosellini, and M. Terzi.** 1983. A new method for the purification of the different stages of carrot embryoids. Plant Cell Rep. **2**:216–218.
25. **Gleddie, S., W. Keller, and G. Setterfield.** 1983. Somatic embryogenesis and plant regeneration from leaf explants and cell suspensions of *Solanum melongena* (eggplant). Can. J. Bot. **61**:656–666.
26. **Green, C. E., R. L. Phillips, and R. A. Kleese.** 1974. Tissue cultures of maize (*Zea mays* L.): initiation, maintenance, and organic growth factors. Crop Sci. **14**:54–58.
27. **Gresshoff, P. M.** 1980. *In vitro* culture of white clover: callus, suspension, protoplast culture, and plant regeneration. Bot. Gaz. **141**:157–164.
28. **Huber, J., F. Constabel, and O. L. Gamborg.** 1978. A cell counting procedure applied to embryogenesis in cell suspension culture of anise (*Pimpinella anisum* L.). Plant Sci. Lett. **12**:209–215.
29. **Kao, K. N.** 1977. Chromosomal behavior in somatic hybrids of soybean-*Nicotiana glauca*. Mol. Gen. Genet. **150**:225–230.
30. **Kartha, K. K., O. L. Gamborg, J. P. Shyluk, and F. Constabel.** 1974. *In vitro* plant formation from stem explants of rape (*Brassica napus* cv. Zephyr). Physiol. Plant. **31**:217–220.
31. **Kartha, K. K., O. L. Gamborg, J. P. Shyluk, and F. Constabel.** 1976. Morphogenetic investigations in *in vitro* leaf culture of tomato (*Lycopersicon esculentum* Mill. cv. Starfire) and high frequency plant regeneration. Z. Pflanzenphysiol. **77**:292–301.
32. **Keller, W. A., and K. C. Armstrong.** 1978. High-frequency production of microspore-derived plants from *Brassica napus* anther cultures. Z. Pflanzenzucht **80**:100–108.
33. **Keller, W. A., G. Setterfield, G. Douglas, S. Gleddie, and C. Nakamura.** 1982. Production, characterization and utilization of somatic hybrids of higher plants, p. 81–114. *In* D. T. Tomes et al. (ed.), Application of plant cell and tissue culture to agriculture and industry. University of Guelph, Guelph, Canada.
34. **Koblitz, H., and D. Koblitz.** 1982. Experiments on tissue culture in the genus *Lycopersicon* mill. Mesophyll protoplast regeneration to plants in *Lycopersicon esculentum* cv. 'Nadja.' Plant Cell Rep. **1**:143–146.
35. **Larkin, P. J., and W. R. Scowcroft.** 1981. Somaclonal variation—a novel source of variability from cell cultures for plant improvement. Theor. Appl. Genet. **60**:197–214.
36. **Linsmaier, E. M., and F. Skoog.** 1965. Organic growth factor requirements of tobacco tissue cultures. Physiol. Plant. **18**:100–127.
37. **Lu, C., V. Vasil, and I. K. Vasil.** 1983. Improved efficiency of somatic embryogenesis and plant regeneration in tissue cultures of maize (*Zea mays* L.). Theor. Appl. Genet. **66**:285–289.
38. **Mroginski, L. A., and K. K. Kartha.** 1981. Regeneration of plants from callus tissue of the forage legume *Stylosanthes guianensis*. Plant Sci. Lett. **23**:245–251.
39. **Murashige, T., and F. Skoog.** 1962. A revised medium for rapid growth and bioassay with tobacco tissue cultures. Physiol. Plant. **15**:431–497.
40. **Nagy, I., and P. Maliga.** 1976. Callus induction and plant regeneration from mesophyll protoplasts of *Nicotiana sylvestris*. Z. Pflanzenphysiol. **78**:453–455.
41. **Nickell, L. G. (ed.).** 1982. Plant growth regulators. Springer-Verlag, New York.
42. **Nitsch, J. P., and C. Nitsch.** 1969. Haploid plants from pollen grains. Science **163**:85–87.
43. **Ohira, K., K. Ojima, and A. Fujiwara.** 1973. Studies on the nutrition of rice cell culture. I. A simple, defined medium for rapid growth in suspension culture. Plant Cell Physiol. **14**:1113–1121.
44. **Phillips, G. C., and G. B. Collins.** 1980. Somatic embryogenesis from cell suspension cultures of red clover. Crop Sci. **20**:323–326.
45. **Rao, A. N. (ed.).** 1982. Tissue culture of economically important plants. Proceedings of an International Symposium in Singapore. COSTED and ANBS, 1982. National University of Singapore, Singapore.
46. **Rechcigl, M.** 1977. CRC handbook series in nutrition and food, sect. G, vol. IV. CRC Press Inc., Boca Raton, Fla.
47. **Roest, S., and G. S. Bokelmann.** 1980. *In vitro* adventitious bud techniques for vegetative propagation and mutation breeding of potato (*Solanum tuberosum* L.). 1. Vegetative propagation *in vitro* through adventitious shoot formation. Potato Res. **23**:167–181.
48. **Schenk, R. U., and A. C. Hildebrandt.** 1972. Medium and techniques for induction and growth of monocotyledonous and dicotyledonous plant cell cultures. Can. J. Bot. **50**:199–204.
49. **Street, H. E., and R. D. Shillito.** 1977. Nutrient media for plant organ, tissue and cell culture, p. 305–359. *In* M. Rechcigl (ed.), CRC handbook in nutrition and food, vol. IV. CRC Press Inc., Boca Raton, Fla.
50. **Thomas, E., S. W. J. Bright, J. Franklin, V. A. Lancaster, and B. J. Miflin.** 1982. Variation amongst protoplast-derived potato plants (*Solanum tuberosum* cv. 'Maris Bard'). Theor. Appl. Genet. **62**:65–68.
51. **Thomas, E., and M. R. Davey (ed.).** 1975. From single cells to plants. Wykeham Publications (London) Ltd., London.
52. **Thorpe, T. A. (ed.).** 1981. Plant tissue culture. Methods and applications in agriculture. Academic Press, Inc., New York.
53. **Vardi, A., and P. Spiegel-Roy.** 1982. Plant regeneration from *Citrus* protoplasts: variability in methodological requirements among cultivars and species. Theor. Appl. Genet. **62**:171–176.
54. **Vasil, I. K. (ed.).** 1984. Cell culture and somatic cell genetics in plants. Laboratory techniques. Academic Press, Inc., New York.
55. **Walker, K. A., and S. J. Sato.** 1981. Morphogenesis in callus tissue of *Medicago sativa*: the role of ammonium ion in somatic embryogenesis. Plant Cell Tissue Organ Culture **1**:109–121.
56. **Weber, G., and K. G. Lark.** 1979. An efficient plating system for rapid isolation of mutants from plant cell suspensions. Theor. Appl. Genet. **55**:81–86.
57. **Widholm, J. M., and S. Rick.** 1983. Shoot regeneration from *Glycine canescens* tissue cultures. Plant Cell Rep. **2**:19–20.

V. Biochemical Engineering

Biochemical Engineering: Introduction

BRUCE K. HAMILTON

This section is intended to facilitate the scale-up and cost estimation of fermentation-based processes. "Scale-Up of Fermentation," by Trilli, discusses the wide range of factors which can affect the performance of a fermentation as it is run at larger and larger volumes, and suggests approaches for achieving favorable scale-up results. Instrumentation and computer systems useful for fermentation development, optimization, and scale-up are described by Wang. How to go about putting together a general-purpose fermentation pilot plant for scale-up and other purposes is covered next by Hamilton et al. Since propagation of a pure culture at large scale is usually a key to successful operation, special attention is devoted to sterilization and prevention of contamination in the chapter by Bader. Finally, Kalk and Langlykke provide a compact guide to fermentation and biotechnology process cost estimation.

Many components must be integrated together in the course of commercializing a bioprocess. The adventure of conceiving a new bioprocess, developing a microorganism capable of carrying it out, creating an efficient operation in which to install that microorganism and recover its product at high yield and sufficient purity, and all the while targeting to meet marketing timetables and budget constraints is always stimulating and usually holds a number of surprises. In scale-up work there are periods of excitement and pressure, but the ultimate achievement of bringing a successful new bioprocess on-stream will make it all worthwhile.

Chapter 22

Scale-Up of Fermentations

ANTONIO TRILLI

Scale-up is the complex of techniques and methodologies employed to transfer to the production scale a fermentation process developed on a smaller scale. However, it is by no means a one-way procedure (i.e., from small to large), as in practice the complex nature of the processes involved often requires going back to a small scale to obtain additional information. Scale-down, therefore, is a basic part of any scale-up procedure (see section 22.2.3). In a wider sense, scale-up is the study and exploitation of the effects of fermentation scale on the various fermentation parameters, both physical and biological. Most of the research carried out in this field in the past has been prompted by the need to rationalize the behavior of aerobic fermentations on different scales of operation. These studies were carried out in agitated vessels designed to achieve the complete mixing of their contents, i.e., stirred tank reactors. The following discussion will be centered mainly upon aerobic fermentations carried out in this type of fermentor, as these processes can pose severe problems with respect to mass transfer and mixing. Scale-up is a highly interdisciplinary task, requiring the combined and integrated use of concepts and methodologies of both chemical engineering and microbial physiology. The quantitative understanding of the chemical engineering aspects of scale-up has proceeded more rapidly, due to the intrinsically simpler nature of physicochemical phenomena.

Such a situation is reflected in the relative abundance of literature reports concerning engineering aspects of scale-up. Actual scale-up problems cannot, however, ignore the complex responses of microorganisms, and a certain degree of empiricism has to be accepted when a complete quantitative understanding of the particular problem is not available. Fermentation development in industrial environments is highly competitive, and time is always a critical factor; this calls for a balanced approach combining established quantitative correlations and empirical results.

There is no unique and universal scale-up procedure, as each process is likely to present its own difficulties and peculiarities. What follows here is therefore aimed at providing a general outline of scale-up procedures together with a description of the relevant techniques, with an emphasis on the practical aspects. Successful and time-efficient scale-up still requires a number of "educated guesses" that require, in turn, an appreciation of the possible problems and pitfalls. Reference is therefore made in the text to the literature (both review articles and original papers) that treats in detail the various aspect of scale-up. Acquaintance with this literature will be of great help to anyone confronted for the first time with a scale-up problem.

Symbols used in the text and equations in this chapter are defined in Table 1.

22.1 EFFECTS OF FERMENTATION SCALE

The scale of fermentation affects a number of physical and biological parameters that may vary either in absolute value, or relative to each other, or both. The term "scale" refers to either the volume or the linear dimensions of a fermentor. Most results concerning scale effects have been obtained with series of geometrically similar fermentors, i.e., vessels whose linear dimensions stand in the same ratios to each other on the different scales. For a series of cylindrical vessels characterized by a height (H_T)-to-diameter (D_T) ratio R, the volume is given by:

$$V_i = R \frac{\pi}{4} D_{T_i}^3 \quad (1)$$

or

$$V_i \propto D_T{}^3{}_i \quad (2)$$

since $R(\pi/4)$ is independent of scale. In such a series the dimensions of the impeller also follow geometric

TABLE 1. List of symbols

Symbol	Definition or explanation	Dimensions
A	Arrhenius constant	s^{-1}
A_c	Area available for heat exchange, or cross-sectional area of vessel $= \frac{\pi}{4} D_i^2$	m^2
a	Specific interfacial area for air-liquid mass transfer, or numerical constant	m^{-1}
a_s	Area of film across which diffusion takes place	m^1
b	Numerical constant	
C	Oxygen concentration in the liquid	$kg\ m^{-3}$
C^*	Saturating oxygen concentration in the liquid	$kg\ m^{-3}$
C_p	Total heat capacity of fermentor	$kcal\ °C^{-1}$
C_c	Dissolved O_2 concentration at cell surface	$kg\ m^{-3}$
C_c^*	Pseudocritical O_2 concentration	$kg\ m^{-3}$
C_B	Steady-state dissolved oxygen concentration	$kg\ m^{-3}$
C_g	Molar fraction of oxygen in gas phase	
c'	Michel and Miller constant	Depend on units used in equation
c_p	Constant relating ND_i^3 with ϕ_p, characteristic of impeller geometry	
c_c	Specific heat of coolant	$kcal\ kg^{-1}\ °C^{-1}$
D_c	Diffusivity constant	$m^2\ s^{-1}$
D_i	Impeller diameter	m
D_T	Tank diameter	m
d_B	Bubble diameter	m
d_s	Sauter bubble diameter $= \Sigma_i f_i(d_i)^3/\Sigma_i f_i(d_i)^2$	m
E	Activation energy	$cal\ mol^{-1}$
F_c	Volumetric flow rate of coolant	$m^3\ h^{-1}$
fCO_2	Molar fraction of CO_2 in gas phase	
F_i	Impeller pumping capacity	$m^3\ s^{-1}$
F_g	Volumetric gas flow	$m^3\ h^{-1}$
g_c	Gravitational constant $= 9.8\ m\ s^{-2}$	
H	Henry constant	$mol^{-1}\ Pa\ m^3$
H_L	Liquid height in fermentor	m
H_T	Tank height	m
h	Thickness of stagnant film in diffusion phenomena	m
h_i	Duration of the ith interval in a sterilization cycle, or film heat transfer coefficient of culture fluid	s $kcal\ m^{-2}\ h^{-1}\ °C^{-1}$
h_o	Film heat transfer coefficient of coolant	$kcal\ m^{-2}\ h^{-1}\ °C^{-1}$
h_b	Impeller blade height/width ratio	
j	Liquid-to-cells volumetric mass transfer coefficient	$mol\ kg^{-1}\ h^{-1}$ (% saturation)$^{-1}$
K	Geometry factor in power consumption relationship, or generic constant	
K_1	Thermal death rate constant	s^{-1}
K_2	Thermal deactivation rate constant	s^{-1}
K_c	Casson viscosity	$kg^{1/2}\ m^{-1/2}\ s^{-1/2}$
K_L	Mass transfer coefficient	$m\ s^{-1}$
K_v	Viscometer constant	m^{-3}
K_w	Thermal conductivity of fermentor wall	$kcal\ m^{-1}\ h^{-1}\ °C^{-1}$
K_La	Volumetric mass transfer coefficient	s^{-1}
L	Height of viscometer cylinder	m
l	Thickness of fermentor wall	m
M	Torque	$kg\ m^2\ s^{-2}$
M_g	Molecular weight of gas	
m	Exponent of correlation between N_p and N_{RE}	
N	Impeller rotation speed, or number of microorganisms	s^{-1}
N_0	Number of microorganisms before sterilization $= n_0V$	
N_g	Number of generations	
N_p	Newton, or power, number	
N_{RE}	Impeller Reynolds number $= ND_i^2\rho/\mu$	
N_{PR}	Prandtl number $= C_p\mu/K$	
n_0	Flow behavior index, in pseudoplastic fluids	
n_0	Concentration of contaminants in medium before sterilization	m^{-3}
OSR	Oxygen supply rate	$mol\ m^{-3}\ h^{-1}$
OTR	Oxygen transfer rate	$mol\ m^{-3}\ h^{-1}$
OUR	Oxygen uptake rate	$mol\ m^{-3}\ h^{-1}$
P_0	Gas absolute pressure	Pa
Po	Nongassed power consumption	$kg\ m\ s^{-1}$, or W
P_g	Gassed power consumption	$kg\ m\ s^{-1}$, or W
P_s	Gas pressure at sparger	Pa
p	Oxygen partial pressure in fermentor	Pa
p_1^*	Oxygen partial pressure in inflowing air	Pa
p_2^*	Oxygen partial pressure in outflowing air	Pa

Continued on following page

TABLE 1—*Continued*

Symbol	Definition or explanation	Dimensions
P_{sp}	Power dissipated for air sparging	$kW\ m^{-3}$
P_v	Gas pressure in vessel	Pa
pCO_2	CO_2 partial pressure	Pa
$pCO_{2(crit)}$	CO_2 partial pressure above which respiration rate or product formation rate decreases with increasing pCO_2	Pa
pO_2	O_2 partial pressure	Pa
$pO_{2(crit)}$	O_2 partial pressure above which q_{O_2} is independent of pO_2	Pa
pO_2^*	O_2 partial pressure in the gas phase	Pa
Q	Mass of diffusing substance,	kg
	or air volumetric flow rate at standard temperature and pressure	$m^3\ s^{-1}$
Q_{acc}	Rate of heat accumulation	$kcal\ m^{-3}\ h^{-1}$
$Q_{agitation}$	Heat equivalent of power of agitation	$kcal\ m^{-3}\ h^{-1}$
$Q_{evaporation}$	Rate of heat loss due to water evaporation	$kcal\ m^{-3}\ h^{-1}$
$Q_{exchange}$	Rate of heat exchange	$kcal\ m^{-3}\ h^{-1}$
Q_{metab}	Metabolic heat evolution rate	$kcal\ m^{-3}\ h^{-1}$
q_{O_2}	Specific rate of O_2 uptake	$mol\ kg^{-1}\ h^{-1}$
R	Gas constant = 1.98 cal $mol^{-1}\ K^{-1}$, or 82.05 cm^3 atm $mol^{-1}\ K^{-1}$, or 8.31 m^3 Pa $mol^{-1}\ K^{-1}$, or aspect ratio of fermentor = H_T/D_T	
S	Linear scale factor	
T	Absolute temperature	K
T_{in}	Inlet temperature of coolant	°C
T_h	Temperature of the maintenance phase in a batch sterilization cycle	K
T_{out}	Outlet temperature of coolant	°C
t	Time	h
t_h	Duration of the maintenance phase in a batch sterilization cycle	s
t_m	Mixing time	s
U	Overall heat transfer coefficient	$kcal\ m^{-2}\ h^{-1}\ °C^{-1}$
V	Tank volume	m^3
V_L	Volume of clear liquid	m^3
V_S	Superficial air velocity = Q/A_c	$m\ s^{-1}$
W	Molar fraction of water in gas stream	
X	Total biomass (=xV)	kg
x	Biomass concentration	$kg\ m^{-3}$
x_m	Biomass concentration of original strain	$kg\ m^{-3}$
x_n	Biomass concentration of variants	$kg\ m^{-3}$
α	Relative rate of temperature increase in a sterilization cycle or numerical constant modifying the effect of the number of impellers in the Richard equation (86)	s^{-1}
β	Numerical constant modifying the effect of the number of impellers in the Richard equation (86)	
γ	Shear rate	s^{-1}
$\bar{\gamma}$	Average shear rate	s^{-1}
ΔT_{in}	Temperature difference between the culture and the coolant, $T - T_{in}$	°C
ΔT_{lm}	Logarithmic mean temperature difference between culture and coolant = $(\Delta T_{in} - \Delta T_{out})/\ln(\Delta T_{in}/\Delta T_{out})$	°C
ΔT_{out}	Temperature difference between the culture and the coolant, $T - T_{out}$	°C
δ	E/RT_0	
θ_1	Duration of an interval in the piecewise linear approximation of a sterilization profile	s
$\wedge$	Nutrient quality criterion = logarithm of the ratio of initial to final concentration of a heat-labile substance in a sterilization cycle	
λ	Mutation rate	No. of variants produced per genome per generation
μ	Specific growth rate,	h^{-1}
	or dynamic viscosity	$kg\ m^{-1}\ s^{-1}$
$\bar{\mu}_a$	Average apparent viscosity	$kg\ m^{-1}\ s^{-1}$
μ_L	Viscosity of suspending liquid	$kg\ m^{-1}\ s^{-1}$
ν	Impeller peripheral velocity = πND_i	$m\ s^{-1}$
ν_0	Gas velocity at sparger = F_g/S_s, where S_s = total surface area of sparger holes	$m\ s^{-1}$
ρ	Density	$kg\ m^{-3}$
ρ_c	Density of coolant	$kg\ m^{-3}$
ρ_g	Gas density	$kg\ m^{-3}$
τ	Shear stress	$kg\ m^{-1}\ s^{-2}$
τ_0	Yield stress	$kg\ m^{-1}\ s^{-2}$
ϕ	Norwood and Metzner's mixing time factor, or fractional gas holdup	
ϕ_c	Volume fraction of cells	
ϕ_p	Impeller pumping capacity = $c_p ND_{i3}$	$m^3\ s^{-1}$
∇	Del factor = logarithm of the ratio of initial to final number of viable organisms in a sterilization cycle	
$\nabla_{cooling}$	Del factor of the cooling phase in a batch sterilization cycle	
∇_h	Del factor of the maintenance phase in a batch sterilization cycle	
$\nabla_{heating}$	Del factor of the heating phase in a batch sterilization cycle	
∇_{tot}	Del factor of the entire batch	

similarity, and if D_i = impeller diameter, then D_i/D_T = constant, and, from equation 2:

$$\frac{V_1}{V_2} = \frac{{D_{T_1}}^3}{{D_{T_2}}^3} = \frac{{D_{i_1}}^3}{{D_{i_2}}^3} \tag{3}$$

for any two vessels within the series. The scale-up factor usually refers to the linear dimensions, and thus a scale-up factor of 10 (i.e., $D_{T_2}/D_{T_1} = 10$) is equivalent to a volumetric scale-up factor of 10^3 = 1,000. The main parameters affected by scale are (i) number of generations, (ii) development of precultures, (iii) sterilization of culture medium, (iv) aeration and agitation, and (v) heat transfer.

22.1.1 Number of Generations

The number of generations required to achieve a given final biomass concentration, x, starting from a given inoculum, is obviously a function of fermentor volume. Assuming for simplicity that exponential growth (59) occurs throughout, we have:

$$X = X_0 e^{\mu t} \tag{4}$$

or,

$$Vx/X_0 = e^{\mu t} \tag{5}$$

where $X = Vx$ and μ = specific growth rate (for symbols, see Table 1). Substitution of the known relationships between μ, the biomass doubling time (t_d), and the number of generations (N_g) (59) into equation 5 yields:

$$Vx/X_0 = e^{N_g \ln 2} \tag{6}$$

or

$$N_g = 1.44\,(\ln V + \ln x - \ln X_0) \tag{7}$$

Equation 7 expresses a linear relation between the number of generations and the logarithm of fermentor volume. This latter condition follows from the initial assumptions of equal inoculum size, X_0, and of identical growth medium, that in turn supports the same final biomass concentration, x. The major implication of equation 7 is that in a fermentation process employing a microorganism characterized by a certain rate of appearance of undesired variants (mutants or otherwise), the final proportion of variants in the population increases with fermentation scale. This will be caused by both the appearance of new variants at each generation and the multiplication of previously appeared variants. The contribution of the latter mechanism is generally greater than that of the former and acquires a dramatic impact if the growth rate of the undesired variants is higher than that of the parent strain. This has been found to be the case with highly developed industrial strains employed in the production of antibiotics (4, 61). Under the simplifying assumption that the variants have the same specific growth rate as the parent strain, the fraction of variants after N_g generations will be:

$$\frac{x_m}{x_m + x_n} = \frac{x_{m_0} + x_{n_0}\,(1 - e^{-0.693\lambda N_g})}{x_{m_0} + x_{n_0}} \tag{8}$$

where x_n = biomass of parent strain; x_m = biomass of variant (x_{n_0} and x_{m_0} are values at inoculation time); and λ = rate of appearance of variants expressed as number of variants produced per genome per generation.

In the more general case in which the parent organism and the variant have different growth rates we have (56):

$$\frac{x_n}{x_m + x_n} = \frac{\alpha + \lambda - 1}{(\alpha - 1) + \lambda 2^{N_g(\alpha + \lambda - 1)}} \tag{9}$$

where α is the ratio of the specific growth rate of the variant to that of the parent organism. Equations 8 and 9 express a simplified view of the actual phenomenon, as in practice it is conceivable that more than one type of non- (or low-) producing variants may appear, each with its own growth characteristics.

Furthermore, in the majority of today's batch fermentation processes, the culture is in a nongrowing state for a certain fraction of the fermentation cycle, called stationary phase, whereas continuing growth is assumed in equations 8 and 9. In practice, in many cases it is more likely that, whereas no net growth takes place in these stationary phases, a process of cell (and thus genome) replication continues at a slow rate, replacing those cells that undergo a lytic process and utilizing the nutrient substances made available by such lysis (cryptic growth). However, provided no gross changes in culture behavior are introduced by scale, the situation described by equations 7 through 9 is likely to occur, namely, an increasing frequency of undesired variants with increasing fermentation scale.

22.1.2 Medium Sterilization

The basic techniques employed for the heat sterilization of culture media are batch sterilization and continuous sterilization. In the former, the entire volume of medium is homogeneously heated up to a preselected temperature, held at that temperature for a predetermined period of time, and then cooled down to the initial temperature of fermentation. This gives rise to a temperature profile (Fig. 1) characterized by three distinct phases, heating up, maintenance, and cooling down, the first and the last being generally scale dependent. This is due to the fact that heating (at least in part) and cooling (always) are obtained by heat exchange through the vessel walls, and the available surface increases with the square of linear dimensions while the volume of liquid increases with the third power of linear dimensions, so that their ratio, and thus the heat exchange rate, decreases with scale (see section 22.1.5).

The main objective of sterilization is to kill all the living organisms present in the medium before inoculation, to eliminate any possible competition or interference with the growth and metabolism of the desired microorganism. In this context the operative definition of death is the loss of ability to reproduce. In practice, the sterilization process is designed to reduce the probability of lack of success to acceptably low values (a typical design value is 10^{-3}, i.e., one accepts a 1 in 1,000 chance that at least one organism will survive the treatment). Such a probabilistic approach to sterilization follows from the first-order kinetic model usually adopted to describe the thermal destruction of microorganisms. It is assumed that the rate of killing is at any time proportional to the number of viable organisms present, i.e.:

$$\frac{dN}{dt} = -K_1N \quad (10)$$

yielding, on integration:

$$N = N_0e^{-K_1t} \quad (11)$$

or:

$$\ln N = \ln N_0 - K_1t \quad (12)$$

Equation 12 allows the experimental determination of the death rate constant, K_1, from a plot of ln N versus t, where N = number of viable organisms at any time t and N_0 = number of viable organisms at $t = 0$. This model predicts that N approaches zero at very high values of t, without actually reaching it at any finite value of t. This deterministic model in fact ceases to represent the real phenomenon when N, a continuous variable, fails to describe the discrete nature of a very small population of microorganisms. In fact, $N = 0$ is achieved in every successful sterilization of a batch of medium. N is thus taken to represent the probability that at least one viable microorganism is still present in the medium after sterilization. For a discussion of this assumption, see text after equation 17. This justifies the otherwise physically meaningless practice of using fractional values of N in designing sterilization cycles (e.g., $N = 10^{-3}$, or 1/1,000 of a microorganism).

To achieve the desired value of N, one must know N_0 and K_1 for use in equation 11. Such an approach is usually not possible in practice because (i) the initial population is extremely variable in size, reflecting the various degrees of contamination of the raw materials employed in the make-up of the medium, and the determination of N_0 is time-consuming and often of doubtful significance; (ii) the population is highly heterogeneous and the various subpopulations have different heat sensitivities (i.e., different K_1 values); and (iii) K_1 changes with temperature in a different fashion for the different members of the population. The problem is simplified assuming that (i) only one population is present; (ii) this population is entirely composed of the most heat-resistant microorganism likely to be encountered (e.g., spores of *Bacillus stearothermophilus* FS1518; 1, 6). The dependence of the thermal death rate of such spores, K_1, on temperature has been determined (21) using moist heat (i.e., saturated steam) and found to follow the Arrhenius equation:

$$K_1 = Ae^{-(E/RT)} \quad (13)$$

where R = gas constant = 1.98 cal mol^{-1} K^{-1} and T = absolute temperature, with the following parameters: $A = 1 \times 10^{36.2}$ s^{-1}; E = 67.7 kcal mol^{-1}. Thus, the value of K_1 can be calculated at any given temperature by substituting the above values in equation 13. Inspection of Fig. 1 shows that K_1 is constant only during the maintenance phase, and therefore equation 12 can be applied as such to calculate the reduction in the number of viable organisms, N/N_0, occurring during this phase. During the heating up and cooling-down periods, the rate of thermal death changes with time,

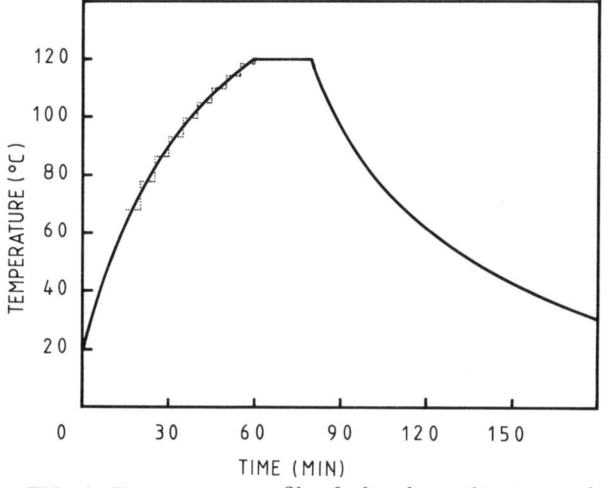

FIG. 1. Temperature profile of a batch sterilization cycle. Solid line, Actual temperature; broken line, stepwise approximation.

and thus the contribution of these phases to sterilization must be estimated by some more elaborate procedure.

It is expedient at this point to rewrite equation 12 as follows:

$$\nabla = \ln \frac{N_0}{N} = K_1t = Ate^{-(E/RT)} \quad (14)$$

where ∇, i.e., the logarithm of the ratio of the initial to the final population, is called the sterilization criterion, Del factor, or design criterion (1, 6). The quantity ∇ is scale dependent, as shown by the fact that N, i.e., the probability of not achieving complete destruction of the contaminants, is fixed at the design stage, whereas:

$$N_0 = Vn_0 \quad (15)$$

where V = volume of medium and n_0 = concentration of contaminants in the nonsterile medium. One expects n_0 to be mainly related to the composition of the medium (i.e., to the degree of contamination of the raw materials employed) and thus to be largely independent of scale. Then

$$\nabla = \ln V + \ln n_0 - \ln N \quad (16)$$

Comparison of equations 14 and 16 shows that if n_0 is scale independent, then ∇, which is a direct measure of the severity of the heat treatment, must increase linearly with the logarithm of culture volume, if the same final probability of not achieving sterility, N, is desired. Equation 16 may be rewritten as:

$$\nabla = \ln V + \ln \frac{n_0}{1 - P} \quad (17)$$

where P = desired probability of achieving complete destruction of contaminants.

Equation 17 is not rigorously correct, as one would expect the probability of an individual batch being sterile (i.e., containing no viable microorganisms) to follow a Poisson distribution. The probability of such an event could be $P(0) = e^{-\mu}$, where μ = average number of contaminants per batch, designated by N

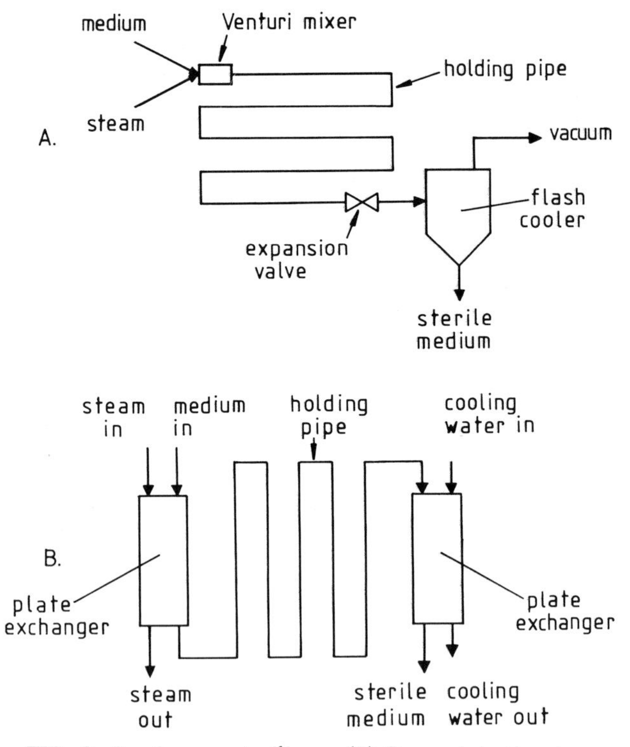

FIG. 2. Continuous sterilizers. (A) Steam injection type; (B) plate exchanger type.

in our notation. The probability of a particular batch being nonsterile is therefore given by $1 - e^{-N}$. In practice this expression approaches N as N approaches zero, and therefore they can be interchanged. For instance, the error introduced by this approximation in the value of ∇ in the example below is less than 0.002%.

The second term in the right-hand side of equation 17 is scale independent and, as a first approximation, expected to depend solely on the composition of the culture medium. The desired value of ∇ can be achieved in several subsequent steps. If, for example, $n_0 = 10^6$ spores per liter, $V = 50{,}000$ liters, and $P = 0.999$, then, from equation 17:

$$\nabla = \ln 50{,}000 + \ln \frac{10^6}{0.001} = 10.82 + 20.72 = 31.54$$

This may be accomplished in three steps, bringing the population first down to 10^2, then to 1, and finally to 10^{-3}. To each step we can associate a ∇_i value, calculated as follows:

$$\nabla_1 = \ln \frac{5 \times 10^{10}}{5 \times 10^6} = 9.21$$

$$\nabla_2 = \ln \frac{5 \times 10^6}{5 \times 10^4} = 4.60$$

$$\nabla_3 = \ln \frac{5 \times 10^4}{10^{-3}} = 17.73$$

$$\nabla_{tot} = 31.54$$

The additivity of ∇ allows the separate estimation of the effects of the heating-up, maintenance, and cooling-down phases of a batch sterilization cycle.

The difference in activation energy between the destruction of microorganisms and that of heat-labile nutrients is exploited to a large extent in the alternative technique of medium sterilization, i.e., continuous sterilization, also known as high-temperature-short-time sterilization. The basic steps of this technique are (i) rapid heating of the culture medium flowing through a pipe via either heat exchangers or live steam injection (Fig. 2); (ii) holding of sterilization temperature for the desired period of time, obtained with a suitable combination of fluid velocity and length of piping; and (iii) rapid cooling via either heat exchangers (with or without heat recovery) or flash-cooling, consisting of the instantaneous evaporation of part of the water in an expansion chamber, with concomitant loss of latent heat. Several configurations of varying degrees of sophistication are possible, and particular attention is paid to the safety aspects to minimize risks of contamination of the whole batch in case of breakdown, drop in steam pressure, etc. From the scale-up point of view, the basic feature of high-temperature-short-time systems is that they can be scaled up with both the time and the temperature of sterilization remaining constant, due to the virtual absence of the heating-up and cooling-down phases. This possibility permits a quantitative approach to the problem of destroying critical, heat-labile substances, provided the Arrhenius relationship for this destruction is known. The technique, discussed in detail by Deindoerfer and Humphrey (21), is based on the definition of a nutrient quality criterion, $\wedge$, analogous to the sterilization criterion, ∇, as:

$$\wedge = \ln \frac{C_0}{C} = K_2 t \tag{18}$$

where C_0 and C are the initial and final concentrations of the critical substance, respectively.

The thermal inactivation rate of heat-labile organic substances, such as vitamins, is less sensitive to temperature changes, as shown by the lower values of the activation energy, E. Typical values are in the range of 15 to 30 kcal mol^{-1}, substantially lower than those for bacterial spores destruction, lying between 50 and 82 kcal mol^{-1}. Such a situation can be exploited to reduce $\wedge$, while maintaining ∇ constant, in a high-temperature-short-time sterilization.

It is clear from the foregoing that the kinetic constant K_2 is less sensitive to temperature changes than K_1, the kinetic constant for spore destruction, as implied by the lower activation energy. This can be visualized by plotting $\ln(t)$ versus $1/T$. In fact, from equation 14:

$$\ln(t) = \ln\left(\frac{\nabla}{A}\right) + \frac{E}{R} \times \left(\frac{1}{T}\right) \tag{19}$$

A plot of $\ln(t)$ versus $1/T$ gives a straight line whose slope is proportional to E and whose elevation is a function of ∇. Different values will produce parallel lines (Fig. 3). The same procedure, if applied to $\wedge$, yields parallel lines with a less steep slope. This fact can be exploited since ∧ can be decreased (moving to a lower line) while maintaining the same ∇ value (moving along the same line), provided both ln(t) (and thus t) and $1/T$ are decreased, which means a higher temperature for a shorter period of time. Ideally, ∧ should be decreased to the point that no further improvements in the performance of the medium are observed because saturating concentrations of the critical substrate are left in the medium. In practice, the objective is often to minimize ∧ compatibly with the engineering constraints placed upon the continuous sterilizer (i.e., maximum allowable temperature, temperature of heating source, maximum flow rate, maximum cooling capacity). The practical application of this technique is somewhat restricted by the fact that the critical substance must be identified, its thermal inactivation kinetics must be determined, and its effect upon medium performance must be quantitated. This is rarely the case with complex organic media (6).

If, however, small-scale experimentation has shown that medium performance decreases with approximately first-order kinetics with increasing sterilization time, and the specific deactivation rate changes with temperature according to equation 13, the construction of a process performance chart can still be attempted by substituting a measure of medium performance (e.g., final titer or production rate) for C in equation 18.

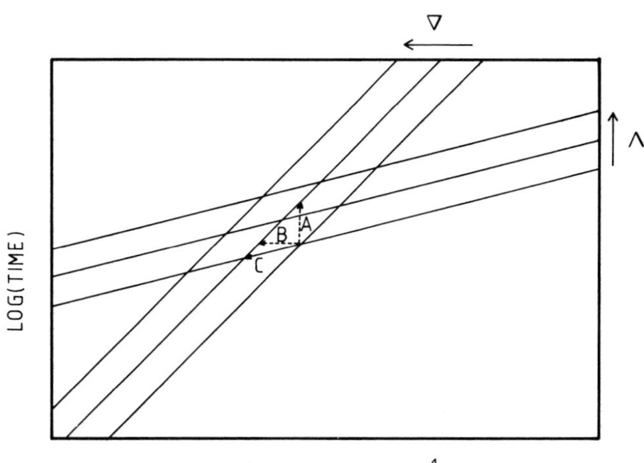

FIG. 3. Time-temperature equivalence curves for the sterilization criterion, ∇, and the nutrient quality criterion, ∧. The increase in ∇ associated with a scale increase can be obtained in three ways: (A) time >, temperature =, and ∧ >; (B) time =, temperature >, and ∧ >; or (C) time <, temperature >, and ∧ =. Only procedure C maintains ∧ constant. (From reference 6.)

22.1.3 Agitation and Aeration

According to a highly simplified representation, a fermentation can be seen as consisting of essentially three dynamic steps, as shown in Fig. 4. The technical objective of fermentation development is to maximize the overall rate of material flow through this sequence. Step 2 comprises all the biochemical events effecting the transformation of substrates into products, while steps 1 and 3 represent all the transport mechanisms that operate to provide the cell with the substrates and to remove its products.

Substrate uptake and product excretion, and of course intracellular metabolism, are under the combined control of the genetic background of the cell and of the environmental conditions (medium composition, pH and temperature of incubation, etc.). The ultimate aim of process development is to achieve rate limitation by the internal factors of the particular strain employed (e.g., intrinsic kinetic characteristics of key enzymes that are specified at the genetic level) through the optimization of the environmental conditions. Further development rests, therefore, with the genetic modification of the microorganism. To achieve this objective, one must ensure that (i) the concentration of substrates at the cell surface is high enough to allow the uptake mechanisms to operate at maximum capacity (likewise, the concentration of products must be kept low enough to avoid possible product inhibition effects), and (ii) the specified optimal conditions are homogeneously attained throughout the culture. The first condition is met by the realization of a sufficient degree of turbulence, and the second is met by a sufficient degree of bulk mixing. The degree of bulk mixing is characterized by the magnitude of the differences of bulk liquid parameters (e.g., temperature, dissolved oxygen, pH) in different parts of the vessel. The term turbulence is used in this context to indicate the existence, and the intensity, of liquid velocity fluctuations in both direction and intensity on a micro-scale, i.e., small compared with the vessel dimensions. A sufficient degree of turbulence minimizes the difference between the concentration of substances at the cell surface, where it really matters, and in the bulk of liquid, where it is measured. Chemical analysis and probes (pO_2, pH, pCO_2) give the bulk liquid concentration, and with this one relates the physiological responses of the microorganism.

The transfer of substrates and products to and from the cells, i.e., mass transfer, is usually visualized as being composed of several steps whose individual behavior can be modeled and analyzed to identify the rate-controlling step in each real situation. A widely adopted model considers any interphase mass transfer phenomenon as being controlled by a series of diffusional resistances, separated by well-mixed stages (1). For instance, the interphase mass transfer of O_2 from the air bubbles dispersed in an aerated fermentation broth to the respiring cells is visualized, according to this theory, as depicted in Fig. 5. Mass transfer across a stagnant film takes place solely by diffusion, and the general expression for the rate of mass flow across the stagnant film is:

$$\frac{dQ}{dt} = \frac{D_c a_s \Delta C}{h} \tag{20}$$

where a_s = area of the surface across which the diffusion takes place; ΔC = concentration difference across stagnant film; h = film thickness; and D_c = diffusivity constant. Since h is generally not known,

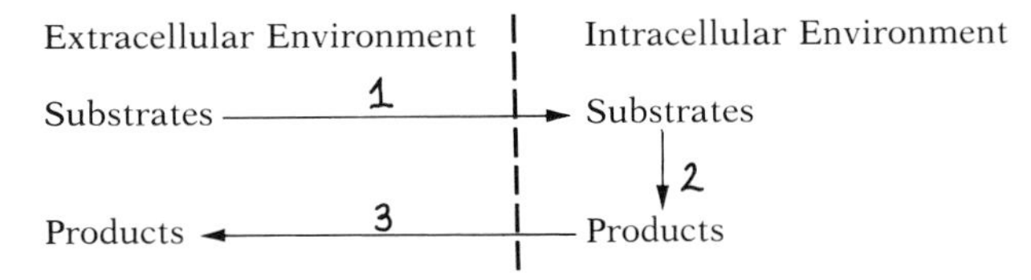

FIG. 4. Three dynamic steps in fermentation.

the usual form of equation 20 is:

$$\frac{dQ}{dt} = K_L a_s \Delta C \tag{21}$$

where $K_L = D_c/h$.

Practical experience has shown that in fermentations the liquid film resistance adjacent to the gas phase is usually larger than that of the other two films and is therefore rate limiting (2, 8, 25). There are, however, some exceptions (58, 75, 87). With respect to the liquid film adjacent to the gas phase, we can express the mass transfer rate as:

$$\frac{1}{V}\frac{dQ}{dt} = K_L a \Delta C \tag{22}$$

where $a = a_s/V$, i.e., the specific interfacial area. If the liquid at the interface with air is saturated with O_2, then $\Delta C = C^* - C$, where $C = O_2$ concentration in bulk liquid and C^* = saturating O_2 concentration in the liquid.

The effect of turbulence is that of decreasing the apparent film thickness by inducing motion of small liquid elements relative to each other, e.g., bringing elements of bulk liquid close to the cell surface. This condition can be difficult to achieve in highly viscous, non-Newtonian broths in which turbulent motion can be brought to a rest very quickly by the internal frictional resistances of the liquid, i.e., viscous forces. It may then happen that shortly after a given portion of the culture broth has left the high-turbulence zone near the impeller, the internal liquid motions die away and the liquid moves in a laminar fashion, with the small liquid elements traveling along parallel trajectories with little or no velocity relative to each other. In such conditions the thickness of the stagnant film around the cells increases considerably (see Fig. 5) and so do the diffusional resistances. In terms of equation 20 this means that h increases, and thus the mass transfer rate decreases, for a given driving force, ΔC. In such conditions the measurement of dissolved oxygen in the bulk of liquid, as performed by oxygen probes, is unsatisfactory. The dissolved oxygen concentration at the cell's surface (the value that really matters) will be lower than that sensed by the probe, and the difference will be a function of the degree of turbulence in the culture. This type of situation will be discussed in greater detail after we examine the basic aspects of agitation, with particular emphasis on their dependence on scale.

In a conventional stirred fermentor, the agitation apparatus has four basic functions: (i) to ensure homogeneity of the culture through macro-mixing; (ii) to promote interface mass transfer through micro-mixing; (iii) to promote heat transfer; and (iv) to provide gas-liquid interfacial area for gas-to-liquid and liquid-to-gas mass transfer. All of these functions require the expenditure of mechanical energy, which is supplied by the motion of various types of impellers and is eventually dissipated as heat. The energy so supplied to the culture broth contributes to the four functions listed above to various degrees, according to the geometry of the apparatus and to the scale of operation. The ability of impellers to transfer energy to a liquid is usually expressed in terms of the Newton number, or power number, N_p, defined as follows (67, 68):

$$N_p = \frac{P_0 g_c}{N^3 D_i^5 \rho} \tag{23}$$

where P_0 = power consumption (kg m^{-1} s^{-1} [not SI units], or W); g_c = gravitational constant (m s^{-2}); N = rotation speed of impeller (s^{-1}); D_i = impeller diameter (m); and ρ = density of liquid (kg m^{-3}). Plots of experimentally determined N_p versus N_{RE}, the impeller Reynolds number (see Fig. 6), show that above a certain value of N_{RE}, N_p remains virtually constant at a value that depends on the impeller geometry. This only applies to baffled systems. In nonbaffled, i.e., free-vortexing systems, turbulent flow ensues at $N_{RE} \simeq 10^2$, and from this point onwards N_p decreases linearly according to the relationship $N_p = K(N_{RE})^{-0.28}$ (13) (Fig. 7). From this point of view the flat-blade turbine is more efficient (higher N_p) than the other types. The Reynolds number is a measure of the turbulence of the system and expresses the ratio of inertial to viscous forces in the liquid:

$$N_{RE} = \frac{N D_i^2 \rho}{\mu} \tag{24}$$

where μ = liquid viscosity (kg m^{-1} s^{-1}). The general

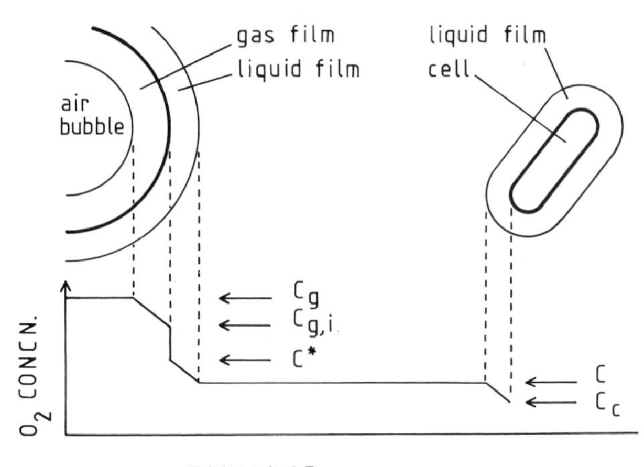

FIG. 5. Three-films theory of gas-liquid O_2 transfer. C_g = O_2 concentration in bulk of well-mixed gas phase; $C_{g,i}$ = O_2 concentration in the gas phase at the gas-liquid interface; C^* = O_2 concentration in the liquid phase at the gas-liquid interface; C = O_2 concentration in bulk of well-mixed liquid phase; and C_c = O_2 concentration at the cell surface.

relationship between N_p and N_{RE} has the form:

$$N_p = K(N_{RE})^m \quad (25)$$

for vessels in which no vortex is formed.

The plot of log N_p versus log N_{RE} reported in Fig. 6 can be subdivided into three regions, according to the value of m in equation 25. Up to $N_{RE} \simeq 10$, $m = -1$, i.e., N_p is inversely proportional to N_{RE}. This gives, from equations 24 and 25:

$$N_p = \frac{K\mu}{N\,D_i^2\,\rho}$$

and thus:

$$P_0 = \frac{K\mu N^2 D_i^3}{g_c} \quad (26)$$

This region is called the viscous flow region, and power consumption is proportional to the culture viscosity, μ. When $N_{RE} > 10^4$, $m = 0$, and then:

$$P_0 = \frac{N_p N^3 D_i^5 \rho}{g_c} \quad (27)$$

i.e., power consumption is independent of viscosity. This is the turbulent flow region. (The region delimited by $10 < N_{RE} < 10^4$ is characterized by a variable value of m and is called the transition region, and no general expression relates N_p to N_{RE}.) The utility of equations 26 and 27 lies in the fact that K is a constant that depends on vessel and impeller geometry, but not on scale. These equations can thus be used to predict power consumption in a larger vessel, provided it is geometrically similar to that used for the experimental determination of K. Their practical utility is significantly reduced by the following facts: (i) geometric similarity in real plants is frequently not maintained; (ii) the above results have been obtained with nongassed Newtonian fluids, whereas most commercially important processes employ aerobic microorganisms and require sparging of the culture content with a suitable O_2 source (usually air); (iii) most cultures change their viscosity with fermentation time, and many of them are non-Newtonian, i.e., they have no unique value of viscosity, and moreover, this is frequently different in the different parts of the fermentor; and (iv) large vessels use multiple impellers, whereas the above results have been obtained in systems with one impeller. A considerable amount of work has therefore concentrated upon the study of gassed fluids, non-Newtonian broths, and geometrically dissimilar fermentors.

The power drawn by impellers rotating in gassed fluids, P_g, in the turbulent regime can be related to the power drawn at the same rotation speed in nongassed conditions, P_0, by the empirical relationship of Michel and Miller (48):

$$P_g = C' \left(\frac{P_0^2 N D_i^3}{Q^{0.56}} \right)^{0.45} \quad (28)$$

where C' is an experimentally determined constant of the given system and Q = volumetric airflow rate.

In an aerated fluid, the gassed power, P_g, is used to promote turbulence and interfacial area for gas ex-

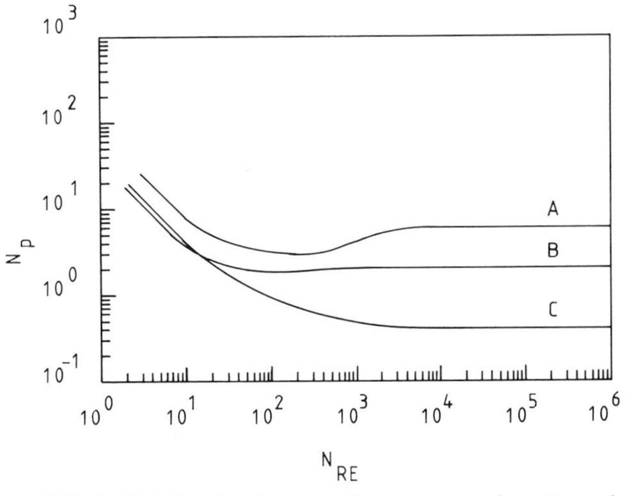

FIG. 6. Relationship between the power number, N_p, and the Reynolds number, N_{RE}, in baffled systems. (A) Flat-blade turbine impeller; (B) paddle impeller; (C) marine propeller. (From reference 1.)

change and can therefore be expected to correlate with the volumetric mass transfer coefficient, K_La (see equation 23). A considerable amount of work has been carried out on this relationship, and a number of different correlations have been proposed (see Table 2). The term P_g/V, the volumetric power input, is present in all correlations, but with widely differing exponents. The main causes of such variation are as follows. (i) Physicochemical properties of the liquid: the presence of electrolytes, for instance, increases the exponent as well as the absolute value of K_La (49, 93). (ii) Fermentation scale: the exponent decreases with increasing scale (7). This is partly due to the decreasing importance of surface aeration with increasing scale (26). (iii) Type and number of impellers: flat-blade turbines (six-bladed) yield steeper correlations than curved-blade turbines or paddle impellers (42). In multiple-impeller systems most of the mass transfer takes place in the region of the impeller situated close to the spargers. The other impellers contribute more to mixing than to mass transfer.

Thus, no general and unequivocal correlation between power consumption and K_La exists, although for moderate variations of operating parameters a particular relationship may apply satisfactorily. The adoption of power/unit volume as a scale-up criterion is no guarantee of success, and in point of fact it is rarely used as such in industrial practice (A. Einsele, Abstr. 5th Int. Ferm. Symp., Berlin, 1976, abstr. no. 4.13). The main reason for this situation probably lies in the fact that the flow behavior of most fermentation broths is different from that of the model systems employed in fundamental studies and usually changes with fermentation time. The relevant fundamental results should therefore be obtained in the conditions prevailing in the critical phase of the fermentation (see section 22.2.2, Characterization of Fermenter Performance). Power dissipation in the liquid is not the only measure of impeller performance. Two other parameters are important in fermentation systems, namely, the shear rate distribution and the pumping capacity of the impeller.

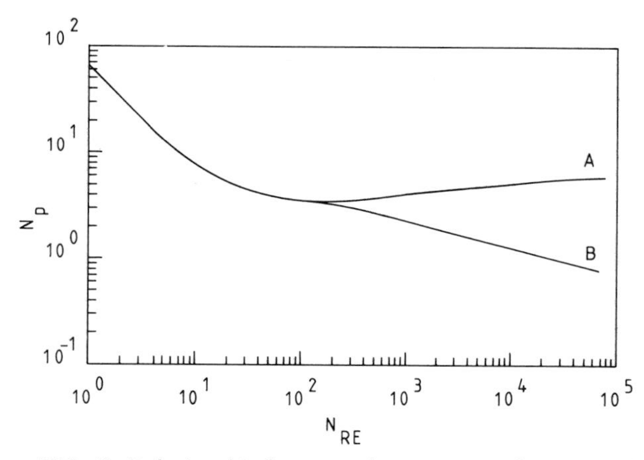

FIG. 7. Relationship between the power number, N_p, and the Reynolds number, N_{RE}, for a flat-blade turbine. (A) Baffled system; (B) free-vortexing system. (From reference 13.)

As discussed previously, the internal liquid friction (i.e., viscosity) tends to slow down the motion of liquid elements relative to each other, and the motion of the impeller is the primary source of these velocity differences. It is therefore evident that the maximum value of shear rate (i.e., the velocity gradient between adjacent liquid elements) occurs at the tip of the impeller blades, and its magnitude is proportional to the peripheral velocity of the impeller, given by $v = \pi N D_i$. Liquid shear has been found to be the major mechanism of physical damage to filamentous organisms, rather than the direct impact of the stirrer blades (79, 84). Tip speed, being proportional to maximum shear rate, can therefore be of primary importance in fermentations employing filamentous microorganisms (e.g., almost all antibiotic fermentations).

The impeller pumping capacity is directly connected to the degree of homogeneity that can be achieved in a fermentor. In fact, a rotating impeller imparts to the liquid a flow motion whose pattern is characteristic of the impeller type. For instance, radial-flow impellers, such as the widely used flat-blade turbines, determine the liquid circulation pattern depicted in Fig. 8. Culture fluid elements experience phases of high turbulence (when they enter the impeller zone), separated by periods of decreasing turbulence (in very viscous broth they can actually move in a streamline fashion for the best part of the cycle). Interphase mass transfer phenomena, such as O_2 transfer from air to liquid, slow down considerably during the low-turbulence phase and, if the recirculation rate is low enough, may limit the biological activity of the culture.

A measure of the degree of homogeneity achieved is given by the mixing time, defined as the time needed to disperse a pulse of tracer uniformly throughout the fermentor. The tracer can be a dye, an acid, an alkali, an electrolyte, or heat (supplied as a portion of medium warmer or colder than the bulk of the fermentor contents) (see Fig. 9). It is clear that the recirculation time and the terminal mixing time are functions of both impeller and tank geometry. Norwood and Metzner (53) defined the mixing time factor, ϕ, as:

$$\phi = \frac{t_m (N D_i^2)^{2/3} g_c^{1/6} D_i^{1/2}}{H_L^{1/2} D_T^{3/2}} \tag{29}$$

and studied its correlation with the impeller Reynolds number, N_{RE}, using a flat-blade turbine rotating in a Newtonian liquid. It was observed that ϕ became independent of N_{RE} at $N_{RE} > 6 \times 10^4$. If the various tank dimensions in equation 29 are expressed in terms of D_i, which is valid if geometric similarity is maintained, it is found that:

$$t_m N^{2/3} D_i^{-1/6} = \text{constant} \tag{30}$$

independent of scale. Inspection of equation 30 shows that t_m is bound to increase with scale (i.e., with D_i) if P/V is kept constant. In fact in the turbulent region, if P_0/V is to be kept constant, then:

$$\frac{P_0}{V} \propto \frac{P_0}{D_i^3} \propto N^3 D_i^2 = \text{constant} \tag{31}$$

from equation 27. Eliminating N from equations 30 and 31, i.e., keeping P_0/V constant, yields:

$$t_m = \text{constant} \cdot D_i^{11/18}$$

or

$$t_{m2} = t_{m1} \left(\frac{D_{i2}}{D_{i1}}\right)^{11/18} \tag{32}$$

for two different scales, indicated by 1 and 2 subscripts.

As far as the impeller peripheral velocity is concerned, keeping P_0/V constant requires, from equation 31:

$$\frac{N_1}{N_2} = \left(\frac{D_{i2}}{D_{i1}}\right)^{2/3}$$

On the other hand,

$$\frac{v_1}{v_2} = \frac{N_1 D_{i1}}{N_2 D_{i2}}$$

Elimination of N_1/N_2 from the above expressions gives:

$$v_2 = v_1 \left(\frac{D_{i2}}{D_{i1}}\right)^{1/3} \tag{33}$$

Therefore the impeller peripheral velocity increases with the cube root of scale if the specific power consumption is kept constant.

It is therefore clear that power/unit volume (and thus $K_L a$), tip speed (i.e., maximum shear rate), and mixing time cannot be kept constant simultaneously as the scale increases in a series of geometrically similar fermentors. The situation is summarized in Table 3, which shows the dependence on the linear scale factor, $S = D_{i2}/D_{i1}$, of the two remaining agitation parameters when one is adopted as a scale-up criterion, i.e., is maintained constant as scale is increased from D_{i1} to D_{i2}.

An additional parameter displaying scale effects is aeration. In aerobic fermentation it has basically two functions, to supply oxygen to the culture and to

TABLE 2. Relationship between K_La and agitation-aeration parameters in stirred tanks

Relationship[a]	Agitation system	Reference
$K_La \propto \left(\frac{P_g}{V}\right)^{0.95} V_s^{0.67}$	Vaned disk	19
$K_La \propto \left(\frac{P_g}{V}\right)^{0.53} V_s^{0.67}$	Paddle impeller	19
$K_La \propto \left(\frac{P_g}{V}\right)^{0.4} V_s^{0.5}$	Six-bladed disk-turbine	15
$K_La \propto \left(\frac{P_g}{V}\right)^{0.7} V_s^{0.3}$	Six-bladed disk-turbine	15
$K_La \propto \left(\frac{P_g}{V}\right)^{0.4} V_s^{0.5} N^{0.5}$	Six-bladed disk-turbine	62
$K_La \propto \left(\frac{P_g}{V}\right)^{0.72} V_s^{0.11}$	Waldhof	33
$K_La \propto \left(\frac{P_g}{V}\right)^{0.56} V_s^{0.7} N^{0.7}$	Turbine	27
$K_La \propto \left(\frac{P_g}{V}\right)^{0.67} V_s^{0.67}$	Turbine	91
$K_La \propto \left(\frac{P_g}{V}\right)^{0.33} V_s^{0.56}$	Turbine	77

[a] K_La = Volumetric mass transfer coefficient; P_g/V = gassed power/unit volume; V_s = superficial air velocity (=volumetric flow rate/cross-sectional area of vessel); N = rotational speed of impeller.

remove CO_2 and other volatile metabolites from the culture. With respect to the first function, the effect of airflow is rather complex and is closely interwoven with agitation. The interfacial surface area for mass transfer provided by bubble formation at the sparger in stirred tanks is usually too small to permit a sufficient oxygen transfer rate. Therefore a fraction of the mechanical energy imparted to the liquid by the impeller is used to increase the interfacial area by breaking large bubbles into smaller bubbles. The total surface area of spherical bubbles per unit volume of liquid, a, is given by:

$$a = \frac{6\,\phi}{d_s(1-\phi)} \tag{34}$$

where ϕ = fractional holdup of gas [$=V_{gas}/(V_{gas} + V_{liquid})$] and d_s = Sauter bubble diameter = $\Sigma_i f_i d_i^3/\Sigma_i f_i d_i^2$. f_i is the frequency of gas bubbles with diameter d_i. During their permanence in the liquid, which is increased by the recirculating flow induced by the impeller (see Fig. 8), bubbles are subjected to shear forces, which tend to break them, and to coalescence, which tends to increase their size. Turbulent forces depend mainly on level of power input and impeller geometry, while coalescence is affected by electrolyte concentration, surface tension of liquid, and viscosity. Electrolytes and surfactants decrease the rate of coalescence, affecting the interfacial area and bubble size. The following relationships apply (49):

$$a = 0.55\left(\frac{P_g}{V}\right)^{0.4} V_s^{0.5} \tag{35}$$

and

$$d_B = 0.27\left(\frac{P_g}{V}\right)^{-0.17} V_s^{0.27} + (9 \times 10^{-4})$$

for air-water (coalescing) systems and

$$a = 0.15\left(\frac{P_g}{V}\right)^{0.7} V_s^{0.3} \tag{36}$$

and

$$d_B = 0.89\left(\frac{P_g}{V}\right)^{-0.17} V_s^{0.17}$$

for air-electrolyte solutions systems, where d_B = bubble diameter. The presence of electrolytes decreases K_L but increases a to a larger extent, so that K_La increases (49). Aeration decreases power consumption, mainly by decreasing the apparent density of the liquid in which the impeller rotates.

The effect of culture viscosity is more complex. On the one hand, growth of large bubbles (large enough to have a substantial rise velocity through the liquid) is enhanced by the capture of small bubbles rising in the long wakes of these large bubbles (69). Their specific interfacial area is small and their permanence in the culture is short, due to their high ascending velocity. Their contribution to oxygen transfer is therefore modest. On the other hand, a high viscosity hinders the contact, and hence the coalescence, of very small bubbles. They remain small and therefore have no appreciable rise velocity in the highly viscous culture. Their residence time in the culture can be very high, and they soon become depleted of oxygen and saturated with CO_2. At this point they become a nuisance in that they decrease the apparent density of the culture, and thus the power input, while contributing nothing to the mass transfer.The air-liquid interfacial area, a, and the volumetric power consumption have been correlated with V_s, the superficial air velocity (Table 2), obtained by dividing the volumetric flow rate by the cross-sectional area of the vessel, A_c: $V_s = Q/A_c$. For geometrically similar fermentors, $A_c \propto D_i^2$ and $V \propto D_i^3$; therefore, if Q/D_i^2 is kept constant, Q/D_i^3 decreases with scale.

It should be stressed at this point that an important function of aeration is that of removing CO_2 and other volatile metabolites from the culture. This function, known as ventilation, is particularly important in processes in which product formation is inhibited by CO_2. Well-known examples of such products are penicillin (54; R. J. Fox, 1st Eur. Congr. Biotechnol.

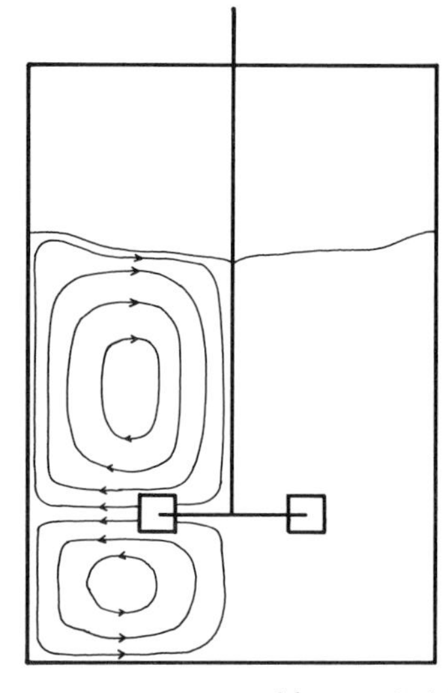

FIG. 8. Flow pattern generated by a radial-flow impeller in a baffled tank.

preprints, part 1, p. 80–83, 1978), erythromycin (51), and inosine (35). The volumetric mass transfer coefficient for CO_2 removal has been shown to agree well with the K_La for O_2 absorption, provided a correction is made for the difference in diffusivity (90). This allows the calculation of pCO_2 in the culture from the exit gas CO_2 concentration (see section 22.2.2) if a pCO_2 electrode is not available.

Maintaining V_s constant decreases the airflow rate/unit volume and therefore increases the CO_2 concentration in the exit gas, if the specific CO_2 evolution rate of the biomass remains constant. This in turn leads to an increase in the pCO_2 in the culture, as is evident from the steady-state CO_2 balance:

$$\frac{(fCO_{2(out)} = fCO_{2(in)})\,Q}{22.4V} = K_La_{CO_2}\,(pCO_{2(b)} - pCO_{2(out)})/H \qquad (37)$$

where fCO_2 = molar fraction of CO_2 in gas; pCO_2 = CO_2 partial pressure (Pa); and for subscripts: in = inflowing gas; out = outflowing gas; b = culture broth. This is an indirect scale effect that must be considered to avoid the danger of supplying enough oxygen to the biomass while allowing the build up of inhibitory CO_2 concentrations.

Actual fermentation broths usually exhibit a more complex flow behavior than the simple model systems employed in early work to establish the basic correlations between operating variables and system response. Recent work has therefore been focused on non-Newtonian culture broths. The general relationship between the shear rate, γ (i.e., the velocity gradient between adjacent layers of liquid), and the shear stress, τ (i.e., the force needed to maintain such gradients), is (6):

$$\tau = \tau_0 + K\,(\gamma)^n \qquad (38)$$

which describes the behavior of Newtonian, pseudoplastic, Bingham plastic, and dilatant fluids, according to the values taken by τ_0, K, and n, as reported in Table 4. An additional type of fluid, known as Casson body, is characterized by the relationship:

$$\tau^{1/2} = \tau_0{}^{1/2} + K_c\gamma^{1/2} \qquad (39)$$

The graphical relationships between τ and γ are illustrated in Fig. 10. The fundamental characteristic of Newtonian fluids is that they have a "true" viscosity that is independent of shear rate and is defined as:

$$\mu = \frac{\tau}{\gamma} \qquad (40)$$

Culture broths of mycelial organisms have been found to belong to almost all these types (20, 22, 38, 44, 74, 78) and to change frequently from one type to another during the course of the fermentation as a result of changes in morphology (22, 44, 83). The effect of the different fluid behaviors on power consumption has been studied mainly in nongassed conditions, which reduces considerably the practical applicability of the relationships developed. On the other hand, air dispersion phenomena in highly viscous liquids are extremely complex. The straightforward application of power curves obtained in the Newtonian fluids (see Fig. 6) is not possible because in general the viscosity is not independent of shear rate in non-Newtonian fluids (see Fig. 10). In fluids described by equation 38, with $n < 1$ (i.e., pseudoplastics), and in Casson bodies, the ratio τ/γ decreases with increasing γ. But γ is maximum at the impeller periphery and decreases (due to internal frictions of the fluid) as the culture moves away from the impeller. The τ/γ ratio will then increase, and therefore the viscosity will not be the same throughout the fermentor. In these cases an average apparent viscosity, $\bar{\mu}_a$, is used to calculate N_{RE} and to build a power curve for the prediction of power consumption (46). The technique, first described for pseudoplastic fluids (46) and later found to apply to other non-Newtonian fluids (14, 15, 28, 45, 47), is based on the finding that in laminar-flow conditions the average shear rate of impellers, $\bar{\gamma}$, is simply given by:

$$\bar{\gamma} = kN \qquad (41)$$

with k having values of 10 to 13. On the other hand, for power law fluids:

$$\tau = K_p(\gamma)^n \qquad (42)$$

and

$$\mu = K_p(\gamma)^{n-1} \qquad (43)$$

If $\bar{\gamma}$ is used in equation 43, the average apparent viscosity is given by:

$$\bar{\mu}_a = K_p(kN)^{n-1} \qquad (44)$$

The value of K_p is obtained from conventional τ versus γ curves determined with a viscometer in connection with equation 42 (see section 22.2.2, Culture Rheology), and k is obtained by the procedure described by Metzner and Otto (46) (see also reference 1), based on the use of a power curve determined in the given fermentor employing a Newtonian fluid or using the widely accepted value of 11 (6, 45, 69). A complete power curve (i.e., extending beyond the viscous flow region) is then constructed by plotting log N_{RE} versus log N_p. At any given impeller speed, N, N_{RE} is calculated from equation 24 using $\bar{\mu}_a$ (equation 44) instead of μ, and N_p is calculated from equation 23 using the measured power consumption, P_0, at the given impeller speed. The power curve is then used to determine, by graphical interpolation, the N_p value for any value of N. First the value of $\bar{\mu}_a$ is calculated by means of equation 44, or is obtained from a rheogram of the culture in correspondence with the value of $\bar{\gamma}$ obtained from equation 41, and then the corresponding value of N_{RE} is found. The resulting N_p is used to predict P_0 from equation 27.

The average apparent viscosity, $\bar{\mu}_a$, can also be calculated from the following equation (14):

$$\bar{\mu}_a = \frac{K_p}{(8N)^{1-n}} \left(\frac{3n + 1}{4n}\right)^n B \qquad (45)$$

where n is the exponent in equation 42, also called the flow behavior index. The value of B has been found to be 1.25 for pseudoplastic fluids ($n < 1$) contained in stirred reactors with $D_i/D_t < 0.67$ (14). Such a technique has extended the nongassed power consumption relationship to most non-Newtonian fluids.

The effect of non-Newtonian viscosity on oxygen transfer is rather complex. Experimental work carried out with either real culture broths (e.g., *Streptomyces niveus* [75], *Penicillium chrysogenum* [58], *Aspergillus niger* [58]) or simulated non-Newtonian fluids (e.g., polymer solutions [88], paper pulp suspensions [10]) has shown that in these conditions K_La is a function not only of power consumption but also of impeller geometry. In the case of *S. niveus* (74), a correlation of the form:

$$\text{OTR} = KP_g^{0.46} \qquad (46)$$

was found to hold for different impeller diameters, but with K increasing with decreasing impeller diameter (6). This phenomenon was explained by assuming that the highly viscous medium prevented bubble coalescence and that therefore the interfacial area generated in the high-turbulence region at the impeller periphery was stable through the period of time taken by the fluid to recirculate to the impeller region. It is therefore possible to achieve the same oxygen transfer rate at a lower power consumption because no (or very little) power is required to redisperse the coalesced bubbles in the bulk of the medium, provided the same maximum shear rate (i.e., impeller tip speed) is achieved. A linear relationship was found between log OTR and log tip speed as illustrated in Fig. 11 (74). The experimental data could be described by the following equation:

$$\log \text{OTR} = 1.6 \log (K\pi ND_i) + K' \qquad (47)$$

Further investigation of this particular system showed that the gas-liquid oxygen transfer was not sufficient to characterize the supply of oxygen to the cells (87). In particular, there was no unique relationship between the O_2 partial pressure in the culture and the respiration rate. At higher stirring speeds, higher respiration rates were observed at the same dissolved oxygen concentration (Fig. 12). The identity of the latter showed that the gas-to-liquid mass transfer was not the controlling step, and pointed to an involvement of the liquid-to-cell mass transfer. The liquid-to-cell mass transfer coefficient, j, was defined by (87):

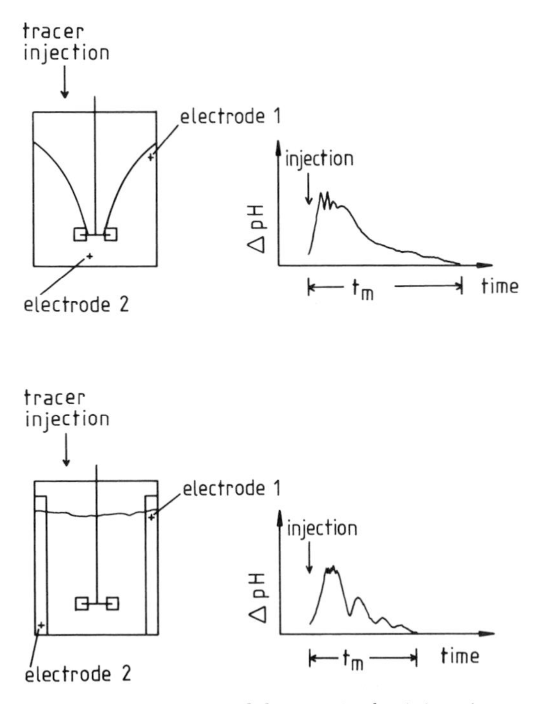

FIG. 9. Determination of the terminal mixing time, t_m, by the tracer injection technique. The pH difference, ΔpH, between the two electrodes is monitored after the tracer injection. (A) Free-vortexing system; (B) baffled system. (From reference 17.)

$$q_{O_2} = j\,(C - C_c) \qquad (48)$$

where q_{O_2} = specific O_2 uptake rate; C = O_2 concentration in bulk of liquid; and C_c = O_2 concentration at the cell's surface (assumed = 0 in these calculations). A unique correlation, valid for different impeller geometries, was found between j and the quantity $ND_i^3h_b$, called the impeller pumping parameter, having the dimensions of a flow rate ($1\ s^{-1}$). h_b was the blade height-to-width ratio in the particular turbine employed.

The effect of impeller geometry on oxygen supply to the cells was also apparent from the correlation between the pseudocritical oxygen concentration, C_c^*, and the shear-to-flow ratio of the impeller. The former was obtained by extrapolating the respiration rate to zero in the q_{O_2} versus DO_2 plots (see Fig. 12), and the latter was defined as N/D_i (87). A linear relationship was found between the shear-to-flow ratio and log C_c^* (see Fig. 13). All of these observations point to the fact that in highly viscous non-Newtonian mycelial cultures, resistance to transport within the mycelial mat can be rate limiting for mass transfer. From the point of view of scale-up, this means that in these instances K_La is probably not the right scale-up parameter, as it measures the gas-to-liquid mass transfer efficiency, which is not rate limiting for the productivity of the culture. Recent work with simulated non-Newtonian broths employing carboxymethyl cellulose solutions (pseudoplastic) has shown that in such solutions K_La can be related to power input, airflow rate, and rheological properties of the culture by expressions that do not contain terms describing impeller characteristics. For a review of this topic, see reference 69.

TABLE 3. Effect of scale on power/unit tip volume, speed, and mixing time[a]

Parameter to be kept constant	Multiplying factor to obtain value of parameter on new scale		
	P/V	t_m	v
P/V	1	$S^{0.67}$	$S^{0.33}$
t_m	$S^{2.75}$	1	$S^{1.25}$
v	S^{-1}	$S^{0.83}$	1

[a] S = Linear scale factor; P/V = power/unit volume (kg m s^{-1} m^{-3}); t_m = mixing time (s); v = peripheral velocity of impeller (m s^{-1}).

The relevance of these results to scale-up will have to be assessed through the application of this type of analysis to real fermentation broths on different scales of operation.

22.1.4 Quality of Medium Ingredients

Different types, or grades, of raw materials are frequently employed on the laboratory scale and the production scale. This practice may reflect the fact that economical considerations become particularly important as the fermentation scale increases, and cheaper or more readily available raw materials are sought at this stage. Such a procedure introduces an unnecessary complication into the scale-up process that can be avoided by employing industrial-grade ingredients when the initial optimization work is carried out. This requires that the economical and logistic aspects be evaluated right at the beginning.

Complex ingredients of agricultural origin, such as corn steep liquor, molasses, soybean and peanut meal, etc., have a highly variable chemical composition and have been found to be critical for the success of a number of fermentation processes. The main reasons for such variability are (i) genetic background of the organism from which the ingredient is derived (e.g., cultivar of soybean or of maize); (ii) cultivation conditions of the organism (e.g., soil composition, addition of fertilizers, and climate); (iii) postharvest treatment, especially in the case of raw materials that undergo subsequent transformation, such as corn steep liquor, in which a lactic acid fermentation takes place. In these cases storage conditions (mainly temperature and time) can be critical.

In a few cases the substance that determines the suitability of a particular raw material for a given fermentation has been identified, but in the great majority of situations, and especially when new processes are being scaled up for the first time, such knowledge is lacking. The most time-efficient procedure seems therefore to be to adopt production-scale raw materials from the first stages of process development. When the fermentation medium contains insoluble ingredients, as is the case with most of today's industrial processes, the surface available for attack by the microorganism controls the rate of substrate utilization and thus the fermentation kinetics. Therefore when the same raw material is available in different granule sizes, it is advisable to use the same type throughout the scale-up procedure. If processs performance seems to depend closely on the composition of one particular ingredient, it is advisable to start an investigation with the objective of identifying the critical substance(s). If successful, this work may eventually lead to better process control through the analytical characterization of raw materials before they are employed in the production process, development of new and simpler fermentation media, and selection of strains that are less sensitive to variations in the level of such critical substance(s) in the raw materials. For a detailed discussion of nutritional factors in process development see chapter 4; substrates for large-scale fermentations are discussed in chapter 10.

A raw material that is frequently overlooked is water. The large amount of water required by fermentation processes is frequently obtained from a number of sources (e.g., wells drilled at various depths) that can yield water of very different characteristics (hardness, metal ions content, etc.). It is common experience that certain types of water give the best results with a particular process. The use of distilled or deionized water is feasible in small-scale fermentations, and indeed a necessity in nutritional studies, but represents a nuisance as well as an additional cost in production fermentors. Unless strictly required by the particular process, it is preferable to use natural, untreated water, obtained from a source as constant and as reliable as possible, for medium make-up. In some geographical areas the level of water-bearing strata is not constant throughout the year, and this can produce changes in the chemical composition of water. As far as scale-up is concerned, it is strongly advisable to use the same type of water throughout.

22.1.5 Heat Transfer

In a fermentation, heat is continuously produced by two main mechanisms, metabolic heat evolution by the culture and mechanical power input (both as agitation and as gas sparging, if used). Fermentations are usually carried out at a predefined temperature, and this requires that a heat balance be maintained in the fermentor. The general form of such a balance is:

$$Q_{metab} + Q_{agitation} - Q_{evaporation} - Q_{exchange} = 0 \quad (49)$$

(where Q_{metab} stands for $Q_{metabolism}$), which neglects the contributions of solution of solids, absorption/desorption of gases, addition of nutrients, and removal of samples. The greatest contributions are made by mechanical agitation and metabolic heat evolution. The former can be easily calculated from the power

TABLE 4. Constitutive equations for various types of fluids[a]

Fluid type	Equation
Newtonian	$\tau = \mu\gamma \ (n = 1, \tau_0 = 0)$
Pseudoplastic	$\tau = K_p\gamma^n \ (n < 1, \tau_0 = 0)$
Dilatant	$\tau = K_p\gamma^n \ (n > 1, \tau_0 = 0)$
Bingham plastic	$\tau = \tau_0 + K_r\gamma, \ (n = 1)$
Casson body	$\tau^{1/2} = \tau_0^{1/2} + K_c\gamma^{1/2}$

[a] τ = Shear stress (kg m^{-1} s^{-2}); γ = shear rate (s^{-1}); K_p = consistency index (kg m^{-1} s^{n-2}); n = flow behavior index; τ_0 = yield stress (kg m^{-1} s^{-2}); K_r = plastic viscosity (kg m^{-1} s^{-1}); K_c = Casson viscosity (kg m^{-1} $s^{-1})^{1/2}$; μ = Newtonian (or true) viscosity (kg m^{-1} s^{-1}).

input, either measured directly or estimated by one of the relationships presented in section 22.1.3 above.

The power dissipated per unit volume (P_{sp}/V) of culture for sparging (3) is given by:

$$\frac{P_{sp}}{V_L} = \frac{F_g \rho_g}{V_L} \left(\frac{\alpha V_0^2}{2} + \frac{RT}{M_g} \ln \frac{P_s}{P_v} \right) \quad (50)$$

where F_g = gas flow rate; ρ_g = gas density; V_0 = gas velocity at the sparger hole; M = molecular weight of gas; P_s = absolute pressure at sparger; P_v = absolute pressure in the vessel; and α = 0.06. For spargers characterized by a sufficiently low air velocity at the holes (i.e., a realistically low pressure drop), the term $V_0^2/2$ can be neglected. The entity of heat loss through evaporation is dictated by the water content of the incoming air and by its temperature, as it can be assumed that the exhaust air is saturated at the fermentation temperature. The total heat change can be calculated by adding the sensible heat variation to the latent heat variation. The contribution of this term is, however, usually small in comparison with those of agitation and metabolism. Metabolic heat is a by-product of the energy-generating mechanisms, and its rate of evolution is proportional to their rate of functioning. In aerobic fermentations the heat evolution rate has been found to be proportional to the oxygen consumption rate for a number of organisms (18, 32, 86).

To satisfy equation 49, it is necessary to remove heat from the system at the rate $Q_{exchange}$, which must be equal to the net accumulation rate obtained by adding together the remaining terms. Heat removal from fermentors is accomplished via exchange surfaces separating the culture from a suitable coolant, usually water. The simplest form of exchange surface is the wall of the vessel itself, which is surrounded by a jacket, either continuous or in the form of half-coils. This arrangement may pose a scale-up problem since the total heat evolution rate at the fermentor increases with the cube of the linear dimensions, while its surface increases with the square of the linear dimensions.

The exchange surface available for removal of heat from the unit volume of culture decreases with scale, and this sets a maximum limit to fermentor size. This is reached when the net rate of heat evolution equals the maximum heat removal rate, i.e.:

$$Q_{exchange} = UA_c \, \Delta T_{lm} \quad (51)$$

where A_c = heat exchange area per unit volume; U = overall heat transfer coefficient; and ΔT_{lm} = logarithmic mean temperature difference between the culture and the coolant. Fermentation scale affects the quantities in equation 51 in various ways. The major effect is on A_c, since:

$$A_c \propto \frac{1}{D_i} \quad (52)$$

ΔT_{lm} is generally independent of scale, provided the same coolant is used throughout, whereas $Q_{exchange}$ may be marginally affected by different specific aeration rates, as happens if the superficial air velocity is maintained constant (see section 22.1.3). In such cases, $Q_{evaporation}$ would decrease with increasing scale

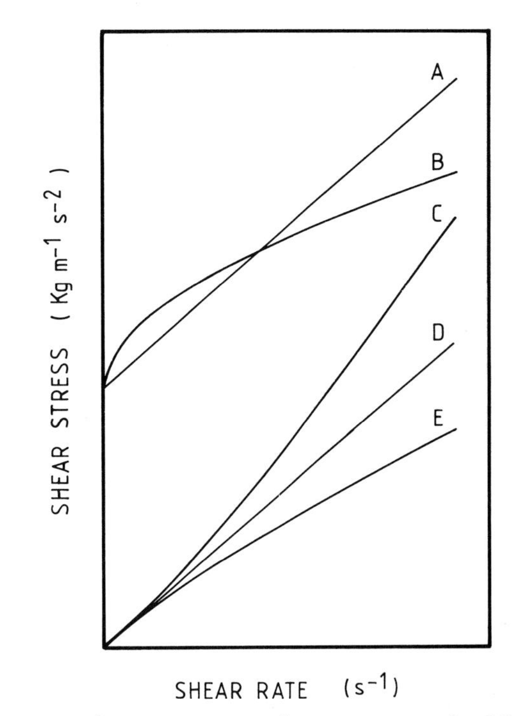

FIG. 10. Shear stress versus shear rate curves for different types of fluids: (A) Bingham plastic; (B) Casson body; (C) dilatant; (D) Newtonian; and (E) pseudoplastic. Constitutive equations can be found in Table 4.

and $Q_{exchange}$ would increase, since, from equation 49:

$$Q_{exchange} = Q_{metab} + Q_{agitation} - Q_{evaporation} \quad (53)$$

The effect on U is more complex. U represents the reciprocal of a resistance to transfer under a driving force, in analogy to K_L (see section 22.1.3). The driving force in this case is the temperature difference between the culture and the coolant. The general expression for U, for heat transfer across plane surfaces, is (3), in the absence of fouling:

$$U = \left(\frac{l}{K_w} + \frac{1}{h_i} + \frac{1}{h_o} \right)^{-1} \quad (54)$$

In the first term within the parentheses, l = wall thickness and K_w = thermal conductivity of wall. Larger fermentors require thicker walls, i.e., larger l, and this decreases U. On the other hand, h_i and h_o are the film heat transfer coefficients relative to the culture and the coolant, respectively. These are analogous to K_L in equation 21 and incorporate heat conductivity and stationary film thickness. In analogy to K_L, they depend on liquid turbulence, as the latter affects the stationary film thickness. The correlation between these variables can be expressed by an equation of the form (3):

$$\frac{h_i D_T}{K} = E N_{RE}^{a} N_{PR}^{b} \left(\frac{\mu}{\mu_m} \right)^{c} X \quad (55)$$

where N_{RE} = Reynolds number = $(ND_i^2\rho)/\mu$; N_{PR} = $(C_p\mu)/K$ (C_p = heat capacity of liquid); μ = viscosity of bulk liquid; μ_m = liquid viscosity at wall; K, E, and X = geometric factors; and a, b, and c are positive

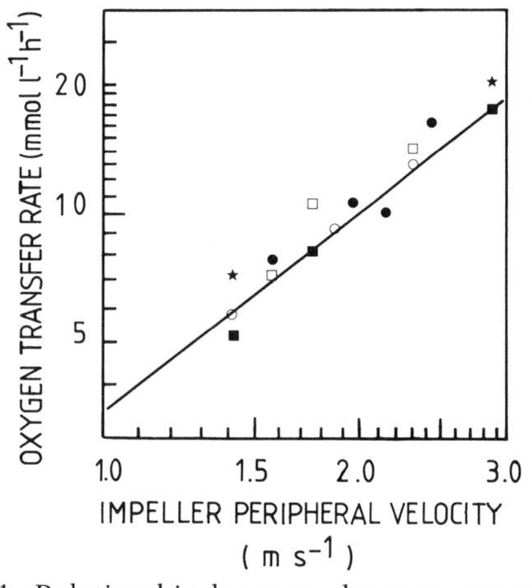

FIG. 11. Relationship between the oxygen transfer rate, OTR, and the impeller peripheral velocity in *S. niveus* cultures. (From reference 76.)

numbers. Equation 55 shows that h_i increases with increasing N_{RE}, and this decreases with increasing viscosity, μ. The situation is even worse with the majority of non-Newtonian fluids, in which the viscosity increases with decreasing shear rate and will thus be maximal in the stagnant film, where $\mu/\mu_m < 1$. Therefore, in addition to the already discussed agitation problems, viscous non-Newtonian culture broths present scale-up difficulties at the level of heat transfer (72). In conclusion, heat transfer capacity decreases with scale and can pose an upper limit to fermentor size. Methods for calculating such size have been discussed by Blenke (11).

22.2 SCALE-UP PROCEDURES

22.2.1 Choice of the Number of Stages

The number of stages, i.e., different fermentor sizes, that should be used in a scale-up procedure has been the subject of considerable debate and is still largely a matter of personal choice. The time factor and the need to contain costs (both investment and operating costs) within reasonable limits both advise against a large number of stages. On the other hand, it is necessary to assess the effects of scale on the particular process, and it must also be possible to carry out extensive experimentation at a reasonable cost in a fairly short time. These requirements can be met by a structure composed of a number of small laboratory fermentors (1 to 20 liters) and fewer fully instrumented pilot plant fermentors. A detailed discussion of pilot plant structure and function can be found in chapter 24. The first and the last stages are more or less fixed; on the one hand there is a wide agreement on the opportunity to start scale-up work in laboratory fermentors, and on the other hand the final stage is the production fermentor, which can be of any size between 20 and 1,000 m^3, according to the particular process and plant.

The main problem is therefore the number and the volume of the intermediate stages. The tendency to proceed through a number of small volume increases can be associated with the largely empirical nature of fermentation development, which tends to avoid gross disturbances of the system as being complex and little understood. As the knowledge of fermentations, and particularly the identification of the rate-controlling steps, develops, the need for such a cautious and slow approach will probably disappear. Today's tendency favors a reduced number of pilot plant stages (one or two), employing a small number of highly instrumented fermentors. It seems desirable to have the larger pilot plant fermentors as big as possible, particularly if mycelial organisms yielding viscous non-Newtonian broths are employed. In such cases larger fermentors can allow an early appreciation of mixing problems that might not occur in smaller apparatus. This applies particularly to fermentations requiring the continuous slow feed of nutrients and to continuous fermentations proceeding at low dilution rates. In such cases a distinct nonhomogeneity of the culture may result, due to the fact that the added nutrients are consumed by only a portion of the culture, as the mixing time is high in comparison with the metabolic rate of the culture (30). A typical pilot plant layout may therefore comprise two sizes of fermentors, e.g., 50 to 100 and 1,000 to 5,000 liters. It is desirable that the two types of vessel be geometrically similar to allow the investigation of scale-up relationships. If the production fermentors already exist, or the design has been decided, then it is of advantage to have the pilot fermentors geometrically similar to them.

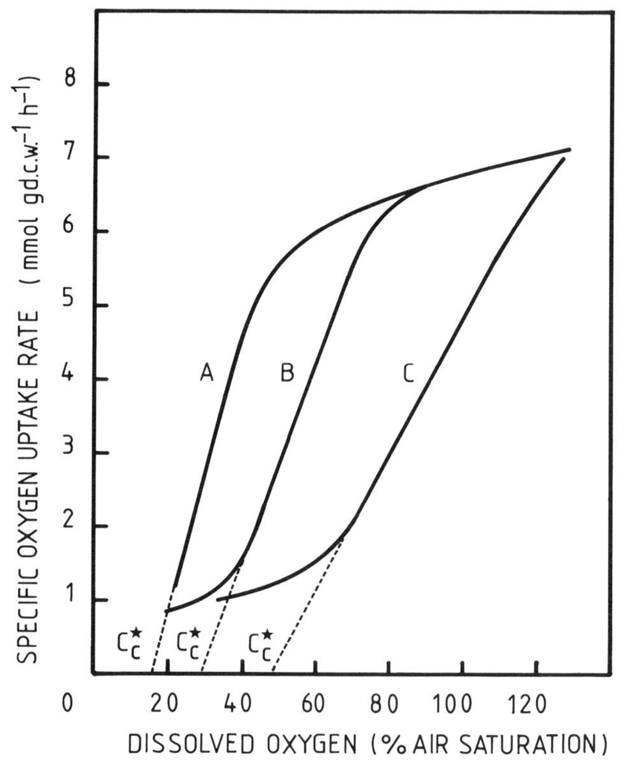

FIG. 12. Effect of impeller speed on the relationship between dissolved oxygen concentration and oxygen uptake rate in *S. niveus* cultures. Speeds: (A) 350 rpm; (B) 250 rpm; (C) 150 rpm. C_c^* = Pseudocritical O_2 concentration. (From reference 87.)

22.2.2 Characterization of the Process

The development of new fermentation processes and the improvement of existing ones frequently involve shake flasks as fermentation vessels (for a detailed discussion see chapter 6). Their main advantage is the possibility of exploring a wide range of strains and culture environments in a short time and at a low cost. The disadvantage is the difficulty of achieving satisfactory environmental control, particularly with respect to aeration-agitation, pH control, and nutrient additions. An early transfer of the process to stirred fermentors is thus highly desirable. This is frequently a critical step, as the characterization of the shake-flask fermentation is usually rather limited. It should, however, be borne in mind that this is not necessarily so, as a number of fermentations perform better in small fermentors than in shake flasks. In general it is advisable to assess the effects of pH, dissolved O_2, dissolved CO_2, and addition of nutrients at this stage. The objective is to optimize the process on what can be considered the starting scale of the scale-up procedure. Transfer onto a larger scale (and finally to production scale) of the complete set of the optimal fermentation parameters is generally not possible, especially for agitation-aeration (see section 22.1.3). On the other hand, one must consider the technical limitations imposed by the production plant (if already existing) and the economical considerations. The rational approach to this problem requires (i) characterization of fermentor performance on the starting scale; (ii) identification of the critical values of fermentation parameters; and (iii) scale down of existing, or future, production scale constraints.

Characterization of fermentor performance

Gas-liquid mass transfer. Gas-liquid mass transfer is measured by the volumetric mass transfer coefficient K_La (see section 22.1.3) and must be correlated with the geometrical and operating conditions of the fermentors on the different scales. Two groups of methods of K_La determination are available, i.e., those applicable to model systems and those applicable to actual fermentations. A detailed discussion of the various methods is beyond the scope of this chapter, and for this purpose the reader is referred to the excellent review of Sobotka et al. (73). From the practical point of view the methods applicable to actual fermentations are the most interesting. Basically two types of methods are available, dynamic methods and steady-state methods.

Dynamic methods consist of the following steps.

1. Reduce the OTR of the culture to zero by either turning off the air supply or substituting a nitrogen stream for the airstream. The general O_2 balance in the culture:

$$\frac{dC}{dt} = K_La\,(C^* - C) - q_{O_2}x \tag{56}$$

will then reduce to:

$$\frac{dC}{dt} = -q_{O_2}x \tag{57}$$

so that $q_{O_2}x$, i.e., the volumetric rate of O_2 consumption, can be estimated from the slope of O_2 concentration versus time. The dissolved oxygen concentration should not be allowed to fall below the critical value at which q_{O_2} becomes a function of C.

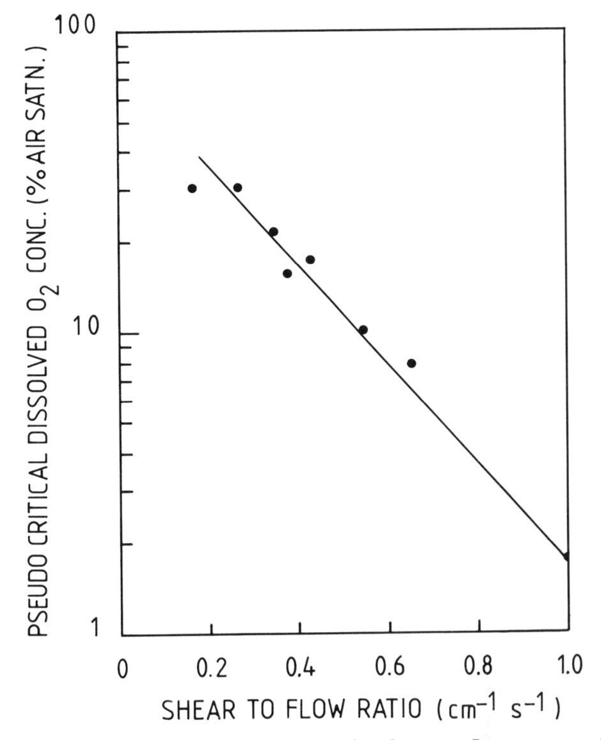

FIG. 13. Relationship between the shear-to-flow ratio and the pseudocritical O_2 concentration in *S. niveus* cultures. (From reference 87.)

2. Resume the original aeration conditions. C will then increase according to equation 56 until the value:

$$C_B = C^* - \frac{q_{O_2}x}{K_La} \tag{58}$$

is reached, representing the stationary value of C in those culture conditions. Equation 56 can be rewritten to give:

$$C = C^* - \frac{\left(\dfrac{dC}{dt} + q_{O_2}x\right)}{K_La} \tag{59}$$

which enables the estimation of K_La as the reciprocal of the slope of a C versus $[(dC/dt) + q_{O_2}x]$ plot. The ideal time course of such an experiment is depicted in Fig. 14.

The main assumptions underlying this method of K_La determination are: (i) the fermentor is perfectly mixed, and there is no lag between the change in aeration and the attainment of the K_La value pertaining to the new aeration conditions; and (ii) the oxygen electrode responds with no lag to changes in dissolved oxygen concentration. The first assumption is unlikely to be fulfilled in large fermentors, and the time needed to attain the new equilibrium conditions (especially the gas hold up and the interfacial area, a) increases with scale. On the other hand, extensive investigations on oxygen electrode dynamics (73) have shown that oxygen diffusion through the membrane and through the electrolyte layer separating the membrane from the cathode introduces a time lag in the electrode response. Models of varying complexity have been proposed to take into account this phenomenon (73). However, the electrode lag poses a limit to the maximum K_La value that can be measured accurately. For the probe delay to be negligible, it must be $K_La << 1/t_p$, where t_p is the probe constant defined as the time needed for the electrode output to reach 63.2% of the equilibrium value after a step change of O_2 concentration. On the other hand, there is a scale limitation in that this determination is not meaningful if the mixing time (see below) is comparable to $1/K_La$. The latter is perhaps the most serious limitation of the technique.

Steady-state methods of K_La determination assume that during the determination all the involved fermentation parameters are constant. If, in particular, the volumetric respiration rate, $q_{O_2}x$, is constant, then, from equation 58:

$$K_La = \frac{q_{O_2}x}{C^* - C_B} \tag{60}$$

The measurement of the volumetric rate of oxygen consumption, $q_{O_2}x$, can be carried out by means of the O_2 balance in the gas phase:

$$Vq_{O_2}x = \frac{P_1F_1}{RT_1}C_{g1} - \frac{P_2F_2}{RT_2}C_{g2} \tag{61}$$

where V = culture volume; P = gas absolute pressure (Pa); F = gas flow rate ($m^3\ h^{-1}$); T = absolute temperature (K); C_g = molar fraction of oxygen (subscripts 1 and 2 refer to inlet and outlet values, respectively); and R = gas constant = 8.31 m^3 Pa mol^{-1} per degree Kelvin.

The group P/RT in equation 61 corrects for temperature and pressure differences in the inlet and outlet gas and converts m^3 to mol. Care must be taken in determining the O_2 molar fraction correctly. In particular, if an O_2 analyzer is used that requires a dry air sample, then the water pressure in the gas streams must be taken into account to correct the O_2 concentration given by the analyzer. Failure to do so produces overestimates of O_2 concentration. The correct O_2 concentration, C_g, is given by:

$$C_g = C_{g(\text{dry})}(1 - W) \tag{62}$$

where W = molar fraction of water in the gas stream. The exhaust air can be assumed to be saturated at the fermentation temperature, and therefore $W = P_W/P_2$, where P_W is the water vapor pressure at the given temperature. P_W in the inflowing air must be measured directly.

The steady-state determination of K_La has the following advantages: (i) it introduces no disturbances in the culture, unlike the dynamic methods; (ii) it can be used at any O_2 concentration, even zero, whereas the dynamic method requires $C_B > O$ to provide the driving force for the transient C increase; and (iii) K_La can be acquired on-line and be used as a stable variable in control strategies, since both C_{g_2} and C_B can be obtained as continuous signals from an O_2 analyzer and an O_2 electrode, respectively. These values can be fed, along with P, T, and F, to a computer for the K_La calculation. Computers are becoming a standard component of fermentation pilot plants as well as of production plants (see chapter 23), and on-line K_La determination is a standard feature.

Throughout the preceding discussion, dissolved oxygen has been expressed as a concentration, whereas the standard method of measurement is by means of oxygen electrodes that produce signals proportional to O_2 partial pressure and not to O_2 concentration (73). The determination of K_La in the units of reciprocal time (s^{-1} or h^{-1}) therefore requires the knowledge of the Henry constant for the particular culture fluids. The O_2 concentration can then be obtained from its partial pressure from the relation:

$$C = \frac{p}{H} \tag{63}$$

H is related in a complex way to the chemical composition of the culture medium (9) and to temperature. In scale-up work one is usually interested in comparing oxygen transfer rates into the same medium on different scales. It is therefore expedient to incorporate the Henry constant into K_La and rewrite equation 60 as follows:

$$\frac{K_La}{H} = \frac{q_{O_2}x}{p^* - p} \tag{64}$$

where the volumetric mass transfer coefficient is in mol $m^{-3}\ h^{-1}\ Pa^{-1}$.

A final point is the determination of p^* and p. Only under the assumption of perfectly mixed gas and liquid phase can a single pair of values describe the true driving force existing in the fermentor. In large vessels the gas phase is better described by a plug flow behavior, and then the correct driving force is given by the logarithmic mean:

$$\Delta p_{\text{mean}} = \frac{p_1^* - p_2^*}{\ln \dfrac{p_1^* - p}{p_2^* - p}} \tag{65}$$

where p_1^* and p_2^* are the partial pressures of O_2 in the gas phase at the inlet and the outlet of the fermentor, respectively. It should be noted that equation 65 assumes perfect mixing of the liquid phase, as only one p value is considered. Recent work carried out in 112-m^3 fermentors (41) has shown that axial profiles of p are obtained that can be interpreted by assuming some degree of top-to-bottom recirculation of the liquid phase. The degree of recirculation could be described by a single dimensionless parameter expressing the ratio of mass transfer to mixing efficiency, a characteristic of a given fermentor geometry. The axial p profiles in actual streptomycete fermentation broths were found to be parallel to the axial profiles of p^*, indicating the existence of a constant driving force throughout the fermentor. If generally applicable to other fermentation systems, this result would greatly simplify the interpretation of K_La in large-scale fermentors.

A relationship must then be established between K_La, or K_La/H, and the operating variables, i.e., essentially power input and aeration rate. Its function is to predict K_La values resulting from given sets of operating conditions on the large scale, and its validity on different scales must therefore be checked. With cultures of unicellular organisms such as yeasts or bacteria, the relationship between power consumption and K_La can be expected to be independent of cell density (44, 89), but with mycelial organisms a dependence on both cell density and hyphal morphology is certain to occur. In such cases it is advisable to study the K_La versus operating parameters, employing the cultures at their most critical age from the point of view of oxygen supply. The use of uninoculated culture medium is generally misleading, as K_La can decrease during the fermentation by a factor of 5 to 7 (36, 83).

It is advisable to use a simple correlation, in view of the many complications that may arise due to variations of the culture properties during the fermentation. For each scale of operation, a number of experiments with different power inputs (measured directly) and airflow rates should be carried out to determine the coefficients K, a, and b of the following equation:

$$\frac{K_La}{H} = K\left(\frac{P_g}{V}\right)^a V_s^{\,b} \tag{66}$$

which can be expected to describe with reasonable approximation even the behavior of mycelial, non-Newtonian broths (77) (Fig. 15). The values of a and b can be quickly obtained as the slopes of log-log plots of K_La versus P_g/V and V_s, respectively, and K is obtained from the plot of equation 66 by using the above values of a and b. This requires two series of K_La determinations, one with P_g/V varying and V_s constant, and the other with V_s varying and P_g/V constant. Alternatively, K, a, and b can be found by multiple regression analysis. If no unique correlation is found to relate K_La to P_g/V and V_s on the different scales, the different number of impellers on the various scales could be responsible. The following relationship could then give a better fit (86):

$$K_La/H = (\alpha + \beta N_i)\left(\frac{P_g}{V}\right)^{0.77}(V_s)^{0.67} \tag{67}$$

where N_i = number of impellers, and α and β are constants. If K_La/H is expressed as mol of O_2 m^{-3} Pa h^{-1}, P_g/V as kW m^{-3}, and V_s as m s^{-1}, then $\alpha = 20$, $\beta = 2.8$, and the right-hand side of equation 67 must be multiplied by the factor 8.06×10^{-3} to convert original units (27) to SI units. Equation 67 was found to apply, irrespective of scale, for up to 42 m^3 of fermentor volume, but K_La/H was measured by the sulfite oxidation technique (27) (Fig. 16). In actual fermentation broths the exponents on P_g/V and V_s are likely to be different and must be determined by experiment. For viscous cultures it is highly desirable to incorporate the effect of viscosity on K_La into an empirical correlation of the form:

$$K_La = (\alpha + \beta N_i)\left(\frac{P_g}{V}\right)^a V_s^{\,b}\, \mu_{app}^{\,c} \tag{68}$$

The preceding equations do not explicitly incorporate the geometrical characteristics of the agitation system, implying that these relationships are valid irrespective of geometry and, in particular, of impeller diameter. This assumption is likely to be invalid for viscous, non-Newtonian broths, where the reduced coalescence of air bubbles and clump formation can render the O_2 transfer mechanism primarily sensitive to the maximum shear rate and therefore to the impeller peripheral velocity (87). Evidence of such a situation can be obtained as described below ("Determination of the critical values of fermentation parameters"). Apart from such a case, a relationship of the type expressed by equations 66 and 67 enables one to predict the gassed power consumption needed on the large scale. The corresponding impeller rotation speed can then be calculated by the combined use of the Michel and Miller relationship for gassed power consumption (equation 28) and equation 27, giving the nongassed power consumption, P_0. The final expression is:

$$N = 544P_g^{\,0.317}\,(nN_p\rho)^{-0.286}\,D_i^{-1.86}\,Q^{0.08} \tag{69}$$

where P_g is HP, N is min^{-1}, D_i is m, Q is m^3 s^{-1}, ρ is kg m^{-3}, and the value of C in equation 28 is 0.08, as obtained by Michel and Miller (48).

This equation is valid for geometrically similar vessels for both Newtonian and non-Newtonian broths (77), provided it is used in the turbulent region, where N_p is essentially independent of N_{RE}. If the rotation speed is fixed, the resulting impeller diameter is given by:

$$D_i = 29.7\,P_g^{\,0.171}\,(nN_p\rho)^{-0.154}\,N^{-0.54}\,Q^{0.043} \tag{70}$$

if the units specified for equation 69 are employed. It is, however, advisable to determine the value of C in each system, as this factor depends on fermentor geometry and broth characteristics. If power consumption is expressed as kW, the numerical constants in equations 69 and 70 become 256.5 and 43.8, respectively.

If the culture broth exhibits non-Newtonian behavior, then the power curve should be determined as described in section 22.1.3. The resulting N and P_0 must be compared with the characteristics of the existing production fermentors (motor, gear box, impeller shaft). P_0 is in this respect more important than P_g because it represents the load on the motor in the absence of aeration. The latter can be either purposeful (e.g., during sterilization) or accidental (e.g., air compressor failure). This problem is overcome by variable-speed agitators.

Mixing. The degree of mixing is usually estimated by the mixing time, defined as the time needed to disperse uniformly a pulse of tracer throughout the fermentor (see section 22.1.3). The mixing time is strongly affected by the rheology of the culture, and it should therefore be measured in actual fermentation conditions. To this purpose the tracer used must be nontoxic to the culture, and the disturbance introduced must be as small as possible. Heat possesses these characteristics, provided small temperature changes are involved (87). Pulses of acid or alkali that give appreciable, but small, pH changes are also suitable and have been widely used. In either case, the desired physical or chemical change is followed by recording the output of a suitable fast-responding sensor (thermoresistance, thermocouple, pH electrode). As the rate of change of the parameter under investigation slows down as the after-pulse equilibrium value is approached, the exact determination of the terminal mixing time is difficult, and the time needed to attain 90 or 95% of the equilibrium value is usually determined (57). This determination of the mixing time was proved to be sensitive to the nature of the tracer employed and to the mode of pulse addition, and therefore a consistent technique should be used on the different scales of operation. The relationship between the mixing time factor, ϕ, and the Reynolds number, N_{RE} (see equation 29 and section 22.1.3), has been found to predict qualitatively the behavior of *Streptomyces niveus* broths (87), although longer mixing times were obtained than those predicted. The modified N_{RE} of Calderbank and Moo-Young (14) was used, incorporating the apparent average viscosity characterizing the particular non-Newtonian broth (see equation 45). The experimental determination of ϕ versus N_{RE} on the various fermentation scales is highly desirable. In particular, the effect of viscosity must be assessed, since this parameter can vary widely during the course of the fermentation.

The use of mixing time as a scale-up parameter is prohibitive in terms of power consumption and requires impeller speeds that may damage mycelial organisms in large fermentors (see Table 3). The degree of nonhomogeneity of the culture is therefore bound to increase with scale. The limit is reached when the overall productivity of the culture starts decreasing as a result of either starvation of nutrients or accumulation of inhibitors in stagnant zones. The time needed to recirculate the culture to the high-turbulence impeller regions controls the alternation between optimal and suboptimal conditions of the biomass. For a given culture volume, the circulation time can be decreased by using multiple impellers, but this introduces the problem of fluid exchange between the various impeller regions. The ratio of inter-impeller fluid exchange flow and circulation flow has been found to be independent of power input in aerated, pseudoplastic suspensions, simulating mycelial cultures, and to be rather low, of the order of 0.13 (50). This means that a given fluid element, on the average, starts a new circulation in the same mixing compartment 87 times out of 100 and changes mixing compartment only 13 times out of 100. This result was obtained with flat-blade turbines and supports the notion that this type of impeller is very efficient in promoting mass transfer, but much less so in promoting bulk mixing.

Culture rheology. The rheological behavior of the culture is an important process parameter, as it affects mixing and mass and heat transfer and, therefore, process productivity. The rheological behavior of a fluid is characterized by the shear stress versus shear rate curve, or rheogram. The various fluids are classified according to the shape of their rheograms and their mathematical description (see Fig. 10 and Table 4). Rotating-cylinder viscometers are usually employed with fermentation broths, since the particulate matter present in such fluids would interfere with capillary or falling-sphere viscometers. In addition, the preparation of rheograms is much easier with rotating-cylinder viscometers, as in most modern apparatus the variation of rotation speed, and therefore of shear rate, is done automatically according to a predefined program. The two main types of rotating-cylinder viscometers are the coaxial-cylinder viscometer and the suspended-bob viscometer. In the former type the gap between the inner (rotating) and the outer (stationary) cylinder is small, so that the ratio of their diameters is close to 1 (usually <1.15). In such conditions the radius of curvature is large compared with the gap, and the two surfaces in contact with the liquid approximate parallel planes. The tangential force that the liquid contained in the gap exerts on the rotating inner cylinder appears as a torque on the cylinder shaft, and therefore the shear stress, τ, is given by:

$$\tau = \frac{M}{2\pi r^2 L} \tag{71}$$

where M = torque (kg m^2 s^{-2}); r = cylinder radius (m); and L = cylinder height (m). The dimensions of τ are kg m^{-1} s^{-2}.

The shear rate, assuming that the liquid in contact with the cylinders' surfaces has the same velocity as the cylinders, is:

$$\gamma = r\frac{\omega}{\Delta r} \tag{72}$$

where ω = angular velocity of rotating cylinder (rad s^{-1}) and Δr = gap between cylinders (m). γ has the dimension of s^{-1}.

For a Newtonian fluid:

$$\mu = \frac{\tau}{\gamma} = K_v \frac{M}{\omega} \tag{73}$$

where K_v is a constant for the given geometry of concentric cylinders. If the gap between the cylinders cannot be considered small compared with the cylinder radius, then the constant K_v in equation 73 becomes:

$$K_v = \left(\frac{1}{R_o^2} - \frac{1}{R_i^2}\right) \bigg/ 4\pi L \tag{74}$$

The relationships between M, ω, τ, and γ for non-Newtonian fluids are more complex.

In suspended-bob viscometers, such as the well-known Brookfield viscometer, the gap between the rotating cylinder and the cup is very large, and therefore the liquid layer can be considered semi-infinite. The shear rate cannot therefore be calculated by equation 72 but can be obtained from the relationship:

$$\gamma = \frac{4\pi N}{n^1} \tag{75}$$

where n^1 = slope of the logarithmic plot of M versus N (72). τ is given by equation 71. A drawback of conventional rotating-cylinder viscometers is that particulate materials such as microorganisms tend to move away from the rotating cylinder so that eventually the latter rotates in a thin layer of clear liquid, experiencing a substantially lower shear stress. This distortion is particularly severe with mycelial broth, whose rheological behavior is dictated by the three-dimensional network formed by the interacting filaments.

This difficulty can be overcome by taking advantage of the relationship between power input and rotation speed of flat-blade turbines operating at low Reynolds numbers (<10). In this range the power consumption of the impeller is given by equation 26 (section 22.1.3) and is proportional to fluid viscosity. Power consumption can be calculated from the torque on the impeller shaft:

$$P = 2\pi NM/g_c \tag{76}$$

Elimination of P from equations 26 and 76 gives:

$$\mu = \frac{2\pi M}{KND_i^3} \tag{77}$$

For non-Newtonian fluids the true viscosity, μ, is substituted for by the average apparent viscosity $\bar{\mu}_a$ (see equation 45). The value of the constant K can be calculated from the slope of an M versus N curve obtained with a Newtonian fluid of known μ. Having thus characterized the particular system, this can be used to measure μ for Newtonian cultures and $\bar{\mu}_a$ for non-Newtonian cultures (12). This particular type of set-up is available from Haake, Berlin, under the denomination "FL measuring device" and covers a range of Newtonian viscosity from 100 to 10^6 mPa s.

The problem of phase separation is also circumvented by the use of the "P measuring devices" of the same company, in which the surface of the coaxial cylinders bear longitudinal indentations acting as small turbine blades. These non-smooth surface measuring devices are strongly recommended for mycelial cultures (44).

There are two points of particular importance in the determination of rheological properties of mycelial culture broths: (i) the sample must be completely deaerated, as small entrapped bubbles can alter its flow behavior, and (ii) torque readings must be taken after a period of intense agitation and within 5 to 10 s after the rotation speed has been regulated to the set value. This minimizes the pronounced effect of sedimentation (44). This technique can only be applied to a fluid in laminar flow, and therefore care must be taken that the impeller speed, N, does not exceed the value above which turbulent flow ensues. In the latter condition equations 76 and 77 do not apply, and in principle power consumption is independent of viscosity. The safe operating range can be determined by a plot of torque versus impeller speed. The transition from laminar to turbulent flow appears as an abrupt increase of the slope of the plot, as a consequence of the fact that torque in the turbulent region is proportional to the square of impeller speed (44).

The next problem is the interpretation and use of rheological data. Cultures of unicellular microorganisms usually exhibit Newtonian behavior, and their viscosity is a function of cell density alone. However, pseudoplastic behavior was observed for a yeast suspension above 100 g of dry biomass per liter (39). The relationship between cell density and viscosity is usually described by the Vand equation:

$$\mu = \mu_L(1 + 2.5\phi_c + 7.25\phi_c^2) \tag{78}$$

where μ_L = viscosity of suspending liquid and ϕ_c = volume fraction of cells. On the other hand, cultures of mycelial organisms generally exhibit non-Newtonian behavior, and often the rheological type changes during the fermentation. The usual reason for rheology changes with culture age is the variation of biomass concentration, morphology, or both, but in cultures producing polymers these also contribute to the overall rheology.

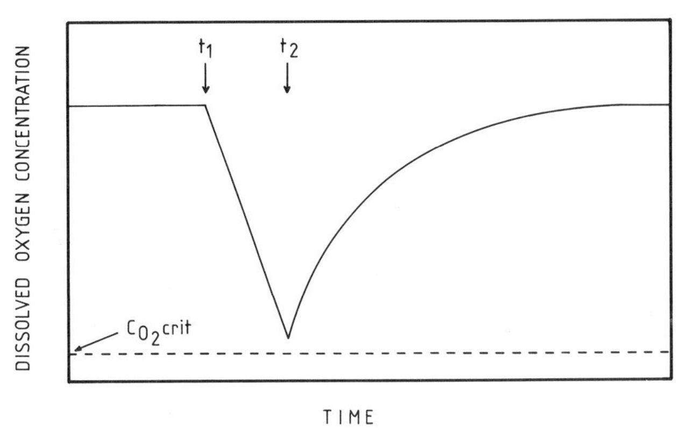

FIG. 14. Ideal dissolved oxygen profile during a K_La determination by the dynamic method. O_2 supply is discontinued at $t = t_1$ and resumed at $t = t_2$. $C_{O_2(crit)}$ = O_2 concentration below which the respiration rate becomes dependent on O_2 concentration.

The most frequent rheological types observed in mycelial cultures are the pseudoplastic, the Bingham plastic, and the Casson type. The corresponding typical rheograms are depicted in Fig. 10, and the constitutive equations are given in Table 4. The appropriate rheological type can be identified by comparison of the experimental rheograms with the typical ones, and the relevant parameters (consistency coefficient, flow behavior index, yield stress, Casson viscosity) can be identified by application of the relevant constitutive equation. From the application point of view it is important to determine the rheological behavior of the culture as a function of culture age, so that one concentrates upon the rheological properties displayed by the culture in the critical phases, when mass (or heat) transfer limitation occurs or is likely to occur in larger fermentors. It is advantageous to describe the rheological behavior of a culture by the power law equation, as this facilitates the prediction of the apparent viscosity from the operating parameters (see section 22.1.3). If the power law does not describe the whole range of shear rate, it is worthwhile to apply it to a limited range of shear rates that includes the values expected in actual fermentation conditions. Pseudoplastic fluids at high shear rates tend to exhibit a linear relationship between shear rate and shear stress, and in this particular region could be treated as Bingham plastics, with the attendant advantage of a constant viscosity. For a detailed discussion of techniques and approaches to culture fluid rheology, the reader should consult references 17, 44, and 72.

Rheological data can be utilized in at least two ways. In the first place, they can be used in connection with the relevant relationships to predict K_La (equation 68), mixing time, and nongassed power consumption through the identification of the correct power number (see section 22.1.3). In the latter case, the apparent viscosity is used to calculate the modified Reynolds number, which is then entered in the exper-

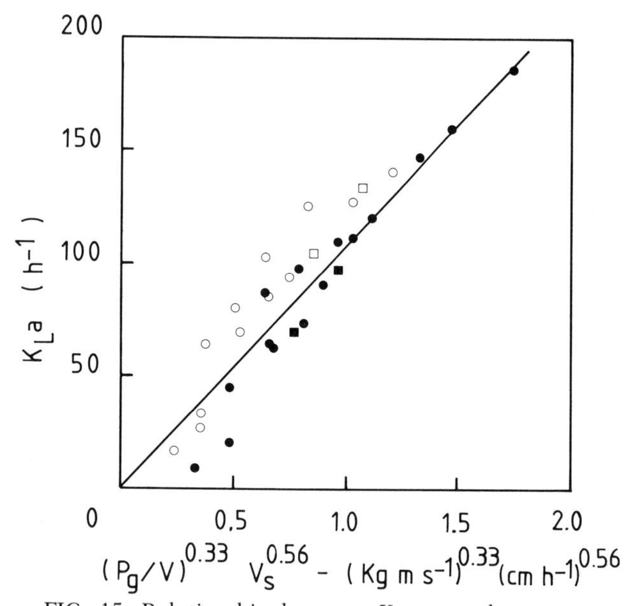

FIG. 15. Relationship between K_La, gassed power consumption, and superficial air velocity in a non-Newtonian broth (*Endomyces* sp.) in vessels of 30 to 50,000 liters. Symbols: (●) 30 liters; (○) 100 liters; (□) 5,000 liters; (■) 50,000 liters. (From reference 78.)

imental θ versus N_{RE} curve or power curve, respectively. Second, the time course of rheological behavior provides valuable information on the development of culture morphology. The latter is strictly related to nutritional and environmental factors in a way that is characteristic of each process and can therefore constitute an indicator of culture performance. In some industrial fermentations operating close to mass transfer and heat transfer limitation, viscosity is used as a state variable on the basis of which control variables, such as feed rate of certain additives, are modulated. In this respect viscosity is highly relevant in the line of research that seeks to quantitatively correlate mycelial morphology with rheological behavior (44, 66).

Determination of the critical values of fermentation parameters

The indentification of the relevant parameters is of necessity somewhat arbitrary, as different levels of process breakdown are possible. Nevertheless, the work carried out in the past has shown that some properties of the fermentation system are particularly important: (i) gas-to-liquid mass transfer; (ii) liquid-to-cell mass transfer; (iii) fluid shear; (iv) macromixing; and (v) heat evolution. These parameters exert their effects on the fermentation by controlling the value of the state variables (temperature, pH, dissolved oxygen, dissolved carbon dioxide, concentration of nutrients) and their homogeneity throughout the culture. The fermentation development work carried out in small fermentors, i.e., the initial stage of scale-up, should lead to the identification of a set of optimal values of state variables. One has then to ensure that these values are achieved in larger scale fermentors, homogeneously throughout the culture.

Gas-liquid mass transfer. A sufficient gas-liquid mass transfer must be ensured to satisfy the oxygen demand of the culture and to remove at a sufficient rate toxic gaseous metabolites such as CO_2. The relevant state variables are therefore $pO_{2(crit)}$, the critical dissolved oxygen partial pressure, and $pCO_{2(crit)}$, the critical carbon dioxide partial pressure (35). These two values must be determined with respect to a relevant measure of process performance that is not necessarily respiration rate. In fact, in several antibiotic fermentations the $pO_{2(crit)}$ for production is higher, and thus more restrictive, than that for respiration (24). The determination of pCO_2 has recently become practicable on the large scale upon the introduction of steam-sterilizable pCO_2 probes. The simplest way to determine $pO_{2(crit)}$ and $pCO_{2(crit)}$ is in steady-state continuous cultures (see chapter 11), preferably at various dilution rates. The transferability of the results to batch systems may not be straightforward, and it is therefore advisable to attempt such determinations in the actual batch process. It is also advisable to do so at different culture ages, as the effects of O_2 and CO_2 on the culture are likely to be dependent on culture age. As a minimum requirement, $pO_{2(crit)}$ and $pCO_{2(crit)}$ should be determined in that phase of the fermentation in which the gaseous exchanges are critical, as suggested by the pO_2 and pCO_2 curves. The critical K_La value can then be calculated using the following equation:

$$OUR_{max}\, H = K_La_{crit}\,(pO_2{}^* - pO_{2(crit)}) \qquad (79)$$

where $pO_2{}^* = O_2$ partial pressure in the gas phase.

As previously discussed, the K_La value obtained by dividing OUR by the driving force expressed as partial pressure incorporates the Henry constant, but this is no problem as temperature and medium composition are assumed to be constant on the various fermentation scales. For the correct determination of $pO_2{}^*$ see Characterization of Fermentor Performance, above. The K_La value so obtained can be used as a scale-up criterion, but it should be remembered that liquid-to-cell mass transfer could be the important step (see below). If this is so, then pO_2 ceases to be a meaningful state variable because it does not measure unequivocally the oxygen partial pressure "sensed" by the cells, and the relation between the measured and the "true" pO_2 is a function of the fluid regime (8, 83) (see section 22.1.3).

Liquid-to-cell mass transfer. Evidence that the liquid-to-cell mass transfer controls the overall gas-to-cell mass transfer can be obtained by examining the relation between pO_2 and q_{O_2}, as determined with a differential O_2 analyzer. If different curves are obtained at different impeller speeds, the liquid-to-cell oxygen transfer is probably the rate-limiting step (87). A quick check during a fermentation run can be carried out by increasing the agitation speed and decreasing the airflow rate or the back pressure so as to maintain pO_2 constant. An increase in OUR constitutes evidence for liquid-to-cell resistance to oxygen transfer. A direct calculation of a mass transfer coefficient is generally not possible, as the pO_2 at the cell surface is not known. The assumption that this is zero is unlikely to be correct, as in this case no respiration should occur. The existence of liquid-to-cell mass transfer limitation is a source of severe problems in scale-up (for instance, K_La ceases to be a good scale-up criterion), and it is perhaps of widespread occurrence in cultures of mycelial organisms. This could in part explain the difficulties connected with scaling up such fermentations. Possible approaches to this problem are discussed in section 22.3.2.

Fluid shear. The relationships between the intensity and the scale of turbulence and process performance are complex, since two opposite effects may coexist. On the one hand, turbulence increases the gas-liquid interfacial area and reduces the stationary film thickness, thus promoting gas-liquid and liquid-cell mass transfer, and on the other hand, excessive shear may damage filamentous organisms or fragile cells. The intensity of such effect is usually correlated with the maximum shear rate in the fermentor, which occurs at the periphery of the impeller and is estimated by the quantity $v = \pi N D_i$.

A series of fermentations carried out at different rotation speeds (maximum peripheral velocity, 12 to 15 m s^{-1}) can be used to assess any possible adverse effect of shear. On the other hand, a positive effect will show as either a decreased $pO_{2(crit)}$ (see above) or a higher productivity if, for instance, the removal of production-inhibiting CO_2 is a rate-influencing step. The objective is to obtain a quantitative relationship between shear intensity and process performance. The latter is better expressed as an instantaneous production rate rather than a final product concentration, as this is a very complex function of physiological and environmental factors, usually difficult to break down.

Macro-mixing. Macro-mixing of fermentor contents is usually characterized by the mixing time, whose experimental determination is described above (Characterization of Fermentor Performance). As already pointed out, the use of mixing time as a scale-up parameter is not practicable, since this would require an increase of power/unit volume almost in proportion to scale. Furthermore, experimental work on pseudoplastic fluids has yielded mixing times even 50-fold higher than predicted (28). In viscous, non-Newtonian cultures, in which turbulence stops soon after the culture has left the impeller region, the relevant parameter may be the impeller pumping capacity:

$$\phi_p = c_p N D_i^3 \quad (80)$$

In a fermentor with a flow pattern such as that depicted in Fig. 17, in steady-state conditions, the culture may be assumed to enter the impeller (i.e., the mass transfer) region with an O_2 concentration of C_1 and to leave it with a saturating concentration of C^*, both constant in time. The O_2 balance in the fermentor is (44):

$$\text{OUR} \times \text{V} = K_L a V(C^* - C_1) \quad (81)$$

where the right-hand side represents the O_2 supply rate to the fermentor under the assumption that C_1 is the O_2 concentration throughout the fermentor. On the other hand, O_2 is added to the culture by the culture flow leaving the impeller at a rate: $V \times \text{OSR} = \phi_p(C^* - C_1)$, and, since in steady state OSR = OUR, we have:

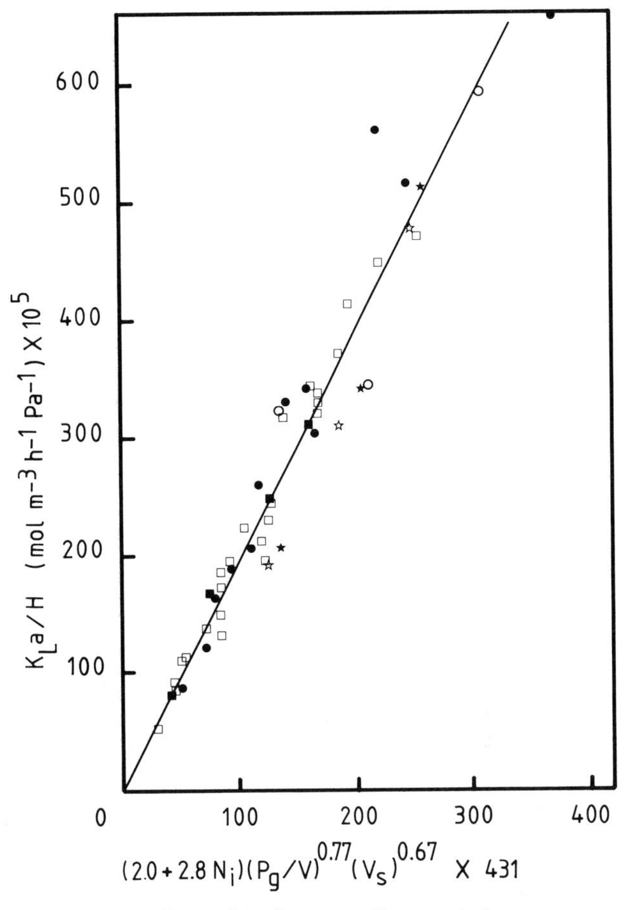

FIG. 16. Relationship between K_La, gassed power consumption, and superficial air velocity in sodium sulfite solutions in vessels of 100 to 40,000 liters (redrawn from data of Fukuda et al. [27]). The equation of the regression line is:

$$K_La/\text{H} = (2.0 + 2.8\ N_i)\ (P_g/V)^{0.77}\ (V_s)^{0.67}\ \text{CF}$$

where N_i = number of impellers; P_g/V = gassed power consumption (kW m^{-3}); V_s = superficial air velocity (m s^{-1}); and CF = factor for conversion of SI units (8.06 × 10^{-3}). Symbols: (●) 100 liters; (○) 500 liters; (■) 3,600 liters; (□) 4,200 liters; (☆) 30,000 liters; (★) 40,000 liters.

$$K_L a = \frac{\phi_p}{V} = \frac{N D_i^3}{V} \quad (82)$$

ϕ_p could therefore be a scale-up parameter, and it is worthwhile to determine its influence on process performance. Wang and Fewkes (87) have obtained a linear relationship between ϕ_p and the liquid-to-cell mass transfer coefficient. It would thus appear that in viscous, non-Newtonian broths, ϕ_p can control the overall process of mass transfer.

Heat evolution. As pointed out previously (see section 22.1.3), metabolic heat constitutes the major contribution to the total heat load of an aerobic fermentation, besides heat of agitation. The determination of heat evolution rate can be carried out by either of two methods: (i) dynamic calorimetry (18) or (ii) heat balance on coolant. In the first method, the fermentor is treated as a calorimeter (i.e., no heat exchange with the environment takes place) whose rate of temperature increase is related to heat evolution rate by the relation:

$$VQ_{acc} = \frac{\Delta T}{\Delta t} C_p \tag{83}$$

where C_p = total heat capacity of the systems. Q_{metab} can be obtained by subtracting from Q_{acc} the heat of agitation. Usually a ΔT of 1°C is employed, and this makes the system suitable for use in actual fermentation conditions. If the heat exchange with the environment cannot be neglected (especially in small fermentors with a high surface/volume ratio), then its contribution can be estimated, together with that of $Q_{agitation}$, by equation 83 used in connection with the noninoculated fermentor, given a certain temperature gradient versus the environment (18).

The heat balance technique is more suitable to larger fermentors (pilot plant and upwards). It is based on the heat balance of the coolant when the fermentor temperature is kept constant by automatic control. In this case $\Delta T/\Delta t = 0$, and therefore $Q_{exchange} = Q_{metab} + Q_{agitation}$, or:

$$VQ_{metab} = F_c \rho_c c_c (T_{out} - T_{in}) - Q_{agitation} \tag{84}$$

where F_c = volumetric coolant flow rate; ρ_c = coolant density; c_c = specific heat of coolant; and T_{in} and T_{out} = coolant temperature at inlet and outlet of jacket or coil, respectively. Q_{metab} should be determined at various times during the fermentation, and the maximum value obtained can be used to predict the heat removal capacity required on the large scale, in connection with the corresponding predicted fluid dynamic parameters (see equation 55). It should be remembered that the rate of metabolic heat evolution is proportional to the rate of energy-yielding metabolism and therefore, in aerobic cultures, to O_2 consumption rate. This, in turn, can be controlled by the rate of feeding of the energy source. The latter is therefore an extremely important control variable that can be used to avoid exceeding the limits of O_2 transfer capacity and heat removal capacity of the particular plant.

Culture stability

When developing a new fermentation process it is strongly advisable to carry out preliminary experiments on the small scale to estimate the likelihood that a culture stability problem will be encountered on a larger scale (see section 22.1.1). The usual procedure consists of scaling down the process that is likely to be the final one in terms of number of preculture stages and of size of inoculum. The economical use of production fermentors, which is essentially characterized by their average production rate, calls for the use of relatively large inoculum sizes to reduce the initial growth phase during which the volumetric productivity is low. As a consequence, the degree of multiplication, x/x_0, in the final stage is fairly low, usually in the range of 10 to 100, corresponding to the number of generations comprised in the interval, three to seven. This is a small fraction of the total number of generations, which can easily reach 40, from the starting stock culture (spore suspension, frozen mycelium, freeze-dried material, or others) to the harvest of the production fermentor. The critical stages are therefore the precultures, and it is their development that must be simulated on the small scale. This is conveniently done by interposing between the stock culture and the production stage a number of precultures giving, at a fixed inoculum ratio, the desired number of generations. For instance, if a fully grown culture developed in a given medium is used to inoculate the same medium at a ratio of 3%, the number of generations needed to attain the same final biomass concentration is (from equation 7) ~5. Evidence of the accumulation of low-producing variants is obtained either by comparing the productivity achieved in the final stage with that observed when only one preculture is used, or by determining the productivity (in the standard test) of several isolates obtained by plating the culture at various stages of the procedure. This is a very cumbersome and time-consuming task if carried out in shake flasks, but it provides direct evidence that a change in the structure of the population is taking place. It is therefore highly desirable to have a quick method for the estimation of the productivity of individual colonies, such as the agar-disk technique (34, 80, 81). Evidence of culture degeneration can also be obtained on a small scale by measuring the productivity of the culture in several successive production stages, each of which is inoculated with a sample of the culture taken from the previous stage (6). This method is quick and widely used, but suffers from the drawback that if the preculture medium is very different in composition from the production medium, different selective pressures may be exerted upon the original strain and the emerging variants in the two culture media. Thus the nonproducing variants might have a selective advantage over the original strain in the vegetative (i.e., preculture) medium but not in the production medium. Such a situation would lead to an underestimation of the phenomenon if the successive transfers are carried out in production medium.

Since the proportion of undesired variants increases with increasing number of generations (equation 8), a lower fraction of such variants can be tolerated in the starting material (i.e., the stock culture) as the scale of fermentation increases. This requires the utmost care in the purification and maintenance of the stock cultures employed in the scale-up procedure, to ensure that possible falls in productivity are indeed due to environmental factors and not to a different composition of the populations in the fermentor on the two scales. Application of equation 8 to a hypothetical process adopting realistic values of x_{m_0} and λ shows that after reasonable numbers of generations (~50) only a very small proportion of the final biomass is made up of the undesired variants. On the other hand, dramatic examples of degeneration of productivity have been described in the literature (5, 60, 70) and are also often encountered in everyday industrial practice.

There are two possible explanations for such discrepancy, namely, either the rate of occurrence of the undesired variants is very high (several orders of magnitude higher than known spontaneous mutation rates), or under the conditions employed, these variants grow considerably faster than the original strain. This has been found to be the case with low-producing variants of highly mutated industrial strains (4, 62). Little, if any, control can be exerted on the appearance of the undesired variants save perhaps searching for a more stable strain. This consideration mainly applies to microorganisms obtained by means of traditional mutation and selection procedures, which generally yield strains carrying unknown genetic modifications involving either chromosomal or extrachromosomal elements, or both, in combinations whose stability is generally unpredictable. Several possible mechanisms of genetic instability have been discussed, e.g., dissociation of heterokaryons (29) and mitotic nonconformity in fungi (52), but no general methods exist for the control of such phenomena. A possible approach could be to search for substances selectively inhibiting the nonproducing variants. The situation is likely to be different with "engineered" strains harboring plasmids. Instability may be caused either by a greater proneness of the plasmids to mutation (23) or by loss of the plasmid. Stability of the plasmids in the latter case is under genetic control and might be improved by means of genetic techniques (31, 37). On the other hand, there seems to be considerable scope for the study and exploitation of the effects of culture environment on the growth rates of different strains, and hence on the outcome of their competition in the culture. Differential effects of temperature (82, 85), growth-limiting substrate in chemostat cultures (65, 71), and pH (64) on the growth rates of different populations coexisting in the same culture vessel have been described in the literature.

Research on plasmid stability in continuous culture has shown in one case that a selective advantage is conferred on plasmidless cells by phosphate limitation but not by carbon or nitrogen limitation (43). In another instance (37), some plasmids were not lost even after 120 generations, but their copy number was decreased. Where gene dosage effects are important, such type of behavior would nevertheless be undesirable. Plasmidless cells have frequently been found to have higher growth rates than their plasmid-bearing counterparts. In such instances growth rates have been found to increase by factors varying from 1.06 to 2.31. A factor of 1.3 is sufficient to reduce the fraction of plasmid-bearing populations to about 13% in 25 generations (37), if $\lambda = 10^{-2}$. The existing evidence points to the fact that the extent of growth rate increase accompanying plasmid loss varies with medium composition.

It is the common experience of industrial microbiologists that certain mutants that are capable of overgrowing the parent strain in, say, the production medium are unable to do so in the preculture medium, or vice versa. Therefore, if the reproduction on the small scale of the number of generations expected to occur on the final scale provides evidence of accumulation of undesired variants, it is worthwile to try several modifications of the preculture medium as well as of the culture conditions (pH and temperature). This is justified by the facts that (i) the largest fraction of the total number of generations occurs in the preparation of the final inoculum, and (ii) the physicochemical environment in which the inoculum is developed is usually (but not always) less critical than that of the production phase (6), thus allowing a certain latitude in the definition of medium composition and of inoculum conditions.

Medium sterilization

The design of a batch sterilization cycle requires the knowledge of the contribution of the different phases of the cycle to the destruction of microorganisms. As discussed previously (see section 22.1.2), the global sterilization criterion, ∇_{tot}, can be broken down into three components: $\nabla_{tot} = \nabla_{heating} + \nabla_{holding} + \nabla_{cooling}$. $\nabla_{holding}$ can be easily calculated by equation 14, using the K_1 value corresponding to the holding temperature. Table 5 contains the tabulated values of K_1 for *B. stearothermophilus* 1518 (6) spores treated with moist heat in the range of 100 to 130°C (=373 to 403 K). $\nabla_{heating}$ and $\nabla_{cooling}$ can be estimated by means of several techniques of varying degrees of complexity and precision, depending upon the available equipment. In principle, heating profiles can be either hyperbolic, linear, or exponential, according to the heating system employed (sparging with live steam, electrical heating, or exchange with jacket or coils containing steam). A knowledge of the relevant geometrical and physical parameters should allow the prediction of the temperature profile (21, 86) and then, through substitution into equation 10 and integration, the calculation of the relevant ∇ value. Cooling is usually effected by heat exchange with a coolant circulated in a jacket or coils, and thus the temperature profile is usually exponential (1). Actual temperature profiles can deviate from the theoretical ones for a number of reasons, and it is thus useful to have approximate methods giving reasonably accurate results. The main techniques are stepwise approximation and piecewise linear approximation.

The first technique (stepwise approximation) assumes that if the period of time under investigation is subdivided into a number of sufficiently small time intervals, the temperature within each interval can be considered constant (Fig. 1). For each interval a ∇ value is calculated using equation 14 with the appropriate K_1 value, and thus the overall contribution to sterilization of the period under examination, ∇_p, is given by:

$$\nabla_p = \sum_{i=1}^{n} h_i K_{1i} \tag{85}$$

where h_i is the duration of the ith interval. If $h = h_1 = h_2 \ldots = h_n$, then:

$$\nabla_p = h \sum_{i=i}^{n} K_{1i} \tag{86}$$

The K_{1i} for each interval is obtained by entering into Table 5 the temperature corresponding to the interval midpoint that is determined by graphical interpolation from the temperature profile. The degree of precision increases with the decreasing size of the time intervals, but this in turn increases the manual work load.

The piecewise linear approximation consists of representing the temperature profile with a succession of segments of a straight line. The contribution of each segment is then calculated using the approximate

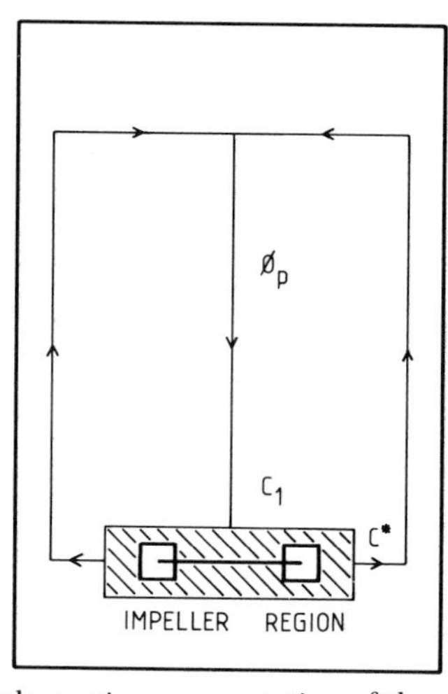

FIG. 17. Schematic representation of the circulation pattern of a viscous culture in a single-impeller tank. ϕ_p = impeller flow (m^3 s^{-1}); C_1 = O_2 concentration of culture entering the impeller region; C^* = saturating O_2 concentration. (From reference 44.)

relationship (1):

$$\nabla = \frac{A\delta}{\alpha}\left[\frac{e^{-\delta/(1+\alpha\vartheta_1)}}{\left(\dfrac{\delta}{1+\alpha\vartheta_1}\right)^2} - \frac{e^{-\delta}}{\delta^2}\right] \quad (87)$$

for heating and:

$$\nabla = \frac{A\delta}{\alpha}\left[\frac{e^{-\delta}}{\delta^2} - \frac{e^{-\delta/(1+\alpha\vartheta_1)}}{\left(\dfrac{\delta}{1-\alpha\vartheta_1}\right)^2}\right] \quad (88)$$

for cooling, where A = Arrhenius constant (see equation 13) (s^{-1}); α = relative rate of temperature increase (or decrease) expressed as degrees K s^{-1} $T_0{}^{-1}$, where T_0 = temperature at the beginning of the time interval (s^{-1}); ϑ_1 = duration of interval (s); $\delta = E/RT_0$, where E = activation energy for spore death (cal mol^{-1}) and R = gas constant (cal mol^{-1} K). δ is dimensionless.

The approximation of the experimental temperature profile can be carried out either manually or by means of a computer, which can be easily programmed to select the interval size giving the desired degree of precision and to carry out the subsequent calculations. The linear approximation method has been greatly simplified by Richards (63), who assumed that only the parts of the cycle above 100°C need to be considered and that during such phases the temperature varies linearly with time. He then computed the cumulative values of ∇ in the temperature range, assuming a rate of temperature variation of 1°C min^{-1} (Table 5). To compensate for different velocities of temperature variation, the appropriate ∇ value is divided by the observed rate of temperature change, expressed as °C min^{-1}. The methods described above allow the determination of the effect on spore destruction of those phases of the sterilization cycle that are largely dictated by the equipment available (i.e., vessel, jacket or coils, pipes, valves, etc.) on that particular scale. This contribution can thus be regarded as an independent variable of the problem. The parameters left for manipulation are therefore the temperature of the maintenance phase and its duration. Inspection of equation 14 shows that the same value of ∇ can be obtained by an infinite number of combinations of time and temperature. A practical approach consists of the following steps.

1. Calculate the desired ∇_h value by difference:

$$\nabla_h = \nabla_{tot} - \nabla_{heating} - \nabla_{cooling} \quad (89)$$

2. Select the sterilization temperature, T_h.
3. Calculate t_h from equation 14 by entering ∇_h and T_h.

Selection of T_h must take into account the dual effect of heat on fermentation medium, i.e., destruction of contaminants and enhancement of chemical reactions involving chemical substances contained in the medium. Many organic substances are destroyed by heating, and the rate of such reactions depends on temperature in a way generally describable by the Arrhenius equation (see equation 13). In this relationship E, the activation energy, measures the sensitivity of reaction rate to temperature. The E values for the majority of chemical reactions are in the range of 10 to 30 kcal mol^{-1}, while those for spore destruction lie in the range of 65 to 85 kcal mol^{-1} (6). The manipulation of sterilization temperature during the maintenance phase, T_h, therefore provides a means of controlling the extent of reactions other than spore destruction while maintaining the latter constant through the adoption of the t_h value giving the desired ∇_h.

From the practical point of view, the basic information that is needed is the relationship between the severity of the heat treatment and the performance of the medium in the actual fermentation process. Media containing critical heat-labile components (e.g., vitamins) (86) can be expected to perform better at lower intensities of the heat treatment, whereas simple, chemically defined media may prove largely insensitive, provided the ingredients that are likely to react together (e.g., glucose and substances containing amino groups) are sterilized separately. Some industrial media containing proteins and polysaccharides may even perform better after a moderate heat treatment, usually due to the breakdown of part of the insoluble nutrients. In the vast majority of cases the nature of the critical heat-labile substance is unknown (there might even be no such substance), and the only practicable approach then consists of carrying out a number of small-scale fermentations employing batches of medium subjected to heat treatments of varying intensity. This is most easily accomplished either by sterilizing flasks at a fixed temperature (e.g., 120°C) for varying periods of time, or by aseptically withdrawing aliquots from a stirred vessel at different times during a long sterilization run and transferring them to sterile shake flasks. In view of the environmental changes associated with scale-up from shake flasks to fermentors, it is highly desirable to conduct this type of experiment directly in small-scale stirred fermentors if preliminary experimentation in shake flasks has suggested that an effect indeed exists. If an adverse effect of heating upon medium performance is confirmed, then the next step is to exploit the above-

TABLE 5. Thermal inactivation constant, K_1, and cumulative sterilization criterion, ∇, between 100 and 130°C for spores of *B. stearothermophilus* 1518 (from reference 6)

Temp (°C)	K_1 (min^{-1})	∇
100	0.019	—
101	0.025	0.044
102	0.032	0.076
103	0.040	0.116
104	0.051	0.168
105	0.065	0.233
106	0.083	0.316
107	0.105	0.420
108	0.133	0.553
109	0.168	0.720
110	0.212	0.932
111	0.267	1.199
112	0.336	1.535
113	0.423	1.957
114	0.531	2.488
115	0.666	3.154
116	0.835	3.989
117	1.045	5.034
118	1.307	6.341
119	1.633	7.973
120	2.037	10.010
121	2.538	12.549
122	3.160	15.708
123	3.929	19.638
124	4.881	24.518
125	6.056	30.574
126	7.506	38.080
127	9.293	47.373
128	11.494	58.867
129	14.200	73.067
130	17.524	90.591

mentioned difference in activation energy and select the highest possible value of maintenance temperature, compatibly with (i) maximum operating pressure (and thus temperature) of vessel (this rarely exceeds 2 kg cm^{-2} [≃132°C for saturated steam]) and (ii) maximum attainable heating rate, which must be sufficient to reach the desired temperature before a $\nabla_{heating}$ level is reached that leaves no time for maintenance, because $\nabla_{heating} + \nabla_{cooling} = \nabla_{tot}$.

It should, however, be considered that $\nabla_{heating}$ is usually higher than $\nabla_{cooling}$ because the rate of change of temperature slows down as the maintenance temperature is approached, while the cooling rate is highest in the early cooling phase (see Fig. 1). In a typical situation, the sum of $\nabla_{heating}$ and $\nabla_{holding}$ may amount to 95% of ∇_{tot}.

If there is an adverse effect of sterilization on medium performance, it is strongly advisable to scale down the production plant sterilization profile in pilot plant vessels and use it as the standard sterilization cycle during scale-up work.

The foregoing discussion is irrelevant if a continuous sterilization technique is employed. Once the optimal combination of holding time and sterilization temperature has been identified, the scale-up is essentially an engineering task. However, this is not without problems, as departure from ideal plug flow in the sterilizer is unavoidable, and a compromise between overheating and chance of contamination must be found (1, 86).

22.2.3 Scale-Up Strategy

After the characterization of the process on the starting scale, a scale-up strategy must be developed and implemented. The sequence of operations, the time to be devoted to the various stages, etc., can vary widely, depending upon the nature of the particular process and the body of knowledge already accumulated. A procedure established a priori is therefore generally of little use, as it may be insufficiently detailed for some processes and unduly long for others. Nevertheless it is useful to consider a generalized procedure to highlight some basic aspects (Fig. 18). The objective of steps 1 to 3 (Fig. 18) is to define a set of values of fermentation parameters (concerning, essentially, mass transfer and mixing aspects) that can be presumably achieved on the production scale.

Such values can be used to predict a fermentation result on the basis of the correlations established in the previous phase of the work (see section 22.2.2, Determination of the Critical Values of Fermentation Parameters). This predicted performance must be

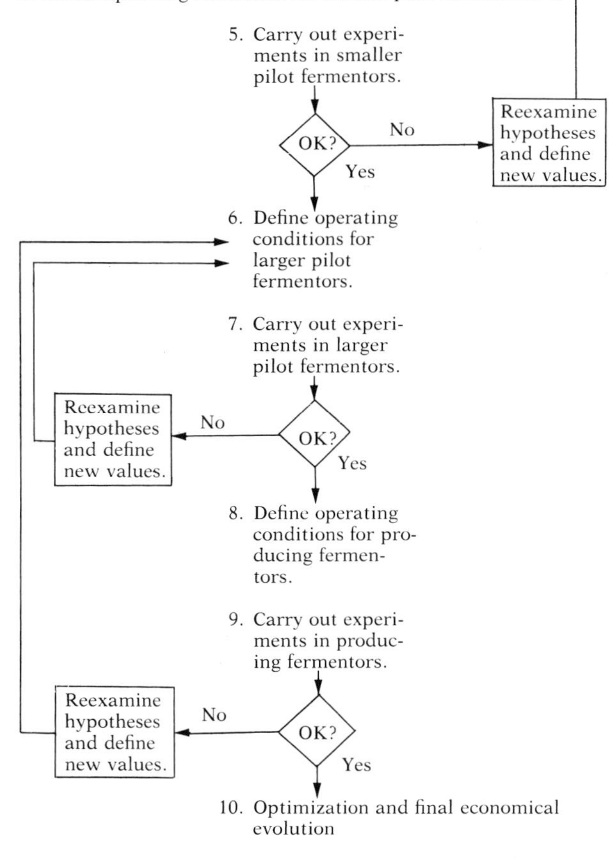

FIG. 18. Generalized flow diagram of a scale-up procedure.

seen as a target to be reached, and suffers from two orders of uncertainties. In the first place the prediction of particular values of, for example, K_La or mixing time on the final scale is usually at best approximate, especially if their relation to geometric and operating variables is taken either from the literature or from model systems. Second, the relationships between, let us say, K_La or impeller peripheral velocity or circulation time and process results are highly empirical and describe very complex systems. This is especially true when the product of interest is a secondary metabolite whose rate of synthesis can be controlled by a large number of interconnected factors.

Step 4 (Fig. 18) is essentially an attempt to scale down production scale conditions, and as such it makes use of published and experimental relationships between operating variables. In step 5 a series of fermentations are carried out in the smaller pilot fermentors (if two sizes are used) with the objective of achieving the target values of fermentation parameters (e.g., K_La, mixing time, etc.). Once this is achieved, and it may take a number of adjustments, the process performance is compared with the one predicted from the laboratory-scale fermentation data. On some occasions the results are satisfactory and the procedure (steps 4 and 5) is repeated in the larger pilot fermentors (if two sizes are used) and then in the production fermentors. On the other hand, some processes will not produce the expected results at some stage. This generally means that the hypotheses underlying the choice of particular values of fermentation parameters, or indeed the choice of the parameters themselves, are incorrect.

Alternatively, new effects may appear that were not apparent, or recognized, on a smaller scale. In such an instance a variety of approaches is possible, and the choice depends on the particular situation. However, the two major attitudes are either to continue the experiments on the same scale or to go back to a smaller scale to obtain more information about the process and test new hypotheses. The first approach could produce quick results if relatively minor modifications are needed, and it is thus advisable to perform a small number of exploratory experiments of this type. If the result does not improve, it is then better to go back to smaller fermentors for more radical work. The physiological state of the inoculum is of paramount importance in a number of batch fermentation and a frequent source of problems in scale-up (6). It is therefore recommended that at an early stage of fermentation development the properties of inocula grown in pilot fermentors be tested in small-scale equipment. This problem is frequently associated with that of medium sterilization and is best treated by the approach called by Banks (6) "the monitor experiment," in which media prepared in larger fermentors, and inocula grown in them, are used on a small scale to identify the source of the problems. An advantage of having two sizes of pilot plant fermentors is that the smaller vessels can be used to provide the inoculum for the fermentations carried out in both the same vessels and the larger ones, thus eliminating the inoculum variable from this step of scale increase and helping to delimit the source of potential problems.

An additional source of problems, frequently encountered with filamentous organisms, is morphology. The rheological properties of filamentous-organism culture broths are influenced by the length of filaments, degree of branching, and filament flexibility (44). On the other hand, these morphological features can be controlled by very complex and subtle mechanisms. In such a way small environmental differences are amplified, giving rise to spectacular changes of the rheological properties. These in turn affect the fermentor performance, e.g., shifting the resistance to mass transfer from the gas-liquid step to the liquid-cell step (87), decreasing the Reynolds number near the heat exchange surfaces and creating stagnant zones (44), entrapping small bubbles that reduce the apparent density (and thus power input) without contributing to mass transfer, etc. Such changes then affect the production rate through oxygen limitation (although not shown by O_2 electrodes), CO_2 accumulation, and starvation of nutrients that are in limited supply in stagnant zones (30). Culture morphology and rheology should therefore be kept under close surveillance by both direct microscopical examination and viscosity measurements (see section 22.2.2, Culture Rheology).

It may happen that during the scale-up procedure no way can be found to obtain the desired result by maintaining the operating conditions within the physical or economical constraints of the production plant. In such an event one has to accept that the particular process cannot be scaled up in that particular plant, and the best that can be obtained must be evaluated in economical terms. On the other hand, a considerable body of knowledge will have accumulated during the work of adjustment and reexamination of ideas and hypotheses prompted by lack of success, as implied by Fig. 18. All this information must be fed back to the process development unit, as it can help it to set targets that are relevant for future scale-up work (see section 22.3.2).

A final point to consider is geometric similarity of the various fermentors. Geometric similarity is important from the practical point of view, in that it simplifies the prediction of fermentor performance on the larger scale. It is interesting in this respect that Taguchi et al. (77), using a non-Newtonian culture, found that the Michel and Miller correlation (equation 28), for gassed power consumption, and a modification of Cooper's correlation for K_La (equation 66) applied in a series of geometrically dissimilar fermentors ranging from 100 to 50,000 liters. In practice one needs to consider two aspects: fermentor geometry and impeller geometry. The latter is easily modified and is therefore open to study and optimization during the scale-up work. The former, apart from small laboratory fermentors, is largely fixed and must be decided in advance. From the point of view of mixing (especially of viscous broths), fermentor geometry is far more critical on the large scale (i.e., long circulation times) than on the small scale. Experience has shown that higher liquid height/tank diameter ratios are required in large fermentors than are typical in laboratory or pilot plant fermentors. In practice, the geometry of the production vessel will be largely dictated by a number of technical and economic constraints (space available, transportation, cost of sparged air as a function of fermentor head, heat exchange, etc.). Whenever possible it is therefore preferable to build the pilot plant vessels as a scale-down of production vessels.

The geometry of impellers is a major variable that can be manipulated to achieve the same values of more than one agitation parameter (55), which is not possible in a strictly geometrically similar series (see Table 3). The most obvious step is to alter impeller diameter while the ratios among the various dimensions remain constant. In this way the power number should not change, and the prediction of power consumption from impeller speed should be straightforward. Oldshue (55) suggested this approach as a means of obtaining the same power absorption (and therefore K_La) and the same impeller peripheral velocity (and therefore maximum shear rate) at the same time. This line of thought enables one to harmonize, within certain limits, various aspects of agitation (mass transfer, shear rates, and mixing efficiency) better than can be obtained in geometrically similar systems.

22.3 RESULTS OF SCALE-UP

22.3.1 Technical and Economic Success

The final objective of scale-up is to bring a fermentation process to its economic fruition, and therefore the technical aspects are important insofar as they can be translated into economic indicators such as unit cost of product, investments required, and output rate, whose combined evaluation eventually decides upon the economic viability of the processes (see chapter 26). Many economic evaluations cannot be done accurately using laboratory equipment, and this therefore is one of the functions of a pilot plant, in which power consumption, mixing requirements, and heat exchange can be more realistically assessed. The same applies to problems connected with feeding nutrients, precursors, etc. A detailed economic analysis must be carried out, however, on the first industrial operations (see Fig. 18, step 10), concomitantly with the final optimization. It frequently happens that a particular technical objective, e.g., final titer, average output rate, etc., is not reached on the final scale but the process is still economically viable or produces a higher profit margin than the previous version. However, from the point of view of process development, it is still essential to investigate the causes of technical failures, as their understanding can contribute to future process improvement. For example, a lower production rate may be balanced on the production scale by the use of cheaper raw materials, but the fact remains that the desired rate could not be obtained. The study of these aspects should form a substantial fraction of subsequent improvement work.

22.3.2 Feedback from Scale-Up Work

As pointed out in the preceding sections, the experimental work associated with scale-up produces a body of information on the physiology of the microorganism employed. In particular, those characteristics will be identified that pose scale-up problems. This information is valuable to the strain improvement units, which can then set precise and relevant targets to their work. The great majority of scale-up problems encountered with filamentous organisms originate from the flow behavior of their cultures, which are usually highly viscous and non-Newtonian. This creates a nonhomogeneity of the culture, as high-shear zones, e.g., around the impellers, have a low apparent viscosity (shear thinning effect) whereas low-shear zones, e.g., behind baffles or near cooling coils, have a high apparent viscosity. The overall effect is that on the one hand air is effectively channeled along the impeller shaft and escapes in large bubbles, contributing little to mass transfer, and on the other hand liquid-cell mass transfer and heat exchange in the low-shear zones become critical. This hydrodynamic behavior is caused by the interactions between filaments that lead to entanglement and floc formation. The effect of filament length can be appreciated by observing the dramatic drop of viscosity caused by fragmentation in streptomycete fermentations, while at the same time cell dry weight changes very little. Therefore, there seems to be wide scope for the selection of morphological mutants having shorter filament length. Hyphal rigidity seems to influence apparent viscosity (44), and this fact could encourage the development of new strain-medium combinations capable of the same productivity but resulting in lower viscosities. Another major area connected with scale-up problem is heat evolution. The heat removal rate is in fact affected by the culture viscosity at the exchange surface (see equation 55) and is proportional to the temperature gradient (see equation 51). The ability to ferment at higher temperatures would then result in a better heat exchange, which would in turn ease the problems created by viscosity and vessel size (i.e., surface-to-volume ratio). The selection of thermotolerant strains appears very interesting in this context.

22.4 LITERATURE CITED

1. **Aiba, S., A. E. Humphrey, and N. F. Millis.** 1973. Biochemical engineering, 2nd ed. Academic Press, Inc., New York.
2. **Aiba, S., and T. Yamada.** 1961. Oxygen absorption in bubble aeration. Part I. J. Gen. Appl. Microbiol. **7:** 100–106.
3. **Atkinson, B., and F. Mavituna.** 1983. Biochemical engineering and biotechnology handbook. The Nature Press, New York.
4. **Backus, M. R., and J. F. Stauffer.** 1955. The production and selection of a family of strains in *Penicillium chrysogenum*. Mycologia **47:**429–463.
5. **Ball, C., and J. L. Azevedo.** 1976. Genetic instability in parasexual fungi, p. 243–251. *In* K. D. McDonald (ed.), Genetics of industrial microorganisms. Academic Press, Inc., New York.
6. **Banks, G. T.** 1979. Scale up of fermentation processes. Top. Enzyme Ferment. Biotechnol. **3:**170–266. Ellis Horwood Ltd., Chichester.
7. **Bartholomew, W. H.** 1960. Scale-up of submerged fermentations. Adv. Appl. Microbiol. **2:**289–300.
8. **Bartholomew, W. H., E. O. Karow, M. R. Sfat, and R. H. Wilhelm.** 1950. Oxygen transfer and agitation in submerged fermentations. Mass transfer of oxygen in submerged fermentations by *Streptomyces griseus*. Ind. Eng. Chem. **42:**1801–1809.
9. **Battino, R., and H. L. Clever.** 1966. Solubility of gases in liquids. Chem. Rev. **66:**395–463.
10. **Blakebrough, N., and K. Sambamurthy.** 1966. Mass

transfer and mixing rates in fermentation vessels. Biotechnol. Bioeng. **8:**25–42.

11. **Blenke, H.** 1979. Loop reactors. Adv. Biochem. Eng. **13:**121–214.
12. **Bongenaar, J. J. T. M., N. W. F. Kossen, B. Metz, and F. W. Meijboom.** 1973. A method for characterizing the rheological properties of viscous fermentation broths. Biotechnol. Bioeng. **15:**201–206.
13. **Brauer, H.** 1979. Power consumption in aerated stirred tank reactor systems. Adv. Biochem. Eng. **13:**87–119.
14. **Calderbank, P. H., and M. Moo-Young.** 1959. The prediction of power consumption in the agitation of non-Newtonian fluids. Trans. Inst. Chem. Engrs. **37:**26–33.
15. **Calderbank, P. H., and M. Moo-Young.** 1961. Power characteristics of agitators for the mixing of Newtonian and non-Newtonian fluids. Trans. Inst. Chem. Engrs. **39:**337–347.
16. **Calderbank, P. H., and M. Moo-Young.** 1961. Continuous phase heat and mass transfer properties of dispersions. Chem. Eng. Sci. **16:**39–54.
17. **Charles, M.** 1978. Technical aspects of the rheological properties of microbial cultures. Adv. Biochem. Eng. **8:** 1–62.
18. **Cooney, C. L., D. I. C. Wang, and R. I. Mateles.** 1968. Measurement of heat evolution and correlation with oxygen consumption during microbial growth. Biotechnol. Bioeng. **11:**269–281.
19. **Cooper, C., G. Fernstrom, and S. Miller.** 1944. Performance of agitated gas-liquid contactors. Ind. Eng. Chem. **36:**504–509.
20. **Deindoerfer, F. H., and E. L. Gaden.** 1955. Effect of liquid physical properties on oxygen transfer in penicillin fermentation. Appl. Microbiol. **3:**253–257.
21. **Deindoerfer, F. H., and A. E. Humphrey.** 1961. Scale-up of heat sterilization operations. Appl. Microb. **9:**134–139.
22. **Deindoerfer, F. H., and J. M. West.** 1960. Rheological examination of some fermentation broths. J. Microbiol. Biochem. Technol. Eng. **2:**165–175.
23. **Dwivedi, C. P., T. Imanaka, and S. Aiba.** 1982. Instability of plasmid-harboring strains of E. coli in continuous culture. Biotechnol. Bioeng. **24:**1465–1468.
24. **Feren, C. J., and R. W. Squires.** 1969. The relationship between critical oxygen level and antibiotic synthesis of capreomycin and cephalosporin C. Biotechnol. Bioeng. **11:**583–592.
25. **Finn, R. K.** 1954. Agitation-aeration in the laboratory and in industry. Bacteriol. Rev. **18:**254–274.
26. **Fuchs, R., D. D. Y. Ryu, and A. E. Humphrey.** 1971. Effect of surface aeration on scale-up procedures for fermentation processes. Ind. Eng. Chem. Process Design Dev. **10:**190–196.
27. **Fukuda, H., Y. Sumino, and T. Kanzaki.** 1968. Scale-up of fermentors. I. Modified equations for volumetric oxygen transfer coefficient. J. Ferment. Technol. **46:**829–837.
28. **Godleski, E. S., J. C. Smith.** 1962. Power requirements and blend times in the agitation of pseudoplastic fluids. Am. Inst. Chem. Eng. J. **8:**617–620.
29. **Haas, F. L., T. A. Puglisi, A. J. Moses, and J. Lein.** 1956. Heterokaryosis as a cause of culture rundown in *Penicillium.* Appl. Microbiol. **4:**187–195.
30. **Hansford, G. S., and A. E. Humphrey.** 1966. The effect of equipment scale and degree of mixing on continuous fermentation yield at low dilution rates. Biotechnol. Bioeng. **8:**85–96.
31. **Hershberger, C. L.** 1982. Recombinant DNA systems for application to antibiotic fermentation in *Streptomyces.* Annu. Rep. Ferment. Processes **5:**101–126.
32. **Ho, L.** 1979. Process analysis and optimal design of a fermentation process based upon elemental balance equations: generalized theoretical equations for estimating rates of oxygen demand and heat evolution. Biotechnol. Bioeng. **21:**1289–1300.
33. **Hospodka, J., Z. Caslavsky, K. Beran, and F. Stross.** 1964. The polarographic determination of oxygen uptake and transfer rate in aerobic steady state yeast cultivation on laboratory and production scale, p. 353–367. *In* I. Malek, K. Beran, and J. Hospodka (ed.), Continuous cultivation of microorganisms, vol. 2. Ceskoslov. Akad. ved., Prague.
34. **Ichikawa, T., M. Date, T. Ishikura, and A. Ozaki.** 1971. Improvement of Kasugamycin-producing strain by the agar piece method and the prototroph method. Folia Microbiol. **16:**218–224.
35. **Ishizaki, A., H. Shibai, Y. Hirose, and T. Shiro.** 1973. Ventilation in submerged fermentations. V. Estimation of aeration and agitation conditions in respect to oxygen supply and ventilation. Agr. Biol. Chem. **37:**107–113.
36. **Jarai, M.** 1972. Oxygen transfer in the fermentations of primary and secondary metabolites, p. 97–103. *In* G. Terui (ed.), Fermentation technology today: Proceedings of the 4th International Fermentation Symposium.
37. **Jones, I. M., S. B. Primrose, A. Robinson, and D. C. Ellwood.** 1980. Maintenance of some ColE1-type plasmids in chemostat culture. Mol. Gen. Genet. **180:**579–584.
38. **Karow, E. O., W. H. Bartholomew, and M. R. Sfat.** 1953. Oxygen transfer and agitation in submerged fermentations. J. Agr. Food Chem. **1:**302–306.
39. **Labouza, T. P., D. B. Santos, and R. N. Roop.** 1970. Engineering factors in single-cell protein production. I. Fluid properties and concentration of yeast by evaporation. Biotechnol. Bioeng. **12:**123–134.
40. **Lehrer, I. H.** 1968. Gas agitation of liquids. Ind. Chem. Eng. Process Design Dev. **7:**226–239.
41. **Manfredini, R., V. Cavallera, L. Marini, and G. Donati.** 1983. Mixing and oxygen transfer in conventional stirred fermentors. Biotechnol. Bioeng. **25:**3115–3131.
42. **Mehta, V. D., and M. M. Sharma.** 1971. Mass transfer in mechanically agitated gas-liquid contactors. Chem. Eng. Sci. **26:**461–479.
43. **Melling, J., D. C. Ellwood, and A. Robinson.** 1977. Survival of R-factor carrying *Escherichia coli* in mixed cultures in the chemostat. FEMS Microbiol. Lett. **2:**87–89.
44. **Metz, B., N. W. F. Kossen, and J. C. van Suijdam.** 1979. The rheology of mould suspensions. Adv. Biochem. Eng. **11:**103–156.
45. **Metzner, A. B., R. H. Feehs, H. L. Ramos, R. E. Otto, and J. D. Toothill.** 1961. Agitation of viscous Newtonian and non-Newtonian fluids. A. I. Ch. E. J. **7:**3–9.
46. **Metzner, A. B., and R. E. Otto.** 1957. Agitation of non-Newtonian fluids. A. I. Ch. E. J. **3:**3–10.
47. **Metzner, A. B., and J. S. Taylor.** 1960. Flow patterns in agitated vessels. A. I. Ch. E. J. **6:**109–114.
48. **Michel, B. J., and S. A. Miller.** 1962. Power requirements of gas-liquid agitated systems. A. I. Ch. E. J. **8:**262–266.
49. **Moo-Young, M., and H. W. Blanch.** 1981. Design of biochemical reactors: mass transfer criteria for simple and complex systems. Adv. Biochem. Eng. **19:**1–69.
50. **Mukataka, S., H. Kataoka, and J. Takahashi.** 1981. Circulation time and degree of fluid exchange between upper and lower circulation regions in a stirred vessel with a dual impeller. J. Ferment. Technol. **59:**303–307.
51. **Nash, C. H.** 1974. Effect of carbon dioxide on synthesis of erythromycin. Antimicrob. Agents Chemother. **5:** 544–545.
52. **Nga, B. H., and J. A. Roper.** 1969. A system generating spontaneous intrachromosomal changes at mitosis in *Aspergillus nidulans.* Genet. Res. **14:**63–70.
53. **Norwood, K. W., and A. B. Metzner.** 1960. Flow patterns and mixing rates in agitated vessels. A. I. Ch. E. J. **6:** 432–437.
54. **Nyiri, L., and Z. L. Lengyel.** 1965. Studies on automatically aerated biosynthetic processes. I. The effect of agitation and CO_2 on penicillin formation in automatically aerated liquid cultures. Biotechnol. Bioeng. **7:**343–354.
55. **Oldshue, J. Y.** 1966. Fermentation mixing scale-up tech-

niques. Biotechnol. Bioeng. **8**:3–24.

56. **Ollis, D. F., and H. Chang.** 1982. Batch fermentation kinetics with (unstable) recombinant cultures. Biotechnol. Bioeng. **24**:2583–2586.
57. **Paca, J., P. Ettler, and V. Gregr.** 1978. Oxygen transfer rate in media used for erythromycin biosynthesis. J. Ferment. Technol. **56**:144–151.
58. **Phillips, D. H., and M. J. Johnson.** 1961. Aeration in fermentations. J. Biochem. Microbiol. Tech. Eng. **3**: 277–309.
59. **Pirt, S. J.** 1975. Principles of microbe and cell cultivation. Blackwell Scientific Publications, Oxford. **15**:1284–1290.
60. **Reusser, F.** 1963. Stability and degeneration of microbial cultures on repeated transfer. Adv. Appl. Microbiol. **5**:189–215.
61. **Reusser, F., H. I. Koepsell, and G. M. Savage.** 1961. Degeneration of *Streptomyces niveus* with repeated transfers. Appl. Microbiol. **9**:342–345.
62. **Richards, J. W.** 1961. Studies in aeration and agitation. Prog. Ind. Microbiol. **3**:141–172.
63. **Richards, J. W.** 1965. Rapid calculations for heat sterilizations. Br. Chem. Eng. **10**:166–169.
64. **Richmond, A.** 1983. Effect of bicarbonate and carbon dioxide on the competition between *Chlorella vulgaris* and *Spirulina platensis*. J. Chem. Technol. Biotechnol. **33B**:197.
65. **Righelato, R. C.** 1976. Selection of strains of *Penicillium chrysogenum* with reduced penicillin yields in continuous cultures. J. Appl. Chem. Biotechnol. **26**:153–159
66. **Roels, J. A., J. Van den Berg, and R. M. Vonchen.** 1974. The rheology of mycelial broths. Biotechnol. Bioeng. **16**:181–208.
67. **Rushton, J. H., E. W. Costich, and H. J. Everett.** 1950. Power characteristics of mixing impellers. I. Chem. Eng. Prog. **46**:395–404.
68. **Rushton, J. H., E. W. Costich, and H. J. Everett.** 1950. Power characteristics of mixing impellers. II. Chem. Eng. Prog. **46**:467–476.
69. **Schugerl, K.** 1981. Oxygen transfer into highly viscous media. Adv. Biochem. Eng. **19**:71–174.
70. **Sermonti, G.** 1969. Genetics of antibiotic producing microorganisms. Wiley Interscience, London.
71. **Sikyta, B. Q., and Z. Fencl.** 1976. Continuous production of enzymes, p. 158–169. *In* A. C. R. Dean, D. C. Ellwood, C. G. T. Evans, and J. Melling (ed.), Continuous culture 6: applications and new fields. Ellis Horwood Ltd., Chichester, U.K.
72. **Skelland, A. H. P.** 1967. Non-Newtonian flow and heat transfer. John Wiley & Sons, Inc., New York.
73. **Sobotka, M., A. Prokop, I. J. Dunn, and A. Einsele.** 1982. Review of methods for the measurement of oxygen transfer in microbial systems. Annu. Rep. Ferment. Processes **5**:127–210.
74. **Steel, R., and W. D. Maxon.** 1962. Some effects of turbine size on novobiocin fermentation. Biotechnol. Bioeng. **4**:231–240.
75. **Steel, R., and W. D. Maxon.** 1966. Dissolved oxygen measurements in pilot and production-scale novobiocin fermentations. Biotechnol. Bioeng. **8**:97–108.
76. **Steel, R., and W. D. Maxon.** 1966. Studies with a multiple-rod mixing impeller. Biotechnol. Bioeng. **8**:109–115.
77. **Taguchi, H., T. Imanaka, S. Teramoto, M. Takatsa, and M. Sato.** 1968. Scale-up of glucamylase fermentation by *Endomyces* sp. J. Ferment. Technol. **46**:823–828.
78. **Taguchi, H., and S. Miyamoto.** 1966. Power requirement in non-Newtonian fermentation broth. Biotechnol. Bioeng. **8**:43–54.
79. **Tanaka, H.** 1976. Studies on the effect of agitation on mycelia in submerged culture. J. Ferment. Technol. **54**:818–829.
80. **Trilli, A., I. Costanzi, F. Lamanna, and N. Di Dio.** 1982. Development of the agar disc method for the rapid screening of strains with increased productivity. J. Chem. Technol. Biotechnol. **32**:281–291.
81. **Trilli, A., J. Michelini, V. Mantovani, and S. J. Pirt.** 1978. Development of the agar disk method for the rapid selection of cephalosporin producers with improved yields. Antimicrob. Agents Chemother. **13**:7–13.
82. **Trilli, A., and S. J. Pirt.** 1976. Permissive and non-permissive conditions for the stable association of *Klebsiella aerogenes* and *Pseudomonas aeruginosa* in a chemostat. J. Appl. Chem. Biotechnol. **26**:328–329.
83. **Tuffile, C. M., and F. Pinho.** 1970. Determination of oxygen-transfer coefficients in viscous streptomycete fermentations. Biotechnol. Bioeng. **12**:849–871.
84. **van Suijdam, J. C., and B. Metz.** 1981. Fungal pellet break-up as a function of shear in a fermentor. J. Ferment. Technol. **59**:329–333.
85. **Veldkamp, H., and H. Jannasch.** 1972. Mixed culture studies with the chemostat. J. Appl. Chem. Biotechnol. **22**:105–123.
86. **Wang, D. I. C., C. L. Cooney, A. L. Demain, P. Dunnill, A. E. Humphrey, and M. D. Lilly.** 1979. Fermentation and enzyme technology. John Wiley & Sons, Inc., New York.
87. **Wang, D. I. C., and R. C. J. Fewkes.** 1977. Effect of operating parameters on the behavior of non-Newtonian, mycelial, antibiotic fermentations. Dev. Ind. Microbiol. **18**:39–56.
88. **Yagi, H., and F. Yoshida.** 1975. Gas absorption by Newtonian and non-Newtonian fluids in sparged agitated vessels. Ind. Eng. Chem. Process Design Dev. **14**:488–493.
89. **Yagi, H., and F. Yoshida.** 1975. Enhancement factor for oxygen absorption into fermentation broth. Biotechnol. Bioeng. **17**:1083–1089.
90. **Yagi, H., and F. Yoshida.** 1977. Desorption of carbon dioxide from fermentation broth. Biotechnol. Bioeng. **19**:801–819.
91. **Yoshida, F., A. Ikeda, S. Imakawa, and Y. Miura.** 1960. Oxygen absorption rates in stirred gas-liquid contactors. Ind. Eng. Chem. **52**:435–438.
92. **Yoshikawa, H., Y. Takiguchi, and T. Michiya.** 1982. Degeneration of herbicidin-A producer, *Streptomyces saganonensis*, and strain improvement. J. Ferment. Technol. **60**:385–391.
93. **Zlokarnik, M.** 1978. Sorption characteristics for gas-liquid contacting in mixing vessels. Adv. Biochem. Eng. **8**:133–151.

Chapter 23

Bioinstrumentation and Computer Control of Fermentation Processes

HENRY Y. WANG

The increasing importance of producing microbial and cellular products has led to an ever-increasing demand for cultivation methods to grow microbial, animal, and plant cells in an artificially created environment. To achieve greater understanding of these cellular and microbial processes and to allow decisions to be made regarding the best operating conditions, process monitoring and control have become essential elements to optimize the fermentation processes.

The basis of understanding and controlling a biological process is the data obtained from the biosensors and instrumentation employed. The information can be only as good as what the sensors and instruments indicate. A key limitation for complete bioprocess monitoring is the inability to measure accurately, on-line and quickly, many key variables such as cell concentration and various substrate and product concentrations.

The use of computers in the fermentation industry is relatively new; the first description of computer-controlled fermentations appeared about 15 years ago (22, 57). In the past 10 years, however, as the value of computer control and monitoring has been recognized, the field has received greater attention. Most new pilot and manufacturing fermentation facilities built in the last decade have incorporated at least some automation, which may range from single-loop digital control to complete plant control. Published work has addressed problems of system design (13, 15, 29, 39, 40), sensor development (14, 21, 44), and control action (7, 34, 48). The work has been well summarized by several recent review articles (23, 24, 26, 50). In the future, complex biological systems will be adequately monitored with new and robust sensors and controlled by means of a computer or some other form of artificial intelligence to maximize the full potential of what will often be "unnatural" producing organisms (i.e, organisms selected or constructed in the laboratory).

23.1 BIOSENSORS AND BIOINSTRUMENTATION IN FERMENTATION PROCESSES

Various physicochemical sensors have been developed to monitor fermentation parameters. A list of these measured parameters is shown in Table 1. Electronic sensors can transmit low-impedance analog or digital signals directly to the computer, or the transmission may be indirect through some form of electronic conversion. Careful shielding and grounding must be installed to eliminate extraneous electrical noise (37). Sensors for fermentation processes can be categorized as those sensing the physical environment, the chemical composition of the environment, and the microbial cell concentration and its intracellular events (Table 1). A list of several manufacturers that make various biosensors and instrumentation is given in Table 2.

23.1.1 Physical Measurement

Temperature

Mercury contact thermometers are often used for temperature measurement and control in laboratory fermentors. The thermometer should directly contact the fluid to minimize time lag. The thermometer must be able to withstand steam sterilization. A typical thermometer covers the range from 0 to 125°C and has an accuracy of ±0.5°C. Thermocouples and thermistors can be used to replace mercury contact thermometers. A metal resistance thermometer made of nickel or platinum has the advantage of stability and linearity of response; a compact form of this sensor is used in most fermentors today.

TABLE 1. Physically and chemically measured parameters for fermentation processes

Parameter	Frequency of use[a] in processes:		
	Bench[b]	Laboratory times pilot plant fermentations[c]	Industrial fermentors[d]
Physical			
Temperature	+++	+++	+++
Pressure	+++	+++	+++
Power consumption	+	+	++
Agitation speed	+	++	+++
Viscosity	++	+	+
Gas flow rate	++	++	+++
Liquid flow rate	++	++	+++
Turbidity	++	+	+
Volume and weight of tank	+++	++	++
Foam detection	++	+++	+++
Chemical			
pH	+++	+++	+++
Dissolved O_2	+++	+++	+++
Exit CO_2 concn	++	++	++
Exit O_2 concn	++	++	++
Redox potential	+++	++	++
Substrate concn	+++	++	+
Product concn	+++	++	++
Biological			
Cell concn	+++	++	++
Enzyme activities	+++	+	+
DNA and RNA	+++	+	+
NADH	+++	++	+
ATP	+++	+	+

[a] +++, Widely used; ++, less frequently used; +, seldom used.
[b] 0.5 to 2 liters.
[c] 2 to 1,500 liters.
[d] Typically >50,000 liters.

In many small fermentors, temperature is controlled by a heating element and cooling water supply, either of which can be on or off depending on whether heating or cooling is required. Large fermentors are controlled using only cooling water, or cooling water and steam as the heating source if necessary.

Pressure measurement

There is not much difference between pressure measurement in fermentors and in other chemical reactors. Nonsterile line pressure can easily be measured by ordinary Bourdon gauges; for asepsis, such a pressure gauge should be located in the exhaust gas filter outlet. The filter should be kept dry if possible, and the mount should avoid foam entrainment. A sterilizable type of pressure gauge can be used if the pressure gauge must be located in the fermentor. The force-balance, transmitter-type diaphragm gauge is widely used in pilot plant fermentors; electronic conversion of the pressure signal is available. Pressure in fermentor vessels can easily be controlled by using regulating valves in the exhaust air line.

Viscosity measurement

The measurement of the rheology of culture broths can be quite difficult. This is primarily due to the non-Newtonian characteristics and solids content of most fermentation broths. There are doubts that any measurement outside of the fermentor can yield a true value. In situ measurement cannot be achieved easily since the shear rate is nonuniform throughout the fermentor. Most apparent viscosity measurements in the literature are performed off-line using a Couette viscometer or Brookfield viscometer.

Gas flow measurement

A common method employed for measuring the inlet and outlet flow of gases to and from a fermentor is the use of variable area flowmeters, or rotameters. These usually consist of a glass tube with a tapered bore, hence a variable cross-sectional area, and a visible float for indication. The usual range of accuracy of these meters is ±2%, and the ratio of maximum to minimum flow rates is usually 10:1. The accuracy of these meters depends upon a constant pressure of the gas supply; therefore, it is desirable to include a pressure indicator with the gas flowmeter.

Large gas flow can be measured by a turbine flowmeter. These are electronic devices that operate by converting the rotary action of a turbine into an electrical signal. Control of flows can be effected manually and electrically through flow regulators, depending on cost considerations and convenience. The thermal mass flowmeter with controller has been demonstrated to be useful for measuring and controlling airflow in pilot plant fermentors (47).

Liquid flow measurement

Some of the liquid flows associated with fermentation systems require sterility. Nonsterile liquid flow, such as cooling water, can be metered easily with a metering pump. The metering of sterile liquids, especially liquids containing solids, is the most difficult to accomplish, and this difficulty increases with small fermentors. Most sterile nutrient flow can be controlled by a sterilizable metering pump. Unfortunately, these pumps are not necessarily accurate, and they deteriorate with time. Therefore, turbine flowmeters or equivalent devices must be used to ensure the accuracy of the metered flow rate. Some of the electromagnetic flowmeters can be used with media containing suspended particles, but a minimum flow rate of 2 ft/s (ca. 0.61 m/s) is usually required for proper operation.

Most of these liquid flow devices are expensive but suitable for pilot and manufacturing plant operations. Laboratory units may resort to the use of load cells to measure the weight of nutrient in feed storage tanks as means to back-calculate the liquid flow rate out of the storage vessel. Accurate and reliable peristaltic pumps must be used in conjunction with load cells to provide accurate liquid flow rate.

TABLE 2. Manufacturers of biosensors and instrumentation for fermentation processes

Apparatus	Manufacturer[a]
Temperature measurement	Omega Instrument Fisher Scientific Co. Gulton Industries LFE Corp. Rosemount, Inc.
Pressure measurement	Duriron Pressure Products Schaevitz Engineering
Viscosity measurement	Brookfield Engineering Automation Products, Inc.
Gas flow measurement	Brooks Instruments Div. Fisher and Porter Co. Foxboro Co. Meriam Instrument Div.
Liquid flow measurement	Brooks Instruments Div. Fisher and Porter Co. Foxboro Co. Interpace Corp. Chemcon, Inc. Milton Roy Co.
Volume and mass measurement	Ranco Controls Dexelbrook Robertshaw Controls Co. BLH Electronics Transducers, Inc. Micro-Strain, Inc. Schaevitz Engineering
Shaft power and agitation speed measurement	Electro Corp. Ohio Semitronics Co. Vibrac Corp.
pH measurement	Activion Glass, Ltd. Ingold Electrodes, Inc. Leeds and Northrop Co. Phoenix Electrode Co. Orion Research, Inc.
Dissolved oxygen measurement	New Brunswick Scientific Co., Inc. Beckman Instruments, Inc. Instrumentation Laboratory, Inc. Yellow Springs Instrument Co.
Exit CO_2 analyzer	Infrared Industries, Inc. Mine Safety Appliances Co. Beckman Instruments, Inc. Anarad, Inc. Perkin-Elmer Corp. (mass spectrometer) VG Instruments (mass spectrometer)
Exit O_2 analyzer	Beckman Instruments, Inc. Mine Safety Appliances Co. Taylor Instrument Co. Perkin-Elmer Corp. (mass spectrometer) VG Instruments (mass spectrometer)
On-line analyzer	New Brunswick Scientific Co., Inc. (turbidity) Technicon Instrument Corp. (wet chemistry analyses)
Highly instrumented fermentors	New Brunswick Scientific Co., Inc. B. Braun AG Chemapec, Inc. LKB Instruments, Inc. L. E. Marubushi Co. Biolafitte

[a] For addresses, see section 23.3.

Weight or volume measurement

Load cells are accepted as an ideal means to monitor the weight of the nutrient feed tanks and fermentors. The load cell is an electromechanical transducer that produces an electrical output proportional to the displacement of a separate movable cone. Through a signal conditioner, an electrical signal can be generated and measured. Most load cells have an accuracy of ±10% and are sensitive to temperature. The load cell is usually mounted in a compression mode such that the reservoir or tank and the platform rest upon the load-bearing shaft. Large fermentors require three or four load cells to generate a more accurate reading. Closed environment control must be built in to minimize any outside disturbance.

Shaft power and agitation speed measurement

There are two common devices used to measure shaft power, the torsion dynamometer and the strain gauge. The torsion dynamometer or torque meter measures the amount of torque which is produced between the motor shaft and the impeller shaft when the former turns the latter. The conductivity of a resistance wire varies with the strain imposed by tension or compression; the strain gauge uses this phenomenon to measure shaft power. The main difficulty is to maintain the electrical connection between the strain gauge and the turning shaft. A strain gauge is frequently used because it gives a more exact reading than a dynamometer. For large fermentors, a wattmeter should be adequate to measure the total power drawn by the agitator motor.

Agitation speed can be measured directly or indirectly. Direct gear or belt-driven tachometers are particularly useful for large fermentors when the speed ranges from 0 to 100 rpm. For higher-speed monitoring, such as in small fermentors, where the agitator speed may range from 0 to 2,000 rpm, indirect monitoring devices such as those based on magnetic and optical methods may be used. Magnetic sensors are noncontact transducers whose output voltage reflects the rate of change of the magnetic field. They are usually mounted on the external drive shaft and can send out a direct electrical signal if needed.

Foam monitoring and control

Foaming is a common problem in many fermentation processes. Large quantities of air passing through the broth can lead to foam formation. Foam can be detected by either a capacitance or conductivity probe. Foaming is controlled by mechanical foam destruction devices, the addition of sterile antifoam agents, or a combination of both. There are two mechanical ways to break up foam or prevent its formation. The first consists of specially designed disks or weir arrangements on a stirrer shaft, which break foam by a beating action. The second method is to destroy the foam using ultrasonic or other energy waves. A range of chemical antifoam agents are available to lower the surface tension at the liquid/air interface. Empirical testing must be done to choose the best. The silicone-based antifoam agents are the most commonly used in the fermentation industries. Foam control systems are usually the on-off type with a time delay to prevent overcharging of the antifoam agent. An extensive review on foam formation, detection, and control was done by Hall et al. (23).

23.1.2 Chemical Measurement

pH measurement

The pH of a fermentation is commonly monitored with a steam-sterilizable pH electrode connected to a pH meter for direct visual readout. Analog control can be achieved by using a titrator. The titrator in turn activates pumps or valves which feed aqueous acid or base solutions from the respective reservoirs into the fermentor if the pH changes from a set point. Computer-based direct digital control has been used to replace the analog titrator in many recent fermentation plant and pilot plant operations (18).

Dissolved oxygen measurement

The partial pressure of dissolved oxygen in the fermentation broth is measured with either polarographic or galvanic oxygen electrodes. These are membrane electrodes designed by Clark (11) and Borkowski and Johnson (8) with subsequent modification to suit fermentation monitoring (19). Most dissolved oxygen sensors operate on the principle of reducing oxygen at the surface of a metal electrode (33):

$$Ag + 1/2\ O_2 + H_2O + 2e^- \rightarrow 2OH^- \quad (1)$$

The electrons are provided from the anode. The value of the current (I) from the sensor depends on the electrons transferred (n), the Faraday constant (F), the area of the cathode (A), the permeability coefficient of the membrane (P_m), the thickness of the membrane (b), and the partial pressure of the oxygen (pO_2):

$$I = nFA(P_m/b)pO_2 \quad (2)$$

The change of the dissolved oxygen concentration, C_L, of a fermentation depends on the oxygen transfer capacity of the fermentor and the metabolic rate of the respiring cells:

$$dC_L/dt = K_La(C^* - C_L) - Q_{O_2}X \quad (3)$$

where C_L = the actual dissolved oxygen concentration in the liquid; C^* = concentration of dissolved oxygen which is in equilibrium with pO_2 in the bulk gas flow; K_La = the overall oxygen transfer coefficient; Q_{O_2} = specific oxygen uptake rate; and X = viable cell mass concentration.

Dissolved oxygen measurement is important in fermentation processes because it can be used to determine K_La and $Q_{O_2}X$ (27) and can also be applied as a sensor to control and maintain a desired dissolved oxygen concentration in the fermentation broth. Agitation speed and gas flow rate are the final control elements in such a control loop.

To obtain meaningful information about dissolved oxygen, the electrode design must be appropriate and proper electronic conversion and amplification must be maintained. The 90% response time should be below 1 min, and membrane fouling by chemical antifoam and other nutrients must be avoided. Dissolved oxygen measurement is also quite sensitive to temperature, and thus a constant temperature must be maintained or temperature compensation must be incorporated to obtain accurate measurements.

Redox potential measurement

The oxidation-reduction potential (ORP, E_h, rH_2, etc.) of a fermentation can be determined with electrodes similar to those used to measure pH. Usually a platinum electrode is employed and the measurement is in millivolts. It has been shown that temperature, dissolved oxygen, pH, and inorganic and organic compounds with oxidizing or reducing capabilities will influence this measurement. However, for anaerobic fermentations, the reading may be used to reflect the relative redox potential of the environment. It is more difficult to interpret the meaning of this variable in aerobic fermentations because of the interference of oxygen and other compounds. The redox potential of a fermentation broth can be controlled by sparging nitrogen gas or adding chemical nutrients such as cysteine, ascorbic acid, sodium thioglycolate, and other reducing agents (31).

Exit O_2 concentration

The oxygen membrane electrode can also be used to monitor oxygen concentration in the exhaust gas stream from a fermentor. The C^* value of the equation is more commonly obtained by measuring oxygen partial pressure in the exhaust gases by means of gas analyzers (14). There are at least two commercially available methods for the continuous analysis of oxygen in gas streams. One depends on the paramagnetic properties of oxygen, and the other one is based on an electrochemical fuel cell. Recent advances in solid-state circuitry have helped to stabilize such devices tremendously. The oxygen uptake rate (OUR) can be calculated from the oxygen concentrations in the gas streams:

$$\text{OUR} = Q_{O_2}X$$
$$= \frac{F_N}{V}\left(\frac{P_{O[\text{in}]}}{P_t - P_{O[\text{in}]} - P_{W[\text{in}]} - P_{C[\text{in}]}} - \frac{P_{O[\text{out}]}}{P_t - P_{O[\text{out}]} - P_{W[\text{out}]} - P_{C[\text{out}]}}\right) \quad (4)$$

where F_N = inert (N_2) gas flow rate; V = liquid volume; and F_t = total pressure of the fermentor. $P_{O[\text{in}]}$, $P_{W[\text{in}]}$, and $P_{C[\text{in}]}$ are the inlet partial pressures of oxygen, water vapor, and carbon dioxide, respectively; similarly, $P_{O[\text{out}]}$, $P_{W[\text{out}]}$, and $P_{C[\text{out}]}$ are the outlet partial pressures of oxygen, water vapor, and carbon dioxide, respectively.

Exact values of OUR can be obtained only by accurate, simultaneous measurement of inlet and outlet oxygen and carbon dioxide concentrations. Special precautions must be taken in treating the sample gas. It should be freed from dust by filtration, and the water-saturated gas stream must be dewatered because water molecules are paramagnetic and introduce error in measurement.

The typical range of oxygen monitored is usually from 21 to 10%. The optimal gas flow is around 0.5 liter/min. Proper piping must be arranged to ensure sterility and to guard against foaming problems (Fig. 1).

There has been an upsurge in the use of on-line process gas chromatographs and process mass spectrometers in the fermentation industry. Some of the problems associated with gas analyzers, such as low specificity and slow response, can be drastically reduced by using these multivariable machines. A process mass spectrometer is best for fermentation monitoring (Fig. 2). The use and maintenance of the mass spectrometer-based gas analyzers have been reviewed elsewhere (25).

Exit CO_2 concentration

There are two methods to analyze specifically CO_2 in a fermentation gas stream. One is infrared analysis, which is specific for CO_2 measurement. The other is the thermal conductivity method, which is less specific to CO_2 monitoring because other gaseous components such as H_2O and H_2 affect its measurement.

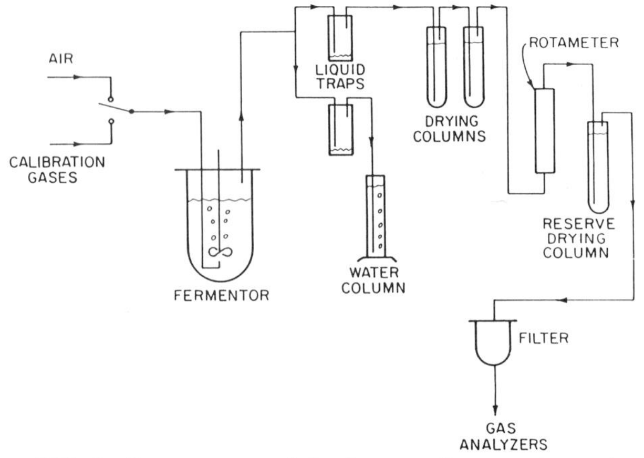

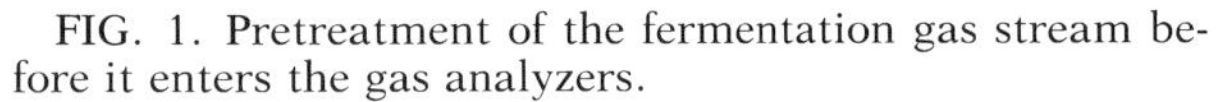

FIG. 1. Pretreatment of the fermentation gas stream before it enters the gas analyzers.

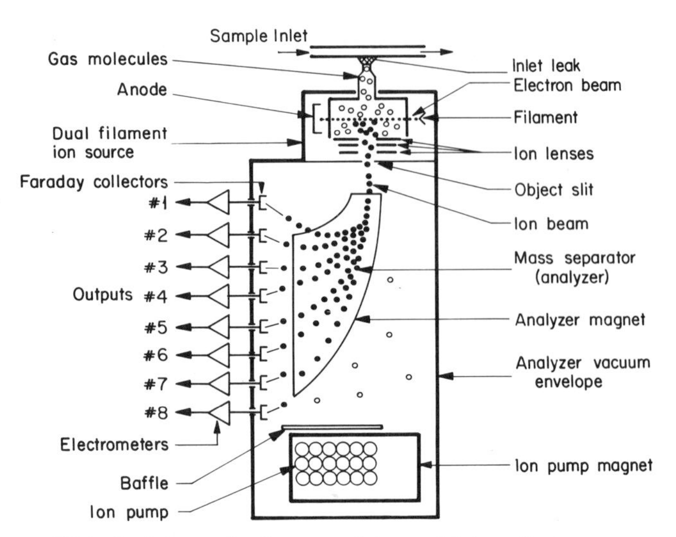

FIG. 2. Schematic diagram of the multiple-collector mass spectrometer for fermentation gas analysis (courtesy of Perkin-Elmer Corp.).

A carbon balance can be made on the gas flow in and out of the fermentor to calculate carbon dioxide evolution rate (CER):

$$\text{CER} = \frac{F_N}{V}\left(\frac{P_{C[\text{out}]}}{P_t - P_{C[\text{out}]} - P_{W[\text{out}]} - P_{O[\text{out}]}} - \frac{P_{C[\text{in}]}}{P_t - P_{C[\text{in}]} - P_{W[\text{in}]} - P_{O[\text{in}]}}\right) \quad (5)$$

Since the solubility of CO_2 in aqueous solution is heavily dependent on pH and buffering agents in the fermentation broth, changes in measured CO_2 value should be compared with pH changes to dissociate the influence of pH on this measurement.

Increasingly, process gas chromatographs and process mass spectrometers are being used to simultaneously measure O_2 and CO_2 contents of the inlet and outlet air streams in fermentations (9, 10).

Measurement of other gases

In some fermentations, the carbon and nitrogen sources may be gases and can be measured with instruments such as the process gas chromatograph and process mass spectrometer (25, 43). For example, methane has been used as a carbon and energy source in single-cell protein production, and gaseous ammonia has been used as a nitrogen source and for pH control purposes. Both of these gases can be continuously measured, and such measurements supply valuable information about the metabolic status of the fermentation. In some cases, the substrate or the product can be quite volatile. For example, ethanol production by yeast fermentation can be adequately monitored by determining the gas-phase ethanol concentration with specific gas analyzers. In general, methods for and use of gas analyses have been advanc-

TABLE 3. Performance of measurement and control instrumentation in a pilot plant fermentor (21)

Performance	Temp (W)[a] (°C)		pH (H)[a] (units)		Dissolved oxygen (R)[a] (% saturation)		Pressure (R)[a]	Vol (R)[a]	Airflow (R)[a]	Exit CO_2 (R)[a]	Exit O_2 (R)[a]
	Measurement	Control	Measurement	Control	Measurement	Control	(psig)	(liters)	(liters/min)	(%, vol/vol)	(%, vol/vol)
Accuracy	±0.1	0.02	±0.14	0.1	6.0	0.52	0.41	0.4	±7.0	±0.1	±0.1
Precision (SD)		0.08		0.01		9.95	0.07	3.4	23.0		
Resolution	0.02		0.001				0.004	0.14	0.02	0.0005	0.0005

[a] Instruments recalibrated weekly (W), hourly (H), or per run (R).

ing steadily in recent years. Their utility in fermentation monitoring and control has been demonstrated and appreciated.

23.1.3 Biological Measurement

Few commercial on-line sensors are available for substrate and product measurements, such as carbon and energy sources, nitrogen content, phosphate content, etc., in fermentation medium. No commercial on-line sensor is available to measure the intracellular events of the microorganisms which dictate the outcome of the fermentation. With the exception of low-molecular-weight volatile compounds such as ethanol, no adequate methods have been developed for on-line liquid-phase measurements. Current methods for examining the process status of industrial fermentations require routine removal of samples from the fermentor through a sampling port for examination and assay. Steam lines and traps are built around the sampling port to avoid microbial contamination. Although this periodic sampling technique is tedious, it is still commonly used by industry for process monitoring. The samples must also be pretreated, to remove suspended solids and microbial cells, and properly diluted before off-line chemical and biological assay. One of the main reasons why most of the commercial chemical- or biochemical-based analyzers are not frequently used for on-line analysis is that a suitable interface device is not available. Dialysis samplers are now available (45, 52). Their complete acceptance still remains to be seen. Membrane fouling on the probe is a key problem that must be solved before any of these membrane-coated electrodes or sensors can be used.

To overcome these problems, discrete measurements based on automatic analyzers can be used. This has already been accomplished in analyzing small molecular components in yeast fermentation (36, 49). Recently an automatic assay for cellulose activity during a fermentation has been developed and used for computer monitoring and control (32). The challenge comes from maintaining these automatic analyzers (such as the Technicon autoanalyzer) in an industrial plant environment.

23.1.4 Performance Evaluation of Biosensors and Bioinstrumentation

Flynn (21) suggested that three important characteristics must be considered in choosing specific measured variables for monitoring control: (i) relevance, (ii) accuracy, and (iii) precision. The relevance of data is judged by the decisions they allow one to make: the greater the number of decisions, the greater the relevance of the data. The accuracy of the measured data is expressed as the difference between the observed value of a variable and its true value. The problem here is to determine the true value of a variable, especially for a fermentation process. The precision of the data relates to the probability that repeated measurements on the same system will produce the same value. Statistics used include the standard deviation, the standard error, and the 95% confidence level. Standard deviation can easily be determined experimentally and is widely used to assess precision. Resolution is another characteristic of specific measured variables; it is the smallest difference between two values of a particular parameter that can be discriminated by normal instrumentation. The performance of some of the stated fermentation variables in a pilot plant fermentor is shown in Table 3.

23.1.5 Conclusion

A key characteristic in automatic monitoring and control of a fermentation is the adequate monitoring of concentrations of substrate, cell mass, and intracellular and extracellular products. With a few exceptions, direct sensors for biochemical and chemical measurements have not been developed sufficiently for general application. Those which are commercially available are not suited to most industrial applications because of interfering characteristics of the fermentations, such as the presence of insoluble substrates or colored medium, nonspecificity of the sensor, and the need to maintain sterility. New means for direct monitoring of substrate and product concentrations are required before these bioprocess parameters can be controlled. We also need information from inside the microbial cells to develop more robust models to describe the process dynamics. Intracellular events such as proteinase activities and product accumulation dictate the eventual outcome of the fermentation. This is especially true of newly developed microbial processes based on recombinant DNA technology and cell fusion techniques.

23.2 PROCESS CONTROL OF FERMENTATION PROCESSES

Before reviewing the specifics of process control of fermentation processes, it is desirable to look at the state of the art in the fermentation industry. Although the application of computers to process control is widely accepted in many chemical and petroleum

TABLE 4. Manufacturers of computer interfaces for data acquisition and control

Manufacturer (model)	Computer[a]
Data Translation Inc. (DT 2801, DT 2083 and other models)	IBM PC, XT, AT
Keithley/DAS (series 500)	IBM PC, XT, AT, Apple II+, IIe
Interactive Microware, Inc. (ADALAB)	Apple IIe Macintosh
Burr-Brown Corp. (PCI-20000)	IBM PC, XT, AT; DEC PDP-11; DEC VAX; HP-3000
Hewlett Packard Co. (3421A and others)	All HP computers
Kirby Lester Inc. (Microlink)	IBM PC, XT, AT; Apple, HP
Gould Electronics (DASA 9000)	IBM PC, XT, AT
Tecmar (Labmaster)	IBM PC, XT, AT
MetraByte (IEEE-488 GPIB)	IBM PC, XT, AT
Cyborg (ISAAC)	Apple II+, IIe; IBM PC, XT, AT
Digital Equipment (ADM)	All PDP computers

[a] For compatibility, check with the manufacturer.

industries, it has been slower to come about in the fermentation industry. Two major reasons for this slowness in development are (i) the biochemical and physical complexity of fermentation processes, for which no generalized process model has been developed, and (ii) the lack of suitable biosensors to measure the important variables (see above).

The two earliest descriptions of computer control of a batch fermentation appeared in the late 1960s (22, 51). Both described the application of direct digital control, which utilized existing sensors and simply eliminated the need for multiple analog controllers, which are still dominant in the existing fermentation industry. In a few cases, a computer is used not only for sequence digital control of the fermentations, but also to form a mathematical model which can be used to monitor the growth of the microbial fermentation. If the actual fermentation pattern as monitored by the computer differs from the model, the computer will immediately warn the operator. Sometimes, a statistical evaluation of past fermentations instead of a mathematical model can be used to derive a "standard" fermentation with which the computer compares the actual process performance. Unfortunately, most of the mathematical models developed so far incorporate one or more assumed constants which may limit their utility.

One of the characteristics of all digital computer control systems is that they are sampled at intervals rather than being continuous in nature. Hence, the control system must select a particular sampling frequency for each control loop. This frequency may differ from one fermentation to another. The desire to sample rapidly (to approximate continuous measurement as closely as possible) usually cannot be satisfied in a computer-controlled system because of either limited memory capacity or instrumentation delay. For example, it does not make much sense to sample more often than the inherent process dead time or delay. For most fermentations, sampling frequency of the order of 0.5 h is adequate. Time-averaged signals are used to reduce measured noise. In addition, a Kalman filter or other digital filtering technique can be used to reduce the buildup of error in the estimated state of the fermentation due to noise in the measurements (28).

23.2.1 Computer Application in Fermentation Processes

Major changes have occurred in the computer industry. Computers are smaller in size, more powerful in memory, and less expensive than in the past. Most microcomputers can now be used to do jobs which required a minicomputer to do just a few years ago. A computer used for fermentation processes must be able to perform the following tasks: (i) log data derived from the bioinstrumentation through an appropriate electronic interface system; (ii) analyze data on the basis of a computer program; (iii) store data and computer programs on a physical memory, such as a disk, which can be recalled; (iv) control process variables based on appropriate process control loops and pumps or valves.

A low-cost microcomputer such as an Apple or IBM PC with more than 64K memory can be used to monitor and control fermentation processes as long as an appropriate data interface system is available for data acquisition and data management to effect control. Various interface systems for microcomputers for process control purposes are commercially available (Table 4).

The central processor board of the microcomputer should contain the following (Fig. 3):

1. A large-scale integrated (LSI) based microprocessor such as M68000.
2. A read-only memory (ROM) device to monitor the program; ROM is not altered by the microprocessor.
3. A system clock for timing orderly functioning of the microcomputer.
4. Bus buffers that buffer all the signals to and from the CPU board.
5. Memory boards that consist of LSI elements called RAMs (random access memory). These circuits enable programs and data to be stored, read, and changed under the control of the CPU board.
6. A software clock which counts the cycles of the main alternating current and provides a pulse to a binary counter after a certain number of cycles.
7. An interface with peripherals that enables the computer to issue and receive commands from a keyboard and type messages on paper. The interfacing between the computer and the bioreactor also consists of analog to digital (A/D) converters and digital to analog (D/A) converters. The A/D converters read measurements relayed by the fermentor's on-line sensors and instrumentation and convert them into digital signals which may then be interpreted by the central processor. Conversely, D/A converters convert digital signals sent by the central processor back into analog signals and relay them back towards the

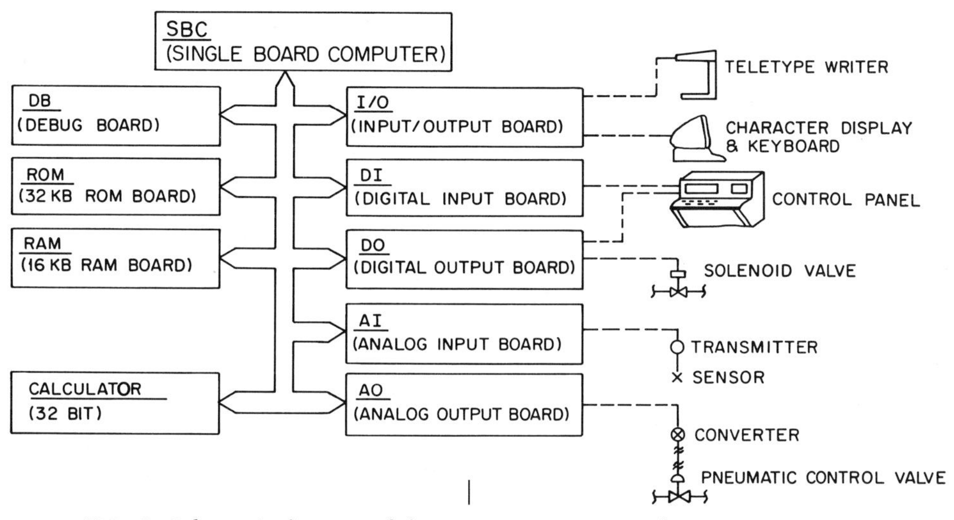

FIG. 3. Schematic diagram of the various components of a microcomputer.

fermentation system to effect control, such as to turn on or off a pump.

The hardware of the computer should also include such items as the program/information processor, terminals, printers, and storage devices, as shown in Fig. 3.

23.2.2 Types of Computer Control in Fermentation Processes

Computer control can be divided into two kinds: direct digital control and set point control. In the case of the first, the computer examines the input variables from the sensors and uses the information to produce a signal that is sent back to effect control. For set point control, the computer only adjusts the set point of existing analog controllers.

Each control loop (Fig. 4) consists of (i) process (the fermentation), (ii) measuring element (e.g., thermometer), (iii) intelligent controller (e.g., computer), and (iv) final control element (e.g., control valve or pump). There is a general trend to replace all analog controllers with digital computer control. After receiving an error signal, the computer sends out another signal to the final control element, such as a control valve, to produce corrective action. The corrective action can be regarded as mathematically related to the output signal from the controller. There are several types of control action frequently used in industrial processes. The kind of control loop most often used is the univariable feedback loop, which contains either pneumatic or electronic circuits with the various control elements.

Proportional control

Proportional control produces an output signal which is proportional to the error ε. The error ε is the difference between the set point and the measured variable from the sensor. This action can be expressed as:

$$T = K_C\varepsilon + T_s \tag{6}$$

The units of the set point and the measured variable must be the same. In a controller having adjustable gain, the value of the gain K_c can be varied accordingly. The error required to move the final control element from fully closed to fully open is called the proportional band, the bandwidth, or the gain. Even though basic proportional action is simplest, other modes of action such as integral and derivative controls are frequently needed to correct process upset (Fig. 5).

A special case of proportional control is the on-off control. If the gain K_c in equation 6 is made very high, the control element will swing from one extreme

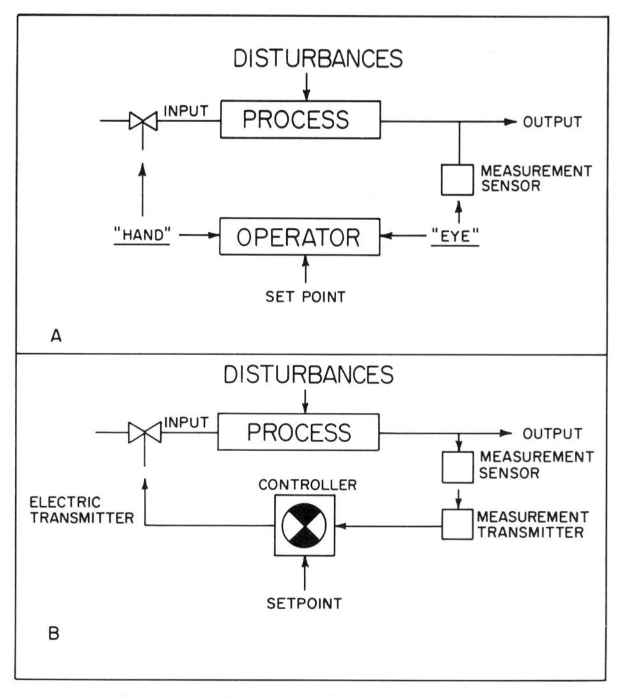

FIG. 4. (A) Manual control of a process variable. (B) Automatic feedback control of a process variable.

position to the other if ε is very small. This very sensitive action is called on-off control because the control element is either fully open (on) or fully closed (off) like a switch. The bandwidth of an on-off controller is approximately zero. Many simple control loops use on-off control action in fermentation processes.

Proportional-integral control

Due to transient responses and time delays, it is often desirable to add other modes of control to the basic proportional action. For example, proportional-integral control can be modeled as follows:

$$T = K_c\varepsilon + \frac{K_C}{\tau_I}\int_0^t \varepsilon\, dt + T_s \qquad (7)$$

Another term which is proportional to the integral of the error has been added to the proportional control. Values of both K_c and T_I can be adjusted to effect control action.

Proportional-derivative control

Proportional-derivative control can be represented as:

$$T = K_c\varepsilon + K_C\,\tau_0\frac{d\varepsilon}{dt} + T_s \qquad (8)$$

In this case, we have added to the proportional control action another term which is proportional to the derivative of the error. Other terms which have been used to describe derivative control are rate control and anticipatory control.

Proportional-integral-derivative control

The fourth mode of control is a combination of the previous modes and is given by the expression:

$$T = K_c\varepsilon + K_c\,\tau_0\frac{d\varepsilon}{dt} + \frac{K_C}{\tau_I}\int_0^t \varepsilon dt + T_S \qquad (9)$$

The curves of Fig. 5 show the behavior of a typical feedback control system using different kinds of control when it is subjected to a disturbance. With proportional control only, the control system is able to

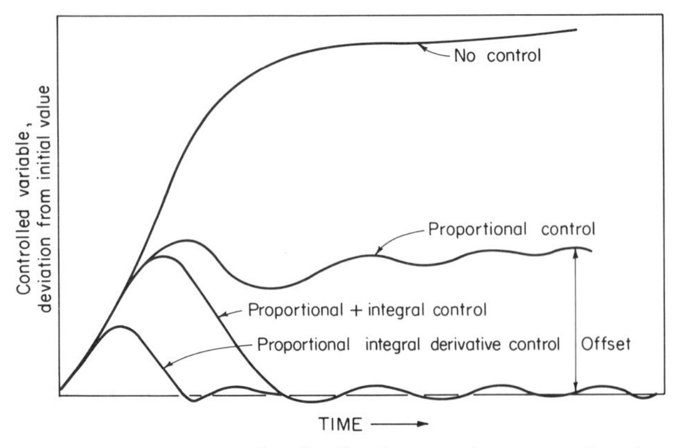

FIG. 5. Response of a feedback control system showing the effect of various modes of control.

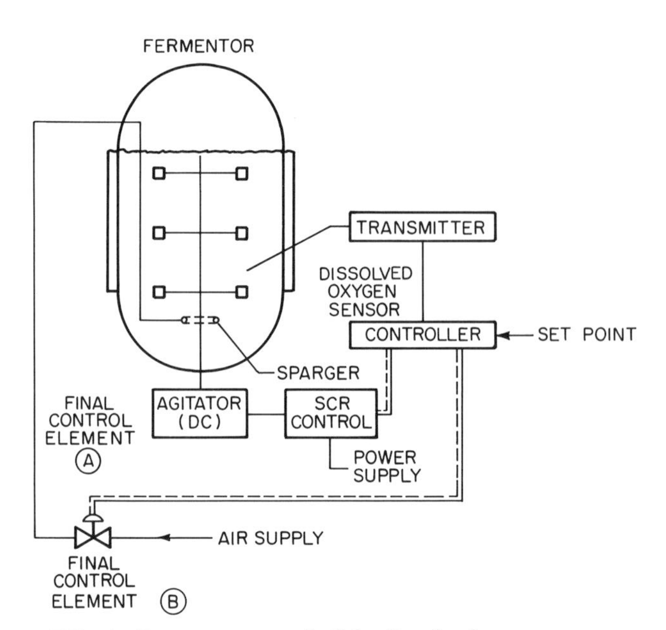

FIG. 6. Computer control of the dissolved oxygen concentration in a fermentor.

arrest the rise of the controlled variable and bring the variable to a new steady-state level which may not be the original value. The difference between this new steady-state value and the original value is called the offset. The addition of integral condition eliminates the offset and brings the controlled variable back to the original value. The advantage of intergal action should be balanced by the disadvantage of a more oscillatory behavior. The addition of derivative action to the proportional-integral action gives a definite improvement in the response. The rise of the controlled variable is arrested more quickly, and it is returned rapidly to the original value with little or no oscillation. Except for economic and other practical considerations, proportional-integral-derivative control is best for steady-state operation. Procedures for tuning the control loop for best performance can be found in most textbooks on process control.

Feedforward and other control algorithms

In feedback control, any disturbance has already passed through the process before the error signal is generated. Feedback control, consequently, is a corrective control aimed at restoring some predetermined state of the process. Another form of control is called feedforward control, in which the sensor is used to monitor disturbances in the input streams. A disturbance is sensed before it enters the process. If the corrective action is fast enough, the upset may be avoided or minimized. The choice between feedback or feedforward control depends on system stability and availability of the sensors. In many cases, a combination of feedforward-feedback control algorithms may provide the best type of control (48).

There are other forms of control such as ratio, ratio with autobias or autoproportion, and so on. Instead of univariable control as discussed previously, multivariable cascade control is becoming the predominant feature for fermentation monitoring and

TABLE 5. Computer control of various fermentation processes

Fermentation process	Type of control	References
Glutamic acid fermentations	Direct digital control; control based on a kinetic model	51
Antibiotic fermentations	Direct digital control; total plant control	22
Baker's yeast fermentations	Nutrient feed control	7, 17, 48
	Temp, pH, dissolved oxygen controls	34, 41
	Statistical control	30
Cycloheximide fermentations	Nutrient feed control	Wang (in press)
Yeast fermentations	Direct digital control	35
	Temp, pH, dissolved oxygen controls	29; Jeffries et al.[a]
Lysine fermentations	Optimization (off-line)	42
Penicillin fermentations	Optimization (off-line)	12
	Nutrient feed control	38

[a] R. P. Jeffries, A. E. Humphrey, and L. K. Nyiri, 164th National ACS Mtg, New York, 1972.

control. This is because many variables of fermentation processes are quite interactive and nonlinear in nature. For example, the control of dissolved oxygen can be achieved by using either airflow or agitation or both as manipulated variables; the set point can also be changed up or down over fixed time intervals according to the demand of the cellular activities (Fig. 6).

23.2.3 Computer Control of Fermentation Processes

There are many advantages and objectives of computer-aided fermentation process control. With monitoring and control of raw material addition and total energy consumption, processes can become far more efficient, resulting in great economic savings. This is especially true in fermentations for which the substrates or raw materials are the major costs. The second objective of computer control is process optimization, which involves the maximization of volumetric productivity, product concentration, and conversion yield of substrate to product. Optimal running conditions are determined through repetitive experimentation and model construction. Computer control can achieve nearly exact reproducibility and maintenance of the optimal conditions. In addition, a computer monitors parameters continuously and much more closely than humanly possible. This allows detection of minute changes in the fermentation over very short intervals. Other potential advantages of computer-aided fermentation control include quick and efficient data management, storage, and reproduction; more plant flexibility in adapting to market demands; and better pollution control and plant safety control. Unfortunately, many of these objectives are not being met at present. The complexity of microbiological systems has made them difficult to model adequately for cause-effect control purposes. There has also been a lack of reliable sensing instruments for measuring essential biological and physicochemical parameters. Even with these handicaps, various forms of computer-aided fermentation monitoring and control systems have been developed for specific fermentations (Table 5).

A computer-controlled fermentation system is shown in Fig. 7. In industrial practice, multifermentor

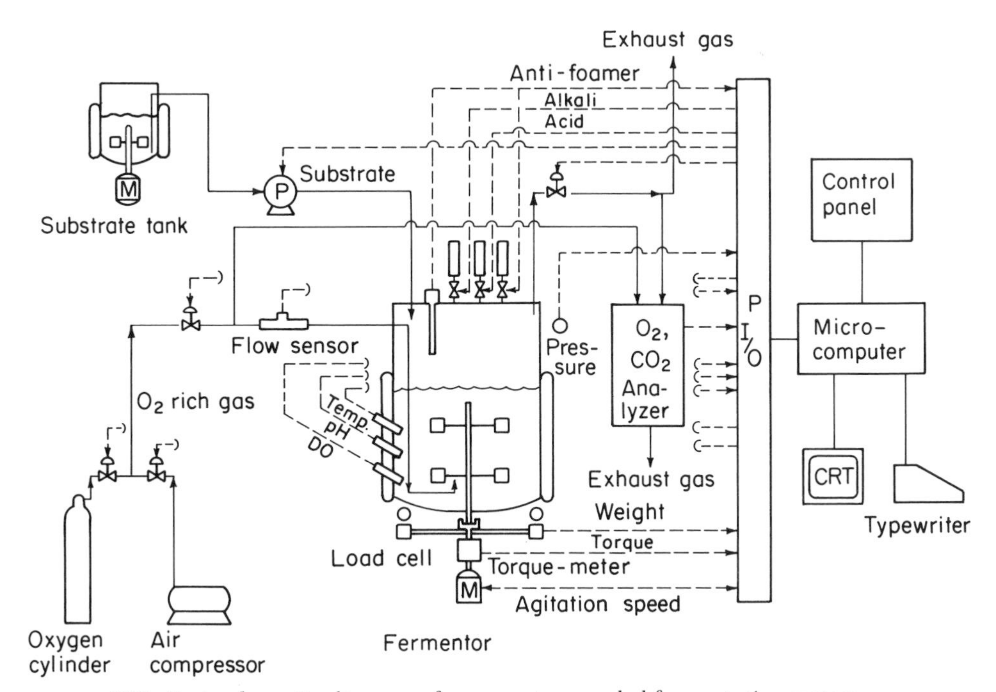

FIG. 7. A schematic diagram of a computer-coupled fermentation system.

monitoring and control is more frequently encountered.

There are two main subsystems of a computer-controlled fermentation system. One is the computer itself with its associated hardware, as discussed previously. The other is the fermentor with associated instrumentation for on-line measurements (Fig. 7).

The software program in the computer commands the fermentation operation. Several different types of software are incorporated in carrying out this function. One type is responsible for enabling the operator to issue commands to a terminal, load and execute programs, and control and monitor information transmitted to and from all system devices. This type of software is called the user-oriented applications program. These programs are written in standard computer languages such as FORTRAN, COBOL, and BASIC. Most of the program has to be custom made according to the computer-coupled fermentation system and the actual fermentation. There is no generalized fermentation monitoring and controlling program available. Other software is responsible for the interfacing between the operator and system programs. This software translates the language of the user-oriented applications programs to the language of the system programs. Such software is termed the assembly language program. The third type of program, system programs, is what the computer actually reads and processes.

One other important function performed by a computer is time sequencing. A computer can perform only so many functions at one time. Therefore, time-sequencing software determines the order in which programs are executed and tasks are completed. Some of the higher priority tasks are starting the scanner, logging data on the disk, generating alarm messages, and actual process control. Among the lowest priority tasks are graphing and report printout; such tasks could easily be delayed without any effect upon the fermentation. Delaying an alarm message or some aspect of the process control, however, could destroy an entire fermentation. Thus, appropriate allocation of computer resources is highly important.

Computer control has been used for various fermentation processes (Table 5). Hundreds of articles have been published concerning computer hardware design, new instrumentation and analyses, and model construction and evaluation as applied to fermentation. Yeast fermentations are the most often cited processes for computer control. This is primarily because the fermentation is well studied and the biochemistry of yeast growth and product formation is well understood. Most other industrial fermentation processes need more fundamental understanding before sophisticated control theory can be applied.

23.2.4 Conclusion

The application of computers to on-line monitoring and control of fermentation processes requires the collection and interpretation of pertinent data. The number of sensors available to measure the relevant key variables such as cell concentration and product concentration are quite limited. Sometimes it may be necessary to deduce the values of these parameters from other measured parameters. It will be important to develop new and robust sensors before more widespread automatic control of fermentation processes can be instituted. The purpose of a control system is to keep the process close to an ideal state. Unfortunately, most of the control strategies developed so far assume steady-state or near-steady-state operation. Fermentation processes, on the other hand, are usually run in batch or fed-batch modes. The measured variables are highly interactive in nature, and process models are not well defined. Multivariable control and dynamic optimization so far have limited utility for most fermentation processes. Recent advances in molecular genetics and cell physiology may help us to develop more accurate process models to define various cause-effect relationships which in turn can be used for process control.

23.3 COMMERCIAL SOURCES

Activion Glass, Ltd., Halstead, Essex, England
Anarad, Inc., Santa Barbara, CA 93103
Automation Products, Inc., Houston, TX 77008
B. Braun Melsungen AG, P.O.B. 346, D-3508 Melsungen, Federal Republic of Germany
Beckman Instruments, Inc., 2500 Harbor Blvd., Fullerton, CA 92634
Biolafitte, 719 Alexander Rd., Princeton, NJ 08540
BLH Electronics, Waltham, MA 02254
Brookfield Engineering, Stoughton, MA 02072
Brooks Instruments Div., Hatfield, PA 19440
Burr Brown, International Airport Industrial Park, P.O. Box 11400, Tucson, AZ 85734
Chemapec, Inc., 230-C Crossway Park Dr., Woodbury, NY 11797
Chemcon, Inc., S. Attleboro, MA 02703
Cyborg, Inc., 55 Chapel St., Newton, MA 02158
Data Translation Inc., 100 Locke Dr., Marlboro, MA 01752
Dexelbrook Engineering, Horsham, PA 19044
Digital Equipment, 77 Reed Rd., Hudson, MA 01749
Duriron Co., 900 Louis Drive, Warminster, PA 18974
Electro Corp., Sarasota, Fla.
Fisher and Porter Co., Warminster, PA 18974
Fisher Scientific Co., 711 Forbes Ave., Pittsburgh, PA 15219
Foxboro Co., Foxboro, Mass.
Gould Electronics, 3631 Perkins Ave., Cleveland, OH 44114
Gulton Industries, Costa Mesa, Calif.
Hewlett Packard, 1820 Embarcadero Rd., Palo Alto, CA 94303
Infrared Industries, Inc., Box 989, Santa Barbara, CA 93102
Ingold Electrodes, Inc., Lexington, Mass.
Instrumentation Laboratory, Inc., Lexington, Mass.
Interactive Microware, Inc., P.O. Box 771, State College, PA 16801
Interpace Corp. (Lapp), Rochester, N.Y.
Keithley/DAS, 349 Congress St., Boston, MA 02110
Kirby Lester, Inc., P.O. Box 43, Riverside, CT 06878
Leeds and Northrup Co., Sumnneytown Pike, North Wales, PA 19454
LFE Corp., Waltham, MA 02254
LKB Instruments, Inc., 12221 Parklawn Dr., Rockville, MD 20852

L. E. Marubushi Co., Koki Bldg., 8-7, 2-chome, Kajicho, Chiyoda, Tokyo, Japan
Meriam Instrument Div. (Scott & Fetzer), Cleveland, Ohio
Metra Byte Corp., 254 Tosca Dr., Stoughton, MA 02072
Micro-Strain, Inc., Spring City, Pa.
Milton Roy Co., St. Petersburg, Fla.
Mine Safety Appliances Co., 600 Penn Center Blvd., Pittsburgh, PA 15235
New Brunswick Scientific Co., Inc., 44 Talmadge Rd., P.O. Box 986, Edison, NJ 08811
Ohio Semitronics Co., Columbus, Ohio
Omega Instrument, Stamford, CT 06907
Orion Research, Inc., Cambridge, MA 02139
Perkin-Elmer Corp., Maine Ave., Norwalk, CT 06856
Phoenix Electrode Co., Houston, TX 77081
Ranco Controls, Columbus, Ohio
Robertshaw Controls Co., Anaheim, CA 92803
Rosemount, Inc., Eden Prairie, MN 55344
Schaevitz Engineering, Pennsauken, N.J.
Taylor Instrument Co., 95 Ames St., Rochester, NY 14601
Technicon Instrument Corp., Tarrytown, NY 10591
Tecmer, Inc., 6225 Cochran Rd., Solon, OH 44139
Transducers, Inc., Whittier, Calif.
VG Instruments, Stamford, CT 06901
Vibrac Corp., Chelmsford, Mass.
Yellow Springs Instrument Co., Yellow Springs, Ohio

23.4 LITERATURE CITED

23.4.1 General References

1. **Aiba, S., A. E. Humphrey, and N. F. Mills.** 1973. Biochemical engineering, 2nd ed. Academic Press, Inc., New York.
   Chapter 11 in this text gives a detailed description of fermentation instrumentation and its use as the basis for computer control.
2. **Halme, A. (ed.).** 1983. IFAC Workshop, 1st, Helsinki, Finland, 1982. Modelling and control of biotechnical processes: proceedings. Pergamon Press, Oxford, England.
   This is the proceedings of the First IFAC Workshop on computer application to fermentation processes. Many interesting articles.
3. **Peppler, H. J., and D. Perlman (ed.).** 1979. Microbial technology, vol. 2. Fermentation technology. Academic Press, Inc., New York.
   Many chapters on fermentation processes. Chapter 14 by Tannen and Nyiri gives a detailed description of various instruments. Chapter 15 is on computer applications in fermentation processes.
4. **Solomons, G. L.** 1969. Materials and methods in fermentation. Academic Press, Inc., New York.
   Several chapters (4–7) describe in detail many instruments and sensors used in fermentation processes.
5. **Staphanopoulos, G.** 1984. Chemical process control. An introduction to theory and practice. Prentice Hall, Englewood, N.J.
   This is a recent book on process control in chemical industries. It describes the state of the art of areas in control algorithms for chemical processes. No discussion of fermentation monitoring and control.
6. **Wang, D. I. C., C. L. Cooney, A. L. Demain, P. Dunnill, A. E. Humphrey, and M. D. Litty.** 1979. Fermentation and enzyme technology. Wiley Interscience, Inc., New York. *A general reference on fermentation technology with some description of bioinstrumentation and computer applications.*

23.4.2 Specific References

7. **Aiba, S., S. Nagai, and Y. Nishizawa.** 1976. Fed batch culture of *Saccharomyces cerevisiae*. A perspective of computer control to enhance the productivity in baker's yeast cultivation. Biotechnol. Bioeng. **18:**1001–1016.
8. **Borkowski, J. D., and M. J. Johnson.** 1967. Long-lived steam-sterilizable membrane probes for dissolved oxygen measurement. Biotechnol. Bioeng. **9:**635–639.
9. **Buckland, B. C., and H. Fastert.** 1982. Analysis of fermentation exhaust gas using a mass spectrometer, p. 119–126. *In* Proceedings of the Third International Conference on Computer Applications in Fermentation Technology. Society of Chemical Industry, London.
10. **Bull, D. N.** 1983. Automation and optimization of fermentation processes. Annu. Rep. Ferment. Processes **6:** 359–375.
11. **Clark, L. C.** 1956. Monitor and control of blood and tissue oxygen tensions. Trans. Am. Soc. Artificial Internal Organs **2:**41–48.
12. **Constantinides, A., and R. R. Vishva.** 1974. Application of the continuous maximum principle to fermentation processes. Biotechnol. Bioeng. Symp. **4:**663–680.
13. **Cooney, C. L.** 1979. Computer application in fermentation technology—a perspective. Biotechnol. Bioeng. Symp. **9:**1–11.
14. **Cooney, C. L., H. Y. Wang, and D. I. C. Wang.** 1977. Computer-sided material balancing for prediction of fermentation parameters. Biotechnol. Bioeng. **19:**55–67.
15. **Corrieu, H., Blachere, and A. Geranton.** 1974. Acquisition and handling of data by computer in fermentation installations. Biotechnol. Bioeng. Symp. **4:**607–611.
16. **Dairaku, K., and T. Yamane.** 1979. Use of the porous teflon tubing method to measure gaseous or volatile substances dissolved in fermentation liquids. Biotechnol. Bioeng. **21:**1671–1676.
17. **Dekkers, R. M.** 1983. State estimation of a fed-batch baker's yeast fermentation, p. 201–211. *In* A. Halme (ed.), IFAC Workshop, 1st, Helsinki, Finland, 1982. Modelling and control of biotechnical processes: proceedings. Pergamon Press, Oxford, England.
18. **Dobry, D. D., and J. L. Jost.** 1977. Computer applications to fermentation operations. Annu. Rep. Ferment. Processes **1:**95–114.
19. **Elsworth, R.** 1972. The value and use of dissolved oxygen measurement in deep culture. Chem. Eng. **258:**63–71.
20. **Evan, L. B.** 1977. Impact of the electronics revolution on industrial process control. Science **195:**1146–1151.
21. **Flynn, D. S.** 1983. Instrumentation for fermentation processes, p. 5–12. *In* A. Halme (ed.), IFAC Workshop, 1st, Helsinki, Finland, 1982. Modelling and control of biotechnical processes: proceedings. Pergamon Press, Oxford, England.
22. **Grayson, P.** 1969. Computer control of batch fermentations. Process Biochem. **4:**43–48.
23. **Hall, M. J., S. D. Dickinson, R. Pritchard, and J. I. Evans.** 1973. Foams and foam control in fermentation processes. Prog. Ind. Microbiol. **12:**170–234.
24. **Hatch, R. T.** 1982. Computer application for analysis and control of fermentation. Annu. Rep. Ferment. Processes **5:**291–311.
25. **Heinzle, E., K. Furukawa, I. J. Dunn, and J. R. Bourne.** 1983. Experimental methods for on-line mass spectrometry in fermentation technology. Biotechnology **4:**181–188.
26. **Humphrey, A. E.** 1972. The use of computers in fermentation systems. Process Biochem. **12:**19–23.
27. **Humphrey, A. E.** 1973. Rationale and economics of computer process control, p. 1–13. *In* Proceedings of the

First European Conference on Computer Process Control in Fermentation. Institut National de la Recherche Agronomique, Dijon, France.
28. **Jazwinskie, A. H.** 1970. Stochastic processes and filtering theory. Academic Press, Inc., New York.
29. **Jeffries, R. P.** 1975. Computer in the fermentation pilot plant. Proc. Biochem. **10:**15–18.
30. **Kishimoto, M., T. Sawano, T. Yosida, and H. Taguchi.** 1984. Application of a statistical procedure for the control of yeast production. Biotechnol. Bioeng. **26:**871–876.
31. **Kjaergaad, L., and B. B. Joergensen.** 1979. Redox potential as a state variable in fermentation systems. Biotechnol. Bioeng. Symp. **9:**85–94.
32. **Leisola, M., J. Virkkunen, E. Karvonen, and A. Meskanen.** 1979. Automatic cellulase assay in computer coupled pilot fermentation. Enzyme Microb. Technol. **1:**117–121.
33. **Mackereth, F. J. H.** 1964. An improved galvanic cell for determination of oxygen concentrations in fluids. J. Sci. Instrum. **41:**38–45.
34. **Melin, C., A. Laanait, F. Delarne, and M. Cordonnier.** 1983. Direct digital control and adaptive control algorithms for a pilot plant fermentor: some applications, p. 283–289. *In* A. Halme (ed.), IFAC Workshop, 1st, Helsinki, Finland, 1982. Modelling and control of biotechnical processes: proceedings. Pergamon Press, Oxford, England.
35. **Mohler, R. D., R. J. Hennigan, H. C. Lim, G. T. Tsao, and W. A. Weigand.** 1979. Development of a computerized fermentation system having complete feedback capabilities for use in a research environment. Biotechnol. Bioeng. Symp. **9:**257–268.
36. **Mor, J. R., A. Zimmerly, and A. Fiechter.** 1973. Automatic determination of glucose, ethanol, amino-nitrogen, and ammonia. Cell counting and data processing. Anal. Biochem. **52:**614–624.
37. **Morrison, R.** 1979. Answers to grounding and shielding problems. Instruments Control Syst. **7:**35–38.
38. **Mou, D. G., and C. L. Cooney.** 1983. Growth monitoring and control through computer-aided on-line mass balancing in a fed-batch penicillin fermentation. Biotechnol. Bioeng. **25:**225–255.
39. **Nyiri, L. K.** 1972. Application of computers in biochemical engineering. Adv. Biochem. Eng. **2:**49–78.
40. **Nyiri, L. K.** 1972. A philosophy of data acquisition, analysis and computer control of fermentation processes. Dev. Ind. Microbiol. **13:**136–144.
41. **Nyiri, L. K., and R. P. Jefferies.** 1974. Process control aspects of single cell production in submerged cultures, p. 105–126. *In* S. R. Tannenbaum and D. I. C. Wang (ed.), Single cell protein. MIT, Cambridge, Mass.
42. **Ohno, H., E. Nakanishi, and T. Takamatsu.** 1976. Optimal control of a semi-batch fermentation. Biotechnol. Bioeng. **18:**847–864.
43. **Reuss, M., H. Piehl, and F. Wagner.** 1975. Application of mass spectrometry to the measurement of dissolved gases and volatile substances in fermentation. Eur. J. Appl. Microbiol. Biotechnol. **1:**323–325.
44. **Swartz, J. R., and C. L. Cooney.** 1978. Instrumentation in computer-sided fermentation. Process Biochem. **2:**3–7.
45. **Valentini, L., and G. Razzano.** 1983. The real time analysis of broth constituents in the fermentation processes: on/off-line fulfillments, p. 253–258. *In* A. Halme (ed), IFAC Workshop, 1st, Helsinki, Finland, 1982. Modelling and control of biotechnical processes: proceedings. Pergamon Press, Oxford, England.
46. **Wang, H. Y.** 1984. Sensor development for fermentation monitoring and control. Biotechnol. Bioeng. Symp. **14:**601–610.
47. **Wang, H. Y., C. L. Cooney, and D. I. C. Wang.** 1977. Computer-sided bakers' yeast fermentations. Biotechnol. Bioeng. **19:**69–86.
48. **Wang, H. Y., C. L. Cooney, and D. I. C. Wang.** 1979. Computer control of bakers' yeast production. Biotechnol. Bioeng. **21:**975–995.
49. **Weibel, K., J. R. Mor, and A. Fiechter.** 1974. Rapid sampling of yeast cells and automated assays of adenylate, citrate, pyruvate and glucose-6-phosphate pools. Anal. Biochem. **58:**208–216.
50. **Weigand, W. A.** 1978. Computer applications to fermentation processes. Annu. Rep. Ferment. Processes **2:**43–72.
51. **Yamashita, S., H. Hoshi, and T. Inagaki.** 1969. Automatic control and optimization of fermentation processes: glutamic acid, p. 441–469. *In* D. Perlman (ed.), Fermentation advances. Academic Press, Inc., New York.
52. **Zabriskie, D. W., and A. E. Humphrey.** 1982. Continuous dialysis for the on-line analysis of diffusable components in fermentation broth. Biotechnol. Bioeng. **20:**1295–1301.

Chapter 24

Establishment of Pilot Plant

BRUCE K. HAMILTON, JEFFREY J. SCHRUBEN, AND J. PATRICK MONTGOMERY

24.1 PILOT PLANT MISSION

Before starting to design a pilot plant facility, it is best to first "scope" the mission of the plant by considering five factors which turn out to be highly interrelated: (i) scope of tasks; (ii) process scope; (iii) scope of the organisms; (iv) scope of containment; and (v) budget and schedule constraints. Elements of each of these five factors must be meshed together. For example, budget constraints might force the scope of tasks and process scope to be narrowed, necessitate elimination of high-level containment, and restrict the range of organisms which the pilot plant would have the capability to handle. By sorting these general factors out at an early stage, a highly useful framework is established to guide design and construction of the pilot plant. The five factors are discussed below.

24.1.1 Scope of Tasks

The initial step is to decide what tasks the pilot plant will be expected to perform. Typical pilot plant tasks include (i) process development and scale-up for new processes, (ii) preparative-scale supply of new products, (iii) process improvement for established processes, (iv) testing of new raw materials, (v) training of operations personnel for full-scale manufacturing plants, and (vi) expansion for on-site manufacturing.

Task i, process development and scale-up for new processes, will probably be the central purpose of most biological pilot plants. Even within this task, it is important to establish what will be the pilot plant's responsibility, and what will not. For example, in the genetic engineering area, development of practical host-vector systems should obviously not be performed by the pilot plant, but instead should be done by genetic and microbial physiology or fermentation technology groups. Ideally, the pilot plant should not be assigned to run a process until a bench-scale version has been run by laboratory staff from beginning (culture storage) to end (product of sufficient quality in a bottle). It is preferable, too, that the bench-scale process be demonstrated using commercial-grade raw materials, as opposed to high-priced reagent-grade laboratory chemicals, before the process goes into the pilot plant. Pilot plant (and engineering) staff should closely monitor the bench-scale phase of process development in the laboratory and should advise laboratory personnel about what is practical and what is not. The pilot plant should then focus on obtaining scale-up data, performing preliminary manufacturing economic evaluations (or supplying data to a separate engineering economics group), solving scale-up problems, and pursuing other tasks (see below) as assigned.

Task ii is preparative-scale supply of new products, typically for applications testing, regulatory clearance, samples for marketing, and test marketing. Typical assignments here include supply of milligram to gram quantities per month of a new recombinant DNA product, or supply of gram to kilogram quantities per month of a new antibiotic. Effective preparative-scale supply of new products is vital for new product development.

Another typical pilot plant task is improvement of established processes (task iii). One important objective is the lowering of manufacturing cost. Another can be reduction of hazardous risks associated with a process, e.g., elimination of a toxic solvent used in recovery.

Task iv is testing of new raw materials. For example, a new vendor may offer what is claimed to be an identical raw material at a lower price; the pilot plant may be assigned to test the performance of that raw material in a pilot-scale run before a substitution is made in the production plant.

The pilot plant is one logical site for training lead operations personnel who will supervise manufacturing work in a new production facility (task v). Finally, in some cases, it may be necessary to expand the pilot plant facility itself so that manufacturing can be performed on-site (task vi).

24.1.2 Process Scope

After defining the scope of tasks for the pilot plant (above), the next step is to decide upon process scope. It may easily be the case that the assigned scope of tasks will require the full range of process capability outlined in Table 1, from fermentation through bioreactors and recovery and ending up with packaging and warehousing. For example, broad process scope is obviously required to accomplish preparative-scale supply of new products.

For each of the process elements listed in Table 1, detailed capability needs must be defined. For example, for fermentation, it must be decided which operational modes (batch, fed batch, continuous) the fermentation section of the pilot plant will need to be able to perform. Likewise, a decision must be made on whether the pilot plant will have capability for continous as well as batch sterilization of fermentation media.

A key decision to make for the recovery section of the pilot plant is whether organic solvents (e.g., acetone, butanol) will be employed. If only aqueous-

TABLE 1. Elements of process scope

Fermentation
Batch, fed batch, continuous
Batch versus continuous sterilization
Bioreactors
Recovery
Aqueous versus solvent systems
Batch versus continuous
Packaging
Warehousing

based recovery systems are used (e.g., perhaps for certain protein products), construction cost of the recovery area will be greatly reduced. However, use of organic solvents will almost certainly be necessary for recovery of products such as antibiotics. It is even possible that the desired product itself may be an organic solvent (e.g., ethanol, acetone, and butanol can be manufactured by fermentation).

A second item to be decided for the recovery section is whether continuous equipment will be used where possible. For example, either a batch vacuum tray dryer or a continous vacuum dryer could be installed. Batch recovery equipment may cost less to acquire and, for a multipurpose, new-product pilot plant, may be more flexible. However, batch equipment is labor intensive, and at the large scale, continuous operations are preferred. Therefore, if batch equipment is employed in the pilot plant, it may well be necessary to use results obtained with batch equipment as the basis for design of large-scale continuous equipment.

24.1.3 Scope of Organisms

Common types of organisms which might be handled by the biological pilot plant include microbes either single celled or multiple celled (mycelial, e.g., actinomycetes, molds), algal organisms, mammalian cells, and plant cells. The organism type influences the design of fermentors, batching equipment, support laboratories, etc.

24.1.4 Scope of Containment

Levels of containment for both biological agents and chemicals must be defined. For biological containment, the large-scale guidelines of the National Institutes of Health (13) are useful, particularly when genetically engineered microorganisms are employed. Pathogenicity of organisms must also be considered, and here the classification system of the Centers for Disease Control is helpful (5). The level of chemical containment necessary depends on what hazards are associated with the products, raw materials, and recovery solvents that the pilot plant will process. For example, one product might be a toxic antitumor compound that will have to be physically contained to protect pilot plant operators as well as the environment. Also, many solvents used in recovery (e.g., butanol) are toxic and must be contained.

24.1.5 Budget and Schedule Constraints

Budgetary limitations may constrain the overall pilot plant mission. For example, installation of a solvent-based recovery area, which is expensive because of the need for "explosion-proof" design, may have to be deferred. The size and number of the fermentors may have to be scaled back. It might not be possible to immediately allocate building space for future expansion. However, as long as sufficient land is acquired at the original site, capabilities can always be upgraded by stages at a future date.

The timetable for launching new products may impose a time framework on the pilot plant design and construction schedule. The design and construction schedule can often be accelerated by use of outside service organizations (see section 24.2.1).

24.2 CONSTRUCTION MANAGEMENT AND EXECUTION

24.2.1 Use of an Engineering Company

An engineering company can be used to supply a wide range of services for design and construction of a biological pilot plant. The desired services might be any of the following: process equipment design or specification; utility equipment specification; site location; building design, construction, or renovation; equipment procurement, layout, and installation; piping, instrumentation, and utility design, procurement, and installation; securing of permits (building, sewage, environmental, etc.); contracting for and supervision of trades (mechanical, electrical, welding, etc.); general consulting; or even complete program responsibility.

Several engineering companies (see partial listing, section 24.10.2) have many years of experience in the biological process industries. The degree to which an engineering company is used will depend on the internal resources, objectives, and timetable of the organization that is funding the pilot plant (the "owner").

24.2.2 Scheduling the Project

A typical project schedule for establishing a biological pilot plant is given in Fig. 1. The trick is to construct a network of tasks so that the project can be completed within time constraints that usually exist (chapter 7 in reference 15). Good scheduling procedures embody the critical-path method in some form. Typically, lead times for equipment or parts delivery are the biggest problem. Use of an expeditor (whose responsibility is to press for rapid deliveries) may be a must.

24.2.3 Project Cost Control

Expenditures will undoubtedly have to be made according to a project budget. Therefore, accurate and prompt cost tracking, reporting, and control is a necessity.

24.2.4 Site and Building Location

Factors important in choosing a site for the pilot plant include: proximity of research and development (R&D) laboratories and manufacturing facility; labor availability, skill, and cost; labor climate; weather; surrounding community; expansion; supply of utilities (water, gas, electricity); waste treatment (sewage, solids, vapors); trucking; local ordinances, including those covering recombinant DNA containment and other building permits; and local construction costs.

Ideally, the pilot plant will be located close to both the R&D laboratories and the manufacturing facility. If these two facilities are distantly separated, a choice

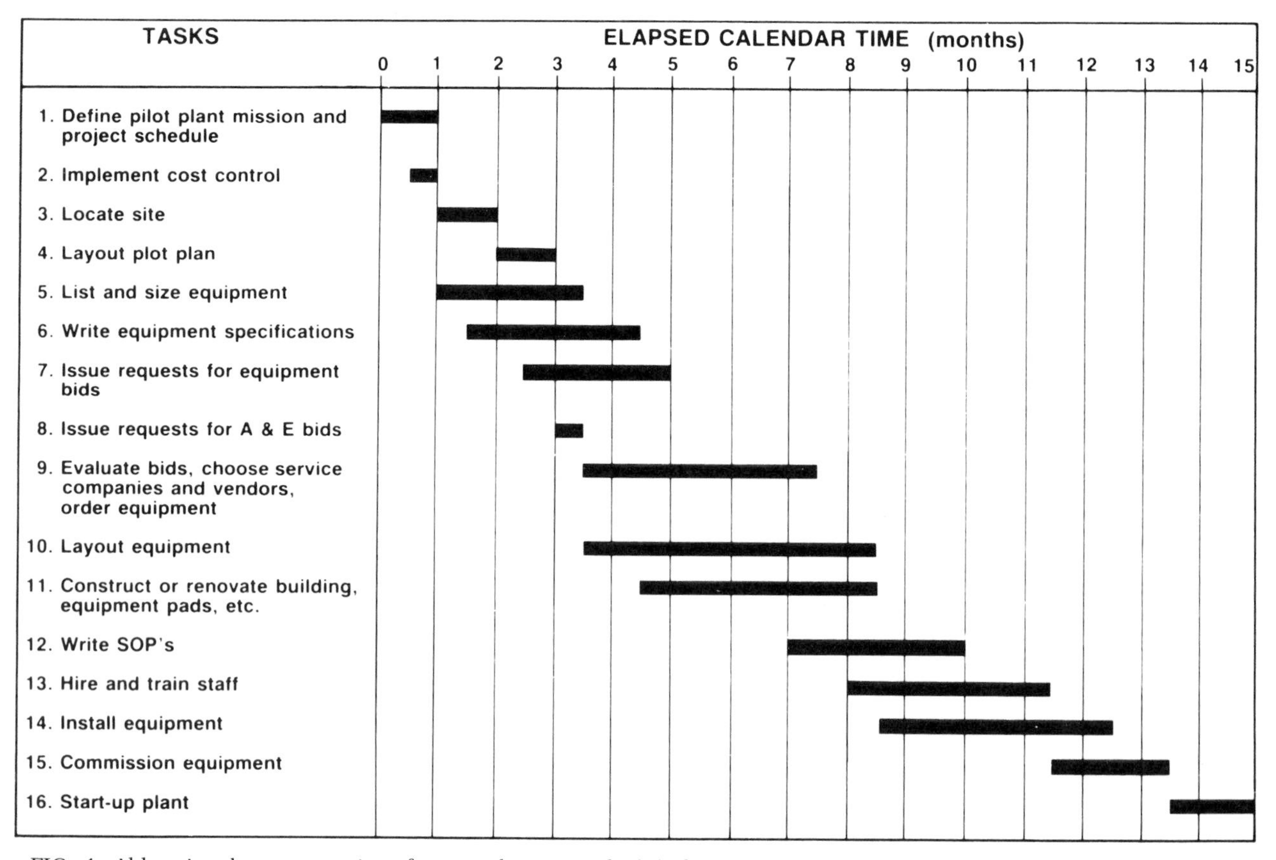

FIG. 1. Abbreviated representation of a typical project schedule for design, construction, and start-up of a biological pilot plant.

will have to be made. Regardless of what location is chosen, it is very important to ensure that information is transferred rapidly and effectively among the R&D laboratory, pilot plant, and manufacturing groups. Development of practical processes suitable for commercialization will then be facilitated.

One disadvantage of separating the pilot plant from the manufacturing facility is that the separation may result in the need to duplicate certain pieces of process equipment (e.g., intermediate-size fermentors which might be used as scale-up vessels in the pilot plant or seed vessels in a manufacturing plant), inefficient use of utilities (e.g., small steam boiler at pilot plant, large one at the manufacturing plant), and redundant manpower.

For a new site, labor availability, skill, and cost should be carefully checked. Weather should obviously not be so harsh that it interferes with construction or operations. If possible, the site should not be too close to residential areas because of the possibility of unpleasant odors occasionally issuing from the pilot plant. It is advantageous for adjacent open land to be available for possible future expansion. Utility (water, gas, electricity) and waste treatment (sewage, solids, fumes) costs must be acceptable. Trucking access for receiving raw materials and shipping product has to be provided. Finally, local ordinances covering recombinant DNA containment, building codes, etc., should be reasonable.

24.2.5 Equipment List and Sizing

A list of the pieces of equipment needed for the pilot plant should be formulated, along with the number and size of each type of equipment. Equipment type, number, and size are determined by several factors: pilot plant mission (see above), the desired throughput rates (kilograms per month), and availability of equipment of desired size. Example sizing calculations will be discussed below in various subsequent sections covering pieces of equipment of particular types.

An example equipment list for a biological pilot plant is given in Table 2. Note the assumptions on pilot plant mission and process scope, in particular: (i) for this example, the plant is intended initially to handle primarily processes for producing amino acids, and (ii) the facility at first will have capability only for aqueous (not organic solvent)-based recovery processes.

24.2.6 Plot Plan and Plant Layout

A list of work areas which the pilot plant may require is given in the legend to Fig. 2. An example layout of the work areas, with assumptions stated, is given in the same figure (for additional discussion on plot plan and layout, see chapter 6 of reference 15). The plant depicted has two stories of vertical headspace to provide room for process equipment,

TABLE 2. Example equipment list for a biological pilot plant[a]

Type of equipment[b]	Description	Size	No.
Process	Mechanically agitated fermentors	14 liters	10
		250 liters	2
		1,500 liters	1
	Continuous cell centrifuge (e.g., Westfalia)		1
	Tubular bowl centrifuge (e.g., Sharples)		1
	Filter press		1
	Batch tank	50 gal	2
	Process tanks:		
	Glass-lined, agitated, jacketed	300 gal	3
	Stainless, agitated, jacketed	300 gal	3
		50 gal	1
	Transfer pump, centrifugal screw		10
			3
	Perforated basket centrifuge	30-in. diameter	1
	Ultrafiltration unit		1
	Column	10 liters	3
	Evaporator		1
	Shelf dryer	150 ft^2	1
	Blender	50 ft^3	1
	Mill		1
Utility	Process air compressor (oil free)	90 scfm	2
	Instrument air compressor (oil free)	15 scfm	1
	Instrument air dryer		1
	Boiler	125 bhpw	1
	Chiller	30 tons	1
	Water purification unit		1
	Cooling tower		1

[a] Mission assumptions: capability for tasks i through vi; expansion for on-site manufacturing (task vi) is limited. This plant is intended initially to handle primarily processes for producing amino acids; floor space is allocated for later expansion into other products. Process scope assumptions (see Table 1): for this plant, fermentation involves (i) batch and fed batch capability; continuous capability by using sister fermentors for medium sterilization and feeding; and (ii) initially, no continuous sterilization. For bioreactors, floor space and tanks are provided. Recovery is aqueous only and by batch. Floorspace is provided for packaging and warehousing.

[b] For laboratory equipment, see section 24.6.2.

including allowance for ease of maintenance (e.g., lifting vessel heads, pulling agitator shafts) and installation of pipe racks. The plant is not intended to handle solvents, so there are no solvent storage or solvent recovery areas. Glass-washing facilities and a lounge area are located elsewhere in the building. Raw materials, parts, and new equipment enter through receiving (area 1); outbound shipments (product, equipment sent out for repair) exit from this same area. Raw materials are stored (areas 2 and 3), in the cold if necessary (area 4). Stored raw materials are first placed in "quarantine," examined for acceptability (identity tests are run if necessary for compliance with current good manufacturing practices), and then released for use. Fermentation media are prepared in a batching area (area 5). Fermentors of three sizes (10, 250, and 1,500 liters) are grouped together (areas 6, 7, and 8). A cell centrifuge is installed nearby (area 9). Space is allocated for bioreactors (area 10) between the fermentor area and the process tank area.

Equipment maintenance is done in a machine shop, where spare parts are also stored (area 11). In-process chemical assays and nonaseptic microbiological process control tests are performed in a laboratory (area 12). Chemical and engineering bench tests occur in a separate laboratory (area 13), as does microbiological benchwork (area 14). Records are stored, and data processing by computer is performed, in the pilot plant supervisors' office (area 15). Electrical switch gear and motor starters are located in a separate room (area 16). Products are recovered from fermentation broths and bioreactor effluents in 50-gal and 300-gal process vessels (area 17). Crystals of final products are separated from their mother liquors in a basket centrifuge (area 18). Wet crystals of final product are dried in a vacuum shelf dryer (area 19). Dried final products are pulverized into fine powder, and batches are blended into uniform lots (area 20). Fourteen-liter fermentor jars and nutrient addition vessels for both the 14-liter and 250-liter fermentors are sterilized by steam in an autoclave (area 21). Utility equipment is separated from the rest of the plant by block walls for both fire prevention and noise reduction (area 22). Quality control assays on finished product lots are performed in a separate laboratory (area 23). The pilot plant staff is administered from the pilot plant manager's office (area 25). Facilities for showering and for changing into and out of work clothes are provided for women (area 27) and men (area 28) of the pilot plant staff. Product is stored in area 29, and safety equipment is kept in area 30. Trash is disposed of in a dumpster (area 31). Sufficient floorspace is available for new equipment (area 32).

For each work area, detailed layouts must be made. Four positioning principles apply for each layout of an individual area as well as for the relative placement of the areas themselves: (i) line of material flow, (ii) gravity feed, (iii) operating and maintenance access,

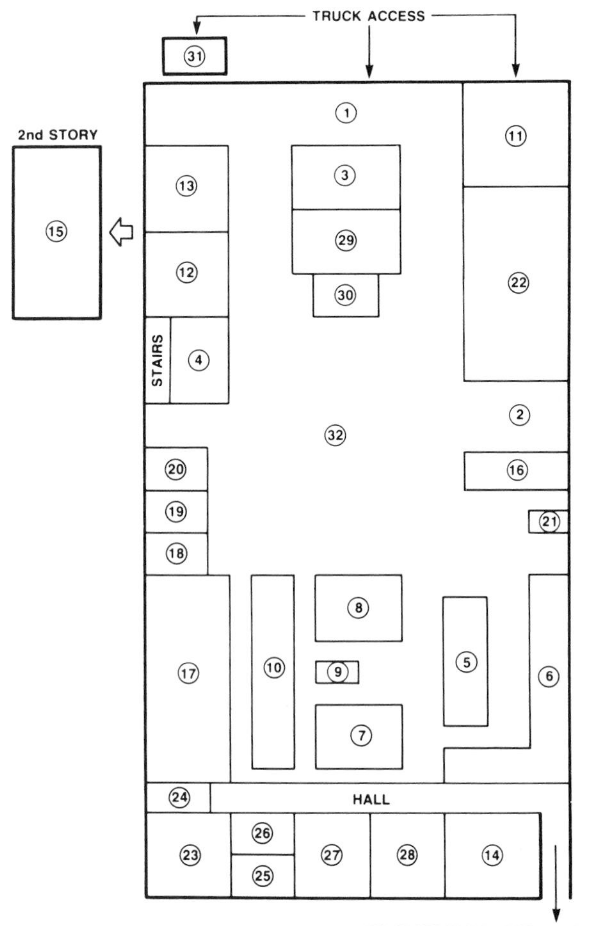

FIG. 2. Example work area layout. This plant is not intended to handle solvents. Glass wash facilities and a lounge ("break room") are located elsewhere in the building. The plant has two stories of vertical headspace. Work areas include: (1) receiving, shipping; (2) fermentation and bioreactor raw materials storage (quarantined, released); (3) recovery raw materials storage (quarantined, released); (4) cold storage; (5) batching area; (6) 14-liter fermentors; (7) 250-liter fermentors; (8) 1,500-liter fermentor; (9) cell centrifuge; (10) bioreactor area; (11) maintenance and spare parts; (12) in-process laboratory; (13) chemistry and engineering laboratory; (14) microbiology laboratory; (15) records storage, computer room, and supervisors' offices; (16) electrical switch gear; (17) process tanks and product isolation and purification; (18) basket centrifuge; (19) dryers; (20) pulverizing, blending; (21) autoclave; (22) utilities; (23) quality control laboratory; (24) quality control office; (25) pilot plant manager's office; (26) secretary's office; (27) women's locker room; (28) men's locker room; (29) product storage (quarantined, released); (30) safety equipment; (31) dumpster; (32) open area for new equipment.

and (iv) emergency (e.g., fire) egress and access. Where possible, equipment should be placed in a sequence so that material flows from one piece of process equipment to a contiguous piece of equipment. Pumping liquid or transporting solid in-process material back and forth between one end of the plant and the other should be avoided. Where gravity flows can be advantageously employed, they should be. Access for operations and maintenance should be carefully considered, and emergency egress and access must not be overlooked. In addition, overhead pipe rack and subfloor piping installation should be taken into account; in particular, do not neglect drains for process and utility equipment, safety showers, and floor washes and spills. Of course, when the pilot plant is to be installed into a pre-existing building, layout may have to be compromised to accommodate the existing structure.

Examples of detailed layouts for various individual work areas (e.g., fermentation) will be given below.

24.2.7 Procurement

Specifications

For each piece of equipment, specifications should be written up. For major pieces, consideration should be given to issuing a request for proposal (RFP). An outline of an example RFP for a 250-liter fermentor is given in Table 3. This example is intended to illustrate, in general, one format for presenting specifications. The particulars for fermentor specifications are covered later in section 24.3.4.

For each piece of equipment, a piping and instrumentation diagram should be obtained. For the case of a complex piece of equipment, it is instructive to break this diagram down into subsystems. For example, for a fermentor, some of the subsystems include: (i) inlet air system, (ii) exhaust gas and pressure control system, (iii) temperature control system, (iv) vessel steam sterilization system, (v) addition systems, etc. (for example piping and instrumentation diagrams, see Bull et al. [3]).

It is recommended that major pieces of equipment be accepted by the owner only after they pass a battery of acceptance tests. For example, for a sterilizable fermentor, one important acceptance test is verification that the fermentor can be sterilized by the standard operating procedure after it has been "challenged" by a contaminant, e.g., viable bacillus spores.

New versus used equipment

Good used equipment, if it is available, can be the key to a pilot plant project constrained by time or funds. In addition to the advantage of lower cost, used equipment can often be delivered and placed into service much more quickly than new equipment, which usually is not kept in stock and must be fabricated from scratch. On the other hand, for new equipment, there is no need to be concerned about use history (e.g., was used equipment in its prior application employed to make a toxic product?), and the new equipment can be tailor-made to meet the owner's particular objectives.

Vendor candidates

Compilations of vendors (e.g., Chemical Engineering Catalog; Chemical Engineering Equipment Buyers Guide; Food Engineering Master: Equipment, Supplies & Services—see section 24.10.1) can be very useful. A brief partial listing of some vendors for various particular services and pieces of equipment is given in section 24.10.

Bids and vendor selection

For competitive purposes, it is advisable to obtain bids from more than a single vendor, typically three. The bids should then be evaluated according to specific criteria, e.g., cost, vendor experience, field performance record, availability of service and spare parts, vendor financial stability and depth of resources, etc. Particularly for expensive equipment, it is recommended to field-inspect representative installations of candidate vendors and to solicit user input on field performance.

24.2.8 Architectural and Engineering

The primary objective of the architectural and engineering (A&E) task of the project is detailed design of a new building or detailed specification of renovations for an old building to house the pilot plant. The plot plan and layouts (section 24.2.6) are key guidelines used during A&E. All plans for piping, wiring, ventilation, heating, air conditioning, lighting, gas supply, etc. must be worked out. Load lists (electrical, air, water, etc.; see section 24.7.2) are important documents. Detailed engineering drawings (e.g., plant piping and instrumentation diagrams) result from the A&E work. In one approach, after A&E is complete, construction bids are solicited and evaluated, a contract is awarded, a budget is set, and construction begins. In another approach, which can save calendar time that is sometimes badly needed, construction is begun before all drawings are complete. For this latter approach, however, costs can be difficult to predict with much precision.

An additional task that can be completed during the A&E phase is procurement of required permits (demolition, building, sewage, environmental, etc.).

24.2.9 Building Construction and Equipment Installation

Construction must proceed in a sequence which allows large pieces of process and utility equipment to be brought inside the building. Concrete mounting pads, floor drains, etc., should already have been set in place. Adequate headroom should be provided for installation and maintenance (e.g., pulling vessel heads and vertical shafts).

24.2.10 Commissioning, Acceptance, and Start-Up

The commissioning period is the time during which the installed equipment is demonstrated to be in good mechanical and electrical working order by the contractor; e.g., motors are checked for rotation in the proper direction; vessels, piping, and pumping systems are checked with water for tightness; instruments are calibrated and verified to function correctly, etc. Any mechanical or electrical malfunctions are corrected during this time.

During the acceptance period, major pieces of process, utility, and laboratory equipment are subjected to test runs. For example, a fermentor sterilization cycle might be validated with an appropriate challenge test (see section 24.3.4), a chiller might be tested under a heat load (e.g., cooling down a hot vessel), a UV-visible spectrophotometer might be used to generate a standard curve for an assay, etc. Formal transfer of the plant from the contractor to the owner should not occur until commissioning and acceptance tests are complete.

During the start-up period, an integrated process is run through the plant. The objective here usually is production of a certain amount of product (e.g., 100

TABLE 3. Outline of an example RFP for a 250-liter fermentor

Section I. General

1. Introduction. Explains who the buyer is, what in general is to be purchased, and what proposals in response to the RFP should cover.
2. Quantity. Enumerates how may units are to be purchased.
3. Proposal preparation, format, and submittal date. Specifies the form that responses to the RFP should take and sets a closing date for submission of proposals.
4. Delivery. Sets a calendar date by which the equipment is to be received by the purchaser and delineates the penalty if the delivery date is not met by the contractor.
5. Testing and acceptance. Makes clear the vendor's responsibilities for verification of performance of the delivered equipment; sets penalty for performance failure.
6. Rejection and acceptance of proposals. States explicitly that the buyer is free to evaluate any and all proposals in terms of the buyer's best interests.
7. Proprietary information. Explains how any proprietary information supplied by the vendor should be so identified and how it will be handled.
8. Oral presentation. At the option of the buyer.
9. Amendments to RFP. Specifies how amendments will be issued if they are necessary.
10. Commitments, warranties, and representations. States that the vendor selected will be held to claims made in its proposal.
11. Shop inspection. States that the buyer shall have the right to inspect the vendor's shop.
12. Drawings and manuals. Specifies what drawings and manuals (including operating and maintenance manuals, with a spare parts list) must be supplied with the equipment.
13. Drawing submittals. Specifies the date by which the vendor must supply drawings to the buyer for purposes of planning the installation of the equipment (e.g., floor mounting, utilities, headspace, getting the equipment into the building, etc.).
14. Method of award. States criteria for choosing a vendor, e.g., quality of construction, operational features, vendor experience, cost, availability of spare parts, service, ease of maintenance, etc.

Section II. Specifications

1. Operation
2. Fermentation vessel
3. Open-frame pipe rack
4. Agitation system
5. Aeration system
6. Instrumentation
7. Addition vessels and injection system
8. Electrical
9. Spare parts
10. Energy conservation
11. Acceptance tests

TABLE 4. Fermentor containment features

Containment
1. Decontaminated exhaust gas (by filtration or incineration)
2. Contained sampling and addition systems
3. Contained seal barrier fluid
4. Hard-piped condensate drains
5. Sterilized condensate
6. Hard-piped rupture disk with vent to contained tank (or else system design to eliminate rupture disk)
7. Check valves and/or automatic positive shut-off on air supply to guard against backflow from fermentor
8. Spill/emergency procedures
9. Chemical, thermal, genetic, or mechanical inactivation of fermentor contents before exposure to operators or environment
10. All welded pipe
11. Validation protocols
12. Floor drains, pitched floor, decontamination sewer

High containment
1. System integrity monitoring device
2. Fail-safe cooling water (if vessel wall is penetrated)
3. Building air decontamination (filters)
4. Restricted facility entrance/exit
5. Shower out protocol
6. Containment suits and headgear
7. Operation at negative gauge pressure

mg of a purified higher eucaryotic protein, such as a human interferon, made through recombinant DNA technology by *Escherichia coli*). For successful start-up, personnel, materials, and equipment must all be brought together in a sufficiently harmonious manner.

24.3 FERMENTORS

24.3.1 Type, Number, and Size

Three basic factors characterizing fermentor type are as follows.

(i) Agitation. Two common types are mechanically agitated fermentors and air-agitated fermentors. Air-agitated fermentors have the virtue of simplicity (e.g., no shaft seals are necessary—for this reason, air agitation has sometimes been used in work with pathogens). However, mechanically agitated tanks are far more common in fermentation laboratories and pilot plants (3).

(ii) Pressure rating. To steam sterilize a large fermentor, the fermentor must be built as a pressure vessel. Pressure vessels are subject to the American Society of Mechanical Engineers code. Many fermentors (e.g., those used for yeast culture) are not pressure vessels and therefore cannot be steam sterilized. Thus, nonpressure vesels are unsuited for generalized pure-culture application.

(iii) Material of construction. While laboratory fermentors are often made of glass, large fermentors are not. The most versatile material for the wetted walls of the fermentor is probably 316 stainless steel. Other grades of stainless (e.g., 304) are sometimes used. For certain applications, e.g., cell culture, "low-carbon" stainless steel is recommended. For stainless-steel fermentors, care must be taken to avoid exposure to concentrated hydrochloric acid, or even to hot sulfuric acid. Carbon-steel fermentors are sometimes employed when iron does not interfere with the fermentation; however, carbon steel is undesirable for food and pharmaceutical applications (9).

Deciding what number of fermentors to install in the pilot plant can involve a trade-off between making many runs per year with numerous basic fermentors versus fewer runs per year with a limited number of highly instrumented fermentors. The reason for the trade-off is that, given a certain limited amount of funds, a larger number of basic fermentors can be purchased than highly instrumented ones. Both approaches have advantages. Typically it is useful to have a bank of at least several identical small bench fermentors so that various values for an important parameter (e.g., temperature, pH, dissolved oxygen [DO], medium composition, etc.) can be evaluated simultaneously after the tanks are inoculated with inoculum split from a single source.

The size of fermentors chosen is usually affected by scale-up and preparative objectives. For scale-up, one rule of thumb calls for mechanically agitated pilot fermentors to be at least 250 liters in volume. Use of smaller fermentors may result in distortions caused by a high surface-to-volume ratio, wall growth, and too much influence by sensor probes on fluid mixing. For preparative objectives, fermentor size is determined by desired throughput. For example, in the case of a pilot plant for development of new antibiotics, it may be desirable to have the capability to prepare a kilogram or so per month of a candidate product. Initially, the product might be produced at a titer of 100 μg/ml by a 6-day fermentation, and perhaps 50% of the product might be isolated in recovery. A 10,000-liter fermentor operated with 7,500 liters of medium would then turn out 375 g of recovered product per run, so that the throughput might be about 1.5 kg/month. Thus, for the case just considered, a 10,000-liter production fermentor would be appropriate to meet preparative objectives. A 10% rule of thumb for the final inoculum tank for the 10,000-liter production fermentor would call for a 1,000-liter seed vessel. This 1,000-liter seed vessel could probably be inoculated adequately by 14-liter bench fermentors (for early stages of seed development, the 10% rule does not apply). These fermentors could also be used for fermentation development studies.

24.3.2 Containment Class

Requirements for containment characteristics of fermentors operated with over 10 liters of working liquid volume are specified by the National Institutes of Health large-scale guidelines (13). These guidelines were written for work with microorganisms containing recombinant DNA molecules. They are also useful, however, for work with pathogens. The Centers for Disease Control have classified pathogens according to hazard level (5).

Table 4 lists several containment features for fermentors used in biohazard work. Which of these features should be incorporated depends on the biohazard level of the work being done.

24.3.3 Instrumentation and Control

Table 5 lists five levels of instrumentation and process control for fermentors. In one approach, the initial installation is kept very simple (encompassing just levels 1 and 2) while design allows for later addition of more sophisticated levels (3, 4, and 5). The advantage of this approach is that with relatively few resources it is possible to get the fermentation section of the pilot plant up and running and turning out good pure-culture fermentations. The danger of attempting simultaneous start-up of all five levels of instrumentation and control is that resources may be spread too thinly to allocate enough time to get an adequate number of fermentations done. Of course, if sufficient resources are available, there are obvious advantages to getting the entire job done at once, all the way through computer control (level 5, Table 5). Finer optimization is then possible, and data collection, analysis, and storage are immensely facilitated.

Control of field valves, solenoids, and motors can be done by distributed control or direct minicomputer control. Distributed control, particularly with manual override options, affords the greatest reliability. Cascade supervision by an appropriate minicomputer then can permit complete control flexibility. This approach, however, is expensive. For pilot plants where reliability is not of top priority, direct minicomputer control may be acceptable and should be cheaper.

24.3.4 Specifications

Typical specifications for a 250-liter mechanically agitated fermentor are as follows.

Operation

The fermentor will be charged through a combination viewing and filling port in the head plate, or else through addition ports in the vessel. The contents will be agitated, and the temperature will be raised to a range of 121 to 125°C and held at this temperature for 20 min to 1 h or longer for sterilization. Most heat energy will be supplied by steam in the vessel jacket rather than by direct injection of steam into the fermentation broth through the sparge line. This is one way to avoid contamination by boiler additives possibly carried in the steam; another is to pass the steam through a sanitary-type activated carbon steam filter. The batch temperature will be lowered with tap water to 40 to 50°C in 1 h or less, assuming maximum tap water temperature of 33°C; then it will be lowered to the operating temperature of 20 to 37°C with chilled water. The batch will be aseptically injected with pure inoculum, agitated, aerated, and held under aseptic conditions for the growth period, after which time it will be cooled to 5°C or lower and harvested. During the growth period, sterile acid, base, antifoam, and nutrients will be added aseptically to the batch through steam-sterilizable ports as required. Periodically, aseptic sampling will be performed through a steam-sterilizable port.

TABLE 5. Levels of instrumentation and control for fermentors

Level	Capability
1	Pure culture fermentation
2	Basic instrumentation (temperature, pH, DO, sparge rate, revolutions per minute, vessel pressure) and addition/feeding capabilities (automatic acid and base, manual antifoam and nutrient feeds)
3	Sophisticated instrumentation (off-gas analysis, load cells, shaft torsion indicators, etc.)
4	Computerized data acquisition
5	Computer control

Fermentor vessel

The vessel will be an all-welded, jacketed tank with a removable head. Welds will conform to Section 8 of the American Society of Mechanical Engineers code and will be subjected to hydrostatic testing and spot checks by X ray. The vessel will be designed with a height-to-diameter ratio of 3:1. The interior shell and head that come in contact with the culture liquid will be fabricated only of low-carbon, type 316 stainless steel. All other surfaces will be fabricated of type 304 stainless steel. The head plate will be sealed to the vessel with "O" rings, preferably in the vessel flange and not in the head (to facilitate assembly). All sealing material should be of ethylene propylene or better. Lifting rings will be provided in the head plate. There will be one or more lights that will illuminate the interior through a light port(s). The light(s) will be protected and mounted on the fermentor head. Design and fabrication of the vessel should meet the criteria of the latest edition of the American Society of Mechanical Engineers unfired vessel code. The complete interior of the vessel will be polished to a no. 4 finish or better. All inside and outside corners within the vessel will be rounded to provide access for cleaning. The internal design pressure of the vessel will be at least 40 psig (lb/in^2, gauge). The jacket design pressure will be at least 35 psig with full vacuum on the inner shell of the vessel (to exclude the possibility of vessel implosion). Openings in the vessel will be provided for the following:

1. At least one charging and view port
2. Light port(s)
3. Five steam-sterilizable addition ports (inoculum, acid, base, chemical antifoam, nutrient feed)
4. Instrumentation ports (pH, DO, foam sensor, temperature [including thermometer well in addition to control sensor], and spare ports for optional sensors)
5. Inlet and outlet air
6. Jacket inlet and outlet water
7. Steam-sterilizable drop valve for harvest
8. Steam-sterilizable sample port, 0.5 in. ID (for mycelial fermentations)
9. Pressure relief device (e.g., rupture disk)
10. Diaphragm-type pressure gauge

Open-frame pipe rack

All piping, plumbing, valving, etc., required to operate the fermentor will be mounted in a rigid, welded-frame, open pipe rack (instrumentation will be mounted in a separate rack). The rack should be spacious for ease of routine maintenance, i.e., filter cartridge changing, accessibility to pumps, automatic valves, steam traps, etc. The rack will be piped for steam, air, tap water, chilled water, condensate, and any other service required for operation. All steam lines will be arranged to eliminate the possibility of trapping condensate. All gauge faces and meters will be positioned so they can be easily read. A sintered, stainless steel cartridge filter or an equivalent will be furnished and installed on the steam supply line. A cartridge-type filter will be furnished and installed on the water supply line. The pipe rack will be completely assembled, requiring the owner to furnish only steam, tap water, chilled water, electricity, drain, and air supply to the rack. All connections will be made to shut-off valves that will be provided and mounted on the rack at the point of entry to the rack. Piping will be painted (color coded, if so specified) to give a strong, resistant finish.

Agitation system

The vessel will be equipped with a top-drive mechanical agitation system. The agitator shaft will pass through the tank head plate and be sealed by a double mechanical seal assembly. Steam must be supplied to the seal chamber for sterilization, and lubrication of the seal faces will be provided by steam condensate. The seals must be commercial, off-the-shelf items available for replacement in a short period of time. The seals will be installed in a bolted-in-place, easily removed housing that can be resealed and replaced in a matter of hours or less.

The impellers will be flat-blade turbine type with three impellers on the shaft as a set, each with six blades per impeller. The impellers will be removable or adjustable to variable heights on the shaft. The ratio of the impeller diameter to the vessel diameter will be 1:3. The impeller asembly itself will be of the Rushton 20:5:4 design (20).

The fermentor vessel will have four removable baffles. The width of each baffle will be 0.10 of the vessel diameter. The baffle mounts will be adjustable up to 1 in. in relation to the edge of the baffle and the inner wall of the vessel.

The shaft will be rotated with a variable speed motor adjustable from 30 rpm to whatever speed is necessary for delivery of a minimum of 8 W of agitation power per liter to water occupying 75% of the vessel total volume. The motor will be provided with a splashproof or waterproof enclosure. A tachometer will be installed on the instrument panel to read revolutions per minute. A meter will be installed on the instrument panel which will read motor power consumption in watts. There will be a nonadjustable clock on the drive unit to indicate total time of operation in hours. Shaft bearings will be ball-bearing type, lubricated and sealed. If an agitator shaft steady bearing is required, it will be designed to resist abrasive medium materials, easily removed and replaced, easily cleaned, and constructed so that visual inspection will indicate degree of wear.

Aeration system

The air delivery system will include necessary piping, valves, a pressure regulator, flow meter, prefilter, pressure gauges, and a steam-sterilizable, cartridge-type filter that will remove all particles greater than 0.22 μm. The flow rate of the system will be adjustable over a range equal to or better than 0.2 to 2.0 vol of air per vol of liquid per min (vvm). The accuracy of the pressure gauges will be equal to or better than ± 0.3 lb/in^2, except in the case of the vessel pressure gauge, where the accuracy will be equal to or better than ± 0.5 lb/in^2. A sparge line will be installed with a multiported ring at the bottom of the vessel. An overlay pipeline will enter at the top of the vessel wall. The overlay pipeline operating ball valve will be installed in a horizontal run to prevent condensate trapping. The sparge line and the overlay line will be supplied from a single line from the pipe rack. The complete air delivery system will be designed to be sterilized manually in place with steam.

The exhaust gases from the fermentor will pass through a steam-sterilizable filter with the capacity to remove all particles greater than 0.22 μm. A back-pressure regulator will be installed in the piping to control the vessel operating pressure. A filter heater will be installed on the exhaust line at or near the filter to ensure that the exhaust gas never reaches its dew point in the filter. This exhaust filter will be equipped with a bypass in case the cartridge is fouled during a fermentation; however, exhaust gas passed through the bypass will exit through an incinerator.

At least one diaphragm-type pressure gauge will be installed to read true vessel pressure. The accuracy of the gauge will be equal to or better than ± 0.5 lb/in^2.

Instrumentation

An instrument panel will be mounted in a separate rack. The rack will be designed to house and protect all of the instrumentation from moisture, heat, and plant dust. The following parameters of the fermentation process will be controlled, recorded, or both.

Temperature control. The temperature of the contents of the vessel will be continually monitored and automatically controlled and recorded over a range of 5 to 45°C. The accuracy of temperature control in the operating range of 28 to 40°C will be equal to or better than ± 0.25°C. The resolution of the recorded temperature will also be ± 0.25°C. The accuracy of temperature control from 5 to 28°C and from 45 to 125°C will be equal to or better than ± 1°C. Temperature control will be accomplished with steam, tap water, or chilled water as conditions dictate.

pH control. The pH of the fermentor broth will be continuously monitored, recorded, and automatically controlled. Control will be accomplished by automatically metered acid or base solutions as required.

DO control. DO will be monitored, recorded, and controlled in four modes of automatic adjustment: (i) agitation only; (ii) aeration only; (iii) agitation and aeration simultaneously; and (iv) agitation and aeration sequentially.

Foam indicator and controller. An indicator probe will be provided to monitor foam level. A foam controller will accurately meter injection of chemical antifoam.

Nutrient feeding. Additions will be made manually as required (see next section).

Addition vessels and injection system

Autoclavable carboys will serve as addition vessels. Peristaltic metering pumps will be used to deliver additions to the fermentation vessel. *Note*: For recombinant DNA work, alternative arrangements are recommended (2).

Electrical

The fermentor, pipe rack, instrument rack, agitation system, addition system, and all components of these items requiring electrical power will be completely factory wired and tested. The vendor will assemble the unit and make all electrical connections required at the job site of the owner. All AC and DC motors over 0.5 horsepower will operate on 460 V, three-phase power. If DC motors are used, all transformers, speed controllers, and any other devices required for optimum operation will be furnished by the fermentor vendor as part of the package. All instrumentation controls and other 115 V-rated items will be wired for 115 V, one-phase power. The owner will supply and install the 460 V, three-phase power and the 115 V, one-phase power to NEMA 1 (National Electrical Manufacturers Association) fused disconnects mounted on the pipe rack or instrument rack. The fused disconnects will be furnished by the fermentor vendor. All electrical equipment and wiring should meet the requirements of the latest editions of the National Electrical Manufacturers Association, Underwriters Laboratories, and national electrical (NEC) codes.

Spare parts

An inventory of spare parts will be furnished with the fermentor as follows: (i) a complete bearing and seal housing, with bearings and seals installed, that will adequately align with the shaft and tank mounting arrangement; (ii) one set of filters for the aeration system supply and exhaust; (iii) one probe each for DO, antifoam, pH, and temperature; and (iv) one impeller shaft and steady bearing assembly. All spare parts will be delivered in the factory cartons and specified as to model, type, etc.

Energy conservation

Fermentor system design shall provide for minimum energy consumption.

Acceptance tests

The following acceptance test will be performed.

Medium (purposely contains insolubles)

Ingredient	*Concn (%, wt/vol)*
Autolyzed yeast	0.75
Defatted cottonseed flour	1.25
Calcium carbonate	1.00
Animal oil	0.50
Fishmeal	1.20
Sodium chloride	0.33
Cerelose	5.00

Procedure

1. Charge fermentor.
2. Inoculate with viable spores of *Bacillus subtilis*.
3. Agitate at 300 rpm for 10 min at 38°C; aeration, 0.5 vvm.
4. Sterilize vessel for 45 min by standard operating procedure (SOP; see section 24.8.3)
5. Set fermentation parameters: aeration, 0.5 vvm; agitation, 300 rpm; temperature, 28°C; vessel pressure, 4 psig. Monitor DO and pH.
6. Samples should be taken (i) before sterilization, (ii) after sterilization at zero hour, and (iii) every 8 h for 120 h.
7. Sample testing: (i) inoculate shake flasks, 250 rpm, 28°C for 72 h; (ii) plate on Trypticase soy agar and incubate for 48 h at 28 and 37°C.
8. During the sterilization test, five addition carboys will be charged with a clear nutrient medium (e.g., 1% glucose, 0.5% yeast extract, and 1.5% peptone mixture) and inoculated with viable spores as specified above. The contents of the five carboys will be sterilized at 121°C for 45 min and then injected into the fermentor through each of the addition ports periodically over the first 24-h period of the test run.

Acceptance criteria. Two sterilization tests will be performed after the equipment is installed and operating. If one test fails (i.e., batches not sterile after each sample evaluation), the system will be rejected until two successful tests are accomplished one after another.

Additional information on fermentor design, specification, and vendors is given by Steel and Miller (20), Richards (16), Sikyta (18), and Soderberg (19).

24.3.5 Fermentation Area Layout

When designing the layout of the fermentation area, the two major considerations that must be kept in mind are: (i) the amount of space allocated to each fermentor unit, and (ii) the arrangement of the fermentor units and auxiliary equipment.

Space

Adequate space must be provided around and above each fermentor unit to allow for operation and maintenance of the equipment. To operate the fermentor efficiently and safely, the operator must have quick and easy access to all parts of the fermentor, including control panels, gauges and indicators, and valves for sterilization, inoculation, sampling, additions, and harvesting. Enough space should be provided in the immediate vicinity of the fermentor to accomodate all accessory equipment, including addition vessels, sample bottles, pumps, and instruments. These accessories should preferably be placed on racks or benches, with easy access from the floor. It is important to

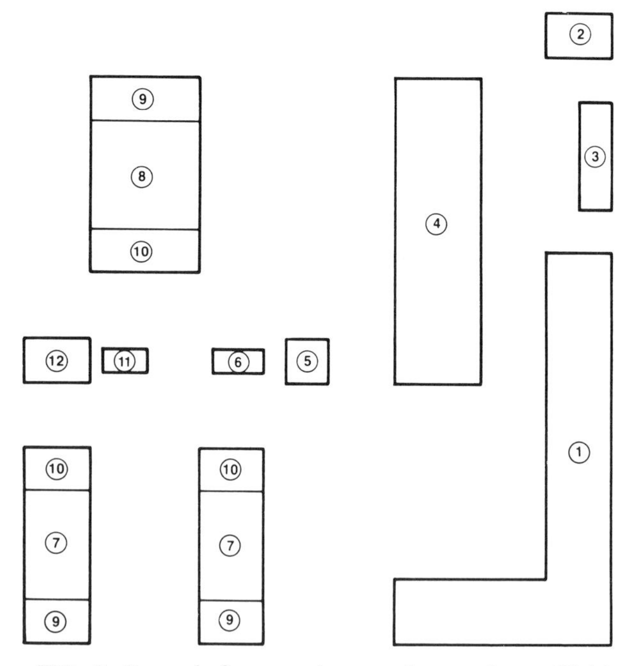

FIG. 3. Example fermentation area layout. Areas: (1) 14-liter fermentors; (2) autoclave; (3) sinks; (4) media batching and supplies storage; (5) agitated batching vessel; (6) media pump; (7) 250-liter fermentors; (8) 1,500-liter fermentor; (9) instrumentation and control panels; (10) addition vessels and pumps; (11) harvesting pump; (12) cell centrifuge.

realize that, in the case of small-scale bench fermentors, the space required by the accessories is often greater than the space occupied by the fermentor itself.

Utility connections to the fermentor unit should be made in a way that interferes as little as possible with access to the fermentor by operators and maintenance workers. These connections include, as a minimum, electricity, steam, process air, instrument (dry) air, tap water, chilled water supply and return, condensate drains, and sewer drains. Usually, the best arrangement is to make all connections with hard pipe running vertically from headers in the ceiling directly above the fermentor unit. Drain lines should also be hard-piped and should run across the floor to the nearest floor drain.

Sufficient space around and above each fermentor must be provided for maintenance work. If fermentors are located at or near a wall, enough space should be left between the fermentor and the wall that a mechanic can repair or remove any components that might fail. In the case of fermentors with top-driven agitators, it is very important to provide enough space above the fermentor to allow the agitator assembly to be pulled completely out of the vessel.

Arrangement

Where a facilty has more than a few fermentors, some thought should be given to the arrangement of the fermentor units. The best arrangement is usually the one that allows the operator(s) to work as efficiently as possible, thereby reducing manpower requirements. In a highly instrumented facilty, it is often possible to locate most of the instrumentation in a single control room, an arrangement that maximizes the productivity of the operator. This cannot be done in facilities where instruments and controls are positioned locally at each fermentor unit. However, it is often possible to arrange the fermentor units in a way that permits a single operator to monitor and control a group of fermentors simultaneously.

Preparation of fermentation media should be done in a centralized area, both for efficiency and to ensure uniformity of media formulations. For fermentors larger than about 40 liters, media should be charged in a separate agitated vessel with a pump and hard-piped connections for distributing the media to the fermentors.

Harvesting of fermentor broths should be done through a single centrifuge clarifier or filter. Again, the most efficient arangement is a system of hard-piped connections leading from the fermentors to a pump, which then feeds the solid-liquid separating unit.

Fermentors larger than about 15 liters should be steam-sterilizable in place. In the case of small-scale jar fermentors, it is often most economical to sterilize the vessels in groups by placing them in an autoclave (as long as the medium employed does not have to be agitated for sterilization). Where this type of system is used, the autoclave should be located as close as possible to both the media charging area and the fermentor units, to minimize the distance over which the jars must be transported.

An example layout of the fermentation area of a pilot plant is shown in Fig. 3. This example is based on a facility having 10 14-liter jar fermentors, two 250-liter fermentors, and one 1,500-liter pilot-scale fermentor.

24.4 BIOREACTORS

24.4.1 Types

Table 6 lists some of the types of bioreactors that may be employed in the pilot plant. Some bioreactors are hybrids of these types, e.g., a circulating-tank cell-free enzyme ultrafiltration bioreactor (4, 22) is a hybrid of tank and membrane types. Some forms of hollow-fiber membrane bioreactors (10, 21) are also packed-bed bioreactors. For each bioreactor type, many novel approaches and designs have already appeared. For example, for packed beds, one simple approach involves using a pressure-leaf filter as a bioreactor (N. E. Lloyd, L. T. Lewis, R. M. Logan, and D. N. Patel, U.S. patent 3,694,314, 1972). For fluidized-bed bioreactors, one novel design employs a tapered geometry (17). Frequently, special designs are necessary due to the biology of the particular system being developed, e.g., photosynthetic bioreactors (14).

24.4.2 Accessories

Substrate and product tanks

Most bioreactors will probably be run continuously, which means that sizeable substrate feed tanks and product receiving tanks are necessary. For example, suppose a fairly small pilot-scale bioreactor, say 10 liters in volume, is fed substrate at a flow rate of 1 space velocity per hour. Then, over the course of a single day, 240 liters of substrate feed tank volume and 240 liters of product receiving tank volume will be necessary. Therefore, if continuous-flow bioreactors of pilot scale are to be employed in the pilot plant, a small in-process tank farm must be installed. The example pilot plant described in sections 24.2.5 and 24.2.6 has six 300-gal tanks, which can all be employed as bioreactor feed or receiving tanks or as tanks used for recovery of bioreactor product solutions (e.g., for use in precipitations).

TABLE 6. Some bioreactor types

Bioreactor type
Tank
Cell-free enzymes (e.g., starch-processing industry use of glucoamylase)
Free cells (e.g., steroid transformations)
Immobilized biocatalyst (e.g., suspended beads)
Membrane
Tubes (e.g., hollow fibers)
Sheets
Packed bed
Fluidized bed

Feed filters

The material that is fed into bioreactors is usually an aqueous solution containing a substrate compound. In an industrial situation, the feed solution will often contain a certain amount of suspended solids, which may originate from any of a variety of sources, e.g., insoluble impurities from the substrate raw material, rust particles from process equipment, etc.

For some types of bioreactors, it is very important that suspended solids be removed from the feed solution before the solution is fed into the bioreactor. Hollow-fiber bioreactors are extremely vulnerable to solid particles that are small enough to enter the hollow fiber, yet large enough to plug the fibers. Packed-bed bioreactors can also be damaged by excessive quantities of suspended solids in the feed solution. The type of filter that should be selected for filtering the feed solution will depend, of course, on the nature of the suspended solids and their concentration. For solids that are relatively large in size (e.g., >10 μm) and relatively low in concentration, a simple cartridge-type filter, using inexpensive fiber filter cartridges, is usually adequate. In more difficult situations, such as where a high concentration of fine particles must be removed, more elaborate filtration systems must be used, e.g., a pressure-leaf filter, precoated with a fine grade of filter aid and with a continuous body feed of filter aid. Precoat filters, when used, should be followed by a membrane cartridge filter, to collect any filter aid particles that might escape the main filter unit.

Feed pumps

The feed solution is usually transported into the bioreactor, and the product stream is carried out of the bioreactor, by means of pumps. It is also possible to move the feed solution by pressurizing the feed solution holding vessel, or even by gravity, but in most cases pumping will be the method of choice. Selection of a feed solution pump will be governed by two separate considerations, namely, the type and the size of the pump. The type of pump required depends primarily on the nature of the feed solution. If the components of the feed solution are not unusually sensitive to mechanical shear stress, then the best choice will probably be a centrifugal pump, due to its low cost, high reliability, and ability to tolerate throttling. If the feed solution does contain substances that might be degraded by a centrifugal pump, then a positive-displacement pump should be considered, such as the rotary, progressing cavity, or diaphragm type. Positive-displacement pumps are also desirable in situations where the feed is a heterogeneous mixture, such as a slurry of solid particles in a liquid.

If the bioreactor is operated in a continuous mode, i.e., a continuous feed and continuous discharge, then it will be very important to have some means of accurately controlling the flow rate of the solutions. The discharge from a centrifugal pump can be easily controlled. A flow-regulating valve installed downstream from the pump can be set automatically by the signal from a flow sensor or adjusted manually by an operator. Positive-displacement pumps should be equipped with variable-speed drives.

After the type of pump has been determined, the next important specification is its size. The size of pump required depends on the flow rate desired, the physical characteristics of the feed solution (e.g., specific gravity, viscosity, etc.), and the total dynamic head of the bioreactor unit. (Total dynamic head is the fluid back pressure that is seen by the pump when it is feeding the bioreactor at the design flow rate and with all in-line filters and control valves in their normal configuration.) After the information above has been determined, the pump manufacturer can recommend the appropriate size.

The pump should, of course, be built of materials that are resistant to corrosion by the feed solution.

Temperature control

Temperature generally has a strong influence on the performance characteristics of biocatalysts. Therefore, an important part of bioreactor design is the control system for regulating the temperature inside the bioreactor.

In the case of continuous-flow bioreactors, temperature variations can occur as the result of heat released or consumed in the biochemical reaction and heat generated by fluid flow in the bioreactor. Since many biochemical reactions are exothermic, the problem of temperature control is often one of removing heat from the bioreactor. If the heat of reaction is not large, or if the temperature rise from the inlet to the outlet is tolerable, then no cooling system will be required. On the other hand, there will be many situations in which a cooling system is desirable or even necessary.

The most widely accepted method of cooling bioreactors is to circulate cooling water through a jacket or some other type of heat transfer surface attached to the bioreactor. An external jacket alone is usually adequate for bioreactors that are relatively small in size, i.e., up to a few hundred gallons. For larger bioreactors, and especially those in which the contents are not mixed or agitated, e.g., packed-bed or hollow-fiber bioreactors, internal heat transfer surfaces may be necessary. The flow of cooling water should be controlled by signals from a flow-through temperature sensor at the outlet of the bioreactor, or sensors inserted into the body of the bioreactor.

If several continuous-flow bioreactors are operated in series, it may be desirable to maintain a different temperature in each reactor. To do this, the effluent from one bioreactor can be passed through a heat exchanger before it is fed to the next. By applying cooling water, hot water, or steam to the heat exchanger, the temperature of the solution can be adjusted to the desired level. In addition, each bioreactor should have an independent internal temperature control system.

In the case of a bioreactor operated in the batch mode, one has the option of either maintaining a constant temperature throughout the reaction cycle, or varying the temperature during the cycle according to a predetermined temperature profile. This latter can be accomplished by applying cooling water, hot water, or steam to the jacket or internal coils of the bioreactor at the appropriate times.

pH control

Many biochemical reactions are accompanied by a change in the pH of the solution. Furthermore, the rates of biochemical reactions are usually strongly dependent on pH. As a result, it is often desirable to have some means of controlling the pH of the solution inside the bioreactor.

Control of pH in a bioreactor is generally accomplished by regulated addition of acids or bases, e.g., sulfuric acid, ammonia, caustic soda, etc. A pH electrode is inserted into the body of the bioreactor, and the electrical signal is sent to a controller which opens and closes the appropriate valves.

In the case of stirred-tank bioreactors, design of the pH control system is simple and straightforward. It is more difficult for immobilized-enzyme bioreactors, such as packed-bed or hollow-fiber reactors. In those systems, measurement of the pH within the bioreactor is difficult, and introduction of concentrated acids or bases in a way that does not create localized extremes of pH is a major problem. One method for dealing with the problem is to add pH-buffering agents to the feed solution. Another solution is to use several bioreactors in series, with small holding tanks in between. In this way, the pH of the solution can be automatically readjusted as it passes from one bioreactor to the next.

Specialized types of pH electrode are often required for use in bioreactors. Feed solutions are often highly concentrated and sometimes contain reactive substances. Bridged electrodes, in which the reference cell is separated from the bulk solution by an extra chamber filled with an inert electrolyte, are generally recommended. In situations where the solution contains substances that would react with the standard Ag/AgCl or calomel reference cells, other types of cells can be supplied.

24.4.3 Special Design Features

Design of bioreactors suitable for supplying gas (e.g., O_2) or removing it (e.g., CO_2) is a focus of current research (7). Also, in some cases vapor containment (e.g., N_2, ammonia) is beneficial. For example, some biocatalysts are inactivated by oxygen, and one approach to protecting such biocatalysts is first to remove oxygen by sparging the bioreactor with nitrogen before adding the biocatalyst, and then to prevent reentry of oxygen by pressurizing with nitrogen during the reaction while the biocatalyst is present. Vapor containment of the nitrogen used for pressurization is then economically advantageous. In other cases, a volatile and possibly noxious component (e.g., ammonia) is a reaction substrate, and so vapor containment is beneficial.

24.4.4 Number and Size

Since the types of bioreactors (Table 6) are numerous, it is probably best to design the pilot plant so that open floorspace is left for the temporary installation of prototype reactors for testing, demonstration, and scale-up purposes. As noted above, a fairly small continuous bioreactor may consume a large volume of substrate solution and produce an equally large volume of product solution. Therefore, a general-purpose biological pilot plant probably will only rarely employ a continuous flow bioreactor larger than 10 liters. Batch or fed-batch tank bioreactors, on the other hand, may be two or three orders of magnitude larger.

24.5 RECOVERY EQUIPMENT

24.5.1 Impact of Materials Handled on Area Construction, Equipment, and Operations

If flammable or explosive materials are to be handled in the recovery area, special construction features, equipment, and operating procedures will probably be necessary. Flammable and explosive materials typically include organic solvents (e.g., ethanol, acetone, ether) and sometimes particulate dusts (e.g., cornstarch or organic product powders).

Special construction features may include explosion venting by blowout panels, grounded equipment and floor grates, and explosion-proof wiring, electrical devices (e.g., perhaps even telephones), and process equipment (particularly electric motors for pumps, etc.). Construction requirements can be defined by review with the owner's insurer.

Most vendors of recovery process equipment provide versions already designed for use in solvent environments.

Special operating procedures for solvent environments include methods designed to eliminate sparking; for example, 55-gal solvent drums must be properly grounded before operators make solvent transfers. Only nonsparking (e.g., brass) versions of tools can be used in solvent environments. Lighted cigarettes, matches, etc., must of course be strictly excluded from solvent handling areas. Likewise, maintenance operations involving welding must be prohibited while flammable solvents are present.

24.5.2 Product Class and Recovery Equipment Type

The category of product recovered (e.g., antibiotics, proteins, amino acids) will likely influence the type of recovery equipment that is necessary; for example, antibiotic recovery will probably require use of solvents in liquid/liquid extractions (in tanks or centrifugal extractors), evaporators, crystallizers, basket centrifuges, and dryers. Recovery of purified proteins, on the other hand, may not involve organic solvents but may require large cold-room operations with chromatographic columns. Amino acid recovery might be achieved with only aqueous systems in vessels having operating temperatures controlled by jacket or coil water circulation.

As an example, aspartic acid made by bioreactor conversion of fumaric acid and ammonia to product can be recovered in the pilot plant depicted in Fig. 2 as follows. The bioreactor product stream is collected in a glass-lined holding tank (Fig. 2, area 17). Sulfuric acid is added to reach the isoelectric point of aspartic acid. The product precipitates to form a slurry. The slurry is fed via a screw-pump-driven pipeline to a basket centrifuge (area 18). The wet product cake is collected and washed in the centrifuge and then transferred to a dryer (area 19). After drying, the product powder is blended (area 20) and loaded into drums for shipping.

24.5.3 Fluid Transfer Systems

Process fluids may be transferred from tank to tank either by permanently installed hard-piped headers and permanently installed pump manifolds, or by temporary flexible lines and portable pumps. Temporary flex lines typically are made of rubber, plastic, stainless steel, Teflon, or combinations of these materials. Temporary lines with quick-fits often appear to offer the greatest degree of versatility. However, a well-designed hard-piped fluid transfer system has the advantage of avoiding a mass of criss-crossed lines running helter-skelter from tank to pump to centrifuge or wherever.

24.5.4 Accessories

It is important to include an adequate number of site glasses, level indicators, flow meters, pressure gauges, temperature indicators, and, especially, sampling systems. It is difficult to have too many of these items and easy to have too few. Also, disposal of waste solids, liquids, and fumes should be carefully planned. If a waste filter cake is generated, a safe procedure and site should be established for disposing of it; the same precautions apply to waste solvents. It is best to recycle as much processing material as possible; e.g., solvent recovery is advisable. Some amount of material unavoidably must be discarded. In many cases, total on-site incineration may be advantageous.

For additional discussion on product recovery, see Calton et al. (chapter 30, this volume).

24.6 LABORATORY SUPPORT

24.6.1 Purpose of the Pilot Plant Laboratory

It is essential for the pilot plant to have access to adequately equipped and staffed laboratory facilities. The laboratory provides a variety of services that are vital to the successful operation of the pilot plant. These services can be divided into three categories, chemical, microbiological, and engineering. Specific examples of laboratory services are the following.

Chemical

1. Analysis and testing of raw materials
2. In-process control assays
3. Analysis of finished product
4. Utilities maintenance support, e.g., water testing

Microbiological

1. Maintenance of cultures
2. Inoculum preparation for fermentors
3. Examination and testing of inocula and fermentor broths for contamination
4. Genetic stability testing, e.g., continuous culture

Engineering

1. Bench-scale testing of physical and chemical unit operations, e.g., filtration, centrifugation
2. Bench-scale testing of microbiological processes, including fermentations and enzymatic transformations
3. Calibration of instruments

24.6.2 Laboratory Equipment Needs

Ideally, the pilot plant might have separate laboratories for chemical, microbiological, and engineering services. The equipment requirements of the three are generally quite different, and any equipment that is common to all three can be shared.

The chemical laboratory should have the basic pieces of equipment required for common types of chemical analysis and, in addition, any specialized instruments necessary for the particular processes and products being tested in the pilot plant. The basic equipment for the chemical laboratory includes an analytical balance (0.1-mg precision), a top-loading balance (200- to 500-g capacity), drying ovens (vacuum and atmospheric), a chemical storage cabinet, an explosion-proof storage cabinet for solvents, a fume hood, a UV-visible spectrophotometer, a pH meter, a muffle furnace, refrigerators, a microcentrifuge, glassware, heated-water baths, magnetic stir plates, and various analytical instruments, e.g., high-pressure liquid chromatograph, gas chromatograph, fluorescence spectrophotometer, etc.

The microbiological laboratory must be designed for aseptic manipulation of cultures. It should have at least the basic equipment needed for routine microbiological work, such as a microbiological hood (biosafety cabinet) with sterile air supply system, microscopes, incubators, (two or more set at different temperatures), laboratory incubator-shakers (two or more set at different temperatures; if large flasks are to be used, include "tall"-type shakers), a refrigerated laboratory centrifuge (Sorvall RC-5B or equivalent), a microcentrifuge, an analytical balance (0.1-mg precision), a top-loading balance (200- to 500-g capacity), a pH meter, a refrigerator (preferably walk-in size), a freezer (−10°C), a liquid-nitrogen freezer, magnetic stir plates, a visible-light spectrophotometer, and an autoclave. In addition, if the pilot plant will be working with organisms that require special containment, e.g., certain categories of genetically engineered microorganisms, more elaborate containment will be necessary.

The engineering laboratory would ideally be equipped with bench-scale versions of as many as possible of the pieces of process equipment in the pilot plant, including fermentors, cell harvest centrifuges, filters, evaporators, stills, dryers, etc. In practice, however, bench-scale process equipment usually is not available. It is often necessary to simulate the conditions encountered in process equipment by using laboratory equipment. The specific types of equipment needed in the engineering laboratory will vary, depending on the mission of the pilot plant.

Much of the work in the engineering laboratory consists of single, small-scale tests to verify or predict the behavior of equipment or materials in larger-scale work. For example, the solubility of a substance under a certain set of conditions can be determined in the laboratory in a small-scale test before proceeding with a crystallization operation on a 1,000-liter batch in the pilot plant. For this type of work, normal laboratory equipment is adequate. Equipment that is generally useful in the engineering laboratory includes top-loading balances (1- and 10-kg capacity), a refrigerated variable-speed centrifuge (Sorvall RC-5B or equivalent), a circulating water bath with heating and refrigeration, vacuum pumps with vacuum regulator (1- to 760-torr control range), a pressure leaf filter, metering pumps (variable speed, 0 to 2,000 ml/min), glass distillation apparatus, glass reactor vessels (jacketed, agitated, 100 to 1,000 ml), chromatography columns (1 to 5 cm in diameter), a rotary evaporator (vacuum), and a mechanical mixer.

24.6.3 Laboratory Staffing and Organization

If the pilot plant operates on a 24-h-a-day, 7-days-a-week basis, it will be necessary to staff the laboratories to handle the workload. It is generally not necessary to have continuous staffing of the engineering laboratory, but the chemical and microbiological laboratories may need to have personnel on duty around the clock. At a minimum, there must be sufficient manpower available to perform in-process control assays and to manage the transfer of cultures and inocula.

The organization of the laboratory staff and its integration with the pilot plant operations staff can be accomplished in a variety of ways. The in-process assay function and the preparation of inocula can be performed by individuals who are part of the pilot plant operations staff. The other activities of the pilot plant laboratory, such as engineering and final product analysis, may also be performed by the pilot plant staff or, alternatively, by supporting technical staff outside of the pilot plant.

24.7 UTILITIES

24.7.1 Utilities Equipment List

Utilities commonly needed in a biological pilot plant (1, 23) include the following.

1. Steam. For fermentor and medium sterilization, vessel temperature control, space heating.
2. Air. For fermentor sparging (oil-free air usually supplied by an oil-free compressor), instrument air if pneumatic controllers or valves are used, and supplied breathing air if it is necessary for certain work areas.
3. Heating (space heating), ventilation, and air conditioning.
4. Vacuum. For vacuum dryers, vacuum evaporators, etc. Supplied by pumps or steam ejectors.
5. Water. Tap water as common supply. Purified water as necessary for process streams. Tower water (evaporative cooling tower) for heat rejection. Chilled (refrigerated) water for low temperature cooling; brine if temperatures of zero or below are necessary. Hot water as necessary.
6. Electricity and gas as necessary.
7. Waste disposal, including sewage and incineration.

24.7.2 Utility Load Lists

Each piece of utility equipment must be sized using a load list formulated on the basis of the service needs of the pilot plant equipment (and also perhaps non-pilot plant equipment, e.g., research laboratory space heaters or autoclaves). In particular, careful attention must be paid to providing sufficient capacity to support peak loads. For example, a steam generator (either a gas- or oil-fired boiler, or perhaps an electric funace) might be used to supply steam simultaneously for sterilization of a number of fermentors, operation of an evaporator and several autoclaves, and wintertime building space and hot-water heating. In addition, when sizing utilities, it may be advantageous to provide excess capacity to support future expansion.

24.8 OPERATIONS

To carry out its function effectively, the pilot plant must have an adequate staff of well-trained personnel, including operators, supervisors, engineers, scientists, and technicians (8). These personnel must be assembled into a coherent organization with clearly defined lines of responsibility and established operating procedures. In these respects, the pilot plant is similar to a full-scale production facility.

24.8.1 Staffing

The most efficient use of a pilot plant is to have it in operation on a 24-h-a-day, 7-days-a-week basis. Therefore, the basic staffing requirement for the pilot plant is to ensure that a sufficient number of adequately trained personnel are on duty at all times. In addition, the staff must be organized in such a way that all

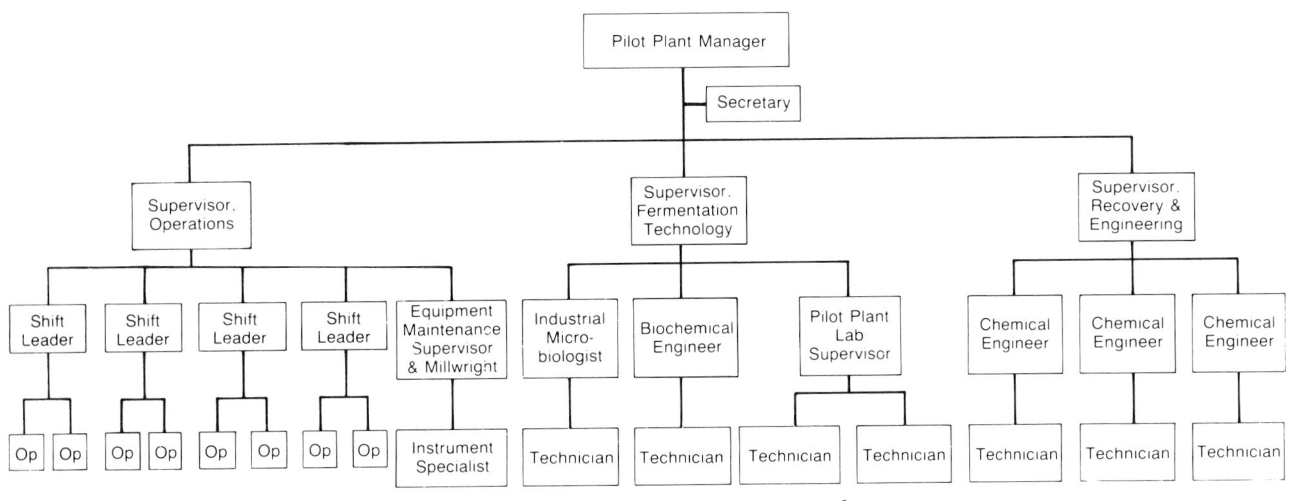

FIG. 4. Example pilot plant organization chart.

parts of the plant's operation are adequately supervised.

The general practice in industry is to divide the 168 h of the week into 21 shifts of 8-h duration. Thus, each day is divided into three shifts: for example, midnight until 8 a.m., the "graveyard shift"; 8 a.m. until 4 p.m., the "day shift"; and 4 p.m. until midnight, the "evening shift" or "swing shift."

A major decision in organizing the pilot plant staff then becomes the choice between permanent ("fixed") shifts or rotating shifts. In a fixed-shift organization, each operator and shift supervisor will remain permanently assigned to one of the three shifts. In a rotating-shift organization, personnel are constantly moved from one shift to another in a regular rotation. The primary advantage of fixed-shift organization is that the personnel on duty are more likely to be alert and well rested at all times. With rotating shifts, the possibility of an operator making an error due to fatigue or drowsiness is significantly greater. Of course, the main disadvantage of the fixed-shift system is the negative impact it may have on the personal lives of those individuals permanently assigned to night shifts.

The size of the pilot plant staff will, of course, depend upon the size of the plant and the workload. The staff should be at least large enough to cover all 21 shifts during the week, with each operator working 5 shifts, to avoid excessive costs for overtime pay. It is also advisable to have at least two individuals on duty at all times, for safety reasons.

The pilot plant staff should be organized in such a way that the lines of authority are clearly established. Thus, the plant should have a manager who oversees all of the operations of the plant. Reporting directly to the pilot plant manager are the area supervisors, e.g., fermentation and bioreactor supervisor, recovery supervisor, maintenance supervisor, laboratory, etc. (For small pilot plants, it is usually not necessary to have supervision of the plant divided into areas). Each shift should have one operator who is designated as shift leader; the shift leaders report to an area supervisor. Ultimate responsibility for ongoing operations in the plant at any given time goes to the senior individual on duty. Thus, if the plant manager is absent, an area supervisor is in charge; if the area supervisor is absent, the shift leader takes over.

The principal activities into which pilot plant operations can be divided are materials handling, fermentation and bioreactor procedures, product recovery, utilities, maintenance, and laboratory procedures. Materials handling includes receiving and storing incoming raw materials, weighing out materials, batching solutions, and packaging and shipping finished products. The fermentation and bioreactor area includes all aspects of fermentor and bioreactor operation, from inoculum to harvest. The product recovery area covers all of the chemical and physical unit operations involved in extracting and purifying products from fluids produced by fermentors or other types of bioreactors. The utilities area covers the operation of all auxiliary equipment in the plant, e.g., boilers, air compressors, water chillers, water deionizers, etc. Maintenance is the area responsible for installing new equipment, modifying existing equipment, and repairing broken equipment. The laboratory area is responsible for handling microbial cultures, preparing inocula for fermentations, and performing in-process chemical and biological assays.

Figure 4 illustrates one possible organization for handling these areas of pilot plant operations activities. This arrangement incorporates three groups, operations, fermentation technology, and recovery and engineering. The operations group consists of equipment-operating specialists who are responsible for round-the-clock shift coverage and equipment operation and repair. The two additional groups, fermentation technology and recovery and engineering, focus on process development, experimental design, and process economic analysis. For any particular process that is run in the pilot plant by the operations group, project staff support is drawn from the fermentation technology group and also from the recovery and engineering group.

24.8.2 Training

The pilot plant operating staff must be well trained. A high level of training is essential not only for maximizing the staff's productivity, but also to protect the safety of the pilot plant personnel and to prevent damage to the plant's equipment.

The training needs of the pilot plant staff may be divided into two distinct categories: training of newly hired staff, and training of existing staff in new procedures, processes, and equipment.

New operating staff training involves, in most cases, individuals hired to fill junior operator positions. Positions above the level of junior operator are usually filled by promoting individuals within the staff; individuals promoted in this way are generally familiar with existing operating procedures and therefore require little or no additional operations training (for new supervisors, training in supervision and management systems is necessary).

When a new operator joins the staff, he or she is first assigned to a shift. It is usually best to assign newly hired operators to the day shift, at least during their training period, so that they can have access to supervisors and technical personnel in departments supporting the pilot plant, most of whom are present during daytime hours. The shift leader is the individual responsible for making sure that the new operator becomes fully trained in all procedures necessary for carrying out his or her duties. Initially, the operator can receive basic instruction by reading written operating procedures or instruction manuals for the equipment and by observing the experienced operators as they go about their regular duties. Later the trainee operator can begin "hands-on" training, under the direct supervision of a senior operator or the shift leader. As the operator demonstrates the ability to carry out procedures efficiently and safely, he or she will be "certified" in those procedures. A record should be kept of the level of training and certification achieved by each operator.

Whenever a new procedure is introduced into the pilot plant, it is usually necessary for the pilot plant staff at all levels to receive training in the new procedure. New procedures may originate from any of a variety of changes in the pilot plant operation, e.g., experimental modification of an existing procedure, acquisition of a new piece of process equipment, or introduction of an entire new process into the pilot plant.

Training in the new procedure should begin at the level of shift leaders and area supervisors. These individuals should obtain all the information about the new procedure that is necessary to implement the new procedure in the pilot plant. It then becomes their responsibility to effectively transfer the information to their subordinates in the form of detailed, written operating procedures. The shift leaders must take an active role in training the operating staff in the new procedure, and it is their responsibility to monitor the implementation of the new procedure, to ensure that the new procedure is being properly carried out, and to identify problems.

It is important for the shift leaders and area supervisors to work closely with any supporting staff from outside the pilot plant that may have had a role in the development of the new procedure. For example, if a new process or unit operation has been developed at the scale of the laboratory bench by a group of scientists and engineers, then the pilot plant staff at the supervisory level must communicate closely with that group to ensure that the new procedure will be effectively implemented, or scaled-up, in the pilot plant. It is also desirable for individuals from the supporting staff to participate directly in the training of operators in the new procedure.

In addition to these normal, or routine, training activities, it is beneficial for the operations staff of the pilot plant to attend organized training sessions periodically. These sessions may cover specialized technical subjects or nontechnical areas such as supervision. Ideally, specialists in areas relevant to the operation of the pilot plant may be brought in from outside the pilot plant to provide the instruction.

24.8.3 Standard Operating Procedures

For every routine activity of the pilot plant operations staff, there must be an established standard operating procedure (SOP). These SOPs should be known and understood by every individual of the operations staff. SOPs should be written down in detail; printed copies should be given to every individual of the operations staff and should be kept at hand in the pilot plant at all times.

The SOP contains detailed, step-by-step instructions for the routine operation of a single piece of equipment. It should be written in the form of a checklist, with spaces provided for the operator to record the completion of each step in the operation, as well as all relevant data, such as gauge readings. As a minimum, the date and hour of the completion of each step and the operator's initials should be recorded. The SOP should contain instructions in at least the inspection, start-up, normal operation, shut-down, cleaning, and routine maintenance of equipment.

Generally, there should be one SOP for each major piece of equipment in the pilot plant. Thus, an SOP should be written for every discrete piece of process equipment, utility equipment, and laboratory equipment.

Each time a piece of equipment is operated, the completed SOP should be retained and stored in a file.

24.8.4 Experimental Operating Procedures

The experimental operating procedure (EOP) differs from the SOP in that the EOP provides specific instructions about the operation of a piece of equipment for application in an experimental process for making a particular product. Much of the equipment in the pilot plant has a considerable amount of flexibility in the way that it can be operated. This is especially true of process equipment, such as fermentors, in which the operating parameters can be varied over wide ranges. For these types of equipment, there are certain aspects of the equipment's operation that are general, in the sense that they are the same regardless of the particular application for which the equipment is being used. These general procedures are covered in the SOP for that piece of equipment. The variable aspects of the equipment's operation, i.e., those that are specific to the particular application involved, are covered in the EOP.

In most other respects, EOPs are very similar to SOPs. Like SOPs, EOPs should be written in a format that provides for the collection and recording of data generated during the operation. EOPs also should be retained once an operation is completed.

24.8.5 Safety Procedures

The pilot plant contains many types of equipment that have the potential for causing serious injury or even death to an operator if proper safety procedures are not followed. Thus, it is mandatory for the pilot plant to have effective safety procedures and to have every member of the operations staff thoroughly trained in all safety procedures.

In most cases, safety procedures for a piece of equipment can be incorporated into the SOP for that piece of equipment, eliminating the need for a separate, written safety procedure. However, there are many aspects of the pilot plant operation that cannot conveniently be covered by an SOP, either because they are too general or because they are not clearly associated with any major piece of equipment. In these cases, written safety procedures are necessary. An example of a situation where a general safety procedure is required might be the use of tie-down wires on "quick-connect" hose fittings on pressurized lines.

All safety procedures must be consistent with pertinent federal, state, and local regulations concerning occupational safety and health. It is the responsibility of the pilot plant supervisory staff, in particular the plant safety officer (or the plant manager if there is no safety officer), to make certain that all pertinent regulations are implemented. One important and detailed set of regulations that must be complied with is the federal regulations issued by the Occupational Safety and Health Administration, which can be found in the Code of Federal Regulations (6) at Title 29, Chapter 17, and especially Part 1910, "Occupational Safety and Health Standards."

24.8.6 Spill Emergency Procedures

In many cases, it will be necessary for the pilot plant to process materials that are hazardous. This type of situation may arise from the use of toxic or explosive solvents, concentrated acids or bases, etc. In some cases, the product being manufactured in the pilot plant may be toxic, pathogenic, or in some way hazardous. In addition, the microbial strain used in the process may be hazardous.

The need to protect workers in the pilot plant from excessive exposure to hazardous substances or microorganisms makes it necessary to have effective procedures in place for dealing with accidential spills. These procedures should be written down in detail, with copies placed in an accessible location in the pilot plant. All pilot plant operations personnel should receive thorough training in the emergency procedures for each type of hazardous material used in the plant.

Spill emergency procedures will vary depending on the properties of the particular material for which the procedure is being developed. There are two basic criteria that must be met by any spill emergency procedure. First, the health and safety of the workers in the pilot plant must be protected at all times. The procedure must be designed to ensure that no worker will receive an exposure in excess of federal, state, or local regulations. Second, release of the hazardous material outside the plant must be prevented. The procedure must be designed to control releases from the plant so that no federal, state, or local regulations are violated.

Whenever an accidental spill of a hazardous material occurs, the shift supervisor should assume responsibility for making sure that the correct emergency procedures are carried out. After the accident is over and the spill is cleaned up, the shift supervisor (or the plant safety officer, if there is one) should investigate the incident and write a complete report, which should be submitted to the plant manager.

24.8.7 Log Book

An important part of the pilot plant operation is the maintenance of a log book, which is a continuous running record of the activities of the operations staff of the pilot plant.

At the end of each shift, the shift supervisor or the senior operator on duty should write in the log book a brief summary of the work that was done during that shift. Included in the report should be a description of the status of each phase of the process; the status of the utility equipment; any significant observations made during the shift; accidents, injuries, or other unexpected occurrences; maintenance work that was done or needs to be done; and any other comments or information that the next shift might need to be aware of. When the next shift reports to the plant for duty, the first order of business for the incoming shift leader or the senior operator is to read the log book, starting from the last shift he or she worked. Then, at the change of shifts, the two shift leaders should discuss the previous shift's activities and the current status of the entire plant, so that the incoming shift will be completely aware of everything that is going on in the plant at that time. Thus, the log book is an essential device for facilitating the transfer of information from one shift to the next.

Logs should be written in a permanent, bound record book. Completed log books should be stored in an accessible but safe location. They are a vital and invaluable source of information about the operations of the pilot plant.

24.8.8 Record Keeping

A pilot plant usually generates very large amounts of information. One of the reasons for this is that a pilot plant is often both a research facility and a production facility. Thus, it is necessary for the pilot plant to process and store all of the technical information produced in the course of its research activities and, in addition to that, maintain all of the record-keeping systems required of a production plant. These record-keeping problems are further compounded in the case of a pilot plant developing a process to manufacture any product for use in foods or pharmaceuticals for human consumption.

To be effective as a process development facility, the pilot plant must keep detailed records of each production run and all experimental work. These records are of vital importance to technical personnel involved in

TABLE 7. Typical pilot plant run records

Record	Types of information recorded
Media batch sheets	Ingredients, lot numbers, volumes, weights, pH
Seed fermentation log sheet	Inoculum strain, fermentation parameters, culture density, culture purity
Production fermentor log sheet	Fermentation parameters, product titer, yield, culture purity
Product recovery log sheet(s)	Process parameters, volumes processed, yields, product purity
Finished product analysis	Purity, contaminants, levels of contaminants

process development, since they are the primary source of scale-up data. In addition, valid pilot plant run records may be required for substantiating patent claims.

Most of the technical information needed for process development is contained in the completed SOPs and EOPs (see sections 24.8.3 and 24.8.4 above). For production runs, the SOPs and EOPs should cover every phase of the process, from media preparation to the finished product. Examples of the types of records that might be required for a typical microbiological production process are listed in Table 7.

In addition to the normal production runs, there is likely to be a considerable amount of small-scale experimental work being done in the pilot plant. Ideally, all experimental work should be covered by individual EOPs. In practice, however, it is often cumbersome to prepare an EOP for some types of experiments. In these cases, data should be recorded in notebooks, which should always be properly signed and dated, to constitute a valid record.

As a production facility, there are certain types of records the pilot plant must keep that are essential for any manufacturing operation. For the most part, these records pertain to the primary factors of production, i.e., labor, utilities, etc. Some examples of manufacturing records are listed in Table 8.

It will often be the case with pilot plants based on microbiological processes that the product being manufactured is intended for use in foods or drugs for human consumption. In these cases, it is important for the manufacturing process and the entire pilot plant operation to be conducted in compliance with federal regulations pertaining to good manufacturing practices. These regulations cover many aspects of the plant operation, including records and record keeping. Good manufacturing practices regulations are issued by the U.S. Food and Drug Administration and can be found in the Code of Federal Regulations (6) at Title 21, Chapter 1. Especially relevant are Part 110 (foods and food ingredients), Parts 210 and 211 (human drugs), and Part 225 (animal drugs).

24.8.9 Maintenance

An effective maintenance program is vital to the successful operation of the pilot plant. This is especially true in the case of complex processes that contain many unit operations, or processes that require highly sophisticated types of equipment.

The basic purpose of a maintenance program is to put equipment in working order and keep it operating for as long as possible. The scope of the maintenance program should include the entire physical plant, i.e., buildings, process equipment, utility equipment, laboratories, vehicles, etc.

It is possible to divide the responsibilities of the maintenance staff into three areas: (i) preventive maintenance; (ii) breakdown maintenance or repair; (iii) new equipment installation.

Preventive maintenance

Ideally, all equipment maintenance would be taken care of in the course of routine, regularly scheduled preventive maintenance. In practice, of course, unexpected equipment failures are inevitable; however, the object of preventive maintenance is to reduce equipment "down time" to a minimum by means of a carefully planned and rigorously executed schedule of repair work. Thus, for every major piece of equipment in the pilot plant, a complete schedule of preventive maintenance should be established at the time the equipment is first installed and put into operation. The preventive maintenance schedule should be designed in accordance with the recommendations of the equipment manufacturer. If, in time, it becomes apparent that the preventive maintenance schedule is inadequate, then the schedule should be changed by increasing the frequency of maintenance or even putting in additional types of maintenance work until the desired level of equipment reliability is achieved.

For any preventive maintenance program, it is necessary to have a shop for doing repair work. The shop should be conveniently and accessibly located in the plant, and it should be equipped with all of the machine tools and instruments required for making repairs on the equipment in the pilot plant. A complete inventory of spare parts should be readily available at all times. A record keeping system should be established to keep track of the spare parts inventory and to ensure that orders for new parts are placed before the supply on hand is exhausted.

Breakdown maintenance

When preventive maintenance fails and a piece of equipment becomes inoperable, breakdown maintenance comes into play. It is difficult to be fully prepared for breakdowns, because they often occur without warning and the cause of the breakdown is often something that could not have been foreseen. The best that can be done is to have a well-equipped maintenance shop and a skilled, experienced maintenance staff. In difficult situations, such as when the cause of the breakdown cannot be determined or the same type of breakdown occurs repeatedly, it is usually best to seek advice from the equipment manufacturer.

TABLE 8. Typical pilot plant operating records

Record	Types of information recorded
Raw materials	Quantities received, results of quality assays of incoming raw materials, quantities accepted and rejected
Utilities	Consumption of water, electricity, fuels; breakdown of utility loads by plant area
Labor	Labor inputs, broken down by plant area
Finished product	Quantities produced, results of quality analyses of finished product, quantities shipped out
Wastes	Quantities and composition of major discharges to air or sewer system

New equipment installation in an existing pilot plant

When a piece of equipment is brought into an existing pilot plant, whether it is a new piece of equipment purchased directly from a manufacturer or used equipment brought in from another facility, the task of installing, testing, and starting up the equipment should be the resonsibility of the maintenance staff. In this task, the maintenance staff should receive guidance and assistance from the technical staff supporting the pilot plant, especially the engineering staff, as well as from the equipment manufacturer.

After a new piece of equipment has been installed, it is very important for the maintenance staff to work closely with the operations staff during the initial phase of operation to make certain that the installation has been done properly and to transfer to the operations staff all of the information about the equipment that is necessary to operate it safely and efficiently. The maintenance staff should participate in writing the SOPs for the new piece of equipment.

In all three of the areas of maintenance described above, it is sometimes difficult to make a clear-cut distinction between the responsibilities of the maintenance staff and those of the operations staff. For example, much of the routine preventive maintenance can easily be incorporated into the equipment SOP and is probably most efficiently done by the equipment operator in the course of his normal duties. Likewise, if equipment breaks down during one of the night shifts, when none of the maintenance staff is on duty, the operator certainly should make every effort to repair it.

The size of the maintenance staff is an important factor to be considered in allocating maintenance duties between the operations staff and the maintenance staff. If the maintenance staff is large enough to cover all of the shifts with at least one full-time maintenance worker, then a more formal division of maintenance duties is both possible and generally desirable.

24.8.10 Housekeeping

The pilot plant should be kept clean at all times, and equipment and supplies should be kept in neat, orderly arrangements, not only to protect the health and safety of the pilot plant personnel, but also to protect the process from contamination. The contamination rate in fermentations will be affected by the state of cleanliness of the fermentation equipment and of the entire plant.

As a minimum, housekeeping procedures should be in accordance with the relevant Occupational Safety and Health Administration and Food and Drug Administration regulations. In pilot plants where recombinant DNA microorganisms are handled, procedures established by the National Institutes of Health should also be followed (13).

Housekeeping duties should be distributed evenly among the operations staff, the only exception being procedures that require specialized training. Ultimate responsibility for the condition of the plant should be assigned to the plant manager.

24.9 COMMERCIAL SOURCES

Note. The list of commercial sources given below is far from comprehensive. The entries made are intended to be illustrative.

24.9.1 General

Chemical Engineering Catalog
(published each year by:
Penton/IPC Reinhold Publishing, Stanford, Conn.)

Chemical Engineering Equipment Buyers' Guide
(published each year by:
McGraw-Hill, New York)

Food Engineering Master: Equipment, Supplies & Services
(published each year by: Chilton Co., Chilton Way, Radnor, PA 19089)

24.9.2 Construction Management and Execution

American BioDesign Inc.
44 Mechanic St.
Newton, MA 02164
(617) 332-1082

The Austin Co.
2001 Rand Rd.
Des Plaines, IL 60016

Badger
One Broadway
Cambridge, MA 02142
(617) 494-7721

Daniel
Daniel Building
Greenville, SC 29602
(803) 298-2500

Davy McKee Corp.
6200 Oak Tree Blvd.
Independence, OH
(216) 524-9300

Foster Wheeler Corp.
110 South Orange Ave.
Livingston, NJ 07039

Jacobs Engineering Group Inc.
251 South Lake Ave.
Pasadena, CA 91101
(213) 681-3781

John Brown Engineers & Construction Inc.
17 Amelia Place
P.O. Box 1432
Stamford, CT 06904
(203) 327-1450

Stone & Webster Engineering Corp.
One Penn Plaza
250 West 34th St.
New York, NY
(212) 290-6000

24.9.3 Fermentors

ABEC
Airport Road
Commercial Park, Box 2606
Lehigh Valley, PA 18001
(215) 837-1871

Bethesda Research Laboratories
Life Technologies Inc.
Gaithersburg, MD 20877
(800) 638-4045

Bilthoven (Netherlands)
U.S. reps: Flow Laboratories, Inc.
7655 Old Springhouse Rd.
McLean, VA 22102
(703) 893-5925

Bioengineering, AG (Switzerland)
Sagenrainstrasse 7
CH8636 Wald, Switzerland

Biolafitte
Alexander Commerce Park
719 Alexander Rd.
Princeton, NJ 08540
(609) 452-7660

BioMed Design
12109 N.E. 95th St.
Vancover, WA 98662
(206) 254-6011

B. Braun
875 Stanton Rd.
Burlingame, CA 94010
(415) 692-6022

Chemapec, Inc.
230 Crossways Park Dr.
Woodbury, NY 11797
(516) 364-2100

Fermac
P.O. Box 15
SF-68601 Pietarsaari, Finland

Fermex, Inc.
Box 2294
Edison, NJ 08818
(201) 238-8984

Gallenkamp
P.O. Box 290, Technico House
Christopher St.
London EC2P 2ER, England

Hitachi
c/o Hitachi America, Ltd.
437 Madison Ave.
New York, NY 10022
(212) 758-5420

Lab-Line Bioengineering Ltd.
Lab-Line Plaza
Melrose Park, IL 60160
(312) 450-2600

LH (England)
U.S. reps: Queue Systems
Box 1901
Parkersburg, WV 26102
(304) 464-5400

LKB Instruments, Inc.
12221 Parklawn Dr.
Rockville, MD 20852

Marubishi, Inc. (Japan)
U.S. reps: Bioengineering Associates, Inc.
44 Mechanic St.
Newton, MA 02154
(607) 890-0018

MBR Bio Reactor AG
Werkstrasse 3
8620 Wetzikon, Switzerland

New Brunswick Scientific Co., Inc.
P.O. Box 986
44 Talmadge Rd.
Edison, NJ 08818
(201) 287-1200

Pegasus Industrial Specialties Ltd. (Canada)
U.S. reps: Biosystems, Inc.
P.O. Box 416
Montville, NJ 07045
(201) 299-9268

Setric Genie Industrial (France)
U.S. reps: Dela Technology Corp.
2263 Lewis Ave.
Rockville, MD 20851
(301) 881-6292

VirTis Co.
Route 208
Gardiner, NY 12525

24.9.4 Bioreactors, Recovery Equipment, and Accessories

Ace Glass, Inc.
1430 N.W. Blvd.
Vineland, NJ 08360
(small pilot-scale bioreactor and recovery vessels, other specialized types of glassware)

Ametek Process Equipment Division
P.O. Box 1406
El Cajon, CA 92022
(pressure leaf filters, basket centrifuges)

Amicon Corp.
Cherry Hill Dr.
Danvers, MA 01923
(800) 343-1387
(hollow-fiber units, column bioreactors, LC columns)

Cordis Dow Corp.
Miami, FL 33145
(hollow-fiber units)

Fulflo Filtration Systems, commercial filters
P.O. Box 1300
Lebanon, IN 46052
(cartridge filters and accessories)

Goulds Pumps Inc.
P. O. Box 330
Fall St.
Seneca Falls, NY 13148
(centrifugal pumps)

Hull Corp.
Davisville Rd.
Hatboro, PA 19040
(215) 672-7800
(dryers)

Ingold Electrodes
One Burtt Rd.
Andover, MA 01810
(pH electrodes)

Mueller
P.O. Box 828
Springfield, MO 65801
(800) 641-2830
(process tanks)

Pharmacia Fine Chemicals Division
810 Centennial Ave.
Piscataway, NJ 08854
(bioreactor and LC columns)

Pfaudler
Rochester, NY 14692
(716) 235-1000
(glass-lined tanks)

Robbins and Myers Inc.
Fluids Handling Division
Moyno Products
1345 Lagonda Ave.
Box 960
Springfield, OH 45501
(progressing cavity pumps)

Romicon, Inc.
100 Cummings Park
Woburn, MA 01801
(hollow-fiber units)

Rosemount, Inc.
12001 W. 78th St.
P.O. Box 35129
Eden Prairie, MN 55344
(temperature sensors, pH electrodes)

Sparkler Filter, Inc.
Box 19
Conroe, TX 77305
(horizontal-plate pressure filters)

Stokes
5500 Tabor Rd.
Philadelphia, PA 19120
(215) 289-0100
(dryers)

Waukesha Foundry Division
Abex Corp.
1300 Lincoln Ave.
Waukesha, WI 53186
(positive-displacement rotary pumps)

24.9.5 Laboratory Support

Bally Case & Cooler, Inc.
Ball, PA 19503
(walk-in refrigerators and freezers)

DuPont Instruments
Sorvall Division
Peck's Lane
Newtown, CT 06470
(refrigerated high-speed laboratory centrifuges)

Fisher Scientific Co.
203 Fisher Bldg.
Pittsburgh, PA 15219
(general laboratory supplies)

New Brunswick Scientific Co.
P.O. Box 986
Edison, NJ 08817
(laboratory incubator-shakers)

Perkin-Elmer Corp.
Instrument Group
Main Ave.
Norwalk, CT 06856
(spectrophotometers and other analytical instruments)

Pharmacia Fine Chemicals Division
810 Centennial Ave.
Piscataway, NJ 08854
(chromatography columns)

Scientific Products
Division of American Hospital Supply Corp.
General Offices
1420 Waukegan Rd.
McGaw Park, IL 60085
(general laboratory supplies)

Sigma Chemical Co.
P.O. Box 14508
St. Louis, MO 63178
(laboratory chemicals)

Waters Associates, Inc.
Maple St.
Milford, MA 01757
(high-performance liquid chromatography apparatus)

24.9.6 Utilities

Illinois Water Treatment Co.
4669 Shepherd Trail
Rockford, IL 61105
(815) 877-3014
(water purification systems)

Joy Manufacturing Co.
Montgomeryville Industrial Center
Montgomeryville, PA 18936
(oil-free air compressors)

Oswego Package Boiler Co., Inc.
Oswego, NY
(steam boilers)

York Div. Borg-Warner Corp.
Richland Ave.
P.O. Box 1592
York, PA 17405
(717) 846-7890
(chillers)

24.10 ACKNOWLEDGMENTS

We thank R. Geoghegen and L. Jacobs for contributions to the section on fermentor specifications.

24.11 LITERATURE CITED

1. **Barrer, P. J.** 1983. Crucial factors for design of a pilot plant. Bio/Technol. **1**:661–666.
2. **Bull, D. N.** 1983. Fermentation and genetic engineering: problems and prospects. Bio/Technol. **1**:847–856.
3. **Bull, D. N., R. W. Thoma, and E. Stinnett.** 1983. Bioreactors for submerged culture. Adv. Biotechnol. Processes **1**:1–30.
4. **Butterworth, T. A., D. I. C. Wang, and A. J. Sinskey.** 1970. Application of ultrafiltration for enzyme retention during continuous enzymatic reaction. Biotechnol. Bioeng. **12**:615–631.
5. **Center for Disease Control.** 1974. Classification of etiological agents on the basis of hazard, 4th ed. July 1974. Center for Disease Control, Atlanta, Ga.
6. **Code of Federal Regulations.** Superintendent of Public Documents, U.S. Government Printing Office, Washington, D.C.
7. **Enfors, S. O., and B. Mattiasson.** 1983. Oxygenation of processes involving immobilized cells, p. 41–60. *In* B. Mattiasson (ed.), Immobilized cells and organelles, vol. 2. CRC Press, Inc., Boca Raton, Fla.
8. **Hockenhull, D. J. D.** 1983. Organization of a pilot plant for the development of new products. Prog. Ind. Microbiol. **17**:191–224.
9. **Kampen, W. H.** 1983. Industrial pilot plant, p. 48–76. *In* H. C. Vogel (ed.), Fermentation and biochemical engineering handbook. Noyes Publications, Park Ridge, N.J.
10. **Kitano, H., and N. Ise.** 1984. Hollow fiber enzyme reactors. Trends Biotechnol. **2**:5–7.
11. **Klibanov, A. M.** 1979. Enzyme stabilization by immobilization. Anal. Biochem. **93**:1–25.
12. **Lloyd, N. E., L. T. Lewis, R. M. Logan, and D. N. Patel.** 1972. Process for isomerizing glucose to fructose. U.S. patent 3,694,314, assigned to Standard Brands Inc.
13. **National Institutes of Health.** 1983. Guidelines for research involving recombinant DNA molecules; June 1983. Fed. Regist. **48**:24556–24581.
14. **Pirt, S. J.** 1981. Fermentation—a question of life, p. 17–23. *In* M. Moo-Young (ed.), Advances in biotechnology, vol. 2. Pergamon Press, New York.
15. **Rase, H. F., and M. H. Barrow.** 1957. Project engineering of process plants. John Wiley & Sons, Inc., New York.
16. **Richards, J. W.** 1968. Design and operation of aseptic fermenters, p. 107–122. *In* Industrial sterilization. Academic Press, Inc., New York.
17. **Scott, C. D.** 1983. Fluidized-bed bioreactors using a flocculating strain of *Zymomonas mobilis* for ethanol production, p. 448–456. *In* K. Venkatasubramanian, A. Constantinides, and W. R. Vieth (ed.), Biochemical engineering, no. 3, vol. 413. New York Academy of Sciences, New York.
18. **Sikyta, B.** 1983. Methods in industrial microbiology, p. 43–74. John Wiley & Sons, New York.
19. **Soderberg, A. C.** 1983. Fermentation design, p. 77–118. *In* H. C. Vogel (ed.), Fermentation and biochemical engineering handbook. Noyes Publications, Park Ridge, N.J.
20. **Steel, R., and T. L. Miller.** 1970. Fermentor design. Adv. Appl. Microbiol. **12**:153–188.
21. **Vickroy, T. B., H. W. Blanch, and C. R. Wilke.** 1983. Microbial hollow fiber bioreactors. Trends Biotechnol. **1**:135–139.
22. **Wandrey, C.** 1983. Enzyme membrane reactor systems, p. 577–588. *In* Biotech 83. Online Publications, Pinner, Middlesex, U.K.
23. **Weisman, E.** 1970. The installation of fermentation pilot plant. Process Biochem. **5**:10–11, 18.

Chapter 25

Sterilization: Prevention of Contamination

FREDRIC G. BADER

To the modern student of microbiology, sterilization is often viewed as a simple technical necessity. Yet in a very broad sense, it is the fundamental technique upon which the whole fascinating world of modern microbiology is based. Pasteur's development of sterilization provided the evidence needed to destroy the theory of spontaneous generation once and for all. Today, we take for granted the alternative theory on which all genetics is based, that life only comes from life. It is, perhaps, ironic that it has been the development of the ability to destroy organisms that has allowed us to understand them well enough to use them for the benefit of humankind. Sterilization is big business. It is fundamental to many major industries.

This chapter is designed to look at sterilization from a practical viewpoint. There are already in existence many good books and articles that review the sterilization literature and the more complicated theories associated with it. There are, however, very few sources of practical information on how to run large-scale equipment with a low probability of contamination. This is the area to which this chapter is addressed: the sterilization and prevention of contamination in fermentation equipment.

25.1 CONCEPTS AND TERMINOLOGY

25.1.1 Definition of Terms

Sterilization has been defined many ways. Reddish (41) defined it as "any process, physical or chemical, which will destroy all forms of life, applied especially to microorganisms, including bacterial and mold spores and the inactivation of viruses." Sykes (8) states that "sterilization is an absolute term and implies the total destruction of all forms of microbial life in terms of their ability to reproduce . . . ." These types of definition are conceptually operative. However, there are a number of points which need further discussion. The emphasis on microorganisms is unwarranted. A sterile system has no life forms, either microscopic or macroscopic. A germfree animal is axenic, not sterile. Both of these definitions rely upon destruction and do not consider removal of forms of life by some other technique such as filtration. It is also difficult to deal in an operational sense with sterilization being an absolute. There is the old laboratory joke that being almost sterile is like being almost pregnant. But one has to recognize that sterility and pregnancy are substantially different. It is possible to prove that one is pregnant. It is never possible to prove that something is sterile. One can only say that no organisms were present that were capable of growing under the specified test conditions. With this in mind, it is also necessary to recognize that a sterilization process is not an absolute, as sterilization kinetics are stochastic and the success of a sterilization is only a probability. Sterility is much like the concept of infinity. It is a concept that is not approachable mathematically, but has an operational function as a limit. In the design of sterilization process conditions, one can only speak of a probability that one will achieve sterility.

Another objection to most common definitions of sterilization is that they do not provide a concept of containment or boundaries. From an engineering perspective, it is important to recognize that one sterilizes systems which have limits or boundaries. It is also important to recognize that there is a difference between open systems and closed systems. In a closed system, the boundaries are all impermeable and nothing passes through them. However, most systems that we sterilize are open systems that are designed to allow at least gases to move into and out of the system. A closed system is sterilized once and maintains sterility by impermeable barriers. Open systems are initially sterilized, but in an operational sense, they are continually being sterilized by the removal capability of the permeable boundary.

For example, consider an open system consisting of a test tube of broth with a cotton plug. It can be sterilized initially in an autoclave so that there is a high probability that no organisms are present either inside or outside the tube. However, on removal from the autoclave, the outside of the tube and the plug will soon become contaminated. In this case, the desired sterile system is the inside walls of the tube, the broth, the gas in the tube, and the surface of the plug on the inside of the tube. Gases will normally move in and out of the tube and are continually being sterilized by the removal capability (filtration) of the plug. In such a system, one could say that the broth was sterilized in the autoclave. However, the gas is continually being sterilized by the cotton as it moves into and out of the tube. The definition of the system that one wants to sterilize and the concept that sterilization can be a continual process are both important when one deals with larger and more complex equipment.

With the above in mind, I propose an operational or engineering definition of "sterilization" as the destruction or removal of all forms of life from a defined system through either a discontinuous or a continuous process.

There are a few other concepts or definitions that should be mentioned, as they often lead to confusion. "Sanitization" is the reduction of microbial contamination to a level which is determined to be safe by public health requirements. Sanitization generally involves a cleaning process which uses chemicals and detergents both to remove dirt and to reduce concomitant microbial levels. However, it should be recognized that sanitary systems are not sterile. This frequently becomes a problem when dealing with equipment vendors, especially from the food and dairy industries. Sanitary valves, fittings, and equipment are designed to be cleanable, especially by clean-in-place processes. Such equipment is frequently also sterilizable, but not necessarily under the conditions specified for sanitary use. In some cases, sanitary equipment is not designed to handle sterilizing conditions.

Since "sterile" defines a system which contains no forms of life, then "sterile technique" would define methods used to maintain sterility. When one transfers a culture to a sterile system, one cannot, by definition, be using sterile technique. "Aseptic" defines a system which contains life forms, but is free from foreign or unwanted life forms which would adversely affect the system. Normal culture transfers are conducted using "aseptic techniques." "Contaminated" defines a system which contains an unwanted or foreign form of life. "Axenic" defines a system which contains only one desired form of life or a pure culture. A pure culture can also be called aseptic, but a defined mixed culture cannot be called axenic.

25.1.2 Sterilization versus Contamination

Sykes' definition (8) that sterilization is an absolute is conceptually correct. From an engineering viewpoint, one is frequently not as much concerned with sterilization as with the actual prevention of contamination. The two are related, but are not always the same. There are many biological processes that operate free from contamination in an operational sense, but are not necessarily axenic or sterile. This is due primarily to the fact that many systems are run under conditions which are inhibitory to many types of contamination. Such conditions need to be taken into account both when designing a process and when evaluating equipment and conditions that others use. Some of the conditions that can affect the sensitivity of a process to contamination are shown in Table 1.

Different industries also tend to be concerned with different microbial contaminants. This is due both to the process conditions and to the potential damage that the contaminant can elicit. In the canning industry, *Clostridium botulinum* is the major organism of concern. It is an obligate anaerobe which can grow well in sealed cans. It is also a heat-resistant sporeformer, which makes it difficult to kill. Moreover, it produces a deadly toxin. A single can contaminated with *C. botulinum* can send a company into bankruptcy. *Clostridium* sp. contamination is a major concern in canning soups, meats, and neutral vegetables. It is not a major concern in catsup (too acid), jams and jellies (too much sugar), or table milk (too cold).

In the production of wine, conditions are anaerobic but too acidic (pH 5) for *Clostridium*. Therefore, producers are more concerned about organisms such as *Lactobacillus*, which produces lactic acid instead of alcohol. In addition, if conditions get to be aerobic, *Acetobacter* can convert the wine to vinegar or molds can give the wine an off flavor.

In the antibiotic industry, typical fermentation media support luxuriant microbial growth, and operating conditions are generally ideal for a wide range of contaminants (molds, yeasts, and many bacteria including *Bacillus* sp.). In such fermentations, prevention of contamination is a major problem, and there are wide ranges in the degree of contamination problems depending upon which antibiotic is being produced. All antibiotic fermentations are highly susceptible to contamination during the first several days of growth (trophophase). Once the culture establishes itself and starts producing metabolites, the antibiotic can help to protect the fermentation environment or even destroy an established contaminant. This, of course, depends upon the antimicrobial spectrum, potency, and titer of the antibiotic. For example, *Bacillus* sp. contamination may be common in erythromycin fermentations, but is likely to be rare in penicillin fermentations.

The most feared contaminants in the fermentation industry are bacteriophages. Fortunately, a bacteriophage can replicate only in a susceptible host,

TABLE 1. Conditions which affect probability of contamination

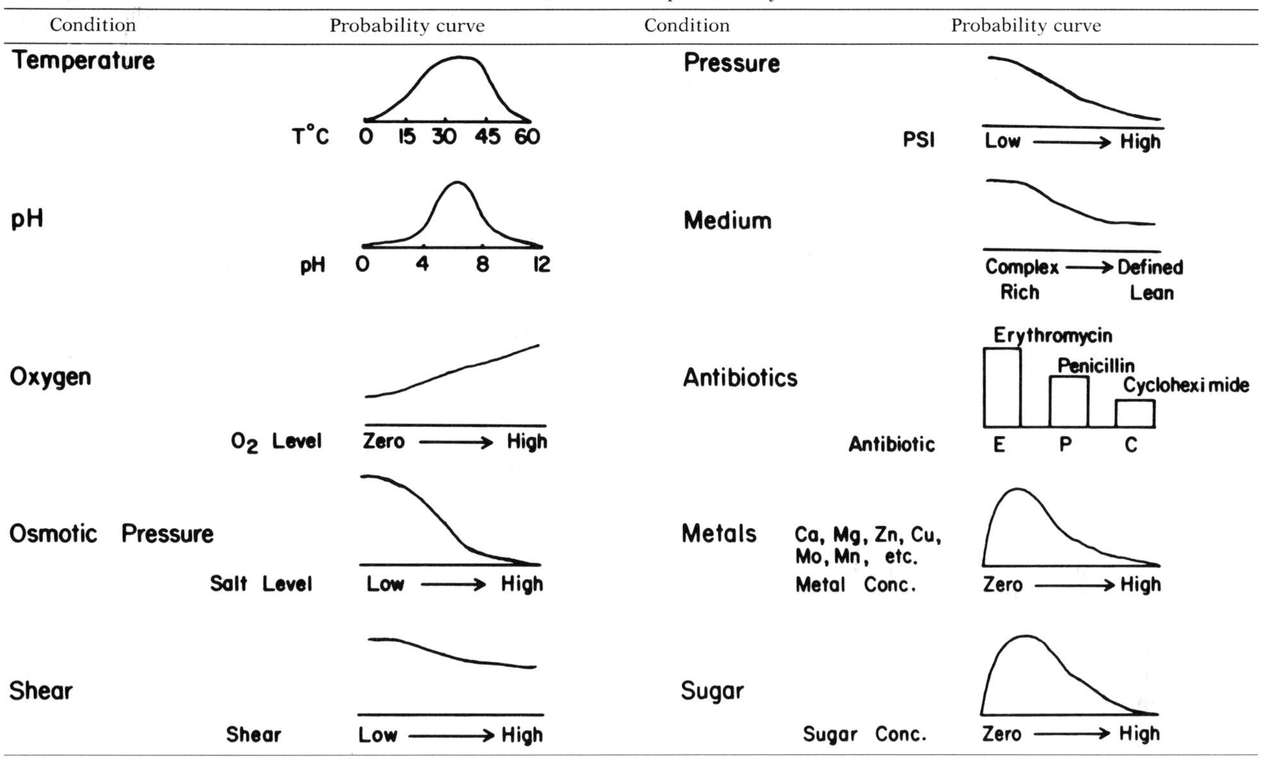

and therefore most phage virions are harmless. However, if a specific phage enters a producing culture during the fermentation run, it can lyse the culture in a matter of hours. Phage contaminants are also difficult to contain and tend to spread rapidly throughout the plant. The only effective method to prevent phage lysis is to develop resistant cultures.

The above comments have been made with two thoughts in mind. First, in developing procedures for sterilization and contamination control, one needs to evaluate the process with respect to susceptibility to contamination and to ascertain the types of contaminants that might be involved. Second, one needs to recognize that what works for one process may not work for others. For example, a liquid medium can be sterilized in an autoclave in 15 min at 121°C, whereas oils generally require a few hours, and concentrated media (10 to 20% solids) cannot be sterilized without agitation. Another example is that a vendor's fermentation equipment that runs uncontaminated with, for example, alcohol fermentations may run with a high contamination rate with a different fermentation process.

Each process is unique and must be evaluated separately with respect to sterilization and contamination control. It is also possible that a particular process may have to be abandoned because of an inability to successfully sterilize the medium or because of a high proclivity for contamination. It is, therefore, essential that the fermentation technologist who is involved in the development and design of new processes be aware of the fundamental principles involved in sterilization and contamination control.

25.2 THEORY OF HEAT STERILIZATION

25.2.1 Kinetics and Terminology of Heat Sterilization

The death of microorganisms occurs under thermal sterilizing conditions according to first-order kinetics as shown in equation 1.

$$dN/dt = -kN \qquad (1)$$

where N = number of viable organisms; k = reaction of death rate (min^{-1}); and t = time. Equation 1 may be integrated to obtain:

$$N = N_0 e^{-kt} \qquad (2)$$

where N_0 is the number of viable organisms at time t = 0. It should be noted that N may be considered as the total number of organisms present, or it may be a concentration of organisms per unit volume. The kinetics are the same. The reaction rate k is a function of the environmental conditions, the organism and its history, and the temperature, as will be discussed later. As indicated in equation 2, the death of organisms occurs exponentially. A large value of k indicates a high or rapid death rate.

It is generally difficult for most people to think in terms of the base e. As a result, equation 2 is most often seen in the base 10 form of:

$$N = N_0 10^{-t/D} \qquad (3)$$

where D is called the D value and usually has units of minutes. It is clear from equation 3 that when $t = D$, $N = 0.1\ N_0$. Therefore, the D value is defined as the time required at temperature to reduce the microbial

TABLE 2. Population of surviving organisms versus time for a D value of 1 min

Time (min)	t/D	No. of organisms surviving	No. of organisms killed	% Killed
0	0	1,000,000	0	0
1	1	100,000	900,000	90.
2	2	10,000	990,000	99.
3	3	1,000	999,000	99.9
4	4	100	999,900	99.99
5	5	10	999,990	99.999
6	6	1	999,999	99.9999
7	7	0.1	999,999.9	99.99999
8	8	0.01	999,999.99	99.999999
9	9	0.001	999,999.999	99.9999999

population by 90% or one order of magnitude. Table 2 more clearly shows the relationship between the population of surviving organisms and time for a D value of 1 min. By comparing equations 2 and 3 it becomes evident that:

$$D = 2.303/k \tag{4}$$

Therefore, the D value is inversely related to k such that a small value of D leads to a rapid kill rate.

Taking the $\log_{10}$ of both sides of equation 3 leads to equation 5:

$$\log_{10} N/N_0 = -t/D = \log_{10} N - \log_{10} N_0 \tag{5}$$

It is evident from equation 5 that if one plots $\log_{10} N$ versus time, one will obtain a straight line where the slope equals $-1/D$ and the intercept equals $\log_{10} N_0$. Figure 1 shows such a plot, which is usually called a survivor curve (37). It becomes evident from Fig. 1 that the number of surviving organisms will never go to zero. It can go to extremely low values; for example, N could equal 10^{-7} organisms per ml. Although it is not possible to have a fraction of a viable organism, one can view fractional populations in two ways. First, if one has 10^{-7} organisms remaining per ml, then this would indicate that one surviving organism would be present in 10^7 ml or 10,000 liters. Second, if one has 10^{-7} organisms per sample, then one might expect that one sample in 10^7 would still have a surviving organism.

It has been shown by Aiba and Toda (10) that the probability that a sample is sterile can be determined by:

$$P = (1 - e^{-kt})^{N_0} \tag{6a}$$

or

$$P = (1 - 10^{-t/D})^{N_0} \tag{6b}$$

where P equals the probability that the sample is sterile and N_0 is the initial number of organisms in the sample (note that N_0 is not a concentration in this calculation).

Conversely, the probability that a sample is not sterile is given by:

$$1 - P = 1 - (1 - e^{-kt})^{N_0} \tag{7a}$$

or

$$1 - P = 1 - (1 - 10^{-t/D})^{N_0} \tag{7b}$$

Table 3 shows the probability that a sample is sterile for the same conditions used in Table 2. As shown in Table 3, there is still a 1% chance of the sample not being sterile, even when 0.01 viable organism is remaining in the sample.

In the above equations, N_0 can be called the bioburden of the sample. It becomes evident from equation 6 that the initial bioburden can strongly affect the probability of achieving the sterilization of a sample. For example, if $N_0 = 10^3$ organisms, a 5-min exposure at 121°C would lead to the same probability of sterilization as an 8-min exposure with $N_0 = 10^6$ organisms. In general, the lower the bioburden or the cleaner the system, the easier it is to achieve sterilization.

Another term that is frequently used in sterilization is the F value. The F value is the time at temperature required to achieve sterilization. Therefore, it is related to the D value, the bioburden N_0, and the required probability of achieving sterilization. The F value is commonly used in the canning industry, but is rarely used in the fermentation industry.

25.2.2 Effect of Temperature on Death Rate

Figure 2 shows a typical plot of log specific death rate ($\log_{10} k$) versus reciprocal absolute temperature in degrees Kelvin for bacterial spores. As can be seen, such a plot is generally linear. The effect of temperature on the death rate k follows the Arrhenius equation as shown in equation 8:

$$k = ae^{-E/RT} \tag{8}$$

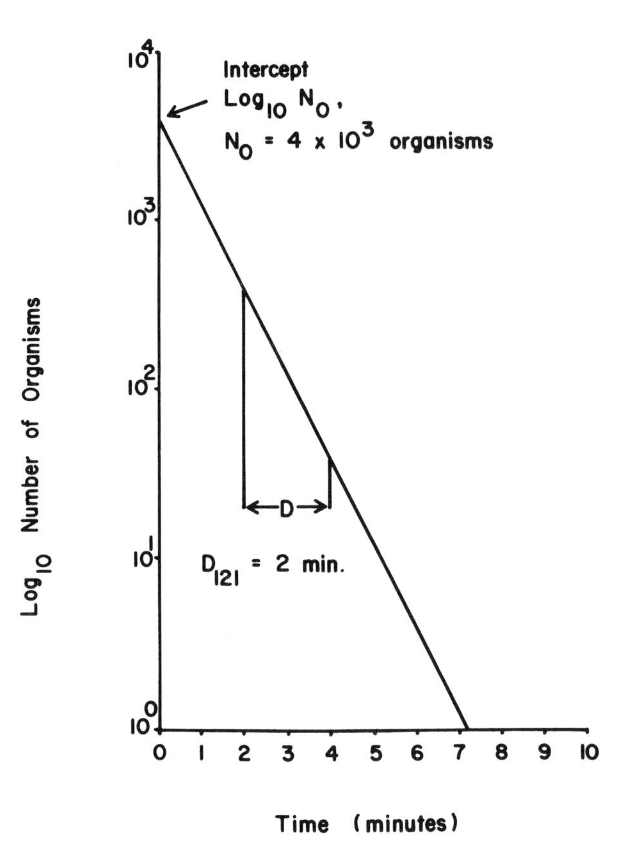

FIG. 1. Survivor curve for a typical sporeformer at 121°C. The $\log_{10}$ of the number of survivors is plotted versus time. The D value is the time required for 1 log reduction.

TABLE 3. Probability of sterilizing a sample ($N_0 = 10^6$; $D_{121} = 1$ min)

Time (min)	t/D	No. of surviving organisms	Probability of sterility	Probability of not being sterile
0	0	10^6	0	1
1	1	10^5	0	1
2	2	10^4	0	1
3	3	10^3	0	1
4	4	10^2	3.70×10^{-44}	1
5	5	10^1	4.54×10^{-5}	0.99995
6	6	10^0	0.36788	0.63212
7	7	10^{-1}	0.90484	0.09516
8	8	10^{-2}	0.99005	0.00995
9	9	10^{-3}	0.99900	0.00100

where E = activation energy (kcal/g-mol); R = gas constant (kcal/g-mol − degrees K); T = absolute temperature (degrees K); a = empirical constant.

Equation 8 can be rearranged to the base 10 as shown:

$$k = a10^{-E/2.303RT} \tag{9a}$$

Taking the $\log_{10}$ of both sides of equation 9a yields:

$$\log_{10}k - \log_{10}a = -E/2.303RT \tag{9b}$$

which is the basis of the plot in Fig. 2. The slope of the line is equal to $-E/2.303R$, and $-\log_{10}a$ can be calculated from an intercept at a designated temperature.

A second method of relating the death rate to temperature is through D values as shown:

$$\log D - \log D_{121} = -(T - 121)/Z \tag{10a}$$

or

$$D = D_{121}10^{-(T - 121)/Z} \tag{10b}$$

In these equations, D_{121} = D value at 121°C; Z = Z value; T = temperature (°C).

A typical plot of $\log_{10}$ of the D value versus temperature, called a thermal resistance curve, is shown in Fig. 3. The Z value equals the number of degrees of temperature required for the D value to change by a factor of 10. The Z value is indicated in Fig. 3.

If one plots the F value versus temperature, the curve is called a thermal death time curve. It has the same shape as the thermal resistance curve and follows the same equations (equation 10).

A third way of relating the death rate to temperature is the Q_{10} value. This is the change in the reaction rate constant for a change in temperature of 10°C. For most chemical reactions, if the temperature is increased by 10°C, the rate increases by a factor of two. For sterilization, Q_{10} values range from 2.2 to 4.6 for dry heat and 10.0 to 18.0 for wet heat (37). The specific death rate is related to Q_{10} as follows:

$$k_{(T + 10°C)} = Q_{10}k_T \tag{11}$$

It should be pointed out that there is a danger in extrapolating any of these curves to higher temperatures. Wang et al. (48) showed a 160% disagreement between extrapolated and measured data at higher temperatures where careful experimental technique was used. Extrapolating low-quality data can easily lead to errors of an order of magnitude.

The above equations and parameters describe most of the important concepts found in the literature that are related to the thermal destruction of microorganisms. Table 4 contains typical values for some of the parameters discussed in the equations above for wet-heat sterilization. Table 5 contains similar values for dry-heat sterilization. These numbers are presented here to serve as a reference point for comparing experimental results. They may also be useful as a first approximation when designing sterilization equipment or cycles. They should be used with some degree of caution, as particular cases can differ significantly from these reported values (Table 4). This is evident in Table 4, where two different strains of *Bacillus stearothermophilus* have death rates that differ by a factor of almost 4 at the same temperature. It is evident from Tables 4 and 5 that the sterilization kinetic parameters vary dramatically between vegetative cells and bacterial spores and between wet-heat and dry-heat sterilization conditions. This will be discussed further in a later section of this chapter.

25.2.3 Abnormal Thermal Resistance Curves

In the previous section, the idealized theory that describes the death of microorganisms was discussed. Actual experimental results frequently do not follow this theory exactly. There are two major types of deviation from the normal linear survival curves. The

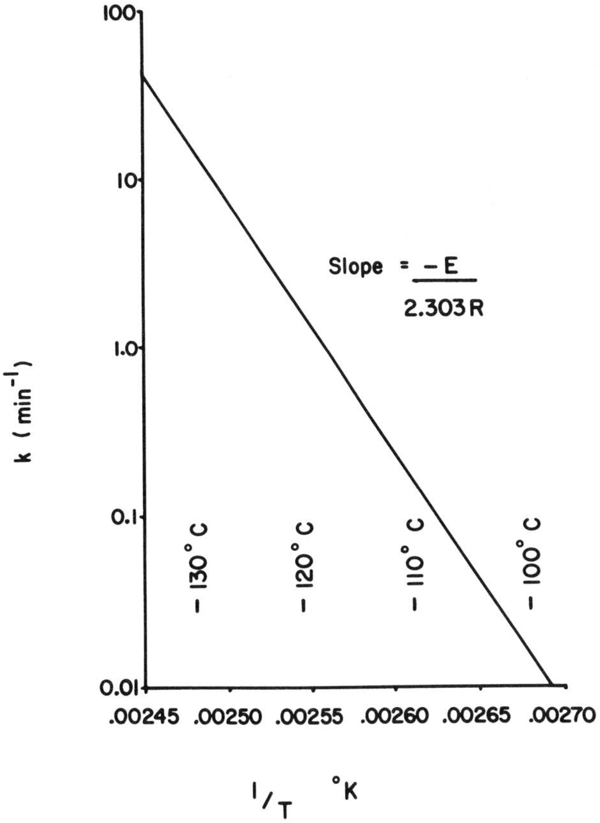

FIG. 2. Arrhenius plot of specific death rate versus reciprocal absolute temperature.

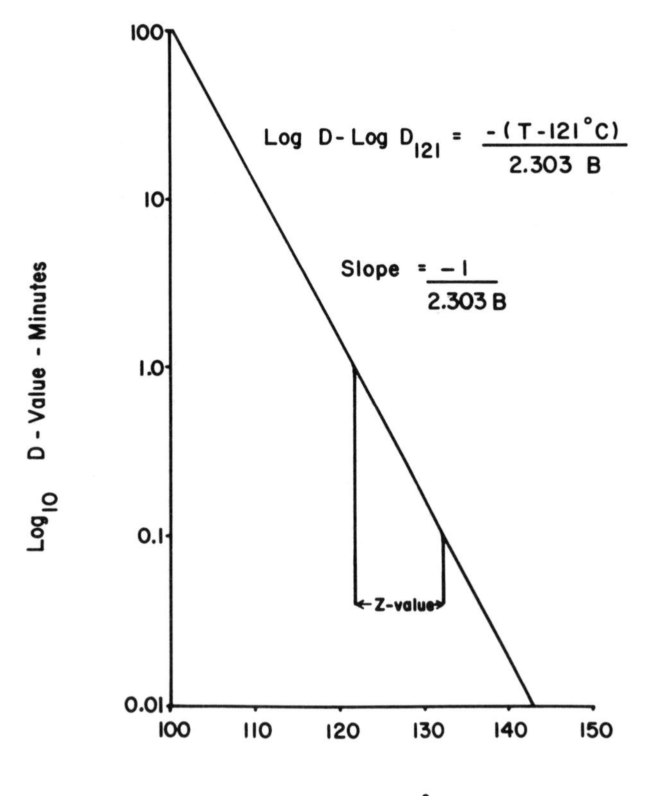

FIG. 3. A typical plot of $\log_{10} D$ value versus temperature for a heat-resistant sporeforming organism. The Z value is the number of degrees required to reduce the D value by 1 order of magnitude.

first type is illustrated in Fig. 4, where one initially sees a rapid kill phase, followed by a slower kill phase. This phenomenon can generally be attributed to the occurrence of a mixed population of spores of different thermal resistances (24).

The plot in Fig. 4 represents the theoretical results that one would expect to see if one mixed equal populations of 2.5×10^6 spores of each of the cultures indicated with the death rates as shown in Table 4. The total initial population for the four cultures would then be 10^7 spores. As shown, the survivor curve for each population would be linear. Since some of the spores die off more rapidly than others, the total combined curve is not linear until all of the rapidly dying spores are gone and only the most resistant remain. Hence, one sees what looks like a rapid kill phase followed by a slower kill phase.

Similar curves are also seen when using a single-spore culture. This is generally indicative of either a nonuniform spore population (i.e., some spores are more resistant than others, possibly due to previous growth conditions) or a nonuniform environment during the sterilization treatment or the outgrowth of the spores.

The second type of common deviation that is seen in spore survival curves is shown in Fig. 5. This type of curve is concave downward and shows very little initial kill, followed by a more rapid kill phase. This type of curve is frequently indicative of spore clumping or the presence of solids in the spore preparation. The effect of clumping of spores can be illustrated mathematically as follows. The normal death of spores was described in equation 2 as $N = N_0 e^{-kt}$. If one has a clump containing C_0 spores, the number of spores surviving in the clump can also be described as:

$$C = C_0 e^{-kt} \tag{12}$$

The probability that all of the spores in a clump are killed is obtained from equation 6a as:

$$P = (1 - e^{-kt})^{C_0} \tag{13}$$

and the probability that at least one spore in the clump is alive is given by:

$$1 - P = [1 - (1 - e^{-kt})^{C_0}] \tag{14}$$

If a total of N_0 spores are present with a uniform number (C_0) of spores in a clump, then the initial number of CFU (T_0) would be given by:

$$T_0 = N_0/C_0 \tag{15}$$

At any given time during the heat treatment or sterilization process, the number of CFU (T) surviving at time t would be given by:

$$\begin{aligned} T &= T_0(1 - P) \\ &= N_0/C_0\,[1 - (1 - e^{-kt})^{C_0}] \end{aligned} \tag{16}$$

TABLE 4. Typical values for sterilization kinetic parameters, wet heat

Species	Death rate constant k at 121°C (min^{-1})	D at 121°C (min)	Reference
Escherichia coli vegetative	10^{13}	2.3×10^{-13}	1
Bacillus subtilis FS5230	3.8–2.6	0.6–0.9	1
B. stearothermophilus FS1518	0.77	3.0	1
B. stearothermophilus FS617	2.9	0.8	1
Clostridium sporogenes PA3679	1.8	1.3	1

Species	E (cal/g-mol)	Q_{10}	Z value (°C)	Reference
Bacillus stearothermophilus FS7954	68,700	8.35	10.85	1
Escherichia coli vegetative	127,000	50.44	5.87	1
Bacillus subtilis subsp. *niger* ATCC 9372[a]	42,400	3.71	17.55	Bader and Boekeloo[a]
B. stearothermophilus FS7954	83,600	14.28	8.71	48

[a] Note: These data are for spores in wet oil (F. G. Bader and M. K. Boekeloo, 184th ACS Annu. Meet., 1982).

TABLE 5. Typical values for sterilization kinetic parameters, dry heat

Species	Death rate constant k at 121°C (min^{-1})	D at 121°C (min)	Reference
Bacillus subtilis 5230	0.0078–0.0118	195–295	38
Clostridium sporogenes PA3679	0.0118–0.0200	115–195	38
Bacillus stearothermophilus FS1518	0.12121	19	18

Species	Activation energy E (cal/g-mol)	Q_{10}	Z value (°C)	Reference
Bacillus subtilis subsp. *niger*	31,600	2.72	23.05	33
B. coagulans	30,200	2.60	24.11	33
B. stearothermophilus	33,200	2.86	21.94	33

Figure 5 shows the survival curve in CFU for spores with a death rate $k = 2$ min^{-1} and the clump sizes of $C_0 = 1, 5$, or 10, i.e., 1, 5, or 10 spores per clump. One can best observe the effect of clumps by comparing the two curves where $N_0 = 10^7$ and $C_0 = 10$, and where $N_0 = 10^6$ and $C_0 = 1$. Both curves start out with the same number of CFU, T_0. In the absence of clumping ($C_0 = 1$), each spore that dies causes a reduction in T. This is not the case with clumped spores, where all 10 spores in a clump have to die before there is a reduction in T. Therefore, sufficient time has to elapse for 90% of the spores in the clump to die before any significant reduction in T will begin to occur.

These types of abnormal curves occur rather frequently in experimental studies. They can be prevented to a certain extent by taking proper precautions in growing uniform spore crops, by washing them well with sterile water plus surfactants like Tween 80 or buffer to remove medium residues and cell fragments, and by sonicating the spore suspension before treatment to break up clumps. These precautions will help, but may not totally eliminate the problem. It is primarily because of these problems that very few researchers use spore survivor curves and kinetics to design sterilization processes. For the most part, sterilization cycles and processes are designed based upon experience, endpoint sampling, and conditions that provide significant overkill.

25.3 PRACTICAL HEAT STERILIZATION

25.3.1 Destruction of Organisms by Dry Heat

The mechanism of the lethal effect of dry heat on microorganisms is not totally clear. The presence of oxygen in the gaseous phase during dry-heat sterilization generally improves the rate, indicating an oxidative reaction. It is also suggested that the lethal action results from heat that is conveyed from the objects with which the organisms are in contact (33). Quesnel et al. (39) have discussed the various factors involved in dry-heat sterilization. It is quite clear that this method requires significantly higher temperatures and longer exposure times than does wet-heat sterilization (compare Tables 4 and 5). For this reason, dry-heat sterilization is generally only used to sterilize materials which are relatively insensitive to higher temperatures.

The exact requirements for dry-heat sterilization are somewhat vague. Table 6 shows some recommended values, of which 180°C for 2 h would be considered the most conservative condition. If the objects to be sterilized are clean, then the shorter times and lower temperatures should be sufficient (see Table 6). If the objects are heavily contaminated with soils, baked medium solids, oils, and the like, then the 180°C–2-h condition may be required.

A dry-heat sterilizer is basically a simple oven. Common laboratory models are electrically heated and use a fan to recirculate air within the oven. These sterilizers are also equipped with sensitive temperature controllers and provide for stable and uniform temperature profiles. It is also possible to use a much

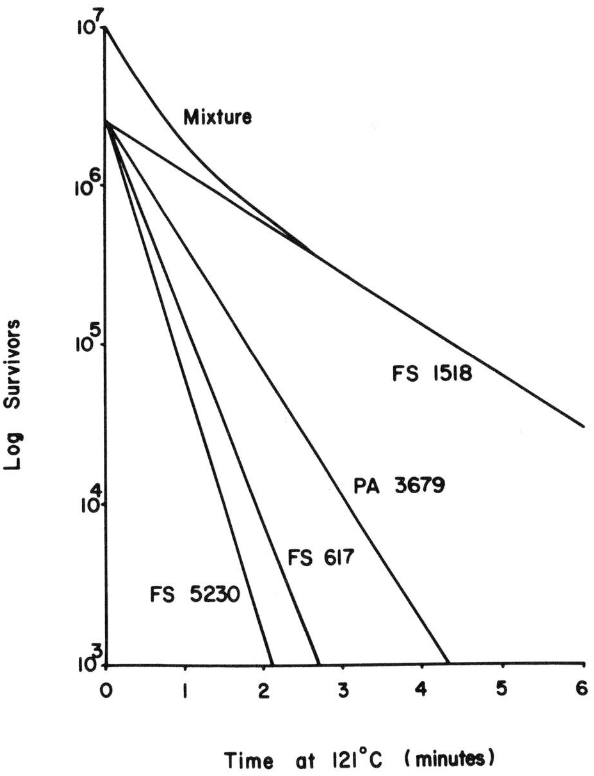

FIG. 4. Mathematically derived survivor curve for a mixture of different spore populations with different death rates. Note that the total survivor curve is no longer linear over the whole time period.

TABLE 6. Times and temperatures required for dry-heat sterilization

Source (reference)	Temp (°C)	Time (min)
Perkins (6)	140	180
Perkins (6)	150	150
Perkins (6)	160	120
Perkins (6)	170	60
British Pharmacopoeia	150	60
U.S. practice	180	120
Frobisher et al. (22)	165	120
Sykes (46)	140	240

TABLE 8. Survival of tetanus spores as a function of temperature and pH (22)

pH	Survival (min) at temp:			
	105°C	100°C	95°C	90°C
1.2	4	5	6	6
4.1	6	11	14	23
6.1	9	14	38	54
7.2	11	29	53	65
10.2	5	11	21	24

cheaper kitchen-type oven for dry-heat sterilization. Generally, electric ovens are preferred to gas because they hold a more constant temperature. The lack of air circulation can lead to substantial temperature variations between different locations in the oven. For more precise work, a specialized sterilization oven would be recommended, but a common kitchen stove will meet most laboratory needs.

Dry-heat sterilization is commonly used to sterilize surgical instruments and laboratory glassware, such as petri dishes and pipettes. It can also be used to sterilize mineral oils, animal and vegetable oils, certain powders, and any other dry materials which are not destroyed or degraded at 180°C. Dry heat cannot be used with liquids that boil unless they are heated under high pressure.

Dry heat is not recommended for paper, cotton, plastics, wood, or other materials which may char or produce toxic residues. It is also not recommended for glass syringes with metal tips because the differences in the coefficients of expansion of the glass and metal can cause breakage.

The proper use of dry-heat sterilization requires that the objects to be sterilized be heated to the proper temperature before the cycle time begins. Hot air has a relatively low heat transfer coefficient, so it can frequently take 1 or 2 h to heat objects to the proper temperature. It is good practice to measure the temperature of objects in the oven with thermocouples to ensure proper sterilization. A more common laboratory practice is to add 2 h to the cycle when sterilizing instruments and glassware, to account for heat-up times.

25.3.2 Destruction of Organisms by Wet Heat

The most common method of destroying microorganisms is by the use of a combination of water and heat, or the wet-heat process. The presence of water substantially reduces the temperature and the exposure time required to achieve a given result (42). It is commonly believed that the lethal effect of moist heat is due to coagulation of cellular protein (40). The effect of moisture on the coagulation temperature of egg albumin protein is shown in Table 7. Clearly, water substantially reduces the temperature required for coagulation.

In general, wet heat is more effective than dry heat in killing organisms for the following two reasons. First, water is a highly reactive chemical species. Many biological polymers (proteins, RNA, DNA, etc.) are produced by reactions which eliminate water between two monomers. The reverse reaction would be damaging to the cell. Many of the constituents of the cell have complicated secondary and tertiary structures which are held in place by hydrogen bonding, which can also be broken by water at high temperatures. In addition, water can react directly with complex biochemicals at higher temperatures and destroy necessary cell components or produce toxic side products. It should also be recognized that the acid and hydroxyl ions that are naturally present in water are more reactive than water. Table 8 shows the effect of pH on the survival of tetanus spores. These data indicate that changes in the concentration of H^+ or OH^- can substantially affect the wet-heat process.

The second way in which water augments the killing process is through an increase in heat transfer rates. Typical heat transfer coefficients for steam, water, and air are roughly 1,000, 100, and 10 BTU/h-ft^2 per degree Fahrenheit. In addition, wet materials conduct heat much more rapidly than do dry materials. L. L. Kemp (University of Michigan) once described these phenomena as follows: a person can quickly pass a finger through an 800°F candle flame without experiencing heat or pain; however, if he passes a finger through a stream of steam from a tea kettle at 212°F, it will hurt.

The overall result that occurs in wet-heat processes is a more highly reactive environment and a more rapid and more penetrating transfer of heat. The

TABLE 7. Effect of water on egg albumen protein coagulation (22)

% Water content	Coagulation temp (°C)
50	56
25	76
15	96
5	149
0	165

TABLE 9. Typical wet-heat sterilization conditions

Temp (°C)	Time (min)	Pressure (lb/in^2 gauge)
121	15	15
126	10	20
134	3	30
140	0.67	38

combination of these leads to the wet-heat sterilization conditions shown in Table 9. These conditions are generally accepted as sufficient to kill all organisms, including heat-resistant bacterial spores, under normal operating conditions. It should be noted that the above conditions require that the materials being sterilized be at the prescribed temperature for the indicated time period. The most common device used to sterilize biological materials is the steam autoclave. The detailed operation of an autoclave has been reviewed in numerous papers and texts (5, 22) and will not be covered in detail here. It is important to point out that a sterilizer time starts its cycle when the chamber temperature or pressure reaches the prescribed set point (usually 121°C). The contents of the autoclave may require a substantially longer time to reach temperature. Table 10 shows the typical autoclave times required to achieve sterilization at 121°C saturated steam for different kinds of materials (34).

The sterilization conditions mentioned above are only required when all microorganisms, including heat-resistant bacterial spores, have to be destroyed. Normal vegetative cells can generally be killed by a much milder heat treatment. In many cases, vegetative cells are destroyed by exposure to 50°C for several minutes. Many organisms produce resting stages, cysts, or spores which are slightly more resistant than the vegetative cells, but these can generally be destroyed by heat treatment at 65°C. Only the heat-resistant spores of the family *Bacillaceae* (*Bacillus* and *Clostridium*) can survive above 100°C for any period of time. These observations have led to the development of a variety of processes for selectively killing only those organisms of concern. The most common of these, pasteurization, involves heating a material to 63°C for 30 min or, for high-temperature, short-time pasteurization, to 80°C for 15 s. Pasteurization destroys all commonly encountered pathogenic organisms, but does not achieve sterilization. Products of pasteurization, such as milk, must be refrigerated and have only limited shelf life. Water can commonly be disinfected by boiling for 5 min, which is sufficient to destroy most pathogenic organisms. Another process that is now generally considered outdated is tyndallization. This process can be used to achieve sterility, but it is not as reliable as steam sterilization. Tyndallization involves heating a material to 100°C (boiling point of water) for a few minutes on three or four separate occasions that are 24 h apart. The theory behind tyndallization is that all vegetative cells are destroyed in the initial heating. During the following 24-h period, heat-resistant spores can germinate and

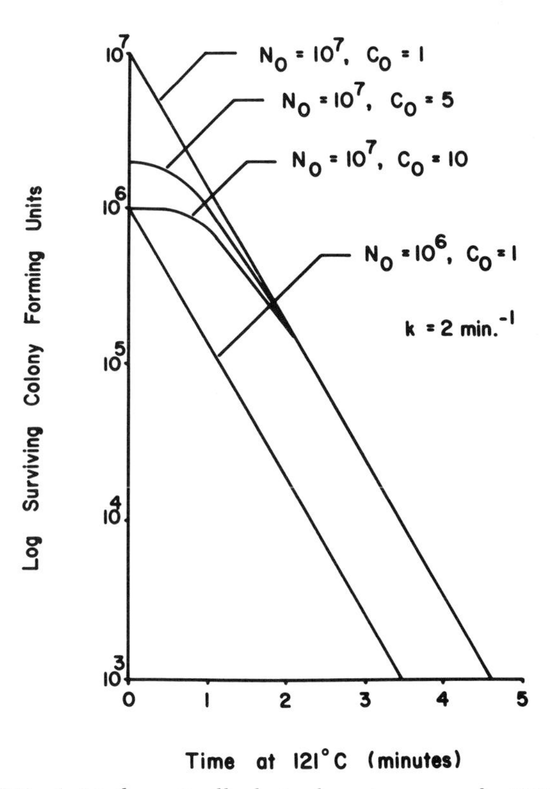

FIG. 5. Mathematically derived survivor curve for CFU as a function of clump size and time. Note that clumping of spores produces an apparent lag in the reduction of the spore population.

TABLE 10. Actual cycle time (121°C, saturated steam) required to sterilize materials in an autoclave (34)

Material	Time (min)
Erlenmeyer flasks	
2,000 ml	30–35
1,000 ml	20–25
500 ml	17–22
200 ml	12–15
125 ml	12–14
"Square-Pak" Pyrex Flask, 1,000 ml	30–35
Serum bottle, 9,000 ml	50–55
Milk dilution bottle, 100 ml	13–17
Test tubes	
18 by 150 mm	12–14
32 by 200 mm	13–17
38 by 200 mm	15–20
Syringes, unassembled, in muslin or paper covers	30
Empty glassware	15
Rubber gloves in muslin wrappers	20
Utensils in muslin covers	15
Instruments in trays, with muslin covers	15
Instruments, wrapped for storage	30

TABLE 11. Probability of detecting contamination in a fermentor[a]

Time (h)	Spore level (no. of spores per 10^8 ml)	Probability of detecting the contaminant in a 1-ml sample
0	1	1×10^{-8}
5	32	3.2×10^{-7}
10	1.024×10^3	1.24×10^{-5}
15	3.277×10^4	3.28×10^{-4}
20	1.049×10^6	1.04×10^{-2}
25	3.355×10^7	0.285
30	1.074×10^9	0.999978

[a] Basis: One spore in 100,000 liters of fermentation broth. Spore germinates and grows with a doubling time of 1 h.

become vegetative. These are then killed in the subsequent heat treatments.

25.3.3 Steam Sterilization Problems

It is not uncommon to run a process through a sterilization cycle in either an autoclave, a fermentor, or a specially designed sterilizer and to find out later that sterility was not achieved. In some cases, much attention should be paid to being certain that sterilizing conditions are achieved at all points within the process chamber. This will be discussed in more detail in the next section for fermentor design, but should also be considered here for the general use of sterilizers.

In a rather novel piece of work, Jacobs et al. (26) showed that medium containing solid materials or particulate matter does not heat up at the same rate as liquid medium containing no solid matter. This is especially the case when using continuous high-temperature short-time processes, where temperatures of 140°C for 2 to 4 min are commonly used (36). Spores that are trapped within the solids may not be exposed to the proper temperature cycles or moisture content to achieve sterility. A similar problem occurs if the equipment being sterilized is dirty or contains scoured medium components from previous runs. It has been shown (12, 27, 29, 32) that oils or grease can also protect spores during a sterilization cycle. Bader et al. (12) showed that the history of the spores was important in that dried spores placed in dry oil were

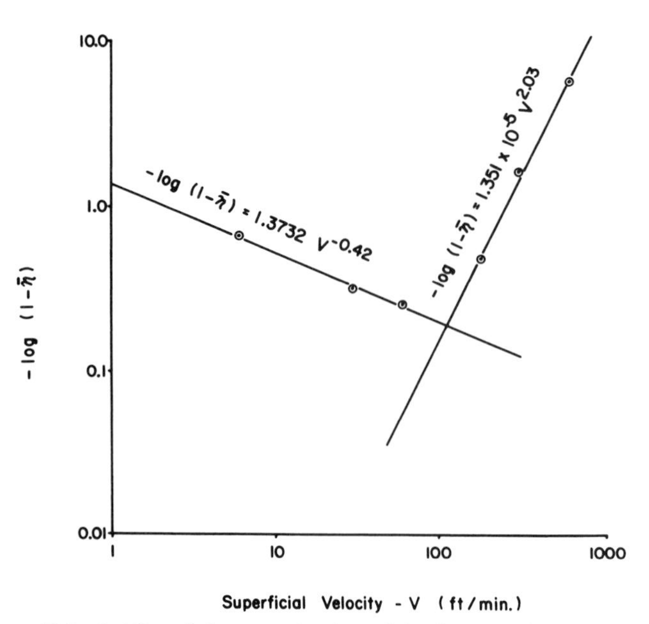

FIG. 7. Plot of the negative log of the fractional penetration of a filter versus the log of the superficial air velocity for the data of Gaden and Humphrey (23). Note that the velocity exponent in equation 24 changes as the velocity increases.

much more resistant than wet spores that were dried in oil.

A common sterilization failure in the laboratory occurs when media containing a high percentage of solids are autoclaved. In the absence of agitation, the solids form a layer on the bottom of a flask or laboratory fermentor jar. The heat transfer rate of these solids can be extremely slow. This problem becomes more severe in the presence of materials like starch that make the liquid highly viscous, which further inhibits mixing. Such problems can be partially alleviated by adding ingredients in the proper order (oils and antifoams should be added last), by mixing them with a high-shear agitator before autoclaving, or by heating the medium to 100°C with agitation immediately before loading into the autoclave. In some cases, lowering the pH to 4.0 will also help (see Table 8). However, there are some media which can only be sterilized in an agitated sterilizer.

In any sterilization process, the lower the initial bioburden, the more effective the sterilization cycle. This is clearly evident from the kinetics section discussed earlier. It is good practice to clean all glassware and equipment immediately after it has been

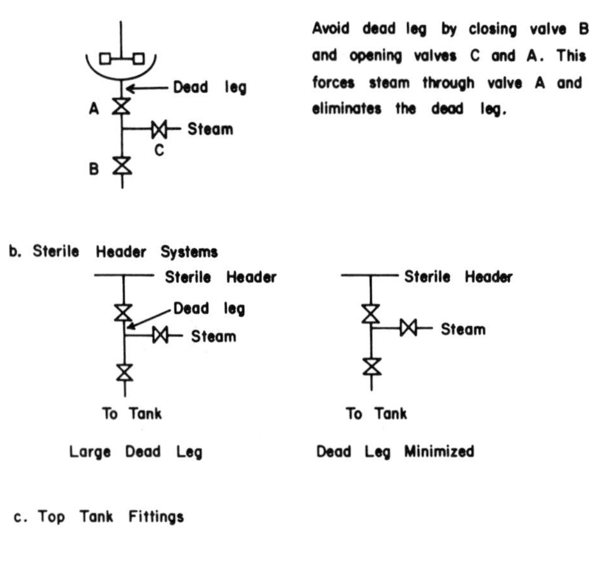

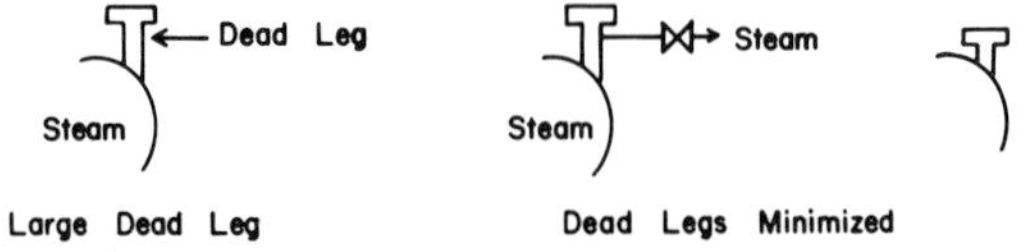

FIG. 6. Examples of how to avoid dead legs in fermentor piping. In (a) the dead leg is avoided by closing valve B and opening valves A and C. This forces steam through the dead leg.

TABLE 12. Efficiency parameters for glass fibers (17)[a]

Fiber diameter (μm)	Parameter			
	a	b	c	d
1.3	4.6	0.8	1.0	−0.2
2.5				−0.25
15	0.085	0.9	1.1	−0.4
30	0.054	0.9	0.9	−0.4
115				−0.5

[a] Equation: $-\log(1 - \bar{\eta}) = aL^b\rho_b{}^cV^d$. Note: V was varied over the range of 3 to 50 ft/min.

used or to keep it soaking in water to prevent old medium from drying onto surfaces. Strong detergents are helpful in cleaning but, in some cases, leave residues that can inhibit the growth of cultures. Ultrasonic cleaning baths are becoming more common to avoid this problem. Good housekeeping can be a great help both in lowering the bioburden and in eliminating residues that can be difficult to sterilize.

A final area of concern is that all surfaces be exposed to saturated steam during a sterilization process. Superheated steam does not transfer heat as well as saturated steam, and it does not saturate surfaces with water to the same degree. As a result, superheated steam performs more like hot air and can approach dry-heat sterilization conditions. Mixtures of air and steam are similar to superheated steam. It is also important to recognize that most autoclaves are controlled by pressure. If an autoclave is set to operate at 15 psig and contains a mixture of one-third steam and two-thirds air, the temperature will only be 108°C; if it contains two-thirds steam and one-third air, the temperature will be 115°C. It is, therefore, necessary to bleed all air from the autoclave before sterilization can begin. This is generally done automatically in a modern autoclave but can become a problem if the steam trap is not functioning properly. It is also important that sealed bottles not be autoclaved, as they may not sterilize properly.

TABLE 13. Effect of superficial velocity on packed filter length, pressure drop, and filter size[a]

V (ft/min)	L	ΔP (lb/in²)	Area (ft²)	Diameter (ft)
6	7.57	0.056	250	15.8
30	16.04	0.595	125	11.2
60	22.17	2.789	25	5.0
180	9.97	2.219	8.34	2.9
300	3.15	1.168	5	2.2
600	0.66	0.489	2.5	1.6

[a] Conditions: $\mu = 0.02\ cp = 1.344 \times 10^{-5}$ lb/s-ft; $(1 - t) = 0.040$; $m = 1.55$; $d_f = 8\ \mu m = 2.6247 \times 10^{-5}$ ft; $\bar{\eta} = 0.9999$. Flow rate = 3,000 SCFM; pressure at filter = 15 psig.

25.4 STERILE OPERATIONS

25.4.1 Testing for Sterility

In the beginning of this chapter, the statement was made that it is never possible to prove that something is sterile. All sterility testing conditions tend to be selective for a certain group of microorganisms, so that a lack of growth only indicates that there are no organisms present that normally grow under these conditions. Because of the selective forces that are naturally applied by the choice of test conditions, it is important to choose conditions that will at least detect those organisms which commonly cause a problem to the process. This may require some development of methodology for each type of process. Beloian (16) provides a good review of many special testing methods, and the United States Pharmacopeia (47) details many standard procedures.

The bulk of my experience with sterility testing has been in the antibiotic fermentation industry, where samples of cultures and fermentation broths are frequently tested for either sterility or asepsis. Common microbiological media used for this purpose are tryptic soy broth (Difco), which detects most bacteria; phenol red broth (Difco), which detects most bacteria and changes color; thioglycolate broth (Difco), which detects anaerobes and other bacteria; tryptic soy agar (Difco), which detects bacteria and fungi; and potato dextrose agar (Difco), which is especially good for fungi.

The incubation temperature for samples will vary. The most common temperature is 35°C, which provides for rapid growth of contaminants. There is a danger in using only 35°C as it may be too high a temperature for some organisms. In many cases, a second set of samples is incubated at 25°C to detect contaminants which have a lower temperature optimum. To eliminate running two samples at different temperatures, some people will choose an intermediate temperature. Since most antibiotic fermentations are carried out between 20 and 37°C, it is not necessary to test for psychrophiles or thermophiles, since they would not survive well under normal process conditions. Testing of fermentation broths for contamination provides a somewhat different problem than testing for true sterility. During the fermentation, the broth contains large concentrations of the fermenting organism. Direct microscopic examination of the broth requires that one be able to distinguish between a contaminant, medium ingredients, and the production culture. This takes considerable experience and is frequently inconclusive. One can also only detect contaminant populations greater than 10^4 organisms per ml under the best conditions with a microscope. Transferring samples of broth to tubes of test medium also suffers from problems of competition between contaminants and culture and from products of the culture that may inhibit the growth of the contaminant. The preferred method of testing fermentation broth for contaminants is by plating a sample on nutrient agar. This allows contaminants to become visible as separate colonies and may also minimize the effect of inhibitory products.

Although it is frequently difficult to detect contaminants reliably in antibiotic fermentations, there are certain aspects of the antibiotic-producing cultures that are helpful. Most antibiotic producers are filamentous streptomycetes or fungi, which can easily be distinguished from contaminating bacteria. The production cultures are also not fast growers as compared with many bacteria. Streptomycetes and antibiotic-producing fungi also produce compact colonies on agar plates, which readily distinguishes them from wild fungal and bacterial colonies.

The type of contaminant in a fermentation may also indicate the source of the contaminant. Mold contaminants generally come from the transfer of cultures, from feed systems, and from air filter failures. Gram-positive sporeformers generally indicate an incomplete sterilization, which may be caused by a failure in the time-temperature sterilization cycle, by the presence of large particles in the medium, or by a pocket of dried material in a crack or crevice in the fermentation vessel. Gram-negative bacteria often indicate a cooling water leak (C. A. Perkowski, G. R. Daronsky, and J. Williams, 184th Natl. ACS Meet.,

Kansas City, Mo., 1982). Contamination by cocci often comes from a technician's hands or clothing during a transfer of the culture. The presence of multiple contaminants indicates a major failure in the sterile system and can frequently be attributed to air filter failures. These indications can frequently be used to start the investigation in the right direction, but should never be considered as absolutes.

Testing for sterility is a statistical problem, as discussed by Pflug (37). Assume that one sterilizes 10,000 tubes, and at the end of the sterilization cycle, 0.1% of the tubes are nonsterile. If 10 tubes were tested for sterility, there would only be a probability of 0.00995 of finding 1 bad tube in the 10 tested. If 1,000 tubes were tested, the probability of finding a bad tube would be 0.632305. Pflug also points out that roughly 1 in 1,000 tubes would become contaminated as a result of the sterility test procedure, even under careful conditions. The probability of detecting a contaminant is given roughly by:

$$P = 1 - e^{-NF} \tag{17}$$

where N is the number of units tested and F is the number of contaminants per unit. As discussed later, if the contamination rate is 0.1%, then F equals 0.001 contaminant per unit. If one tests 1,000 units, then the probability of detecting the contaminant is 0.632 or 63%.

In the fermentation industry, the concern is to detect contaminants in fermentors. Table 11 shows a hypothetical case where one spore survives sterilization in a 100,000-liter fermentor. If the spore germinates and begins to grow with a doubling time of 1 h and the fermentor is sampled every 5 h by transferring 1 ml of broth to a test plate, then it is not likely that the fermentor contamination would show up until the 25-h plate. Since the spore germinates and grows, it will eventually be detected. A more frustrating problem occurs when one tries to detect contaminants in feed streams (like sugars, oils, or air) where the contaminant may be dormant. Low levels of contaminants may exist in these streams and are virtually impossible to detect. It is for this reason that the sterility of a product should be guaranteed by the process itself. Pflug (37) states this rather well: "It is only through careful design and diligent execution of the sterilization process that a sterility probability level of the order of 1 in 10^{-6} can be achieved. The sterility level must be built into the product; it cannot be tested into the product given reasonable economic considerations."

25.4.2 Equipment Design for Sterile Operations

This section will primarily focus upon equipment design rules for fermentors, although the same principles apply to any type of equipment for sterile operations. It is necessary to recognize that in any design problem, one faces trade-offs between the world of ideality and reality. For example, in the design of a fermentor, it is best to have all penetrations into the tank above the liquid level, as any penetration below the liquid level is more prone to contamination. This becomes quite impractical when designing a large production fermentor. It is also a necessary part of any design to keep in mind the people who will be operating and maintaining it. The more difficult the operation and maintenance, the more likely errors become. With this in mind, the following design guidelines should be taken as the ideal case, and not as absolute laws.

1. All surfaces within the system to be sterilized should be smooth and rounded. Crevices, cracks, rough welds, threads, bolts and nuts, square corners, and the like can all provide places for media to concentrate and create an environment in which contaminants can escape sterilization.
2. Equipment should be designed to be easily cleanable. Automatic cleaning systems are preferred, but are not often practical.
3. Equipment should be designed for easy disassembly and cleaning of any parts that pose a potential contamination problem. An example of this is the commonly used InGold fitting for below-liquid-level probes. This fitting contains an O ring that can get medium behind it and create a potential pocket for contamination. However, the fitting is designed with a threaded cap that can be easily removed by hand for cleaning between runs, making it a good design.
4. It is safest to make penetrations into a fermentor through the top, above the liquid level.
5. It is best to use as much machine-welded pipe as possible to avoid pipe threads and potentially sloppy hand welds.
6. Avoid the presence of dead legs in the system. A dead leg is an area that does not have steam flowing through it and is not well mixed. See Fig. 6 for examples.
7. Design equipment with simplicity in mind. Minimize the number of valves and fittings required.
8. Always maintain equipment under positive pressure with sterile air. Leakage should always be out and never in.
9. Where possible, keep connections to the vessel under a steam seal. A typical steam seal is shown in Fig. 6a.
10. Use valves that are simple in design with a minimum of crevices. All valves should have seals that can withstand steam service. The preferred valves are ball valves, plug valves, and diaphragm valves. Diaphragm valves were once preferred but are now being used less frequently due to high maintenance cost and their tendency for catastrophic failure.
11. Stainless steel, type 304 or 316, is the preferred material of construction for sterile equipment. It is cleanable, and it does not readily corrode. Rust in iron or carbon steel equipment provides a good environment for harboring contaminants.
12. It is best to sterilize equipment empty rather than full. The equipment can then be sterilized for longer times and at higher temperatures and pressures to ensure that all cracks and crevices have had sufficient steam penetration. Many fermentation plants have separate sterilizers for media; the sterilized medium is then transferred to a sterile fermentor.
13. It is preferred to sterilize fermentors in place and not in an autoclave. Heavy industrial-type media, which frequently contain 10 to 20% solids, cannot be sterilized in an autoclave.
14. Bacteria can grow through valves, through filters, and up a vent line. At no time should there be a

direct connection between sterile and nonsterile sections of the system.

It is generally not useful to try to design and build small laboratory fermentors. Numerous manufacturers of laboratory fermentors offer a wide line of equipment which is quite acceptable for normal use. All manufacturers have their own style of equipment with specialized fittings and valves. The choice can generally be made on the basis of serviceability, availability of spare parts, and particular process needs. It is also wise to talk to someone who has used a vendor's equipment before purchasing it. Most researchers are willing to discuss their experience with different types of laboratory equipment.

Pilot plant equipment may be purchased from vendors or be designed in house. In general, large fermentation companies design and build their own pilot fermentors based upon their own experience. Such companies also have access to the piping and welding skills that are necessary to build fermentor systems. Smaller and newer companies generally buy pilot fermentors from vendors.

Production-scale equipment is generally either designed in house or with the aid of a qualified consulting firm to meet a company's specifications. Most production equipment is too large and specialized to be provided by a vendor. A typical antibiotic production fermentor will have a 100,000-liter operating capacity.

It has been mentioned briefly that many companies use continuous sterilizers for sterilizing media or products. Continuous sterilization is a touchy process and requires a great deal of experience to handle. Most companies purchase continuous sterilizers from qualified vendors who have the experience to design effective systems (11, 36).

Additional information on fermentor system design can be found in the literature (1, 15, 43, 45).

25.5 STERILIZATION OF AIR

25.5.1 General Methods of Air Sterilization

Over the years, many different methods have been used to sterilize air for numerous different applications. The first method that was used was gravity settling in Pasteur's original experiments. This method is applicable when extremely low airflow rates are used. Gravity settling is the principle behind the use of the test tube caps that are still used today. Other methods that have been used with limited success are incineration and heat of compression (dry-heat sterilization), electrostatic precipitation (used in home furnaces), steam jets, UV lights (used in sterile rooms), and various methods of filtration. In the production of large quantities of sterile air, only heat of compression and filtration have been used with any degree of reliability.

Air that is used for fermentation must first pass through a compressor, where its pressure is increased to a level that is high enough to provide the motive force to push the air through pipe lines and into the fermentor. Typically, production fermentor air is compressed to a pressure between 35 and 60 psig, depending upon the type and size of fermentor used and the operating back pressure on the vessel. The temperature increase of the air due to compression can be approximated by (35):

$$T_2 = T_1(P_2/P_1)^{0.286} \quad (18)$$

where T_2 = exit temperature of air (degrees Kelvin); T_1 = inlet temperature of air (degrees Kelvin); P_2 = outlet pressure (psia); P_1 = inlet pressure (psia).

From equation 18, one can readily calculate that if air enters at 14.7 psia and 20°C (293 K), the exit temperature will be 415 K (142°C) if the air is compressed to 35 psig or 435 K (161°C) if compressed to 60 psig. Frictional heat from the compressor can also increase this exit temperature. Using the heat of compression, Decker et al. determined that bacterial spores in air could be killed by exposure to 218°C for 24 s (20). This type of process was used successfully by Bartholomew et al. (14) as a method to produce phage-free air for glutamic acid fermentation. It should be pointed out that bacterial spore contamination was not generally a problem with the glutamic acid fermentation.

Heat of compression operates as a method to reduce the bioburden in the air whenever compressors are used. As a result, the number of organisms per cubic foot of air is lower after compression than it is at the air intakes. However, heat of compression has rarely been used as a method to sterilize air for fermentors for a number of reasons: (i) a large holding chamber is required to provide sufficient residence time at the higher temperature; (ii) as indicated above, normal compression ratios are not high enough to provide sufficient temperature rise; (iii) most compressors used today are multistage centrifugal compressors which use intercoolers to reduce temperatures and horsepower requirements; and (iv) it is difficult to maintain sterile conditions throughout the air header downstream from the compressor.

25.5.2 Filter Sterilization of Air by Depth Filters

The primary and only generally acceptable method to sterilize large quantities of air is by filtration. Over the years, many different types of filter materials have been tried, including porous ceramic filters, cotton fibers, steel wool, granular carbon, glass fibers, and a number of specially designed filter media or membranes. Of these, porous ceramic filters have generally only been useful for small fermentors due to size limitations. Granular carbon and steel wool are generally ineffective. Cotton fibers produce high-pressure drops and tend to support bacterial growth. At the present, only glass fibers and special membrane-type filters are commonly used.

In the process of removing particles from air with a fiber filter, the five principal mechanisms involved are diffusion, interception, impaction, gravitation, and electrostatic collection (13). It has been shown that electrostatic forces increase the efficiency of collection of organisms (30, 44) and that bacterial spores do tend to have positive or negative charges (25). It has been difficult to quantify the effects of electrostatic forces, so this mechanism will not be discussed further. Yao et al. (49) show the following mathematical expressions for single-fiber collection efficiencies for the other four mechanisms:

Diffusion/Brownian motion: $E_B = 0.9 (k_B T/\pi d_f d_p V)^{2/3}$ (19)

Interception: $E_I = 1.5 (d_p/d_f)^2$ (20)

Impaction: $E_{IM} = 0.075 (\rho_p V d_p^2/g\mu_a d_f)^{1/2}$ (21)

Gravitation: $E_G = \Delta_\rho d_p^2/18\mu_a V$ (22)

where k_B = Boltzmann constant; V = upstream fluid velocity; ρ_P = particle density; d_f = diameter of fiber; d_p = diameter of particle; T = temperature; μ_a = viscosity of air; Δ_ρ = density difference between particle and air; g = gravitational constant; and E = collection efficiencies.

From equations 19 through 22, the following generalizations can be made: (i) at low air velocity, gravitation and diffusion may be significant mechanisms; (ii) at high air velocity, impaction will be a significant mechanism; (iii) larger particles (large d_p) should be easier to filter than smaller particles; and (iv) the use of small-diameter filter fibers (d_f) will increase filter efficiency.

From these generalizations, one might expect that the filter efficiency might be relatively high at low airflow rates due to diffusion and perhaps gravitation. As the airflow rate increases, the filter efficiency would decrease to a point where interception predominates. As the airflow rate is increased to higher levels, the filter efficiency might increase due to impaction. Gaden and Humphrey (23) showed that this is generally the case.

To design a filter for fermentor air, one must first determine the required efficiency of the filter, which will be designated here as $\bar{\eta}$. Typical ambient air contains roughly 50 to 400 organisms per ft^3, with a typical design value of 100 organisms per ft^3. This number can generally be considered conservative as compressor heat will substantially reduce this number before the air gets to a filter. The acceptable number of organisms that may pass through the filter is more prone to debate. Some engineers design filters based upon one organism per batch as being acceptable. Others prefer to design based upon one organism per 10^6 batches, or a 10^{-6} probability (31). In practice, one frequently compromises somewhere between these two limits. Once a penetration number has been decided upon, then the efficiency of the filter can be determined by equation 23:

$\bar{\eta} = 1 -$ (acceptable penetration/total challenge) (23)

Consider, for example, that an acceptable penetration for the filter is 0.1 organisms per year. If the flow rate through the filter is 100 SCFM, containing 100 organisms per ft^3, then the total challenge is given by: 100 organisms per $ft^3 \times 100\ ft^3/min \times 60\ min/h \times 24\ h/day \times 365$ days/year $= 5.256 \times 10^9$ organisms per year. The efficiency required would then be: $\bar{\eta} = 1 - [0.1/(5.256 \times 10^9)] = 0.999999999981$. This type of number is difficult to deal with as it requires more significant figures than are available on most calculators. For that reason, it is generally easier to work with the term $(1 - \bar{\eta})$, which for the previous example would be: $1 - \bar{\eta} = 1.9 \times 10^{-11}$.

The details for the design of a depth filter have been described in detail by Aiba et al. (1). A simplified design method will be used here. Blazewitz and Judson (17) found that the efficiency of a glass fiber filter could be described by the following expression:

$$1 - \bar{\eta} = 10^{-aL^b P_b^c V^d} \quad (24)$$

where a, b, c, and d are empirical constants, L is the depth of the filter (in.), P_b is the packing density of the filter (lb/ft^3), and V is the superficial air velocity through the filter (ft/min). The values for the empirical constants are shown in Table 12 and are dependent upon the fiber diameter as shown. In general, $b = 0.9$, $c = 1.0$, and the coefficient of velocity (d) is negative. This indicates that increased velocity decreases the filter efficiency over the range of velocities tested (3 to 50 ft/min). This is what would be expected from equations 19 through 22 if the velocity were not high enough for impaction to become significant.

Gaden and Humphrey (23) tested IMF filters over an air velocity range of 6 to 600 ft/min. Since they worked with a single type of filter, their data shed no light on the effect of packing density or fiber diameter. However, their data can be correlated in terms of equation 24 if one assumes the length of the filter exponent (b) remains 0.9 and one ignores the packing density term. The effect of air velocity then is shown in Fig. 7. At air velocities in the range of 6 to 60 ft/min, equation 24 becomes:

$$-\log(1 - \bar{\eta}) = 1.373 L^{0.9} V^{-0.42} \quad (25)$$

and in the range of 120 to 600 ft/min, the correlation becomes:

$$-\log(1 - \bar{\eta}) = 1.351 \times 10^{-5} L^{0.9} V^{2.03} \quad (26)$$

For practical purposes, one could assume that the collection efficiency is relatively independent of velocity in the range of 60 to 120 ft/min.

These data correlate well with what one would expect from the mechanisms given in equations 19 through 22. Apparently, at superficial air velocities above 120 ft/min, the impaction term starts to dominate as the mechanism of removal. It is worth noting that the velocity exponent for low velocities is roughly identical with that of Blazewitz and Judson. However, the Gaden-Humphrey study indicates that it is dangerous to extrapolate data beyond the actual range tested. This fact is further indicated by the data of Esumi and Ashida (21) on polyvinyl alcohol filters, which shows that at high air velocities, reentrainment of spores occurs. It is assumed that at high air velocities, the air supplies sufficient force on spores to strip them away from the fiber medium.

The pressure drop across a fiber filter can be correlated to the Reynolds number (N_{RE}) via a modified drag coefficient, C_{DM} (28). Aiba (9) showed that the relationship between the drag coefficient and Reynolds number for glass fiber filters could be expressed by the following equation:

$$C_{DM} = 48.50\, N_{RE}^{-1.0} \quad (27)$$

where:

$$N_{RE} = d_f V \rho/\mu \quad (28)$$

$$C_{DM} = 864\pi g_c d_f \Delta P/\rho L V^2 (1 - \varepsilon)^m \quad (29)$$

and where d_f = fiber diameter (ft); V = velocity (ft/s); ρ = density of air (lb/ft^3); μ = viscosity (lb/s-ft); g_c =

32.17 lbm ft/lb$_f$ s^2; ΔP = pressure drop (lb/in^2); L = filter length (in.); ε = void fraction of filter; and m = 1.35 for 19-μm fibers or 1.55 for 8-μm fibers.

By substituting equations 28 and 29 into equation 27 and rearranging, one can solve for the pressure drop across a filter:

$$\Delta P = 9.257 \times 10^{-6}\ V(1 - \varepsilon)^m \mu L/d_f^2 \qquad (30)$$

where V is in ft/min.

From equations 25, 26, and 30, one can then calculate the length required to obtain a specified filter efficiency and the pressure drop across the filter. The results of a set of sample calculations are presented in Table 13. Increasing the velocity through the filter initially produces an increase in the length and the pressure drop. Above a velocity of 180 ft/min, the length and the pressure drop begin to decline. The size of the filter decreases as the velocity increases, as indicated. The final design of a filter depends upon a balance between the cost of the initial filter housing, the cost of the filter packing, and the energy cost of the pressure drop across the filter. These items need to be optimized according to local costs and requirements.

Packed depth filters are sterilized with flowing steam at 15 to 30 psig for 30 min. Air is then passed through the filter immediately after sterilization to dry the filter bed. The drying air is generally bled to the atmosphere. During operation, the inlet air to the filter must be below saturation to prevent the fibers from becoming wet. Depth filters will not properly sterilize air when wet. During operation, the collection of dust and dirt on the filter will generally increase the filter efficiency, but will also increase the pressure drop. Since bacteria penetrate into the depth of the filter, there will eventually be a point where the bacteria may break through the filter. To avoid excessive pressure drop and break-through, depth filters are generally repacked every 6 months.

25.5.3 Sterilization of Air by Filter Cartridges

In recent years, there has been a rather strong trend to replace depth filters with specially designed air filtration cartridges. There are a number of reasons for this change, some of which are presented here. (i) Depth filters tend to be inconsistent, unpredictable, and difficult to validate. (ii) Depth filters cannot handle wet air. (iii) The pressure drop across a depth filter can be high. (iv) Depth filters are large, require a lot of plant space, and are labor intensive to repack. (v) The state of the art of filter cartridge design has improved dramatically in recent years. (vi) Filter cartridges have been reported to significantly reduce contamination rates (C. F. Bruno and L. A. Szabo, Div. Microbiol. Biochem. Technol. Symp., ACS Annu. Meet., New York, N.Y., 1981).

The term filter cartridge can be applied to any of a wide range of manufactured filters which fit into specially designed filter housings. Such filters have existed for a long time, but were generally not suitable for the sterilization of air. During the past decade, a number of manufacturers have developed specialized technology which has enabled them to produce a filter medium or membrane that is capable of removing micrometer- and sub-micrometer-sized particles with a high degree of efficiency. Some of these filter media have been capable of withstanding steam sterilization conditions with a relatively low pressure drop at high airflow rates. The combination of these properties and the development of filter support and packaging systems has led to the production of a number of cartridges that are suitable for the sterilization of air.

All filter cartridges have generally the same type of overall design; in fact, most manufacturers design their cartridges to be interchangeable with their competitors' cartridges. The basic construction of the cartridge consists of an inner cylindrical core which supports a sheet of filter material which is contained within an outer protective cage. The top of the filter has an end cap which also frequently supports a locating fin. The bottom of the filter generally has a end cap with a tube with one or more O rings that slip into the filter housing and seal the cartridge. Air always moves from the outside of the cartridge cylinder towards the center and exits through the bottom tube connector.

The primary difference between types of filter cartridges lies in the type of filter medium or membrane used.

(i) Absolute membranes generally consist of a solid sheet of a polymer through which small holes of defined size are cut, generally by a process of nuclear bombardment. In such filters, all of the pores in the membrane are of the same relative size and are generally small enough to prevent bacterial penetration. Absolute membranes are not commonly used for formentor air sterilization because of cost, high pressure drop, and rapid plugging. Absolute membrane filters can be used to sterilize both gases and liquids.

(ii) Statistical membranes consist of small-diameter fibers which are bonded into a thin mat. The pore size in such membranes statistically varies over a wide range. However, by control of the manufacturing process, manufacturers are able to develop membranes which statistically are capable of retaining all particles above a certain size, even though surface pores may exist which are larger than the particles retained. It is likely that statistical membranes operate on a tortuous-path principle and can be thought of as something in between an absolute membrane and a depth filter. These filters can be used for sterilizing gases and liquids.

(iii) Depth medium filters use a thicker sheet of a specially designed micro fiber material that is compressed to an optimum pack density for removal of microorganisms. The depth medium does not have a controlled pore size but uses a filter medium thickness that is roughly 30 times that of statistical membrane filters. The use of micro fibers and the controlled packing density provides a high-efficiency filter, as would be predicted by the data of Blazewitz and Judson (17) (see Table 12). The depth medium filters offer a low pressure drop and high dirt-loading capacity, but are not capable of handling moisture.

The basic principles of operation of the filter cartridge are identical with those of depth filters, with the additional mechanism of sieving for the absolute and statistical membrane filters. Sieving would not apply significantly to depth medium cartridges.

There is very little involved in the design of systems that use cartridge filters. Both the cartridges and the filter housings are generally purchased from the manufacturer. The only design involved is the selection of the number of filter cartridges required for a given application, allowing for an appropriate pressure drop, and the piping required for the sterilization and drying of the filter. The most commonly used statistical membrane filters for air sterilization are the 0.45-μm and 0.2-μm type filters. Typical economical air flow rates for a common 10-in. cartridge are between 50 and 100 SCFM of air at 30 psig. Absolute and statistical membrane filters can only be steamed in the forward direction and cannot generally withstand a reverse pressure drop. Depth media filters can be sterilized in either direction within limitations and can pass air in either direction. In any filter application, the use of prefilters tends to extend the life and reliability of the filters. For the actual design of a cartridge filter application, it is best to contact a supplier or manufacturer for the appropriate literature and guidance.

One of the advantages of cartridge filters is that they can be validated to ensure that they are working effectively. A difficulty that occurs is that the validation procedures developed by the different manufacturers are not comparable. Absolute and statistical membrane filters generally use a forward flow test for validation (see, e.g., the Validation Guide for Pall 0.2-μm nylon 66 membrane cartridges; Pall Corp., 1980). The forward flow test involves wetting the filter with a solvent or solvent-water mixture. The solvent solution is then drained from the filter housing and a controlled pressure of air is applied to the filter in the normal direction of flow. The airflow rate is then monitored, and if it does not exceed a prescribed test value, the filter is considered validated. The mechanism of this test relies upon the solvent solution filling the pores of the filter. The surface tension of the solvent tends to plug the pores and prevent air from passing, except by diffusion. If pores or manufacturing defects are present that are larger than the normal pore size, then the surface tension of the solvent is not sufficient to plug them and an excessive airflow occurs. The validity of the foward flow test has been correlated to filter challenges with liquid suspensions of *Pseudomonas diminuta* (R. S. Conway, 184th ACS Annu. Meet., 1982). In general, the sterile filtration of liquids is more difficult than that of gases, so this test should be sufficient for any application. The difficulty of the forward flow test is that it is hard to conduct in the plant environment on filters that have already been sterilized. Testing of filters before sterilization does not ensure against defects that may occur during the sterilization process.

The forward flow has no meaning when applied to depth medium filters. The manufacturers of depth medium filters generally recommend the use of an oil smog such as DOP or a sodium chloride aerosol to challenge the filter (K. R. Domnick, 184th ACS Annu. Meet., 1982). The basic methodology used is to produce an aerosol of fine particles or droplets in the range of 0.1 to 0.3 μm. This aerosol is supplied upstream of the filter, and a photometer is placed downstream of the filter. The photometer records the number of particles which pass through the filter. Aerosol generators exist which produce up to 5×10^{11} particles per ft^3. Photometers are available which can detect particles in the range of 10^5 particles per ft^3. As a result, a lack of response by the downstream photometer would ensure a filter efficiency of 99.99998%. The drawback of the aerosol test is that a filter exposed to a 60-SFM airflow rate containing 100 organisms per ft^3 would challenge a filter with 8.6×10^6 organisms in a single day. If one organism passed the filter per day, the efficiency of the filter would be equivalent to the limit of detection of the aerosol challenge test. The advantage of the aerosol test method is that it can be conducted on line in the plant without affecting the sterility of the filter. It is worth noting that it is not uncommon for filters to fail during the sterilization cycle.

The ultimate test of any air filter is the use of a bacterial challenge. This is generally not a test that is routinely applied in the field. Rather, it is a test used to validate the ultimate capability of the filter in the laboratory and to supply data for correlation with field test methods. Robertson and Frieben (J. H. Robertson and W. R. Frieben, 184th ACS Annu. Meet., 1982) challenged a number of statistical membrane filter cartridges with *Bacillus subtilis* spores with total biochallenges of 10^9 spores per filter. They found no passage of spores on any filter that passed the manufacturer's forward flow test criteria. Leahy and Gabler (T. J. Leahy and R. Gabler, 184th ACS Annu. Meet., 1982) challenged statistical membrane filter cartridges over a 4-day period with 10^{10} *P. diminuta* organisms and found no penetration. Conway (19) challenged 0.2-μm hydrophobic statistical membrane filters with 10^{10} T1 bacteriophage under conditions of 100% relative humidity and also found no penetration. The T1 bacteriophage has dimensions of 0.05 by 0.1 μm, smaller than any bacteria. Domnick (184th ACS Annu. Meet., 1982) reported tests for depth medium filters in which challenges of 10^{12} organisms of *B. subtilis*, *Serratia marcescens*, and *Escherichia coli* showed no penetration. In separate tests, he also showed that depth medium filters could withstand a 10^{13} challenge with bacteriophage with no penetration, provided that the relative humidity of the air was maintained below 60%.

These tests all prove that the common cartridge-type filters available today are capable of sterilizing air for a significant period of time. It has generally been my experience that all of the commercially available sterile air filters perform well in fermentation application, provided they are free from manufacturing defects. It is becoming more accepted that they are superior to conventional depth filters, and most new fermentation installations are being designed with cartridge filters. It should also be recognized that packed depth filters still predominate in production facilities and are an acceptable alternative where other constraints or conditions exist.

Recently, hydrophobic statistical membrane cartridges have become available in the market places. These filters have a natural water shedding capability and can therefore be used to filter air at 100% relative humidity. In general, it would be poor practice to supply saturated air to a fermentor, but such filters may prove of significant value as vent or off-gas filters on fermentations that require high containment, such as recombinant DNA fermentations.

25.6 SUMMARY

In the text above, only heat sterilization and air filtration have been discussed. Filters may also be used to sterilize liquids, but only under special conditions and with careful quality control. There are many other methods that can be used to achieve sterilization. Block (3) provides an excellent source of information on many other methods. The methods discussed above are by far the most commonly used in the biotechnology industry. Other methods are generally much more expensive, require careful monitoring, and tend to be less reliable.

In recent years, there has been a trend to look for methods of sterilization that would use less energy than heat sterilization. A number of novel methods have been discussed at technical meetings and in the literature. In light of this, it is important to recognize that a 100,000-liter production fermentor, including medium, can be sterilized with steam at a cost of $200 to $400. It is hard to anticipate a new technology that could reduce this cost or even be competitive with it. It is likely that heat sterilization will be around for some time, and it is my hope that the above discussion will prove useful to those in the field who find a need to understand this process better. In contrast to heat sterilization, the filter sterilization of air is an area that is under rapid development. Significant improvements in filter cartridge design have been developed within the last couple of years. At this time it is questionable as to whether cartridges will be further improved, or whether the competition will lead to a reduction of cost.

25.7 LITERATURE CITED

25.7.1 General References

1. **Aiba, S., A. E. Humphrey, and N. F. Millis.** 1973. Biochemical engineering, 2nd ed. Academic Press, Inc., New York.
2. **Ball, C. O., and F. C. W. Olson.** 1957. Sterilization in food technology. McGraw-Hill Book Co., New York.
3. **Block, S. S. (ed.).** 1977. Disinfection, sterilization, and preservation, 2nd ed. Lea & Febiger, Philadelphia.
4. **Lawrence, C. A., and S. S. Block (ed.).** 1968. Disinfection, sterilization, and preservation. Lea & Febiger, Philadelphia.
5. **Meynell, G. G., and E. Meynell.** 1970. Theory and practice in experimental bacteriology, 2nd. ed. Cambridge University Press, London.
6. **Perkins, J. J.** 1956. Principles and methods of sterilization. Charles C. Thomas Publisher, Springfield, Ill.
7. **Stumbo, C. R.** 1973. Thermobacteriology in food processing, 2nd ed. Academic Press, Inc., New York.
8. **Sykes, G.** 1965. Disinfection and sterilization, 2nd ed. E. and F. N. Spon Ltd., London.

25.7.2 Specific References

9. **Aiba, S.** 1962. Design of fibrous air sterilization filters. J. Gen. Appl. Microbiol. **8:**169–179.
10. **Aiba, S., and K. Toda.** 1965. An analysis of bacterial spores thermal death rate. J. Ferment. Technol. **43:**527–533.
11. **Ashley, M. H. J., and J. G. Mooyman.** 1982. Design optimization of continuous sterilizers. Biotechnol. Bioeng. **24:**1547–1553.
12. **Bader, F. G., M. K. Boekeloo, J. W. Cagle, and H. E. Graham.** 1984. Sterilization of oils: data to support the use of a continuous point-of-use sterilizer. Biotechnol. Bioeng. **26:**848–856.
13. **Bailey, J. E., and D. F. Ollis.** 1977. Biochemical engineering fundamentals, p. 465–473. McGraw-Hill, New York.
14. **Bartholomew, W. H., D. E. Engstrom, N. S. Goodman, A. L. O'Toole, J. L. Shelton, and L. P. Tannen.** 1974. Reduction of contamination in an industrial fermentation plant. Biotechnol. Bioeng. **16:**1005–1013.
15. **Bartholomew, W. H., E. O. Karow, and M. R. Sfat.** 1950. Design and operation of a laboratory fermentor. Ind. Eng. Chem. **42:**1827–1830.
16. **Beloian, A.** 1977. Methods of testing for sterility, p. 11–48. *In* S. S. Block (ed.), Disinfection, sterilization, and preservation, 2nd ed. Lea & Febiger, Philadelphia.
17. **Blazewitz, A. G., and B. F. Judson.** 1955. Filtration of radioactive aerosols by glass fibers. Chem. Eng. Prog. **51:**6J–11J.
18. **Bruch, C. W., M. G. Koesterer, and M. K. Bruch.** 1963. Dry-heat sterilization: its development and application to components of exobiological space probes. Dev. Ind. Microbiol. **4:**334–342.
19. **Conway, R. S.** 1984. State of the art in fermentation air filtration. Biotechnol. Bioeng. **26:**844–847.
20. **Decker, H. M., F. J. Citek, J. B. Harstad, N. H. Gross, and F. J. Piper.** 1954. Time-temperature studies of spore penetration through an electric air sterilizer. Appl. Microbiol. **2:**33–38.
21. **Esumi, S., and K. Ashida.** 1966. Experimental studies on 'PVA.' II. Effect of pore size and thickness. J. Ferment. Technol. **44:**529–535.
22. **Frobisher, M., R. D. Hinsdill, K. T. Crabtree, and C. R. Goodheart.** 1974. Fundamentals of microbiology, 9th ed. W.B. Saunders Co., Philadelphia.
23. **Gaden, E. L., Jr., and A. E. Humphrey.** 1956. Fibrous filters for air sterilization, design procedure. Ind. Eng. Chem. **48:**2172–2176.
24. **Hodges, N. A., J. Melling, and S. J. Parker.** 1980. A comparison of chemically defined and complex media for the production of *Bacillus subtilis* spores having reproducible resistance and germination characteristics. J. Pharm. Pharmacol. **32:**126–130.
25. **Humphrey, A. E., and E. L. Gaden, Jr.,** 1955. Air sterilization by fibrous media. Ind. Eng. Chem. **47:**924–930.
26. **Jacobs, R. A., L. L. Kempe, and N. A. Milone.** 1973. High temperature-short time processing of suspensions containing bacterial spores. J. Food Sci. **38:**168–172.
27. **Jensen, L. B.** 1954. Microbiology of meats. Garrard Press, Champaign, Ill.
28. **Kimura, N., and G. Iinoya.** 1959. Pressure-drop characteristics of fiber mats. Chem. Eng. **23:**792–799.
29. **Lang, O. W.** 1935. Thermal processes for canned marine products. Univ. Calif. Pub. Public Health **2:**1–182.
30. **Lundgren, D. A., and K. I. Whitby.** 1965. Effect of particle eleotrostatic charge on filtration by fibrous filters. Ind. Eng. Chem. Process Des. Dev. **4:**345–349.
31. **Maxon, W. D., and E. L. Gaden, Jr.** 1956. Fibrous filters for air sterilization: experimental studies with a pilot scale filter. Ind. Eng. Chem. **48:**2177–2179.
32. **Molin, N., and B. G. Snygg.** 1967. Effect of lipid materials on heat resistance of bacterial spores. Appl. Microbiol. **15:**1422–1426.
33. **National Aeronautics and Space Administration.** 1968. Biological handbook for engineers, NASA CR-61237, June 1968. National Technical Information Service, Springfield, Va.
34. **Opfell, J. B., C. E. Miller, N. S. Kovar, P. E. Naton, and R. D. Allen.** 1964. Sterilization handbook, final report, NASA Contract NASw-777. Dynamic Science Corp., Pasadena, Calif.
35. **Perry, R. H., and C. H. Chilton.** 1973. Chemical engineers handbook, p. 6–16. McGraw-Hill, New York.
36. **Peterson, A. L.** 1974. Continuous sterilization of ferment-

ing broth for the pharmaceutical industry. Biotechnol. Bioeng. Symp. **4**:879–890.
37. **Pflug, I. J.** 1972. Heat sterilization, p. 239–282. *In* G. B. Phillips and W. S. Miller (ed.), Industrial sterilization. Duke University Press, Durham, N.C.
38. **Pheil, C. G., I. J. Pflug, R. C. Nicholas, and J. A. L. Augustin.** 1967. Effect of various gas atmospheres on destruction of microorganisms in dry heat. Appl. Microbiol. **15**:120–124.
39. **Quesnel, L. B., J. M. Hayward, and J. W. Barnett.** 1967. Hot air sterilization at 200°. J. Appl. Bacteriol. **30**: 518–528.
40. **Rahn, O.** 1945. Physical methods of sterilization of microorganisms. Bacteriol. Rev. **9**:1–47.
41. **Reddish, G. F.** 1957. Antiseptics, disinfectants, fungicides, and chemical and physical sterilization, 2nd ed., p. 23–29. Lea & Febiger, Philadelphia.
42. **Reyes, A. L., R. G. Crawford, A. J. Wehby, J. T. Peeler, J. C. Wimsatt, J. E. Campbell, and R. M. Twedt.** 1981. Heat resistance of *Bacillus* spores at various relative humidities. Appl. Environ. Microbiol. **42**:692–697.
43. **Shepherd, P. G., B. Fraissignes, and W. A. Peet.** 1974. Fermentation process design aspects of single cell protein from hydrocarbons. Biotechnol. Bioeng. Symp. **4**: 721–732.
44. **Silverman, L., E. W. Conners, Jr., and D. M. Anderson.** 1955. Mechanical electrostatic charging of fabrics for air filters. Ind. Eng. Chem. **47**:952–960.
45. **Steel, R., and T. L. Miller.** 1970. Fermentor design. Adv. Appl. Microbiol. **12**:153–188.
46. **Sykes, G.** 1969. Methods and equipment for sterilization of laboratory apparatus and media. Methods Microbiol. **1**:77–121.
47. **United States Pharmacopeia.** 1970. United States Pharmacopeia, 18th rev. Sterility tests, p. 851–857. Mack Publishing Co., Easton, Md.
48. **Wang, D. I. C., J. Scharer, and A. E. Humphrey.** 1964. Kinetics of death of bacterial spores at elevated temperatures. Appl. Microbiol. **12**:451–454.
49. **Yao, K. M., M. T. Habibian, and C. R. O-Melia.** 1971. Water and waste water filtration: concepts and applications. Environ. Sci. Technol. **5**:1105–1112.

Chapter 26

Cost Estimation for Biotechnology Projects

JAMES P. KALK AND ASGER F. LANGLYKKE

Cost analysis generally endeavors to answer the questions, How much will it cost to produce a product? what investment will be required? and what return on the investment can be expected? In addition, cost analysis attempts to identify and quantify the interrelation between various design parameters or process operations and manufacturing and investment costs (e.g., optimum design, sensitivity analysis, etc). As a subdivision of cost analysis, cost estimation specifically applies itself to answering the first two questions, regarding product cost and capital investment.

The estimation of capital and operating costs for chemical processes is the subject of many texts and articles (see Literature Cited) and is generally a major part of any undergraduate program in chemical engineering. By necessity, this article will highlight only briefly the important aspects of such estimates and refer the reader to more extensive studies. As a focus, specific examples of cost estimation related to biotechnology projects will be presented.

26.1 TOTAL PRODUCT COST, CAPITAL INVESTMENT, AND PROFITABILITY

26.1.1 Elements of Total Product Cost

Table 1 presents a breakdown of the elements of total product cost as they are typically defined. The table also gives some rough approximations of the percentage contribution of each category to product cost. Reference is also made to the specific sections of this chapter which deal with the individual items. The most common accounting basis for figuring product cost is the annual basis. The most obvious reason for this is for tax purposes, but it also has the effect of leveling out day-to-day variations in output and making various profitability calculations simpler. Total product cost is usually divided into two parts, manufacturing cost and general expenses.

Manufacturing cost

Those costs which can be considered directly required for manufacture of the product are assigned to the category of manufacturing cost (also sometimes called production or operating cost). There are three major subsections of manufacturing cost: (i) direct operating (or variable) costs; (ii) fixed costs; and (iii) plant overhead costs.

Charges for manpower, materials, and utilities specifically associated with production operations are considered direct operating costs. Detailed information regarding the individual accounts which make up direct operating costs, and how to estimate them, are given below in section 26.2.1.

Direct operating costs are sometimes referred to as variable costs. This is because these costs are assumed to be directly related to production capacity; thus, as the level of output of a plant varies, direct operating expenses also vary. This is an oversimplification, however, because some of the items included in direct operating costs do not vary substantially with production volume (e.g., supplies and labor).

Fixed costs are charges for capital investment in the production facility. They vary little, if any, with production volume changes. Details of the items making up the fixed costs category are given in section 26.2.2.

The third category of manufacturing cost is plant overhead costs (see section 26.2.3). This is often a catch-all category for items not detailed or itemized in the estimation of variable and fixed charges. These charges are those which are necessary for day-to-day operation of the plant, but which are not directly involved in production of the product (e.g., janitorial

TABLE 1. Elements of total product cost and typical values for estimation

I. Direct operating costs (section 26.2.1)
- A. Raw materials and supplies
  - 1. Raw materials (30 to 80% of manufacturing cost)
    - a. Primary
    - b. Secondary
    - c. Freight
  - 2. Supplies
    - a. Operating (3 to 5% of direct operating labor and supervision)
    - b. Maintenance (100% of maintenance labor and supervision; 2 to 5% of fixed capital investment)
    - c. Laboratory (20 to 40% of laboratory labor)
    - d. Other (10 to 20% of plant overhead; see III below)
  - 3. By-product credits/debits
- B. Labor and supervision (10 to 40% of manufacturing cost)
  - 1. Base salaries and wages from detailed manpower estimate
  - 2. Overtime (6% of base [hourly] wages)
  - 3. Fringe benefits (30 to 40% of salaries and wages)
- C. Utilities (5 to 20% of manufacturing cost)
  - 1. Steam (or fuel)
  - 2. Electricity
  - 3. Water
  - 4. Waste treatment

II. Fixed costs (section 26.2.2)
- A. Depreciation and interest (8 to 12% of fixed capital for depreciation, 10 to 15% for interest; as alternative, use capital recovery factor)
- B. Taxes (1 to 4% of fixed capital)
- C. Insurance (1 to 3% of fixed capital)
- D. Rent (highly variable)

III. Plant overhead (section 26.2.3)
(10 to 70% of labor and supervision depending upon level of detail of manpower estimate)

IV. Administration (section 26.1.1)

V. Marketing (section 26.1.1)

VI. R&D (section 26.1.1)

For cost estimation:
Manufacturing cost = I + II + III
General expenses = IV + V + VI
Total product cost = manufacturing cost + general expenses

services, personnel, accounting, surveillance, etc). As a support function to actual production operations, plant overhead costs are somewhat related to direct operating manpower. As such, these costs are usually estimated as a percentage of the salaries and wages for direct operations.

General expenses

Included in the category of general expenses are charges for administration (i.e., corporate offices), marketing, and research and development (R&D). (Sometimes interest or the cost of financing is included under general expenses [33]. This is often a reasonable approach from an accounting standpoint when funds for investment may come from a variety of sources. For the purpose of preliminary estimation of manufacturing cost for a single product, however, it is more straightforward to assign interest cost as a fixed charge.) General expenses are highly variable, depending upon the industrial setting and corporate history, priorities, and business philosophy (3–5, 13, 14). It is generally true that the parties responsible for estimation of manufacturing cost and capital investment have little knowledge of and no responsibility for assessing the costs for general expenses. As such, this chapter will not attempt to detail or quantify these costs. This should not be taken as an indication that general expenses are a minor part of product cost and are to be ignored. They can, in fact, be a major contributor to product cost. For example, development costs for a pharmaceutical product can be about $90 million per product (e.g., for clinical trials, government approval, etc.) (40).

26.1.2 Capital Investment

All of the funds required to build, start, and operate a production facility make up the capital investment. The investment in purchase of land, construction of buildings, and purchase and installation of all equipment detailed in the final engineering design of the facility are all parts of the fixed capital investment. In addition, after initial construction is complete, it is often necessary to make physical modifications of the original design layout during the initial period of operation of the plant. The costs associated with these modifications are known as start-up costs and are also considered a part of the fixed capital. To operate the plant, an inventory of raw materials and supplies, work in progress, finished product, and operating cash (to pay salaries, vendors, etc.) is required. The total value of these items is called working capital. Together, fixed capital and working capital are considered the total capital investment. Table 2 briefly outlines and describes the costs associated with the total capital investment. More detailed, "checklist" type versions of Table 2 are available elsewhere (9, 11, 12, 28, 33, 44). Techniques for estimating capital investment are detailed and referenced below in section 26.3.

26.1.3 Profitability Measures

There are a variety of methods which are used to evaluate whether or not a potential investment will be profitable. The simplest technique is a calculation of return on investment, which is simply the ratio of annual profit to capital investment expressed as a percentage. Also called rate of return on investment, this figure can be expressed on a before- or after-tax basis. Other investment assessment methods are based upon cash flow analysis, which relates net cash influx (profits plus depreciation) to cash outflow (initial investment). Various techniques used to evaluate cash flow are payback time, present worth, and internal rate of return (discounted cash flow). As with the assignment of general expenses, the method of calculation and definition of acceptable profit are dependent upon corporate business philosophy. Detailed discussions of profitability evaluation are available

TABLE 2. Outline of total capital investment

Item	% of fixed capital
I. Fixed capital	
A. Direct costs	
1. Land: property, surveys, recording fees, commissions, taxes, etc.	2–3
2. Site development: utility hook-ups, site clearing, grading, excavating, roads, walkways, landscaping, railway	4–6
3. Buildings: foundations, offices, shops, warehouses, processing and utility areas, laboratories, locker rooms, cafeteria, services to buildings (plumbing, electrical, heating, ventilation, air conditioning, phones, painting, fire protection, etc.)	10–15
4. Processing: equipment, installation, piping, electrical, instrumentation, special foundations and structural supports, insulation, paint, freight, insurance, taxes (sometimes the last three items are considered indirect costs [25])	40–70
5. Services (utilities): As for (4) above, plus distribution systems; also nonprocessing equipment and machinery (laboratory and office equipment, etc.)	20–30
B. Indirect costs	
1. Engineering: design, piping and instrumentation drawings, procurement, administration, cost control, travel, etc.	5–15
2. Construction: field supervision, temporary facilities, tools, equipment, etc.	5–15
3. Contingency: dependent on level of detail of estimate	3–50
4. Fees: engineering, construction, contractor	4–6
C. Start-up costs: equipment and construction modifications required to bring plant up to a specified capacity; personnel training, technical support, operating expenses	5–20
II. Working capital	15
A. Inventory: raw materials and supplies in warehouse, raw materials and product(s) in process, finished product(s)	
B. Accounts receivable: shipped finished product awaiting payment	
C. Accounts payable (offsets A and B): salaries and wages due, raw material and supply payments due, utilities payments due	

elsewhere (12, 17, 33, 38, 39, 44) and will not be reviewed here.

26.2 MANUFACTURING COST ESTIMATION

26.2.1 Direct Operating Costs

Raw materials and supplies

Raw materials directly consumed in the process of producing a product or used in its recovery constitute a major part of the manufacturing cost. This is more true for biologically based production systems (where raw materials can represent 30 to 80% of the production cost) than for conventional chemical production plants (10 to 50% of the production cost [33]).

There are many possible sources to use for determining specific costs for individual raw materials. The weekly publication *Chemical Marketing Reporter* is by far the most up-to-date and comprehensive source. Because this publication usually presents price information for bulk purchase, the volume of raw material actually required should be calculated before relying on this source. Industrial gases and many small-volume specialty chemicals are noticeably absent from the *Chemical Marketing Reporter*. *Chemical Week* and *Chemical and Engineering News* present periodic updates on these and other materials and are useful publications. Of course, the best possible sources of information are vendor quotes. Source books for particular raw-material vendors include the *Chemical Buyers Guide*, *Chemical Engineering Catalog*, and *Chemcyclopedia*. In particular, confirmation of prices should be obtained from vendors when the overall process economics appear to be sensitive to an individual raw-material cost.

Prices for raw materials are generally quoted f.o.b. (free on board), which usually means the price includes loading the material at its site of production onto some mode of transportation. Thus, freight charges are not usually included in the raw material price quote, but should be included in the cost analysis. Freight charges are highly variable, depending on the type of material, load, destination, and mode of transportation (7, 8, 45). The more defined the scope of the project (i.e., site selected, vendors selected, etc.), the better is the estimate of freight charges. Trucking rates can be estimated at $1 to 2/mile for 40,000-lb loads (200-mile minimum). A range of $0.01 to 0.05/lb represents typical rates for all types of freight.

Supplies are those materials not directly used in the production process but still necessary to carry out manufacturing operations. The specific charges for supplies are difficult to itemize individually and are usually accounted for by some percentage of other costs. (For example, Peters and Timmerhaus [33] recommend estimating these costs as 15% of the total cost for maintenance and repairs.) Costs for supplies are usually related to the personnel or equipment involved in a particular aspect of the overall manufacturing effort. Operating supplies are those materials required by direct operations, such as log sheets, copier materials, batteries, etc. These can be conveniently estimated as 3 to 5% of salaries and wages for direct operating labor and supervision. Maintenance supplies represent the spare parts and tools required to keep all plant equipment in satisfactory working condition. Total cost for maintenance (supplies plus labor) generally represents between 5 and 10% of the fixed capital investment for the plant (see below) (1, 6). Supplies and labor are roughly 50% each of total maintenance cost (20), so maintenance supplies can also be estimated as 100% of maintenance labor. (For operating plants, maintenance costs are usually based on a percentage of capital replacement value, which more adequately integrates the effects of inflation [1, 6, 20]). Laboratory supplies include costs for reagents, disposable equipment, etc., and can be estimated at 20 to 40% of associated personnel salaries and wages. (Laboratory supplies include those for process control and quality control. Although these two functions are usually organizationally separate, they often share space, equipment, and materials.) Other supplies not specifically included in the previous categories can be estimated as 10 to 20% of plant overhead (see below).

A manufacturing process for a specific product will often generate by-products or side products. Generally there will be some charge for disposing of these materials. Occasionally they may be sold, in which case the charge becomes a credit. It is a matter of convention, but often by-product debits are considered in the utilities account as a waste treatment charge, whereas by-product credits are listed as a (positive) line item in the raw materials and supplies account.

Labor and supervision

Next to raw material costs, salaries and wages usually represent the largest fraction of total manufacturing costs (10 to 40%). There are a number of ways to estimate the manpower charges associated with manufacturing a product (9, 12, 28, 33). Other than direct experience with a similar process, no simple technique can be expected to give very accurate results. The best system is to assign labor and supervision on the basis of the actual operations expected to be performed.

Often, only the labor and supervision directly associated with production are estimated, and the rest of the costs associated with paid personnel at the plant are lumped into a rather large category of "plant overhead" (see section 26.2.3). Plant overhead is then estimated at 50 to 70% of the direct labor, supervision, and maintenance costs (33). Because many of the charges put in this category (e.g., fringe benefits, surveillance, quality control, shipping and receiving, packaging, general maintenance, etc.) can be detailed quite adequately in a full labor estimate, it seems more appropriate to do so rather than to rely on a large and uncertain percentage item.

Once a detailed manpower estimate is made, allocations for overtime and fringe benefits can be made. Obviously, overtime should be added only for nonexempt (hourly) labor. Typical overtime is assessed at 6% of the base pay (4% actual time × 1.5 hourly rate). Fringe benefits are then estimated at 30 to 40% of the direct payroll (45).

Actual pay scales for both exempt and nonexempt employees are quite variable, dependent mainly upon the geographical location of the plant (45). If the project scope is defined sufficiently to include plant location, fairly accurate data for wage rates may be obtained from local authorities or from national publications like the *Monthly Labor Review* or *Engineering News Record*. If reliable data are not available, a range of labor costs should be considered so that the sensitivities of the process to this variable can be adequately understood.

Utilities

Usually included in the category of utilities are the costs for steam, electricity, water, and waste treatment. For biologically based production processes, utilities costs can range from less than 5% to more than 20% of the production cost. An accurate estimate of utility consumption can only be obtained from a detailed process flow sheet and plant layout. Once specific utility usage is determined, the costs can be arrived at by multiplying by the appropriate rate. Utility rates are also highly variable and heavily dependent on geographical location of the plant. Typical rates for utilities are listed in Table 3.

When substantial amounts of steam and electricity are used directly in the process, building utilities (lighting, heating, ventilation, air conditioning, water, phones, etc.) will usually be covered by the 10 to 25% contingency that should be added to each utility account. However, in those cases where process utilities consumption is quite low, building utilities can represent more than 25% of the total utility account. Direct estimates of building utilities can be obtained by making a sketch of the plant layout and using cost-per-unit-space factors given elsewhere (9, 23).

Steam. Unless the new production facility being estimated is located next to a plant with excess steam capacity, steam is usually not "purchased" per se. Rather, some other form of energy is purchased, which is used to produce steam in some type of boiler unit. Sometimes, for the purpose of rapid manufacturing cost estimates, it is convenient to lump all of the costs associated with producing steam (fuel, water treatment, labor, distribution, and equipment) together as a single steam cost-production rate. This total production rate is typically about $13 to 16/1,000 kg of steam. Depending upon the fuel type and cost, however, total costs can fall outside this range (Table 3). If a lump-sum steam cost rate is used, one should be careful to include the cost of equipment for steam production and distribution in the capital estimate of the production facility. Also, when analyzing the economics of below-capacity operation, steam costs will not be directly proportional to capacity if the steam cost rate includes fixed charges (i.e., equipment and, to some extent, labor).

If not required for recovery operations (e.g., evaporation), the major steam usage in a typical microbiologically based production facility is for sterilization. Steam consumption for batch sterilization is usually about 0.2 to 0.4 kg/liter of fermentation medium. Continuous sterilization can cut this usage by as much as 75%. With a steam cost of $15/1,000 kg of steam, batch fermentor sterilization costs are about $3 to 6/1,000 liters. Although steam for sterilization is not a negligible cost, the major utility charges associated with fermentations are electrical charges for aeration and agitation.

TABLE 3. Typical utility rates

Utility	Fuel	Rate	Cost ($/1,000 kg, at 80% boiler efficiency)	
			Fuel	Other charges[a]
Steam[b]	Natural gas	$4–5/1,000 ft^3	11–14	2–4
	Oil (no. 2)	75–85¢/gal	14–17	2–4
	Coal	$2.50–3.50/10^6 BTU	7–10	4–7

Utility	Region	¢/kW-h	
		High load (40,000 kW)	Low load (<5,000 kW)
Electricity	Pacific Northwest	2.5	5.0
	West Coast	5.0	9.0
	Southwest	4.0	7.0
	Southeast	3.0	6.0
	Central	4.0	7.0
	East Coast	4.0	7.0

Utility	Type	Cost
Water	Well, river	$0.005–0.03/m^3
	Municipal (in/out charge)	$0.30–0.75/m^3
	High purity[c]	
	Reverse osmosis	$1–5/m^3
	Distillation	$15–20/m^3
Waste treatment	Landfill	2–4¢/kg
	Wet oxidation	$10–20/m^3[d]
	Biological	10–20¢/m^3[e]

[a] Labor, waste treatment, building and equipment depreciation.
[b] Total production cost, $13–16/1,000 kg.
[c] Includes capital depreciation and operating expenses for on-site production.
[d] 5×10^4 to 10×10^4 ppm COD (chemical oxygen demand).
[e] 500 to 1,000 ppm BOD (biological oxygen demand).

Electricity. Even more than other utilities, electricity rates vary widely with geographical location of the plant and load requirements (35, 45) (Table 3).

Electric power consumption is usually the largest part of the utility charges for a microbial production process. The three major line items in the electrical utility account are aeration and agitation for fermentation and pumping for the entire process. Another item which can be a major load item, if required, is refrigeration.

Typical aeration rates in a fermentor range from 0.5 to 2 vvm (volume of air per volume of fermentor broth per min). For air delivered at 100 psia (lb/in^2 absolute), 1 vvm represents power consumption of about 5 kW/1,000 liters of fermentation. At $0.05/kW-h, the cost for aeration in a 60-h fermentation is about $15/1,000 liters. Agitation power requirements are usually 1.0 to 3.0 kW/1,000 liters of fermentation. For a 60-h fermentation using 2 kW/1,000 liters, the cost for agitation would be $6/1,000 liters (again, at $0.05/kW-h). Thus, the total electrical charge for fermentation can be more than $20/1,000 liters.

The electrical costs for pumping are highly process dependent and can be accurately estimated only from a fairly detailed process design. For production processes based on fermentation, electricity costs for total process pumping usually run about $0.50 to 1.00/1,000 liters of fermentation broth produced (10 to 20 kW-h/1,000 liters).

Water. Water costs are usually a relatively small fraction of total utility costs, although water is often used in tremendous quantities (for cooling) in the production process. The two major uses are process water and cooling water. Unless a lake or well is nearby to supply cooling water, the source of both process and cooling water may be the same: the municipal water supply. This is usually not a cheap source, with costs sometimes as high as $1/1,000 liters (in/out charge). Process water consumption is not usually high enough to have a significant impact on production cost even with a high water charge. (Exceptions to this rule may occur when very high-quality water is required in the process; e.g., production of parenteral products.) On the other hand, cooling water consumption can be quite high, and indeed prohibitive, unless water conservation is practiced by use of an open or closed cooling water recirculation system.

Waste treatment and disposal. If no on-site waste treatment facility is required, waste disposal is sometimes listed as a line item under raw materials as a by-product debit. Regardless of how it is accounted, it is important that the waste treatment involved in a process be adequately addressed during the development and final design of a plant. Failure to take into account environmental discharge regulations in the final design and construction of a facility can result in substantial delays in start-up of the plant and in reduced plant capacity after start-up (31).

26.2.2 Fixed Costs

Depreciation and interest

Capital invested in construction of the plant and associated buildings is normally recovered by use of the accounting procedure termed depreciation. By this system, the value of the plant is assumed to decrease over time, and this decrease in value is written on the books as an expense of manufacture. There are many procedures for depreciation (straight line, sum of the digits, declining balance, etc.) which are discussed in detail elsewhere (38, 39). Generally, the taxing authorities have the final say in determining the lifetime and depreciation formula that will be used. In United States practice, typical lifetimes are 5 to 15 years for process equipment and 30 years for buildings (land is considered to have an infinite lifetime and therefore cannot be depreciated).

Some or all of the capital invested in building a plant is often borrowed money, and some interest charge may be involved. This is also considered a fixed charge. For the purposes of a preliminary design estimate, a more convenient way to account for the capital investment as part of the manufacturing cost is to use an amortized or capital recovery cost in place of the depreciation and interest charge. Using this technique, an annual charge which will recover the capital invested at a fixed interest rate is entered for equipment and buildings. This simplifies the economic analysis by making the process directly responsible for paying back all the capital associated with construction of the plant. The result is a "break-even" manufacturing cost which considers the value of money (based on current interest rates) rather than the source (i.e., internal versus borrowed funds). (This method is essentially a modification of the discounted cash flow method of profitability analysis.)

Taxes

All annual and one-time taxes paid to state or local authorities which are based upon the value of property (i.e., machinery, equipment, buildings, and land) are included in the fixed-cost category. Property tax rates are highly variable, ranging from 1 to 4% of the fixed capital investment.

Other taxes are normally accounted for in the fringe benefits estimate (i.e., unemployment insurance, social security, etc.) or in the plant overhead estimate. Income taxes are not included in the cost of manufacture, but are accounted for when calculating net income from gross income.

Insurance

Insurance against the hazards of operating a manufacturing process will obviously relate to the type of operation and potential loss from accident. Rates are thus highly variable, though usually between 1 and 3% of the fixed investment.

Rent

Rent is highly variable and, of course, related to the property (land and buildings) value. Unless the scope is detailed enough to include definite site selection (in which case rent can be estimated as 10% of the site value), it is best to assume that land and buildings will be purchased and constructed and to account for their cost in the capital estimate.

26.2.3 Plant Overhead

As previously mentioned, plant overhead often is estimated as a very large percentage (50 to 70%) of salaries and wages for direct labor and supervision and maintenance. By general definition, overhead costs are those which, though required for efficient operation of the plant, are not directly assignable to some part of the manufacturing operation (e.g., medical, cafeteria, janitorial, personnel, plant protection, etc). Often also included in this account, however, are items which could be directly assignable to manufacturing operations (e.g., fringe benefits, warehousing, shipping and receiving, packaging, and in-process and quality control laboratories). Therefore, if the manpower estimate is sufficiently detailed to include these and other ancillary manufacturing operations or costs, the percentage used for plant overhead estimation should be reduced accordingly (to 10 to 20% of salaries and wages).

26.3 CAPITAL INVESTMENT ESTIMATION

26.3.1 Elements of Total Capital Investment

As detailed in Table 2, the amount of money required to build a manufacturing facility can be broken down into two categories according to the use of the funds. Thus, the total capital investment is made up of fixed capital and working capital. Fixed capital refers to all of the funds necessary to design, construct, and start up the plant. Fixed capital (except land and sometimes start-up costs) is also the normal basis used for calculation of depreciation. Working capital refers to the cash and goods that must be on hand to keep the plant operating once it has been started up. These two parts of capital investment can be further subdivided into numerous categories. There is no single system for delineating these categories, although some engineering and construction firms do use the same standard cost codes (44).

Fixed capital

Three components make up the fixed capital investment: direct costs, indirect costs, and start-up costs.

Direct costs cover all of the equipment and materials which compose the manufacturing facility and the labor required to construct the facility. The major categories of direct costs apply to land, site development, buildings, processing, and services (utilities).

The cost of land is included as part of the capital investment estimation, but should be excluded when calculating depreciation for estimation of manufacturing cost. Land costs are highly variable and generally not known until the project scope is fairly detailed. For preliminary estimation (see below), land cost can be estimated as 2 to 3% of the fixed capital investment.

Site development includes whatever physical preparation of the property is necessary to allow construction of the foundations, buildings, and structures required for all plant operations. Site development costs are related to the characteristics of the property purchased (i.e., improved or unimproved). As such, site development costs can only be detailed once the project has proceeded to the point at which it is possible to identify a specific plant location. For cost estimates in the preliminary stage, site development costs are typically estimated as 4 to 6% of fixed capital investment.

Building costs relate to construction of the foundations and structures which house or support all of the plant activities. Building costs can be estimated from a spatial layout of the manufacturing facility by using published unit space cost factors (9, 23, 34). Alternatively, these costs can be preliminarily estimated as 10 to 15% of the fixed capital investment.

The processing and services (utilities) categories of fixed capital apply to the purchase and installation of all the necessary equipment for, or in support of, manufacturing operations. These items are the largest contributors to capital investment and are, of course, dependent upon the size and type of process envisioned. The itemization of the equipment and machinery required for these two categories forms the starting point for most capital investment estimates (see section 26.3.2 below).

Indirect costs are made up of engineering, construction, contingency, and fees. Engineering costs include design and layout of the plant, procurement, scheduling, and supervision of the project. These costs are most closely related to the direct materials required for the project, so they can be estimated as 15 to 20% of the value of these materials (25). Engineering costs are generally "home office" expenses of the engineering contractor, as opposed to "on-site" expenses which are considered part of construction costs. In addition to field supervision, construction costs include temporary facilities, equipment rental, tools, and related overhead. Usually included in construction overhead are the fringe benefits for all field labor. Because most of the construction expenses are related to the direct construction labor, they are often estimated as 60 to 80% of the costs for direct field labor (25). Because the separation of engineering and construction expenses is somewhat arbitrary, it is usually more convenient to estimate these indirect costs together as 20 to 40% of the direct fixed capital cost. Contingency and commission, although considered indirect costs, are estimated separately for reasons discussed below.

Contingency is not an estimatable cost, but rather represents a subjective impression that the cost estimator has of the reliability of the estimate. This reliability is related to the extent of detail used in compiling the estimate and the adequacy of the fundamental design assumptions. The further along in development a project is, the more justifiable detail can be given in the project estimate, and the more reliable the results. Contingency costs can range from as high as 100% of direct and indirect costs, at the conceptual stage, to as low as 5% in the final design stage. (One might expect the contingency allowance to eventually be eliminated as the project approaches actual initiation of construction. In fact, even the most detailed estimate can be expected to vary from the actual dollars spent due to unpredictable factors [work force productivity, scope changes, new regulatory requirements, weather, accidents, etc.].)

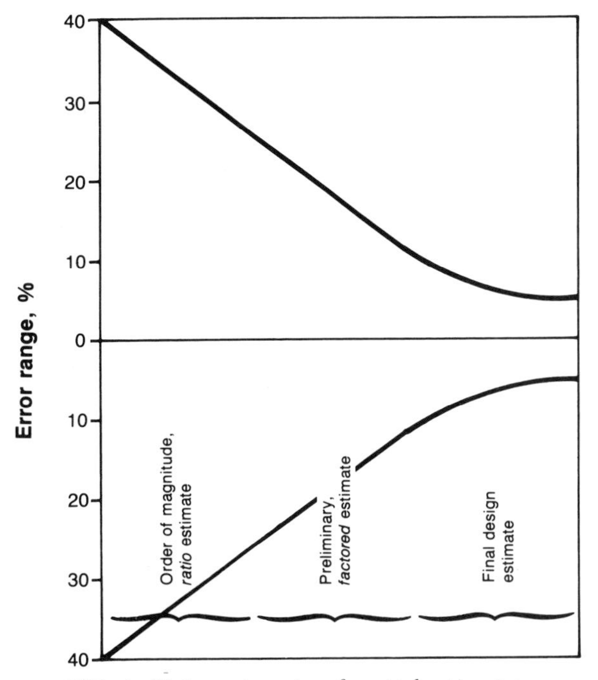

FIG. 1. Major categories of capital estimates.

Commissions or fees charged by the engineering and construction contractors are most often calculated as a percentage of the total direct capital and indirect capital (engineering, construction, and contingency). Thus, they are normally calculated as the last item in the estimate, at about 5%.

Although start-up costs are considered part of the fixed capital, not all of the items which come under start-up are truly "fixed" capital (i.e., depreciable assets). Direct and indirect costs incurred making structural or equipment modifications to bring the plant up to design capacity are an example of true fixed capital costs. Costs for training personnel, extra engineering and technical support, and actual operating expenses during the start-up period are legitimate start-up expenses but nondepreciable. Even though part of the start-up costs can be considered depreciable, company accounting procedures and policies will determine whether they are depreciated or "expensed." Start-up costs are difficult to estimate because it is generally impossible to predict the length of time start-up will take (usually 3 to 6 months) or the extent of process modifications that may be necessary. Start-up costs can range from less than 5% of the fixed capital investment, for a plant based on well-established technology and previous experience, to more than 50% of the fixed capital for new technology with no experience base (2, 31) (e.g., biotechnology projects!).

Working capital

Even after the plant has been built and started up and is running at or near design capacity, a certain amount of capital will always be tied up in inventory and accounts receivable and payable. Although working capital is sometimes estimated at 15% of the fixed investment, a better approach is to use the manufacturing cost estimate to calculate working capital. For example, a reaonable allowance might be 3 months worth of raw materials and supplies (this accounts for 1 month's unfinished, in-progress, and finished goods) plus 1 to 2 months' operating expenses (less raw materials and supplies).

26.3.2 Estimation Types and Techniques

Capital estimates have been classified into five types by the American Association of Cost Engineers (12). The five categories are distinguished more or less by the level of detail and the expected accuracy of the estimate. For simplicity in this chapter, the types of capital estimates are divided into three broader categories, which overlap the Association's five categories. They are order-of-magnitude, preliminary, and final design estimates. These categories can be grouped according to the techniques used in arriving at the estimate (as described below) or according to the accuracy of the estimate. Figure 1 (adapted from reference 44) is an illustration of the expected accuracy of these estimates. Although with more detail the accuracy of the estimate is typically projected to move toward the actual cost within a symmetrical confidence limit, it is very infrequent that the actual capital cost will be less than the estimated cost. This is due to the fact that improvement of detail almost always entails addition of costs due to oversight in the initial estimate rather than elimination of excesses. For this reason, a contingency account is always added to the capital investment estimate.

One objective of this chapter is to illustrate the ways in which cost analysis can be used to direct R&D. Within this context, the major types of estimates used are the order-of-magnitude and preliminary estimate. This is because at the development stage some or all of the fundamental design criteria will be based, at best, upon very preliminary experimental work. The accuracy of a more detailed estimate will be more than offset by the inaccuracies in the design assumptions. Preliminary estimates are used to screen potential projects, to set goals for projects, and to monitor and direct R&D. Both preliminary and final design estimates are used for the variety of profitability analyses required for financial decisions on capital expenditures.

Order-of-magnitude estimates

Also called a "ratio" or "scaled" estimate, the order-of-magnitude estimate is made by deriving the capital estimate for a new facility from a previous estimate or cost of construction of another facility. As implied by its name, the order-of-magnitude estimate is usually not very accurate. There are situations, however, in which the name is a misnomer and the estimate can be quite accurate.

Example 1. The AA company built a fermentation-based production facility in 1980 to produce 2,000,000 kg of amino acid X per year for a total capital investment of $30,000,000. The market demand for the amino acid has continued to rise dramatically. The company plans to build another facility of identical size in 1986 and would like to estimate how much it will cost.

Because the required plant is identical, the new capital investment can be estimated by adjusting the original investment to account for the effect of inflation between the years 1980 and 1986. The first step is to adjust the 1980 cost to a more current (1983) cost, by using one of the construction indices found in the *Engineering News Record* or *Chemical Engineering* magazines. Thus:

$$\text{1983 cost} = \text{1980 cost}\,\frac{\text{(cost index 1983)}}{\text{(cost index 1980)}}$$

$$= \$30{,}000{,}000\,\frac{(316.9)}{(261.2)}$$

$$= \$36{,}400{,}000$$

(Using CE Plant Cost Annual Index from *Chemical Engineering*. An even better estimate could be obtained by splitting the total investment into its parts [i.e., equipment, labor, buildings, engineering, etc.] and using the corresponding itemized cost indices.)

The current cost is then adjusted for the expected inflation over the next 3 years. Assuming 5% annual inflation rate:

$$\text{1986 cost} = \text{1983 cost}\,(1 + \text{inflation rate})^n \quad (n = \text{years})$$
$$= 36{,}400{,}000\,(1.05)^3$$
$$\text{1986 cost} = \$42{,}100{,}000$$

Unless inflation gets out of hand, the accuracy of the estimate for new capital investment in the above example is probably quite good (i.e., within 10 to 15%), and certainly better than order-of-magnitude estimates. As the size and type of new facility becomes less and less similar to the one used to make the estimate, confidence in the estimate will decrease.

Example 2. The AA company would also like to consider building the new amino acid X facility with a capacity one and a half times that of their first plant. To estimate the cost for this facility, one can make use of the fact that capital investment typically varies with plant capacity according to a power function. That is:

$$\text{Plant A cost} = \text{plant B cost}\left[\frac{\text{(plant B capacity)}}{\text{(plant A capacity)}}\right]^x$$

where x is a fractional power almost always between 0.5 and 1.0. For fermentation processes a power of 0.75 has been suggested (11). Using the "power rule," the 1986 cost estimated in example 1 can be scaled to account for the difference in capacities. Thus:

TABLE 4. Lang factors

Plant type	Original:[a] fixed capital	Current[b]	
		Fixed capital	Total capital
Solids processing	3.10	3.87	4.55
Solids-fluids processing	3.63	4.13	4.87
Fluids processing	4.74	4.83	5.69

[a] Reference 29.
[b] Reference 33.

Larger plant cost

$$= \$42,100,000 \left[\frac{(3,000,000 \text{ kg/year})}{(2,000,000 \text{ kg/year})}\right]^{0.75}$$

$$= \$57,100,000$$

The accuracy of the estimate in example 2 may still be fairly good (within 20 to 25%), but as one might expect, the accuracy will decrease as the difference in plant capacity increases. Scaled estimates based on capacity ratios higher than 2 or 3 to 1 should only be used for very preliminary economic evaluation.

Scaled estimates can be useful even when the process is not identical, if good judgment is used.

Example 3. The AA company also has an R&D project under consideration to produce amino acid Y by fermentation. However, the anticipated market is expected to be smaller, 1,000,000 kg/year. Also, preliminary expectations of the R&D group are that the recovery process will be the same as for X but that the titer of amino acid Y at the end of fermentation will be half that of X.

The initial inclination in estimating the capital investment for a facility to produce Y might be to use the power rule with the two plant capacities. However, because the concentration of Y in the fermentation is half that of X, the volume of fermentation broth that must be handled in process Y is the same as that in the process X plant. Because the plant X equipment was mostly sized on the basis of liquid volume handled, the power rule should be applied here using the ratio of liquid handling capacity as opposed to actual product capacity. Thus:

Plant Y cost

$$= \text{plant X cost} \left[\frac{(\text{liquid in process Y})}{(\text{liquid in process X})}\right]^{0.75}$$

$$= \$30,000,000\ (1)^{0.75}$$

Plant Y cost

$$= \$30,000,000$$

Other ratioed techniques have been proposed for chemical plant cost estimation which can give quite good results (i.e., accuracy within 15 to 20%) (15, 41, 43). However, our experience with these techniques indicates that they will need to be modified to give reliable results for biotechnology-based facilities.

Preliminary estimates

Preliminary estimates are also called feasibility, study, and factored estimates. "Factored estimates" is a more descriptive term of the actual technique used for this type of capital estimate. The starting point of all factored estimates is a list of all the major equipment required to conduct the manufacturing process. The delivered price (f.o.b. cost + freight) for each major piece of process equipment is determined. The fixed or total capital investment in the plant is then estimated by multiplying the delivered equipment cost by a factor dependent upon the materials handling characteristics of the plant. This factor concept was first presented by Lang (29), and the phrase "Lang factor" has become a common part of the jargon of cost estimating. Table 4 presents values for the original Lang factors and a more up-to-date set of values. These factors can be broken down into percentages based on the different activities involved in installation of the equipment (e.g., foundations, piping, electrical, buildings, etc.) (Table 5). This is useful from an informational standpoint, but the detail in no way improves the accuracy of the estimate.

TABLE 5. Breakdown of Lang factors for fluid process plant[a]

Item	% of delivered equipment cost
Direct costs	
Purchased equipment—delivered (including fabricated equipment and process machinery)	100
Purchased-equipment installation	47
Instrumentation and controls (installed)	18
Piping (installed)	66
Electrical (installed)	11
Buildings (including services)	18
Yard improvements	10
Service facilities (installed)	70
Land (if purchase is required)	6
Total direct plant cost	346
Indirect costs	
Engineering and supervision	33
Construction expenses	41
Total direct and indirect plant costs	420
Contractor's fee (about 5% of direct and indirect plant costs)	21
Contingency (about 10% of direct and indirect plant costs)	42
Fixed capital investment	483
Working capital (about 15% of total capital investment)	86
Total capital investment	569

[a] Adapted from reference 33. These percentage factors (as well as most published factors) are based on purchase of carbon-steel equipment. Because most biologically based production facilities use more expensive stainless-steel equipment (2- to 2.5-fold higher), the percentages will be lower relative to delivered stainless-steel equipment costs.

TABLE 6. Factored cost estimates for vertical pressure vessel (<50 lb/in^2; carbon steel) field installation modules

Item	Factor (% of equipment cost) for base year:			
	1969[a]		1983[b]	
Equipment cost f.o.b. (E)		100.0		100.0
Field materials (m)				
Piping	60.0		63.6	
Concrete	10.0		10.7	
Steel	8.0		8.6	
Instruments	11.5		10.1	
Electrical	5.0		4.7	
Insulation	8.0		8.6	
Paint	1.3		1.4	
Total m		103.8		107.7
Direct materials (E + m = M)		203.8		207.7
Direct field labor (L)				
Material erection	84.0		61.7	
Equipment setting	15.2		11.2	
Total L		99.2		72.9
Direct cost (M + L)		303.0		280.6

[a] From reference 25.
[b] Guthrie (25) values adjusted to December 1983 using cost indices from *Chemical Engineering*.

To improve the accuracy of the factored estimate, individual Lang-type factors for different types of equipment can be used. This technique has been presented in most detail as a "module" approach by Guthrie (25). Guthrie's technique uses various multipliers to derive direct costs (material and labor) and indirect costs (engineering and construction) from the f.o.b. equipment cost. The sum of these direct and indirect costs is called the "bare module" cost. (The bare module cost can also be derived directly from f.o.b. equipment cost by multiplying by the bare module factor [bmf], which is a composite of the direct and indirect factors.) Once all of the individual bare module costs are added together, other elements of fixed capital (contingency and fees; also land, site development, and buildings if these have not been estimated separately) can be added to arrive at a total figure. This technique is probably the most accurate of the factored techniques (9, 26), although some modifications have been suggested (18) (also see below).

The major problem with any of these factored estimates is the tendency of the factors to become outdated. As an example (Table 6), if one estimates the direct material and labor cost of a $40,000 (f.o.b.) vertical pressure vessel (50 lb/in^2 maximum; carbon steel) using the Guthrie method, the cost would be $121,200 ($40,000 × 3.03 factor). However, if the individual components used to derive this factor are adjusted, according to the itemized indices in *Chemical Engineering*, from 1969 (date of the Guthrie publication) to 1983, the direct material and labor factor should be 2.81. In this case the associated costs have decreased relative to the equipment cost.

The preceding should illustrate the necessity of keeping up-to-date records on costs, as well as the inherent inaccuracies involved in the factor methods. However, excessive efforts to further improve the accuracy of preliminary estimation techniques are almost always unsuccessful due to lack of detail in the scope of the project, inaccuracies in the fundamental design assumptions used for the estimate, or both.

Final design estimates

Once a project has reached the process development stage and certain important technical milestones have been passed, the initiation of a fairly detailed design of a process facility should begin. This work will take as a starting point whatever preliminary cost estimates have already been performed, refining the mass and energy balances to reflect the most current results of process development. The improved design data should be sufficient to begin getting "serious" quotes for equipment from vendors and to allow development of a fairly detailed plant layout. This refinement of detail should bring the estimate within a closer confidence limit and make profitability analysis meaningful. If the product passes whatever economic criteria the corporation requires at this point, a final design estimate will be authorized. This estimate will be based upon a specific selected site, a detailed plant layout, and complete piping and instrumentation drawings. All materials will be completely itemized, as well as the specific craft labor associated with installation. All indirect charges will also be estimated in detail, based upon a fully described schedule of construction operations.

It should be readily apparent from the brief description in the preceding paragraphs that a significant amount of effort from a variety of engineering disciplines is required for a final design estimate. Most frequently, a company involved in manufacturing or development operations will not have sufficient engineering staff to carry out such an assignment. When this is the case, it is most prudent to go to an outside engineering contractor for the final design estimate. The *Engineering News Record* periodically publishes lists of the major engineering design and construction companies around the world. These lists should be explored with regard to particular expertise relating to the project of interest. The engineering contractor's expertise should complement that of the contracting company. This experience is invaluable not only in accurately assessing how much a project will cost, but also in identifying important items which may have been overlooked during the preliminary estimate or development efforts.

26.3.3 Specific Equipment Costs

All factored and detailed estimates depend on specific information regarding the purchase cost of major equipment. Sources for these costs are published charts and graphs (9, 25, 26, 32, 33), annual equipment and construction cost publications (e.g., *Richardson Rapid System*, *Means Building Construction Cost Data*, Lee Saylor's *Current Construction Costs*), previous equipment purchases, and, of course, actual equipment vendors. Given sufficient detail for a quote, the vendor will be the most reliable. On the other hand, when only sketchy design details are available, current published data are often more quickly available and are as accurate as the level of detail warrants.

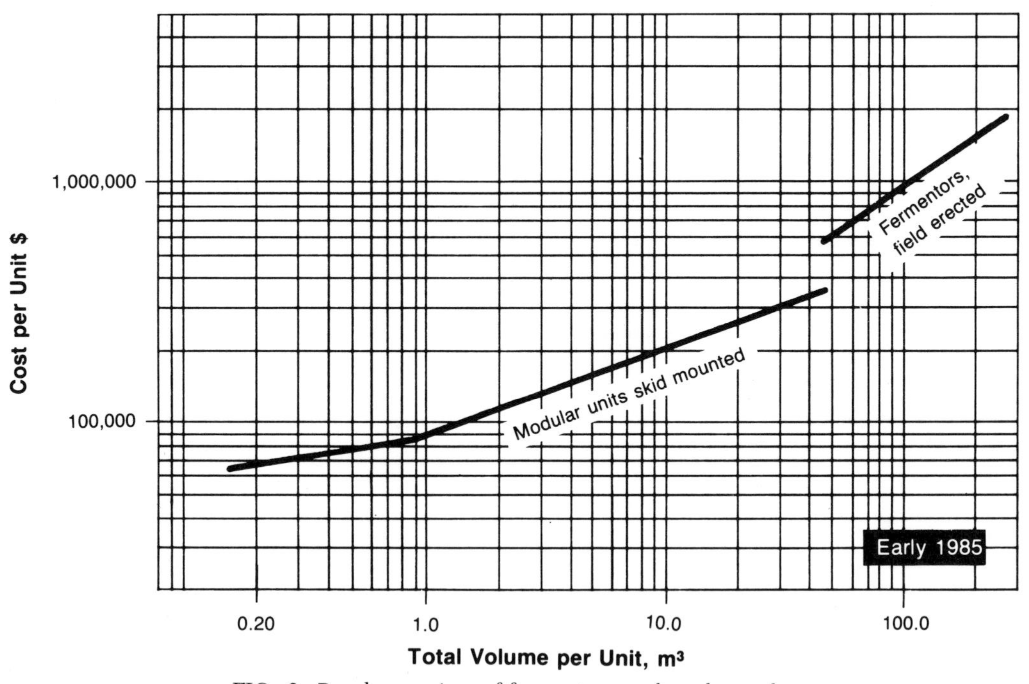

FIG. 2. Purchase prices of fermentors and seed vessels.

Previous equipment purchases or vendor quotes can be updated by using the published cost indices. Updating equipment costs using indices more than 5 years apart may lead to serious inaccuracies.

When price quotes have been obtained for specific pieces of equipment, they can be scaled to other capacities by using the power rule (see example 2 above) and an exponent identified with that particular type of equipment (9, 12, 28, 33, 44). Capacity exponents for the individual elements involved in field construction and installation are also available (30). Again, caution should be taken when using this technique to extrapolate beyond capacity ratios of 2 or 3 to 1.

Figures 2 through 5 present some biotechnology equipment cost charts not generally seen elsewhere. Details concerning these figures are given below. It should be noted when using these charts, and other similar ones, that equipment and machinery are usually sold in standard sizes. Thus, although the solid lines indicate a continuous function of capacity, in actual practice the function is discontinuous with capacity.

Fermentors

Figure 2 presents unit costs (courtesy of Jack Wilson, ABEC, Inc.) for seed vessels and fermentors equipped for accurate, automatic control and recording of all basic operational variables. Each unit is constructed to standard code requirements, and the material of construction is generally 316 stainless steel, particularly in areas of contact with process fluids. In the unit size range of 1.5 to 45 m^3, costs plotted are for complete modular units mounted on transportable skids f.o.b. factory. The cost of installation for each unit would include transportation cost, location at the point of use on a suitable foundation, and attachment of the necessary utility services. For the skid-mounted units, installation costs (excluding indirect charges) would probably fall in the range of 10 to 20% of the f.o.b. modular unit cost.

The short section of the graph plotted to the right provides estimates for units too large to be conveniently produced in complete skid-mounted modular units; instead, this section may be used to estimate the cost of field-erected fermentors over the size range of 70 to 250 m^3. To establish this section of the curve, the vendor's estimate of $1,000,000 for a field-erected fermentor, 113.6 m^3 (30,000 gallons) in size, was extrapolated using the vendor's recommended exponent of 0.7. For derivation of the bare module cost, it should be understood that there is no separate charge to be made for installation of field-erected units (indirect charges should be added, however).

The costs presented here are for basic, "no-frills" fermentors. Special requirements (e.g., biosafety level 1 containment, complete computerized control, extra addition vessels and nozzles, special instruments, etc.) can raise the base cost substantially (by as much as 50 to 100%).

Disk centrifuges

Disk centrifuges are commonly rated for capacity in terms of hydraulic throughput, or in terms of the volume of water which may be transported through a centrifuge machine in a unit of time. Functionally, however, the purpose of the centrifuge is generally to separate a heavier particulate suspended solid from the suspending fluid. Depending upon the nature of the mixture, the throughput may vary over a considerable range. Obviously the separation of large microorganisms (e.g., yeasts) from a fermented broth will proceed more rapidly or more efficiently than the separation of a bacterium such as *Escherichia coli*.

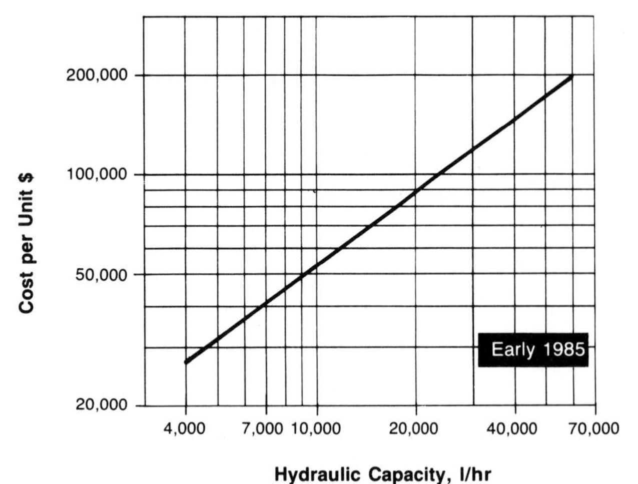

FIG. 3. Purchase prices of disk centrifuges.

The functional capacity of a given model of centrifuge will have to be established for each required separation. For the recovery of yeast cells, the functional rate for effective recovery of a cell paste would be expected to be about 50 to 100% of the hydraulic rate capacity. For effective recovery of an *E. coli* cell paste or concentrated slurry, a feed rate of 5 to 10% of the hydraulic capacity would probably be satisfactory.

Figure 3 charts the hydraulic capacity of continuous (desludging) disk centrifuge units against the purchase cost. Cost estimates derived from the use of this chart may be regarded as base costs; special features or capabilities added by modification of standard units may be expected to add to ultimate costs. Furthermore, cost estimates based on this chart are for centrifuges, f.o.b. fabricator. To estimate total installed cost, it would be necessary to add additional cost load as defined, for example, by the Guthrie factors (see example 5, below).

Homogenizers

When the contents of the microbial cell must be recovered for processing or fractionation, it generally becomes necessary to break or disrupt the cell at an early stage in the course of operations. Cells may be broken by chemical means, by enzymatic attack, or by mechanical rupture of the cell. One mechanical method which has found some favor is the use of a bead mill; another mechanical process which is quite effective is the high-pressure homogenizer, which involves the pumping of a slurry or suspension of cells through a restricted passageway at very high pressures.

Figure 4 presents the cost for high-pressure homogenizers operated at 8,000 lb/in^2 versus operating rate in liters per hour. The chart gives equipment cost f.o.b. fabricator; estimates of total installation cost may be derived from application of the Guthrie factors (see example 6).

FIG. 4. Purchase prices of cell homogenizers.

Ultrafilters

Ultrafiltration is a recent development of great utility in recovering and purifying biologically active substances. By selection of filtration membranes of appropriate permeability, a certain amount of separation can be achieved, but more effectively, concentration of product can be obtained without significant loss.

There are a number of manufacturers of ultrafiltration equipment, each with different types of modular (cartridge) membrane designs (e.g., hollow-fiber, spiral-wound, and plate type). Variation in operating capacity is generally obtained by arranging a number of cartridge systems in parallel.

The membrane area required is established from the average permeate flux rate. Flux rate will vary over time depending upon the system and extent of concentration. Typical average flux rates are 5 to 50 liters/m^2 per h.

Figure 5 relates membrane area to f.o.b. equipment cost, exclusive of membranes. The cost information in Fig. 5 has been derived from vendors' quotes for ultrafiltration units of various membrane surface areas. As such the cost line represents an average cost, with actual prices probably falling within about 25% depending upon the application. Membrane costs can vary from \$150 to 250/m^2. Depending upon the useful

TABLE 7. Target protein case study: production conditions

Variable	Quantity
Annual production	5,000 kg
Recovery efficiency	40%
Annual production (protein in broth)	12,500 kg
Yeast, broth concentration, dcw	40 g/liter
Total intracellular protein, broth concentration	16 g/liter
Fermentor schedule	
Fermentation	24 h
Turnaround	6 h
Operating days/year	350
Weight yield, yeast (per kg of glucose)	0.4 kg
Target protein concentration in broth	0.8 g/liter
Annual broth volume	15,625,000 liters
Fermentors	
Number	2
Size	40,000 liters
Working volume	27,902 liters
Weight yield (recovered target protein per kg of glucose)	0.0032 kg

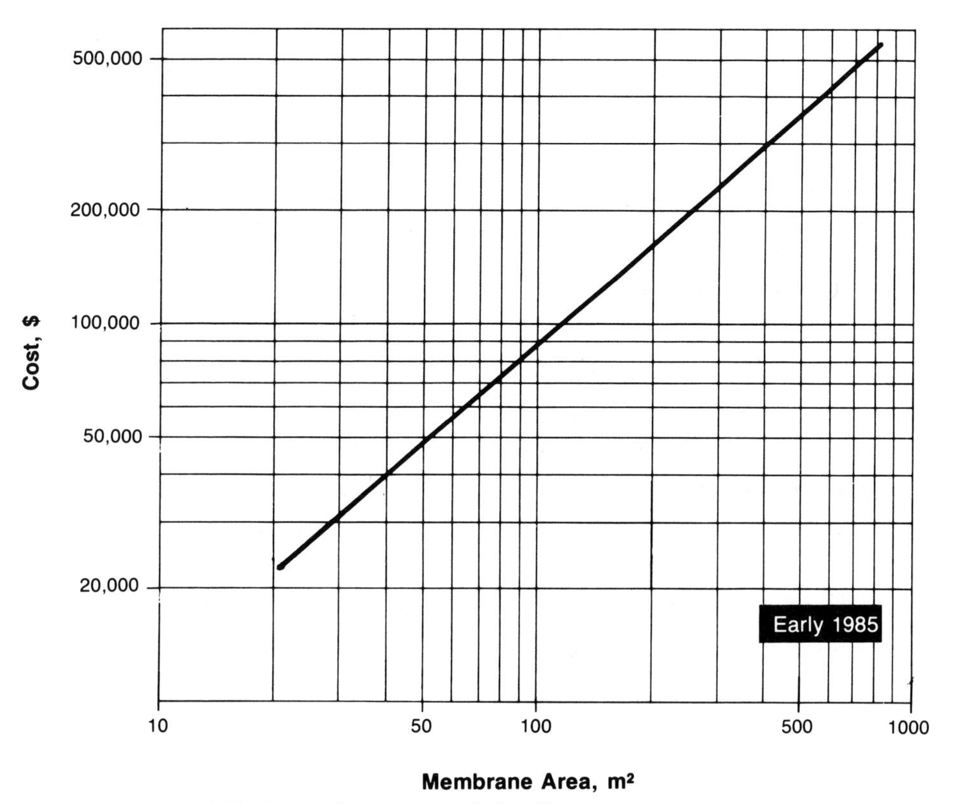

FIG. 5. Purchase prices of ultrafiltration equipment.

life of the membranes, the cost can be accounted for as a direct operating expense under raw materials and supplies, or as a fixed cost under depreciation.

26.4 CASE STUDY: MANUFACTURING COST ESTIMATE FOR A PROTEIN PRODUCED INTRACELLULARLY BY A RECOMBINANT MICROORGANISM

As an illustration of a manufacturing cost estimate, a preliminary cost estimate prepared for a 40,000-dalton protein (target protein) projected to be produced intracellularly by a genetically engineered yeast strain is presented. The purpose of the analysis is to determine whether the R&D required will result in an economically attractive process. It is expected that if 5,000 kg of the product can be manufactured per year for less than $500/kg, a significant profit can be made.

26.4.1 Process Design

After discussions with the molecular biologists, microbial physiologists, and biochemical engineers expected to be involved with the development, a conceptual process design is arrived at. From this design a process description can be written, flow charts can be developed (Fig. 6, 7, and 8), and energy and material balances can be derived.

Process parameters

The general conditions projected for the protein's production are listed in Table 7. Specific points of interest are as follows.

Containment. The National Institutes of Health (NIH) *Guidelines for Research Involving Recombinant DNA Molecules* (Fed. Regist. 1 June 1983) recommend that large-scale production using a recombinant microorganism be conducted in a manner that eliminates release of viable organisms from the process. In general, compliance with these guidelines will warrant either development of a procedure to inactivate the host organism before processing, or construction of a processing facility which is physically contained so as to prevent release. The former solution may involve a difficult development effort to kill the microorganism and still maintain active, recoverable product. On the other hand, physical containment, though certainly possible from an engineering standpoint, may be prohibitively expensive. No provision for compliance with the NIH Guidelines is made in this estimate. If physical containment were required, it would probably result in an increase in fixed capital investment of at least 10 to 20% and an increase in manufacturing cost of 5 to 10%.

Cell concentration: 40 g (dry cell weight [dcw])/liter. Since the protein is produced intracellularly, its production per unit volume of fermentor broth is related to cell density. High cell density is, of course, desirable, and that proposed here is reasonably optimistic.

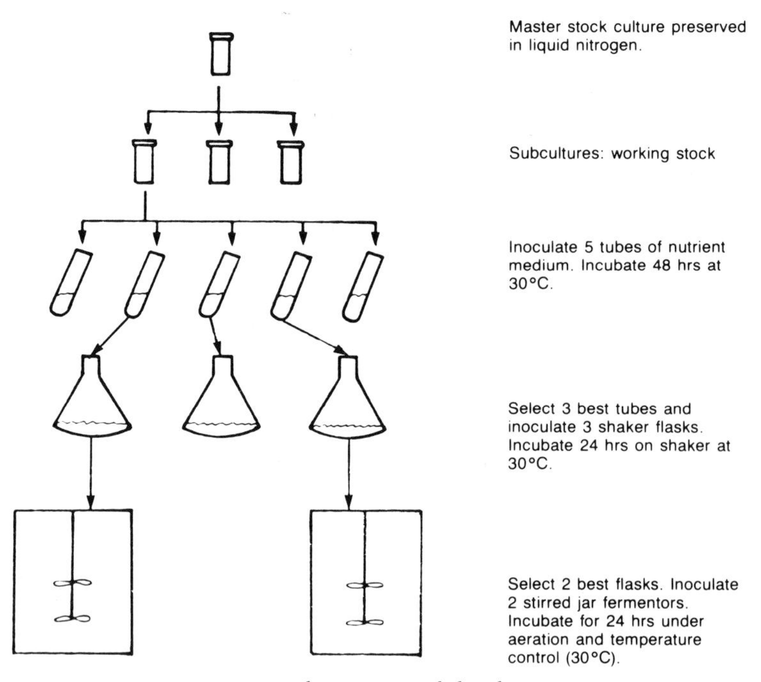

FIG. 6. Laboratory seed development

Fermentation time: 24 h with 6 h of turn-around time. The time required is only that necessary to take the cells up to a high density and is therefore shorter than required for normal fermentations. The value chosen is considered to be optimistic, but has been achieved in other cases.

Total intracellular protein concentration: 40% of dcw (i.e., 16 g/liter). The projected total intracellular protein concentration is typical of yeasts (10).

Raw materials and cell yield. The cell yield is assumed to be 0.4 kg of cells per kg of carbohydrate. A "clean" carbohydrate source (96-dextrose-equivalent corn syrup) and a high-value, complex nitrogen source (casein hydrolysate) are chosen. These choices are based upon the assumptions that a carbohydrate source free of other dissolved materials would permit better protein recovery and that good growth of the genetically engineered organism and production of the target protein would be enhanced by a nitrogen supplement containing peptides and amino acids, such as casein hydrolysate.

The nitrogen required is estimated at 10% of total cell mass and is supplied at twice that amount. About one-tenth of this would be supplied by the complex nitrogen source, with the remainder supplied by a simple nitrogen source, ammonium sulfate.

Assumed target protein concentration. The concentration of the target protein is assumed to be 5% of total intracellular protein. This value represents a level of production considered easily obtainable by genetic engineering.

Product recovery efficiency: 40%. Recovery efficiencies in excess of 50% for purified intracellular protein are uncommon. (Note: to limit the possibility of losses associated with denaturation, all processing in the present study is projected to be done at low temperatures, around 5°C.)

Cell development

A 5% inoculum is assumed for seed development stages. Culture development at the first level (from frozen seed cultures) is expected to take 48 h; that at all other levels is projected at 24 h. The procedures for cell development and production are illustrated in Fig. 6 and 7.

Recovery

Because no R&D work has even been started, the recovery process must necessarily be a generalized one, not specific for the protein of interest. The projected process is therefore based upon information obtained from various general references on protein purification (16, 21, 22, 24, 42) and on industrial procedures used in blood fractionation (19, 36). As such, the processing steps envisioned, although probably not the same as those which might ultimately be developed, serve as "place-holders" which can be used to develop an economic estimate. Generally, the accuracy of this approach will be surprisingly good, provided the recovery efficiency is accurately estimated (see also section 26.5).

The recovery process for the protein is diagrammed in Fig. 8. The whole broth is rapidly discharged from the fermentor, through one or more plate exchangers which cool it to around 5°C, and into a holding vessel. From the hold tank the broth is continuously pumped to a high-speed disk centrifuge which continuously discharges the recovered cell cream (50% solids) and clarified broth at separate ports. The liquid is discarded, and the cell cream is pumped to a set of two high-pressure homogenizers in series which serve to disintegrate the cells mechanically. The high-pressure action also causes a temperature rise, so the homogenizers are equipped with cooling coils. The homoge-

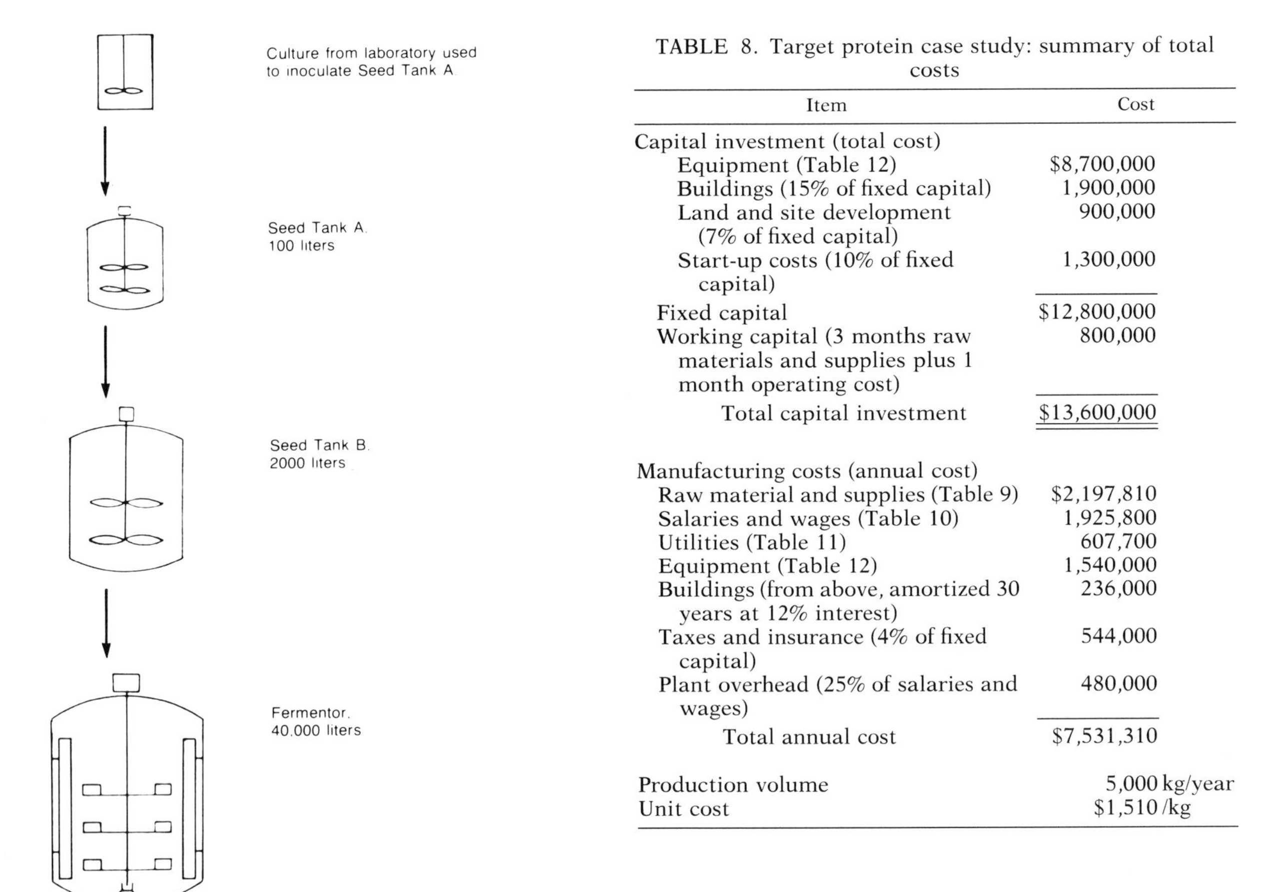

FIG. 7. Factory cell development.

TABLE 8. Target protein case study: summary of total costs

Item	Cost
Capital investment (total cost)	
Equipment (Table 12)	$8,700,000
Buildings (15% of fixed capital)	1,900,000
Land and site development (7% of fixed capital)	900,000
Start-up costs (10% of fixed capital)	1,300,000
Fixed capital	$12,800,000
Working capital (3 months raw materials and supplies plus 1 month operating cost)	800,000
Total capital investment	$13,600,000
Manufacturing costs (annual cost)	
Raw material and supplies (Table 9)	$2,197,810
Salaries and wages (Table 10)	1,925,800
Utilities (Table 11)	607,700
Equipment (Table 12)	1,540,000
Buildings (from above, amortized 30 years at 12% interest)	236,000
Taxes and insurance (4% of fixed capital)	544,000
Plant overhead (25% of salaries and wages)	480,000
Total annual cost	$7,531,310
Production volume	5,000 kg/year
Unit cost	$1,510 /kg

nate is pumped to a mixing vessel where it is made 0.05 M with $MnSO_4$ to precipitate free nucleic acids. Filter aid is also added at this point, and the entire slurry is pumped to a vacuum rotary drum filter for removing cell debris and the precipitated nucleic acids. The filtrate is a crude, dilute (about 1 wt%) solution of intracellular proteins.

The crude protein solution next undergoes a rough fractionation by ultrafiltration. In this step, very large-molecular-weight and micrometer-size debris which passed the rotary drum filtration is removed by passing the solution through ultrafiltration membranes permeable to about 100,000 molecular weight (MW) or lower (300,000 nominal MW membrane). The extent of losses at this point will depend upon a balance of ultrafiltration operating costs and the extent of high-molecular-weight concentration. The permeate from the high-molecular-weight cut is then pumped through a low-molecular-weight (30,000 nominal MW) ultrafilter. This membrane retains the proteins of higher weight and passes smaller molecules. This step is thus a simultaneous fractionation and concentration. The permeate is discarded, and the fivefold concentrate (about 5 wt%) is subjected to further processing.

After ultrafiltration the protein concentrate proceeds through a cold ethanol fractionation. Ethanol concentration is altered and pH is varied to selectively precipitate proteins other than the protein of interest.

The solution is next purified by selective binding and elution using an ion-exchange/gel chromatography system. The exact conditions and performance of this system cannot be predicted with any accuracy, but the equipment is included to account for its contribution to manufacturing costs. The purified protein solution resulting from the chromatographic procedures is finally concentrated by ultrafiltration. After addition of buffers or stabilizers, the solution is sterile filtered and bottled.

26.4.2 Economic Estimate

The capital and operating costs for the process can be calculated during the process of developing the material and energy balances, or later. The cost accounts for the items discussed in sections 26.1 and 26.2 are summarized in Tables 8 through 12. Some specific examples of part of the capital cost estimate for the case study are detailed below.

As summarized in Table 8, the unit cost of manufacture (COM) for the target protein is estimated to be $1,510/kg. Because the target COM for the protein is $500/kg (as stated at the beginning of this section), R&D effort on this project, which results in a production system with the conditions assumed in Table 7, is of questionable value. (See example 8 below for further discussion.)

Example 4. Fermentation area (main fermentor[s])

1. (5,000 kg of protein recovered per year) ÷ 0.4 recovery factor = 12,500 kg produced per year

TABLE 9. Target protein case study: raw materials and supplies

Materials and supplies	Quantity	Unit cost	Annual cost
Fermentation			
95 Dextrose equivalent corn syrup	2,550,000 kg	$0.30/kg	$765,000
Casein hydrolysate	13,750 kg	$7.00/kg	96,250
$(NH_4)_2SO_4$ (food grade)	123,750 kg	$0.52/kg	64,350
K_2HPO_4	51,560 kg	$1.25/kg	64,450
$MgSO_4 \cdot 7H_2O$	8,600 kg	$0.75/kg	6,450
NaCl	17,190 kg	$0.10/kg	1,720
$CaCO_3$	17,190 kg	$0.022/kg	380
Antifoam	17,190 kg	$0.75/kg	12,890
Miscellaneous (trace elements, etc.)		ca. 5%	53,000
			$1,064,490
Recovery			
$MnSO_4 \cdot H_2O$	118,000 kg	$1.50/kg	$177,000
Filter aid	82,800 kg	$0.33/kg	27,320
95% Ethanol	35,000 liters	$0.50/liter	17,500
Ultrafilters	350 m^2 × 300,000 MW	$200/m^2	70,000
	150 m^2 × 30,000 MW	$200/m^2	30,000
Cation IX resin	1,400 liters	$205/liter	287,000
Anion IX resin	170 liters	$225/liter	38,250
Gel media	225 liters	$190/liter	42,750
			$689,820
Supplies			
Operating (3% of operating salaries and wages)			$26,200
Laboratory (40% of laboratory salaries and wages)			59,900
Maintenance (100% of maintenance salaries and wages)			204,400
Other (10% of plant overhead)			48,000
			$338,500
Subtotal			2,092,810
5% Contingency			105,000
Total			$2,197,810

2. 12,500 kg produced ÷ 0.05 fraction of total protein = 250,000 kg of total protein
3. 250,000 kg of total protein ÷ (0.4 fraction of yeast dcw) = 625,000 kg dcw
4. 625,000 kg dcw ÷ 0.04 kg dcw per liter of fermentation = 15,625,000 liters of fermentation
5. (350 days × 24 h) ÷ 30 h per fermentor cycle = 280 fermentor cycles per year
6. Say two fermentors, then:
   fermentor working volume
   = 15,625,000 ÷ 2 ÷ 280
   = 27,902 liters
   Figuring working volume at 70%, then:
   actual fermentor volume = 39,860 liters (say 40,000 liters)
7. From Fig. 2 the purchase price for a skid-mounted 40,000-liter fermentor = $330,000
8. From Guthrie (25), the bare module factor (bmf) for a vertical process vessel is 3.18; however, because the fermentor is skid mounted and of stainless-steel construction, the factor must be adjusted.
   a. Adjustment for skid mount: assume skid price includes all direct material and labor except concrete, insulation, and equipment setting; then the bmf, including indirect costs, is 2.12.
   b. Adjustment for stainless steel: if the unit were carbon steel, then the total module cost would be f.o.b. cost × bmf. Assuming installation and indirect costs for a stainless-steel vessel are about the same as for a carbon-steel vessel, and the material cost ratio of stainless steel to carbon steel is 2.25, then:

total fermentor module cost

$$= \text{f.o.b. stainless-steel unit} + \left(\frac{\text{f.o.b. stainless-steel unit}}{2.25}\right) \times (\text{bmf} - 1)$$

$$= 350{,}000 + [(350{,}000/2.25)(2.12 - 1)]$$

$$= \$494{,}270 \text{ (say } \$495{,}000)$$

9. Enter two 40,000-liter fermentors at $495,000 on equipment list (Table 12).

Example 5. Cell recovery

1. Assume cells harvested from drop tank over 8-h period: 27,902 liters ÷ 8 h = 3,488 liters/h (say two continuous disk centrifuges with 2,000-liters/h functional rate)
2. For yeasts, assuming a hydraulic-to-functional-capacity ratio of 2:1; from Fig. 3, f.o.b. cost for a 4,000-liters/h hydraulic capacity centrifuge = $27,000.
3. From Guthrie (25), the direct materials and labor factor for a stainless-steel solid-bowl centrifuge is 1.6. (Actually the term "solid-bowl" is an inadequate adjective because it is commonly used to describe both disk- and scroll-type units. Because the centrifuge required here is a "dropping-bowl" type, it is assumed that the direct materials and labor factor will be higher, 2.0. With indirect costs estimated at 29% of direct materials and labor, the bmf is 2.0 × 1.29 = 2.6. Then total centrifuge module cost

TABLE 10. Target protein case study: salaries and wages

Category	Personnel required	Wage rate ($/h)	Annual cost
Salaries			
Production manager	1		$42,000
Maintenance supervisor	1		38,000
Shift supervisors	4		140,000
Secretaries	4		60,000
			$280,000
Wages			
Shipping and receiving	2	8	$33,280
Batching	4	8	66,560
Fermentation	12	10	249,600
Recovery	20	10	416,000
Laboratory (quality control and in-process)	8	9	149,760
Maintenance	8	10	166,400
			$1,081,600
Overtime (6%)			64,900
			$1,146,500
Subtotal			$1,426,500
Fringe benefits (35%)			499,300
Total			$1,925,800

TABLE 11. Target protein case study: utilities

Utility	Annual consumption	Annual cost
Steam (@ $15/1,000 kg)		
Sterilization	5,156,250 kg	$77,340
Solvent recovery	875,450 kg	13,130
Contingency (25%)		30,000
		$120,470
Water		
Process ($0.50/1,000 liters)	25,000,000 liters	$ 12,500
Cooling ($0.50/1,000 liters)	45,000,000 liters	22,500
High purity ($2.00/1,000 liters)	8,000,000 liters	16,000
Contingency (10%)		6,000
		$ 57,000
Electricity (@ $0.07/kW-h)		
Batching	7,700 kW-h	$ 540
Fermentation: Aeration	2,062,500 kW-h	144,380
Agitation	825,000 kW-h	57,750
Centrifuges	35,840 kW-h	2,510
Homogenizers	268,800 kW-h	18,820
Drum filters	134,400 kW-h	9,410
Agitators	67,200 kW-h	4,700
Pumps	140,000 kW-h	9,800
Chiller	336,000 kW-h	23,520
Cooling tower	12,900 kW-h	900
General load (ventilation, air conditioning, lighting, etc.) @ 10%		30,000
		$302,330
Waste treatment		
Landfill (@ $0.03/kg)	4,129,000 kg	$123,870
Biological (@ $0.20/1,000 liters)	20,000,000 liters	4,000
		$127,870
Total utilities		$607,700

= 2.6 × $27,000
= $70,200 (say $71,000)

4. Enter two 4,000-liters/h hydraulic capacity continuous disk centrifuges at $71,000 on equipment list (Table 12).

Example 6. Cell lysis

1. Assume cell slurry from centrifuges is 50% (wet weight cells per volume) and that slurry undergoes two passes (each 4 h) through a high-pressure homogenizer.
2. Fermentor cell titer
   = 0.04 kg of dry cells/liter × 5
   = 0.20 kg of wet cells/liter (20%)
3. Cell cream flow rate
   = 27,902 liters × (20%/50%) ÷ 4 h
   = 2,790 liters/h
   (say two 1,500-liters/h machines)
4. From Fig. 4, purchase price is $38,000.
5. Guthrie (25) does not list a bmf specific for the homogenizer; therefore, assume a bmf for a reciprocating pump (which the homogenizer essentially is), i.e., 3.38. Because this bmf is for a carbon-steel unit and the homogenizer is stainless steel, the factor must be corrected as in example 4. Thus, total homogenizer module cost:
   = 38,000 + (38,000/2.25) × (3.38 − 1)
   = $78,200 (say $79,000)
6. Enter two 1,500-liters/h high-pressure homogenizers at $79,000 on equipment list (Table 12).

Example 7. Ultrafiltration

(i) High-molecular-weight and micrometer-size debris removal

1. Assume 8-h batch operation and 50-fold concentration (removal) of high-molecular-weight debris at average flow rate of 10 liter/m^2 per h:
   Total permeate = 27,900 liters (49/50)
   = 27,342 liters
   Required membrane area = (27,342 liters/8 h)/(10 liters/m^2 per h)
   = 341.8 m^2 (say 350 m^2)
2. From Fig. 5, purchase price for a 350-m^2 ultrafiltration system is $270,000.
3. There is no Guthrie factor for ultrafiltration systems. Because these units are normally packaged systems, assume a small installation factor (1.1) and normal indirect factor (1.29). Thus, the bmf = 1.42, and installed cost is $270,000 (1.42) = $383,400 (say $385,000).
4. Enter one 350-m^2 ultrafiltration unit at $385,000 on Table 12.
5. Figure membrane life at 1 year and cost at $200/m^2.
6. Enter 350 m^2 of 300,000-MW filters at $200/m^2, or $70,000 annual cost on raw material and supplies list (Table 9).

TABLE 12. Target protein case study: installed equipment (including indirect charges)

Equipment	Cost
Batching area	
1 15,000-liter glucose syrup storage tank (agitator, stainless steel, coils, insulated)	$77,000
1 5,000-liter batching tank (stainless steel, with load cells loading, agitator)	42,500
5 1,000-liter blending tanks @ $12,000	60,000
Fermentation area	
4 20-liter seed fermentors @ $20,000	80,000
2 100-liter seed fermentors (stainless steel, with addition vessels, instruments, jacket, baffles, and agitator) @ $75,000	150,000
2 2,000-liter seed fermentors (stainless steel, equipped as above) @ $180,000	360,000
2 40,000-liter production fermentors (stainless steel, equipped as above) @ $495,000	990,000
Recovery	
2 100-m² plate heat exchangers (broth chilling) @ $60,000	120,000
1 40,000-liter chilled broth holding tank (stainless steel, with agitation, insulation, coils)	106,000
2 Desludging disk centrifuges (4,000 liters/h, stainless steel) @ $71,000	142,000
2 High-pressure homogenizers (stainless steel, with cooling coils, 1,500 liters/h) @ $79,000	158,000
1 10,000-liter $MnSO_4$ solution mixing/holding tank (stainless steel, agitated, insulated, coils)	30,000
1 40,000-liter mixing vessel (stainless steel, agitated, insulated, coils)	120,000
1 4,000-liter filter aid slurry tank (agitator, stainless steel)	23,000
1 50-m² vacuum rotary drum filter (stainless steel)	285,000
1 Ultrafiltration unit (300,000 MW, 350 m²)	385,000
1 Ultrafiltration unit (30,000 MW, 150 m²)	185,000
3 10,000-liter protein precipitation vessels (stainless steel, agitated, insulated, cooled) @ $30,000	90,000
1 Desludging disk centrifuge (protein precipitate removal; 4,000 liters/h)	71,000
1 10,000-liter 95% ethanol receiving vessel (stainless steel, with insulation and cooling coils)	33,000
1 3,500-liters/day 95% ethanol recovery distillation unit	152,000
2 Ion-exchange systems @ $250,000	500,000
1 Gel filtration system	250,000
1 Sterile filtration and bottling system	100,000
Ancillary equipment	
85 Pumps and motors, various sizes and types	637,500
1 40-ton chiller (20°F [ca. −7°C], with coolant holding vessel)	100,000
1 1,500-kg/h boiler, with water treatment, distribution, and condensate return systems	70,000
1 40-m³/min centrifugal air compressor (100 lb/in², 225 kW)	277,000
1 Cooling tower, 125 gal/min, 25°F range, 2-horsepower fan	20,000
Laboratory equipment	500,000
Office equipment	100,000
Subtotal	$6,214,000
+ Contingency and fee (40%)	2,486,000
Total	$8,700,000

Amortized cost, 10 years @ 12% interest = $1,540,000

(ii) Concentration and low-molecular-weight debris removal

1. Assume 8-h batch operation and 50-fold concentration with 30,000-MW membrane at average flux of 20 liters/m² per h.
   Permeate = 27,342 liters (16/20) = 21,874 liters
   Membrane area = (21,874 liters/8 h)/(20 liters/m² per h) = 137 m² (say 150 m²)
2. As before,

System cost = 130,000 (from Fig. 5) × 1.42 (bmf)
= $184,600 (say $185,000 installed cost)

   Enter on Table 12.
3. Enter 150 m² of 30,000-MW membranes at $200/m², or $30,000 per year, on Table 9.

(iii) Final product concentration

1. Assume final product at 30 g/liter.
2. Assume dilution during chromatographic procedures results in product solution of 1 g/liter. Thus, 30-fold concentration is required.
3. With 30,000-MW membranes for concentration, 4-h batch time, and flux rate of 50 liters/m² per h:
   Per batch final product solution = (5,000 kg)/(0.03 kg/liter)/(2) (280 batches per year) = 298 liters/batch
   Preconcentrate solution = 298 liters (15)
   Permeate = 8,940 − 298 = 8,172 liters
   Membrane area = (8,642 liters/4 h)/(50 liters/m² per h) = 43 m² (say 50 m²)
4. Because this requirement is small, assume that the ultrafiltration unit sized above will serve dual duty.

26.5 USING COST ANALYSIS FOR R&D DECISION MAKING

Biotechnology offers many opportunities for exciting new products and elegant production techniques. These new opportunities for product and process development will support extensive new efforts in basic scientific research. Early in the course of such scientific studies, the likelihood that the product can be made economically by the new process should be determined. The economic potential of a project should be one of the bases for choice among the many product opportunities available.

While the undertaking of a development program may, in part, depend upon a cost analysis for justification, cost analysis has other useful applications. Alternative development programs may be chosen on the basis of cost comparison, and indeed, prospective cost of operations and prospect for profit should always be a major consideration during the course of a process or product development program. Formula-

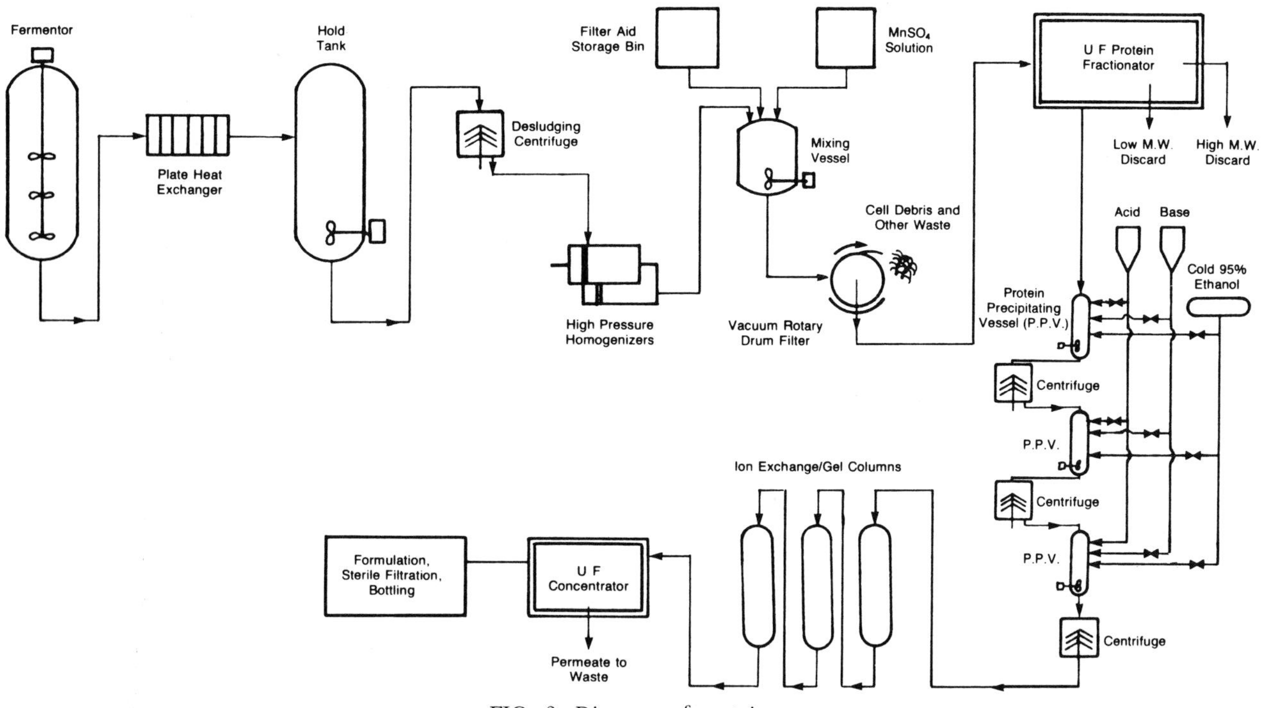

FIG. 8. Diagram of protein recovery.

tion of cost estimates should, in modern practice, be computerized so that continual cost analysis will accompany the course of development studies. As a product or process development program approaches the point of application, process optimization studies will become relatively important, and such optimization procedures should, of course, be keyed to the process economics. Even when commercial operation of the program is undertaken, continual process optimization directed toward minimizing operating costs should be pursued (37).

Cost analysis is particularly important for those cases in which the potential project will be in competition with already established chemical, enzymatic, or fermentative synthesis. It is important to maintain close cost surveillance on a new development project to avoid the dissipation of funds and effort on noncompetitive ventures. Obviously, cases will occur in which the cost of a unit of production is of less importance than some other product quality, and in such cases, much process development will undoubtedly be undertaken without careful cost analysis. (Traditionally, pharmaceuticals have fallen in this category.) In the end, however, it would appear to be prudent to maintain continual awareness of all of the elements of cost which a new product, whether simple or complex, will have to meet in the quest for profit.

In general, the very first step in development of a process is the identification of a need, desire, or ability to make a specific product. This need, desire, or ability can be analyzed from an economic perspective as to its potential for attracting investment and returning that investment plus some required amount of profit. (It will be assumed that this is the basic requirement for justifying any development effort.) From the per-

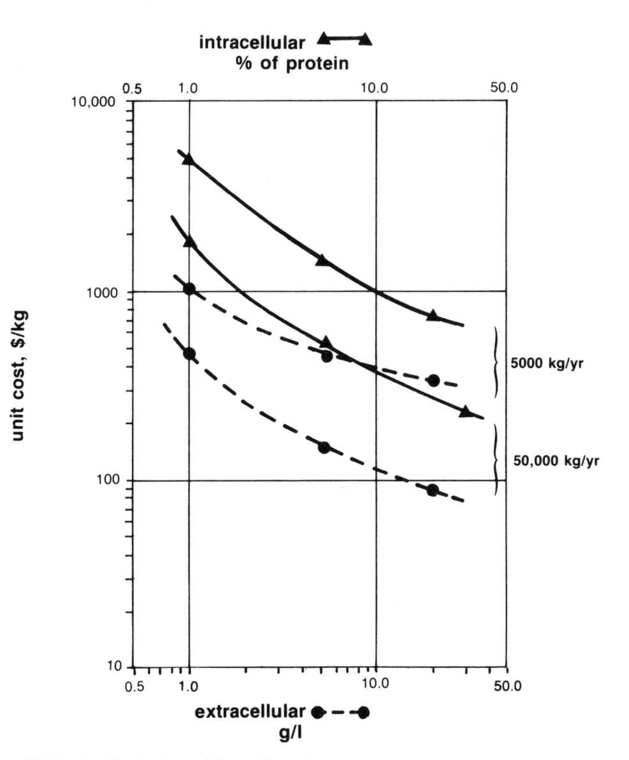

FIG. 9. Relationship of unit cost to protein expression level (based on assumptions detailed in section 26.4 and examples 8 and 9).

spective of biotechnology, the tremendous accomplishments of genetic engineering have given us alternative ways to synthesize established products as well as establishing needs and desires for totally new products.

Once the concept of making a specific product has been identified, the process of exploring its economic feasibility can begin. This starts by conceptualizing a reasonable set of process parameters, from which a plant design can be established and a cost analysis generated as shown in section 26.4. The conceptualization of a process does not require cost analysis, but the economics associated with individual unit operations are an important consideration in the conceptualization process.

The first set of parameters that must be established relate to marketing considerations. What is the value of the product of interest (i.e., dollars per unit), and how much can reasonably be expected to be sold (i.e., number of units)? The value of good marketing information cannot be overemphasized. Many research dollars may be wasted if the sales price or production volume of a product is overestimated, and potentially valuable research projects may never be initiated because of similar underestimates. The production volume is an absolute necessity to estimate the size and cost of capital equipment and the associated operating costs. The product value is not of absolute necessity to an economic estimate; however, it can greatly reduce the amount of time spent on unfeasible projects by raising an economic warning flag.

One of the first economic criteria upon which to base a decision to commit research funds is the cost of the raw materials needed to produce the product of interest. Obviously, a company would be foolish to spend time and money trying to develop a process to make a less valuable product from a more valuable one. Such analysis may be intuitively obvious, but it represents an often used tool for screening new product ideas or alternative raw material sources.

In biotechnology-based production processes, not only the primary raw materials (those directly converted to product) have a significant impact on economic feasibility. Secondary raw materials (e.g., cofactors, enzymes, inducers, auxotrophic requirements, and recovery materials) can tremendously influence the costs of production. The high costs of these materials will not necessarily cause a project to be terminated (or not to be initiated), but they may serve to focus aspects of the research on minimizing or eliminating prohibitive costs.

It is generally true that the total raw material costs associated with production and recovery serve as very good directors and monitors of development efforts aimed at optimization. This is especially the case when raw material costs represent between 50 and 80% of the total manufacturing cost of a product. For this reason, the raw material costs can be used to give quick insight into the potential of proposed projects or the importance of process developments.

When considering direct fermentation products, it is convenient to classify them according to "value groups" (as suggested specifically for proteins by Hamilton [27]). High-value products include most of the scarce, "glamor" therapeutic proteins that generally are produced by the powerful recombinant DNA techniques (e.g., interferon, antihemophilic factors, etc). These products are currently worth from a million to as much as a billion dollars per kilogram. This "worth" is based not so much on their actual value in use (most have not yet gone through clinical trials), but more on the difficulty in isolating minute quantities from very scarce natural tissue sources. The techniques of genetic engineering are allowing much higher concentrations of these products to be produced in alternative host systems. Products of more moderate value, produced by recombinant DNA, will take their value from currently produced equivalent or substitute products (insulin from bovine/porcine sources, human serum albumin from human blood, rennin from calves or microbial sources). These products are worth thousands of dollars per kilogram, a price which reflects the fact that they are present in low to moderate concentrations in moderately scarce tissues. These products are also amenable to recombinant DNA production, but the levels of expression required for success are substantially higher than for high-value products. Low-value products include commercial enzymes, (some) antibiotics, amino acids, etc. These generally sell for no more than $100 to 200 per kilogram and are often already produced by microbial fermentations in relatively high concentrations. Here too, recombinant DNA technology can make contributions, but those improvements will generally represent incremental improvements in manufacturing cost. As a rule the market requirement for production volumes of these products is inversely proportional to their value (a fact without which few low-value products would be produced).

Another general rule for these fermentation products is that the impact of raw materials on their production cost is also inversely proportional to their value. That is, raw material cost, as a percentage of product cost, tends to increase as the value of the product decreases. This is due not so much to the fact that raw material costs change in any way (in fact, they are quite similar for the same volume of fermentation broth), but that as the volume of product required decreases, the amount of labor required to produce the product does not decrease proportionately and the required technical skills of the personnel involved tend to increase.

It is possible to arrive at rough estimates of production costs strictly from raw material costs and production volume. (For many specialty organic chemicals made by a biological process, for example, a better estimate of the production cost [than as a factor of 1.5 to 2.0 times raw material cost] is to estimate the processing [non-raw material] cost at $1 to 5/kg depending inversely upon production volume [10^6 to 10^7 kg/year].) Certainly, however, much more confidence can be obtained by more closely examining the proposed process for production and estimating all of the costs in more detail. This kind of (preliminary) estimate is always justifiable before any significant amount of capital is invested in research.

To do a preliminary cost estimate, it is necessary to have a process description with all of the important process parameters established. For a well-developed process this should not be a difficult task. A more-difficult task is to develop a process description and assign parameters for a process which has not yet been demonstrated. This is generally the task faced when doing an economic feasibility study for a proposed research project. The biological research strategy may have been identified, but the actual results of that research, such as product titer, rate of synthesis, yield, etc., cannot be predicted in advance of the experimental work. In addition, the process by which the product is isolated from the reaction vessel (fermentor or bioreactor) may be even less certain.

Regardless of the uncertainties, some reasonable attempt to quantify the possible results of projected research must be made. A combination of thorough literature research, practical experience, and intuition can often give surprisingly accurate estimates of how much, how long, and how successful the R&D procedure might be. Economic analysis can further enhance the estimates of time and effort for R&D by helping to define what the milestones and goals of the work should be. For example, by doing a number of cost estimates which differ only in terms of the value of a key process parameter, the goal for research on that key parameter can be set on the basis of economic necessity.

Example 8. The economic estimate made for the case study presented in section 26.4 indicated that the chosen set of process assumptions would not result in an economically attractive operation. Before making the decision to terminate R&D for this protein, the management and scientific staff would like to know what level of expression might result in hitting the target cost of manufacture of $500/kg. To do this, the economic estimate exercise for the target protein is repeated, but with the protein expression level varied. So that a COM-versus-expression-level chart can be developed for future reference, the expression levels chosen for the new estimates are 1 and 20% of the total intracellular protein. Also, to account for the fact that recovery efficiency will be influenced by initial concentration of product, the recovery efficiencies are assumed to be 30 and 50%, respectively (as opposed to the 40% value used in the original case study).

As a result of the new estimates, the following figures are established: at expression levels of 1, 5, and 20%, COM will be $4,840, $1,510, and $740, respectively.

These figures indicate that an expression level higher than 20% will be required to achieve a COM of $500/kg. A plot of COM versus expression level developed from these estimates (as shown in Fig. 9) indicates that between 40 and 50% expression would be required. Because this level of expression, though not impossible, is beyond reasonable expectations, the decision is made to terminate the program.

Example 9. Some time after R&D on intracellular expression of the target protein has been terminated, a proposal is made to develop a microbial expression system which will cause secretion of selected proteins to outside the cell. Such a system should not have the inherent limitations of product concentration that intracellular systems have.

To evaluate the economic potential of such an extracellular production system, a series of economic estimates for different concentrations of a specific protein are developed. To directly compare intracellular versus extracellular expression, the target protein evaluated in the case study and example 8 is chosen as the specific protein for the estimates. As in example 8, a range of values for the key process parameter, product titer, are selected: 1, 5, and 20 g/liter of fermentation. Also, the recovery efficiencies are varied slightly: 40, 50, and 60%, respectively.

The results of the new set of economic estimates are plotted in Fig. 9. It is clear that to achieve a production cost of less than $500/kg, research on a production strain capable of extracellular production should be pursued. (The decision is not really that simple, of course. The research effort required to develop an extracellular expression system might in fact warrant discontinuation of the project depending upon the cost of research, the long-term profit potential, the availability of funds, staff expertise, and other project priorities.)

The economics of recovery processes have an interesting sort of tautology associated with product value. As has already been mentioned, the higher the product value, generally the higher the costs associated with recovery of the product. On the other hand, it can also be stated that the higher the product value, the more it can support expensive recovery operations. Does the difficulty of isolating the product make it expensive, or does the high value of the product obviate the need for a cheap recovery process? The answer to this question usually depends upon an historical perspective of the product in question. If the product is new and has a new use, the cost of producing it may be secondary to establishing a supply to meet demand, and its sales price will be relatively insensitive to the actual production process economics. Unless the new product is protected (e.g., by patent), this situation can probably be expected to deteriorate rapidly as competition gives rise to "second-generation" production systems. On the other hand, a product trying to break into an established market will be immediately subject to economic pressure to establish the most cost-effective recovery process.

With regard to recovery, the most cost-effective sequence of process operations may be impossible to establish while trying to develop a preliminary economic feasibility analysis. Because recovery of a purified product almost always involves at least four or five unit operations, it would seem that to determine the economic potential of a process before any development work has been started would border on the impossible. Fortunately, this is usually not the case. For many products, the exact sequence and type of operations required have little impact on the economics of the process. What is important, however, is the efficiency of the recovery process. This is the key variable in assessing the economics, and it is generally a very straightforward procedure to address its impact on the economic viability of a process.

Any plausible set of recovery operations can be postulated for a process in the preliminary stages of development. By developing a number of scenarios (all identical except for the key parameter, recovery efficiency), a relatively good feel for the potential economics can be achieved. The unit operations act as "place-holders" for those that may actually be developed in the course of the project, while varying the recovery efficiency gives a picture of the sensitivity of the overall process economics to recovery.

The foregoing discussion of the importance of recovery efficiency versus specific unit operations is, of course, an oversimplification. The specific unit operations to be used and their optimum sequence of application are what determine efficiency, and not the other way around (22). However, for medium- to low-value products which are expected to be recovered by more or less conventional operations (centrifugation, homogenization, extraction, filtration, ion exchange, evaporation, etc.), the technique of varying recovery efficiency in the calculations required for feasibility estimates is quite useful.

On the other hand, products which are of medium to high value are often recovered or proposed to be recovered by more exotic operations (isoelectric focusing, affinity chromatography [e.g., using monoclonal antibodies], high-pressure liquid chromatography, etc.). In this type of situation, the operation is often highly efficient and specific for the product of interest. As such, the recovery efficiency can be a less important parameter than the cost of purchasing and operating the highly specific separation equipment. Unfortunately, the high operating costs for many of these selective techniques make them useful only for very high-value products.

26.6 LITERATURE CITED

1. **Anonymous.** 1982. A tightening rein on maintenance costs. Chem. Week, 28 July, p. 20.
2. **Anonymous.** 1982. Bringing plant cost estimates closer to reality. Chem. Week, 4 August, p. 30–33.
3. **Anonymous.** 1983. Between the plans for R&D and reality. Chem. Week, 1 June, p. 22–24.
4. **Anonymous.** 1983. Borrowing costs bite into earnings. Chem. Week, 19 January, p. 42–43.
5. **Anonymous.** 1983. The budget-makers bet on marketing. Chem. Week, 30 March, p. 38–39.
6. **Anonymous.** 1983. Maintenance spending levels off. Chem. Week, 6 July, p. 25.
7. **Anonymous.** 1983. Ocean shipping rates: a bargain for chemicals. Chem. Business, 22 August, p. 17–18.
8. **Anonymous.** 1983. Rail, trucks, barges: lots of space. Chem. Business, 22 August, p. 20–21.
9. **Baasel, W. D.** 1977. Preliminary chemical engineering plant design. Elsevier, New York.
10. **Bailey, J. E., and D. F. Ollis.** 1977. Biochemical engineering fundamentals. McGraw-Hill, New York.
11. **Bartholomew, W. H., and H. B. Reisman.** 1979. Economics of fermentation processes, p. 463–496. *In* H. J. Peppler and D. Perlman (ed.), Microbial technology, vol. II. Academic Press, Inc., New York.
12. **Bauman, H. C.** 1964. Fundamentals of cost engineering in the chemical industry. Reinhold, New York.
13. **Brown, A. S.** 1983. Keeping the books on chemicals. Chem. Business, 25 July, p. 17–23.
14. **Burkhart, D.** 1983. Tough choices for chemical R&D chiefs. Chem. Business, 25 July, p. 26–29.
15. **Cevidalli, G., and B. Zaidman.** 1980. Evaluate research projects rapidly. Chem. Eng. **87:**145–152.
16. **Charm, S. E., and C. C. Matteo.** 1971. Scale-up of protein isolation. Methods Enzymol. **22:**476–556.
17. **Chemical Engineering Magazine.** 1979. Modern cost engineering: methods and data. McGraw Hill, New York.
18. **Cran, J.** 1981. Improved factored method gives better preliminary cost estimates. Chem. Eng. **88:**65–79.
19. **Curling, J. M., and J. M. Cooney.** 1982. Operation of large scale gel filtration and ion-exchange systems. J. Parenteral Sci. Technol. **36:**59–63.
20. **de Matteis, U.** 1982. Planning and budgeting for maintenance. Chem. Eng. **89:**105–112.
21. **Dunnill, P.** 1972. The recovery of intracellular products, p. 187–194. *In* G. Terui (ed.), Fermentation technology today. Proceedings of the IV International Fermentation Symposium.
22. **Fish, N. M., and M. D. Lilly.** 1984. The interactions between fermentation and protein recovery. Bio/Technol. **2:**623–627.
23. **Godfrey, R. S. (ed.).** 1982. Means building construction cost data 1982, 40th ed. Robert Snow Means Co. Inc., Kingston, Mass.
24. **Gray, P. P., P. Dunnill, and M. D. Lilly.** 1972. The continuous-flow isolation of enzymes, p. 347–351. *In* G. Terui (ed.), Fermentation technology today. Proceedings of the IV International Fermentation Symposium.
25. **Guthrie, K. M.** 1969. Capital cost estimating. Chem. Eng. **76:**116–142.
26. **Hall, R. S., J. Matley, and K. J. McNaughton.** 1982. Current costs of process equipment. Chem. Eng. **89:**80–116.
27. **Hamilton, B. K.** 1983. Design factors for constructions of competitive production strains and manufacturing processes, p. 109–143. *In* D. Wise (ed.), Organic chemicals from biomass. Benjamin-Cummings Publishing Co., Menlo Park, Calif.
28. **Humphreys, K. K., and S. Katell.** 1981. Basic cost engineering. Marcel Dekker, Inc., New York.
29. **Lang, H. J.** 1948. Simplified approach to preliminary cost estimates, p. 12–13. *In* C. H. Chilton (ed.), Cost engineering in the process industries. McGraw-Hill, New York.
30. **Lunde, K. E.** 1984. Capacity exponents for field construction costs. Chem. Eng. **91:**71–74.
31. **Merrow, E. W., K. E. Phillips, and C. W. Myers.** 1981. Understanding cost growth and performance shortfalls in pioneer process plants. Rand Corp., Santa Monica, Calif.
32. **Perry, R. H., and C. H. Chilton (ed.).** 1973. Chemical engineers' handbook. McGraw-Hill, New York.
33. **Peters, M. S., and K. D. Timmerhaus.** 1980. Plant design and economics for chemical engineers. McGraw-Hill Book Co., New York.
34. **Saylor, L. S.** 1981. LSI current construction costs 1981, 18th ed. Lee Saylor, Inc., Walnut Creek, Calif.
35. **Short, H.** 1984. Shocking is the word for CPI's electric bill. Chem. Eng. **91:**30–35.
36. **Stryker, M. H., and A. A. Waldman.** 1978. Blood fractionation, p. 25–61. *In* R. E. Kirk and D. F. Othmer (ed.), Encyclopedia of chemical technology, 3rd ed. John Wiley & Sons, Inc., New York.
37. **Swartz, R. W.** 1979. The use of economic analysis of penicillin G manufacturing costs in establishing priorities for fermentation process improvement, p. 75–110. *In* D. Perlman (ed.), Annual reports on fermentation processes, vol. 3. Academic Press, Inc., New York.
38. **Taylor, G. A.** 1975. Managerial and engineering economy: economic decision-making. D. Van Nostrand, New York.
39. **Thuesen, H. G., and W. J. Fabrycky.** 1964. Engineering economy. Prentice-Hall, Inc., Englewood Cliffs, N.J.
40. **Tokay, B.** 1985. Antibiotic fever grips drugmakers again. Chem. Business **7:**40–43.
41. **Viola, J. L.** 1981. Estimate capital costs via a new, shortcut method. Chem. Eng. **88:**80–86.

42. **Wang, D. I. C., C. L. Cooney, A. L. Demain, P. Dunnill, A. E. Humphrey, and M. D. Lilly.** 1979. Fermentation and enzyme technology. John Wiley & Sons, Inc., New York.
43. **Ward, T. J.** 1984. Predesign estimating of plant capital costs. Chem. Eng. **91:**121–124.
44. **Weaver, J. B., and H. C. Bauman.** 1973. Cost and profitability estimation, p. 25/1–25/47. *In* R. H. Perry and C. H. Chilton (ed.), Chemical engineers handbook, 6th ed. McGraw-Hill Book Co., New York.
45. **Winton, J. M.** 1980. Plant sites: spreading out from Houston. Chem. Week, 3 December, p. 54–64.

VI. Special Topics

Special Topics: Introduction

ALLEN I. LASKIN

As indicated by its title, this section includes chapters that do not fit readily into the foregoing sections, but present important material relevant to the subject of this manual.

A topic that has received a great deal of attention has been the question of legal protection for practitioners of biotechnology. The Supreme Court ruling on the Chakrabarty case, which went a long way toward clarifying the issue of patenting living microorganisms, stimulated great activity in the field. Saliwanchik's chapter describes inventions and conceptions, discusses trade secrets, and outlines the procedures for obtaining a United States patent. It details the necessary components of a patent application and describes aspects peculiar to biotechnology inventions. The chapter concludes with a brief discussion on marketing a patented biotechnology invention.

Rarely has the development of a scientific field stimulated as much public debate and controversy as has the field of genetic engineering and recombinant DNA. The history of the self-imposed moratorium on certain kinds of recombinant DNA research, the subsequent Asilomar Conference, and the formation and activities of the Recombinant DNA Advisory Committee of the National Institutes of Health has been extensively documented. Liberman et al. present a concise treatment of the biosafety considerations and practices related to research and development in biotechnology.

It is almost axiomatic in the field of antimicrobial substances that a major key to progress in developing an agent is the availability of an assay method that is simple, rapid, and accurate. The chapter by Isaacson and Kirschbaum describes the classic agar diffusion and turbidimetric assays and presents tables on the microorganisms, media, and reagents used in these microbiological assays. More recently, high-pressure liquid chromatography (HPLC) has become an important methodology for the assay of antimicrobial agents; this chapter describes the HPLC apparatus and the techniques used.

After the initial flush of excitement following the development of recombinant DNA technology and other tools of genetic engineering and their application to the production of useful compounds came the realization that product recovery is an area requiring much attention. The chapter by Calton et al. outlines principles that are applicable to the recovery of compounds from plants, natural fluids, and fermentation broths, but concentrates discussion on the latter. The topics covered include product distribution, cell separation, cell disruption, and liquid-liquid extraction, with an extensive coverage of various aspects of chromatographic separations.

Chapter 27

Legal Protection for Biotechnology

ROMAN SALIWANCHIK

Members of the American Society for Microbiology are becoming increasingly aware of the fact that patent and other lawyers are taking a greater interest in their science. This is not accidental. Chances are that lawyer interest in the microbiological sciences will increase rather than diminish as time goes on. It is not that lawyers have an inherent desire to learn this science; it is more that lawyers are "protectors of rights," and, quite clearly, microbiologists have rights which need to be protected.

We can't say that microbiologists' rights have not been protected over the past 40 years or so, however. Microbiological technology has been the focus of much legal work since the late 1940s in the United States. Perhaps most prevalent have been patents and legal decisions relating to the production of antibiotics.

The so-called "antibiotic era" has been long-lived. Beginning intensively in the 1940s, antibiotic production remains highly visible today as a microbiological science with a tremendously broad impact on the medical sciences. Much of the stability of the antibiotic business is due to several significant and proper legal decisions rendered during the years. Without such decisions, which either upheld existing patent rights or allowed new ones, the impetus and staying power for industrial antibiotic production would have suffered immeasurably.

Thus, lawyers and scientists have demonstrated that in microbiology, as well as in other sciences, law and science are advantageously miscible.

The microbiological sciences are now broadening further into biological and biochemical areas. Numerous advancements rely on microbiology to a large extent to bring the larger sciences to a practical focus. Biology used to be primarily a laboratory science, but this is no longer the case, with such developments as the successful cloning of genes that can be used to make useful polypeptides. There is a new perspective in the general science of microbiology. The field now concerns itself with biology, biochemistry, and genetics at a depth never before considered necessary for "good microbiology." Today, a microbiologist must know classic microbiology, biology, biochemistry, and genetics as a minimum. Additionally, a good microbiologist will be cognizant of the legal system that has been established and developed to enable scientists to legally protect the fruits of their scientific labors.

Being cognizant does not mean that scientists should act as their own lawyers. On the contrary; there is an oft-stated truism that a person who is his or her own lawyer has a fool for a client. There are absolutely too many facets to the practice of law to allow for anyone but a trained professional to assume the responsibility of handling legal matters. Yet the lawyer who does the lawyering will be more successful if the client has an appreciation of the need, and the means, to preserve facts for legal review. Stated otherwise, a client totally ignorant in legal affairs will need luck more than a good lawyer to preserve his legal rights.

The purpose of this chapter is to give the microbiological scientist an insight into the means that can be used to legally protect biotechnology in the United States. Key points from a scientist's perspective are given prime consideration. Again, this is not intended to make lawyers of scientists. But if the result is that some scientists become interested in law to the extent that they go to law school, so be it!

27.1 RESEARCH ENVIRONMENT

27.1.1 Conception of Invention

"Invention" as used in this section covers all inventions whether patentable or not. Basically, an invention is something that is created or produced for the first time. If a person skilled in the science produces a compound that, to his or her knowledge, has never been produced previously, then it is reasonable to assume that an invention exists. Whether this particular invention is patentable will depend upon the law of the country where a patent application is filed for the invention. Although there are many similarities in the patent laws of countries throughout the world, differences exist that can block patentability of an invention in one country but not in others.

A conception of an invention is a complete mental image of the invention. For example, if the invention is a new antibiotic compound, then the mental image must be such as to enable a skilled person to make the antibiotic. It is not enough to have a mental image of a desired chemical compound and not have a process for making it.

An invention need not be conceived in a laboratory. Quite often, history shows, significant inventions have been conceived at odd hours and in various environments quite apart from a recognized research environment. Thus, the time and locale of the conception have no legal significance aside from use as evidence in an interference proceeding, which is a hearing held to determine which party among two or more claiming first inventorship is, in fact, the first inventor.

It should be noted that conception does not require the actual making of the invention. For example, a novel antibiotic need not be made, nor a gene cloned and expressed in a novel fashion. The conception must, however, include the means for making the compound or cloning the gene.

Additionally, a complete conception of the invention must include a utility for it. If it is a new antibiotic, then the conception must envision conditions in which the antibiotic would be useful. This should not be difficult if the conceived antibiotic is analogous to a well-known and useful one. For example, if the proposed drug is a new analog of penicillin G, then it would be reasonable to assume that the analog would have an antibacterial spectrum comparable to that of penicillin G. The new compound need not have a comparable degree of activity; any degree of activity would be sufficient for utility under the law. However, if the conceived antibiotic represents a new class of compounds, there would be little basis for presuming a utility without actually making the compound first and then testing it against various bacteria.

27.1.2 Co-Researchers

A conception of an invention often is the act of a single person. In such a case, if the conception is sufficient as discussed above, the situation is termed sole inventorship. However, since it is customary in various research environments for researchers to collaborate with one another and discuss the nature of their ideas, it becomes likely that conception of new inventions arises from the contributions of more than one person. When the contributions are essential to perfect the conception, the researchers making those contributions become co-inventors of the invention. This aspect of inventorship is not cut and dried, but is a legal determination that ultimately will have to be reviewed and decided upon by a patent lawyer using guidelines set forth in various legal decisions over the years.

27.1.3 Record of Conception

Whether a conception is that of a sole inventor or joint inventors, it is extremely important that it be reduced to writing as soon as possible. This should be done preferably in a bound laboratory notebook, signed and dated by the person(s) making the conception. Additionally, the page(s) containing the conception should be reviewed and signed by a person who is capable of understanding the nature of the conception. All entries and signatures should be written in ink and dated.

The record of the conception is the beginning, in a sense, of the recording of the research directed toward the actual reduction to practice of a conception. Preferably, research work subsequently conducted should be entered into the bound laboratory notebook on the pages following the conception. All entries should be written in ink. As with the conception, all pages of the notebook containing laboratory results should be signed and dated by the person doing the work and by a witness who understands the nature of the work.

Understandably, researchers are reluctant to sit down and record their research results in a laboratory notebook. This takes time away from the bench, and it is much easier to have a stack of papers which can be stuffed into a folder or a pocket. However, if the researcher wants to preserve legal rights in an invention, it is absolutely essential that research results be transcribed into a bound laboratory notebook routinely and conscientiously. Anything less than this can leave the researcher without legal recourse after having spent countless hours in the laboratory working on an invention. Unless the researcher can prove his or her right of first inventorship by competent legal evidence, he cannot expect to prevail in a legal procedure even though he may actually be the first and true inventor. Having proper records and maintaining research results in the bound laboratory notebook is as important to a researcher as having a deed to a house or the title to a car. Without these legal bits of evidence of ownership, there is no reliable mechanism in the law to unravel the complexities of the assertions of two or more individuals professing to be inventors of the same invention.

27.1.4 Secrecy

It should go without saying that all research work concerning an invention should be maintained with a suitable amount of secrecy until the inventor(s) is ready to disclose the invention to others. Prematurely discussing an invention with persons outside the inventive circle can lead to loss of potential patent rights as well as unauthorized disclosure and use of the invention. If there is the need or desire to discuss work surrounding the invention with an outsider, a legal confidential relationship should first be established with the person to whom the invention will be disclosed. This can be established by use of a confidential disclosure agreement, as follows.

Confidential Disclosure Agreement

Dr. S. Aureus, residing in Grand Rapids, Michigan, is willing to accept a disclosure from Dr. E. Coli, residing in Battle Creek, Michigan, relating to a novel process for promoting the transfer of genetic material between gram-positive and gram-negative organisms. Dr. Aureus will hold the novel aspects of Dr. Coli's disclosure in confidence for five (5) years from the date of receipt unless sooner released from this obligation by Dr. Coli. The novel aspects of Dr. Coli's disclosure are those which:

(1) are not already known to Dr. Aureus from his own research or from a third party who has a right to disclose the information to Dr. Aureus, or
(2) are not or do not become subject of a patent issued to a third party, or
(3) are not or do not become open to the public by any means including disclosure in patents or publications, or public use or sale.

This disclosure to Dr. Aureus is for the sole purpose of permitting Dr. Aureus to evaluate his interest in Dr. Coli's process. Dr. Aureus will not make any other use of this process without the prior written consent of Dr. Coli. Further, upon completion of Dr. Aureus' evaluation, within sixty (60) days following receipt of the disclosure of Dr. Coli, Dr. Aureus shall return the disclosure to Dr. Coli, but can retain one copy solely for the purpose of having a copy of his legal responsibility under this Agreement.

AGREED TO:

S. Aureus, Ph.D. E. Coli, Ph.D.

Date ______________________

All records pertaining to an invention should be maintained as carefully as any other legal document, such as the deed to a house. Invention records are useful primarily to establish first inventorship. They also allow the scientist to review the work done at various stages of a scientific endeavor.

27.2 WHAT TO DO WITH RESEARCH RESULTS

27.2.1 Trade Secret Protection

Trade secret protection is available for patentable and nonpatentable inventions. When an invention is determined to be nonpatentable, then trade secret protection would be most appropriate. It is also appropriate when the invention is patentable but would be difficult to "police" if a patent were granted. For example, if the invention concerned an alternative process for preparing a compound by fermentation, such as by the use of a new mixture of nutrients, then the use of the process by another would not be easily ascertainable. True, we like to think that patents are respected and that they are not infringed (practiced without permission from the patentee) intentionally. However, this is not always the case, and it is not unexpected when instances arise showing that a patented invention has been practiced for years without the patentee's knowledge. These kinds of inventions are described as hard to police. On the other hand, when an invention concerns a unique process for making a compound, then it is much more easily policed. Such an invention receives maximum protection from the patent system, because patent infringement is readily noticed in the commercial trade.

Another consideration as to whether an invention might best be maintained as a trade secret is whether the invention can be ascertained by the public after it is practiced. If the invention concerns a process which is not self-evident from the product that is marketed after being produced by the process, then it would be difficult, if not impossible, to reverse-engineer and determine the invention process. In such a case, the invention process, advantageously, could be maintained as a trade secret.

A desirable feature of trade secret protection is that there is no time limit, as opposed to the 17-year term of a United States patent. If a trade secret remains secret, it can afford protection indefinitely for its owner. However, once the secret becomes known to the public, trade secret protection ceases to exist. Thus, maintaining a trade secret requires considerable effort.

The basic characteristics of a trade secret are that it is a tangible entity and gives the trade secret owner an advantage over competitors. A trade secret can include new processes, new microbes, assay techniques, medium ingredients, or the like. It is evident, therefore, that a large amount of laboratory work is generally maintained as trade secrets. This is particularly true in industry, where process variables are constantly being reworked, media and conditions of fermentation are being changed, and new production strains are being developed. Most of these advantageous developments are better kept as trade secrets to provide a competitive advantage for the trade secret owner.

There is one drawback to maintaining a particular development as a trade secret, and that is that if a trade secret is used commercially for more than a year, then the user of the trade secret is thereafter foreclosed by law from going back and applying for patent rights. In that case, if another person were to come upon the trade secret development independently and apply for and ultimately obtain a patent, then the original trade secret user would be unable to practice his development without obtaining a license from the patent holder. This fact causes many industrial managers to go the patent route, particularly when a development is such that (i) it is clearly patentable, (ii) it is significant to the operation of the production unit, and (iii) it would be difficult to circumvent.

Reflecting upon the crux of the trade secret, i.e., secrecy, it is obvious that publication of the trade secret is out of the question. In a strictly commercial setting, such a prohibition is not difficult to accept. Rewards for developing a trade secret often take the form of promotions or bigger paychecks. Comparable rewards, however, are not always available in the academic environment and in fact are often closely linked with publication ("publish or perish"). Accordingly, trade secret maintenance is not a favored form of legal protection in academia. The choice would then have to be made whether to apply for a patent, after which the invention can be published, or to dedicate the invention to the public by publishing outright without acquiring patent protection.

27.2.2 Patent Protection in General

The word "patent" means different things to different people. At one time, a person obtaining a patent was looked upon as being akin to a genius. It is now clear that for an invention to be patentable it does not necessarily have to be the work of a genius. This may be unfortunate in some respects, but it is a condition that makes the patent system available to a broader group of inventors.

A patent is a document given by the United States Government in recognition that an invention has passed the statutory tests for patentability. This patent document is an official pronouncement that the owner of the patent has a legal right to prevent anyone within the confines of the United States and its territories from practicing the claimed invention for 17 years from the date the patent is granted. This legal right possessed by the owner of the patent can be licensed to others, exclusively or nonexclusively.

Aside from giving the patent owner a legal right to prevent someone from making, using, or selling the claimed invention, a patent itself offers no further legal rights. In other words, a patent does not necessarily entitle the patent owner to practice his invention. An example of this fact is a situation in which a patent is obtained for a new antibiotic. The claims in the patent would be to the antibiotic per se and, probably, to a process for preparing the antibiotic. This antibiotic, however, could not be marketed upon the issuance of the patent until suitable permission had been given by the Food and Drug Administration.

The entire patent system in the United States is made possible by legislative action. Patents are the creatures of a statute (United States Code, Title 35) enacted to implement the constitutional provision of Article 1, Section 8, which states, in part, that "Congress shall have power. . . To promote the progress of science and useful arts, by securing for limited times to authors and inventors the exclusive right to their respective writings and discoveries." The requirements for patentability of an invention under the United States patent laws are clearly set forth in the patent statute. The meanings of the words in the statute have been the subject of numerous court cases over the years. Thus, to really know what is patentable under the United States patent law, it is essential that the statute be read in the light of the many legal decisions interpreting it.

27.2.3 Dedicate to the Public, i.e., Publish

Certain legal rights are obtained by maintaining research results as a trade secret or by patenting them. Should there be no need for either protective mechanism (or should there be a desire to prevent another from obtaining legal protection), there remains the avenue of dedicating research results to the public, which is done by publishing them. The publication of subject matter effectively eliminates the possibility for trade secret protection. On the other hand, patent protection in the United States is still possible if a patent application is filed within 1 year of the publication date. If no patent application is filed within a year, then the publication is a statutory bar to the future patenting, by anyone, of the subject matter disclosed in the publication. Thus the publication of a trade secret can prevent another party from obtaining patent rights.

There are several other facets to this situation, but this chapter is not intended to go into these areas, which realistically fall within the bailiwick of the patent lawyer. One of these points is that the publication must be an enabling disclosure of the subject matter that might be patentable. That is, if upon reading the publication one skilled in the art (a hypothetical person who is trained in the particular science to which the publication pertains) cannot reproduce the published subject matter without undue experimentation, the publication is not an enabling disclosure. In this case, publication cannot be considered a statutory bar to subsequent patenting of the subject matter, even if the publication date is more than 1 year prior to a patent application filing date. Numerous legal decisions that have dealt with various aspects of these situations must be understood to fully comprehend the substance of a particular problem.

27.3 BRIEF REVIEW OF THE PROCEDURE FOR OBTAINING A UNITED STATES PATENT

27.3.1 Hire a Competent Patent Lawyer

A patent lawyer is a person trained in science or engineering and the law. To be a patent lawyer, it is necessary to have a degree in one of the technical sciences as well as a law degree and to have passed the examination given by the Patent and Trademark Office (PTO). A patent agent is also scientifically trained and must also pass the PTO examination, but does not have to have a law degree. A patent agent is restricted to practice before the PTO, whereas a patent lawyer can practice before the PTO as well as before the various courts to which he or she has been admitted.

It is important for the inventor of a microbiological invention to retain patent counsel having, preferably, both education and experience in the microbiological sciences. This will ensure that the scientist will obtain the maximum protection under the law for the invention being considered.

At the time a patent lawyer is contacted, it is wise to have a frank discussion of the fees, charges, and total costs that could be expected in the patenting procedure. It is best to settle this matter initially, with the parties making every attempt to honor any obligations that have been agreed upon. Patent lawyers are generally flexible with regard to the financial problems in pursuing patent protection. For example, should cash be a problem for the scientist, some patent lawyers will render services in return for a percentage interest in the patent when granted. There are other methods that can be employed, and these should be explored during the initial contact with the patent lawyer.

27.3.2 Disclose the Invention to the Patent Lawyer

After hiring a competent patent lawyer, the scientist should disclose the entire invention to the lawyer. Although the research results can be classified as an invention, in that they are new to the researcher, these results may still not be of the caliber required for patentability. The patent lawyer will determine this. If the lawyer believes the research is inadequate for patent protection, then trade secret protection may be a viable alternative. Also, the patent lawyer frequently can suggest further work that could be done to obtain the necessary data for a suitable patent application. In this manner, a knowledgeable patent lawyer can be invaluable in helping to develop the invention.

In addition to the research results, it is imperative that the scientist also disclose all the pertinent prior art relating to the invention. "Prior art" means all publications available to the public at the time of the invention. Since the patentability of the invention will be determined against the background of the prior art, the patent lawyer must have this complete picture right from the beginning. The scientist should show the patent lawyer how the invention differs from the closest prior art known to the scientist. When a problem recognized by those skilled in the art is solved by the invention, this should be disclosed to the patent lawyer.

27.3.3 Patent Application

Once a patent lawyer decides that the invention disclosed by the client has patentable features, the all-important task of preparing a patent application will begin. The patent application is a legal document that must be prepared in a certain manner to meet statutory requirements and the requirements resulting from legal decisions interpreting the statute. It is extremely important that this document be prepared by a patent lawyer fully knowledgeable in the science to which the invention pertains. The patent lawyer will prepare an initial draft of the patent application and submit it to the inventor for review and comment. Once the inventor has reviewed it, the patent lawyer will redraft the application to incorporate any changes proposed. It is not unusual for this drafting and review procedure to go back and forth several times before a final draft is prepared and ready for submission to the PTO.

27.3.4 File the Patent Application

The final draft of the patent application will be filed with the PTO in Washington, D.C. A complete patent application, as defined by the patent statute, consists of the specification and claims which disclose the invention; the filing fee prescribed by the statute; and an oath or declaration by the inventor(s) attesting that he or she is the inventor of the subject matter claimed. Until recently, all of these separate papers had to be filed at the same time in the PTO to receive a filing date for the patent application. Because of a rule change by the PTO, it is now possible to file the specification and claims and the required fee without an oath or declaration, to receive a filing date. The oath or declaration must be filed subsequently, accompanied by a late-payment fee of $100. This rule change was instituted primarily to preserve the legal right to an invention when an inventor was not available to sign the oath or declaration and it was critical that the application be filed before a given date.

27.3.5 Examination of the Patent Application

When the patent application arrives at the PTO, it is first handled in the receiving branch. Here the application is stamped with an application serial number and filing date, and it is then assigned to one of a number of examining groups in the PTO on the basis of the nature of the subject matter of the invention and the skills of the patent examiners in the groups. For example, a patent application pertaining to a microbiological invention will be assigned to an examining group specializing in such inventions. This examining group will be staffed by persons trained in microbiology as well as in PTO procedures. Some examiners have law degrees; others do not. Those who do not have law degrees, however, are skilled in applying the patent law requirements in determining the patentability of inventions.

Initially the patent examiner examining a particular patent application will determine whether the disclosure of the invention in the application meets all the requirements of the patent law. More often than not, the application will be rejected and returned to the inventor's lawyer.

27.3.6 Rejections

The written rejection by a patent examiner will be based upon various sections of the patent statute. The patent lawyer will be given a period of 3 months from the date of the rejection in which to reply to the patent examiner. During this 3-month period, the patent lawyer will consult with the inventor to determine the best manner in which to overcome the examiner's rejection. Among the most frequent grounds for rejection of microbiological inventions are that (i) the invention is obvious from what is known in the prior art, or (ii) the patent application does not contain a sufficient disclosure of the claimed invention. Both of these rejections are based on statute requirements, and both have been the subject of numerous legal decisions by patent courts. The knowledgeable patent lawyer will be cognizant of the legal decisions and, with the help of the inventor, will develop an appropriate rebuttal to the examiner's rejection. This rebuttal will be reduced to writing and sent to the PTO before the 3-month period ends.

27.3.7 Final Rejection of a Patent Application

If the examiner accepts the patent lawyer's rebuttal of a rejection, then the rejection will be withdrawn and the claims allowed. The formal procedure for the granting of a patent will then follow.

When the patent examiner does not accept the patent lawyer's rebuttal, the application is given a final rejection, again transmitted in writing to the patent lawyer. In many cases, the patent lawyer can overcome the examiner's rejection by canceling certain claims from the application. Such a procedure is feasible when the claims remain sufficient to adequately protect the invention. However, if such an alternative is not desirable in a patent application, then the next step for the patent lawyer, should the decision be made to continue to try to obtain a patent, would be to file an appeal to the PTO Board of Appeals.

27.3.8 Appeal to the PTO Board of Appeals

The PTO maintains a number of three-person appeal panels to hear appeals concerning final rejections given by patent examiners. These panels are considered administrative courts, and hearings are not open to the public. The participants before the PTO Board of Appeals in a hearing are generally the Board members and the patent applicant's (now the appellant's) patent lawyer. It is possible for the examiner also to appear, but generally this is not done. Before appearing before the PTO Board to argue the appellant's case, the patent lawyer will prepare and submit to the PTO a written brief of the arguments that he or she deems sufficient to overcome the final rejection. The patent examiner will also submit a brief in the form of an Answer to the appellant's brief. These briefs, along with the other correspondence between the patent examiner and the patent lawyer (patent file), will become the record that the PTO Board will consider in judging the appeal.

The appellant's patent lawyer will be notified of the hearing date. At the hearing, 20 min will be allowed in which to present the appellant's case orally before the PTO Board. Generally the PTO Board will render a decision in writing within 6 months of the Board hearing. The decision can be that the final rejection is overruled; in such a case, the application will go back to the patent examiner, who will then follow a more or less formal procedure in allowing the application and placing it in the formal process of a subsequent patent grant. If the PTO Board affirms the examiner's final rejection, then the appellant can either abandon the patent application before the PTO or appeal to the Court of Appeals for the Federal Circuit (CAFC), which is located in Washington, D.C.

27.3.9 Appeal to the CAFC

An appeal to the CAFC of a PTO Board decision opens the subject matter to the public. In preparing such an appeal, the patent lawyer must address a host of technical requirements, including the preparation of a suitable appellant's brief.

The PTO solicitor represents the PTO before the CAFC, aided by several associates who are assigned to various patent appeals. The associate assigned to a particular case will also prepare a written brief supporting the PTO Board's decision in the case. Upon completion of the brief-writing in a patent case, the Court will set a hearing date. At the designated time, the appellant's patent lawyer and the representative from the solicitor's office will argue their cases before the Court. In this open-court setting (as opposed to the closed PTO Board hearing), it is possible for the public to read the briefs and listen to the proceedings. Upon the completion of the hearing, the Court will generally render a decision within 6 months to 1 year. If the Court's decision is to reverse the PTO Board, the application is remanded back to the patent examiner for appropriate action leading to the grant of a patent. If the CAFC affirms the PTO Board, the only recourse remaining for the appellant, aside from abandoning the application, is to seek a review by the Supreme Court of the United States via a writ of certiorari.

(It should be recognized that this seeking of a writ can also be done by the PTO when the CAFC reverses a PTO Board decision. A recent case in which the PTO sought and obtained a writ to the Supreme Court of the United States was the case styled *Parker* v. *Bergy* and *Chakrabarty* [In re Bergy et al.; In re Chakrabarty, CCPA, 596 F.2d 952, 201 USPQ 352, 1979]. This case concerned the patentability of a living microorganism. Subsequently Bergy dropped out and the case then concerned only the Chakrabarty application. [*Diamond, Commissioner of Patents and Trademarks* v.

A. M. Chakrabarty, Supreme Court of the United States, 100 S.Ct. 2204, 206 USPQ 193, 1980].)

27.3.10 Appeal to the Supreme Court of the United States

When a writ of certiorari is filed to the Supreme Court of the United States, there again is a host of legal technicalities that must be followed by the patent lawyers in the procedure. If the Supreme Court grants the writ, the pertinent parties will then submit required written briefs and the case ultimately is heard and decided by the Court. The decision by the United States Supreme Court, of course, is final. However, the legislature can come in subsequently and, in effect, modify or overcome some United States Supreme Court decisions.

27.3.11 Granting of a Patent

Most patents (almost 90 to 99% of those granted by the PTO) do not have to pass through the appellate procedure discussed above. The usual situation is to file the patent application, receive a rejection, and overcome the final rejection to gain an allowance of the application. Once a Notice of Allowance is received by the patent applicant, there is a period of 3 months in which to pay the statutory fee for the issuance of a patent. A patent will be issued approximately 3 months after the payment of this fee to the PTO.

27.4 REQUIRED PARTS OF A PATENT APPLICATION DISCLOSURE

The above section gives a brief review of the process of obtaining a patent in the United States. Key to this process is the initial step of preparing the patent application for filing in the PTO. It thus behooves the scientist to become familiar with the required parts of a patent application disclosure. It should be recognized that such a disclosure is not accomplished by merely filling in blanks or checking appropriate boxes. Rather, it is a narrative description of the invention, presented in a manner that will enable a person skilled in the pertinent field to reproduce (i.e., practice) the invention once the application is made public. This test of enablement is critical, since an enabling disclosure is the consideration which the patent applicant gives in return for the government grant of a patent. If this disclosure is insufficient, then obviously the patent applicant has not "paid the price" for the patent grant and the application will be doomed to failure from the start.

Though the test of enablement is applied to the patent application disclosure, this should not be construed in any way to imply that the public has a right to practice the invention upon the publication of the patent document. On the contrary, upon the publication of the patent, the public cannot practice the invention without license from the patent owner. The caveat to this is that the law has recognized the legality of, or at least has not enforced the patent right against, minimal experimental practicing of a patented invention.

We can now progress to the various substantive parts of a patent application disclosure. These parts will be exemplified by reference to patents for microbiological and molecular biology inventions.

27.4.1 Title of the Invention

When a patent application is first filed in the PTO, it will generally contain one of the statutory titles representing classes of subject matter that are patentable in the United States. For example, a microbiological process would be titled "Process." A novel microbe, plasmid, and the like would be titled "Composition of Matter" or "Manufacture." When the invention concerns both a process and a product, for example, a process for preparing a novel antibiotic and the novel antibiotic itself, then the title would be "Composition of Matter and Process." There is nothing critical about the title on the application as filed with the PTO. The title can be amended while the application is pending before the PTO. In fact, before giving an Allowance, the patent examiner will request a more specific descriptive title for the application; for example, where the initial title may have been "Composition of Matter and Process," the final title, which would then appear on the patent, could be "Antibiotic X and Process for Preparing the Same."

27.4.2 Abstract of the Disclosure

The next part of a patent application disclosure is termed the "Abstract of the Disclosure." The Abstract is a concise statement of the salient points of the invention. By way of illustration, below are abstracts from five different types of microbiological inventions, including one relating to molecular genetics.

Patent no. U.S. 4,250,256
Title: Microbiological Test Device
Abstract: Microbiological test device having a nutrient card or cards and a covering layer thereover which is permeable to nutrients but impermeable to bacteria. The covering layer is made from an aqueous dispersion of a water-insoluble resin and a pore former.

Patent no. U.S. 4,250,259
Title: Microbiological Production of Ketones from C_3-C_6 Secondary Alcohols
Abstract: A process is disclosed for the microbiological production of ketones from C_3-C_6 secondary alcohols by contacting C_3-C_6 secondary alcohols under aerobic conditions with resting microbial cells derived from a microorganism of [sic; or] enzyme preparation derived from said cells, wherein the microorganism has been previously grown under aerobic conditions in a nutrient medium containing a methyl radical-donating compound. Typical methyl radical-donating compounds are: methane, methanol, dimethyl ether, methylamine, methyl formate, methyl carbonate, ethanol, propanol, butanol, etc. Isolation and purification of a novel C_3-C_6 secondary alcohol dehydrogenase enzyme is also disclosed.

Patent no. U.S. 4,267,274
Title: *Streptomyces mediterranei* Mutant Capable of Production of Rifamycin B

Abstract: A biologically pure culture of *Streptomyces mediterranei*, mutant strain 18, ATCC 21789.

Patent no. U.S. 4,306,021
Title: Composition of Matter and Process
Abstract: Novel antibiotic U-60,394 producible in a fermentation under controlled conditions using a biologically pure culture of the microorganism *Streptomyces woolenses*, Dietz and Li sp. n., NRRL 12113. This antibiotic is strongly active against various gram-positive bacteria, for example, *Staphylococcus aureus* and *Streptococcus hemolyticus*. It is also strongly active against the gram-negative bacterium *Streptococcus pneumoniae*. Thus, antibiotic U-60,394 can be used in various environments to eradicate or control such bacteria.

Patent no. U.S. 4,393,137
Title: Cloning Plasmid for *Streptomyces*
Abstract: Novel chemical compound plasmid pUC1061 obtained by deletion of ~2.0 kilobases of DNA from the *Streptomyces espinosus* plasmid pUC6. This plasmid is useful as a cloning vehicle in recombinant DNA work. For example, using DNA methodology, a desired gene, for example, the glucose isomerase gene, can be inserted into the plasmid and the resulting plasmid can then be transformed into a suitable host microbe which, upon culturing, produces the desired glucose isomerase.

(Copies of patents are available upon writing to the U.S. Department of Commerce, Patent and Trademark Office, Washington, D.C. 20231. Patents should be requested by their number, and each patent copy costs $1.50. Personal checks will be accepted and should be made to the Commissioner of Patents and Trademarks. Allow 2 to 4 weeks for receipt.)

27.4.3 Background of the Invention

After the Abstract, the patent application disclosure delineates the "Background of the Invention." The Background of the Invention is designed to apprise the reader, initially the patent examiner, of the general area of the invention. In this section there is a discussion of prior art (publications that are in existence at the time of filing the application) that is most closely related to the invention. The patent lawyer will highlight the closest prior art in a manner showing clear differences between this prior art and the invention. The Background of the Invention can be a short paragraph, or it can be rather extensive. The length will depend primarily upon the nature of the art to which the invention is related most closely.

27.4.4 Brief Summary of the Invention

The "Brief Summary of the Invention" is comparable to the Abstract discussed above, but is generally more extensive and detailed. Again, the amount of detail will depend somewhat on the nature of the particular invention. An example of a relatively brief summary of the invention is found in U.S. 4,267,274, cited above. The summary in this patent reads as follows:

> The present invention relates to the isolation of a mutant strain of *Streptomyces mediterranei* and the elaboration by it of essentially rifamycin B only, irrespective of the presence or absence of sodium diethyl barbiturate in the fermentation medium, so that an improved process for making rifamycin B results. The rifamycin B is recovered from the fermentation medium in a conventional manner.

There is no standard concerning the actual length of the Brief Summary of the Invention. The summary of the invention in U.S. 4,250,259 is approximately eight times longer than the one above. It will clearly depend upon the scope and complexity of the invention.

27.4.5 Reference to the Drawings

Many microbiology patent application disclosures contain drawings that are used to further describe the invention. It is desirable to give a brief description of the drawings to aid the reader in understanding them. For example, U.S. 4,250,256 contains the following portion relating to the description of the drawings:

> The present invention will be more fully understood in view of the following description taken in conjunction with the accompanying drawing wherein:
>
> FIG. 1 is a side view of a test device according to the invention wherein nutrient cards (1) mounted on a handle/carrier (3) are covered with a film (2); and
>
> FIG. 2 is a cross-sectional view of the device shown in FIG. 1 also showing a package (4) surrounding the device.

U.S. 4,393,137, which concerns a genetic engineering invention, contains the following reference to the drawings:

> FIG. 1—This shows the construction scheme for making pUC1061 from pUC6.
>
> FIG. 2—Restriction endonuclease cleavage map for pUC1061.
>
> The map is constructed on the basis of plasmid pUC1061 having a molecular length of ca. 7.2 kilobases. The restriction endonuclease abbreviations are as follows: (1) *Bgl*II is an enzyme from *Bacillus globigii*; (2) *Bcl*I is an enzyme from *Bacillus caldolyticus*; (3) *Pvu*II is an enzyme from *Proteus vulgaris*; and (4) *Xho*I is an enzyme from *Xanthomonas holicola*.

27.4.6 Detailed Description of the Invention

It is necessary, in the disclosure, to give a description in sufficient detail to enable a person skilled in the art to practice the invention. When a novel microbe is involved, then it will be necessary to describe the microbe taxonomically, using standard tests. Additionally, a specimen of the novel microbe must be deposited with a recognized culture repository before the patent application is filed with the PTO. The accession number the repository gives the culture must appear in the patent application when it is filed. Further details on this procedure will be presented in section 27.5.1.

When the invention concerns the preparation of a novel antibiotic via a microbiological fermentation, sufficient detail must be disclosed concerning the fermentative preparation of the antibiotic, its isolation, and its purification to a final claimed product. In making this disclosure, the patent applicant is required to disclose the "best mode" for practicing the claimed invention. Basically, this means that the best fermentation conditions must be disclosed as well as the best culture available at the time of filing the patent application.

In the process of isolating and purifying novel antibiotics, it is essential for a person skilled in the field to understand the physical and chemical characteristics of the antibiotic. These characteristics must be disclosed in the patent application and form the basis for a claim to the antibiotic entity itself. When the structure of an antibiotic is known, this also would be included in the patent application disclosure. Similar disclosure requirements are required for any invention. To claim title to a new invention, the inventor must establish that the findings are in fact new by the various identifying characteristics available in the field.

A patent application disclosure also requires the statement of a utility for the invention. This utility must be something reasonable, but it does not have to be of a commercial nature. The utility for a novel microbe can be the production of a useful antibiotic; that for a plasmid can be to create a recombinant plasmid, containing a gene coding for a useful protein, which can then be inserted into a host microbe. See the patent abstracts cited above (section 27.4.2) for examples.

27.4.7 Claims

The claims are based on the patent disclosure and are the means by which the patent applicant delineates that which is considered to be the invention. The drafting of proper claims is something that cannot be approached lightly. If the applicant is negligent in claiming the invention, it is possible to lose valuable patent rights. A patent application can be filed with certain claims which can then be amended during prosecution of the patent application before the patent examiner. Claims are frequently amended during the prosecution to attempt to overcome examiner rejections or to better define the invention. The structure of a claim can be quite simple, or it can be extremely complex. Below are examples from the above-cited patents.

Patent no. U.S. 4,250,256

Claim 1: Microbiological test device comprising a nutrient card and a covering layer which is impermeable to bacteria but permeable to nutrients, said covering layer comprising a water-insoluble synthetic resin film modified with a pore former which is a water-soluble or water-swellable compound, said film having been produced from an aqueous dispersion of the synthetic resin and wherein the pore former and resin film have no negative effect on bacterial growth.

Patent no. U.S. 4,250,259

Claim 1: A process for the microbiological conversion of a C_3-C_6 linear secondary alcohol to the corresponding methyl ketone comprising oxidizing said alcohol by contacting a C_3-C_6 linear secondary alcohol, under aerobic conditions, with microbial cells derived from a methylotrophic microorganism or an enzyme preparation derived from said microorganism, wherein said secondary alcohol is converted to the corresponding methyl ketone by said cells or enzyme preparation, wherein said microorganism has been previously grown under aerobic conditions in a nutrient medium containing a carbon-containing compound which provides the carbon and energy source for growth of the cells of the microorganism and induces C_3-C_6 secondary alcohol dehydrogenase enzyme activity in said cells and wherein said methyl ketone is produced in insoluble amounts.

Patent no. U.S. 4,250,259

Claim 13: A secondary alcohol dehydrogenase enzyme which converts C_3-C_6 secondary alcohols to the corresponding methyl ketones under aerobic conditions and further characterized as being derived from cells of a methylotrophic microorganism which has been previously grown under aerobic conditions in a nutrient medium containing a carbon-containing compound which provides the carbon and energy source for growth of the cells of the microorganism and induces C_3-C_6 secondary alcohol dehydrogenase enzyme activity in said cells.

Patent no. U.S. 4,267,274

Claim 1: A biologically pure culture comprising a new microorganism mutant identified as *Streptomyces mediterranei* M 18, ATCC 21789, and a nutrient medium consisting essentially of an assimilable carbon source, an assimilable nitrogen source, and essential mineral salts, said culture being capable of producing rifamycin B in a recoverable quantity upon fermentation.

Patent no. U.S. 4,306,021

Claim 1: Antibiotic U-60,394, which is active against gram-positive bacteria, and which in its essentially pure crystalline form has the following characteristics:

(a) molecular weight of 391.08014 (high-resolution mass spectrometry);
(b) color and form of crystals: yellowish-green needles;
(c) is insoluble in water, soluble in methanol and acetone with difficulty, and easily soluble in ethyl acetate, methylene chloride, and dimethyl sulfoxide;
(d) a characteristic ^{13}C-NMR spectrum as shown in FIG. 1 of the drawings;
(e) a characteristic ^{1}H-NMR spectrum as shown in FIG. 2 of the drawings;
(f) a characteristic UV spectrum as shown in FIG. 3 of the drawings;
(g) a characteristic infrared absorption spectrum when dissolved in a mineral oil mull as shown in FIG. 4 of the drawings;
(h) a melting point of 265 to 266°C with decomposition;

and base addition salts thereof.

Patent no. U.S. 4,393,137
Claim 2: Plasmid pUC1061, characterized as shown by the restriction map in FIG. 2 of the drawings and having the following additional restriction sites: *Sal*I (five to six sites) and *Sst*I (two sites); lacking restriction sites for the following restriction endonucleases: *Bam*HI, *Hin*dIII, *Kpn*I, *Hpa*I, *Eco*RI, and *Xba*I; and having a molecular length of ca. 7.2 kilobases.

27.5 SPECIAL CHARACTERISTICS OF THE DISCLOSURE FOR A BIOTECHNOLOGY INVENTION

Biotechnology inventions have been around for a long time. Perhaps the greatest surge in the patenting of such inventions occurred in the late 1940s and early 1950s with the dawning of the antibiotic era. Since then many patents have been granted for new antibiotics and microbiological processes for preparing them. In addition, many patents have been granted for microbiological methods, including fermentation processes and equipment, laboratory procedures, assays, and the like. Many of these patents were so important that they spawned new industries and permitted old ones to remain in business. It is well known that a patent covering a useful antibiotic that makes it to the marketplace is indeed a valuable asset. The assurance of a secure market for such an antibiotic, by virtue of the patent position, will enable the assignee to realize significant profits over the years of the patent duration for the antibiotic. This story has been repeated many times with many companies. Thus, patents function in the development of these technologies and in bringing them to the marketplace for the use of the public, with a suitable profit to the marketer.

Most recently, biotechnology has extended its scope into the exciting area of molecular biology. For years molecular biology was primarily an academic curiosity, unknown to industrial concerns. With the successful transformation of plasmids into bacteria, this academic curiosity suddenly became a breakthrough in the biotechnology field. Work in this section of biotechnology has accelerated greatly in the last 5 years. Many new genetic engineering companies have been formed, and existing industries either have bought research from these companies or have instituted their own in-house molecular biology units. All of this activity signals the prospect of a large number of inventions arising from this work, and many of these inventions have, and will continue to have, patentable significance. Perhaps the most well known at this stage is the basic Cohen-Boyer patent, which has generated an enormous amount of income from licenses granted throughout the industry ("Process for Producing Biologically Functional Molecular Chimeras;" S. N. Cohen and H. W. Boyer [Board of Trustees of the Leland Stanford Jr. University, assignee], U.S. patent 4,237,224, December 1980).

This new biotechnology will present new problems in patenting. Many of these problems will be solved by reliance on tested procedures and legal decisions that have been in place to service the already existing body of biotechnology inventions. In this section I will highlight some of the critical characteristics of a patent application disclosure for a biotechnology invention. These characteristics, though developed primarily with antibiotic-type inventions, are clearly applicable to the newer molecular biology inventions.

27.5.1 Cultures: Deposit and Availability

For any biotechnology invention which utilizes a novel microbe, it is essential to recognize certain established procedures concerning the disclosure of these microbes in a patent application. These procedures were initially approved in the legal decision that is known as the *Argoudelis* case (In re Argoudelis et al., 58 CCPA 769,434 F.2d 1390, 168 USPQ 99, 1970). Subsequently, the substance of the *Argoudelis* case was incorporated into the rules of the PTO, where it remains today and where it is applied routinely to patent application disclosures (Manual of Patent Examining Procedure, §608.01(p)C., Deposit of Microorganisms; available from the Superintendent of Documents, U.S. Government Printing Office, Washington, D.C.). Basically, the requirement concerning a novel microbe disclosure in a patent application is that a specimen of the microbe be deposited in a recognized culture repository before the patent application is filed with the PTO. The accession number given by the culture repository for the deposit must be in the patent application when it is filed. Additionally, in the United States, the deposit must be made with the understanding that the culture will be available to the public upon the grant of a patent disclosing the culture by its accession number.

Two well-known culture repositories in the United States are the Agricultural Research Service Culture Collection (Northern Regional Research Laboratory; NRRL) in Peoria, Ill., and the American Type Culture Collection (ATCC) in Rockville, Md.

NRRL charges a user fee of $500 for each patent culture strain, payable at the time of deposit. A $20 fee is charged to the deposit requestor for the distribution of an available (released) patent culture deposited after September 30, 1983. Released patent cultures deposited before this date are available to requestors without charge. Fee payments should be made by check payable to the Agricultural Research Service, U.S. Department of Agriculture. Further information on types of microorganisms accepted and conditions for deposit may be obtained by contacting A. J. Lyons, Curator, ARS Patent Culture Collection, Northern Regional Research Center, USDA-ARS-NCR, 1815 N. University Street, Peoria, IL 61604; (309) 685-4011.

ATCC also charges fees for patent cultures. If the deposit is in connection with patent application filing in the United States and other countries, a one-time fee of $380 for 20 years, $475 for 25 years, or $570 for 30 years is required. The 30-year maintenance period is the time generally required by patent systems in key foreign countries. In other words, the depositor must make provision with the repository to maintain the deposit for 30 years. If the deposit is in connection with a United States patent application only (not other countries), an annual fee may be paid. The present fee for deposits in connection with United States patents is $145 as the initial fee and $100 annual fee thereafter until the United States patent is issued. Deposits in connection with the European Patent Office must be maintained for 30 years and require that the ATCC inform the depositor of all requests for the cultures in question. The one-time fee for European Patent Office deposits for 30 years and the informing fee is $870 per culture. The contact for additional information concerning deposits and fees, and also the person to whom cultures should be sent, is B. A. Brandon, American Type Culture Collection, 23402 Parklawn Dr., Rockville, MD 20852.

Although the culture becomes available to the public upon the grant of a patent, this does not mean that the public can practice the claimed invention commercially. The public can, however, use the culture experimentally and possibly make improvements on the culture by, for example, mutation procedures. Such mutants at times can possess advantageous properties over the parent which make them, in turn, suitable candidates for patenting.

Along with depositing the culture and making it available at the proper time, the patent applicant should also disclose a suitable and standardized taxonomic description of the microorganism.

The above discussion does not distinguish between any particular biotechnological inventions. The only requirement is that a novel microbe be involved. Thus, the requirement for culture deposit would extend to inventions arising from molecular biology research. Perhaps the most extensive use of culture deposits in molecular biology inventions is for recombinant plasmids in suitable host microbes. The host microbe containing the recombinant plasmid is a novel microbe and, as such, can be claimed in a patent application.

U.S. patent 4,393,137, referred to in section 27.4, discloses the deposit of a *Streptomyces* culture hosting a recombinant plasmid. The patent contains claims to the host microbe and also to the recombinant plasmid per se. Additionally, there are process claims directed to the preparation of the recombinant plasmid.

27.5.2 Best-Mode Requirement

Another characteristic of a biotechnology invention disclosure (also required for any other invention disclosure) is that the best mode of practicing the claimed invention must be disclosed in the patent application when filed in the PTO. This would mean best mode known to the inventor(s) at the time the application is filed. This has special significance in biotechnology inventions because it seems clear that this best mode would also apply to the culture that is deposited. There have been no legal decisions directly on this point, but it would be reckless for an inventor to deposit anything but the best culture at the time of filing the application. Should the invention become commercially important, for example, the patent owner might find it necessary to enforce his patent rights against a person practicing the claimed invention, by an infringement suit in an appropriate federal court. As a defense, the alleged infringer could argue that the patent was invalid on the grounds that the patent applicant did not disclose the best mode of practicing the claimed invention, i.e., he or she did not deposit the best culture at the time the application was filed. If the alleged infringer has sufficient proof of this, a federal district court is highly likely to find it repugnant, which could possibly affect the patent owner's case.

27.5.3 Antibiotic Inventions

Since antibiotics constitute a large proportion of biotechnology inventions, it is appropriate to discuss details for disclosing such inventions in patent aplications. The producing culture must be deposited as described above, but it is also necessary to provide an enabling disclosure of a process to make the antibiotic. Generally, for a new antibiotic, this will be a microbial fermentation process. The method disclosed need not be on the level of a commercial process; one using shake flasks, for example, would be suitable. The procedure should be on a scale large enough so that recovery of the antibiotic is feasible. The means for recovering the antibiotic must be disclosed, along with the various physical and chemical parameters that identify it. Occasionally the structure of the antibiotic is known at the time the application is filed; if so, this should be provided as it is considered to be the highest order of identification of the antibiotic.

27.5.4 Molecular Biology Inventions

Molecular biology inventions have characteristics that are quite different from those encountered in other types of biotechnology inventions. Patent applications for such inventions often disclose restriction endonuclease cleavage sites on plasmids constructed to contain foreign genes, promoters, and the like by means of schematic outlines and maps depicting the cleavage sites and description of the procedures used to obtain the recombinant plasmid. Such inventions also record nucleotide sequences that encode the production of useful proteins. If the nucleotide sequence depicted is the first known sequence for encoding the particular useful protein, it is also possible to deduce and claim the novel amino acid sequence from the nucleotides.

Molecular biology inventions can have claims directed to (i) DNA having a specific nucleotide sequence, (ii) amino acid sequence deduced therefrom, (iii) a recombinant plasmid containing DNA having a specific nucleotide sequence, and (iv) a host microbe modified to contain a specific sequence. Additionally, process claims can cover various methods employed in molecular biology, including processes for making desired proteins by the use of engineered microbes.

27.5.5 Recent Patents Issued for Biotechnology Inventions

In addition to the United States patents discussed previously, some recent biotechnology patents are listed in Table 1.

The Office of Technology Assessment and Forecast, U.S. Department of Commerce, issues the publication *Patent Profiles*, which deals with patent activity in various technological areas. Two issues of *Patent Profiles* (July 1982 and 1982 Update—September 1983) dealt exclusively with biotechnology patents. *Patent Profiles* is published by the U.S. Government Printing Office, Washington, D.C.

TABLE 1. Some recent biotechnology patents

Patent no.	Title/description
U.S. 4,182,860	Modified heteropolysaccharides useful in oil recovery
U.S. 4,184,917	Process for producing a structurally modified interferon
U.S. 4,190,495	Modified microorganisms (*E. coli* K-12 × 1776 and 1976) and method of preparing and using same
U.S. 4,237,224	Process for producing biologically functional molecular chimeras
U.S. 4,241,184	Enzymatic conversion of C_3-C_6 secondary alcohol dehydrogenase derived from novel methylotrophs
U.S. 4,247,641	Enzymatic production of epoxides and glycols from alkenes using a halogenating enzyme, oxidizing agent, and halide source
U.S. 4,257,774	Intercalation inhibition assay for compounds that interact with DNA or RNA
U.S. 4,259,451	Pure culture of *Agrobacterium radiobacter* ATCC 31643 capable of producing heteropolysaccharides
U.S. 4,264,731	DNA joining method by pretreatment to remove 5′-terminal phosphate groups
U.S. 4,273,875	Pure culture of *Streptomyces espinosus* NRRL 11439 and plasmid pUC6 produced therefrom
U.S. 4,287,303	Production of ethanol by continuous fermentation
U.S. 4,311,639	Immunogenic interferon peptides
U.S. 4,317,811	Herpes simplex type I subunit vaccine
U.S. 4,320,109	Immunoradiometric assay employing terminal radiomelide labeling and synthesis of conjugates for such assay
U.S. 4,322,497	Method of transducing genetic markers into strains of *E. coli* K-12 × 1776
U.S. 4,322,499	Recombinant DNA plasmid or bacteriophage transfer vector comprising a cDNA sequence comprising the endorphine gene cDNA sequence
U.S. 4,326,358	Process for producing hybrid seeds by genetic crossing of two parent plants to produce a uniform phenotype hybrid and cloning same
U.S. 4,332,900	Construction of cointegrate plasmids from plasmids of *Streptomyces* and *Escherichia*; includes plasmid pUC1024 from pUC1019
U.S. 4,334,017	Method for detecting cancer in mammalian tissue by detecting beta-linked galactose moieties associated with cells
U.S. 4,345,030	Microorganism mutant conversion of sterols to androst-4-ene-3,17-dione
U.S. 4,349,528	Monoclonal hybridoma antibody specific for high-molecular-weight carcinoembryonic antigen
U.S. 4,360,597	*Streptomyces* plasmid and culture
U.S. 4,361,549	Complement-fixing monoclonal antibody to human T cells and method of preparing same

27.6 MARKETING A BIOTECHNOLOGY INVENTION

Making an invention is mentally satisfying to the inventor. It is the accomplishment of something that has never been done before and which, it is hoped, will solve a problem and be of significant use in society. Beyond this mental satisfaction as well is the thought that the invention might make money for the inventor. But making the invention and then taking it to the marketplace are two quite different things. To come up with an invention requires scientific knowledge and skills, but to market it requires a knowledge of the marketplace and of legal procedures that can be used to introduce the invention to potential developers.

27.6.1 Disclosure of the Invention to Others before a Patent Application Is Filed

It is generally not advisable to disclose an invention to a third party for the purpose of ascertaining an interest in developing the invention before a suitable patent appliation has been filed. The preferred course of action once an invention has been made is to contact a competent patent lawyer as detailed above in section 27.3.1. The filing of a patent application in the PTO gives the inventor a record date that can be relied upon in various subsequent legal proceedings. When no patent application has been filed for an invention, a disclosure to a third party can be accomplished, with risk, by use of a confidential disclosure agreement. One example of such an agreement is set out in section 27.1.4. Another, more detailed, example is as follows.

Confidential Disclosure Agreement Letter

Dear Mr. Receiver:

During our meeting on 1 December 1983 you expressed an interest in learning details about an invention that I broadly indicated to you I had made, related to high-level expression of human growth hormone in a lower eucaryote. I will provide you with sufficient details of my invention for you to evaluate your interest in either purchasing my invention or collaborating with me in the development of the invention. I will divulge the details to you under the following terms and conditions.

1. Promptly after receipt of this agreement signed by you I will forward to you sufficient data concerning my invention, which shall hereinafter be referred to as "Confidential Information."
2. You agree to the following in return for receipt of this Information and the opportunity to evaluate your interest as stated above:
   a. to receive and hold the Confidential Information in strict confidence,
   b. to exercise all reasonable precaution to prevent the disclosure of the Confidential Information to others, and
   c. to use the Confidential Information only for the evaluation as described above.
3. Your obligation described above shall be of three years' duration from the date of your signature on this agreement. However, you shall not have any obligations of confidentiality with regard to any portion of the Confidential Information wherein you can establish that:

a. It was known to you prior to receipt of the same, or
b. it becomes generally known or available to the public hereafter through no act or failure to act on the part of you, or
c. it is disclosed at any time to you by a third party who did not violate a confidential agreement with me.

4. You will, promptly after receipt of the Confidential Information from me, evaluate said Information and within sixty days advise me of your interest in negotiating a business relationship with me.

If the foregoing is agreeable to you, would you please so indicate by executing this letter and returning it to me, whereupon this letter shall constitute an agreement between us. The duplicate copy is for your records.

Very truly yours,
SUBMITTER

by ______________________
Mr. T. T. Submitter Date

Accepted:

____
Mr. C. H. Receiver Date

The use of the above agreement or one comparable to it gives the submitter a legal cause of action for breach of the agreement should the receiver disclose the information to another party. However, it is possible for the receiver to have been working on essentially the same invention and, upon receiving the disclosure of the invention, to be stimulated to file a patent application on his or her version of the invention as soon as possible. Such a filing could well precede the submitter's filing, which would give the receiver a procedural advantage in the PTO in a subsequent interference procedure. This is one of the hazards in disclosing an invention, even with a suitable confidential disclosure agreement, before filing a patent application in the PTO.

27.6.2 Disclosure of the Invention to Others after a Patent Application is Filed

Once a patent application is filed in the PTO, the legal risks of disclosing the invention to others for evaluation of interest are minimized. A convenient way of making this disclosure at this point is to submit a copy of the application along with a disclosure agreement along the lines of that given above. Some parties advocate submitting to the other party only the specification of a patent application and not the claims. This may have some advantages in terms of not divulging in an initial contact the scope of the claimed invention. However, before a receiving party can realistically determine its interest in an invention, it is virtually necessary to have the complete application.

In ascertaining an interest in an invention, the receiver of the invention disclosure should be satisfied with the total validity of the application. Accordingly, the patent application should be evaluated not only by a scientist competent in the field and by a business person experienced in marketing, but also by a patent lawyer knowledgeable in the demands of the patent laws. A desirable invention can turn out to be no more than an interesting curiosity if the patent application is rejected.

27.6.3 Licensing of the Patent Right

Once a receiving party shows an interest in the invention, licenses to the invention can be negotiated. This license procedure may occur before the patent application is filed; in this case the licensee, as a part of the license agreement, often may undertake suitable patent action to protect the invention. This would be taken into consideration in the license agreement.

When a patent application has been filed in the PTO, a license can be structured on it and any patents that may issue on the application or on continuations, divisionals, or reissues of it. The license agreement can be either exclusive or nonexclusive. Generally when a licensee wants an exclusive license, the terms of the license agreement are more favorable to the licensor, and problems of patent infringement then become the province of the exclusive licensee. On the other hand, a nonexclusive licensee, or a series of nonexclusive licensees, allows for perhaps a broader scope of development for the invention. A case in point is the nonexclusive licensing of the Cohen-Boyer Stanford patent, wherein over 60 licensees have taken nonexclusive licenses. Generally, a nonexclusive license is sufficient for a licensee when the licensed entity covers a procedure in which no exclusivity is needed or desired by the licensee, such as the invention in the Cohen-Boyer patent. In comparison, when the invention concerns a new antibiotic, then a licensee would generally prefer to have an exclusive license to establish a strong marketing position. The stronger the license, the higher the fee the licensor can command from the licensee.

ADDENDUM IN PROOF

A decision by the United States Court of Appeals for the Federal Circuit, dated 16 September 1985, in the case styled "In re Robert L. Lundak," holds that a culture deposit need not be made before a patent application is filed. The deposit only has to be made before a patent is granted.

Chapter 28

Biosafety and Biotechnology

DANIEL F. LIBERMAN, RICHARD FINK, AND FREDERICK SCHAEFER

Biotechnology can be defined as the application of biological systems to technical and industrial processes. This implies the integration of all biology, including molecular biology, genetics, microbiology, cell biology, and biochemistry, with chemical and process engineering in a way that develops the full potential of each of these systems (1, 12).

From a practical standpoint, academic and industrial applications of biotechnology can be divided into two general categories: research/development and production. Traditionally, the differences between research and production have involved both the goal and the scale of operation for these activities.

Production is the manufacture of specific materials (organisms or metabolites) by carefully developed and controlled procedures. Research, on the other hand, represents an investigative effort which may result in the development of a procedure, in new information concerning the properties of an existing procedure, or in process improvement. Production is restricted to industrial activity, whereas research is within the domain of both academia and industry.

Special hazards of biotechnology are associated with three properties of microorganisms: (i) the potential of a few strains to cause disease; (ii) the potential for undetected genotypic or phenotypic changes to alter a tested and approved process; and (iii) the ubiquity of organisms which can contaminate the system.

While it is unlikely that dangerous human and animal pathogens will be used in very large-scale fermentation, the industrial use of plant pathogens is increasing (6, 15, 45). There is a fundamental lack of information concerning plant diseases and the agents which cause them (see reference 15 for a review). The aerosol dissemination of fungal spores (3), bacteria (21), and viruses which are pathogenic for plants (4, 31) has been described. The economic consequences of the inadvertent release of a pathogen in an area where there are susceptible hosts are potentially catastrophic (15). Therefore it is crucial that efforts be made to prevent the escape of these organisms to the environment.

There is little opportunity for new genotypes to become established in batch production. This is due to the relatively few generations of the organisms involved. However, in continuous fermentations, which are becoming more and more popular and important, e.g., in single-cell protein production, the content of the bioreactor may need to be maintained in a homogeneous state for hundreds of days. This would involve many thousands of generations. With natural mutation rates as high as 1 in 10^7 (42, 45) and with the additional chance of contamination from outside the system, the potential for genetic change would seem enormous. One should remember, however, that the majority of these mutational events would be selected against by the fermentation conditions and that unless some specific advantage was conferred, the mutant or contaminant probably would not become established (45).

On the other hand, if a contaminant entered which was able to maintain itself, it could disrupt the system in several ways. The contaminant could directly inhibit or interfere with the biocatalyst (enzyme, cell, or microorganism). It could destroy the catalyst or destroy the product by using it as an energy source. In addition, the contaminant could introduce substances that are difficult to separate from the product, thereby rendering the product unusable.

Unlike mutation, phenotypic change is a response to a changed environment and is maintained only while the new conditions persist. It should therefore be prevented by the close process control used to maintain maximum efficiency.

In practice, an industrial fermentation process is extremely unlikely to become contaminated with a highly pathogenic microorganism because the environment inside the fermentor is so different from the human body that pathogenicity confers no advantage upon the organism. On the other hand, unless steps are taken to preclude the introduction of environmental microorganisms, they can and will get in and disrupt the system.

While the orientation of this chapter is toward worker safety, it must be kept in mind that the steps taken to protect the worker are equivalent to the steps one takes to protect the integrity of the system. Whenever there is a significant potential for introducing undesirable organisms into the bioreactor, there is an equivalent opportunity for organisms to escape into the environment.

It is therefore appropriate to examine more closely the various stages involved in manipulating organisms in a process and to describe the practices which historically have been shown to minimize or eliminate contamination of personnel, product, and the environment. We first consider overall containment strategy and subsequently will examine various stages of manipulation to see how these containment strategies apply.

28.1 BIOSAFETY CONSIDERATIONS

Analysis of surveys (18–20, 28, 37–41) of laboratory-associated infections reveals that the actual causes for most laboratory-acquired illnesses are not known. Fewer than 20% of documented infections can be attributed to accidental contact, ingestion, or injection with infectious material. The remaining 80% have been attributed to unknown or unrecognized causes (19, 20, 28, 37–41, 48–50). These analyses suggest that personnel engaged in research are at a higher risk than personnel associated with diagnostic, educational, or industrial activities (41). The risk to personnel who are in direct contact with the agent is higher than for personnel who are only remotely involved. Eighty percent of the 300 illnesses at Fort Detrick reviewed by Wedum et al. (50) involved trained laboratory personnel. Support personnel such as janitors, dishwashers, and maintenance and clerical workers were at lower risk. These results serve to confirm previous estimates reported by Sulkin and Pike (46).

Laboratory studies on the potential sources of infection have focused on the hazard potential of routine laboratory techniques (2, 26, 43). Since these studies suggest that most laboratory techniques create aerosols, inhalation exposure to undetected infectious aerosols may contribute significantly to occupational illness among laboratory workers who handle infectious material (5, 19, 24, 28, 42, 44, 47).

Based upon available data, preventive measures have been developed which provide safeguards for the protection of scientific and support personnel, the experiment, and the environment. These safeguards are collectively referred to as containment practices (5, 29, 35, 44, 47).

28.2 TYPES OF CONTAINMENT

Practical planning for safety is hampered by the fact that safety cannot be measured directly. The words "safe" and "safety" represent ideal concepts which, while desirable, are unattainable in absolute terms. Practical planning for safety is therefore performed by evaluating its opposite, risk. Safety in laboratory research activity then becomes an exercise in recognizing what the risks are and then introducing procedures, practices, etc., to control the identified hazards or reduce them to acceptable levels.

The purpose of physical containment is to reduce the potential for exposing laboratory workers, workers outside the laboratory environment, and the public to infectious agents contained within the laboratory. The elements of physical containment include the scrupulous use of safe laboratory techniques, well-designed facilities, and equipment appropriate for the given activity (29, 32, 35, 36). Safety equipment provides primary and secondary physical containment to control aerosols and to maintain the integrity of the system under study or the fermentation in progress.

The most important elements of physical containment are the procedures and equipment used by the worker to control the hazards associated with these agents. The laboratory facility itself provides an important barrier to protect the public and the product from exposure to potentially detrimental material (hazardous organisms and their metabolites as well as saprophytic contaminants).

It would be difficult to describe in detail all the specific rules and procedures designed to prevent or control laboratory-acquired illness or product contamination. These procedures vary and depend to a great extent on the agent, the type of experiment, the equipment, the facilities, and the proficiency of the personnel. Fortunately, there are a number of excellent reviews on this subject, and the reader is encouraged to consult them (5, 19, 23, 24, 28, 29, 35–44). It is incumbent on each laboratory that is involved with or contemplates the use of hazardous or potentially hazardous agents to examine these references and to develop protocols which meet the associated safety concerns. There are, however, several basic practices which apply that should be mentioned. They constitute the basic recommended practices which have been shown to be effective for the protection of personnel, the product, and the environment.

28.3 PERSONNEL PRACTICES

Laboratory supervisors should prohibit eating, smoking, drinking, food storage, and the application of cosmetics in the laboratory. They should not permit pipetting by mouth; the use of mechanical devices must be required. These devices are more accurate than standard pipettes and clearly eliminate aspiration as a source of laboratory-acquired illness (37–41, 44, 50). These personnel protective practices are designed to eliminate ingestion as a mode of exposure.

A second group of personnel protective practices includes the use of appropriate gloves when working with hazardous material. Gloves function to prevent the worker's hands or fingers from being contaminated. This helps to further reduce the hazards associated with ingestion (hand-to-mouth transfer) or the penetration of material through broken or unbroken skin. Although there are relatively few organisms which can penetrate unbroken skin, one should not lose sight of the fact that experiments performed today also involve chemicals (some of which are toxic) and radionuclides. Therefore, gloves should be regarded as a minimum requirement.

Laboratory clothing, which is designed to protect one's personal clothing, should be worn. The purpose of laboratory clothing is to keep street clothing and the worker's forearms free of contamination. It is important to realize that wearing these coats to the cafeteria, to the library, to meetings, and to other buildings provides a mechanism for spreading contamination to others as well as to oneself. Laboratory coats should never be worn outside the laboratory area.

Work surfaces must be decontaminated both daily and immediately after spills. This will reduce the spread of contamination to the worker and at the same time reduce the potential for contaminating the experiment or process at hand or elsewhere in the laboratory.

As indicated previously, considerable information has been accumulated which clearly indicates that nearly all routine laboratory procedures are capable of producing aerosols, including particles of respirable size (2, 8, 14, 22).

Any operation which involves an animal or plant pathogen or an agent which could disrupt the environment if inadvertently released should be contained within safety equipment or facilities. These containment systems must be subjected to periodic inspection and certification to ensure proper function (35). All biological waste should be decontaminated or inactivated before disposal. This is especially true if known pathogens are involved. Contaminated materials such as bioreactors, glassware, laboratory equipment, etc. should be decontaminated before washing, reuse, or disposal. This again will help protect personnel who are not directly associated with the laboratory activity (e.g., glassware workers, janitors, repair personnel, technical support, etc.)

Finally, all employees working with or who may be potentially exposed to hazardous organisms or substances must receive sufficient information and training to enable them to work safely and to understand or appreciate the relative importance of the potential hazards. This instruction should include: a thorough review of operations and procedures with emphasis on material transfer and other possible sources of exposure; adverse health effects (and, where defined, early symptoms); specific acceptable and unacceptable work practices; engineering controls (hazard control ventilation, contained centrifuges, safety cabinets, etc.) in use or being considered for use to limit personnel or environmental exposure; proper disposal of contaminated waste; decontamination of surfaces; and specific emergency procedures to be followed in the event that there is an accident or spill. Each member of the work team must be familiar with the biology of the system or process under way.

While these basic practices may appear to represent a commonsense approach to safety, it must be emphasized that they are based on over three decades of experience with laboratory-associated illness. There are risks associated with activity involving infectious agents. It is possible to minimize the risk if appropriate attention is paid to the safety practices described above as well as primary and secondary containment barriers.

28.4 PRIMARY CONTAINMENT BARRIERS

The selection of specific items of equipment is dictated by the hazard potential of the organism (10, 34, 47). Safety equipment used as primary barriers includes biological safety cabinets, safety blenders, safety centrifuge cups, and a variety of enclosed containers (e.g., fermentors). These items are designed to prevent the escape of aerosols into the laboratory environment.

Biological safety cabinets are used as a primary barrier to prevent the escape of aerosols from the point of origin into the laboratory environment. Three types of cabinets (Class I, II, and III) are used in laboratories (35). Open-fronted Class I and II biosafety cabinets are partial-containment cabinets which offer protection to laboratory personnel and to the environment when used in conjunction with good microbiological technique. Since the inward air velocities (face velocities) are similar for both classes, they provide an equivalent level of personnel protection. Class II cabinets offer the additional capability of protecting materials contained within them from extraneous airborne contaminants. This is provided by a laminar stream of a high-efficiency particulate air (HEPA)-filtered air mass which passes vertically within the workspace (including the front opening). Neither of these cabinet classes is appropriate for the containment of the highest-risk infectious agents because of the chance for inadvertent escape of aerosols across the open front (5, 44, 47). The Class III cabinet, commonly referred to as a glove box, provides this level of personnel and material protection. Protection is provided by the physical isolation of the space in which the infectious agent is manipulated.

When Class III cabinets are required, all procedures involving infectious agents must be conducted within them. These cabinets are frequently designed as a system of interconnected units which contain all the equipment required by the laboratory program, e.g., incubators, refrigerators, centrifuges, and even animal storage cages (35).

28.5 SECONDARY CONTAINMENT BARRIERS

Secondary barriers are the features of the facility which surround the primary barriers. These barriers provide a separation of the laboratory from the outside environment as well as from other facilities in the same building. Examples of secondary barriers are floors, walls, ceilings, airlocks, doors, differential pressures, exhaust air filters, and provisions for treating contaminated wastes. These and other design features provide barriers which serve to prevent the escape of hazardous microorganisms in the event of a failure in a primary barrier and can be designed to prevent external organisms from contaminating the workspace (33).

Actually, the more effective the primary barriers are, the less need there is for emphasis on secondary barriers. In the design of a research laboratory, it is both important and economically necessary first to determine and select the primary containment to be used, thereby reducing the complexity and cost of the secondary barriers (11).

The primary function of the facility is to provide a physical environment in which work can be performed efficiently and safely. A well-designed facility will facilitate good laboratory practice, contain equipment designed to protect the worker from the potential hazards associated with the activity or the system, and ensure the protection of people and the environment outside of the laboratory or building.

Three categories of facility safeguards have been established for hazardous or potentially hazardous research projects. These categories can be classified as the basic facility, the containment facility, and the maximum containment facility. The levels of protection increase with the risks associated with the agents to be used and, in general, correspond with levels of containment required for low-, moderate-, and high-risk agents or procedures. The design criteria for these facilities are published elsewhere (29, 33, 35, 47).

It should be noted, however, that while the facility can provide an environment in which work can be conducted safely, the facility itself does not ensure that the work environment is safe. Well-designed facilities primarily protect the environment and cannot be considered as a replacement for good laboratory procedures and practices.

These comments on containment are consistent with those put forth by the National Institutes of Health for research with hazardous materials such as carcinogens (34), oncogenic viruses (23), and recombinant DNA molecules (36). The Centers for Disease Control, in conjunction with the National Institutes of Health, has developed a similar approach in its recommendations for microbiological and biomedical laboratories (47). The bottom line in all these documents is that hazardous organisms or materials must be contained in a fashion that is commensurate with the level of hazard associated with the organisms or materials (47).

The principles which underlie the containment practices for large-scale activity are based upon the recognition that there may be risks associated with the activity. The best way to control these risks is through the use of techniques and equipment designed to contain aerosols and prevent their release to the work and general environment. The requirements to treat effluent air and liquid to render them biologically inactive before release are as important as the requirements to use aseptic technique (35, 36, 45).

28.6 BIOTECHNOLOGY

Biotechnology was defined above as the application of biological systems to basic and applied technical and industrial processes. These processes are based on the use of some form of organic catalyst in the conversion of a substrate to a desired product. These biocatalysts may be highly complex, e.g., microorganisms or cells, or they may be "simple," e.g., enzymes (27).

The selection of the appropriate biocatalyst depends on the process and product of interest. In the production of beer or wine, yeasts are selected for the conversion of sugars into ethanol and carbon dioxide. In the production of a biologically active protein like interferon, the catalyst can be a bacterium, a yeast, or a cultured cell which has been engineered to express proteins.

A concern common to nearly all biotechnologies is the need to maintain aseptic conditions (11). The reason is that most products of such processes are made by a pure culture, a population of cells derived from a single strain or clone. Introduction of another type of organism into this culture could affect the efficiency of the process or even destroy the culture altogether. To avoid contamination, all phases of the process must be carefully evaluated and monitored.

As one can see from its definition, biotechnology encompasses a rather wide range of activities. While each activity is associated with its own starting material (bacterial, fungal, yeast, plant, or animal cells or enzymes), technical manipulation (cell fusion, cell or enzyme immobilization, genetic engineering, etc.), processes (continuous or batch fermentation), and scale (small to very large) which are unique to that activity, there are common features or stages which lend themselves to a generic treatment. Regardless of the goal, biotechnology will involve (i) isolation and preservation of the organism or cell which is the biocatalyst or the source of the specific catalyst of interest; (ii) preparation of biocatalyst; (iii) process scale-up or large-scale growth; (iv) separation of the desired product from the biocatalyst; (v) purification of the product; and (vi) waste treatment.

28.6.1 Culture Preservation

Organisms to be preserved are usually cultivated in small volumes (less than 100 ml), and the resultant mixture is transferred in quantities of up to 1 ml into small tubes with a cryopreservative. These tubes are sealed and stored in or above liquid nitrogen (−180 or −196°C), in a low-temperature freezer (−80°C), or in a standard freezer (−20°C) (16). Alternatively, samples can be lyophilized (freeze-dried) and stored between room temperature and 0°C (16, 25).

These samples serve as the starting material for all future cultivations. Each sealed tube may contain a few micrograms (dry weight) of microorganism. The number of organisms per tube may vary from 10^8 for yeast and molds to 10^{10} for bacteria (45).

Even though the volume of medium or the number of organisms in the starting sample is comparatively small, the biology of the organism (pathogen or saprophyte, mechanism of pathogenicity, identification of suitable hosts at risk, type of potential illness, etc.) must be understood before work begins. Appropriate work practices, in an appropriate biological safety cabinet, should protect both the worker and the biological system from gross contamination.

When a lyophilizer is used, care should be taken to prevent the vacuum lines from being contaminated. The insertion of a suitable filter or trap is usually sufficient. All lines, tubing, or glass connectors which come in contact with the primary vessel or tube should be decontaminated by chemical disinfection, followed by extensive washing to remove residual disinfectant, or by steam sterilization before reuse.

28.6.2 Preparation of the Biocatalyst

The next stage is to provide the required number and quality of organisms or enzymes to initiate a larger-scale cultivation or process.

Cells

A tube of preserved material is opened, and its contents are transferred into a flask or bottle containing an appropriate volume of a sterile nutrient solution. The culture is capped or closed with a filter or plug which permits the passage of gases but not of organisms. The culture is then maintained at an appropriate growth temperature until satisfactory growth is achieved. This temperature varies, depending on the organism involved. The culture is then used to seed a similar but larger vessel. This flask is maintained at the appropriate temperature and is often placed on an automatic shaker, roller, or stirrer to increase gaseous interchange across the liquid-air interface (16). After a suitable growth period, the culture is transferred to a sterile container from which it is further transferred into a larger culture vessel (bioreactor).

Up to this point, it is frequently necessary to limit the amount of growth at each of these amplification phases to prevent the medium from becoming too acid or alkaline, which might result in a cessation of growth or even kill the organism. Once the seed culture has been transferred to the culture vessel, these restrictions no longer apply. Typically, there may be a few grams (dry weight) of organism at this point, which means that the increase in the number of organisms present is on the order of 10^3 to 10^4 (45).

Enzymes

If the biocatalyst is a purified enzyme, then sufficient enzyme must be isolated from the producing organisms. The preparation of the starting inoculum would be similar to that described above.

When a sufficient number of organisms are attained, the enzyme is isolated and attached to a suitable carrier. By definition, immobilization is the conversion of enzymes from a water-soluble, mobile state to a water-insoluble, immobile state (27).

All plastic or glassware which comes in contact with the organism should be decontaminated or disposed of promptly after use and not allowed to accumulate. Unattended bottles, flasks, plates, etc. containing medium (or even a residual film of medium) can support the growth of saprophytes which may result in unnecessary contamination of the system. Keep in mind that many of these environmental organisms are sporeformers and therefore are capable of extensive contamination.

If pathogens are to be used, precautions to prevent accidental breakage of flasks during shaking are necessary. All glassware should be examined before use; any item which is chipped or cracked should not be used.

When a dry air shaker or a shaker platform is used with pathogens (especially for organisms known to be respiratory tract pathogens), the unit should be enclosed in a chamber which operates under a negative pressure, and the exhaust air should be cleaned by filtration. If a vessel should break or a cotton plug or stopper come off, then any aerosol that might be generated would be trapped in the filter.

If such an accident does occur, then personnel who are suitably trained, clothed, and provided with appropriate respiratory protection should remove the undamaged vessels, clean up the debris, and decontaminate the shaker and chamber surfaces appropriately. A listing of appropriate disinfectants and decontaminants and procedures for their use may be found in reference 35.

28.6.3 Scale-Up of Process

To maximize the yield of a desired product, microorganisms are grown in bioreactors under rigorously controlled conditions. Often the final process is preceded by growth in successively larger vessels, for example, 20, 200, and 2,000 liters (7), usually constructed of stainless steel.

There are three basic fermentor-bioreactor designs: (i) small, portable units which are filled with medium and sterilized in an autoclave; (ii) portable units which are sterilized in an autoclave and filled aseptically with sterile medium; and (iii) fixed units which are sterilized in place (7, 9, 11). In a batch process, most of the constituents of the medium are combined with the biological catalyst at the start. The process vessel is sterilized before the medium is introduced. Depending on the scale of operation, the starting materials are added by means of tubes and pipes (7, 11). In the vessel, the catalyst and constituents of the medium are mixed by a rotating control shaft that carries several impeller rotor blades. As the biological conversion proceeds, any additional nutrients needed can be added via additional tubes and lines. This requires that conditions in the reaction vessel be monitored during the conversion and that sensors be inserted through the vessel wall at various locations. A vessel for a continuous process is similar, except that nutrient is continuously added and the products of the reaction are continuously removed (7).

In either the batch or continuous mode of operation, the design of fermentation equipment and facilities is central to the containment strategy for biological control. The term biological control is used because both the worker and the integrity of the culture must be protected. To avoid contamination of the culture, all materials that come in contact with it, as well as the medium constituents, must be sterilized. Air should be filtered through a deep bed of glass wool or by special high-efficiency particulate air filters. Reactor vessels, pipelines, and other surfaces with which the medium comes in contact are steam sterilized. The mechanical system must be designed and operated so that the opportunities for contaminants to enter, or for organisms to escape, are eliminated or minimized. Maintaining the integrity of various entry and exit points in the system is both crucial and often difficult. The potential for human error and mechanical failure is great, and culture contamination unfortunately does occur.

Massive contamination is usually caused by mistakes, whereas slight contamination which develops gradually is frequently due to inadequate sterilization. If the contamination is chronic, the whole system should be given a detailed, critical inspection.

In the initial installation and during contamination review, care must be taken to ensure that all piping and fittings are welded properly and that the use of connecting fittings is minimized (7, 9). Dead spaces, crevices, and nondraining portions of lines should be avoided. Sterile and nonsterile segments of the system should be separated by steam blocks. Valves should be examined for dead spots or crevices before being installed.

In dealing with pathogens, exhaust gases must be either filter sterilized or incinerated to remove or destroy any organisms suspended in them. An effluent stream with a high moisture content will reduce HEPA filtration efficiency, and in the event of a "foam out" it will reduce containment efficacy. In such a situation, hydrophobic prefilters in conjunction with a catch basin or tank to collect condensate or foam should be included in the design.

Agitator seals must be examined for leaks. The possibility that leaks can occur is not trivial. The recommendation for top-mounted agitation for large-scale processes involving recombinant DNA organisms is an attempt to reduce the impact of a seal malfunction (36). If the impeller shaft is located at the bottom, then a seal leak can result in the vessel's contents contaminating the motor, the floor, etc. If the system is located at the top, then the extent of contamination is reduced substantially.

Three types of seals (packed seal, lip or oil seal, and mechanical seal) are commonly used (9, 11). The current trend in design of contained fermentors is toward double mechanical seals.

Sampling ports must also be considered as possible routes of contamination. Samples should be taken via a closed system to avoid the generation of aerosols (36). Flush-mounted valves which eliminate dead spaces should be used where possible. Piston and ball valves can be designed to include a steam block, in which the sample flows from the vessel to the sample port when the valve is open and steam flows back through the valve toward the sample port and the vessel when the valve is closed. Butterfly valves are also very popular.

Ports must also be provided for probes to monitor the progress of the fermentation. Probes which can be sterilized along with the reaction vessel should be used. It is important to avoid crevices on the ports and to incorporate both internal and external seals in the design of probe ports.

28.6.4 Recovering the Product

The specific method of product recovery and purification depends on the properties of the product, such as its location (intracellular, dissolved in medium, or both) and its stability to heat and chemical disinfectants (reagents used in the extraction and purification may inactivate the product). Yeasts and fungi are usually harvested by continuous filtration, whereas bacteria, viruses, and cells are routinely harvested by centrifugation. It is possible to recover on the order of 100 kg of wet cell material from 1,000 liters of culture (9, 45).

Separation and subsequent processing always involve the breaking of air-liquid interfaces and therefore result in the release of aerosols. If the biocatalyst is a pathogen, it is desirable to inactivate the organism first. This can be accomplished by "in place" sterilization within the reactor vessel (batch sterilization). Alternatively, the reactor contents can be sterilized by being passed through a heat exchanger (continuous-flow sterilization) (36). Either way is acceptable if the product of interest is stable under the conditions of sterilization.

If the product is heat labile, then it is necessary to explore the possibility of chemical inactivation of the organism. This requires that the product be unaffected by the disinfectant selected.

In the event that the product is not stable, then it must be separated from the cells before further processing. Such bulk processing must be performed under "closed conditions" to minimize personnel and environmental exposures (36).

The vessel contents must be delivered to the separation tank, the filtration column, or centrifuge by a series of lines, the integrity of which should be examined in the same fashion as that of the vessel itself. The columns and centrifuge should be closed systems (primary containment) or placed in a chamber or room which is specifically designed to control the spread of aerosols (secondary containment). Such rooms or chambers are maintained under negative pressure (more air is exhausted than supplied) to prevent dissemination of organisms to neighboring areas or to the environment (33).

28.6.5 Processing the Product

Subsequent processing methods vary widely and depend on the purpose of the fermentation. In some cases, the whole culture may be used without separation of cells. Separated microorganisms may be used without further processing beyond drying, such as the yeasts, fungi, and bacteria used as single-cell protein (30). In yet other instances, the cells may be subjected to some form of chemical or physical disruption to release the desired product (9).

Cells may be disrupted by nonmechanical methods, such as osmotic shock or enzymatic or chemical lysis, or by mechanical methods. Because of difficulties in scaling up the former procedures, mechanical methods are more popular on the industrial scale.

Two types of mechanical disrupters are widely used: high-pressure homogenizers and high-speed agitator mills. Since both have the potential for generating aerosols, the system design must ensure either that any aerosol generated be retained within the unit itself or, as mentioned above for filtration systems and centrifuges, that the unit be placed within a suitable chamber or room which will provide the necessary aerosol control.

It is also often essential to protect the product from contamination. In this case, efforts must be directed toward keeping external organisms out rather than keeping process organisms in. This requires facilities and equipment which are operated under positive-pressure conditions. Containment facilities then are transformed into barrier facilities (33). In such circumstances it is essential that any pathogenic organism used as the biocatalyst be inactivated before processing to avoid personnel exposures.

28.7 WASTE TREATMENT

Although some of the activities associated with biotechnology are comparatively new (e.g., genetic engineering, enzyme and cell immobilization, cell fusion), applied microbiology (fermentation), which is the cornerstone of this technology, is not (11–13). The practice of industrial fermentation since World War II has provided us with sufficient background to anticipate the kinds of waste that will be generated and, of equal import, the methods available to treat them.

The important point to remember is that if pathogens are used, they must be rendered inactive before being introduced into the environment. This also applies to genetically engineered organisms which are regulated by the National Institutes of Health Guidelines (36). Such cells can be inactivated by steam sterilization or chemical treatment. Clearly, the scale of operation and the stability of the product are essential parameters in the method(s) used.

More often than not, the product is heat labile and the use of a disinfectant is not feasible or possible due to product instability. In such cases, the biocatalyst is separated from the product and the waste that is generated is processed by an appropriate method. Frequently, the product is intracellular and the cells are disrupted to release the product into the milieu. This disruption should render the organism nonviable. Since it is not certain that all the cells will be disrupted, however, care should be exerted whenever pathogens are used to avoid the introduction of viable organisms into the environment. This will necessitate the treatment of all biowaste generated in any step which has not been subjected to previous treatment.

In principle, the disposal of waste is fairly straightforward whether the waste is subsequently processed on-site or off-site. In sewage treatment, mixed microbial populations are used to reduce the amount of oxidizable organic material. The main components of a biological waste stream are organic chemicals containing nitrogen, phosphorus, and carbon and inorganic chemicals which include buffers, acids, and alkalis.

Several monographs deal with the disposal or processing of biological waste; that by Grady and Lim (17) is comprehensive and useful.

28.8 POSSIBLE HAZARDS FROM INDUSTRIAL PRODUCTION AND USE OF PATHOGENS

A pathogen that is still virulent after growth in a fermentor cannot be handled safely on a very large scale. Nevertheless, if the kinds of precautions described above are employed, there is little reason to expect that industrial processing of pathogens on a scale not exceeding a few thousand liters will lead to their release, either as bulk liquid or, much more importantly, as an aerosol. Therefore, at this scale, we need only consider the possible consequences of a system failure.

The possibility of such a failure can be reduced to a very low level, but never eliminated. Accidental release might follow a mechanical failure, filter failure, or foam-out of the culture. The negative consequences of such a failure would depend on a number of factors: (i) the strain of microorganism used (10, 47); (ii) the quantity of microorganism released (24, 34, 45, 47); (iii) the physiological state of the microorganism, including its pathogenicity, infectivity, spore production, and aerostability as determined by growth conditions (15, 50); (iv) the local environmental conditions, which will determine the organism's dispersal, dilution, survival, and ability to enter the ecosystem; (v) the location of the facility with respect to susceptible hosts (37); and (vi) the susceptibility of possible hosts to infection, as determined by a variety of factors (34).

All of these are interrelated and subject to variation. At present, it is quite impossible to predict their relative importance. The best way to eliminate the hazard of a pathogenic organism is not to use that organism at all, but instead to employ a substitute.

Less hazardous organisms that would fulfil the same purpose as the undesirable pathogen can be discovered, selected, or created. Industry for many years has employed selection techniques to screen wide fields of related organisms for nonpathogens, as well as more complex methods involving mutation and selection of desirable strains. Genetic engineering is also used to remove from the pathogen the genetic components responsible for its useful properties and to incorporate these elements into nonpathogenic organisms such as *Escherichia coli*, *Bacillus subtilis*, or *Saccharomyces cerevisiae* (36).

28.9 CONCLUSIONS

Biotechnology is, at this time, a rapidly developing field. Its potential is so vast and diverse that it is impossible to predict all the areas of academic or industrial activity that will benefit from its application. However, it is certain that biotechnology will play an increasingly important role in future industrial processes. Therefore, it will be necessary to make certain that each individual is properly trained, that each program of activity is adequately reviewed, and that equipment necessary to protect personnel, the product, and the environment is provided and used correctly, so that the promise of biotechnology does not become the reality of a biological catastrophe.

The widely divergent activities associated with biotechnology make it difficult if not impossible for safety specialists to anticipate the use of potentially hazardous materials and to monitor appropriately each and every operation that involves these materials. It is obvious, then, that the user must have some degree of knowledge of the hazard potential presented by these materials so that the precautions necessary to conduct the activity safely can be instituted.

There must be a continuous dialogue between the various parties associated with the project to ensure that the safety of personnel and the environment is not compromised.

28.10 LITERATURE CITED

1. **Abelson, P. H.** 1983. Biotechnology: an overview. Science **219:**609.

2. **Anderson, R. E., L. Stein, M. L. Moss, and N. H. Gross.** 1952. Potential infectious hazards of common bacteriological techniques. J. Bacteriol. **64:**473–481.
3. **Aylor, D. E., and P. E. Waggoner.** 1980. Aerial dissemination of fungal spores. Ann. N.Y. Acad. Sci. **353:**116–122.
4. **Banttari, E. E., and J. R. Venette.** 1980. Aerosol spread of plant viruses: potential role in disease outbreaks. Ann. N.Y. Acad. Sci. **353:**167–173.
5. **Barkley, W. E.** 1981. Containment and disinfection, p. 487–503. *In* P. Gerhardt (ed.), Manual of methods for general bacteriology. American Society for Microbiology, Washington, D.C.
6. **Barton, K. A., and W. J. Brill.** 1983. Prospects in plant genetic engineering. Science **219:**671–676.
7. **Blakebrough, M.** 1969. Design of laboratory fermenters, p. 473–504. *In* Methods in microbiology, vol. 1. Academic Press, Inc., New York.
8. **Brown, J. H., K. M. Cook, F. G. Ney, and T. Hatch.** 1950. Influence of particle size upon the retention of particulate matter in the human being. Am. J. Public Health **40:**450–458.
9. **Calam, C. T.** 1969. Culture of microorganisms in liquid medium, p. 255–326. *In* Methods in microbiology, vol. 1. Academic Press, Inc., New York.
10. **Center for Disease Control.** 1974. Classification of etiologic agents on the basis of hazard. Center for Disease Control, Atlanta.
11. **Cooney, C. L.** 1983. Bioreactors: design and operation. Science **219:**728–733.
12. **Demain, A. L.** 1981. Industrial microbiology. Science **214:**987–995.
13. **Demain, A. L.** 1983. New applications of microbial products. Science **219:**709–714.
14. **Druett, H. A., D. W. Henderson, L. Packman, and S. Peacock.** 1953. Studies on respiratory infection. I. The influence of particle size on respiratory infection with anthrax spores. J. Hyg. **51:**359–371.
15. **Evans, C. G. T., T. F. Preece, and K. Sargeant.** 1981. Microbial plant pathogens: natural spread and possible risks in their industrial use. Commission of the European Communities, United Kingdom.
16. **Gherna, R. L.** 1981. Preservation, p. 208–217. *In* P. Gerhardt (ed.), Manual of methods for general bacteriology. American Society for Microbiology, Washington, D.C.
17. **Grady, C. P. L., and H. C. Lim.** 1980. Biological waste treatment: theory and applications. Marcel Dekker, Inc., New York.
18. **Grist, N. R.** 1980. Hepatitis in clinical laboratories. J. Clin. Pathol. **33:**471–473.
19. **Grist, N. R.** 1982. Epidemiology and control of virus infections in the laboratory. Yale J. Biol. Med. **55:**213–218.
20. **Harrington, J. M., and H. S. Shannon.** 1977. Survey of safety and health care in British medical laboratories. Br. Med. J. **1:**626–628.
21. **Harrison, M. D.** 1980. Aerosol dissemination of bacterial plant pathogens. Ann. N.Y. Acad. Sci. **353:**94–104.
22. **Hatch, T. F.** 1961. Distribution and deposition of inhaled particles in respiratory tract. Bacteriol. Rev. **25:**237–240.
23. **Hellman, A. (ed.).** 1969. Biohazard control and containment in oncogenic virus research. National Cancer Institute, Bethesda, Md.
24. **Hellman, A., M. N. Oxman, and R. Pollack (ed.).** 1973. Biohazards in biologic research. Cold Spring Harbor Laboratory, Cold Spring Harbor, N.Y.
25. **Kennett, R. H.** 1980. Freezing of hybridoma cells, p. 375. *In* R. H. Kennett, T. J. McKearn, and K. B. Bechtol (ed.), Monoclonal antibodies. Plenum Press, New York.
26. **Kenny, M. T., and F. L. Sabel.** 1968. Particle size distribution of *Serratia marcescens* aerosols created during common laboratory procedures and simulated laboratory accidents. Appl. Microbiol. **16:**146–150.
27. **Klibanov, A. L.** 1983. Immobilized enzymes and cells as practical catalysts. Science **219:**722–727.
28. **Liberman, D. F.** 1979. Occupational hazards: illness in the microbiology laboratory. Public Health Lab. **37:**118–129.
29. **Liberman, D. F.** 1980. Facility description and personnel practices for research activity of comparable hazard. Public Health Lab. **37:**118–129.
30. **Litchfield, J. H.** 1983. Single-cell proteins. Science **219:**740–746.
31. **Maramorosch, K.** 1980. Spread of plant viruses and spiroplasmas through airborne vectors. Ann. N.Y. Acad. Sci. **353:**179–185.
32. **Medical Research Council of Canada.** 1977. Guidelines for the handling of recombinant DNA molecules and animal viruses and cells. Medical Research Council of Canada, Ottawa.
33. **National Cancer Institute.** 1979. Design of biomedical research facilities. National Institutes of Health, Bethesda, Md.
34. **National Institutes of Health.** 1974. Biohazards safety guidelines. National Institutes of Health, Bethesda, Md.
35. **National Institutes of Health.** 1979. Laboratory safety monograph. National Institutes of Health, Bethesda, Md.
36. **National Institutes of Health.** 1983. Guidelines for research involving recombinant DNA activity. Fed. Regist. **48:**24555–24581.
37. **Phillips, G. B.** 1969. Control of microbiological hazards in the laboratory. Am. Ind. Hyg. Assoc. J. **30:**170–176.
38. **Pike, R. M.** 1976. Laboratory-associated infections: summary and analysis of 3,921 cases. Health Lab. Sci. **13:**105–114.
39. **Pike, R. M.** 1978. Past and present hazards of working with infectious agents. Arch. Pathol. Lab. Med. **102:**333–336.
40. **Pike, R. M., and S. E. Sulkin.** 1952. Occupational hazards in microbiology. Sci. Mon. **75:**222–229.
41. **Pike, R. M., S. E. Sulkin, and M. L. Schulze.** 1965. Continuing importance of laboratory infections. Am. J. Public Health **55:**190–199.
42. **Powell, E. O.** 1958. Criteria for growth of contaminant and mutants in continuous culture. J. Gen. Microbiol. **18:**255–268.
43. **Reitman, M., and A. G. Wedum.** 1956. Microbiological safety. Public Health Rep. **71:**659–665.
44. **Richardson, J. N., and R. H. Huffaker.** 1980. Biological safety in the clinical laboratory, p. 960–964. *In* E. H. Lennette, A. Balows, W. J. Hausler, Jr., and J. P. Truant (ed.), Manual of clinical microbiology, 3rd ed. American Society for Microbiology, Washington, D.C.
45. **Sargeant, K., and C. G. T. Evans.** 1979. Hazards involved in the industrial use of micro-organisms. Commission of the European Communities, Brussels.
46. **Sulkin, S. E., and R. M. Pike.** 1951. Survey of laboratory acquired infections. Am. J. Public Health **41:**769.
47. **U.S. Department of Health and Human Services.** 1984. Biosafety in microbiological and biomedical laboratories. HHS Publication no. (CDC)84-8395. U.S. Government Printing Office, Washington, D.C.
48. **Wedum, A.** 1961. Control of laboratory airborne infection. Bacteriol. Rev. **25:**210–216.
49. **Wedum, A. G.** 1964. Laboratory safety in research with infectious aerosols. Public Health Rep. **79:**619–633.
50. **Wedum, A. G., W. E. Barkley, and A. Hellman.** 1972. Handling of infectious agents. J. Am. Med. Assoc. **161:** 1557–1567.

Chapter 29

Assays of Antimicrobial Substances

DAVID M. ISAACSON AND JOEL KIRSCHBAUM

Until recently, growth inhibition of selected susceptible microorganisms was the criterion used to determine the qualitative and quantitative properties of most antimicrobial substances. The microorganisms were utilized to determine the potency of an antimicrobial substance in a wide variety of media, substances, and products. These microorganisms were frequently the first means of detecting low-level antimicrobial activity in soil samples or the presence of inhibitory substances in crude fermentation broths. They were used to determine the potency of the earliest crystallized preparations of a new antimicrobial substance, often prepared only in milligram quantities. These microorganisms were also used to quantify the amount of an antibiotic or other antimicrobial substance contained in animal tissues and body fluids during investigations on the pharmacokinetic properties of an antimicrobial substance.

Many of these samples required extensive extraction and separation procedures before they were suitable for microbiological testing. The experiences gained from the applications of these separation and cleanup procedures were applicable later to chromatographic and immunologic procedures developed for the assay of antimicrobial substances. These procedures provide an excellent measure of the chemical specificity of an antimicrobial substance. In addition, chromatography provides superior quantitative measurements of the purity and the amounts and kinds of impurities in an antimicrobial preparation. However, unlike the microbiological method, neither the chromatographic nor the immunologic method provides a measurement of the antimicrobial activity of a substance.

This chapter on assays of antimicrobial substances has been divided into two broad sections. In the first section (Assays by Microbiological Methods), some procedures developed for the microbiological assay of the potencies of antimicrobial substances are described. These procedures can be found described in greater detail in several of the references (10, 23, 32). In the second section (Assays by HPLC), high-pressure liquid chromatography (HPLC) procedures used for the assay of the potencies and purities of antimicrobial substances are described. We have not included procedures for immunologic assays. Also excluded are procedures for the assay of mixtures of antimicrobial substances by microbiological methods and the assay of antimicrobial residues in tissues by either microbiological or HPLC methods. These excluded methods involve techniques that are extremely specialized and cannot be adequately described in this limited space.

29.1 ASSAYS BY MICROBIOLOGICAL METHODS

Microbiological assays of antimicrobial substances principally employ either agar diffusion or turbidimetric (photometric or broth) techniques. In each method, a specific microorganism is placed in contact with the antimicrobial agent and the extent of growth inhibition of the organism is quantified. Quantitative microbiological assays are relative (not absolute) assay methods. Quantitation is obtained by comparing the degree of inhibition in the presence of the test antimicrobial agent versus the degree of inhibition in the presence of a known quantity of a similar or identical antimicrobial substance of known purity.

When agar diffusion assays are used, the areas or diameters of circular zones of inhibition of microbial growth on a seeded agar plate are the raw data that are quantified.

When turbidimetric assay procedures are used, the changes in the number or size or both of the microbial test cells that are produced by an antimicrobial substance in a suspension result in a change in turbidity of the microbial suspension. The turbidity changes are the raw data that are quantified.

Because of the variability of biological systems, a large number of replicate results at various concentrations are necessary to yield useful data. In practice, the data can be measured with great accuracy and are sufficient, under controlled conditions, to produce valid quantitative results.

29.1.1 Official Methods

Many countries have organizations that establish and publish official methods for the assay of pharmaceutical and agricultural products. For example, in the United States, official methods are described in detail in *Code of Federal Regulations, Title 21, Food and Drugs, Part 436, Subpart D* (32); *Pharmacopeia of the United States of America (USP)*, 20th revision (165); and *Official Methods of Analysis of the Association of Official Analytical Chemists* (AOAC) (68). In the United Kingdom, methods are described in the *British Pharmacopoeia* (26). In other European nations, methods are published in the *European Pharmacopoeia*, 2nd ed. (41). In Japan, methods are found in *Minimum Requirements of Antibiotic Products* (115). These official methods of assay must be used by both parties when an assay is in dispute and one party is a regulatory agency.

Official methods may not be the best or the most convenient assay methods available. This fact is acknowledged in the United States and Great Britain, which permit the use of methods of proven equivalent accuracy. That is, any method can be used by manufacturers for internal assay procedures as long as equivalency with official methods can be established. If the assays are free from bias, then the official method and the equivalent method should produce equivalent results in any laboratory performing the assays.

29.1.2 Agar Diffusion Assay

Principle

For agar diffusion assays, properly diluted test solutions are applied to the surface of a carefully selected agar medium that has been inoculated with the appropriate test culture (129). During incubation of the test system, the antimicrobial agent diffuses in an ever-widening circle around the point of application, resulting in a circular area in which growth has been prevented in an otherwise solid lawn of test culture. As noted earlier, the diameters or areas of the zones of inhibition are the raw data. These measurements are related to the concentration of the antimicrobial agent. When the $\log_{10}$ values of these concentrations are plotted against zone diameters or areas, a linear dose-response curve is produced. Quantification of results is made from the dose-response curve by procedures described below.

Method

The methods for performing agar diffusion assays are described below. It is assumed that the assayer is familiar with basic microbiological methodologies. The basic designs of the tests are described in more detail in references 10, 32, 129, and 165. Examination of the methods described in reference 32 will show that with modifications of the kinds of test organism, agar systems, diluents, or other cultural conditions, it is possible to determine the potencies of 52 approved antibiotic agents.

Step-by-step procedures for assays are outlined below. Table 4, which appears at the end of this section and contains references to Tables 1, 2, and 3, provides necessary information for each procedure. The assay designs found in references 10, 32, and 165 are incorporated into the information in Table 4.

In the United States, solutions of antimicrobial agents are usually tested by being added to stainless steel cylinders (6-mm inner diameter, 8-mm outer diameter, and 10-mm length) that have previously been placed on the surface of an agar medium contained in a 100-mm petri plate. The agar medium usually consists of two separate layers. The lower layer is uninoculated and may lack one or more of the ingredients found in the upper layer. The upper layer usually is thinner than the lower layer and contains the test organism (Table 4).

An assay by agar diffusion is performed as follows.

(i) Prepare two-layer plates appropriate for the antimicrobial agent to be assayed (Table 4). Add six cylinders to each plate (at 60° intervals and on a 2.8-cm radius).

(ii) Prepare standard (Table 4) and appropriate sample solutions.

(iii) Prepare 12 plates for use in determining the standard curve responses. These plates include three for each standard curve concentration except the reference concentration solution. The reference concentration solution is included on every plate. On each set of three plates, fill alternate cylinders with the reference concentration solution. Fill the other three cylinders with the concentration of the standard solution under test. In total there will be 36 reference concentration zones of inhibition and 9 zones of inhibition for each of the other four concentrations of the standard curve solutions.

(iv) Prepare three plates for each sample to be tested. Fill three alternate cylinders on each plate with the standard reference concentration solution, and fill the other three cylinders with the sample solution. (The reference concentration is placed on each plate to aid in correcting the inhibition zone changes that are due to uncontrolled conditions which produce plate-to-plate variations in zone size for each replicate sample or standard curve concentration.)

A complete standard curve (Table 4) should be used each time an assay is performed, especially with samples of unknown potencies. But if repetitive samples of known potencies (as with pharmaceutical preparations) are tested, it is appropriate to construct a dose-response curve with only two standard concentrations prepared to approximate closely the anticipated sample potency.

(v) After samples and standard solutions have been added to the test plates, incubate the plates at the temperatures noted in Table 4. Incubation time is usually 16 to 18 h. After incubation, measure the diameters of the inhibition zones, and calculate the potencies of the samples.

Calculation of potency of sample

Potency of a sample is calculated manually as follows.

(i) Prepare a standard curve response. To prepare the standard curve, average the diameters of the zones of inhibitions of the standard reference concentration, and average the diameters of the zones of inhibition of the other standard concentrations tested on each set of three plates. Average also all 36 diameters of the zones of inhibition of the reference concentration for all four sets of plates. The average of the 36 diameters of the reference concentration is the correction point of the standard curve line (32). Correct the average diameter obtained for each concentration to the value it would be if the average reference concentration for that set of three plates were the same as the correction point. For example, if, in correcting the highest concentration of the standard curve line, the average of the 36 diameters of the reference concentration is 16.5 mm and the average of the reference concentration of the set of three plates (the set containing the highest concentration of the standard curve concentration) is 16.3 mm, the correction to be applied is 0.2 mm. If the average reading of the highest concentration of the standard curve line of these same three plates were 16.9 mm, then the corrected diameter would be 17.1 mm.

(ii) Plot these corrected diameters, including the average of the 36 diameters of the reference concentration, on two-cycle semilogarithm paper. Plot the concentration of the antimicrobial agent, expressed as micrograms or units per milliliter, on the logarithmic scale. Plot the diameter of the zones of inhibition on the arithmetic scale.

(iii) The standard curve line is drawn either through these points by inspection or through points plotted for the greatest and smallest zone diameters obtained by means of the following equations:

$$L = \frac{3a + 2b + c - e}{5}$$

$$H = \frac{3e + 2d + c - a}{5}$$

where L is the calculated zone diameter for the lowest concentration of the standard response line, H is the calculated zone diameter for the highest concentration of the standard response line, c is the average zone diameter of all readings of the reference concentration solution, and a, b, d, and e are the corrected average values for the other standard solutions, from the lowest to the highest concentrations, respectively.

(iv) To estimate the potency of the sample, separately average the zone diameters of the standard and the zone diameters of the sample on the three plates used. If the average zone diameter of the sample is larger than that of the standard, add the difference between them to the reference concentration diameter of the standard curve line. If the average zone diameter of the sample is lower than that of the standard, subtract the difference between them from the reference concentration diameter of the standard curve line. From the standard curve line, read the concentrations corresponding to these corrected values of zone diameters. Multiply the concentration obtained by the appropriate dilution factor to obtain the content of the antimicrobial agent in the sample.

Since the above procedures are simply manual programs with a subset of statistical routines, the calculations are readily amenable to microprocessor-controlled calculation.

29.1.3 Turbidimetric Assay

Principle

Turbidimetric or photometric assays are performed by adding test solutions to liquid media previously inoculated with the test organism. After an incubation period, the effect of the antimicrobial agent on the resulting growth is measured as a change in turbidity of the test system. Most turbidimetric antimicrobial assay systems contain relatively high concentrations of inoculum and employ incubation periods as short as 0.5 to 4 h. Because of the large inoculum and the short incubation intervals, aseptic test conditions are less important than in other microbiological assay procedures. (Vitamin turbidimetric assays, for example, use smaller numbers of cells in the inocula and have an overnight incubation. Therefore, aseptic conditions are critically important to avoid possible interference or errors due to contaminants.)

Method

The basic design of the turbidimetric test is described in detail in reference 32. With this design it is possible to determine the potency of 23 approved antibiotics.

In this chapter, the necessary information for the performance of turbidimetric assays is outlined in Tables 1, 2, 3, and 4.

A tube turbidimetric assay is performed as follows.

(i) For each standard and sample concentration, prepare three replicate sterile test tubes in a test tube rack. Each rack must have its own standard curve (81). The test tubes should be large enough to give satisfactory results and be of uniform lengths and diameters (32). A most useful size is a tube 16 by 150 mm. Normally, a five-point standard concentration curve is employed. Therefore, 15 tubes are needed to prepare a standard response curve.

(ii) Prepare standard curve concentration solutions (Table 4) and appropriate sample solutions.

(iii) Prepare the growth medium that is appropriate for the antimicrobial agent under test (Table 4), and inoculate it as directed in Table 4 just before it is used in the test.

(iv) Add 1 ml of sample solution or standard curve concentration solution to the appropriate tubes, and then add 9 ml of the inoculated broth to each tube.

(v) Immediately place the tube in a water bath at the proper temperature, and keep the tube in the bath for 0.5 to 4 h. It is most important that constant incubation temperatures be maintained throughout the assay period. The exact length of the incubation period is determined by observing the growth in the tubes containing the standard reference concentra-

TABLE 1. Preparation of suspensions of the test organisms[a]

Method	Procedure
A	Maintain the culture by weekly transfers on slants of medium 1. Suspend a 24-h slant culture (grown at 35°C) with 3 ml of sterile diluent N. Transfer the suspension onto the surface of 250 ml of medium 1 in a Roux bottle. Incubate the culture for 24 h at 32 to 37°C. Wash the resulting cell growth from the agar surface with 50 ml of sterile diluent N. Store the cell suspension at 5°C.
B	Maintain the culture by weekly transfers on slants of medium 1. Suspend a 24-h culture (grown at 35°C) with 3 ml of sterile diluent N. Transfer the suspension onto the surface of 250 ml of medium 32 in a Roux bottle. Incubate the culture for 24 h at 32 to 37°C. Wash the resulting cell growth from the agar surface with 50 ml of sterile diluent N. Transfer the suspension to a sterile centrifuge bottle. Centrifuge the suspension at 2,000 × *g* for 10 min. Decant the supernatant. Resuspend the cells in 60 ml of sterile diluent N. Heat the suspension for 30 min at 70°C. Store the spore suspension at 5°C.
C	Maintain the culture by weekly transfers on slants of medium 1. Suspend a 24-h slant culture (grown at 35°C) with 3 ml of sterile diluent N. Transfer the suspension onto the surface of 250 ml of medium 1 in a Roux bottle. Incubate the culture for 24 h at 32 to 37°C. Wash the resulting cell growth from the agar surface with 50 ml of sterile diluent N. Transfer the suspension to a sterile centrifuge bottle. Heat the suspension for 30 min at 70°C. Centrifuge it at 2,000 × *g* for 10 min. Decant the supernatant. Suspend the cells with 30 ml of sterile diluent G. Repeat the centrifugation, and wash the cells two additional times with sterile diluent G. Store the spore suspension at 5°C.
D	Follow the procedure for method A, except incubate the slants for 24 h at 30°C, and incubate the Roux bottles for 48 h at 30°C.
E	Maintain the organisms in 100-ml portions of medium 3. To prepare the test broth culture, transfer a loopful of the stock culture to 100 ml of medium 3. Incubate the culture for 18 h at 37°C. Store the test broth culture at 5°C.
F	Follow the procedure for method C, except incubate the Roux bottle for 7 days at 37°C. Suspend the spores in 100 ml of sterile diluent G.
G	Maintain the test culture by weekly transfers on slants of medium 6. Incubate the culture for 24 h at 37°C. Suspend a 24-h slant culture with 10 ml of medium 39. Adjust the cell density to 80% light transmission at 660 nm.
H	Maintain the test culture by weekly transfers on slants of medium 1. Incubate the culture for 20 h at 37°C. To prepare the test inoculum, transfer a loopful of the stock culture to 200 ml of medium 3. Incubate the culture for 18 h with shaking at 30°C.
I	Maintain the test culture on slants of medium 36. Transfer the culture weekly. To prepare the test inoculum, incubate the slant for 2 days at 37°C. Wash the slant with 3 ml of sterile diluent N, and transfer it to 100 ml of medium 34. Add 50 g of sterile glass beads. Incubate the culture on a rotary shaker (130 cycles per min, radius of 3.5 cm) for 5 days at 27°C. Store the test broth culture for 2 weeks at 5°C.
J	Maintain the test culture by weekly transfer on slants of medium 36. Suspend a 24-h slant culture (grown at 37°C) with 3 ml of sterile diluent N. Transfer the culture to 250 ml of medium 36 in a Roux bottle. Incubate the culture for 24 h at 37°C. Wash the resulting growth from the agar surface with 50 ml of medium 37. Store the cell suspension at 5°C for 1 week.
K	Maintain the test culture by weekly transfer on slants of medium 16. Incubate the culture for 18 h at 37°C. To prepare the test inoculum, transfer a loopful of the stock culture to 100 ml of medium 39. Incubate the culture for 18 h at 37°C. Store the test broth culture at 5°C.
L	Follow the procedure for method A, except use medium 27 and incubate the culture for 48 h at 37°C.

[a] Descriptions of the media cited in this table are in Table 2. Descriptions of the solutions cited are in Table 3.

tion. Terminate the assay when the growth appears to have increased by 1 to 2 logs.

(vi) To terminate the assay, remove the tubes from the water bath, and add 0.5 ml of a 12% formaldehyde solution to each tube.

(vii) Determine the A_{530} of each tube in a suitable photoelectric colorimeter. Set the instrument to zero absorbance with an uninoculated blank composed of the same amounts of broth and formaldehyde as are used in the assay.

Calculation of potency

Potency is calculated manually as follows.

(i) Average the absorbance values obtained for each concentration of standard solution used.

(ii) Plot the average absorbance values for each concentration of standard on 1-cycle semilogarithmic graph paper. Plot absorbance on the arithmetic scale and concentrations on the logarithmic scale.

(iii) Draw the response line either through these points by inspection or through points plotted for the highest and lowest absorbance values obtained by means of the following equations:

$$L = \frac{3a + 2b + c - e}{5}$$

$$H = \frac{3e + 2d + c - a}{5}$$

TABLE 2. Selected microbiological culture media[a] for microbiological assays of antimicrobial substances

Ingredient	Amt (g/liter) in medium no. (pH ± 0.05):																		
	1	2	3	5	8	9	10	11	15	16	19	27	32	34	35	36	37	38	39
Beef extract	1.5	1.5	1.5	1.5	1.5			1.5	3.0	3.0	2.4	1.5	1.5	10.0	10.0				3.0
Yeast extract	3.0	3.0	1.5	3.0	3.0			3.0		2.0	4.7	3.0	3.0						
Peptic digest of animal tissue	6.0	6.0	5.0	6.0	6.0			6.0	5.0	5.0	9.4	6.0	6.0	5.0	5.0			15.0	5.0
Pancreatic digest of casein	4.0					17.0	17.0	4.0				4.0	4.0			15.0	17.0		
Papaic digest of soybean meal				5.0	3.0	3.0										5.0	3.0	5.0	
Glucose	1.0		1.0			2.5	2.5	1.0			10.0	1.0				2.5	5.5		
Glycerol														10.0	10.0				
L-Cystine																		0.7	
Potassium dihydrogen phosphate			1.32																
Dipotassium phosphate			3.68			2.5	2.5										2.5		
Sodium chloride			3.5			5.0	5.0				10.0			3.0	3.0	5.0	5.0	4.0	
Sodium sulfite																		0.2	
Neomycin												0.1							
Polysorbate 80 (ml)							10.0												
Agar	15.0	15.0		15.0	15.0	20.0	12.0	15.5	15.0	15.0	23.5	15.0	15.0		17.0	15.0		15.0	
pH ± 0.05 (after sterilization)	6.55	6.55	7.0	7.9	5.9	7.25	7.25	7.9	7.9	6.8	6.1	6.55	6.55	7.0	7.0	7.3	7.3	7.0	6.8

[a] See Tables 2 and 4 for specific applications of these media, most of which are available commercially as complete preparations.

where L is the calculated absorbance value for the lowest concentration of the standard response curve point, H is the calculated absorbance value for the highest concentration of the standard response curve point, and a, b, c, d, and e are the average absorbance values for each concentration of the standard response curve, from the lowest to the highest values, respectively.

(iv) To estimate the potency of the sample, average the absorbance values for the sample, and determine the antibiotic concentration from the standard curve line.

(v) Multiply the concentration by the appropriate dilution factor to obtain the content of the antimicrobial agent in the sample.

These statistical procedures are readily amenable to microprocessor-controlled calculation, and the reader is encouraged to pursue that means of calculation.

29.1.4 Sources of Error

Sampling

A nonhomogeneous sample is a common cause of error. This condition can occur in pharmaceutical powders if mixing processes are incomplete and in liquid samples if viscous samples are improperly mixed prior to dilution. Extraction procedures have inherent errors that also affect assay results. For example, errors may be due to (i) incomplete extraction with solvents, (ii) the extraction solution reacting with and perhaps destroying the active antimicrobial agent when the pH of the solvent is inappropriate, or (iii) the use of a filtering medium that adsorbs some of the active material.

To detect destruction or loss during sample preparation, it is necessary to add a known amount of a substance to a second sample and process the two samples in a manner as nearly identical as possible. The amount of the recovery of the added substance is an indication of the destruction associated with the method.

Bias

A bias is a nonrandom error caused by an asymmetry in the assay system (80). A bias is most easily detected when the potency of a sample varies with the sample dilution. The magnitude of the bias may differ for agar diffusion and turbidimetric assays of the same substance. Some of the biases or drifts are inherent, as is noted with an antibiotic such as bacitracin, which can contain variable mixtures of nine related active components. Certain samples of antimicrobial agents may have several biologically active degradation forms present which will produce an assay bias when tested against a pure substance.

Another important source of error is the variability in potencies of samples because they were assayed by unproven or nonequivalent methods. Different batches of growth medium and of standards can increase the likelihood of bias.

Systematic errors

Because many antimicrobial agents are somewhat unstable in aqueous solutions and lose potency with time, the repeated use of such a degraded compound as a standard can introduce a systematic error into an assay system. Another systematic error can occur with repeated use of an antimicrobial agent when the solvent in which it has been stored has volatilized. If undetected, this situation will result in an apparent increase in potency of the standard compound. Ideally, the standard concentration solutions should be prepared fresh on each test day. How much degradation or solvent evaporation can be tolerated depends on the accuracy desired.

TABLE 3. Solvents, buffers, and diluents for microbiological assays of antimicrobial substances

Solution or diluent	Solution name[a]	Ingredients	Amt	
A	1% Phosphate buffer, pH 6.0 ± 0.05	Dipotassium phosphate	2.0	g
		Potassium dihydrogen phosphate	8.0	g
		Diluent G	to make 1,000	ml
B	0.1 M phosphate buffer, pH 7.9 ± 0.1	Dipotassium phosphate	16.73	g
		Potassium dihydrogen phosphate	0.523	g
		Diluent G	to make 1,000	ml
C	0.1 M phosphate buffer, pH 4.5 ± 0.05	Potassium dihydrogen phosphate	13.6	g
		Diluent G	to make 1,000	ml
D	10% Phosphate buffer, pH 6.0 ± 0.05	Dipotassium phosphate	20.0	g
		Potassium dihydrogen phosphate	80.0	g
		Diluent G	to make 1,000	ml
E	0.2 M phosphate buffer, pH 10.5 ± 0.1	Dipotassium phosphate	35.0	g
		10 N potassium hydroxide	2.0	ml
		Diluent G	to make 1,000	ml
F	0.1 M phosphate buffer, pH 7.0 ± 0.1	Dipotassium phosphate	13.6	g
		Potassium dihydrogen phosphate	4.0	g
		Diluent G	to make 1,000	ml
G	Distilled water	Distilled or deionized water		
H	Dimethyl sulfoxide	Dimethyl sulfoxide		
I	95% Ethyl alcohol	95% Ethyl alcohol		
J	Methanol	Methyl alcohol		
K	0.01 N hydrochloric acid	1.0 N hydrochloric acid	10.0	ml
		Diluent G	to make 1,000	ml
L	0.01 N methanolic hydrochloric acid	1.0 N hydrochloric acid	10.0	ml
		Diluent J	to make 1,000	ml
M	0.05 M Tris buffer	Tris(hydroxymethyl)nitromethane	7.56	g
		Diluent G	to make 1,000	ml
N	0.9% Saline	Sodium chloride	9.0	g
		Diluent G	to make 1,000	ml
O	80% Isopropyl alcohol	Isopropyl alcohol	800	ml
		Diluent G	to make 1,000	ml

[a] Adjust the phosphate buffer to the correct pH with 18 N phosphoric acid or 10 N potassium hydroxide.

Other causes of error

Poor laboratory techniques can produce differences in results from assay to assay. These errors may stem from nonuniform temperatures in an incubator, nonuniform inoculation, incorrect design of the assay, deviation from a time schedule for agar diffusion assays, and interaction of the antimicrobial agent and the test organism in the assay tubes before the start of incubation in a turbidimetric assay. Both agar diffusion and turbidimetric assays are sensitive to nonuniformity of interim and final incubation temperatures. Erythromycin assays, for example, are extremely temperature sensitive. In a turbidimetric test for erythromycin against a 1,000-μg/ml standard, the erythromycin sample can measure 840 μg/ml at the beginning of a test and 1,460 μg/ml at the end of the test. This difference is due to the temperature lag that can occur between tubes containing a standard and tubes containing a sample.

Nonuniform inoculum is another major source of error, especially in turbidimetric assays. If tube dilution turbidimetric tests are used and if the organism grows in the broth as the set of tubes constituting a test are filled, the tube filled last will receive a heavier inoculum than did the tube filled first. How this difference affects the test depends on the position of the standards in the test (80).

Incorrect assay design puts bias into assays, especially in the petri plate agar diffusion method, because the organisms in the seeded layer are subject to time lags that affect their growth rate. For example, the top or seeded layer is usually prepared in a large volume and maintained at about 43 to 45°C. Equal samples are then removed and placed onto the base layer. During this time, the inoculum may either grow

TABLE 4. Methods for the microbiological assay of selected antimicrobial substances

Antibiotic	Test method[a]	Test microorganisms: Name[c]	Test microorganisms: Preparation[d]	Test microorganisms: Amt (ml) of inoculum per liter of medium	Agar diffusion, Base layer: Medium no.[f]	Agar diffusion, Base layer: Amt (ml) per 90-mm petri plate	Agar diffusion, Seed layer: Medium no.[f]	Agar diffusion, Seed layer: Amt (ml) per 90-mm petri plate	Turbidimetric: Medium no.[f]	Turbidimetric: Amt (ml) per tube	Standard solution: Initial concn and solvent[e]	Standard solution: Diluent for additional dilution[e]	Standard solution: Concn range of solution or μg/ml (reference standard in parentheses)	Incubation temp (°C)[b]	Reference
Amoxicillin	Agar	8	A	5.0	11	21	11	4			1,000 μg/ml in G	B	0.064–0.156 (0.100)	32–37	32
Amphomycin	Agar	11	L	5.0	2	21	1	4			100 μg/ml in B	B	6.4–15.6 (10.0)	37	32
Amikacin sulfate	Turb.	18	A	1.0					3	9	1,000 μg/ml in G	G	8.0–12.5 (10.0)	37	32
Amphotericin B[g]	Agar	17	D	2.0			19	8			1,000 μg/ml in H	E	0.64–1.56 (1.0)	30	10, 20
Ampicillin	Agar	8	A	10.0	11	21	11	4			100 μg/ml in G	B	0.064–0.156 (0.100)	32–35	10, 32
Bacitracin zinc	Agar	6 or 7	A	2.0	2	21	1	4			100 U/ml in A	A	0.64–1.56 (1.0)	32–35	10, 32, 165
	Agar	9 or 10	A	2.0	2	21	1	4			100 U/ml in A	A	64–156 (100)	32–35	32, 165
Bleomycin	Agar	12	I	10.0	35	10	35	6			2.0 U/ml in F	F	0.01–0.16 (0.04)	32–35	32
Candicidin	Turb.	17	A	2.0					13	9	1,000 μg/ml in H	G	0.03–0.12 (0.06)	25	32, 165
Capreomycin	Turb.	5	A	1.0					3	9	1,000 μg/ml in G	G	80.0–125 (100)	37	10, 23, 32, 145
Carbenicillin	Agar	14	A	0.5	9	21	10	4			1,000 μg/ml in A	A	12.8–31.2 (20.0)	37	10, 32
Carbomycin	Agar	8	A	4.0	1	21	1	4			1,000 μg/ml in J	B	0.6–1.5 (1.0)	30–32	35, 58
Cefaclor	Agar	8	A	0.2			1	10			1,000 μg/ml in C	C	0.025–0.20 (0.1)	30	44
	Turb.	19	H	30.0					3	10	1,000 μg/ml in C	C	0.05–0.25 (0.15)	37	32
Cefadroxil	Agar	18	A	0.5	2	21	1	4			1,000 μg/ml in A	A	12.8–31.2 (20.0)	37	32
Cefazolin	Agar	18	A	0.5	2	21	1	4			1,000 μg/ml in D	A	0.64–1.56 (1.0)	32–35	32
Cephacetrile	Agar	18	A	5.0	2	21	1	4			1,000 μg/ml in A	A	6.4–15.6 (10.0)	32–35	32
Cephalexin	Agar	18	A	0.5	2	21	1	4			1,000 μg/ml in A	A	12.8–31.2 (20.0)	32–35	10, 32
Cephaloglycin	Agar	18	A	2.0	2	21	1	4			100 μg/ml in G	C	6.4–15.6 (10.0)	32–35	10, 32
Cephaloridine	Agar	18	A	3.0	2	21	1	4			1,000 μg/ml in A	A	0.64–1.56 (1.0)	32–35	10, 32
Cephalothin	Agar	18	A	1.0	2	21	1	4			1,000 μg/ml in A	A	0.64–1.56 (1.0)	32–35	10, 32, 165
Cephapirin	Agar	18	A	0.8	2	21	1	4			1,000 μg/ml in A	A	0.64–1.56 (1.0)	32–35	25, 32
Chloramphenicol	Turb.	4	A	1.0					3	9	10,000 μg/ml in I	A	2.0–3.12 (2.5)	37	10, 32, 165
Chlortetracycline	Turb.	18	A	1.0					3	9	1,000 μg/ml in L	C	0.048–0.075 (0.060)	37	10, 32
Clindamycin	Agar	8	A	15.0	11	21	11	4			1,000 μg/ml in G	B	0.64–1.56 (1.0)	37	10, 32
Cloxacillin	Agar	18	A	2.0	2	21	1	4			1,000 μg/ml in A	A	3.2–7.81 (5.0)	32–35	10, 32, 165
Colistimethate sodium	Agar	3	A	0.4	9	21	10	4			10,000 μg/ml in G	D	0.64–1.56 (1.0)	37	10, 32
Colistin	Agar	3	A	0.4	9	21	10	4			10,000 μg/ml in G	D	0.64–1.56 (1.0)	37	10, 32
Cycloserine	Turb.	18	A	1.0					3	9	1,000 μg/ml in G	G	40–62.5 (50.0)	37	10, 32
Dactinomycin	Agar	2	B	0.2	5	10	5	4			10,000 μg/ml in J	B	0.5–2.0 (1.0)	37	10, 32, 165
Demeclocycline	Turb.	18	A	1.0					3	9	1,000 μg/ml in L	C	0.08–0.125 (0.10)	37	10, 32
Dicloxacillin	Agar	18	A	2.0	2	21	1	4			1,000 μg/ml in A	A	3.2–7.81 (5.0)	32–35	10, 32
Dihydrostreptomycin	Agar	2	B	0.5	5	21	5	4			1,000 μg/ml in B	B	0.64–1.56 (1.0)	37	10, 32
	Turb.	5	A	1.0					3	9	1,000 μg/ml in G	G	24.0–37.5 (30.0)	37	10, 32
Doxycycline	Turb.	18	A	1.0					3	9	1,000 μg/ml in L	C	0.08–0.125 (0.10)	37	10, 32
Erythromycin	Agar	8	A	10.0	11	21	11	4			10,000 μg/ml in J	B	0.64–1.56 (0.10)	37	10, 32, 165
Gentamicin	Agar	20	A	0.3	11	21	11	4			1,000 μg/ml in B	B	0.064–0.156 (0.10)	37	10, 32, 165
Gramicidin	Turb.	21	E	10.0					3	9	1,000 μg/ml in I	I	0.032–0.050 (0.040)	37	10, 32

Kanamycin B	Agar	2	B	0.5	5	21	5	4			1,000 μg/ml in B	B	0.64–1.56 (1.0)	32–35	10, 32
Lincomycin	Turb.	18	A	1.0					3	9	1,000 μg/ml in G	G	0.400–0.625 (0.06)	37	10, 32, 165
Methacycline	Agar	1	C	4.0	8	21	8	4			1,000 μg/ml in L	C	0.064–0.156 (0.10)	30	10
	Turb.	18	A	1.0					3	9	1,000 μg/ml in L	C	0.048–0.075 (0.066)	37	10, 32
Methicillin	Agar	18	A	2.0	2	21	1	4			1,000 μg/ml in A	A	6.4–15.6 (10.0)	32–35	10, 32, 165
Mithramycin	Agar	18	A	1.0	8	10	8	4			1,000 μg/ml in G	A	0.5–2.0 (1.0)	32	10, 32
Mitomycin[g]	Agar	2	B	5.0	2	21	15	4			200 μg/ml in G	F	0.5–4.0 (1.5)	32–35	115
Nafcillin	Agar	18	A	2.0	2	21	1	4			1,000 μg/ml in A	A	1.28–3.12 (2.0)	32–35	10, 32, 165
Neomycin	Agar	18	A	0.4	11	21	11	4			1,000 μg/ml in B	B	6.4–15.6 (10.0)	32–35	10, 32
	Agar	20	A	2.0	11	21	11	4			1,000 μg/ml in B	B	0.64–1.56 (1.00)	37	10, 165
Novobiocin	Agar	20	A	20.0	2	21	1	4			10,000 μg/ml in I	D	0.32–0.781 (0.5)	35	10, 32
Nystatin[g]	Agar	16	D	10.0			19	8			1,000 μg/ml in H	D	12.8–31.2 (20.0)	30	10
Oleandomycin	Agar	20	A	2.0	11	21	11	4			10,000 μg/ml in I	B	3.2–7.81 (5.0)	37	10, 32
Oxacillin	Agar	18	A	1.0	2	21	1	4			1,000 μg/ml in A	A	3.2–7.81 (5.0)	32–35	10, 32
Oxytetracycline	Turb.	18	A	1.0					3	9	1,000 μg/ml in L	C	0.192–0.30 (0.240)	37	10, 165
Paromomycin	Agar	20	A	20.0	11	21	11	4			1,000 μg/ml in B	B	0.64–1.56 (1.0)	37	10, 32
Penicillin G	Agar	18	A	2.0	2	21	1	4			1,000 U/ml in A	A	0.64–1.56 (1.0)	37	10, 32, 165
Phenethicillin	Agar	8	A	4.0	11	21	11	4			1,000 U/ml in G	A	0.064–0.156 (0.1)	32–35	10, 32
Phenoxymethylpenicillin	Agar	18	A	2.0	2	21	1	4			1,000 U/ml in J	A	0.64–1.56 (1.0)	37	10, 32, 165
Phosphonomycin	Agar	13	K	30.0			16	5			1,000 μg/ml in N	N	0.22–7.0 (1.75)	37	144
Polymyxin B	Agar	3	A	0.4	9	21	10	4			20,000 U/ml in G	D	6.4–15.6 (10.0)	37	10, 32, 165
Rifampin	Agar	2	B	1.0	2	21	2	4			1,000 μg/ml in J	A	3.2–7.81 (5.0)	30	10, 32
Rolitetracycline	Turb.	18	A	1.0					3	9	1,000 μg/ml in J	A	0.192–0.30 (0.240)	37	10, 32, 38
Spectinomycin[h]	Turb.	4	A	0.2					3	9	1,000 μg/ml in B	G	24.0–37.5 (30.0)	37	10, 32
Streptomycin	Agar	2	B	0.5	5	21	5	4			1,000 μg/ml in B	B	0.64–1.56 (1.0)	30	10, 32, 165
Tetracycline	Agar	1	C	4.0	8	21	8	4			1,000 μg/ml in L	C	0.64–1.56 (1.0)	32–35	10
	Turb.	18	A	1.0					3	9	1,000 μg/ml in L	C	0.192–0.300 (0.240)	37	10, 23, 32
Ticarcillin	Agar	15	J	15.0	38	21	38	4			1,000 μg/ml in A	A	3.20–7.81 (5.00)	37	32
Tobramycin	Turb.	18	A	1.0					3	9	1,000 μg/ml in G	G	2.00–3.125 (2.50)	37	93
Troleandomycin	Turb.	9	A	0.5					3	9	1,000 μg/ml in O	A	20.0–31.25 (25.0)	37	32
Tyrothricin	Turb.	21	E	10.0					3	9	1,000 μg/ml in I	I	0.032–0.050 (0.040)	37	10, 32
Vancomycin	Agar	2	B	2.0	8	10	8	4			400 μg/ml in G	C	6.4–15.6 (10.0)	30	10, 32, 165

[a] Turb., Turbidimetric assay.

[b] Incubation times were 18 to 24 h for agar diffusion assays and 3 to 6 h for turbidimetric assays.

[c] Microorganisms used for the assay of selected antimicrobial substances were as follows: 1, *Bacillus cereus* subsp. *mycoides* ATCC 11778; 2, *B. subtilis* ATCC 6633; 3, *Bordetella bronchiseptica* ATCC 4617; 4, *Escherichia coli* ATCC 10536; 5, *Klebsiella pneumoniae* ATCC 10031; 6, *Micrococcus luteus* ATCC 7468; 7, *M. luteus* ATCC 7468D; 8, *M. luteus* ATCC 9341; 9, *M. luteus* ATCC 10240; 10, *M. luteus* ATCC 10240a; 11, *M. luteus* ATCC 14452; 12, *Mycobacterium smegmatis* ATCC 607; 13, *Proteus mirabilis* ATCC 21100; 14, *Pseudomonas aeruginosa* ATCC 25619; 15, *P. aeruginosa* ATCC 29336; 16, *Saccharomyces cerevisiae* ATCC 2601; 17, *S. cerevisiae* ATCC 9763; 18, *Staphylococcus aureus* ATCC 6538P; 19, *S. aureus* ATCC 9144; 20, *S. epidermidis* ATCC 12228; and 21, *Streptococcus faecium* ATCC 10541.

[d] See Table 1.

[e] See Table 3.

[f] See Table 2.

[g] Store in low-actinic glassware.

[h] Prediffuse plates for 5 h at 5°C before plates are incubated.

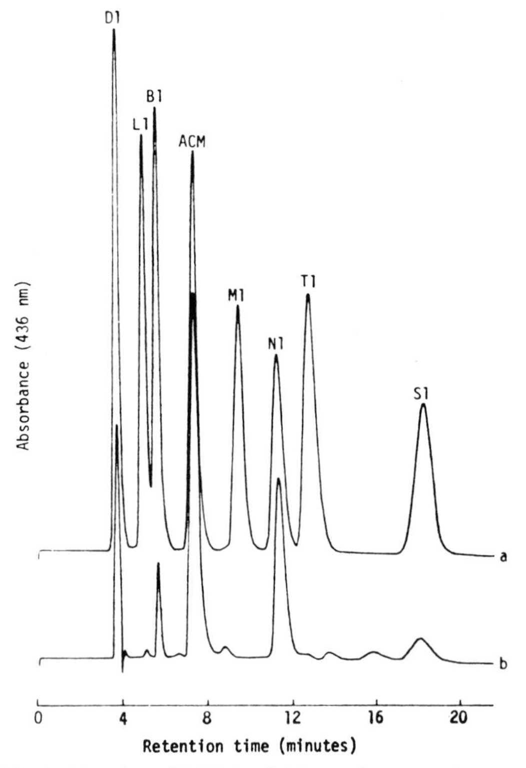

FIG. 1. Results of HPLC of (a) a reference mixture containing the anthracycline antitumor antibiotic aclacinomycin A (ACM) and its analogs containing different sugars and (b) an extract from the fermentation broth. See Table 5 for the chromatographic conditions. Reproduced courtesy of the *Journal of Antibiotics* and the authors (123).

slightly or, if it is heat sensitive, begin to lose viability. Thus the physiological state of the inoculum can differ markedly between the initially prepared and last-prepared plates.

29.1.5 Explanation of Tables for Microbiological Assays of Antimicrobial Substances

The tables that follow this section are modifications of tables found in references 10, 27, and 165. Table 4 is the master table. For a particular antimicrobial assay procedure, the reader should use Table 4 to select the appropriate test organism, the correct procedure for preparing the test organism (Table 1), and the correct volume of test organism to add to the test medium (Table 4). Table 4 directs the reader to the appropriate solvent system for a particular antimicrobial agent (Table 3). In Table 4 are listed the correct volumes of agar to use for an agar diffusion assay and of liquid media to use for a turbidimetric assay. Table 4 also lists the range of concentrations of standards to use in an assay; the reference concentration, in parentheses, for a particular standard antimicrobial agent; and the incubation temperatures to be used. The references cited describe in much greater detail procedures for the assay of specific antimicrobial substances. The reader is encouraged to review the appropriate references.

29.2 ASSAYS BY HPLC

Figure 1 illustrates typical data obtainable by HPLC from a fermentation broth (123). The antibiotic aclacinomycin A and its analogs were quantified from the peak responses by use of a reference mixture. Such data collected over a span of 100 h gave the data represented in Fig. 2 for the kinetics of the production of these compounds. Figures 1 and 2 summarize many of the strengths of HPLC, since this method is selective, sensitive, quantitative, and applicable to a wide variety of compounds, making it now the preferred method for the assay of fermentation broths and other biological matrices for natural products. HPLC is sufficiently sensitive to analyze small quantities of impurities in final, purified products, which is the reason that both governmental regulatory agencies and industrial laboratories generally consider HPLC

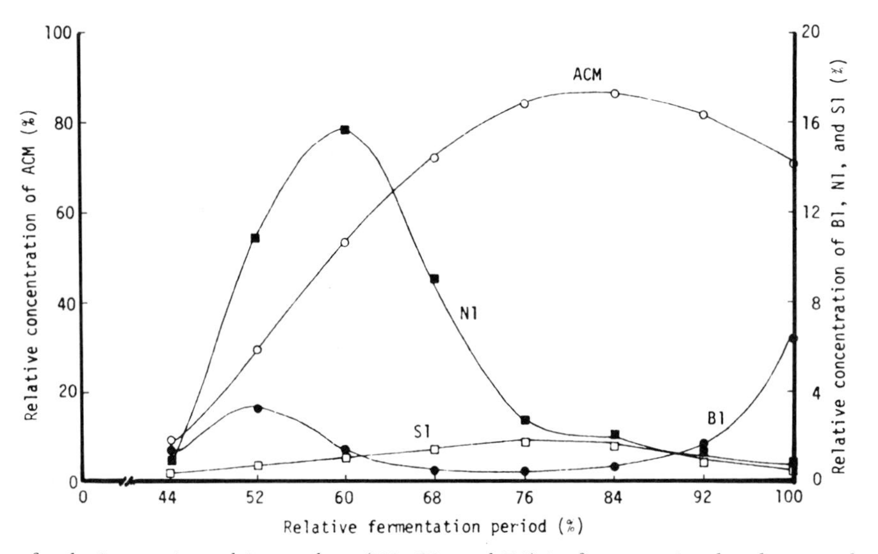

FIG. 2. Contents of aclacinomycin and its analogs (N1, S1, and B1) in fermentation broth, reproduced courtesy of the authors (123) and the *Journal of Antibiotics*.

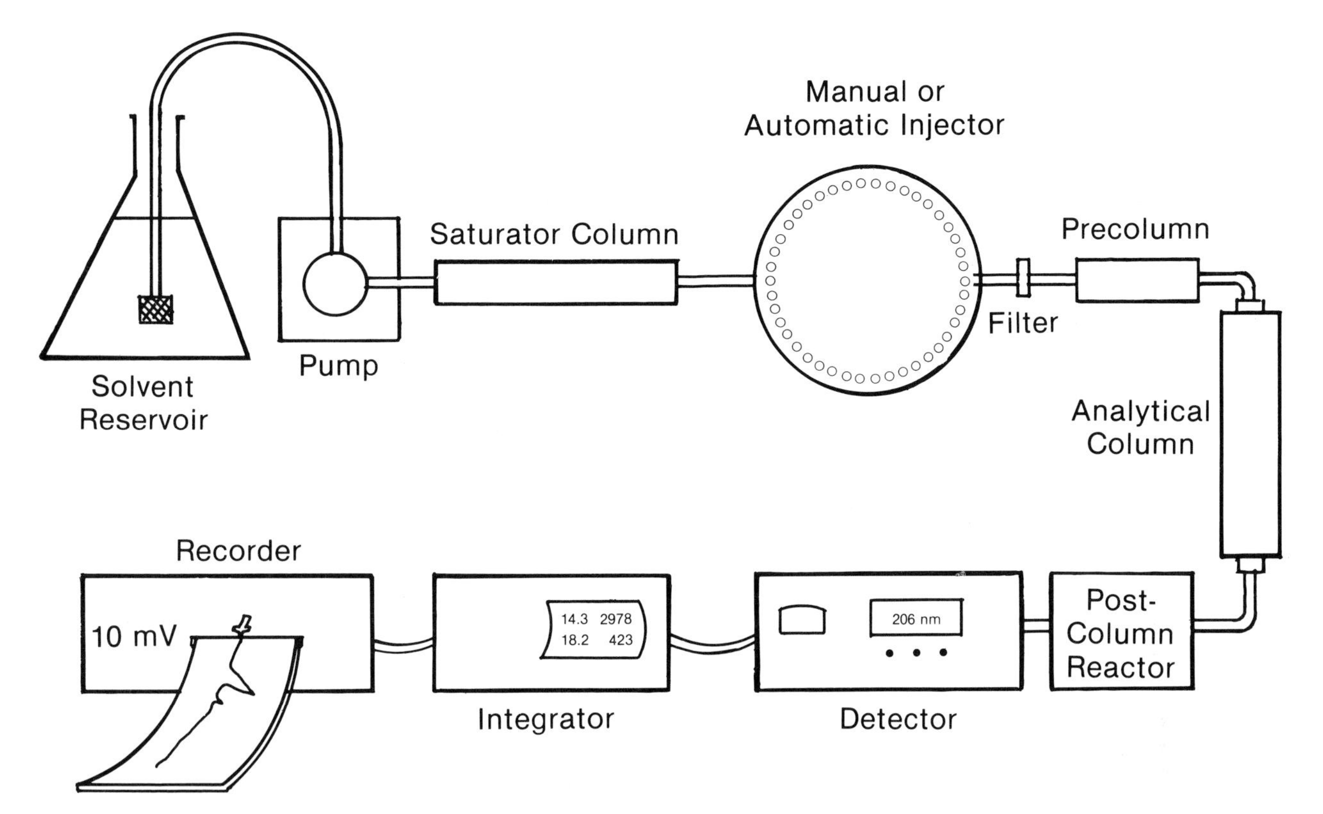

FIG. 3. Schematic diagram of a liquid chromatograph. See the text for a discussion of the components.

the premier method. Additionally, technicians do not require extensive training to prepare and analyze samples. With time, most technicians can learn to do most of the maintenance and repair of the instruments. The disadvantages of this technique include the relatively high cost of equipment ($5,000 to $38,000), maintenance of the equipment, and the need to correlate biological and chemical results.

29.2.1 Description of the Equipment

Figure 3 is a schematic illustration of the HPLC equipment generally used. A constant-volume pump pressurizes a liquid that may range in polarity from hexane or chloroform through a mixture of methanol, tetrahydrofuran, or acetonitrile with aqueous buffers to pure water. The flow rate is generally 1 to 2 ml/min, resulting in a pressure of 400 to 3,000 lb/in^2, although semipreparative and preparative scale systems utilize rates of 100 to 1,000 ml/min. At the other extreme, narrow-bore (microbore) analytical columns may require flow rates of 0.1 ml/min or lower. Occasionally two or more pumps or solvent outlets will have the streams varied and merged according to a programmed and timed sequence to obtain a mobile-phase composition that produces the desired elution of components (gradient elution).

The flowing liquid typically passes through stainless-steel tubing packed with silica particles 37 μm in diameter (saturator column). This passage saturates the mobile phase with silica and thus diminishes dissolution in aqueous mobile phases of the silica-based packings that are usually used in the analytical columns.

Samples of untreated, centrifuged, or highly purified material are injected, usually in a volume of 10 to 20 μl (sample preparation is discussed below in section 29.2.2), although volumes as small as 0.1 μl (for microbore HPLC columns) and as large as >50 ml (for preparative liquid chromatography) have been injected. The injection can be manual, requiring an operator for each injection, or automated, with up to 100 samples and standards introduced via a programmable, microprocessor-controlled autoinjector. A filter (0.45-μm pore diameter) traps small particles, and a precolumn can be used to remove unwanted particles before the analytical column is used.

There are many types of analytical columns. So-called normal-phase columns, containing silica particles 3 to 10 μm in diameter, are used as the stationary phase with organic solvents like methylene chloride, chloroform, and hexane (with controlled or isohydric water contents) or with aqueous methanol or acetonitrile mixtures. 1,2-Dihydroxy hydrocarbons (diol) can be bonded to silica to give a column with properties akin to those of the silica column. Intermediate-polarity columns include amino, cyano, and nitro groups bonded to silica via a hydrocarbon linkage. Reversed-phase columns contain hydrocarbons 1 to 28 carbons long, phenyl, cyclohexyl, fluorocarbon, and related structures bonded to silica, which are generally used with methanol, acetonitrile, or tetrahydrofuran-aqueous buffer mobile phases. The single most commonly used column of this polarity is octadecylsilane, which can be 5 to 15% covered with hydrocarbon. (Residual silanol groups can be shielded by another reaction with a low-molecular-weight hydrocarbon.)

TABLE 5. Selected HPLC methods for antibiotics and related compounds

Compound	Type of column[a]	Mobile phase[b]	Detection (nm)[c]	Comments	References
Aclacinomycin	C18	CH_3CN-CH_3OH-H_2O (40:90:70) + H_3PO_4, 1.5 ml/100 ml	254	Resolves daunorubicin	40
	Si	CH_3Cl-CH_3OH-CH_3CO_2H-$H_2O(C_2H_5)_3N$ (68:20:10:2:0.01)	436	Resolves many constituents	123
Actinomycin	Phenyl C18	H_2O-CH_3CN (1:1)	254	Resolves three components	136
	C18	CH_3CN-H_2O (6:4)	254	Resolves related compounds	Aszalos[d]
Adriamycin	C18	0.01 M *d*-10-camphorsulfonic acid–CH_3CO_2H–CH_3OH-H_2O (4:50:46)	254	Versus micro.[e]	113
	C18	0.01 M $NH_4H_2PO_4$ (pH 4)–CH_3OH (35:65)	254	Also daunorubicin and aglycones	13
6-Aminopenicillanic acid	C8	(0.01 M Na_2HPO_4, 0.2% CH_3CO_2H, 0.01 M tetrabutylammonium hydroxide) (pH 2)–CH_3CN (23:2)	220 or D[f] (357/449)	0.5 μg/ml of broth	132
	NH_2	CH_3CO_2H-CH_3OH-CH_3CN-H_2O (2:4:7.5:86.5)	254	Resolves penicillin V and *p*-OH–penicillin V	113
Amipurimycin	C18	CH_3OH–0.01 M phosphate (pH 5.7) (1:4)	270	Resolved from mildiomycin	61
Amphotericin B	C18	H_2O-CH_3OH-THF (420:90:45)	350	Resolves other polyenes	108
	C18	0.05 M phosphate (pH 2.6)–CH_3OH (3:7)	313	Resolves amphotericin B, A, and X	14
Anisomycin	C18	CH_3CN–0.05 M KH_2PO_4 (pH 6) (1:7)	225	Resolves vancomycin	85
Antimycin A	C18	CH_3OH–0.25 M acetate buffer (pH 5) (3:1)	254, EC + 1V; F (365/481)	Preparative LC and mass spectrometry of 10 isolated components	1
Antracyclines	Si	30–45% CH_3CN in citrate (0.1 M citric acid, 0.2 M Na_2HPO_4), pH 2.2	F (474/590)	Resolves seven components	124
Apramycin	RP8	CH_3CN-H_2O-H_3CCO_2H (55:45:0.15)	D[f] (350)	Kinetics of production	62
Aristeromycin	C18	CH_3OH–0.01 M phosphate (pH 5.7) (1:4)	270	Resolved from mildiomycin	61
Asparenomycins	C18	0.02 M phosphate (pH 7)	220 or 254	Resolves A, B, and C	142
Aureofungin	C18	CH_3CN–0.05 M citrate (pH 5.3–6 gradient) (35:65)	350	Resolves other heptaenes	109
Avermectins	C18	CH_3CN-CH_3OH-H_2O (53:35:7)	245	Resolves isomers	128
	C18	CH_3OH-H_2O (85:15), 40°C	246	Resolves eight components in complex	114
Avoparcin	RP8	0.1 M ammonium formate–acetonitrile (9:1) (pH 7.3)	254	Resolves vancomycin	150
Ayfactin	C18	CH_3CN–0.05 M citrate (pH 5.3–6 gradient) (35:65)	350	Resolves other heptaenes	109
Bacitracin	C18	CH_3OH-CH_3CN-H_2O-0.1 M phosphate (pH 4.5 gradient)	254	Many components; A, B, and B_2 have equal antimicrobial activity	88 163
	C18	CH_3CN–CH_3OH–phosphate-EDTA buffer–H_2O (24:7:20:49) (pH 6.8)	254	Analyzed in feeds; LC method more accurate than micro.	51
Benzoic acid	C18	CH_3OH–0.5% acetate (pH 5.2) (1:9)	230	In foods	48
Blasticidin S	C18	CH_3OH–0.01 M phosphate (pH 5.7) (1:4)	270	Resolved from mildiomycin	61
Bleomycins	C18	10–40% CH_3OH gradient in 0.005 M pentanesulfonic acid (pH 4.3)	254	Resolves related constitutents	14
	C18	+0.005 M 1-heptanesulfonic acid–CH_3OH (1:1) (pH 8.3)	280	Separates A_2 and B_2 components	138
Cadequomycin	C18	H_2O-CH_3OH-CH_3CO_2H (70:30:1)	254	Preparative LC	179
Candicidin	C18	H_2O-CH_3OH-THF (420:90:75)	350	Resolves other polyenes	108
	C18	CH_3CN–0.05 M ammonium acetate (pH 4.6) (3:5)	380	Resolves other polyenes	64

Continued on following page

TABLE 5—*Continued*

Compound	Type of column[a]	Mobile phase[b]	Detection (nm)[c]	Comments	References
Candidin	C18	H_2O-CH_3OH-THF (420:90:60)	350	Resolves other polyenes	108
Candimycin	C18	CH_3CN–0.05 M citrate (pH 5.3–6 gradient) (35:65)	350	Resolves other heptaenes	109
Carboxyristomycin	C8	0.1 M citrate (pH 6.4)–methyl cellosolve (88:12)	254	Resolves vancomycin	150
Cephalosporin C	C18	0.01 M tetrabutylammonium hydroxide (pH 7)–H_3CO_2H-CH_3OH (4:1)	254	Results similar to micro.	82
	C8	(0.01 M NaH_2PO_4, 0.2% CH_3CO_2H, 0.01 M tetrabutylammonium hydroxide) (pH 2)–CH_3CN (23:2)	220 D[f] (352/452)	1 μg/ml of broth	132
	NH_2	CH_3CO_2H-CH_3OH-CH_3CN-H_2O (2:4:7.5:86.5)	254	Also resolves penicillin N	112, 113
	Si	2% Aqueous adipic acid	254	Also resolves deacetyl and deacetoxy forms	77
Cephalosporins	C18	0.005 M tetrabutylammonium hydroxide + 1% ammonium phosphate (pH 7) + 1% CH_3CN in H_2O	254	Mixture of cephalosporins and metabolites in broth	113
	NH_2	CH_3CO_2H-CH_3OH-CH_3CN-H_2O (2:4:7.5:86.5)	254		
Cephamycin C	C8	(0.01 M Na_2HPO_4, 0.2% CH_3CO_2H, 0.01 M tetrabutylammonium hydroxide) (pH 2), (NaOH)–CH_3CN (23:2)	220 D[f] (347/450)	0.5 μg/ml of broth	132
Chlortetracycline	C18	0.001 M EDTA (pH 6.6)–CH_3OH (82:18)	380	Resolves impurities	14
	C8	0.1 M NaH_2PO_4 + 0.2 M *N,N*-dimethyloctylamine–CH_3CN (7:3) (pH 8)	280	Resolves other impurities	65
Circulin	C18	CH_3CN-H_2O-phosphate (pH 2 gradient) + CH_3OH	Flame ionization	Resolves impurities	163
Clavulanic acid	RP18	(0.010 M tetrabutylammonium bromide, 0.004 M Na_2HPO_4, 0.006 M NaH_2PO_4)–CH_3OH (10:1) (pH 7.02)	201, F (375/445)	Resolved from amoxicillin	60
Colistin (polymyxin E)	C18	CH_3CN–H_2O–0.2 M phosphate (pH 2) (2:7:1) to CH_3CN–CH_3OH–H_2O–0.02 M phosphate (pH 2) (50:20:29:1) in 30 min	Flame ionization	Separates many components	163
	C18	CH_3OH-0.2 M chloride buffer (pH 2) (1:1)	210	Resolves multiple constitutents	84
	C18	CH_3CN–0.023 M H_3PO_4 + 0.01 M CH_3CO_2H + 0.05 M Na_2SO_4 at pH 2.5 with $(C_2H_5)_3N$	220	Resolves many components	39
Cyclosporin	C18	0.1% F_3CCO_2H–CH_3CN (35:65 to 5:95)	215	Results compared with radioimmunoassay	181
Daunorubicin (daunomycin)	C8	CH_3CN–0.025 M camphorsulfuric acid (pH 3.8) (1:1)	254	Separates adriamycin and carminomycin	176
	C18	Water (pH 2, H_3PO_4)–CH_3OH (4:6)	254	Find new related compounds	126
	C18	0.01 M $NH_4H_2PO_4$ (pH 4)–CH_3OH (35:65)	254	Also adriamycin and aglycones	14
Deflectins	C18	CH_3OH-triethylammonium formate (pH 6) (8:2)	339	Resolves eight components	8
Demeclocycline	C8	0.1 M citric acid–CH_3CN (76:24) (0.33 ml/min)	350	Resolves impurities	121
	C18	0.001 M EDTA (pH 6.6)–CH_3OH (82:18)	380	Resolves impurities	15
	C8	(0.1 M NaH_2PO_4 + 0.2 M *N,N*-dimethyloctylamine)–CH_3CN (7.3) (pH 8)	280	Resolves other tetracyclines	65
Deoxycycline	C8	0.1 M citric acid–CH_3CN (76:24) (0.33 ml/min)	350	Resolves impurities	121
	C18	0.001 M EDTA (pH 6.6)–CH_3OH (82:18)	380	Resolves impurities	14
	C8	(0.1 M NaH_2PO_4 + 0.2 M *N,N*-dimethyloctylamine)–CH_3CN (7:3) (pH 8)	280	Resolves other tetracyclines	65
Erythromycin	C18	CH_3CN–CH_3OH–0.2 M $NH_4CH_3CO_2$–H_2O (45:10:10:35) (pH 7.8)	215	Discussed in sample preparation section	160

Continued on following page

TABLE 5—*Continued*

Compound	Type of column[a]	Mobile phase[b]	Detection (nm)[c]	Comments	References
	C8	CH_3OH-H_2O-NH_3 (28 Be) (80:19.9:0.1)	215	Resolves A from B; comparison of LC, chemical, and micro. assays	103
	C18	CH_3CN–0.025 M KH_2PO_4–$C_2H_5NH_2$ (400:600:1)	RI	Preparative scale; isolates new erythromycin, F	127
Etoposide	C18	CH_3OH-H_2O (6:4)	215	Resolves teniposide	146
Eurocidin	C18	H_2O-CH_3OH-THF (420:90:70)	350	Resolves other polyenes	108
Filipin	C18	H_2O-CH_3OH-THF (420:90:60)	350	Resolves other polyenes	108
Fredericamycin	C18 Si	CH_3OH-H_2O-CH_3CO_2H (70:30:1) or $CHCl_3$-CH_3OH-CH_3CO_2H (87:3:3)	254 254	Purify components by preparative LC	14 14
Fungimycin	C18	H_2O-CH_3OH-THF (420:90:75)	350	Resolves other polyenes	108
Fusidic acid	C18	CH_3OH–0.01 M KH_2PO_4 (pH ~6.6) (3:1)	254	Good agreement, HPLC versus micro.	66
Gentamicin	C18	5 g of sodium heptanesulfonate in 250 ml of H_2O + 50 ml of H_3CCO_2H diluted to 1 liter with CH_3OH	D[f] (330); D, F[f] (340/418 filter)	Resolves four components versus micro.	46 174
	C18	CH_3OH-H_2O (79:21) + 0.2% tripotassium EDTA	D[f], F (345/430)	Resolves eight components; correct micro., summary components. See introduction	91
	C18	0.015 M sodium pentanesulfonate, 0.2 M Na_2SO_4, and 0.1% CH_3CO_2H	EC + 1.3V	16-μg limit; resolves C_{1a}, C_2, C_1^+	52
Gilvocarcin	C18 C18 Si, 2×	CH_3OH-H_2O (7:3) CH_3OH-H_2O-THF (40:45:15) Ethyl acetate-isopropanol-H_2O (87:13:0.5), 100 ml/min	254 245 400	Resolves complex Preparative LC Purify 300 mg	173 173 18
Gougerotin	C18	0.005 M citrate (pH 5.8)	270	Resolved from mildiomycin	61
Gramicidin	C18	0.005 M $(NH_4)_2SO_4$–CH_3OH (26:74)	220	Resolves many components	16
	Phenyl	CH_3OH-H_2O (73:27)	254	Semipreparative LC of many components	90
	Si	*n*-Hexane–ethanol (75:25)	282	Agrees with micro. and thin-layer chromatography methods	118
Griseofulvin	CN	CH_3OH-H_2O (3:2)	254	Appears to indicate stability	158
Hamycin	C18	H_2O-CH_3OH-THF (420:90:75)	350	Resolves other polyenes	108
	C18	CH_3CN–0.05 M citrate (pH 5.3–6 gradient) (35:65)	350	Resolves other heptaenes	109
p-Hydroxyphenoxyacetic acid	C18	CH_3CN–1.5% H_3PO_4 adjusted to pH 3 with NH_4OH (3:7)	220 EC + 0.8V	+ Penicillin V	70
Interferon	RP8 Diol	*n*-Propanol O–20% for 30 min, 20–40% for 3 h in 1 M sodium acetate (pH 7.5) and then 72.5–50% *n*-propanol in buffer	Fluorescamine microbore analyzer	Resolves three major and several minor constituents	135
	Cyclohexyl RP8	Pyridine-formic acid (8%:8%) containing 20% isopropanol to 8% pyridine–8% formic acid–25% isopropanol–20% *n*-butanol	1 bid	Resolves different molecular weight and activity, other LC systems	47
Kijanimicin	SCX	0.01 M $(NH_4)_3PO_4$ (pH 9.1)	RI	Resolves A from B and C	104
	SCX	0.01 M potassium EDTA (pH 9.3)	D[f], F (320/450)	High sensitivity	104
	RP8	CH_3CN-H_2O-H_3CCO_2H (55:45:0.15)	D[g] (350)	Follow kinetics production	62

Continued on following page

TABLE 5—*Continued*

Compound	Type of column[a]	Mobile phase[b]	Detection (nm)[c]	Comments	References
	Si, 2×	CH_3OH-CH_2Cl_2 (8:92, 16:84, 50:50)	254	Preparative LC	172
Lavendamycin	Si	CH_2C_2-CH_3OH-H_2O-NH_3 (30%) (80:20:1:1)		Related to streptonigrin	17
	C18	0.01 M KH_2PO_4–CH_3CN (8:2)			
Lincomycin	C8	2.9 g of sodium dodecyl sulfate + 10 ml of H_3PO_4 in 660 ml of H_2O + 330 ml of CH_3CN, adjusted to pH 6 (NH_4OH)	214	Resolves lincomycin A from B	12
Lymecycline	C8	0.1 M NaH_2PO_4 0.2 M *N*,*N*-dimethyloctylamine–CH_3CN (7.3) (pH 8)	280	Resolves other tetracyclines	65
Maridomycins	C18	CH_3CN-H_2O-$(C_2H_5)_2NH$ (500:499:1)	203	Resolves polyene family	21
Mediocidin	C18	H_2O-CH_3OH-THF (420:90:75)	350	Resolves other polyenes	108
Merulinic acids	Si	*n*-Hexanes + 1% ethyl acetate + 0.02% *n*-propanol	254	Resolves A, B, and C	54
Methacycline	C18	0.001 M EDTA (pH 6.6)–CH_3OH (82:18)	380	Resolves impurities	14
	C8	0.1 M NaH_2PO_4 and 0.2 M *N*,*N*-dimethyloctylamine–CH_3CN (7.3) (pH 8)	280	Resolves other tetracyclines	65
Mildiomycin	C18	CH_3OH–0.01 M phosphate (pH 5.7) (1:4)	270	For crude and purified	61
Mitomycin	C18	0.01 M phosphate (pH 7), 0–50% CH_3OH gradient [semipreparative LC with 0.01 M $(NH_4)_2CO_3$]	D[h] (313) 336	Resolves 15 constituents	7
	Si, 2×	$CHCl_3$-CH_3OH (9:1)	245	Separates A from B plus others	143
Monensin	C18	CH_3OH-H_2O (88:12)	215	Resolves A and B plus unknown and preparative scale	22
Natamycin	C18	CH_3OH-H_2O (54:35)	303	Resolves related compounds	45
Neocarzinostatin	C18	0.01 M $NH_4CH_3CO_2$ (pH 4)–50 to 84% CH_3OH gradient	245 F (340/418)	Resolves three compounds	120
	C18	35–100% CH_3OH + 0.1% triethanolamine and 0.1% CH_3CO_2H	254	Preparative scale	79
Neomycin	Si	$CHCl_3$–THF-water-CH_3CO_2H (590:400:8:2); microbore, 0.2 μl/min	190	Resolves B from C	159
	Si	$CHCl_3$-THF-H_2O (600:392:8)	D[g] (350), 254	Resolves B from C	161
Nikkomycins	C8	0.03 M ammonium formate (pH 4.7) containing 0.01 M heptanesulfonic acid (also gradient)	290	Resolves five components	42
Nocardin	NH_2	CH_3CO_2H-CH_3OH-CH_3CN-H_2O (2:4:7.5:86.5)	254	Activity versus three microorganisms	113
Novobiocin	Si	Water-saturated butyl chloride–butyl chloride-THF-methanol–CH_3CO_2H (44:44:5:4:3); microbore, 0.3 μl/min	354	Resolves many impurities	166
	C18[i]	CH_3OH–0.02 M phosphate buffer (pH 7) (15:85)	254	Resolves 7 constituents	162
	C18	CH_3OH-H_2O (4:1) containing 0.005 M sodium 1-heptanesulfonate	254	Resolves 11 constituents	67
	ECTEOLA-Cellulose	0.025 M perchlorate plus 0.01 M phosphate (pH 6.8)	254	Resolves 16 constituents	166
Nystatin	C18	0.005 M sodium acetate (pH 5.8)–CH_3OH (55–67%)	304	Resolves impurities; no correlation with micro.	156
	C18	CH_3OH-H_2O-THF (35:43:22) or 0.05 M phosphate (pH 2.3) (3:1)	313	Attempt microassay correlation	Aszalos[d]
	C18	0.005 M acetate–methanol (gradient, 55–67%) (pH 6.8)	304	Resolves 8 components; can separate amphotericin B	157
	C18	H_2O-CH_3OH-THF (420:90:75)	350	Resolves other polyenes	108

Continued on following page

TABLE 5—*Continued*

Compound	Type of column[a]	Mobile phase[b]	Detection (nm)[c]	Comments	References
Octapeptins	C18	0.005 M tartrate (pH 5) + 0.005 M sodium 1-butane sulfonate + 0.05 M Na_2SO_4–CH_3CN (77.5:22.5)	220	Resolves many components	155
Oligostatin	NH_2	CH_3CN-H_2O (6:4)	RI	Complex resolved into 5 components	73
Partricin	C18	CH_3CN–0.05 M citrate (pH 5.3–6 gradient) (35:65)	350	Resolves other heptaenes	109
Paulomycin	C18	CH_3CN–0.5 M phosphate buffer (pH 7)(38:62)	320	Separates 4 components	9
Penicillin G	C8	0.016 M phosphate + 0.1% $(C_2H_5)_3N$ (pH 5.5)–CH_3OH (7:3)	216	Separates anionic from cationic forms	117
	C18	CH_3OH–PIC[j] reagent B-7 (Waters) (6:4)	260 254	Separates dibenzylethylenediamine	147
	C18	0.008 M tetrabutylammonium chloride, 0.008 M phosphate buffer (pH 7.5)–CH_3CN (7:3)	254	Separates analyte from degradation products	53
Penicillin N	C8	0.01 M Na_2HPO_4–0.2% CH_3CO_2H, 0.01 M tetrabutylammonium hydroxide (pH 2)–CH_3CN (23:2)	220	1 μg/ml of broth	132
	C18	CH_3OH-0.05 M KH_2PO_4 (pH 4, H_3PO_4) (5:95)	220	Resolves isopenicillin N and α-(L-α-aminoadipyl)-L-cysteinyl-D-valine	
	C18	Pyridine—CH_3CO_2H-H_2O (0.4:0.4:99.2)	RI		
Penicillin V	NH_2	CH_3CO_2H-CH_3OH-CH_3CN-H_2O (2:4:7,5:86.5)	254	Resolves 6-APA[k] and *p*-OH–penicillin V	
	C8	0.015 M $(NH_4)_2HPO_4$ (pH 6)–CH_3OH (14:1)	D[f] (340/450)	1-ng limit	27
	C8	0.016 M phosphate (pH 4.5)–CH_3OH (7:3), 50°C	216	Separates cationic from anionic forms	117
	C18	CH_3CN–1.5% H_3PO_4 adjusted to pH 3 with NH_4OH (3:7)	EC + 0.8V 220	Resolves *p*-hydroxypenicillin V, phenoxyacetic acid, and *p*-hydroxyphenoxyacetic acid	70
Phenoxyacetic acid	C18	CH_3CN–1.5% H_3PO_4 adjusted to pH 3 with NH_4OH (3:7)	220	+ Penicillin V	70
Polymyxin	PS-DVB	CH_3OH–0.2 M chloride buffer (pH 2) (1:1)	210	Analytic plus preparative methods, separates many components	84
	C18	0.05 M triethanolammonium phosphate (pH 2.2) + 0.025 M 1-butanesulfonic acid + 25% CH_3CN	220	Preparative plus analytic LC, separates polymyxins from colistins	78
	C18	CH_3CN–0.023 M H_3PO_4, 0.01 M CH_3CO_2H, 0.05 M Na_2SO_4 and 0.05 M $(C_2H_5)_3N$ (pH 2.5) (23:72)	220	Polymyxin B and E each resolved into ~11 components plus preparative LC	37
	C18	H_2O-CH_3OH-methanesulfonic acid (50:50:1)	215	In pharmaceuticals	43
Polyoxin	C18	0.005 M citrate (pH 5.8)	270	Analyte resolved from mildiomycin	61
	C18	H_2O-F_3CCO_2H (99.75:0.25), 3 ml/min	254	Resolves mixture	141
Rifamycin (rifampin)	C18	THF–0.5% CH_3CO_2H (4:6)		Resolved from other drugs	55
	NH_2	$CHCl_3$-CH_3OH (97:3)	334	Separates derivatives	170
Ristomycin A	C8	0.1 M citrate (pH 6.4)–methyl cellosolve (88:12)	254	Resolves avoparcin	150

Continued on following page

TABLE 5—*Continued*

Compound	Type of column[a]	Mobile phase[b]	Detection (nm)[c]	Comments	References
Rosaramicin	C18	CH_3CN–0.01 M acetate (pH 4) (3:1)	254	Also resolves desepoxyrosaramicin	99
Sorbic acid	C18	CH_3OH–0.5% acetate (pH 5.2) (1:9)	230	In foods	48
Spectinomycin	C8	0.02 M sodium heptanesulfonate, 0.2 M Na_2SO_4	D; F (350/450)	Resolves related compounds	116
Spiramycin	C18	CH_3CN-H_2O-$(C_2H_5)_2NH$ (500:499.5:0.5)	232	Resolves polyene family	21
Streptomycin (dihydrostreptomycin)	C18	0.02 M sodium hexanesulfonate + 0.025 M Na_3PO_4 in CH_3CN-H_2O (8:92) (pH 6)	195	Resolves 8 streptomycin sulfate components; LC agrees with micro.	175
Teniposide	C18	CH_3OH-H_2O (6:4)	215	Resolves etoposide	146
Tetracycline	C18	CH_3OH–H_2O–0.2 M phosphate (pH 2.5) (3:6:1) to CH_3OH–CH_3CN–H_2O–0.2 M phosphate (pH 2.5) (5:2:2:1) in 15 min	280	Follow kinetics	160
	C18	H_2O-CH_3CN-CH_3CO_2H (71:18.5:10.5)	272	Resolves impurities	89
	C8	0.1 M citric acid–CH_3CN (76:24) (0.33 ml/min)	350	Resolves impurities	121
	C18	0.001 M EDTA (pH 6.6)–CH_3OH (82:18)	380	Resolves impurities	13
	C8	0.1 M NaH_2PO_4 + 0.2 M *N,N*-dimethyloctylamine–CH_3CN (7:3) (pH 8)	280	Resolves other tetracyclines	65
Thiostrepton	C8	H_2O-THF-CH_3OH (2:2:1)	245	Can follow kinetics	Aszalos[d]
Tobramycin	C8	CH_3CN-H_2O-H_3CCO_2H (55:45:0.15)	D[f] (350)	Follow kinetics production	62
Trichomycin	C18	H_2O-CH_3OH-THF (420:90:75)	350	Resolves other polyenes	108
	C18	CH_3CN–0.05 M citrate (pH 5.3–6)	350	Resolves other heptaenes	109
Tunicamycin	C18	CH_3CN-H_2O (38:62)	254	Resolves 10 constituents	72
	C18	CH_3OH-H_2O (68:32 or 4:1)	260	Resolves 16 components	102
Turimycins	Si	Diisopropyl ether-isooctane-H_2O (500:500:0.04) to diisopropyl ether-CH_3OH-H_2O (500:497.4:2.6)	232	Resolves polyene family	21
Tylosin	C18	CH_3CN–2% aqueous $HOCH_2CH_2NH_2$ (2:3)	290	Resolves A, B, C, and D	82
Tyrothricin	C18	CH_3OH-H_2O (4:1)	220	Resolves impurities in drugs	33
Vancomycin	C18	CH_3CN–0.05 M KH_2PO_4 (pH 6) (1:7)	225	Resolves danisomycin	85
	C18	0.1 M citrate (pH 6.4)–methyl cellosolve (88:12)	254	Resolves impurities	150
Vermiculine	C18	CH_3OH-H_2O (2:3) or CH_3CN-H_2O (35:65)	235	Resolves 3 substances	50
Vincristine	Phenyl	CH_3CN–0.02 M ammonium acetate and 0.005 M sodium 1-pentanesulfonate (pH 2) (HNO_3) (3:7)	254	Resolved from related products	24
	RP8	CH_3CN–0.01 M $(NH_4)_2CO_3$ (47:53)	298	Separates 25 alkaloids	57
	2RP-18	CH_3CN–0.01 M Na_2HPO_4 (pH 7.4) (1:1)	254	Purify fractions	94
Virantmycin	C18	CH_3OH-H_2O-H_3CCO_2H (80:20:5)	300	Kinetics of synthesis followed	119

[a] C18, Octadecylsilane; Si, silica; C8, octylsilane; NH_2, amino; RP8, reversed phase; CN, cyano; SCX, strong cation exchange; PS-DVB, poly(styrene-divinylbenzene).

[b] THF, Tetrahydrofuran; LC, liquid chromatography. Rate of flow of the mobile phase is usually 1 to 2 ml/min, unless stated otherwise. Inorganic salts and buffers are in water, unless noted otherwise.

[c] Detection is usually via UV absorbance. EC, Electrochemical; F, fluorescence (excitation, emission); D, derivative (reagent is given in a footnote; numbers indicate the detector, which is usually UV light or fluorescence); RI, refractive index.

[d] Personal communication.

[e] Micro., Microbiological assay.

[f] *O*-Phthalaldehyde or similar derivative.

[g] 1-Fluoro-2,4-dinitrobenzene derivative.

[h] Acetic acid-pyridine (5:1).

[i] Originally hydrocarbon polymer. Mobile phase may need modification.

[j] PIC, Paired ion chromatography solution (Waters Associates, Inc.).

[k] APA, 6-Aminopenicillanic acid.

Ion-exchange groups may also be bonded to silica to enable compounds to be resolved on the bases of charge and affinity to the hydrocarbon backbone holding the anionic or cationic group. Aqueous buffer-organic or aqueous buffer mixtures of solvents are usually used.

For the separation of low-molecular-weight compounds on the basis of polarity and high-molecular-weight compounds on the bases of size and polarity, poly(styrene-divinylbenzene) columns are recommended. Such non-silica-based packings can be manufactured with specified pore sizes. Molecular entities as large as viruses can be separated. Such a size exclusion technique, however, is not always sensitive to minor changes in structure, such as the substitution in a protein of one amino acid for another of similar polarity.

Although there are many types of detectors, the most widely used is UV light (Tables 5 and 6). The analyte usually possesses one or more groups that absorb UV light, making possible the use of a simple UV detector. Sensitivities range as low as $\pm 0.1\% \times 10^{-3}$ absorbance units, full scale. Colored compounds require only a visible-light detector. Fluorescence detection, which is often very sensitive (depending on the compound), refractive index (which is based on differences in the light-carrying [refractive] properties of the solute in mobile phase versus mobile phase above without solute), and infrared (long-wavelength) light are listed in decreasing order of use. Mass spectrometry is increasing in popularity, since it also can provide structural information (140).

Electrochemical detectors are becoming increasingly popular because of their selectivity and sensitivity, especially for biologically active compounds in body fluids. The major limitation of these detectors is that the analyte must be electrochemically active, possessing groups like amine or sulfhydryl.

Occasionally, compounds lacking chromophores must be analyzed at low concentrations. In such cases, if end UV absorption at 190 to 206 nm is inadequate, derivatization may be required. In these instances, a stable, easily detected group can be introduced via reaction with such reagents as *o*-phthalaldehyde and 1-fluoro-2,4-dinitrobenzene. A pre- or postcolumn reactor is often useful in these instances. Alternatively, reactive modification of the analyte in a biological matrix prior to assay is often useful.

The modification, absorbance, or emission of energy by the analyte, as measured by the detector, can be manipulated several ways. First, the measurement is displayed on a strip chart type of recorder. Second, the peak responses are electronically measured by an integrator or some other type of microprocessor, and frequently, calculated results are produced for both quantification of the analyte and determination or estimation of impurities.

The typical cost of a basic HPLC system, in 1986 prices, is as follows. The pump costs from $1,000 to $6,000, depending mainly on solvent-mixing capabilities, with a four-solvent pump costing ~$13,000. Injectors cost from $350 to $700 for a manual valve, $5,000 to $7,000 for an unsophisticated autoinjector, and $10,000 to $12,000 for a programmable type with up to 100 positions. A simple, fixed-wavelength, mercury vapor lamp UV detector operating at 254 nm costs less than $2,000. Variable-wavelength UV detectors cost $4,000 to $18,000, with those in the upper part of the range capable of recording a full spectrum eight times per second. A dedicated minicomputer is required to handle the rapid flow of data. An electrochemical detector costs about $2,500, fluorescence detectors cost $8,000 to $12,000, and simple filter monitors cost half that amount or less. Infrared detectors costing $28,000 are too insensitive at this time; as much as $90,000 is required to buy a sensitive Fourier transform instrument. The most versatile detector is the mass spectrometric detector, which is sensitive and selective and may provide information on structure, including that of isomers. However, the cost is more than $100,000, and the technique is not sufficiently quantitative. A two-pen recorder is $1,000. An unsophisticated laboratory integrator is below $2,500, although a 45-input computer network system costs $400,000 or more. The cost per station, including terminals and printer-plotters, is reasonable. Breakdowns are inevitable, especially as the computer system becomes more complex, thus requiring use of the recorders as a back-up to limit the loss of data.

29.2.2 Sample Preparation

The analysis of extracellular products is usually simple. Microorganisms are separated by filtration or centrifugation, and the aqueous solution is injected without any treatment other than dilution with water or mobile phase. Intracellular products usually require additional sample preparation (5). Sometimes pH adjustment and organic solvent extractions are necessary. For example, to assay erythromycin (160), the fermentation broth must be mixed for uniform sampling (especially if it has been frozen and thawed or kept at 4°C). A portion is centrifuged at 20,000 rpm, and the supernatant is passed through a filter (0.45-μm diameter) into a flask. Heptane is added, the flask is shaken, and the aqueous layer is collected. The heptane layer is washed twice with water, and the washings are combined in a centrifuge tube. Chloroform (10 ml) is added, the tube is shaken for 5 min and then centrifuged at 5,000 rpm, and the aqueous layer is removed. The chloroform is evaporated, and the residue is redissolved in mobile phase and then injected into the LC system. To analyze tetracycline in broth, the pH must be adjusted to 1.7 with HCl or H_3PO_4, and then the liquid is centrifuged, the oily layer is removed, and after filtration, the liquid is directly injected.

However, the antibiotics tobramycin, kanamycin B, and apramycin, which are produced by fermentation of the nebramycin complex, require derivatization for detection (62). First, protein is removed from the broth by heat treatment under acidic conditions, followed either by filtration of the precipitated proteins and suspended particles or by treatment with Tris and centrifugation for 10 min at 3,000 rpm, which gives a solution to which is added 3 ml of 0.15 M methanolic 1-fluoro-2,4-dinitrobenzene. The mixture is refluxed for 45 min, and the final volume is adjusted to 4.0 ml. These examples show that relatively little sample preparation is necessary, compared with that required for some traditional techniques. The chro-

TABLE 6. Selected HPLC methods for some other compounds

Compound or class	Column[a]	Mobile phase[b]	Detection (nm)[c]	Comments	References
Aflatoxins	C18	H_2O-CH_3CN-CH_3OH (15:3:2)	F (365/450), 365	Analyze 6 aflatoxins	151
Amines	C18	0.08 M H_3CCO_2H–CH_3OH–CH_3CN (45:52:3), and then gradient to 100% CH_3OH	D[d], F (340/440)	In must and wine	28
Amino acids	SCX	Citrate buffers (pH 3.2, 4.3, or 4.9) and ethanol gradient	Ninhydrin (570/440)	Many applications	49
	C8, C18	0.1 M ammonium acetate (pH 6.8)–CH_3CN (1:1)–0.05 M ammonium acetate gradient	D[e] (254)	Also peptides	63
Anthocyanidins	C18	CH_3OH-H_3CCO_2H-H_2O (4:1:15)	280	In wines	2
β-Asarone	C18	CH_3OH-H_2O (62:38)	254; F (310/355)	In liquors	111
Berberine	C18	0.005 M sodium lauryl sulfate–CH_3CN (1:19)	254	Gives pressure and accuracy	4
Betaine	NH_2	CH_3CN-H_2O (3:1)	RI	In sugar and wine	169
	SCX	0.05 M KH_2PO_4–CH_3OH (19:1)	195	+Sulphin analogs	56
2,3-Butanediol	Ion-exchange	H_2O at 85°C	RI	Various sugars	171
Caffeine	SCX	0.1 M $NH_4H_2PO_4$ (pH 5)–CH_3OH (3:2)	254	Studied pH and I[f]	164
Carbohydrate	Si	15–40% H_2O in CH_3CN and 0.1% amine modifier	RI	Resolves simple and complex sugars	3
Cellodextrins	Ion-exchange	H_2O at 85°C	RI	Resolves many sugars	92
Cinchona alkaloids	Si	Hexane-CH_2Cl_2-CH_3OH-$(C_2H_5)_2NH$ (66:31:20:0.65)	312	Resolves 8 components	105
Cobamides	C18	0.1 M LiCl in 25–48% aqueous methanol	254	Resolves 5 constituents	177
Cocaine	SCX	0.5 M $NH_4H_2PO_4$ (pH 5)–CH_3OH (3:2)	254	Studied pH and I	164
Coenzyme	C18	0.025 M acetate (pH 6)–CH_3OH (1:1)	F (400/470)	Coenzyme F_{420} assayed	20
Colchicines	C18	0.1 M triethylamine hydrochloride in 0.2 M phosphate (pH 2.2)–CH_3CN–CH_3OH (715:200:85)	370	Also followed kinetics	69
Coumarins	C18	HCO_2H-H_2O (5:95) to CH_3OH	280	Many different components resolved	30
Dicarboxylic acids	SCX	Ethylacetate-*i*-butyl ester–isobutanol–water (150:100:12)	RI	In grape must and wine	130
DNA	C18	0.1 M ammonium acetate (pH 6.6)–CH_3CN–H_2O	Radioactivity	Resolves single- from double-stranded DNA	98
Ellipticines	Phenyl	CH_3CN–0.1% $(NH_4)_2CO_3$ (1:1)	254	Enhance yields with additives	31
Ergot	C18	CH_3OH-H_2O-NH_4OH (60:40:0.1)	Mass spectrometry	Search for new clavines	36
	Si	Iso-octane-CH_2Cl_2-CH_3OH (5:4:1)			
Fatty acids	C18	CH_3CN-H_2O (3:1 or 17:3)	D[g] (242)	Resolves 10+ isomers	178
	C18	CH_3OH-H_2O-CH_3CO_2H (89:11:0.2) or 80:20:0.2)	213	Resolves 6 fatty acids, not isomers	19
Flavonoids	C18	2.5% CH_3CO_2H in water + methanol (gradient)	254, 280, 365, or 546	In hops and barley	107
	C8, C18	CH_3OH-H_3CCO_2H-H_2O gradient	312	Resolve aglycones and glycosides	34
Flavonols	C18, 2×	CH_3OH-CH_3CO_2H-H_2O (6:1:14)	280	In wine	2
Furfurals	C18	CH_3OH-H_2O (1:9)	285	+ Hydroxymethyl furfural	76
Furocoumarins	Si	$CHCl_3$-CH_3OH (100:0 to 99:1)	270	Resolves 6 components	39
Histamines	C18	K_2HPO_4 (to pH 7)–CH_3CN (3:2)	D[d] (200 or 220)	In wines and musts	148
Hypochloresteremic agents	C8	CH_3CN-CH_3OH-H_2O (69:2:29)	D[d] (260 or 237)	In broth	59
Insulin	C18	0.1 M $(NH_4)_2SO_4$, 0.005 M tartaric acid (pH 3)–CH_3CN (3:1) + cetyltrimethylammonium bromide to 0.014 mM	225 or 280	Resolves bovine from porcine + impurities	101
Iso-α-acids	C18	CH_3OH-H_2O (4:1) + 1.3% 40% tetrabutylammonium hydroxide + 1.7% H_3PO_4	270	In beers	6

Continued on following page

TABLE 6—*Continued*

Compound or class	Column[a]	Mobile phase[b]	Detection (nm)[c]	Comments	Reference
		60–83% gradient, CH_3OH + 1% H_3PO_4–H_2O	280	Direct injection of beers	168
Lactic acid	SCX	0.05% EDTA	210	In foods	11
Nicotine	SCX	0.5 M NH_4PO_4 (pH 3–5)–CH_3OH (3:2)	254	Studied pH and I	164
Nucleotides	C18	0.01 M potassium phosphate–CH_3OH (9:1)	254	For ATP, GTP, etc.	74
Oligosaccharides	SCX, Ca	Water	RI	From enzymatic hydrolysis	139
Peptides	CN C3	0.01–0.08% CF_3CO_2H in 0–70% CH_3CN 0.12% CF_3COH in 0–70% CH_3CN	280 colorimetry	Resolves many constituents	154
Pergolide	C18	CH_3CN–0.01 M ammonium carbonate (pH 8.4) (3.75:2)	290	+ Metabolites in extracts	83
Phenolic acids	C18	Water-CH_3CO_2H (984:20) to H_2O–CH_3OH–1-propanol–CH_3CO_2H (815:140:25:20) gradient + 0.02 M sodium acetate	270 (260–304)	8 Nonvolatile compounds analyzed	97
	C18	1-Propanol–H_3CCO_2H–CH_3OH–0.036 M ammonium acetate (2.18:1.98:8.71:87.13)	EC + 1V	13 Compounds analyzed	134
Phenolic	C18	H_2O-CH_3CO_2H (95:5), H_2O-CH_3CO_2H-CH_3OH (80:5:15, 75:5:20, or 65:5:30)	254 280	In grapes	180
	C18	1-Propanol–H_3CCO_2H–CH_3OH–H_2O (2.2:2:8.7:87.1)	EC + 0.95V	In beverages	133
Phenols	C18	HCO_2H-H_2O (5:85) to CH_3OH	280	Many different components resolved	30
Procyanidins	Hexyl	CH_3OH-H_2O (25–98% gradient)	280	In ciders and wines	96
Proteins	Size exclusion	0.1% Sodium dodecyl sulfate in 0.05 M phosphate (pH 6.8)	254	Separates 5 components	149
	Alumina	0.2 M phosphate buffer (pH 9)	RI (220 or 280)	Investigate mobile phases	95
Quassinoids	CN	0.5% CH_3CN in $CHCl_3$	256	Separates 5 terpenes; also preparative LC	131
Quinine	SCX	0.5 M $NH_4H_2PO_4$ (pH 3)–CH_3OH (3:2)	254	Studied pH and I	164
Rotenone	C8	CH_3CN-H_2O (65:35)	229	Resolves rotenone	106
Steroidal sapogenins	Si C18	Hexane-ethanol (98:2 or 93.5:6.5) CH_3OH-H_2O (88:15)	RI or 205	30 Sapogenins examined	100
Steroids	C8	2-Propanol-CH_3CN (3:7), 95–100 gradient CH_3CN-H_2O	210 D[d] (210)	Extracted by resin	29
Sugars	CN	H_2O-CH_3CN (3:17)	RI	In hydrolysates	125
	Si	8–20% C_2H_5OH–0.5% water–$CHCl_3$	D[h] F (350/500)	Reducing sugars	152
Tannic acids	Si	Hexane-CH_3OH-THF (80:15:5 to 35:50:15 gradient)	280	From nuts	167
	C18	H_2O-CH_3OH (90:10 to 0:100) + 0.5% H_3PO_4			
Tubocurarine	SCX	0.5 M $NH_4H_2PO_4$ (pH 5)–CH_3OH (3:2)	254	Studied pH and I	164

[a] See Table 5, footnote *a*.
[b] See Table 5, footnote *b*.
[c] See Table 5, footnote *c*.
[d] *O*-Phthalaldehyde or similar derivative.
[e] Phenylthiocarbamyl.
[f] I, Ionic strength.
[g] α-Bromoacetophenone.
[h] Dimethylaminonaphthalene-5-sulfonic acid hydrazine.

matography provides the selectivity and ability to quantify.

Newer procedures used to prepare samples prior to HPLC are becoming increasingly chromatographic in nature. These procedures include affinity chromatography (182), immobilizing resins (137), and bonded silica separations (71). The purification of interferons from human cell cultures is a good illustration (110), involving the techniques described above and others, such as use of monoclonal antibodies.

29.2.3 Preparation of Standards

Since quantification requires the use of a standard, reference standards must be carefully purified and characterized to obtain the degree of accuracy that

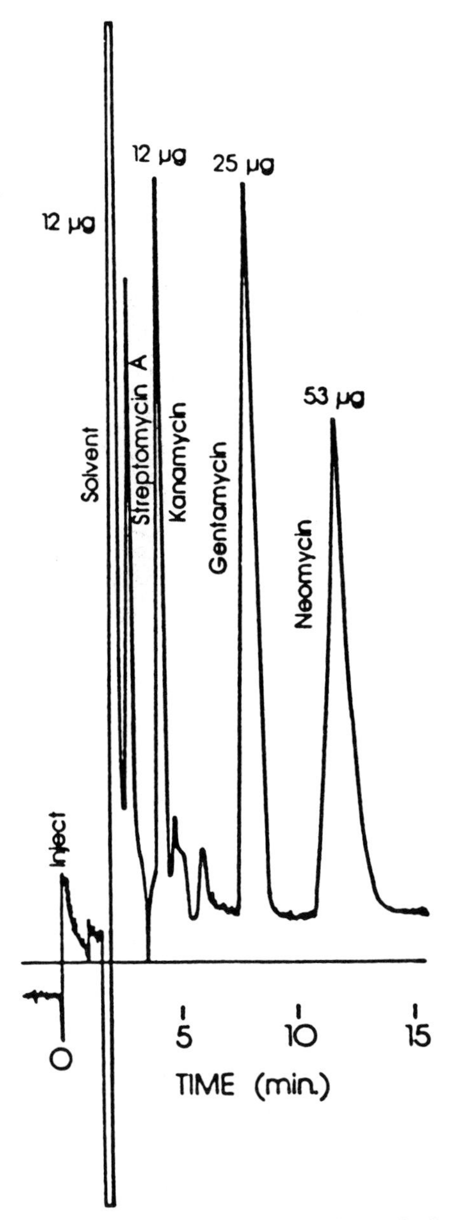

FIG. 4. HPLC separation of the aminoglycosides kanamycin, streptomycin A, gentamicin, and neomycin. See the text for details. Reproduced with the permission of the *Journal of Antibiotics* and the authors (176).

HPLC is capable of producing. Purity, including content, should be determined by several techniques. The vial of standard, usually stored at 4 to 20°C, is allowed to equilibrate to room temperature and is dried, if necessary, and 10 to 25 mg is carefully weighed (unless little is available, in which case a microbalance is used) and, finally, transferred to a volumetric flask (protected from light, if necessary). Solvents like water, buffer, methanol, or dimethyl sulfoxide are used, and frequently, a second dilution with mobile phase is necessary. In our laboratory, we prefer to weigh two portions of standards from two bottles, dissolve and dilute each separately, and then compare the chromatographic results. Agreement within a 1.5% experimental error for purity and resolution and estimation of an inevitable, minor constituent indicates that the liquid chromatography system is suitable for use in the analysis of unknown samples. Solutions are prepared fresh daily. Samples are manipulated like standards as far as possible.

For compounds for which no standards are available, two assays with two polarity systems (87) (like silica- and octadecylsilane-based separations) are preferred. Results obtained with these opposite polarity systems should agree within experimental error. Impurities are estimated with equal molar absorptivity assumed. Calculations must include solvent, salt, and water contents.

29.2.4 Typical HPLC Results

Resolution of related antibiotics

HPLC can usually resolve related antibiotics by similar chromatographic conditions. Figure 4 illustrates the separation of related aminoglycosides with an octadecylsilane column packed with particles 5 μm in diameter and a mobile phase of methanol–0.05 M camphorsulfonic acid (pH 1.7) in the proportion 3:2 flowing at 1 ml/min into a refractive index detector (176). Eluting prior to gentamicin are several minor components which may be other species of gentamicin.

Resolution of a multicomponent antibiotic

Many natural products are composed of a mixture of related, active constituents. Figure 5 depicts a chromatogram of a commercial preparation of bacitracin. Components A_1, B_1, B_2, C, D, E, F_1, F_2, G, X, and a large number of additional components are visible. The procedure (163) used an octadecylsilane column with the mobile phase increasing from 5 to 50% methanol and 20% acetonitrile in 0.01 M phosphate buffer (pH 4.5). Detection was at 254 nm.

Resolution of impurities

Nystatin in impure and purified (A. Aszalos, personal communication) pharmaceutical preparations is resolved into various constituents (Fig. 6) with an octadecylsilane column with a mobile phase of either methanol-water-tetrahydrofuran (35:43:22) or methanol–0.05 M phosphate buffer (pH 2.3) (3:1) flowing at 0.4 to 1.5 ml/min into a UV-light detector adjusted to 313 nm. HPLC was used to monitor further purification.

Resolution of mixtures of antibiotics

Unrelated compounds can be resolved by adjusting column polarity, solvent composition, and pH. If necessary, gradient elution can be used to provide greater flexibility.

29.2.5 Correlation of HPLC with Biological Results

Most of the molecules produced in fermentation broths of commercial importance are antimicrobial agents. In general, the analysis of simple, purified

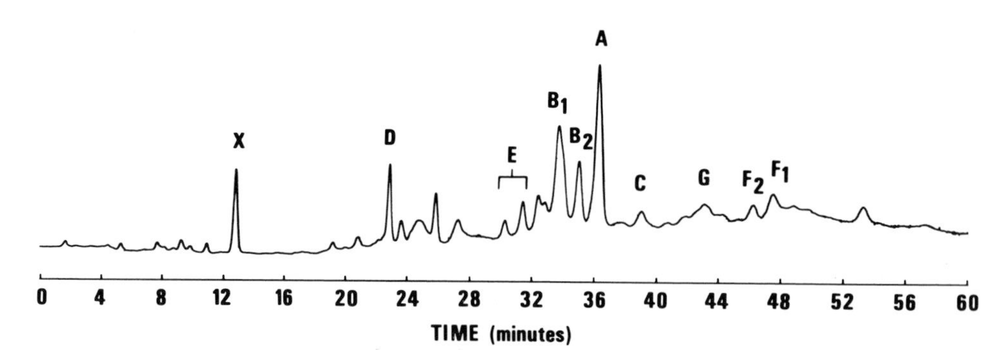

FIG. 5. HPLC of a commercial preparation of bacitracin powder, demonstrating the resolution of a large number of components. See the text for details. Reproduced from the *Journal of Chromatography* with the permission of the editor and the authors (163).

compounds give results similar to those of biological assays. However, multifunctional compounds often yield strange results. For example, amorphous nystatin contents (determined from peak height by the HPLC procedure described above) showed a good correlation with microbiological assays. Inexplicably, crystalline nystatins give no such correlation, although the samples are 90 to 95% one component, based on countercurrent distribution (86; A. Aszalos, personal communication).

If an antibiotic is composed of more than one molecular species, microbiological activity depends on the relative potency of each constituent. Perhaps the best known example is that of gentamicin (Fig. 3), which consists mainly of the C_1, C_{1a}, and C_2 components with smaller quantities of C_{2a} and C_{2b}. For agreement with the microbiological method (174), which apparently does not distinguish between C_2 and C_{2a}, the HPLC value must include the sum of the constituents.

29.2.6 Sources of Error

Each stage of the HPLC assay is subject to error, as is the case with any other assay. Initially, extraction of the analyte from the biological matrix may be incomplete. This error may be minimized for biologically active constituents by re-extracting the residue and reassaying it. Since analyses all depend, eventually, on reference standards, variations in the standard are reflected in the analytical results. In HPLC, the average errors are generally below 2%; however, as for all chemical assays of material active in vivo, at some point a biological correlation must be performed, at least periodically, to ensure validity of the results. To ensure accuracy, results with two (preferably opposite) polarity columns should agree, recoveries of added material should be close to 100%, and each liquid chromatography system should be linear, reproducible, and capable of resolving related compounds.

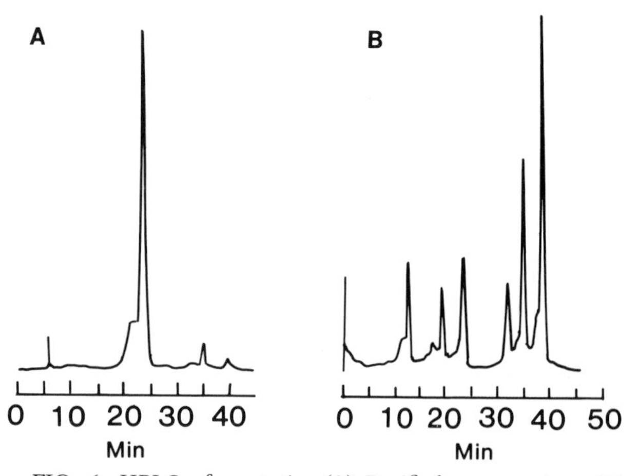

FIG. 6. HPLC of nystatin. (A) Purified preparation; (B) impure sample. See the text for details of the chromatography. Illustration courtesy of the publishers, Elsevier and Marcel Dekker, Inc., and the authors (86, 157; A. Aszalos, personal communication).

29.2.7 Preparative HPLC

One major strength of HPLC is that the analytical technique frequently can be scaled up from micrograms to grams by switching to a larger-diameter column and a greater-volume pump head and, perhaps, by modifying the mobile phase. These changes permit the collection of large amounts of material for structure elucidation and biological testing, if the material is biologically active. For example, five antibiotics were resolved and obtained in nearly pure form with preparative liquid chromatography (153).

29.3 AFTERWORD

In many laboratories, HPLC is rapidly displacing classical chemical and microbiological analytical techniques. The ease of sample preparation, relative simplicity of operation, and selective and accurate results have caused this shift in techniques. Microbiological assays remain an important tool because they are the only device for assessing the actual antimicrobial activity of the chemical under study.

29.4 ACKNOWLEDGMENTS

We thank K. Florey, G. Brewer, J. McGuire, M. Rosenthale, and E. Tolman for their helpful comments; S. Perlman for his helpful comments, proofreading, and thiostrepton data; C. Herrick for typing this manuscript; M. George and L. Moore for obtaining many of the references; and our employers, Ortho Pharmaceutical Corporation and the Squibb Institute for Medical Research, for their support. A. Aszalos of the

Food and Drug Administration supplied considerable information on nystatin and amphotericin B, for which we are grateful.

29. 5 LITERATURE CITED

1. **Abidi, S. L.** 1982. High-performance liquid chromatographic resolution and quantification of a dilactonic antibiotic mixture (antimycin A). J. Chromatogr. **234:**187–200.
2. **Adamovics, J., and F. R. Stermitz.** 1976. High-performance liquid chromatography of some anthocyanidins and flavonoids. J. Chromatogr. **129:**464–465.
3. **Aitzetmüller, K.** 1978. Sugar analysis by high-performance liquid chromatography using silica columns. J. Chromatogr. **156:**354–358.
4. **Akada, Y., S. Kawano, and Y. Tanase.** 1980. High-speed liquid chromatographic analysis of drugs. X. Simultaneous determination of acrinol and berberine chloride in pharmaceutical preparations. Yakugaku Zasshi **100:**766–770.
5. **Alemanni, A., and M. Riedmann.** 1979. HPLC routine analysis of biosynthetic active compounds in fermentation media. Chromatographia **12:**396–398.
6. **Anderson, B. M.** 1983. Separation of α-, β- and iso-α-acids in hop products and beer by high-performance liquid chromatography. J. Chromatogr. **262:**448–450.
7. **Andrews, P. A., S.-S. Pan, and N. R. Bachur.** 1983. Liquid chromatographic and mass spectral analysis of mitosane and mitosene derivatives of mitomycin C. J. Chromatogr. **262:**231–247.
8. **Anke, H., T. Kemmer, and G. Höfle.** 1981. Deflectins, new antimicrobial azaphilones from *Aspergillus deflectus*. J. Antibiot. **34:**923–928.
9. **Argoudelis, A. D., T. A. Brinkley, T. F. Brodasky, J. A. Buege, H. F. Meyer, and S. A. Mizsak.** 1982. Paulomycins A and B: isolation and characterization. J. Antibiot. **35:**285–294.
10. **Arret, B., D. P. Johnson, and A. Kirshbaum.** 1971. Outline of details for microbiological assays of antibiotics, 2nd revision. J. Pharm. Sci. **60:**1689–1694.
11. **Ashoor, S. H., and J. Welty.** 1984. Determination of organic acids in foods by high-performance liquid chromatography: lactic acid. J. Chromatogr. **287:**452–456.
12. **Asmus, P. A., J. B. Landis, and C. L. Vila.** 1983. Liquid chromatographic determination of lincomycin in fermentation beers. J. Chromatogr. **264:**241–248.
13. **Aszalos, A.** 1984. Analysis of antitumor antibiotics by high-performance liquid chromatography. J. Liq. Chromatogr. **7:**69–125.
14. **Aszalos, A., T. Alexander, and M. Margosis.** 1982. High-performance liquid chromatography in the analysis of antibiotics. Trends Anal. Chem. **1:**387–393.
15. **Aszalos, A., C. Heneke, M. J. Hayden, and J. Crawford.** 1982. Analyses of tetracycline antibiotics by reversed-phase high-performance liquid chromatography. Chromatographia **15:**367–373.
16. **Axelsen, K. S., and S. H. Vogelsang.** 1977. High-performance liquid chromatographic analysis of gramicidin, a polypeptide antibiotic. J. Chromatogr. **140:**174–178.
17. **Balitz, D. M., J. A. Bush, W. T. Bradner, T. W. Doyle, F. A. O'Herron, and D. E. Nettleton.** 1982. Isolation of lavendamycin, a new antibiotic from *Streptomyces lavendulae*. J. Antibiot. **35:**259–265.
18. **Balitz, D. M., F. A. O'Herron, J. Bush, D. M. Vyas, D. E. Nettleton, R. E. Grulich, W. T. Bradner, T. W. Doyle, E. Arnold, and J. Clady.** 1981. Antitumor agents from *Streptomyces anandii*: gilvocarcins V, M and E. J. Antibiot. **34:**1544–1555.
19. **Batta, A. K., V. Dayal, R. W. Coleman, A. K. Sinka, S. Shefer, and G. Salen.** 1984. Separation of underivatized C_{20} fatty acids by reversed-phase high-performance liquid chromatography. J. Chromatogr. **284:**257–260.
20. **Beelen, P., A. C. Dijkstra, and G. D. Vogels.** 1983. Quantitation of coenzyme F_{420} in methanogenic sludge by the use of reversed-phase high-performance liquid chromatography and a fluorescence detector. Eur. J. Appl. Microbiol. Biotechnol. **18:**67–69.
21. **Bens, G. A.** 1982. Modern techniques for the determination of macrolide antibiotics. Verh. K. Acad. Geneeskd. Belg. **44:**131–165.
22. **Beran, M., J. Tax, V. Schön, Z. Vaněk, and M. Podojil.** 1983. Preparative reversed-phase high-performance liquid chromatography of monensin A and B sodium salts. J. Chromatogr. **268:**315–320.
23. **Black, H. R., R. S. Griffith, and A. M. Peabody.** 1966. Absorption, excretion and metabolism of capreomycin in normal and diseased states. Ann. N.Y. Acad. Sci. **135:**974–982.
24. **Bodnar, J. E., J. R. Chen, W. H. Johns, E. P. Mariani, and E. C. Shinal.** 1983. High-performance liquid chromatographic determination of vincristine sulfate in performulatin studies. J. Pharm. Sci. **72:**535–537.
25. **Bran, J. L., M. E. Levison, and D. Kaye.** 1972. Clinical and in vitro evaluation of cephapirin, a new cephalosporin antibiotic. Antimicrob. Agents Chemother. **1:**35–40.
26. **British Pharmacopoeia, vol. 2, appendix XIVA.** 1980. Biological and biochemical tests, p. A121–126. Department of Health and Social Security, Her Majesty's Stationery Office, London.
27. **Buchberger, W., K. Winsauer, and F. Nachtmann.** 1983. Spurenbestimmung von Penicillin V mittels Hochdruckflüssigkeits-chromatographie. Z. Anal. Chem. **315:**525–527.
28. **Buteau, C., C. L. Duitschaever, and G. C. Ashton.** 1984. High-performance liquid chromatographic detection and quantitation of amines in must and wine. J. Chromatogr. **284:**201–210.
29. **Byrne, K. J., G. H. Reazin, and A. A. Andreasen.** 1981. High-pressure liquid chromatographic determination of β-sitosterol and β-sitosterol-D-glucoside in whiskey. J. Assoc. Off. Anal. Chem. **64:**181–185.
30. **Casteele, K. V., H. Geiger, and C. F. van Sumere.** 1983. Separation of phenolics (benzoic acids, cinnamic acids, phenylacetic acids, quinic acid esters, benzaldehydes and acetophenones, miscellaneous phenolics) and coumarins by reversed-phase high-performance liquid chromatography. J. Chromatogr. **258:**111–124.
31. **Chien, M. M., and J. P. Rosazza.** 1980. Microbial transformations of natural antitumor agents: use of solubilizing agents to improve yields of hydroxylated ellipticines. Appl. Environ. Microbiol. **40:**741–745.
32. **Code of Federal Regulations, title 21, Food and Drugs, part 436, subpart D.** 1983. Microbiological assay methods, p. 242–259. Office of the Federal Register, National Archives and Records Service, General Services Administration, Washington, D.C.
33. **Cox, G. B., C. R. Loscombe, and K. Sugden.** 1977. Some applications of bonded-phase high-performance liquid chromatography to the analysis of pharmaceutical formulations. Anal. Chim. Acta **92:**345–352.
34. **Daigle, D. J., and E. J. Conkerton.** 1983. Analysis of flavonoids by HPLC. J. Liq. Chromatogr. **65:**105–118.
35. **Dony, J.** 1954. Considerations techniques sur le dosage de la magnamycine. Ann. Pharm. Fr. **12:**307–313.
36. **Eckers, C., D. E. Games, D. N. B. Mallen, and B. P. Swann.** 1982. Studies of ergot alkaloids using high-performance liquid chromatography-mass spectrometry and mass spectrometry. Anal. Proc. **19:**133–137.
37. **Elverdam, I., P. Larsen, and E. Lund.** 1981. Isolation and characterization of these new polymyxins in polymyxin B and E by high-performance liquid chromatography. J. Chromatogr. **218:**653–661.
38. **English, A. R., T. J. McBride, and R. Riggio.** 1962. Biological studies of 6-methylene oxytetracycline, a

new tetracycline, p. 462–473. Antimicrob. Agents Chemother. 1961.

39. **Enríques, R. G., M. L. Romero, L. I. Escobar, P. Joseph-Nathan, and W. F. Reynolds.** 1984. High-performance liquid chromatographic study of *Casimiroa edulis.* II. Determination of furocoumarins. J. Chromatogr. **287:**209–214.
40. **Erttmann, R.** 1983. Determination of aclacinomycin A by reversed-phase high-performance liquid chromatography. J. Chromatogr. **277:**433–435.
41. **European Pharmacopoeia, 2nd ed., part 1.** 1980. Microbiological assay of antibiotics, sect. VIII. 4. (Replacement text, 1984.) Council of Europe, Maison Neuve, Sainte-Ruffine, France.
42. **Fielder, H. P.** 1981. Quantitation of nikkomycins in biological fluids by ion-pair reversed phase high-performance liquid chromatography. J. Chromatogr. **204:**313–318.
43. **Fisher, B. V., and R. B. Raja.** 1982. High-performance liquid chromatography of polymyxin B sulfate and gramicidin. Anal. Proc. **19:**137–140.
44. **Foglesong, M. A., J. W. Lamb, and J. V. Dietz.** 1978. Stability and blood level determinations of cefaclor, a new oral cephalosporin antibiotic. Antimicrob. Agents Chemother. **13:**49–52.
45. **Frede, W.** 1977. Detection of pimaricin on the surface of cheese. Milchwissenschaft **32:**66–67.
46. **Freezman, M., F. A. Hawkins, J. S. Loran, and J. A. Stead.** 1979. The analysis of gentamicin sulfate in pharmaceutical specialties by high-performance liquid chromatography. J. Liq. Chromatogr. **2:**1305–1317.
47. **Friesen, H.-J., S. Stein, M. Evinger, P. C. Familletti, J. Moschera, J. Meinenhofer, J. Shively, and S. Pestka.** 1981. Purification and molecular characterization of human fibroblast interferon. Arch. Biochem. Biophys. **206:**432–450.
48. **Fröhlich, D. H.** 1982. Optimized separation of benzoic and sorbic acid as preservatives in food stuffs by reversed phase HPLC. J. High Resolut. Chromatogr. Commun. **5:**158–160.
49. **Fujita, K., S. Takeuchi, and S. Ganno.** 1979. Fast separation of amino acids using ion exchange chromatography, p. 81–97. *In* G. Charalambous (ed.), Liquid chromatographic analysis of food and beverages, vol. 1. Academic Press, Inc., New York.
50. **Fuska, J., and B. Proksa.** 1983. Chromatographic determination of vermiculine. Pharmazie **38:**634–635.
51. **Gallagher, J. B., P. W. Love, and L. L. Knotts.** 1982. High-pressure liquid chromatographic determination of bacitracin in premix feeds and finished feeds: collaborative study. J. Assoc. Off. Anal. Chem. **65:**1178–1185.
52. **Getek, T. A., A. C. Haneke, and G. B. Selzer.** 1983. Determination of gentamicin sulfate C_{1a} and C_2 and C_1 components by ion-pair liquid chromatography with electrochemical detection. J. Assoc. Off. Anal. Chem. **66:**172–175.
53. **Ghebre-Sellassie, I., S. L. Hem, and A. M. Knevel.** 1982. Separation of penicillin and its major degradation products by ion-pair reversed-phase high-pressure liquid chromatography. J. Pharm. Sci. **71:**351–353.
54. **Giannetti, B. M., W. Steglich, W. Quack, T. Anke, and F. Oberwinkler.** 1978. Murulinsäuren A, B and C, neue Antibiotika aus *Merulius tremellosus* Fr. und *Phelebia radiata* Fr. Z. Naturforsch. **33C:**807–816.
55. **Gidoh, M., S. Tsutsumi, and S. Takitani.** 1981. Determination of the main antileprosy drugs and their main metabolites in serum by high-performance liquid chromatography. J. Chromatogr. **223:**379–392.
56. **Gorham, J.** 1984. Separation of plant betaines and their sulphur analogues by cation-exchange high-performance liquid chromatography. J. Chromatogr. **287:** 345–351.
57. **Görög, S., B. Herenyi, and K. Javánovics.** 1977. High-performance liquid chromatography of *Catharanthus* alkaloids. J. Chromatogr. **139:**203–206.
58. **Grove, D. C., and W. A. Randall.** 1955. Assay methods of antibiotics, a laboratory manual. Medical Encyclopedia, New York.
59. **Gullo, V. P., R. T. Goegelman, I. Putter, and Y.-K. Lam.** 1981. High-performance liquid chromatographic analysis of derivatized hypocholesteremic agents from fermentation broths. J. Chromatogr. **212:**234–238.
60. **Haginaka, J., T. Nakagawa, T. Hoshino, and T. Uno.** 1982. HPLC determination and pharmacokinetic investigation of urinary excretion of clavulanic acid. J. Pharmacobio-Dyn. **5:**s–31.
61. **Harada, S., and T. Kishi.** 1978. Isolation and characterization of mildiomycin, a new nucleoside antibiotic. J. Antibiot. **31:**519–524.
62. **Harangi, J., M. Deák, and P. Nánási.** 1984. Determination of the major factors of fermentation of the nebramycin complex by high-performance liquid chromatography. J. Liq. Chromatogr. **7:**83–93.
63. **Heinrikson, R. L., and S. C. Meredith.** 1984. Amino acid analysis by reverse-phase high-performance liquid chromatography: precolumn derivatization with phenylisothiocyanate. Anal. Biochem. **136:**65–74.
64. **Helboe, P., M. Thomsen, and S. H. Hansen.** 1980. Improved high-performance liquid chromatographic method for the comparison of heptaene macrolide antibiotics. J. Chromatogr. **189:**249–254.
65. **Hermansson, J., and M. Andersson.** 1982. Reversed-phase ion-pair chromatography of tetracycline, tetracycline analogs, and their potential impurities. J. Pharm. Sci. **71:**222–229.
66. **Hikal, A. H., A. Shibl, and S. El-Hoofy.** 1982. Determination of sodium fusidate and fusidic acid in dosage forms by high-performance liquid chromatography and a microbiological method. J. Pharm. Sci. **71:**1297–1298.
67. **Hornish, R. E.** 1982. Paired-ion high-performance liquid chromatographic determination of the stability of novobiocin in mastitis products sterilized by ^{60}Co irradiation. J. Chromatogr. **236:**481–488.
68. **Horwitz, W. (ed.).** 1980. Official methods of analysis of the Association of Official Analytical Chemists, 13th ed., p. 719–731. Association of Official Analytical Chemists, Washington, D.C.
69. **Hughes, J. T., and P. J. Davis.** 1981. High-performance liquid chromatographic determination of *N,N*-dimethylcolchiceinamide and its metabolites, *N*-methylcolchiceinamide and colchiceinamide, in microbial culture. J. Chromatogr. **219:**321–324.
70. **Hussey, R. L., W. G. Mascher, and A. L. Lagu.** 1983. Determination of *p*-hydroxypenicillin V, *p*-hydroxyphenoxyacetic acid, phenoxyacetic acid and penicillin V in production fermentation broth. J. Chromatogr. **268:**120–124.
71. **Hyde, P. M.** 1984. Sample preparation symposium. LC Mag. **2:**320–322.
72. **Ito, T., A. Takatsuki, K. Kawamura, K. Sato, and G. Tamura.** 1980. Isolation and structures of components of tunicamycin. Agric. Biol. Chem. **44:**695–698.
73. **Itoh, J., S. Omoto, T. Shomura, H. Ogino, K. Iwanatsu, and S. Inouye.** 1981. Oligostatins, new antibiotics with amylase inhibitory activity. I. Production, isolation and characterization. J. Antibiot. **34:**1424–1428.
74. **Jahngen, E. G. E., and E. F. Rossomando.** 1984. Intermediary purine-metabolizing enzymes from the cytosol of *Dictyostelium discoideum* monitored by high-performance liquid chromatography. Anal. Biochem. **137:**493–504.
75. **Jensen, S. E., D. W. S. Westlake, and S. Wolfe.** 1982. High performance liquid chromatographic assay of cyclization activity in cell-free systems from *Streptomyces clavuligerus*. J. Antibiot. **35:**1026–1032.
76. **Jeuring, H., and F. J. E. M. Kuppers.** 1980. High

performance liquid chromatography of furfural and hydroxyfurfural in spirits and honey. J. Assoc. Off. Anal. Chem. **63:**1215–1218.

77. **Jost, W., H. E. Hauck, and F. Eisenbeiss.** 1983. Simple method for the separation of cephalosporins using silica gel 60 as the stationary phase in thin-layer chromatography and in high-performance liquid chromatography. J. Chromatogr. **256:**182–184.
78. **Kalász, H., and C. H. Horváth.** 1981. Preparation-scale separation of polymyxins with an analytical high-performance liquid chromatography system by using displacement chromatography. J. Chromatogr. **215:**295–302.
79. **Kappan, L. S., M. A. Napier, and I. H. Goldberg.** 1980. Roles of chromophore and apo-protein in neocarzinostatin action. Proc. Natl. Acad. Sci. USA **77:**1970–1974.
80. **Kavanagh, F.** 1972. Photometric assaying, p. 44–121. *In* F. Kavanagh (ed.), Analytical microbiology. Academic Press, Inc., New York.
81. **Kavanagh, F.** 1972. Introduction, p. 1–12. *In* F. Kavanagh (ed.), Analytical microbiology. Academic Press, Inc., New York.
82. **Kennedy, J. H.** 1978. High performance liquid chromatographic analysis of fermentation broths: cephalosporin C and tylosin. J. Chromatogr. Sci. **16:**492–495.
83. **Kerr, K. M., R. V. Smith, and P. J. Davis.** 1981. High-performance liquid chromatographic determination of pergolide and its metabolite pergolide sulfoxide in microbial extracts. J. Chromatogr. **219:**317–320.
84. **Kimura, Y., H. Kitamura, T. Araki, K. Noguchi, M. Baba, and M. Hori.** 1981. Analytical and preparative methods for polymyxin antibiotics using high-performance liquid chromatography with a porous styrene-divinylbenzene copolymer packing. J. Chromatogr. **206:**563–572.
85. **Kirchmeier, R. L., and R. P. Upton.** 1978. Simultaneous determination of vancomycin, anisomysin and trimethoprim lactate by high pressure liquid chromatography. Anal. Chem. **50:**349–351.
86. **Kirschbaum, J., and A. Aszalos.** 1986. HPLC of antibiotics, p. 239–322. *In* A. Aszalos (ed.), Modern analysis of antibiotics. Marcel Dekker, Inc., New York.
87. **Kirschbaum, J., S. Perlman, J. Joseph, and J. Adamovics.** 1984. Ensuring accuracy of HPLC assays. J. Chromatogr. Sci. **22:**27–30.
88. **Kiyoshi, T., J. H. Robertson, and J. A. Bach.** 1974. Quantitative high-pressure liquid chromatographic analysis of bacitracin, a polypeptide antibiotic. J. Chromatogr. **1974:**597–608.
89. **Knox, J. H., and J. Jurana.** 1979. Mechanism of reversed-phase separation of tetracyclines by high-performance liquid chromatography. J. Chromatogr. **186:** 763–782.
90. **Koeppe, R. E., II, and L. B. Weiss.** 1981. Resolution of linear gramicidins by preparative reversed-phase high-performance liquid chromatography. J. Chromatogr. **208:**414–418.
91. **Kraisintu, K., R. T. Parfitt, and M. G. Rowan.** 1982. A high-performance liquid chromatographic method for the determination and control of the composition of gentamicin sulfate. Int. J. Pharm. **10:**67–75.
92. **Ladisch, M. R., A. L. Huebner, and G. T. Tsao.** 1978. High-speed chromatography of cellodextrins and other saccharide mixtures using water as the eluent. J. Chromatogr. **147:**185–193.
93. **Lamb, J. W., J. W. Mann, and R. J. Simmons.** 1972. Factors influencing the microbiological assay of tobramycin. Antimicrob. Agents Chemother. **1:**323–328.
94. **Langone, J. J., M. R. D'Onofrio, and H. Van Vunekis.** 1979. Radioimmunoassays for the *Vinca* alkaloids, vinblastine and vincristine. Anal. Biochem. **95:**214–221.
95. **Laurent, C. J. C. M., H. A. H. Billiet, L. De Galan, F. A. Buytenhuys, and F. P. B. van der Maeden.** 1984. High-performance liquid chromatography of proteins on alumina. J. Chromatogr. **287:**45–54.
96. **Lea, A. G. H.** 1980. Reversed-phase gradient high-performance liquid chromatography of procyanidins and their oxidation products in ciders and wines, optimised by Snyder's procedures. J. Chromatogr. **194:**62–68.
97. **Lehtonen, M.** 1983. High-performance liquid chromatographic determination of nonvolatile phenolic compounds in matured distilled alcoholic beverages. J. Assoc. Off. Anal. Chem. **66:**71–78.
98. **Liautard, J. P.** 1984. Rapid separation of single-stranded DNA from double-stranded DNA by reversed-phase high-performance liquid chromatography. J. Chromatogr. **285:**221–225.
99. **Lin, C., H. Kim, D. Schuessler, E. Olden, and S. Symchowicz.** 1980. High-pressure liquid chromatographic method for determination of rosaramicin in humans. Antimicrob. Agents Chemother. **18:**780–783.
100. **Lin, J.-T., and C.-J. Xu.** 1984. High-performance liquid chromatography of steroidal sapogenins. J. Chromatogr. **287:**105–112.
101. **Lloyd, L. F.** 1982. Analysis of insulin preparation by high-performance liquid chromatography. Anal. Proc. **19:**131–133.
102. **Mahoney, W. C., and D. Duksin.** 1980. Separation of tunicamycin homologues by reversed phase high-performance liquid chromatography. J. Chromatogr. **198:**506–510.
103. **Martin, J. R., R. L. DeVault, A. C. Sinclair, R. S. Stanaszek, and P. Johnson.** 1982. A new naturally occurring erythromycin: erythromycin F. J. Antibiot. **35:**426–430.
104. **Mays, D. L., R. J. van Apeldoorn, and R. G. Lauback.** 1976. High-performance liquid chromatographic determination of kanamycin. J. Chromatogr. **120:**93–102.
105. **McCalley, D. V.** 1983. Analysis of cinchona alkaloids by high-performance liquid chromatography. J. Chromatogr. **260:**184–188.
106. **McCown, S. M.** 1984. Determination of rotenone and rotenonine in fresh water by HPLC. LC Mag. **2:**318–319.
107. **McMurrough, I.** 1981. High-performance liquid chromatography of flavonoids in barley and hops. J. Chromatogr. **218:**683–693.
108. **Mechlinski, W., and C. P. Schaffner.** 1974. Separation of polyene antifungal antibiotics by high speed liquid chromatography. J. Chromatogr. **99:**619–633.
109. **Mechlinski, W., and C. P. Schaffner.** 1980. Characterization of aromatic heptaene macrolide antibiotics by high-performance liquid chromatography. J. Antibiot. **33:**591–599.
110. **Menge, U., and M. R. Kula.** 1984. Purification techniques for human interferons. Enzyme Microb. Technol. **6:**101–111.
111. **Micali, G., P. Curro, and G. Calabro.** 1980. Reversed-phase high-performance liquid chromatography for the determination of β-asarone. J. Chromatogr. **194:** 245–250.
112. **Miller, R. D., C. Affolder, and N. Neuss.** 1981. HPLC of cephalosporins and their oxa-derivatives. Experientia **37:**928–930.
113. **Miller, R. D., and N. Neuss.** 1978. High performance liquid chromatography of natural products. II. Direct biological correlation of components in the fermentation broth. J. Antibiot. **31:**1132–1136.
114. **Miller, T. W., L. Chaiet, D. J. Cole, L. J. Cole, J. E. Flor, R. T. Goegelman, V. P. Gullo, H. Joshua, A. J. Kempf, W. R. Krellwitz, R. L. Monaghan, R. E. Ormond, K. E. Wilson, G. Albers-Schönberg, and I. Putter.** 1979. Avermectins, new family of potent anthelmintic agents: isolation and chromatographic properties. Antimicrob. Agents Chemother. **15:**368–371.

115. **Ministry of Health and Welfare.** 1961. Minimum requirements of antibiotic products. Ministry of Health and Welfare, Government of Japan, Tokyo.
116. **Myers, H. N., and J. V. Rindler.** 1979. Determination of spectinomycin by high-performance liquid chromatography with fluorometric detection. J. Chromatogr. **176:**103–108.
117. **Nachtmann, F., and K. Gstrein.** 1982. Simultaneous determination of the cationic and anionic parts in repository penicillins by high-performance liquid chromatography. J. Chromatogr. **236:**461–468.
118. **Nachtmann, F., K. Gstrein, and B. Wallnöfer.** 1983. Quantitative Bestimmung von Gramicidin mittels Hochdruckflüssigkeits-chromatographie. Fresenius Z. Anal. Chem. **315:**241–242.
119. **Nakagawa, A., Y. Iwai, H. Hashimoto, N. Miyazaki, R. Ōlwa, Y. Takahashi, A. Hirano, N. Shibukawa, Y. Kojima, and S. Ōmura.** 1981. Virantmycin, a new antiviral antibiotic produced by a strain of *Streptomyces*. J. Antibiot. **34:**1408–1415.
120. **Napier, M. A., B. Helmquist, D. J. Strydom, and I. H. Goldberg.** 1981. Neocarzinostatin chromophore: purification of the major active form and characterization of its spectral and biological properties. Biochemistry **20:**5602–5608.
121. **Nelis, H. J. C. F., and A. P. De Leenheer.** 1980. Retention mechanisms of tetracyclines on a C_8 reversed-phase material. J. Chromatogr. **195:**35–42.
122. **Neuss, N., D. M. Berry, J. Kupka, A. L. Demain, S. W. Queener, D. C. Duckworth, and L. L. Huckstep.** 1982, High performance liquid chromatography (HPLC) of natural products. V. The use of HPLC in the cell-free biosynthetic conversion of α-aminoadipyl-cysteinyl-valine (LLD) into isopenicillin N. J. Antibiot. **35:**580–584.
123. **Ogasawasa, T., S. Goto, S. Mori, and T. Oki.** 1981. High performance liquid chromatographic determination of aclacinomycin A and its related compounds. I. Normal phase HPLC for monitoring fermentation and purification processes. J. Antibiot. **34:**47–51.
124. **Oosterbaan, M. J. M., R. J. M. Dirks, T. B. Vree, and E. van der Kleijn.** 1984. Rapid quantitative determination of seven anthracyclines and their hydroxy metabolites in body fluids. J. Chromatogr. **306:**323–332.
125. **Palmer, J. K.** 1975. A versatile system for sugar analysis via liquid chromatography. Anal. Lett. **8:**215–224.
126. **Pandey, R. C., and M. W. Toussaint.** 1980. High-performance liquid chromatography and thin-layer chromatography of anthracycline antibiotics. Separation and identification of components of the daunorubicin complex from fermentation broth. J. Chromatogr. **198:** 407–420.
127. **Pellegatta, G., G. P. Carugati, and G. Coppi.** 1983. High-performance liquid chromatographic analysis of erythromycins A and B from fermentation broths. J. Chromatogr. **269:**33–39.
128. **Pivnichny, J. V., J.-S. K. Shim, and L. A. Zimmerman.** 1983. Direct determination of avermectins in plasma at nanogram levels by high-performance liquid chromatography. J. Pharm. Sci. **72:**1447–1450.
129. **Platt, T. B., and D. Isaacson.** 1973. Microbiological assays, p. 1017–1051. *In* A. I. Laskin and H. A. Lechevalier (ed.), Handbook of microbiology, vol. 3. Microbial products. CRC Press, Inc., Cleveland.
130. **Rapp, A., and A. Ziegler.** 1976. Trennung von Dicarbonsäuren von Traubenmost and Wein mit Hilfe der Hochdruckflüssigkeits-Chromatographie an einen Kationenaustauscher. Chromatographia **9:**148–150.
131. **Robins, R. J., and M. J. C. Rhodes.** 1984. High-performance liquid chromatographic methods for the analysis and purification of quassinoids from *Quassia amara* L. J. Chromatogr. **283:**436–440.
132. **Rogers, M. E., M. W. Adlard, G. Saunders, and G. Holt.** 1983. High-performance liquid chromatographic determination of β-lactam antibiotics, using fluorescence detection following post-column derivatization. J. Chromatogr. **257:**91–100.
133. **Roston, D. A., and P. T. Kissinger.** 1981. Identification of phenolic constituents in commercial beverages by liquid chromatography with electrochemical detection. Anal. Chem. **53:**1695–1699.
134. **Roston, D. A., and P. T. Kissinger.** 1982. Series dual-electrode detector for liquid chromatography/electrochemistry. Anal. Chem. **54:**429–434.
135. **Rubinstein, M., W. P. Levy, J. A. Moschera, C.-Y. Lai, R. D. Hershberg, R. T. Bartlett, and S. Pestka.** 1981. Human leukocyte interferon: isolation and characterization of several molecular forms. Arch. Biochem. Biophys. **210:**307–318.
136. **Rzeszotarski, W. J., and A. M. Mauger.** 1973. Reversed-phase high-pressure liquid chromatography of actinomycins. J. Chromatogr. **86:**246–249.
137. **Sacco, D., and E. Dellacherie.** 1983. Liquid exchange chromatography of cephalosporin C on polystyrene resins containing copper complex of lysine derivatives. J. Liq. Chromatogr. **6:**2543–2556.
138. **Sakai, T. T.** 1978. Paired-ion high-performance liquid chromatography of bleomycins. J. Chromatogr. **161:**389–392.
139. **Schmidt, J., M. John, and C. Wandrey.** 1981. Rapid separation of malto-, xylo- and cello-oligosaccharides (DP 2-9) on cation-exchange resin using water as eluent. J. Chromatogr. **213:**151–155.
140. **Schulten, H.-R.** 1982. Off-line combination of liquid chromatography and field desorption mass spectrometry: principles and environmental, medical and pharmaceutical applications. J. Chromatogr. **251:**105–128.
141. **Shenbagamurthi, P., H. A. Smith, J. M. Becker, and F. Naider.** 1982. Purification of polyoxin D by reversed-phase high-performance liquid chromatography. J. Chromatogr. **245:**133–137.
142. **Shoji, J., H. Hinoo, R. Sakazaki, N. Tsuji, K. Nagashima, K. Matsumoto, Y. Takahashi, S. Kozuki, T. Hattori, E. Kondo, and K. Tanaka.** 1982. Asparenomycins A, B and C, new carbapenen antibiotics. II. Isolation and chemical characterization. J. Antibiot. **35:**15–23.
143. **Srivastava, S. C., and U. Hornemann.** 1978. High-pressure liquid chromatography of the antibiotics mitomycin A, B and C and of polar mitomycin C conversion products. J. Chromatogr. **161:**393–395.
144. **Stapley, E. O., D. Hendlin, J. M. Mata, M. Jackson, H. Wallick, S. Hernandez, S. Mochales, S. A. Currie, and R. M. Miller.** 1970. Phosphonomycin. I. Discovery and in vitro biological characterization, p. 284–290. Antimicrob. Agents Chemother. 1969.
145. **Stark, W. M., C. E. Higgins, R. N. Wolfe, M. M. Hoehn, and J. M. McGuire.** 1963. Capreomycin, a new antimycobacterial agent produced by *Streptomyces capreolus* sp. n., p. 596–606. Antimicrob. Agents Chemother. 1962.
146. **Strife, R. J., I. Jarine, and M. Colvin.** 1981. Analysis of the anticancer drugs etoposide (VP 16-213) and teniposide (VM 26) by high-performance liquid chromatography with fluorescence detection. J. Chromatogr. **224:**168–174.
147. **Stuber, B.** 1982. Hochdruckflüssigkeitschromatographie von Dibenzylethylendiamin in Benzathin-Penizillinen. Pharm. Acta Helv. **57:**293–294.
148. **Subden, R. E., R. G. Brown, and A. C. Noble.** 1978. Determination of histamines in wines and musts by reversed-phase high-performance liquid chromatography. J. Chromatogr. **166:**310–312.
149. **Sun, S. F.** 1984. Effect of concentration on size-exclusion liquid chromatography of protein polypeptides. LC Mag. **2:**314–316.

150. **Sztaricskai, F., J. Borda, M. M. Puskás, and R. Bognár.** 1983. High performance liquid chromatography (HPLC) of antibiotics of vancomycin type comparative studies. J. Antibiot. **36:**1691–1698.
151. **Takahashi, D. M.** 1977. Reversed-phase high-performance liquid chromatographic analytical system for aflatoxins in wine with fluorescence detection. J. Chromatogr. **131:**147–156.
152. **Takeda, M., M. Maeda, and A. Tsuji.** 1982. Fluorescence high-performance liquid chromatography of reducing sugars using dns-hydrazine as pre-labeling reagent. J. Chromatogr. **244:**347–355.
153. **Taylor, R. L., and M. F. Mallette.** 1978. Purification of antibiotics from *Physarum gyrosum* by high-performance liquid chromatography. Prep. Biochem. **8:**241–257.
154. **Tempst, P., M. W. Hunkapiller, and L. E. Hood.** 1984. Separation of peptides by reverse-phase high-performance liquid chromatography using propyl- and cyanopropylsilyl supports. Anal. Biochem. **137:**188–195.
155. **Terabe, S., R. Konaka, and J. Shoji.** 1979. Separation of polymyxins and octapeptins by high-performance liquid chromatography. J. Chromatogr. **173:**313–320.
156. **Thomas, A. H., P. Newland, and G. J. Quinlan.** 1981. Identification and determination of the qualitative composition of nystatin using thin-layer chromatography and high-performance liquid chromatography. J. Chromatogr. **216:**367–373.
157. **Thomas, A. H., P. Newland, and N. R. Sharma.** 1982. The heterogeneous composition of pharmaceutical grade nystatin. Analyst **107:**849–854.
158. **Townley, E., and P. Roden.** 1980. High-performance liquid chromatographic analysis of griseofulvin in drug substance and solid dosage forms: separation of impurities and metabolites. J. Pharm. Sci. **69:**532–536.
159. **Tsuji, K., and R. Binns.** 1982. Micro-bore high-performance liquid chromatography for the analysis of pharmaceutical compounds. J. Chromatogr. **253:**227–236.
160. **Tsuji, K., and J. F. Goetz.** 1978. HPLC as a rapid means of monitoring erythromycin and tetracycline fermentation processes. J. Antibiot. **31:**302–308.
161. **Tsuji, K., J. F. Goetz, W. VanMeter, and K. A. Gusciora.** 1979. Normal-phase high-performance liquid chromatographic determination of neomycin sulfate derivatized with 1-fluoro-2,4-dinitrobenzene. J. Chromatogr. **175:**141–152.
162. **Tsuji, K., and J. H. Robertson.** 1974. High-pressure liquid chromatographic analysis of novobiocin. J. Chromatogr. **94:**245–253.
163. **Tsuji, K., and J. H. Robertson.** 1975. Improved high-performance liquid chromatographic method for polypeptide antibiotics and its application to study the effects of treatments to reduce microbial levels in bacitracin powder. J. Chromatogr. **112:**663–672.
164. **Twitchell, P. J., A. E. P. Gorvin, and A. C. Moffat.** 1976. High-pressure liquid chromatography of drugs. II. An evaluation of a microparticulate cation-exchange column. J. Chromatogr. **120:**359–368.
165. **U.S. Pharmacopeia, 20th revision.** 1980. Antibiotics—microbial assays, p. 882–888. U.S. Pharmacopeial Convention, Inc., Rockville, Md.
166. **van der Wal, S., and J. F. K. Huber.** 1977. Improvement of the efficiency of ion-exchange cellulose columns and their application. J. Chromatogr. **135:**287–303.
167. **Verzele, M., and P. Delahaye.** 1983. Analysis of tannic acids by high-performance liquid chromatography. J. Chromatogr. **268:**469–476.
168. **Verzele, M., C. Dewaele, and M. V. Kerrebroeck.** 1982. Analysis of beer iso-α-acids by high-performance liquid chromatography without sample pre-treatment. J. Chromatogr. **244:**321–326.
169. **Vialle, J., M. Kolosky, and J. L. Rocca.** 1981. Determination of betaine in sugar and wine by liquid chromatography. J. Chromatogr. **204:**429–435.
170. **Vlasáková, V., J. Beneš, and K. Živný.** 1978. Analysis of rifampicin and of its hydrogenated derivatives by high-performance liquid chromatography. J. Chromatogr. **151:**199–205.
171. **Voloch, M., M. R. Ladisch, V. W. Rodwell, and G. T. Tsao.** 1981. Separation of *meso* and *racemic* 2,3-butanediol by aqueous liquid chromatography. Biotechnol. Bioeng. **23:**1289–1296.
172. **Waitz, J. A., A. C. Horan, M. Kalyanpur, B. K. Lee, D. Loebenberg, J. A. Marquez, G. Miller, and M. G. Patel.** 1981. Kijanimicin (Sch 25663), a novel antibiotic produced by *Actinomadura kijaniata* SCC 1256. J. Antibiot. **34:**1101–1106.
173. **Wei, T. T., J. A. Chan, P. P. Roller, U. Weiss, R. M. Stroshane, R. J. White, and K. M. Byrne.** 1982. Detection of gilvocarcin antitumor complex by a biochemical induction assay (BIA). J. Antibiot. **35:**529–532.
174. **Weigand, R., and R. J. Coombes.** 1983. Gentamicin determination by high-performance liquid chromatography. J. Chromatogr. **281:**381–385.
175. **Whall, T. J.** 1981. Determination of streptomycin sulfate and dihydrostreptomycin sulfate by high-performance liquid chromatography. J. Chromatogr. **219:** 89–100.
176. **White, E. R., and J. E. Zarembo.** 1981. Reverse phase high speed liquid chromatography of antibiotics. III. Use of ultra high performance columns and ion-pairing techniques. J. Antibiot. **34:**836–844.
177. **Whitman, W. B., and R. S. Wolfe.** 1984. Purification and analysis of cobamides of *Methanobacterium bryantii* by high-performance liquid chromatography. Anal. Biochem. **137:**261–265.
178. **Wood, R.** 1984. High-performance liquid chromatography analyses of isomeric monoenoic and acetylenic fatty acids. J. Chromatogr. **287:**202–208.
179. **Wu, R. T., T. Okabe, M. Namikoshi, S. Okuda, T. Nishimura, and N. Tanaka.** 1982. Cadeguomycin, a novel nucleoside analog antibiotic. II. Improved purification, physicochemical properties and structure assignment. J. Antibiot. **35:**279–284.
180. **Wulf, L., and C. W. Nagel.** 1976. Analysis of phenolic acids and flavonoids by high-pressure liquid chromatography. J. Chromatogr. **116:**271–279.
181. **Yee, G. C., D. J. Gmur, and M. S. Kennedy.** 1982. Liquid-chromatographic determination of cyclosporine in serum with use of a rapid extraction procedure. Clin. Chem. **28:**2269–2271.
182. **Zebin, I., and A. V. Fowler.** 1984. Purification of thiogalactoside transacetylase by affinity chromatography. Anal. Biochem. **136:**493–496.

Chapter 30

Product Recovery

GARY J. CALTON, CARRINGTON S. COBBS, AND JOHN P. HAMMAN

Since the cost of product recovery is often greater than 50% of the manufacturing cost, high-yield recovery is essential for a product to compete successfully in the marketplace. Very few schools teach courses in this important area, and thus most of the expertise available is found in industrial patents and in papers describing the isolation of natural products and proteins. This chapter, limited by constraints of space, offers general information on product recovery.

The principles referred to in this chapter are applicable to recovery of compounds from plants, natural fluids, and fermentation broths, but the discussion is limited to isolation from fermentation broths. It is to be expected that the concentration of the desired product in the broth will usually be relatively low. In addition, there will be numerous contaminants, many of which may be quite similar in size, charge, and structure. Thus, a multistep isolation scheme is usually required, and the imagination of the investigator is called for in assembling a practical isolation scheme based on the fundamental principles of separation science. The investigator should recognize before the initiation of a project that many possible permutations of an isolation scheme may have to be explored, requiring a considerable investment of time, effort, and resources.

The isolation of any product may be represented as a fractionation "tree" (Fig. 1) presenting the logical choices available to the investigator. This tree allows others to follow the rational design of the purification. Such an approach is also helpful for comparison of the purification achieved when reversing steps in the procedure. Further steps in the isolation of compounds are shown in Fig. 2.

30.1 PRODUCT DISTRIBUTION AND ASSAY

In fermentation processes, the desired product may be either excreted into the medium or retained within the cells. In either case, the product must be recovered from a solution in which a substantial number of undesirable compounds are present. A specific analysis for the product will be required. In the event that the molecule to be isolated has been identified before extraction (e.g., an amino acid), development of an analysis method is simplified. However, care should be exercised since a complex mixture may contain more than one molecule with the same functional groups or chromatographic retention characteristics.

Thin-layer chromatography is the cheapest and easiest method for this analysis. Many samples can be examined on a single plate, and the separated compounds can be visualized with a sulfuric acid spray followed by heating to char the organic fraction. Stahl gives an exhaustive treatise on thin-layer chromatographic techniques (5).

If a specific biological activity is sought in the metabolite to be isolated, a rapid pharmacological or antibiotic assay should be developed as soon as possible. Isolation may then be followed by assay of the activity. For example, an assay of antimicrobial activity may be carried out with samples of the fermentation broth, cell extract, or chromatography fractions by placing a small quantity (5 to 20 μl) on an agar plate previously inoculated with a bacterium or fungus. Growth inhibition will be observed if antimicrobial activity is present.

30.1.1 Cell Separation

The fractionation scheme shown in Fig. 1 is an example of natural product isolation and represents a standard established in our laboratory for isolation of low-molecular-weight (less than 1,000) antineoplastic agents. The initial step uses a low-speed centrifuge to sediment the microbial cells. Fungi may be centrifuged, or, depending on the character of the mycelial mat, the broth may be clarified by filtration through coarse cloth (e.g., bandage gauze), followed by filtration through filter paper. On an industrial scale, this can be accomplished with a plate-and-frame or rotary press filter. In some streptomycete fermentations, genetic manipulation has been used to obtain an organism having filterable mycelia and high-level production of a secondary metabolite. Laboratory centrifuges made by Beckman or Sorvall may be equipped with continuous-flow rotors which will hold up to an 800-ml pellet. The flow is limited, however, to approximately 4 liters/h. The cost of this rotor is not trivial to the academic laboratory, as it may exceed

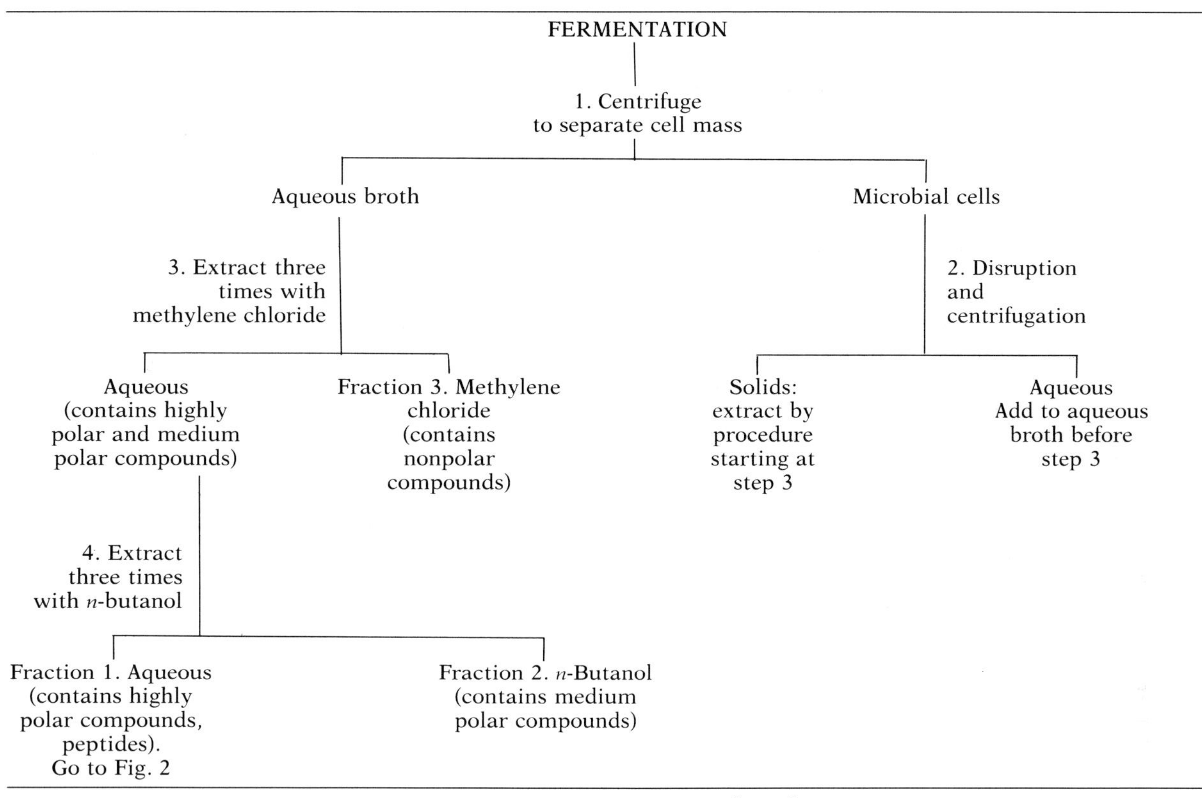

FIG. 1. Fractionation "tree."

$9,000. The centrifugation speed and the flow of the fermentation broth are adjusted to obtain a clarified broth. On a larger scale, DeLaval and Sharples make continuous centrifuges which have a solid containment capability. Laboratory models of either of these centrifuges have a capacity of approximately 200 g of cell paste. We believe the DeLaval bowl is considerably easier to use. The cost of either of these is considerably higher than the rotors used in common refrigerated laboratory centrifuges.

30.1.2 Cell Disruption

Cells may be ruptured (Fig. 1, step 2) after dilution with buffer to approximately 10% solids (consistency of heavy cream). There are several devices available for this purpose. An ultrasonic disrupter (available from most laboratory supply houses) is suitable for small-scale (1 to 10 g) disruption of cells. Power may be varied by tuning, and the probe size may also be varied. A considerable amount of heat may be generated; therefore, short bursts of vibration (1 to 2 min) followed by cooling are recommended for heat-sensitive materials. Some cells will not be ruptured by the sonic vibrations.

Grinding the cells with a fine grade of sand in a mortar will often give results superior to sonic vibration. Although this is inexpensive, it is also labor intensive.

An inexpensive automated cell disrupter based on the same principle is the Bead Beater (Biospec Products, Bartlesville, Okla.). This device, which looks like a blender, stirs the mixture of cells and sand or glass beads with a Teflon agitator. After disruption, the liquid is filtered through a coarse fritted-glass filter and processed as shown in Fig. 1. The manufacturer states that suitable disruption of most types of cells will occur. This is certainly the most inexpensive piece of equipment available for cell disruption. Cooling can also be applied with certain models.

A larger, continuous device (Dyno-Mill, W.A. Bachofen Mfg. Eng., Basel, Switzerland) is also available in stainless steel with temperature control. The principle is the same as that of grinding cells with sand, but the equipment is considerably more sophisticated in construction and operation.

Disruption of fragile cells may be accomplished in a pressure bomb (Parr Instrument, Silver Spring, Md.). These devices are generally available in chemistry departments, where they are used for hydrogenation. The cells to be ruptured are placed in a beaker small enough to sit easily inside the cavity of the bomb. Pressure is applied with compressed nitrogen. Once the pressure is at the maximum cylinder pressure, the valve to the bomb is closed and the cylinder is disconnected. The pressure is then released rapidly. The cell walls are ruptured due to internal gas expansion in the cell. The entire procedure may be carried out at low temperature by placing the apparatus in a cold room or in a bath. Temperature is controlled more easily with this apparatus than by any other cell disruption method. In all of the above methods, quantitation of the disruption efficiency can be made by microscopic examination. Product yield may not correlate with maximum cell disruption due to time and heat variation.

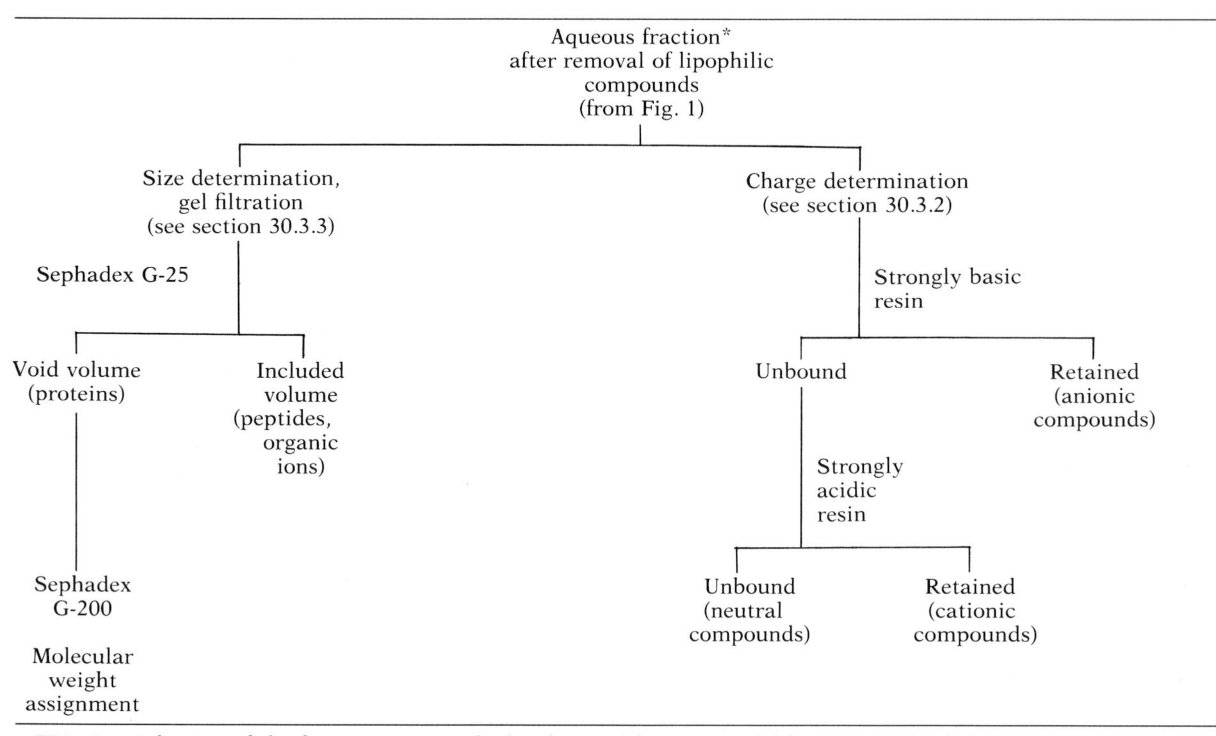

FIG. 2. Selection of the fractionation method to be used for water-soluble compounds. * If organic solvents destroy the activity of the desired compound, clarified broth should be used at this point.

30.2 LIQUID-LIQUID EXTRACTION

In solvent extractions, ionic compounds are more water soluble than nonionic. Both neutral and nonionic lipophilic compounds will be extracted by the organic solvent. Ionized compounds may be extracted by reducing the charge via pH adjustment. By reversing the pH, the neutral compounds will be extracted with the ionized compounds. This often results in substantial purification of the desired chemical.

For both disrupted cells and fermentation broths, initial solvent volumes of one-third the volume of the aqueous layer are recommended for steps 1 to 3 in Fig. 1. Each of the extractions should be repeated three times, giving a total extraction volume equal to the volume of the initial aqueous layer. A separatory funnel is normally used in the laboratory, but equivalent separations may be obtained on an industrial scale with a Podbielniak liquid-liquid contactor.

The extraction procedure shown in Fig. 1 will divide the compounds into three classes: (i) highly polar (in 1, the aqueous fraction); (ii) moderately polar (in 2, the *n*-butanol fraction); and (iii) nonpolar (in 3, the methylene chloride fraction). Examples of compounds in fraction 1 are amino acids, organic acids, and proteins. Examples of compounds in fraction 2 are penicillins and cephalosporins. Fraction 3 compounds include tetracyclines, griseofulvin, etc.

Emulsion formation can be a severe problem. The use of a centrifuge to break the foam on a small laboratory scale is of value in planning for larger scale isolation. Foambreakers compatible with the product can be selected. Of primary importance in this selection is a knowledge of the phase in which the foambreaker and the product will finally reside. Otherwise, the foambreakers may prove difficult to remove from the product. Chemical foambreakers such as polysiloxanes and polypropylene glycol P-2000 are available from a number of manufacturers (e.g. Dow-Corning, BASF), as are oils from natural sources (e.g., olive oil, lard oil). Small quantities of these materials are required, and the minimum effective quantity should be used. Very often, changing the ratio of solvent to aqueous solution will break an emulsion. Stirring with a glass rod is also of value in some cases.

Concentration of the extract should be carried out at low temperature (usually less than 40°C) and reduced pressure to lessen the possibility of chemical changes in the desired molecule. In the laboratory, a rotary evaporator is recommended for rapid removal of the solvent while maintaining even heat distribution.

After solvent removal, the organic extracts and aqueous layers may be assayed for the molecule of interest. If partition of the molecule into both layers has occurred, the solvent polarity (Table 1) should be changed. The polarity of the molecule may also be changed by varying the pH of the aqueous layer as shown in Fig. 3. The extraction scheme illustrated in Fig. 3 will separate the compounds in a manner different from that shown in Fig. 1, since the polarity will be shifted by the changes in pH. For instance, tryptophan, which has a high solubility in water (11.4 g/liter), is insoluble in methylene chloride and is slightly soluble in *n*-butanol. In Fig. 3, step 1, the polarity of tryptophan will be changed such that there is virtually no solubility in water and the tryptophan

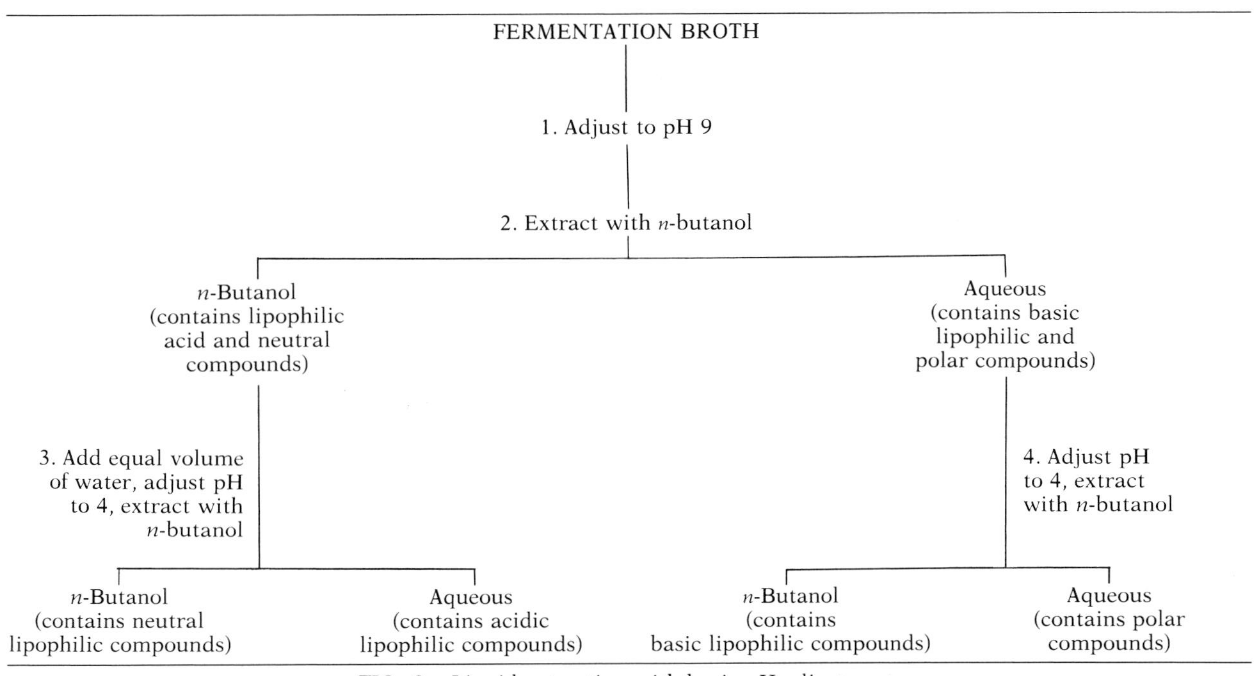

FIG. 3. Liquid extraction with basic pH adjustment.

may be extracted. It is wise to carry out this scheme in reverse (i.e., pH 4 before pH 9) (Fig. 4) as well, since degradation of the desired compound may occur in either acid or base.

If a partial extraction is obtained with organic solvents, the investigator should attempt to refine the final choice of solvents (Table 1) and aqueous buffer used in the initial product extraction.

30.3 CHROMATOGRAPHY

An extracted product is normally a mixture of a large number of compounds. To isolate the desired compound, it will be necessary to carry out one or more chromatographic steps. The partitioning of a solute between the solvent and a solid matrix packed in a column is one of the most versatile methods for the fractionation of biological products. The partitioning can be based on adsorption (chromatography), molecular size (gel filtration chromatography), ionic charge (ion-exchange chromatography), surface activity (hydrophobic chromatography), enzyme specificity (affinity chromatography), or molecular recognition (immunosorbent chromatography). Major advantages of chromatographic methods of product isolation are that (i) the methods are gentle and do not involve heat generation or shear; (ii) recoveries are high, often approaching 100%; (iii) high degrees of purity can be obtained in a single step; (iv) often several solutes can be isolated at one time; (v) it is relatively simple to scale up; (vi) a wide range of chromatographic media are commercially available; and (vii) the equipment is simple.

30.3.1 Adsorption Chromatography

The selection of chromatographic medium is based on initial fractionation results (Table 2). If the compounds have medium or low polarity (extracted in fraction 3, Fig. 1), silica is the chromatography support of choice. Initial separation conditions are best determined by thin-layer chromatography. Screening is carried out with three solvent systems of widely differing polarity. Based upon the initial thin-layer chromatography results, column chromatography

TABLE 1. Relative polarity of organic solvents

Solvent	Solubility of water in solvent (% by wt)	Relative solvent strength	Refractive index	UV cutoff for HPLC[a] (nm)
Pentane	0.010	0.00	1.358	<210
Hexane	0.010	0.01	1.375	<200
Cyclohexane	0.012	0.05	1.406	210
Carbon tetra-chloride	0.008	0.18	1.466	265
Toluene	0.046	0.29	1.496	285
Benzene	0.058	0.32	1.501	280
Diethylether	1.3	0.38	1.353	220
Chloroform	0.072	0.40	1.443	245
Methylene chloride	0.17	0.42	1.424	235
Tetrahydrofuran	Miscible	0.45	1.408	215
Dioxane	Miscible	0.56	1.422	220
Ethyl acetate	9.8	0.58	1.370	260
Acetonitrile	Miscible	0.65	1.344	190
Ethanol	Miscible	0.88	1.361	<210
Methanol	Miscible	0.95	1.329	<205
Acetic acid	Miscible	6.2	1.370	
Water		80	1.333	

[a] HPLC, High-pressure liquid chromatography.

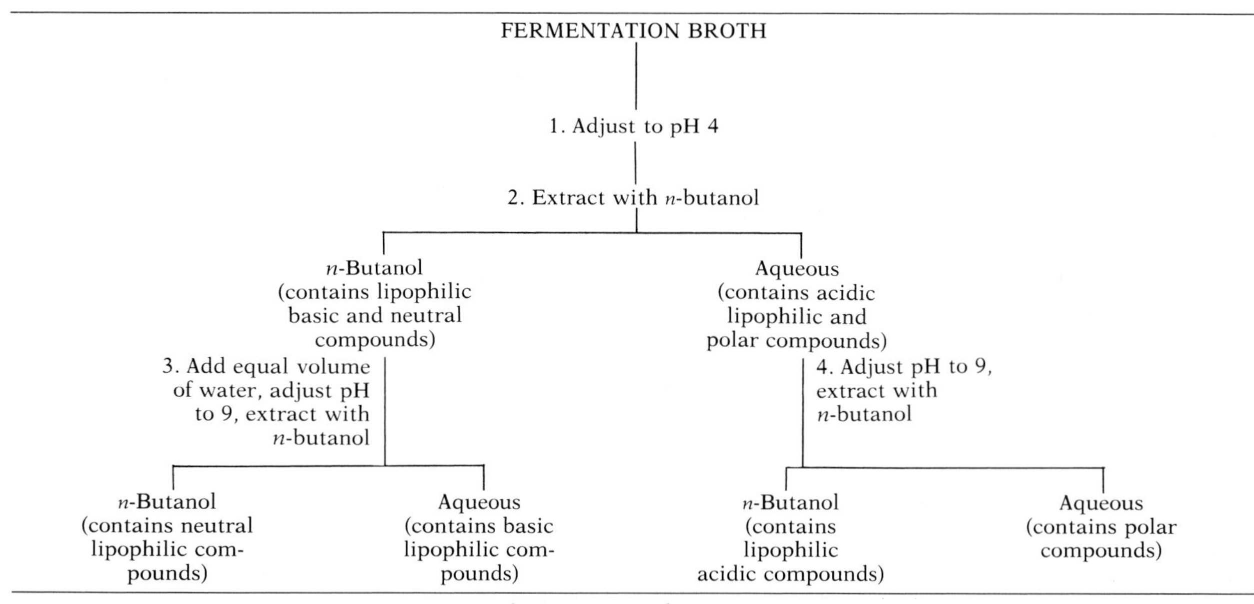

FIG. 4. Liquid extraction with acidic pH adjustment.

with silica or organic-bonded silica may be done with the solvent system (or gradient) giving the clearest separation. These data may be used to scale up directly to semipreparative high-pressure liquid chromatography columns which can separate up to 1 g of sample. Waters Associates and other companies make large-scale separation systems for the laboratory which are capable of handling up to 25-g samples. On a larger scale, Waters has also developed pilot-plant-scale operations which can purify kilogram quantities of specific compounds. Reversed-phase columns are also available for these systems.

30.3.2 Ion-Exchange Chromatography

An ion-exchange resin consists of an insoluble matrix to which charged groups have been covalently attached. In an ion-exchange procedure, the bound counterions are replaced by sample ions that have the same charge and which, in turn, are bound to the functional group on the resin. Neutral molecules and those having the same charge as the functional group pass through the column and are separated from the ionically attached compound. Another ion is then used to displace and elute the ionically attached compound of interest. There are four fundamental types of synthetic ion-exchange resins. Each type is analogous to a common acid or base except that the ionizable group is covalently attached to an insoluble matrix. The four types are strongly acidic cations, weakly acidic cations, strongly basic anions, and weakly basic anions. Ion-exchange resins remove ions from solutions to form insoluble resin salts.

Applications

In product recovery, ion-exchange chromatography is useful in two areas: concentration and fractionation.

In the first application, dilute solutions of product, usually in an aqueous medium, are removed by the ion-exchange resin until saturation is reached. An effective elution ion is then passed through the resin, rapidly exchanging the product ion in a small volume.

The second application, fractionation, utilizes the differing affinities that charged molecules have for a particular cation- or anion-exchange resin. Thus, positively charged molecules are not bound by an anion-exchange resin, whereas negatively charged molecules are retained. Compounds of like charge in solution often differ in their selectivity of binding to the functional group of a resin. Such differences, although sometimes difficult to predict for a particular group of ions, allow one to achieve effective separations by using this technique. By careful pH adjustment of the sample before application to the column, organics that are weak acids or bases can usually be separated by ion exchange.

Selection of correct ion-exchange resin

The choice of a suitable resin for a particular separation will depend on the properties of the compound of interest. Resins based on polystyrene or other hydrophobic polymers are used for isolation of organic compounds. When resins of this type are substituted with ionic groups, they have high capacities for small ions. Rarely, however, are biopolymers compatible with these hydrophobic supports. Such biomaterials are normally fractionated on hydrophilic supports prepared from polysaccharide polymers.

Strongly acidic or strongly basic exchange resins can be used to separate or adsorb all ionic species. Under conditions where the biological activity of a compound (e.g., an enzyme) is important, a weak ion exchanger should be used. With such resins, elution conditions do not require strong base or acid because the interaction is weak. Elution can normally be

TABLE 2. Selection of chromatography medium based on polarity of the compound to be isolated

Compound polarity	Suggested media
Nonpolar	Alumina, silica, charcoal
Medium polar	Silica, organic-bonded silicas, hydrophobic resins
Polar	Gel filtration media, ion-exchange resins, affinity matrices, immunosorbents

carried out under conditions of altered pH or ionic strength without changing the counterion.

Selection of a resin with a high degree of cross-linking will enable the investigator to exclude larger molecules from binding and thus to remove an organic material from a stream on the basis of molecular weight and charge. The physical porosity of the resin does affect adsorption and elution kinetics, requiring slower loading and elution rates to utilize the full capacity of highly cross-linked resins.

In industrial applications of ion exchange, a large resin particle size is normally selected (16/50 mesh). High flow rates can be achieved with large-mesh resins, allowing many cycles daily. Also a minimum pressure drop across a column is achieved. A chromatographic grade of resin (100/200 mesh) is sometimes necessary when separating similar ions or proteins. Finer-mesh resins give higher resolution across the same bed volume; however, the increased pressure drop requires slower flow rates.

Bulk or analytical grades of resins

Analytical grades of ion-exchange resins are available for specialized application where a prewashed resin of a narrow mesh range is important. Bulk grades of resin still contain soluble prepolymers and other impurities that must be removed from the resin before its use in most applications. The cost of analytical-grade resins is high, and such resins are not practical when large quantities of product are being recovered, unless the value of the product is correspondingly high.

Ion-exchange separations of biopolymers

Most resins suitable for separations of small organic molecules are highly cross-linked for mechanical stability and are normally hydrophobic, rendering them unsuitable for separation of biopolymers. Hydrophilic resins are used for these separations. Matrix materials include derivatized polysaccharide polymers such as cellulose, dextran, and agarose. These polymer "beads" exhibit hydrophilic properties important for maintaining "native" proteins. Cross-linked forms of agarose (e.g., Sepharose CL-6B) are reasonably rigid resins when subjected to moderate flow rates. Operation of hydrophilic resins at high flow rates can result in loss of resolution due to bead distortion and "channeling."

Pilot-plant-scale high-pressure liquid chromatography equipment has become increasing available for separations of proteins. In the area of ion exchange, ligands such as DEAE (diethylaminoethyl), carboxymethyl, and sulfopropyl are bound to rigid forms of silica. The theory of operation remains the same, but separation times are greatly reduced. Biopolymers are normally separated by gradient elution with increasing salt concentrations. Many commercial suppliers of standard analytical high-pressure liquid chromatography columns now provide prepacked columns for ion-exchange purposes. Truly preparative columns are expected to be available in the future.

Ion-exchange operation protocol

The following general protocol illustrates the use of ion-exchange resins for separation of a given compound.

1. The resin is first equilibrated with the counterion of choice. The number of column volumes necessary is dependent upon the "selectivity" of the ion for the exchanger. There is an inverse relationship between the column volume of buffer needed and selectivity.
2. The sample is loaded. The amount of sample is determined by the absolute capacity of the resin for the compound or by the complexity of the mixture. Unbound ions are removed by washing the column with water.
3. Desorption (elution) is initiated by changing the counterion, the ionic strength, the pH, or a combination of these parameters.
4. After elution, the resin is returned to its original counterion form by a regeneration step. The relative selectivity of the ion used for regeneration will determine the number of column volumes of liquid needed. Conversion is complete when the first ion is no longer detected in the effluent. At this point, the resin is normally rinsed with a column volume of water.

The capacity of a particular resin for the compound must be determined empirically. The presence of competing salt ions or organics of like charge will decrease capacity in relation to relative concentration and selectivity of binding. If a high-value product is to be recovered, an ion-exchange column should be loaded with 5 to 10% less product than absolute capacity to minimize losses.

As stated above, elution is normally accomplished by introduction of a suitable counterion solution that will rapidly elute the organic ion of interest in the salt form. If strong acids are used to elute, then the organic ion would elute as the free acid. The organic ion of interest must also exhibit reasonable solubility in the eluting solution. If a hydrophobic interaction between an eluted compound and the resin exists, the inclusion of dimethyl sulfoxide, dioxane, or an alcohol will reduce such an interaction and minimize broadening of the elution peak. The use of elevated temperatures for the elution solution will weaken ionic interactions, increase the solubility of the eluted compound, and usually decrease elution cycle length.

As a rule, the elution profile of a product peak yields a highly concentrated portion followed by a dilute "trailing" portion that can continue for several column volumes. Again, the value of the product will dictate the pool that is saved for a downstream process step. Often dilute streams can be used to prepare the next batch of elution solution and thus minimize losses. The choice of a volatile elution ion will allow subsequent separation from the compound of interest and recovery of elution ions for reuse. The selection of an ion-exchange system is ultimately based on maximum product recovery at minimum cost.

Column design is beyond the scope of this chapter, but important aspects include proper flow distribution to minimize channeling of liquid and a proper bed support that allows a high flow rate but is fine enough to retain the minimum size of resin particle. Columns meeting these criteria are available from a number of equipment firms such as Pharmacia, Amicon, LKB, and Bio-Rad.

The proper resin can be used for hundreds of cycles of loading, elution, and regeneration. Loss of capacity can be due to particulate matter in process streams. An increase in back pressure is a symptom of trapped particulates. Back washing the resin will usually drive such material out of the resin pores and restore the capacity of the system. Resin attrition can be caused by bead fracturing or oxidation. If such attrition occurs, milder elution conditions should be sought, including the removal of oxidizing agents from process streams before the exchange resin.

Ion-exchange chromatography examples

Amino acid separations can be achieved by exploiting the fact that they may be classified into three groups: acidic, neutral, and basic. The acidic amino acids have isoelectric points at approximately pH 3; the isoelectric points of neutral and basic amino acids lie at approximately pH 6 and pH 8 to 9, respectively. Since amino acids exist as cations or anions depending upon the pH, it is possible to separate them by pH adjustment and proper resin choice.

As an example, the separation of glutamic acid and alanine from a fermentation broth may be achieved by adjusting the broth to pH 7.0. The broth is loaded on a weak anion-exchange resin (free base form) such as Amberlite IRA-68 (Rohm and Haas Co.) at two to five column volumes per h. The glutamic acid binds, and the alanine passes through. The glutamate (sodium salt) is eluted with two column volumes of acetate buffer (pH 4.0) at two column volumes per h. The resin is then regenerated to the free base form using two to three column volumes of 1.0 N NaOH. Finally, the column is rinsed with a column volume of water.

The alanine fraction, which is not retained, is loaded on a strong cation exchanger such as Amberlite IR120 (Rohm & Haas Co.), hydrogen form, at two to five column volumes per h. The alanine is eluted as the sodium salt using approximately two column volumes of 1.0 N sodium hydroxide at one to two column volumes per h. The column is regenerated with two to three column volumes of 1.0 N H_2SO_4, followed by a column volume of water.

Gel filtration chromatography

Gel filtration chromatography is a separation technique based on the distribution of a molecule between the external and internal solvent in a porous, insoluble matrix. The sample is introduced as a narrow band at one end of a column packed with the insoluble matrix, and the column is developed with solvent. Molecules larger than the pore size of the matrix pass through the column and exit as a peak (broadened by diffusion) immediately after the external volume (V_0). Molecules smaller than the pore size of the matrix distribute equally between the external and internal volumes (V_i) of the matrix and exit as a peak immediately after the sum of the external and internal volumes. Molecules of intermediate size distribute between the external and internal volumes of the matrix in proportion to their molecular size. These molecules will be retarded and exit the column as a peak somewhere between the void volume and the sum of the external and internal volumes.

30.3.3 Gel Filtration Chromatography

Principles

Gel filtration matrix. The total volume (V_t) of the space occupied by a gel matrix packed in a column can be determined by calibration with water or calculated from the column dimensions. The external volume or void volume (V_O) is determined empirically as the elution volume of a solute larger than the pores in the matrix. The internal volume or salt volume (V_i) is most accurately determined empirically as the elution volume of deuterated or tritiated water. The matrix volume (V_m) is difficult to determine, but if V_t, V_O, and V_i are known it can be calculated from the relationship $V_t = V_O + V_i + V_m$.

Solute distribution. The elution volume of the solute (V_e) is the volume of solvent leaving the bed from the time the sample enters the bed until the leading edge of the solute peak appears at the exit. The migration of the solute through the bed is characterized by its distribution coefficient, given by $K_d = V_e - V_0/V_i$. Because of the difficulty of determining V_i, it is convenient to neglect V_m and substitute $V_t - V_0$ for V_i, to give $K_{av} = (V_e - V_0)/(V_t - V_0)$.

Selectivity. It has been found that for solutes of similar molecular shape and density, a linear relationship exists between K_{av} and the logarithm of the molecular weight of the solute over a certain range of molecular weights, depending on the porosity of the matrix. The greater the slope of the selectivity curve, the greater the resolution capability of the matrix for solutes of similar molecular weights.

Resolution. Resolution of two solutes is defined as the difference in their elution volumes divided by half the sum of the base line intercepts of the two peaks:

$$R_s = (V_{e2} - V_{e1})/{}^1/_2\,(W_1 + W_2)$$

The degree of resolution is directly proportional to the difference in elution volumes, which is determined by the selectivity of the matrix, and inversely proportional to the widths of the peaks. The widths of the peaks, assuming the solutes are loaded on the column in a small band relative to the column height, are a function of zone spreading. Zone spreading is caused by longitudinal diffusion of solute in the bed, lack of equilibrium, and eddy diffusion. Longitudinal diffusion is proportional to the time the solute remains in the bed and can therefore can be reduced by increasing the flow rate. The lack of equilibrium can be alleviated by increasing the time for equilibrium to occur, i.e., decreasing the flow rate. Consideration of these two opposing factors can result in an optimum flow rate. Eddy diffusion results from a distribution of

TABLE 3. Properties of gel filtration matrices

Product (composition)	Particle size (μm)	Fractionation range (kilodaltons)	Flow rate (cm/h)
Sephadex (cross-linked dextran)[a]			
G-25F	20–80	1–5	>80
G-50F	20–80	1.5–30	45
G-75	40–120	3–80	25
G-100	40–120	4–150	20
G-150	40–120	5–300	15
G-200	40–120	5–600	10
Bio-Gel (polyacrylamide)[b]			
P-2F	40–80	0.1–1.8	
P-4F	40–80	0.8–4	
P-6F	40–80	1–6	22
P-10F	40–80	1.5–20	18
P-30F	80–150	2.5–40	
P-60F	80–150	3–60	12
P-100F	80–150	5–100	
P-150F	80–150	15–150	10
P-200F	80–150	30–200	
P-300F	80–150	60–400	5
Ultragel (acrylamide-agarose)[c]			
ACA 202	60–140	1–15	4–6
ACA 54	60–140	5–70	3–6
ACA 44	60–140	10–130	3–6
ACA 34	60–140	20–350	3–5
ACA 22	60–140	100–1,200	2–3
Sephacryl (allyldextran cross-linked with bisacrylamide)[a]			
S-200	40–105	5–250	30
S-300	40–105	10–1,500	25
S-400	40–105	20–8,000	
Sepharose (agarose)[a]			
2B	60–200	70–40,000	15
4B	60–140	60–20,000	26
6B	45–165	10–4,000	30
Bio-gel (agarose)[b]			
A-0.5M M	80–150	10–500	30
A-1.5M M	80–150	10–1,500	30
A-5M M	80–150	10–5,000	25
A-15M M	80–150	40–15,000	12
A-50M F	80–150	100–50,000	8
A-150M F	80–150	1,000–150,000	5
Trisacryl (*N*-acryloyl-2-amino-2-hydroxymethyl-1, 3-propane diol polymer)[c]			
GF 2000	40–80	10–15,000	60
Fractogel (hydrophilic gel from vinyl polymers)[d]			
TSK HW 40F	32–63	0.1–10	
TSK HW 50F	32–63	0.5–80	
TSK HW 55F	32–63	1–700	20
TSK HW 65F	32–63	50–5,000	
TSK HW 75S	25–40	500–50,000	30

[a] Pharmacia, Piscataway, N.J.
[b] Bio-Rad, Richmond, Calif.
[c] LKB Instruments, Inc., Gaithersburg, Md.
[d] EM Sciences, Cherry Hill, N.J.

matrix particle size and imperfections in matrix particle packing. It can be reduced by decreasing the length of channels between matrix particles which have approximately the same dimensions as the particle diameter.

For two different solutes moving with different velocities through the column bed, the difference in elution volume will increase with increasing column bed height. However, zone spreading increases with the square root of the column bed height. Since resolution is the ratio of separation volume to zone spreading, it increases as the square root of the column bed height.

Practical aspects

Gel filtration matrices. Hundreds of gel filtration matrices are commercially available, differing in material of construction, porosity, and particle size. Materials of construction include dextrans, polyacrylamides, agarose, cross-linked agarose, copolymers of acrylamide and agarose, beaded cellulose, styrene divinyl-benzene copolymer, *N*-acryloyl-2-amino-2-hydroxymethyl-1,3-propanediol polymer, and controlled-pore glass. For analytical separations (such as molecular weight determinations) and laboratory-scale purifications, a matrix is chosen for its porosity, particle size, and lack of interaction with the solute to be separated. For large-scale separations, particle size and particle rigidity are more important criteria.

The porosity of the matrix determines the fractionation range and selectivity for the separation. Larger pores allow larger solutes to partition between the external and internal volumes of the matrix. Manufacturers publish selectivity data which allow an intelligent choice to be made, assuming the size of the molecule to be separated is known. From these data, a porosity is chosen such that the molecule elutes in a volume between V_0 and $V_0 + V_i$. Matrix particle size is chosen on the basis of application. For analytical- or laboratory-scale chromatography, the small particle sizes reduce eddy diffusion with an increase in resolution. However, as particle size decreases, the hydrostatic pressure required to maintain a constant flow rate increases. For matrices of low rigidity, the increase in pressure can cause a collapse of the matrix, further increasing the pressure until the column stops flowing. For large-scale chromatography, throughput demands larger particle sizes, rigid matrices, or both. In Table 3, the properties of some commercially available gel filtration matrices are given.

Column size and design. The column bed height is determined by the degree of resolution required. This is determined in laboratory-scale equipment. The column diameter needed is specified by the throughput required, which in turn is determined by (i) the amount of product required per unit time, (ii) the size of the batch to be processed, and (iii) the time allowed for product processing.

Once the sample volume to be processed per cycle is determined, the column diameter is scaled up from laboratory experiments by keeping constant the ratio of sample starting band width to column bed height. Assuming it is possible to obtain a uniformly packed bed and to distribute the sample uniformly across the bed cross-section, the resolution will be independent of column diameter. The limit to maximum column bed height is the compressibility of the matrix. The force exerted on any cross-section of bed is equal to the weight of the bed material and the drag force acting on the bed above it. The densities of most matrices used are close enough to the eluant density that the force due to gravity can be ignored. The drag force is proportional to the total surface area of the matrix, the velocity of flow past the surface, and the viscosity of the eluant. Reducing the total surface area of the matrix by increasing the particle size and reducing the flow rate to decrease the drag force adversely affects resolution (see section 30.3.3). One solution to this problem is to decrease the drag force by decreasing the bed height. Resolution is maintained by using several short columns in series. This requires carefully designed columns to minimize the zone broadening that occurs at the inlet and outlet of any column. Another solution is to use matrices with low compressibility. The use of a rigid matrix of suitable selectivity may permit sufficient bed height in a single column to yield the desired resolution without excessive back pressure. Other options in column design are available, such as internal lateral bed supports or internal longitudinal rods or sections. The latter devices support the matrix by wall effects.

Column packing. The required bed volume of the matrix is calculated from the size of the columns(s), or, if the column(s) contains inserts, the volume of water required to fill the column(s). Preswollen matrices are diluted with eluant and mixed gently. After the matrix has settled, the eluant over the matrix is decanted or siphoned off. A volume of eluant approximately equal to the settled volume is added to make a slurry that is thin enough to allow gas bubbles to escape easily. A vacuum is then applied to deaerate the matrix completely. After further settling, enough eluant is removed to give a thick slurry that will still allow air bubbles to escape. If a dry matrix is used, the weight of dry material required to give the desired volume is calculated from the manufacturer's literature. The matrix is swollen in approximately twice the calculated bed volume of eluant for the required time, and the swollen matrix is then deaerated as above.

The column should be securely mounted and exactly vertical. Enough eluant is pumped through the bottom distributor to remove all the air in the line and under the bed support. An extension is attached to the top of the column to allow all of the matrix slurry to be added at one time. The slurry, at the intended operating temperature, is added by pouring down the side of the column at a rate slow enough to avoid introduction of air into the matrix. After the matrix has settled, the excess eluant is drained off, the column extension is removed, and the top distributor is secured. Care must be taken to remove any air from the top of the column when the top distributor is attached. The column is then equilibrated with two to three column volumes of eluant at or 10% above the intended operating flow rate.

Sample application and elution. For process-scale chromatography, the only practical method of sample application is through the inlet line directly to the top distributor. This requires that the top distributor be directly above the surface of the packed matrix bed. The homogeneity of packed bed, V_0, and $V_0 + V_i$ are determined by application of a mixture of blue dextran 2000 (2 mg/ml) and potassium ferricyanide made up in a volume of the eluant equal to the intended sample volume.

Sample preparation. Samples to be chromatographed are freed of particulate matter by filtration or centrifugation. If the sample is a complex mixture, treatment before chromatography to simplify the composition of the mixture may be warranted. Such procedures as precipitation with ammonium sulfate or polyethylene glycol to remove unwanted proteins, precipitation with dextran sulfate to remove lipids and lipoproteins, and so on can improve the chromatography and the resulting purity of the eluted material. The maximum sample volume to allow the required resolution is determined from laboratory-scale experiments and used to calculate the column size. For good resolution, the sample volume should be usually not more than 1 to 4% of the column bed volume. For process separations, the sample size may be 30% of the bed volume, depending on the resolution. The viscosity of the sample should not be more than two times the viscosity of the eluant. For dilute eluants, this corresponds to a protein concentration of about 70 mg/ml. If the viscosity is high due to a high concentration of low-molecular-weight components, desalting by gel filtration or dialysis is a simple solution.

The eluant is chosen based on the nature of the component to be fractionated. The eluant pH and ionic strength are adjusted to maintain the activity of the material to be separated. When using compressible matrices, it is important to keep the viscosity of the eluant as low as possible to reduce drag force (see section 30.3.3).

30.3.4 Typical Approach to Protein Isolation

As an aid in designing an appropriate isolation scheme for a given protein, it is always helpful to know its molecular weight and isoelectric point (pI). This information can be obtained by straightforward analytical techniques. Polyacrylamide gel electrophoresis will yield an approximate molecular weight for a protein. Isoelectric focusing will yield the pI, an important property necessary to predict the net charge of a protein at a particular pH. If the protein of interest is an enzyme, knowledge of a specific cofactor or inhibitor is helpful for designing an affinity chromatography procedure based on supports where small molecules are present in an immobilized form.

A general approach to isolation and purification of proteins begins with an extract or fermentation broth. Concentration of the protein is usually necessary before a chromatographic step. Concentrated solutions of ammonium sulfate or polyethylene glycol are often used to effect precipitation. An appropriate concentration of ammonium sulfate to "bring down" the bulk of the desired soluble protein must be determined empirically. Alternatively, isoelectric precipitation can sometimes by effected by proper pH adjustment.

Gel filtration is usually the first chromatography step used in an isolation procedure because this technique has a high capacity for protein. With the proper choice of medium, proteins with molecular weights in the range of 5,000 to 500,000 can be separated. A fraction will be obtained containing a heterogenous group of proteins in a narrow molecular weight "window."

Ion-exchange chromatography is usually the next step. Before a sample is applied, the material is often dialyzed to effect a buffer change. For process-scale buffer change, gel filtration is used. Ion exchange is an excellent second step, in part because it can serve to concentrate the sample from the previous gel filtration step. Unless the original protein mixture was composed of only a few proteins, a third step such as affinity chromatography would be needed to achieve final purification. This step might utilize immobilized cofactors, dyes, antibodies, or lectins. The greater the degree of interaction specificity between the affinity ligand and the protein of interest, the higher the purity of the final isolated protein is likely to be. The purity of the final product should be checked by sodium dodecyl sulfate-polyacryamide gel electrophoresis using a sensitive protein stain. Before storage, a protein should be concentrated by ultrafiltration or lyophilized in the presence of a volatile buffer.

30.4 LITERATURE CITED

1. **Bio-Rad Laboratories.** 1975. Gel chromatography. BioRad Laboratories, Inc., Richmond, Calif.
2. **Fiechter, A. (ed.).** 1982. Advances in biochemical engineering, vol. 25. Chromatography. Springer-Verlag AG, Berlin.
3. **Pharmacia Fine Chemicals.** 1979. Gel filtration, theory and practice. Pharmacia Fine Chemicals, Piscataway, N.J.
4. **Pharmacia Fine Chemicals.** 1980. Ion exchange chromatography. Pharmacia Fine Chemicals, Piscataway, N.J.
5. **Stahl, E.** 1969. Thin layer chromatography, a laboratory manual. Translated by M. R. F. Ashworth. Springer-Verlag AG, Berlin.

Author Index

Subject Index